# CALCULUS
# & ITS
# APPLICATIONS

# CALCULUS & ITS APPLICATIONS

### ELEVENTH EDITION

## Larry J. Goldstein
Goldstein Educational Technologies

## David C. Lay
University of Maryland

## David I. Schneider
University of Maryland

## Nakhlé H. Asmar
University of Missouri

PEARSON

Prentice Hall

Upper Saddle River, NJ 07458

**Library of Congress Cataloging-in-Publication Data**

Goldstein, Larry Joel.
  Calculus & its applications / Larry J. Goldstein, David C. Lay, David I. Schneider.—11th ed. / with the
assistance of Nakhlé H. Asmar.
      p. cm.
  Includes index.
  ISBN 0-13-191963-6
  1. Calculus–Textbooks. I. Title: Calculus and its applications. II. Lay, David C. III. Schneider, David I. IV.
Title.
  QA303.2.G66  2007
  515--dc22                                                      2005048737

Executive Aquisitions Editor: *Petra Recter*
Editor in Chief: *Sally Yagan*
Production Editor: *Lynn Savino Wendel*
Senior Managing Editor: *Linda Mihatov Behrens*
Executive Managing Editor: *Kathleen Schiaparelli*
Manufacturing Buyer: *Maura Zaldivar*
Manufacturing Manager: *Alexis Heydt-Long*
Director of Marketing: *Patrice Jones*
Marketing Assistant: *Jennifer De Leeuwerk*
Project Manager: *Michael Bell*
Editorial Assistant/Supplements Editor: *Joanne Wendelken*
Art Director: *John Christiana*
Interior Designer: *Kenny Beck*
Cover Designer: *John Christiana*
Art Editor: *Thomas Benfatti*
Creative Director: *Juan R. López*
Director of Creative Services: *Paul Belfanti*
Cover Art: *Origami by Ilene Kahn*
Art Studio: *Interactive Composition Corporation/Laserwords*
Compositor: *Dennis Kletzing*

© Copyright 2007, 2004, 2001, 1999, 1996, 1993, 1990, 1987, 1984, 1980, 1977
by Pearson Education, Inc.
Pearson Prentice Hall
Pearson Education, Inc., Upper Saddle River, NJ 07458

Pearson Prentice Hall™ is a trademark of Pearson Education, Inc.

Printed in the United States of America
10  9  8  7  6  5  4  3  2  1

ISBN 0-13-191963-6

Pearson Education LTD., *London*
Pearson Education Australia PTY, Limited, *Sydney*
Pearson Education Singapore, Pte. Ltd
Pearson Education North Asia Ltd, *Hong Kong*
Pearson Education Canada, Ltd., *Toronto*
Pearson Educación de Mexico, S.A. de C.V.
Pearson Education—Japan, *Tokyo*
Pearson Education Malaysia, Pte. Ltd

# Contents

*Sections preceded by a * are optional in the sense that they are not prerequisites for later material.

# Preface

WE have been very pleased with the enthusiastic response to the first ten editions of *Calculus & Its Applications* by teachers and students alike. The present work incorporates many of the suggestions they have put forward.

Although there are many changes, we have preserved the approach and the flavor. Our goals remain the same: to begin the calculus as soon as possible; to present calculus in an intuitive yet intellectually satisfying way; and to illustrate the many applications of calculus to the biological, social, and management sciences.

The distinctive order of topics has proven over the years to be successful—easier for students to learn, and more interesting because students see significant applications early. For instance, the derivative is explained geometrically before the analytic material on limits is presented. To reach the applications in Chapter 2 quickly, we present only the differentiation rules and the curve sketching needed for those applications. Advanced topics come later when they are needed. Other aspects of this student-oriented approach follow below.

## Applications

We provide realistic applications that illustrate the uses of calculus in other disciplines. See the Index of Applications on the inside cover. Wherever possible, we have attempted to use applications to motivate the mathematics.

## Examples

The text includes many more worked examples than is customary. Furthermore, we have included computational details to enhance readability.

## Exercises

The exercises comprise about one-quarter of the text—the most important part of the text in our opinion. The exercises at the ends of the sections are usually arranged in the order in which the text proceeds, so that the homework assignments may easily be made after only part of a section is discussed. Interesting applications and more challenging problems tend to be located near the ends of the exercise sets. Supplementary exercises at the end of each chapter expand the other exercise sets and include problems that require skills from earlier chapters.

## Practice Problems

The practice problems have proven to be a popular and useful feature. Practice Problems are carefully selected questions located at the end of each section, just before the exercise set. Complete solutions are given following the exercise set. The practice problems often focus on points that are potentially confusing or are likely to be overlooked. We recommend that the reader work the practice problems and study their solutions before moving on to the exercises. In effect, the practice problems constitute a built-in workbook.

## Minimal Prerequisites

In Chapter 0, we review those concepts that the reader needs to study calculus. Some important topics, such as the laws of exponents, are reviewed again when they are used in a later chapter. Section 0.6 prepares students for applied problems that appear throughout the text. A reader familiar with the content of Chapter 0 should begin with Chapter 1 and use Chapter 0 as a reference, whenever needed.

## New in this Edition

Among the many changes in this edition, the following are the most significant:

1. *Over 25% New and Revised Exercises and Examples*

   - Over 400 new and updated exercises that have a wide range including those designed to aid students with a limited math background to exercises that challenge the more capable students.

   - Chapters 1–3 include over 30 new examples. The examples are completely worked out, with an emphasis on how to write solutions with appropriate sentences and units for the answers (see Examples 10–11, Section 1.1; Examples 1, 3, Section 1.2; Examples 3–7, Section 1.3).

   - Engaging applications based on up-to-date real world data: health expenditures, federal deficit, debt per capita, effect of Mad Cow disease on beef exports, price of gasoline (see Exercises 47–50, 53–55, Section 1.1; Exercises 15–18, Section 1.2; Exercises 57–64, Section 1.6).

   - New exercises involve using two data points for the quantity of a commodity sold at different prices, assuming it is linear, and then finding and maximizing the revenue and profit functions (see Exercises 45–46, Section 2.3).

2. *Revised presentation of crucial topics*

   a. The Derivative

   - Early introduction of the derivatives as rate of change (starting in Section 1.2 and continuing in Sections 1.6 to 1.8).

   - In Chapter 1, repeated applications of the derivative as a rate of change and slope of a curve, designed to drill these important topics through the minds of students that are exploring this material for the first time.

   - Applications of the derivative are included throughout Chapters one and two. In Chapter 1, most applications are included at the end of sections, allowing instructors to use them at their own discretion.

   - Examples have been added to Section 3.2 to facilitate the discussion of the role of the variables in differentiation problems (distinctions between time rate of change and marginal concepts).

   b. Optimization

   - Section 2.3 on optimization has been revised to provide an even clearer presentation of the First and Second Derivative Tests.

   - Section 2.5 has been revised to incorporate more details and figures to facilitate a deeper understanding of optimization.

   c. Careful Discussion of Integration

   - In Section 6.1 there is a careful and thorough treatment of the integration constant when dealing with discontinuous functions.

3. *Enhancements to Pedagogy*

   - Many examples and exercises have been labeled to signify the type of application.
   - Over 25 new graphs have been added to facilitate comprehension of the examples and exercises.
   - New four color design used to visually offset important pedagogical features, such as the examples, the art program, the definition and theorem boxes. The use of color will help students identify the elements they are looking for when studying.

4. *"Warnings!"* which provide tips on common pitfalls and mistakes appear at relevant times throughout the text.

This edition contains more material than can be covered in most two-semester courses. Optional sections are starred in the table of contents. In addition, the level of theoretical material may be adjusted to the needs of the students. For instance, Section 1.4 may be omitted entirely, if the instructor does not wish to present the notion of limit beyond the required material that is contained in Section 1.3.

## Acknowledgments

The following is a list of reviewers from this and previous editions. We apologize for any omissions. While writing this book, we have received assistance from many persons. And our heartfelt thanks goes out to them all. Especially, we would like to thank the following reviewers, who took the time and energy to share their ideas, preferences, and often their enthusiasm with us.

**ARIZONA** W.E. Conway, University of Arizona; Donald E. Myers, University of Arizona, Tempe   **ARKANSAS** Dennis Brewer, University of Arkansas   **CALIFORNIA** Robert Brown, University of California, Los Angeles; Larry Gerstein, University of California, Santa Barbara; Shirley A. Goldman, University of California, Davis; Harvey Greenwald, California State Polytechnic University; W.R. Hintzman, San Diego State University; T.Y. Lam, University of California, Berkeley; Melvin D. Lax, California State University, Long Beach; Russell Lee, Allan Hancock College; John H. Mathews, California State University, Fullerton; H. Suey Quan, Golden West College; Ronald Rose, American River College; Edward Spanier, University of California, Berkeley; Bruce Swenson, Foothill College   **COLORADO** Howard Frisinger, Colorado State University   **DELAWARE** Georgia B. Pyrros, University of Delaware   **DISTRICT OF COLUMBIA** Der-Chen Chang, Georgetown University   **FLORIDA** James Brewer, Florida Atlantic University; Bruce Edward, University of Florida   **ILLINOIS** Biswa Datta, Northern Illinois University; Karabi Datta, Northern Illinois University; James L. Heitsch, University of Illinois, Chicago; Roy Lowman, University of Illinois at Chicago   **IOWA** Brenda Diesslin, Iowa State University   **KANSAS** Robert Brown, University of Kansas; Charles Himmelberg, University of Kansas   **LOUISIANA** E. John Hornsby, Jr., University of New Orleans   **MARYLAND** Arlene Sherburne, Montgomery College, Rockville   **MASSACHUSETTS** James Kaplan, Boston University; Arthur Rosenthal, Salem State College; Robert Seeley, University of Massachusetts, Boston   **MICHIGAN** Claude Schochet, Wayne State University; Arthur J. Schwartz, University of Michigan   **MINNESOTA** Dennis White, University of Minnesota   **MISSOURI** James A. Huckaba, University of Missouri; William McCord, University of Missouri; H. Keith Stumpff, Central Missouri State

University    **NEBRASKA** Dana Nimic, Southeast Community College-Lincoln
**NEW JERSEY** Alan Candiotti, Drew University    **NEW MEXICO** W.T.
Kyner, University of New Mexico; Carla Wofsky, University of New Mexico
**NEW YORK** Jack R. Barone, Baruch College, CUNY; Albert G. Fadell,
SUNY Buffalo; Jack E. Graves, Syracuse University; Shujuan Ji, Columbia
University; Lawrence J. Lardy, Syracuse University; Robert A. Miller, City
University New York; Geraldine Taiani, Pace University    **NORTH CAR-
OLINA** James E. Honeycutt, North Carolina University; Gabriel Lugo, Univer-
sity of North Carolina, Wilmington; W.R. Wilson, Central Piedmont Community College    **OHIO** Samuel Jasper, Ohio University; Carl David Minda, Uni-
versity of Cincinnati    **OKLAHOMA** Dennis Bertholf, Oklahoma State Univer-
sity    **OREGON** Betty Fein, Oregon State University; Joan M. Thomas, Univer-
sity of Oregon    **PENNSYLVANIA** Dennis DeTurck, University of Pennsylvania;
David Harbater, University of Pennsylvania; Donald Hight, Kansas State College
of Pittsburg; Joyce Longman, Villanova University; Murray Schechter, Lehigh University; Frankl Warner, University of Pennsylvania    **RHODE ISLAND** Wallace
A. Wood, Bryant University    **SOUTH DAKOTA** Charles Clever, South Dakota
State University    **TENNESSEE** James V. Balch, Middle Tennessee State University    **TEXAS** Janice Epstein, Texas A&M University; Gordon Lukesh, University
of Texas, Austin; Ann McGaw, University of Texas, Austin    **VERMONT** Heath
K. Riggs, University of Vermont    **VIRGINIA** Michael J. Berman, James Madi-
son University; Judy B. Kidd, James Madison University; David W. Penico,
Virginia Commonwealth University; James Sochacki, James Madison University
**WISCONSIN** Fred Brauer, University of Wisconsin; Albert J. Milani, University of Wisconsin-Milwaukee; Denise A. Widup, University of Wisconsin-Parkside
**CANADA** Jens Rademacher, University of British Columbia

The authors would like to thank the many people at Pearson Education who
have contributed to the success of our books over the years. We appreciate the
tremendous efforts of the production, art, manufacturing, and marketing depart-
ments. Special thanks go to Lynn Savino Wendel, who managed production of this
book. The expert skills of our typesetter, Dennis Kletzing, have once again eased
the burden of preparing this new edition. Many thanks to Janet Crandell Tremain
for her superb work on revising the electronic test bank.

Thanks also to Gracia Nabhane, M.D., for her help with the applications in
biology and medicine.

The authors would like to thank our editor, Petra Recter, who helped us plan
this edition.

Larry J. Goldstein
larrygoldstein@comcast.net

David C. Lay
lay@math.umd.edu

David I. Schneider
dis@math.umd.edu

Nakhlé H. Asmar
nakhle@math.missouri.edu

# Student and Instructor Resources

## Student Resources

### Printed Resources

**Student Study-Pack**

- Text and Student Study Pack Bundle ISBN: 0-13-174625-1
- Student Study Pack Stand alone ISBN: 0-13-174588-3

Everything a student needs to succeed in one place. Provided at no charge when packaged with the book or available for purchase stand-alone. Student Study-Pack contains:

➢ **Student Study Guide**

Includes study tips, strategies for studying, warnings of common mistakes, assistance with calculator operations, hints for exercises and solutions to every third odd-numbered exercise.

➢ **CD Lecture Videos**

A comprehensive set of topical videos, in which examples for each topic are worked out by an instructor. The videos provide excellent support for students who require additional assistance, for distance learning and self-paced programs, or for students who missed class.

➢ **Pearson Tutor Center**

Tutors provide one-on-one tutoring for any problem with an answer at the back of the book. Students access the Tutor Center via toll-free phone, fax, or e-mail.

➢ **Visual Calculus**

Built from the ground up by David Schneider (University of Maryland), Visual Calculus enhances the visualizations and therefore the understanding of the *ideas* behind the calculus. With this software students can: analyze a function and its derivatives, examine solutions of differential equations, graph functions, calculate definite integrals, evaluate finite and infinite sums, graph Taylor approximations, plot cosine and sine functions and much more. Find Visual Calculus in MyMathLab and the CD Lecture Videos.

### Internet Resources

**MyMathLab**

- Text and MyMathLab Bundle ISBN: 0-13-223231-6
- Student Study Pack Stand alone ISBN: 0-13-147894-X

MyMathLab is a text-specific, online course that provides unlimited tutorial exercises correlated to the exercises in the text, a multimedia textbook with links to learning aids, such as videos and student supplements in electronic format—the student solutions manual, PowerPoint Presentations and much more! Since 2001, over one million students have done better in math with MyMathLab's dependable and easy-to-use online homework, guided solutions, multimedia, tests, and e-books.

## MathXL®

- 12-month MathXL and Text Bundle ISBN: 0-13-238723-9
- 24-month MathXL and Text Bundle ISBN: 0-13-238722-0
- 12-month Stand-alone ISBN: 0-13-147892-3
- 24-month Stand-alone ISBN: 0-13-147897-4

MathXL offers unlimited self-paced tutorials; practice guided mathematical instructions, and multimedia learning aids. MathXL allows students to learn and practice the concepts so they will succeed. MathXL also creates personalized study plans based on the students test results and offers unlimited (algorithmically generated) practice exercises correlated directly to the exercises in the textbook.

# Instructor Resources

## Instructor Resource Center

All instructor resources can be downloaded from www.prenhall.com (then search for Goldstein, *Calculus*, and click on "Instructor Resources" in left column). This is a password protected site that requires instructors to set up an account or, alternatively, instructor resources can be ordered from your Prentice Hall sales representative.

**Instructor's Solutions Manual** (ISBN: 0-13-191964-4)
Fully worked solutions to every textbook exercise.

**TestGen** (ISBN: 0-13-191968-7)
Test-generating software—create tests from textbook section objectives. Questions are algorithmically generated, allowing for unlimited test versions. Instructors may also edit problems or create their own.

**Test Item File** (ISBN: 0-13-188671-1)
A printed test bank derived from TestGen.

**PowerPoint Lecture Slides**
Fully editable and printable slides that follow the textbook. Use during lecture or post to a Web site in an online course. For download from the Instructor Resource Center and available within MyMathLab.

**MyMathLab**
Powered by CourseCompass, the course management system, and MathXL, the online homework, tutorial and assessment system, MyMathLab is a text-specific, customizable course that allows you to deliver all or a portion of your course online. You can create and assign online homework and tests that are automatically graded and tightly correlated to the text. Manage you students' results in the online gradebook—designed specifically for mathematics—which automatically tracks students' homework and test results and gives you control over how to calculate final grades. You can customize your course to meet your course objectives. In addition to the e-book, you gain access to all instructor and student supplements electronically—PowerPoint Presentations, solutions manuals, videos, applets, etc. For added flexibility, instructors can import TestGen tests. Request an access code from your local PH Sales Representative.

## MathXL®

Offered as a stand-alone product or as a component of MyMathLab, MathXL is the powerful online homework, tutorial and assessment system designed to make teaching your course easier and help your students succeed! With MathXL, instructors can create, edit, and assign online homework and tests and track all student work in the online class gradebook.

## Visual Calculus

Built from the ground up by David Schneider (University of Maryland) the Visual Calculus software enhances the visualization and therefore the understanding of the *ideas* behind the calculus. With this software students can: analyze a function and its derivatives, examine solutions of differential equations, graph functions, calculate definite integrals, evaluate finite and infinite sums, graph Taylor approximations, plot cosine and sine functions and much more. This software adds further value for you for constructing graphs for exams. Find Visual Calculus in MyMathLab and the CD Lecture Videos.

# Introduction

OFTEN it is possible to give a succinct and revealing description of a situation by drawing a graph. For example, Fig. 1 describes the amount of money in a bank account drawing 5% interest, compounded daily. The graph shows that as time passes, the amount of money in the account grows. In Fig. 2 we have drawn a graph that depicts the weekly sales of a breakfast cereal at various times after advertising has ceased. The graph shows that the longer the time since the last advertisement, the fewer the sales. Figure 3 shows the size of a bacteria culture at various times. The culture grows larger as time passes. But there is a maximum size that the culture cannot exceed. This maximum size reflects the restrictions imposed by food supply, space, and similar factors. The graph in Fig. 4 describes the decay of the radioactive isotope iodine 131. As time passes, less and less of the original radioactive iodine remains.

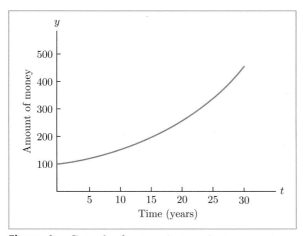

**Figure 1.**  Growth of money in a savings account.

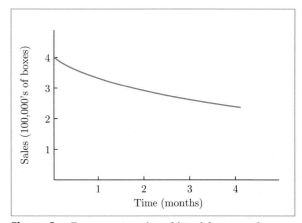

**Figure 2.**  Decrease in sales of breakfast cereal.

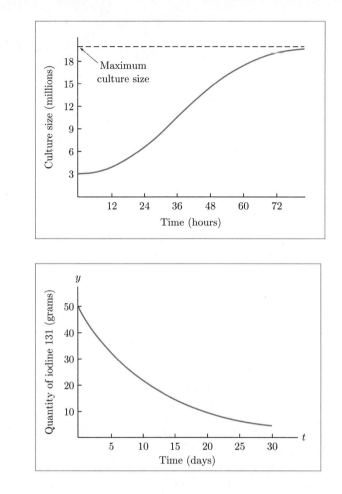

**Figure 3.** Growth of a bacteria culture.

**Figure 4.** Decay of radioactive iodine.

Each of the graphs in Figs. 1 to 4 describes a change that is taking place. The amount of money in the bank is changing as are the sales of cereal, the size of the bacteria culture, and the amount of the iodine. Calculus provides mathematical tools to study each of these changes in a quantitative way.

# Functions

ACH of the graphs in Figs. 1 to 4 of the Introduction depicts a relationship between two quantities. For example, Fig. 4 illustrates the relationship between the quantity of iodine (measured in grams) and time (measured in days). The basic quantitative tool for describing such relationships is a *function*. In this preliminary chapter, we develop the concept of a function and review important algebraic operations on functions used later in the text.

## 0.1 Functions and Their Graphs

### Real Numbers

Most applications of mathematics use real numbers. For purposes of such applications (and the discussions in this text), it suffices to think of a real number as a decimal. A *rational* number is one that may be written as a finite or infinite repeating decimal, such as

$$-\frac{5}{2} = -2.5, \qquad 1, \qquad \frac{13}{3} = 4.333\ldots \qquad \text{(rational numbers)}.$$

An *irrational* number has an infinite decimal representation whose digits form no repeating pattern, such as

$$-\sqrt{2} = -1.414214\ldots, \qquad \pi = 3.14159\ldots \qquad \text{(irrational numbers)}.$$

The real numbers are described geometrically by a *number line*, as in Fig. 1. Each number corresponds to one point on the line, and each point determines one real number.

**Figure 1.** The real number line.

We use four types of inequalities to compare real numbers.

$$x < y \qquad x \text{ is less than } y$$
$$x \leq y \qquad x \text{ is less than or equal to } y$$
$$x > y \qquad x \text{ is greater than } y$$
$$x \geq y \qquad x \text{ is greater than or equal to } y$$

The double inequality $a < b < c$ is shorthand for the pair of inequalities $a < b$ and $b < c$. Similar meanings are assigned to other double inequalities, such as $a \leq b < c$. Three numbers in a double inequality, such as $1 < 3 < 4$ or $4 > 3 > 1$, should have the same relative positions on the number line as in the inequality (when read left to right or right to left). Thus $3 < 4 > 1$ is never written because the numbers are "out of order."

Geometrically, the inequality $x \leq b$ means that either $x$ equals $b$ or $x$ lies to the left of $b$ on the number line. The set of real numbers $x$ that satisfy the double inequality $a \leq x \leq b$ corresponds to the line segment between $a$ and $b$, including the endpoints. This set is sometimes denoted by $[a, b]$ and is called the *closed interval* from $a$ to $b$. If $a$ and $b$ are removed from the set, the set is written as $(a, b)$ and is called the *open interval* from $a$ to $b$. The notation for various line segments is listed in Table 1.

The symbols $\infty$ ("infinity") and $-\infty$ ("minus infinity") do not represent actual real numbers. Rather, they indicate that the corresponding line segment extends infinitely far to the right or left. An inequality that describes such an infinite interval may be written in two ways. For instance, $a \leq x$ is equivalent to $x \geq a$.

| TABLE 1 | Intervals on the Number Line | |
|---|---|---|
| **Inequality** | **Geometric Description** | **Interval Notation** |
| $a \leq x \leq b$ | | $[a, b]$ |
| $a < x < b$ | | $(a, b)$ |
| $a \leq x < b$ | | $[a, b)$ |
| $a < x \leq b$ | | $(a, b]$ |
| $a \leq x$ | | $[a, \infty)$ |
| $a < x$ | | $(a, \infty)$ |
| $x \leq b$ | | $(-\infty, b]$ |
| $x < b$ | | $(-\infty, b)$ |

**EXAMPLE 1**   Describe each of the following intervals both graphically and in terms of inequalities.

(a) $(-1, 2)$     (b) $[-2, \pi]$     (c) $(2, \infty)$     (d) $(-\infty, \sqrt{2}\,]$

**Solution**   The line segments corresponding to the intervals are shown in Figs. 2(a)–(d). Note that an interval endpoint that is included (e.g., both endpoints of $[a, b]$) is drawn as a solid circle, whereas an endpoint not included (e.g., the endpoint $a$ in $(a, b]$) is drawn as an unfilled circle. ◼

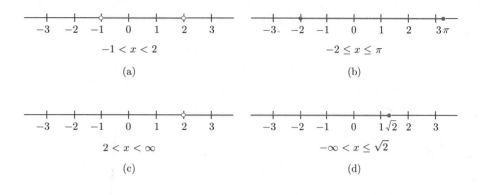

**Figure 2.**   Line segments.

---

**EXAMPLE 2**   The variable $x$ describes the profit that a company is anticipated to earn in the current fiscal year. The business plan calls for a profit of at least 5 million dollars. Describe this aspect of the business plan in the language of intervals.

**Solution**   The phrase "at least" means "greater than or equal to." The business plan requires that $x \geq 5$ (where the units are millions of dollars). This is equivalent to saying that $x$ lies in the infinite interval $[5, \infty)$. ◼

**Functions**   A *function* of a variable $x$ is a *rule* $f$ that assigns to each value of $x$ a unique number $f(x)$, called *the value of the function at $x$*. [We read "$f(x)$" as "$f$ of $x$."] The variable $x$ is called the *independent variable*. The set of values that the independent variable is allowed to assume is called the *domain* of the function. The domain of a function may be explicitly specified as part of the definition of a function or it may be understood from context. (See the following discussion.) The *range* of a function is the set of values that the function assumes.

The functions we shall meet in this book will usually be defined by algebraic formulas. For example, the domain of the function

$$f(x) = 3x - 1$$

consists of all real numbers $x$. This function is the rule that takes a number, multiplies it by 3, and then subtracts 1. If we specify a value of $x$, say $x = 2$, then we find the value of the function at 2 by substituting 2 for $x$ in the formula:

$$f(2) = 3(2) - 1 = 5.$$

---

**EXAMPLE 3**   Let $f$ be the function with domain all real numbers $x$ and defined by the formula

$$f(x) = 3x^3 - 4x^2 - 3x + 7.$$

Find $f(2)$ and $f(-2)$.

**Solution** To find $f(2)$ we substitute 2 for every occurrence of $x$ in the formula for $f(x)$:

$$f(2) = 3(2)^3 - 4(2)^2 - 3(2) + 7$$
$$= 3(8) - 4(4) - 3(2) + 7$$
$$= 24 - 16 - 6 + 7$$
$$= 9.$$

To find $f(-2)$ we substitute $(-2)$ for each occurrence of $x$ in the formula for $f(x)$. The parentheses ensure that the $-2$ is substituted correctly. For instance, $x^2$ must be replaced by $(-2)^2$, not $-2^2$.

$$f(-2) = 3(-2)^3 - 4(-2)^2 - 3(-2) + 7$$
$$= 3(-8) - 4(4) - 3(-2) + 7$$
$$= -24 - 16 + 6 + 7$$
$$= -27$$ ■

**EXAMPLE 4**    **Temperature scales**  If $x$ represents the temperature of an object in degrees Celsius, then the temperature in degrees Fahrenheit is a function of $x$, given by $f(x) = \frac{9}{5}x + 32$.

(a) Water freezes at $0°$C (C = Celsius) and boils at $100°$C. What are the corresponding temperatures in degrees Fahrenheit?

(b) Aluminum melts at $660°$C. What is its melting point in degrees Fahrenheit?

**Solution** (a) $f(0) = \frac{9}{5}(0) + 32 = 32$. Water freezes at $32°$F.
$f(100) = \frac{9}{5}(100) + 32 = 180 + 32 = 212$.
Water boils at $212°$F.

(b) $f(660) = \frac{9}{5}(660) + 32 = 1188 + 32 = 1220$. Aluminum melts at $1220°$F. ■

**EXAMPLE 5**    **A voting model**  Let $x$ be the proportion of the total popular vote that a Democratic candidate for president receives in a U.S. national election (so $x$ is a number between 0 and 1). Political scientists have observed that a good estimate of the proportion of seats in the House of Representatives going to Democratic candidates is given by the function

$$f(x) = \frac{x^3}{x^3 + (1-x)^3}, \qquad 0 \le x \le 1,$$

whose domain is the interval $[0, 1]$. This formula is called the *cube law*. Compute $f(.6)$ and interpret the result.

**Solution** We must substitute .6 for every occurrence of $x$ in $f(x)$:

$$f(.6) = \frac{(.6)^3}{(.6)^3 + (1 - .6)^3} = \frac{(.6)^3}{(.6)^3 + (.4)^3}$$

$$= \frac{.216}{.216 + .064} = \frac{.216}{.280} \approx .77.$$

This calculation shows that the cube law function predicts that if .6 (or 60%) of the total popular vote is for the Democratic candidate for president, then approximately .77 (or 77%) of the seats in the House of Representatives will be won by Democratic candidates; that is, about 335 of the 435 seats will be won by Democrats. Note that for $x = .5$, we have $f(.5) = .5$. Is this what you would expect? ■

In the preceding examples, the functions had domains consisting of all real numbers or an interval. For some functions, the domain may consist of several intervals, with a different formula defining the function on each interval. Here is an illustration of this phenomenon.

---

**EXAMPLE 6**  A leading brokerage firm charges a 6% commission on gold purchases in amounts from $50 to $300. For purchases exceeding $300, the firm charges 2% of the amount purchased plus $12.00. Let $x$ denote the amount of gold purchased (in dollars) and let $f(x)$ be the commission charge as a function of $x$.

(a) Describe $f(x)$.

(b) Find $f(100)$ and $f(500)$.

**Solution**  (a) The formula for $f(x)$ depends on whether $50 \leq x \leq 300$ or $300 < x$. When $50 \leq x \leq 300$, the charge is $.06x$ dollars. When $300 < x$, the charge is $.02x + 12$. The domain consists of the values $x$ in one of the two intervals $[50, 300]$ and $(300, \infty)$. In each of these intervals, the function is defined by a separate formula:

$$f(x) = \begin{cases} .06x & \text{for } 50 \leq x \leq 300 \\ .02x + 12 & \text{for } 300 < x. \end{cases}$$

Note that an alternate description of the domain is the interval $[50, \infty)$. That is, the value of $x$ may be any real number greater than or equal to 50.

(b) Since $x = 100$ satisfies $50 \leq x \leq 300$, we use the first formula for $f(x)$: $f(100) = .06(100) = 6$. Since $x = 500$ satisfies $300 < x$, we use the second formula for $f(x)$: $f(500) = .02(500) + 12 = 22$. ■

In calculus, it is often necessary to substitute an algebraic expression for $x$ and simplify the result, as illustrated in the following example.

---

**EXAMPLE 7**  If $f(x) = (4 - x)/(x^2 + 3)$, what is $f(a)$? $f(a + 1)$?

**Solution**  Here $a$ represents some number. To find $f(a)$, we substitute $a$ for $x$ wherever $x$ appears in the formula defining $f(x)$:

$$f(a) = \frac{4 - a}{a^2 + 3}.$$

To evaluate $f(a + 1)$, substitute $a + 1$ for each occurrence of $x$ in the formula for $f(x)$:

$$f(a + 1) = \frac{4 - (a + 1)}{(a + 1)^2 + 3}.$$

The expression for $f(a + 1)$ may be simplified, using the fact that $(a + 1)^2 = (a + 1)(a + 1) = a^2 + 2a + 1$:

$$f(a + 1) = \frac{4 - (a + 1)}{(a + 1)^2 + 3} = \frac{4 - a - 1}{a^2 + 2a + 1 + 3} = \frac{3 - a}{a^2 + 2a + 4}.$$ ■

**More about the Domain of a Function**  When defining a function, it is necessary to specify the domain of the function, which is the set of acceptable values of the variable. In the preceding examples, we explicitly specified the domains of the functions considered. However, throughout the remainder of the text, we will usually mention functions without specifying domains. In such circumstances, we will understand the intended domain to consist of all numbers for which the defining formula(s) make sense. For example, consider the function

$$f(x) = x^2 - x + 1.$$

The expression on the right may be evaluated for any value of $x$. So in the absence of any explicit restrictions on $x$, the domain is understood to consist of all numbers. As a second example, consider the function

$$f(x) = \frac{1}{x}.$$

Here $x$ may be any number except zero. (Division by zero is not permissible.) So the domain intended is the set of nonzero numbers. Similarly, when we write

$$f(x) = \sqrt{x},$$

we understand the domain of $f(x)$ to be the set of all nonnegative numbers, since the square root of a number $x$ is defined if and only if $x \geq 0$.

**Graphs of Functions**   Often it is helpful to describe a function $f$ geometrically, using a rectangular $xy$-coordinate system. Given any $x$ in the domain of $f$, we can plot the point $(x, f(x))$. This is the point in the $xy$-plane whose $y$-coordinate is the value of the function at $x$. The set of *all* such points $(x, f(x))$ usually forms a curve in the $xy$-plane and is called the *graph of the function* $f(x)$.

It is possible to approximate the graph of $f(x)$ by plotting the points $(x, f(x))$ for a representative set of values of $x$ and joining them by a smooth curve. (See Fig. 3.) The more closely spaced the values of $x$, the closer the approximation.

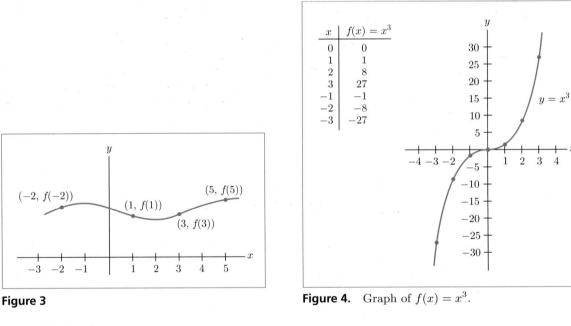

**Figure 3**

**Figure 4.**   Graph of $f(x) = x^3$.

---

**EXAMPLE 8**    Sketch the graph of the function $f(x) = x^3$.

**Solution**   The domain consists of all numbers $x$. We choose some representative values of $x$ and tabulate the corresponding values of $f(x)$. We then plot the points $(x, f(x))$ and draw a smooth curve through the points. (See Fig. 4.)    ∎

**EXAMPLE 9**  Sketch the graph of the function $f(x) = 1/x$.    $x$ ✓ cannot be zero

**Solution**  The domain of the function consists of all numbers except zero. The table in Fig. 5 lists some representative values of $x$ and the corresponding values of $f(x)$. A function often has interesting behavior for $x$ near a number not in the domain. So when we chose representative values of $x$ from the domain, we included some values close to zero. The points $(x, f(x))$ are plotted and the graph sketched in Fig. 5.  ■

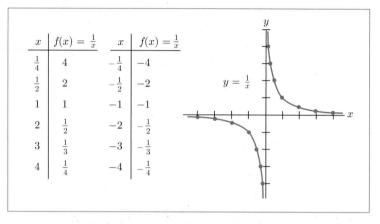

| $x$ | $f(x) = \frac{1}{x}$ | $x$ | $f(x) = \frac{1}{x}$ |
|---|---|---|---|
| $\frac{1}{4}$ | 4 | $-\frac{1}{4}$ | $-4$ |
| $\frac{1}{2}$ | 2 | $-\frac{1}{2}$ | $-2$ |
| 1 | 1 | $-1$ | $-1$ |
| 2 | $\frac{1}{2}$ | $-2$ | $-\frac{1}{2}$ |
| 3 | $\frac{1}{3}$ | $-3$ | $-\frac{1}{3}$ |
| 4 | $\frac{1}{4}$ | $-4$ | $-\frac{1}{4}$ |

**Figure 5.**  Graph of $f(x) = \frac{1}{x}$.

Now that graphing calculators and computer graphing programs are widely available, one seldom needs to sketch graphs by hand-plotting large numbers of points on graph paper. However, to use such a calculator or program effectively, one must know in advance which part of a curve to display. Critical features of a graph may be missed or misinterpreted if, for instance, the scale on the $x$- or $y$-axis is inappropriate.

An important use of calculus is to identify key features of a function that should appear in its graph. In many cases, only a few points need be plotted and the general shape of the graph is easy to sketch by hand. For more complicated functions, a graphing program is helpful. Even then, calculus provides a way of checking that the graph on the computer screen has the correct shape. Algebraic calculations are usually part of the analysis. The appropriate algebraic skills are reviewed in this chapter.

Note also that analytic solutions to problems typically provide more precise information than graphing calculators and can provide insight into the behavior of the functions involved in the solution.

The connection between a function and its graph is explored in this section and in Section 0.6.

**EXAMPLE 10**  Suppose that $f$ is the function whose graph is given in Fig. 6. Notice that the point $(x, y) = (3, 2)$ is on the graph of $f$.

(a) What is the value of the function when $x = 3$?

(b) Find $f(-2)$.

(c) What is the domain of $f$?

**Solution**  (a) Since $(3, 2)$ is on the graph of $f$, the $y$-coordinate 2 must be the value of $f$ at the $x$-coordinate 3. That is, $f(3) = 2$.

(b) To find $f(-2)$ we look at the $y$-coordinate of the point on the graph where $x = -2$. From Fig. 6 we see that $(-2, 1)$ is on the graph of $f$. Thus $f(-2) = 1$.

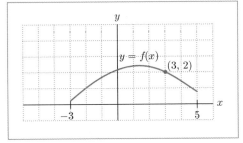

**Figure 6**

(c) The points on the graph of $f(x)$ all have $x$-coordinates between $-3$ and $5$ inclusive; and for each value of $x$ between $-3$ and $5$ there is a point $(x, f(x))$ on the graph. So the domain consists of those $x$ for which $-3 \le x \le 5$. ■

To every $x$ in the domain, a function assigns one and only one value of $y$, namely, the function value $f(x)$. This implies, among other things, that not every curve is the graph of a function. To see this, refer first to the curve in Fig. 6, which *is* the graph of a function. It has the following important property: For each $x$ between $-3$ and $5$ inclusive there is a *unique* $y$ such that $(x, y)$ is on the curve. The variable $y$ is called the *dependent variable*, since its value depends on the value of the independent variable $x$. Refer to the curve in Fig. 7. It cannot be the graph of a function because a function $f$ must assign to each $x$ in its domain a *unique* value $f(x)$. However, for the curve of Fig. 7 there corresponds to $x = 3$ (for example) more than one $y$-value, namely, $y = 1$ and $y = 4$.

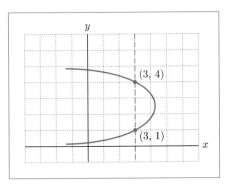

**Figure 7.**  A curve that is **not** the graph of a function.

The essential difference between the curves in Figs. 6 and 7 leads us to the following test.

**The Vertical Line Test**    A curve in the $xy$-plane is the graph of a function if and only if each vertical line cuts or touches the curve at no more than one point.

**EXAMPLE 11**    Which of the curves in Fig. 8 are graphs of functions?

**Solution**  The curve in (a) is the graph of a function. It appears that vertical lines to the left of the $y$-axis do not touch the curve at all. This simply means that the function represented in (a) is defined only for $x \ge 0$. The curve in (b) is *not* the graph of a function because some vertical lines cut the curve in three places. The curve in (c) is the graph of a function whose domain is all nonzero $x$. [There is no point on the curve in (c) whose $x$-coordinate is 0.] ■

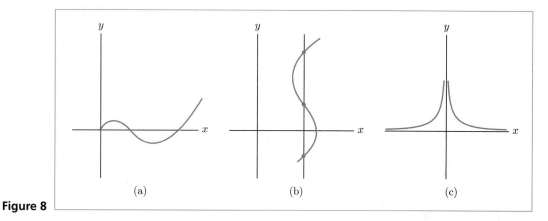

(a)    (b)    (c)

**Figure 8**

There is another notation for functions that we will find useful. Suppose that $f(x)$ is a function. When $f(x)$ is graphed on an $xy$-coordinate system, the values of $f(x)$ give the $y$-coordinates of points of the graph. For this reason, the function is often abbreviated by the letter $y$, and we find it convenient to speak of "the function $y = f(x)$." For example, the function $y = 2x^2 + 1$ refers to the function $f(x)$ for which $f(x) = 2x^2 + 1$. The graph of a function $f(x)$ is often called *the graph of the equation $y = f(x)$*.

**Graphing Functions Using Technology**  As you might have already inferred from the preceding examples, sketching the graphs of functions by plotting points can be a tedious procedure. However, it is one that can be carried out using technology—graphing calculators and mathematical software.* Any of these tools will produce accurate graphs of functions. Most of the technology discussed in this text will concern graphing calculators, the most widely available tool for students. Although graphing calculators are not required for study from this text, we will show how they can be used to simplify computations and enhance understanding of the fundamental topics of calculus. Helpful information about calculators and technology will appear at the ends of most sections in subsections entitled "Incorporating Technology."

Actual keystrokes for calculators are not discussed in the main body of the text, but Appendices A through D provide a detailed reference for the keystrokes needed to carry out most calculus tasks on the TI-82, TI-83[†], TI-85, and TI-86 calculators. A TI-83 generated most of the calculator screens shown in the text.

To graph a function using a graphing calculator, you must perform the following steps:

1. Enter the expression for the function.

2. Enter the specifications for the viewing window.

3. Display the graph.

Let's examine each step in turn.

Functions are normally defined in the function editor (sometimes referred to as the **Y=** editor) and often are given names such as $\mathbf{Y_1}, \mathbf{Y_2}, \mathbf{Y_3}, \ldots$. Figure 9 shows the

---

*Powerful software for calculus includes Derive, Maple, *Mathematica*, MathCAD, and Visual Calculus. A customized copy of Visual Calculus is supplied with the *Study Guide* for this book.

[†]The designation TI-83 refers to the TI-83, TI-83+, and the TI-84+ family of calculators.

calculator expressions for the functions

$$\mathbf{Y_1} = \mathbf{x}^2 - 2\mathbf{x} - 1, \quad \mathbf{Y_2} = \mathbf{x}^{1/3}, \quad \mathbf{Y_3} = \frac{\mathbf{x}-1}{\mathbf{x}+1}, \quad \text{and} \quad \mathbf{Y_4} = -\sqrt{\mathbf{x} - \frac{1}{\mathbf{x}}}.$$

**Figure 9**

Powers are denoted by a caret (^). With some calculators, multiplication must be explicitly indicated by *. Also, you must use parentheses wherever necessary to clarify the mathematical meaning of an expression. The two minus signs in $\mathbf{Y_4}$ are entered with different keys—the first signifies negation and the second signifies subtraction. In Fig. 9, the highlighted equals sign for $\mathbf{Y_1}$ indicates that this function has been selected to be graphed.

The parameters for the viewing window are set from a screen known as the WINDOW or RANGE screen. Figure 10(a) shows the WINDOW screen for the TI-83 graphing calculator and Fig. 10(b) shows the meanings of the parameters. We write **[a,b]** *by* **[c,d]** to denote the viewing window with setting **Xmin = a**, **Xmax = b**, **Ymin = c**, and **Ymax = d**. One of the most important tasks in using a graphing calculator is to determine the viewing window that shows the features of interest. To determine these values often requires calculus, as we shall learn in later chapters. For now, we will either use the default settings or determine appropriate settings by experimentation.

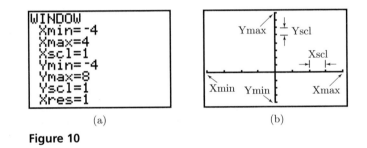

(a)                    (b)

**Figure 10**

After the viewing window has been specified, the graph can be displayed with the press of a single key. Figure 11(a) shows the graph of $\mathbf{Y_1}$ using the settings of Fig. 10(a). The trace feature (or the value or **EVAL** commands) can then be used to determine the coordinates of points on the graph. See Fig. 11(b). If desired, the settings for the viewing window can be altered to see more of the graph or to zoom in on a portion of the graph.

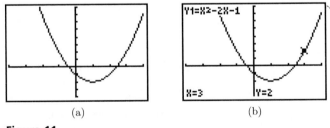

(a)                    (b)

**Figure 11**

### Graphs of Equations
The equations arising in connection with functions are all of the form

$$y = [\text{an expression in } x].$$

However, not all equations connecting the variables $x$ and $y$ are of this sort. For example, consider these equations:

$$2x + 3y = 5$$
$$x = 3$$
$$x^2 + y^2 = 1.$$

It is possible to graph an equation by plotting points just as for functions. The only difference is that the resulting graph may not satisfy the vertical line test. For example, the graphs of the three preceding equations are shown in Fig. 12. Only the first graph is the graph of a function.

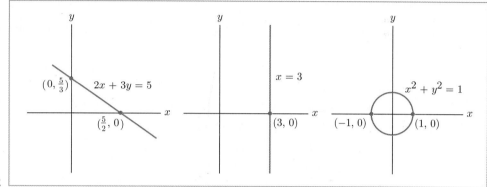

**Figure 12**

Graphing calculators cannot display the graphs of all equations. They can only display graphs of equations you can solve for $y$ in terms of $x$. For instance, in order to graph the first of the preceding equations, it is necessary to first solve for $y$ in terms of $x$:

$$2x + 3y = 5$$
$$3y = 5 - 2x$$
$$y = \frac{5}{3} - \frac{2}{3}x.$$

The second equation cannot be solved for $y$ in terms of $x$ and so cannot be graphed using a graphing calculator. For the third equation, we may solve for $y$ in terms of $x$ to obtain

$$x^2 + y^2 = 1$$
$$y^2 = 1 - x^2$$
$$y = \pm\sqrt{1 - x^2}.$$

The $\pm$ indicates that the last equation is really two equations, one corresponding to each choice of sign. On a graphing calculator, we must enter each of these equations:

$$y = \sqrt{1 - x^2}, \qquad y = -\sqrt{1 - x^2}.$$

The calculator will draw the graphs of both equations, which together comprise the graph of the original equation.

Letters other than $f$ can be used to denote functions and letters other than $x$ and $y$ can be used to denote variables. This is especially common in applied problems where letters are chosen to suggest the quantities they depict. For instance, the revenue of a company as a function of time might be written $R(t)$.

## PRACTICE PROBLEMS 0.1

**1.** Is the point $(3, 12)$ on the graph of the function $g(x) = x^2 + 5x - 10$?

**2.** Sketch the graph of the function $h(t) = t^2 - 2$.

## EXERCISES 0.1*

*Draw the following intervals on the number line.*

**1.** $[-1, 4]$    **2.** $(4, 3\pi)$    **3.** $[-2, \sqrt{2})$

**4.** $[1, \frac{3}{2}]$    **5.** $(-\infty, 3)$    **6.** $(4, \infty)$

*Use intervals to describe the real numbers satisfying the inequalities in Exercises 7–12.*

**7.** $2 \le x < 3$    **8.** $-1 < x < \frac{3}{2}$

**9.** $x < 0$ and $x \ge -1$    **10.** $x \ge -1$ and $x < 8$

**11.** $x < 3$    **12.** $x \ge \sqrt{2}$

**13.** If $f(x) = x^2 - 3x$, find $f(0)$, $f(5)$, $f(3)$, and $f(-7)$.

**14.** If $f(x) = 9 - 6x + x^2$, find $f(0)$, $f(2)$, $f(3)$, and $f(-13)$.

**15.** If $f(x) = x^3 + x^2 - x - 1$, find $f(1)$, $f(-1)$, $f(\frac{1}{2})$, and $f(a)$.

**16.** If $g(t) = t^3 - 3t^2 + t$, find $g(2)$, $g(-\frac{1}{2})$, $g(\frac{2}{3})$, and $g(a)$.

**17.** If $h(s) = s/(1 + s)$, find $h(\frac{1}{2})$, $h(-\frac{3}{2})$, and $h(a + 1)$.

**18.** If $f(x) = x^2/(x^2 - 1)$, find $f(\frac{1}{2})$, $f(-\frac{1}{2})$, and $f(a + 1)$.

**19.** If $f(x) = x^2 - 2x$, find $f(a + 1)$ and $f(a + 2)$.

**20.** If $f(x) = x^2 + 4x + 3$, find $f(a - 1)$ and $f(a - 2)$.

**21.** An office supply firm finds that the number of fax machines sold in year $x$ is given approximately by the function $f(x) = 50 + 4x + \frac{1}{2}x^2$, where $x = 0$ corresponds to 1990.

(a) What does $f(0)$ represent?

(b) Find the number of fax machines sold in 1992.

**22.** (*Response of a Muscle*) When a solution of acetylcholine is introduced into the heart muscle of a frog, it diminishes the force with which the muscle contracts. The data from experiments of the biologist A. J. Clark are closely approximated by a function of the form

$$R(x) = \frac{100x}{b + x}, \qquad x \ge 0,$$

where $x$ is the concentration of acetylcholine (in appropriate units), $b$ is a positive constant that depends on the particular frog, and $R(x)$ is the response of the muscle to the acetylcholine, expressed as a percentage of the maximum possible effect of the drug.

(a) Suppose that $b = 20$. Find the response of the muscle when $x = 60$.

(b) Determine the value of $b$ if $R(50) = 60$—that is, if a concentration of $x = 50$ units produces a 60% response.

*In Exercises 23–26, describe the domain of the function.*

**23.** $f(x) = \dfrac{8x}{(x - 1)(x - 2)}$    **24.** $f(t) = \dfrac{1}{\sqrt{t}}$

**25.** $g(x) = \dfrac{1}{\sqrt{3 - x}}$    **26.** $g(x) = \dfrac{4}{x(x + 2)}$

*In Exercises 27–32, decide which curves are graphs of functions.*

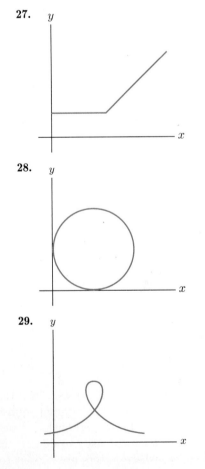

**27.**

**28.**

**29.**

---

**30.**

**31.**

**32.**

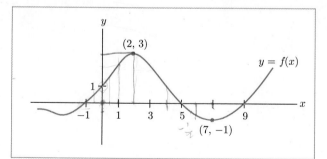

**Figure 13**

**33.** Find $f(0)$.

**34.** Find $f(7)$.

**35.** Find $f(2)$.

**36.** Find $f(-1)$.

**37.** Is $f(4)$ positive or negative?

**38.** Is $f(6)$ positive or negative?

**39.** Is $f(-\frac{1}{2})$ positive or negative?

**40.** Is $f(1)$ greater than $f(6)$?

**41.** For what values of $x$ is $f(x) = 0$?

**42.** For what values of $x$ is $f(x) \geq 0$?

*Exercises 43–46 relate to Fig. 14. When a drug is injected into a person's muscle tissue, the concentration $y$ of the drug in the blood is a function of the time elapsed since the injection. The graph of a typical time-concentration function $f$ is given in Fig. 14, where $t = 0$ corresponds to the time of the injection.*

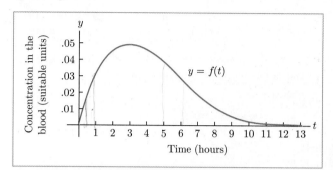

**Figure 14.** Drug time-concentration curve.

**43.** What is the concentration of the drug when $t = 1$?

**44.** What is the value of the time-concentration function $f$ when $t = 6$?

**45.** Find $f(5)$.

**46.** At what time does $f(t)$ attain its largest value?

**47.** Is the point $(3, 12)$ on the graph of the function $f(x) = (x - \frac{1}{2})(x + 2)$?

**48.** Is the point $(-2, 12)$ on the graph of the function $f(x) = x(5 + x)(4 - x)$?

**49.** Is the point $(\frac{1}{2}, \frac{2}{5})$ on the graph of the function $g(x) = (3x - 1)/(x^2 + 1)$?

**50.** Is the point $(\frac{2}{3}, \frac{5}{3})$ on the graph of the function $g(x) = (x^2 + 4)/(x + 2)$?

**51.** Find the $y$-coordinate of the point $(a + 1, \blacksquare)$ if this point lies on the graph of the function $f(x) = x^3$.

**52.** Find the $y$-coordinate of the point $(2 + h, \blacksquare)$ if this point lies on the graph of the function $f(x) = (5/x) - x$.

*In Exercises 53–56, compute $f(1)$, $f(2)$, and $f(3)$.*

**53.** $f(x) = \begin{cases} \sqrt{x} & \text{for } 0 \leq x < 2 \\ 1 + x & \text{for } 2 \leq x \leq 5 \end{cases}$

**54.** $f(x) = \begin{cases} 1/x & \text{for } 1 \leq x \leq 2 \\ x^2 & \text{for } 2 < x \end{cases}$

**55.** $f(x) = \begin{cases} \pi x^2 & \text{for } x < 2 \\ 1 + x & \text{for } 2 \leq x \leq 2.5 \\ 4x & \text{for } 2.5 < x \end{cases}$

**56.** $f(x) = \begin{cases} 3/(4 - x) & \text{for } x < 2 \\ 2x & \text{for } 2 \leq x < 3 \\ \sqrt{x^2 - 5} & \text{for } 3 \leq x \end{cases}$

**57.** Suppose that the brokerage firm in Example 6 decides to keep the commission charges unchanged for purchases up to and including $600 but to charge only 1.5% plus $15 for gold purchases exceeding $600. Express the brokerage commission as a function of the amount $x$ of gold purchased.

**58.** Figure 15(a) shows the number 2 on the $x$-axis and the graph of a function. Let $h$ represent a positive number and label a possible location for the number $2 + h$. Plot the point on the graph whose first coordinate is $2 + h$ and label the point with its coordinates.

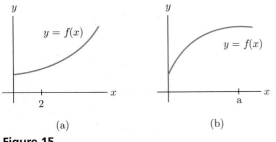

(a)                (b)

**Figure 15**

**59.** Figure 15(b) shows the number $a$ on the $x$-axis and the graph of a function. Let $h$ represent a negative number and label a possible location for the number $a + h$. Plot the point on the graph whose first coordinate is $a + h$ and label the point with its coordinates.

### Technology Exercises

**60.** What is wrong with entering the function $f(x) = \dfrac{1}{x+1}$ into a graphing utility as $\mathbf{Y_1 = 1/X + 1}$?

**61.** What is wrong with entering the function $f(x) = x^{3/4}$ into a graphing utility as $\mathbf{Y_1 = X^\wedge 3/4}$?

*In Exercises 62–65 graph the function with the specified viewing window setting.*

**62.** $f(x) = x^3 - 33x^2 + 120x + 1500;$
$[-8, 30]$ *by* $[-2000, 2000]$.

**63.** $f(x) = -x^2 + 2x + 2;$ $[-2, 4]$ *by* $[-8, 5]$

**64.** $f(x) = \sqrt{x+1};$ $[0, 10]$ *by* $[-1, 4]$

**65.** $f(x) = \dfrac{1}{x^2 + 1};$ $[-4, 4]$ *by* $[-.5, 1.5]$

*If your graphing utility can display tables of function values, in Exercises 66 and 67 create a table similar to the one shown. Then use the table to determine the value of the function when* $\mathbf{X} = 7$*. (Appendices A, B, and D explain how to create tables with the TI-82, TI-83, and TI-86 calculators.)*

**66.**

$Y_1 = X^2 - 2X - 1$

**67.**

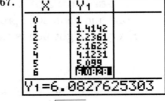

$Y_1 = \sqrt{1 + x^2}$

---

### SOLUTIONS TO PRACTICE PROBLEMS 0.1

**1.** If $(3, 12)$ is on the graph of $g(x) = x^2 + 5x - 10$, then we must have $g(3) = 12$. This is not the case, however, because

$$g(3) = 3^2 + 5(3) - 10$$
$$= 9 + 15 - 10 = 14.$$

Thus $(3, 12)$ is *not* on the graph of $g(x)$.

**2.** Choose some representative values for $t$, say $t = 0$, $\pm 1$, $\pm 2$, $\pm 3$. For each value of $t$, calculate $h(t)$ and plot the point $(t, h(t))$. See Fig. 16.

## SOLUTION TO PRACTICE PROBLEMS 0.1 (CONTINUED)

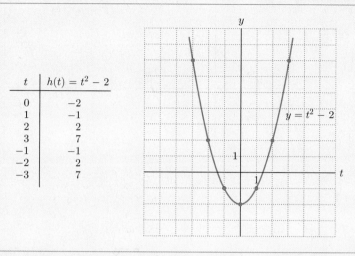

| $t$ | $h(t) = t^2 - 2$ |
|-----|------------------|
| 0   | −2 |
| 1   | −1 |
| 2   | 2 |
| 3   | 7 |
| −1  | −1 |
| −2  | 2 |
| −3  | 7 |

**Figure 16.**  Graph of $h(t) = t^2 - 2$.

## 0.2  Some Important Functions

In this section we introduce some of the functions that will play a prominent role in our discussion of calculus.

**Linear Functions**  As we shall see in the next chapter, a knowledge of the algebraic and geometric properties of straight lines is essential for the study of calculus. Every straight line is the graph of a linear equation of the form

$$cx + dy = e,$$

where $c$, $d$, and $e$ are given constants, with $c$ and $d$ not both zero. If $d \neq 0$, then we may solve the equation for $y$ to obtain an equation of the form

$$y = mx + b, \tag{1}$$

for appropriate numbers $m$ and $b$. If $d = 0$, then we may solve the equation for $x$ to obtain an equation of the form

$$x = a, \tag{2}$$

for an appropriate number $a$. So every straight line is the graph of an equation of type (1) or (2). The graph of an equation of the form (1) is a nonvertical line [Fig. 1(a)], whereas the graph of (2) is a vertical line [Fig. 1(b)].

The straight line of Fig. 1(a) is the graph of the function $f(x) = mx + b$. Such a function, which is defined for all $x$, is called a *linear function*. Note that the straight line of Fig. 1(b) is not the graph of a function, since the vertical line test is violated.

An important special case of a linear function occurs if the value of $m$ is zero, that is, $f(x) = b$ for some number $b$. In this case, $f(x)$ is called a *constant function*, since it assigns the same number $b$ to every value of $x$. Its graph is the horizontal line whose equation is $y = b$. (See Fig. 2.)

Linear functions often arise in real-life situations, as the first two examples show.

**Figure 1**

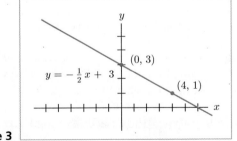

**Figure 2.** Graph of the constant function $f(x) = b$.

**EXAMPLE 1** When the U.S. Environmental Protection Agency found a certain company dumping sulfuric acid into the Mississippi River, it fined the company $125,000, plus $1000 per day until the company complied with federal water pollution regulations. Express the total fine as a function of the number $x$ of days the company continued to violate the federal regulations.

**Solution** The variable fine for $x$ days of pollution, at $1000 per day, is $1000x$ dollars. The total fine is therefore given by the function

$$f(x) = 125{,}000 + 1000x.$$ ◼

Since the graph of a linear function is a line, we may sketch it by locating any two points on the graph and drawing the line through them. For example, to sketch the graph of the function $f(x) = -\frac{1}{2}x + 3$, we may select two convenient values of $x$, say 0 and 4, and compute $f(0) = -\frac{1}{2}(0) + 3 = 3$ and $f(4) = -\frac{1}{2}(4) + 3 = 1$. The line through the points $(0, 3)$ and $(4, 1)$ is the graph of the function. (See Fig. 3.)

**Figure 3**

**EXAMPLE 2**    **Cost**   A simple cost function for a business consists of two parts—the *fixed costs*, such as rent, insurance, and business loans, which must be paid no matter how many items of a product are produced, and the *variable costs*, which depend on the number of items produced.

Suppose a computer software company produces and sells a new spreadsheet program at a cost of $25 per copy, and the company has fixed costs of $10,000 per month. Express the total monthly cost as a function of the number of copies sold, $x$, and compute the cost when $x = 500$.

**Solution**   The monthly variable cost is $25x$ dollars. Thus

$$[\text{total cost}] = [\text{fixed costs}] + [\text{variable costs}]$$
$$C(x) = 10{,}000 + 25x.$$

When sales are at 500 copies per month, the cost is

$$C(500) = 10{,}000 + 25(500) = 22{,}500 \ (\text{dollars}).$$

See Fig. 4.  ■

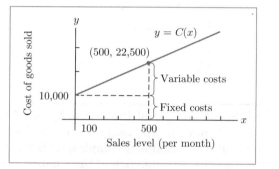

**Figure 4.**   A linear cost function.

The point at which the graph of a linear function intersects the $y$-axis is called the *y-intercept* of the graph. The point at which the graph intersects the $x$-axis is called the *x-intercept*. The next example shows how to determine the intercepts of a linear function.

**EXAMPLE 3**   Determine the intercepts of the graph of the linear function $f(x) = 2x + 5$.

**Solution**   Since the $y$-intercept is on the $y$-axis, its $x$-coordinate is 0. The point on the line with $x$-coordinate zero has $y$-coordinate

$$f(0) = 2(0) + 5 = 5.$$

So the $y$-intercept is $(0, 5)$. Since the $x$-intercept is on the $x$-axis, its $y$-coordinate is 0. Since $f(x)$ gives the $y$-coordinate, we must have

$$2x + 5 = 0$$
$$2x = -5$$
$$x = -\tfrac{5}{2}.$$

So $\left(-\tfrac{5}{2}, 0\right)$ is the $x$-intercept. (See Fig. 5.)  ■

The function in the next example is described by two expressions. Functions described by more than one expression are said to be *piecewise defined*.

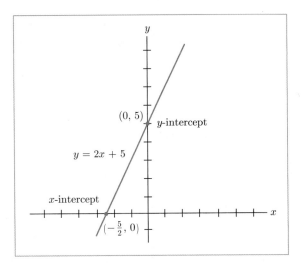

**Figure 5.** Graph of
$f(x) = 2x + 5$.

---

**EXAMPLE 4**    Sketch the graph of the following function.

$$f(x) = \begin{cases} \frac{5}{2}x - \frac{1}{2} & \text{for } -1 \leq x \leq 1 \\ \frac{1}{2}x - 2 & \text{for } x > 1. \end{cases}$$

**Solution**    This function is defined for $x \geq -1$. But it is defined by means of two distinct linear functions. We graph the two linear functions $\frac{5}{2}x - \frac{1}{2}$ and $\frac{1}{2}x - 2$. Then the graph of $f(x)$ consists of that part of the graph of $\frac{5}{2}x - \frac{1}{2}$ for which $-1 \leq x \leq 1$ plus that part of the graph of $\frac{1}{2}x - 2$ for which $x > 1$. (See Fig. 6.)    ■

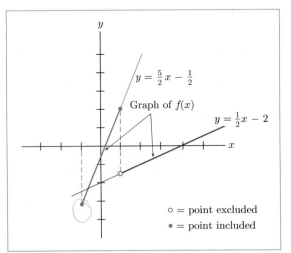

**Figure 6.** Graph of a function specified by two expressions.

**Quadratic Functions**    Economists utilize average cost curves that relate the average unit cost of manufacturing a commodity to the number of units to be produced. (See Fig. 7.) Ecologists use curves that relate the net primary production of nutrients in a plant to the surface area of the foliage. (See Fig. 8.) Each of the curves is bowl-shaped, opening either up or down. The simplest functions whose graphs resemble these curves are the quadratic functions.

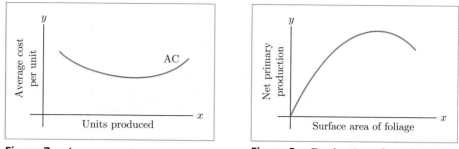

**Figure 7.**   Average cost curve.

**Figure 8.**   Production of nutrients.

A *quadratic function* is a function of the form

$$f(x) = ax^2 + bx + c,$$

where $a$, $b$, and $c$ are constants and $a \neq 0$. The domain of such a function consists of all numbers. The graph of a quadratic function is called a *parabola*. Two typical parabolas are drawn in Figs. 9 and 10. We shall develop techniques for sketching graphs of quadratic functions after we have some calculus at our disposal.

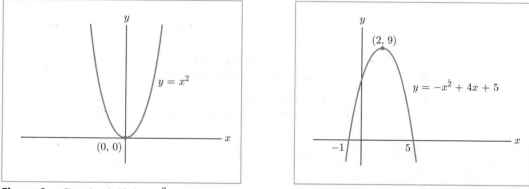

**Figure 9.**   Graph of $f(x) = x^2$.

**Figure 10.**   Graph of $f(x) = -x^2 + 4x + 5$.

### Polynomial and Rational Functions

A *polynomial function* $f(x)$ is one of the form

$$f(x) = a_n x^n + a_{n-1} x^{n-1} + \cdots + a_0,$$

where $n$ is a nonnegative integer and $a_0, a_1, \ldots, a_n$ are given numbers. Some examples of polynomial functions are

$$f(x) = 5x^3 - 3x^2 - 2x + 4$$
$$g(x) = x^4 - x + 1.$$

Of course, linear and quadratic functions are special cases of polynomial functions. The domain of a polynomial function consists of all numbers.

A function expressed as the quotient of two polynomials is called a *rational function*. Some examples are

$$h(x) = \frac{x^2 + 1}{x}$$

$$k(x) = \frac{x + 3}{x^2 - 4}.$$

The domain of a rational function excludes all values of $x$ for which the denominator is zero. For example, the domain of $h(x)$ excludes $x = 0$, whereas the domain of $k(x)$ excludes $x = 2$ and $x = -2$. As we shall see, both polynomial and rational functions arise in applications of calculus.

Rational functions are used in environmental studies as *cost-benefit* models. The cost of removing a pollutant from the atmosphere is estimated as a function of the percentage of the pollutant removed. The higher the percentage removed, the greater the "benefit" to the people who breathe that air. The issues here are complex, of course, and the definition of "cost" is debatable. The cost to remove a small percentage of pollutant may be fairly low. But the removal cost of the final 5% of the pollutant, for example, may be terribly expensive.

---

**EXAMPLE 5**   **A cost-benefit model**   Suppose a cost-benefit function is given by

$$f(x) = \frac{50x}{105 - x}, \qquad 0 \le x \le 100,$$

where $x$ is the percentage of some pollutant to be removed and $f(x)$ is the associated cost (in millions of dollars). See Fig. 11. Find the costs to remove 70%, 95%, and 100% of the pollutant.

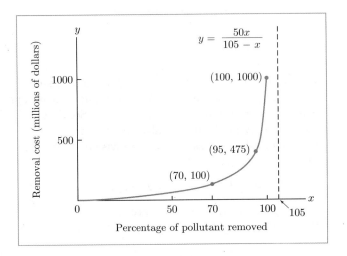

**Figure 11.**   A cost-benefit model.

**Solution**   The cost to remove 70% is

$$f(70) = \frac{50(70)}{105 - 70} = 100 \text{ (million dollars)}.$$

Similar calculations show that

$$f(95) = 475 \quad \text{and} \quad f(100) = 1000.$$

Observe that the cost to remove the last 5% of the pollutant is $f(100) - f(95) = 1000 - 475 = 525$ million dollars. This is more than five times the cost to remove the first 70% of the pollutant! ∎

**Power Functions**   Functions of the form $f(x) = x^r$ are called *power functions*. The meaning of $x^r$ is obvious when $r$ is a positive integer. However, the power function $f(x) = x^r$ may be defined for any number $r$. We delay until Section 0.5 a discussion of power functions. In that section, we will review the meaning of $x^r$ in case $r$ is a rational number.

**The Absolute Value Function**  The absolute value of a number $x$ is denoted by $|x|$ and is defined by

$$|x| = \begin{cases} x & \text{if } x \text{ is positive or zero,} \\ -x & \text{if } x \text{ is negative.} \end{cases}$$

For example, $|5| = 5$, $|0| = 0$, and $|-3| = -(-3) = 3$.

The function defined for all numbers $x$ by

$$f(x) = |x|$$

is called the *absolute value function*. Its graph coincides with the graph of the equation $y = x$ for $x \geq 0$ and with the graph of the equation $y = -x$ for $x < 0$. (See Fig. 12.)

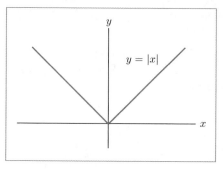

**Figure 12.**  Graph of the absolute value function.

**Incorporating Technology**  Inequalities such as $X \leq 1$ are either true or false depending on the value of $X$. Graphing calculators evaluate the inequality as 1 when it is true and 0 when it is false. See Fig. 13. (*Note*: On many calculators, inequality symbols are usually entered from the TEST menu.) Inequalities can be used to obtain expressions for piecewise-defined functions. For instance, the function in Example 4 can be defined by

$$\mathbf{Y_1 = (-1 \leq X) * (X \leq 1) * ((5/2)X - (1/2)) + (X > 1) * ((1/2)X - 2)}.$$

Piecewise-defined functions are usually best graphed in **Dot** (as opposed to **Connected**) mode. The preceding function is graphed in **Dot** mode in Fig. 14.

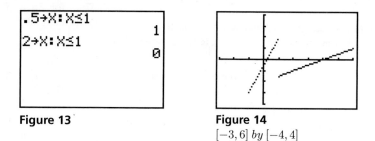

**Figure 13**

**Figure 14**
$[-3, 6]$ *by* $[-4, 4]$

If a rational function is undefined at $x = a$, an extraneous nearly vertical line might appear when the function is graphed. Figure 15(a) shows the cost-benefit function from Example 5. This extraneous line can be suppressed by choosing the window setting so that $a$ either appears in the middle or at one end of the screen. See Figs. 15(b) and 15(c).

The absolute value function is displayed as **abs(X)**. The function **abs** is found in the MATH/NUM menu of most TI calculators. (The TI-82 has an ABS key.)

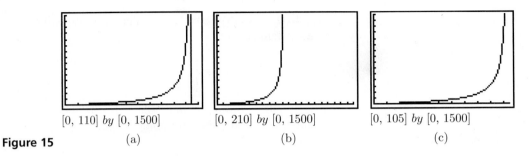

$[0, 110]$ *by* $[0, 1500]$          $[0, 210]$ *by* $[0, 1500]$          $[0, 105]$ *by* $[0, 1500]$

**Figure 15**          (a)                    (b)                    (c)

## PRACTICE PROBLEMS 0.2

1. A photocopy service has a fixed cost of $2000 per month (for rent, depreciation of equipment, etc.) and variable costs of $.04 for each page it reproduces for customers. Express its total cost as a (linear) function of the number of pages copied per month.

2. Determine the intercepts of the graph of

$$f(x) = -\tfrac{3}{8}x + 6.$$

## EXERCISES 0.2

*Graph the following functions.*

1. $f(x) = 2x - 1$
2. $f(t) = 3$
3. $f(x) = 3x + 1$
4. $f(x) = -\tfrac{1}{2}x - 4$
5. $f(x) = -2x + 3$
6. $f(u) = \tfrac{1}{4}$

*Determine the intercepts of the graphs of the following functions.*

7. $f(x) = 9x + 3$
8. $f(x) = -\tfrac{1}{2}x - 1$
9. $f(v) = 5$
10. $f(x) = 14$
11. $f(x) = -\tfrac{1}{4}x + 3$
12. $f(w) = 6w - 4$

13. (*Enzyme Kinetics*) In biochemistry, such as in the study of enzyme kinetics, one encounters a linear function of the form $f(x) = (K/V)x + 1/V$, where $K$ and $V$ are constants.

    (a) If $f(x) = .2x + 50$, find $K$ and $V$ so that $f(x)$ may be written in the form $f(x) = (K/V)x + 1/V$.

    (b) Find the $x$-intercept and $y$-intercept of the line $y = (K/V)x + 1/V$ (in terms of $K$ and $V$).

14. The constants $K$ and $V$ in Exercise 13 are often determined from experimental data. Suppose that a line is drawn through data points and has $x$-intercept $(-500, 0)$ and $y$-intercept $(0, 60)$. Determine $K$ and $V$ so that the line is the graph of the function $f(x) = (K/V)x + 1/V$. [*Hint:* Use Exercise 13(b).]

15. (*Cost of Car Rentals*) In some cities you can rent a car for $18 per day and $.20 per mile.

    (a) Find the cost of renting the car for one day and driving 200 miles.

    (b) Suppose that the car is to be rented for one day, and express the total rental expense as a function of the number $x$ of miles driven. (Assume that for each fraction of a mile driven, the same fraction of $.20 is charged.)

16. (*Right to Drill*) A gas company will pay a property owner $5000 for the right to drill on the land for natural gas and $.10 for each thousand cubic feet of gas extracted from the land. Express the amount of money the landowner will receive as a function of the amount of gas extracted from the land.

17. (*Medical Expense*) In 1998, a patient paid $300 per day for a semiprivate hospital room and $1500 for an appendectomy operation. Express the total amount paid for an appendectomy as a function of the number of days of hospital confinement.

18. (*Velocity of a Baseball*) When a baseball thrown at 85 miles per hour is hit by a bat swung at $x$ miles per hour, the ball travels $6x - 40$ feet.* (This formula assumes that $50 \le x \le 90$ and that the bat is 35 inches long, weighs 32 ounces, and strikes a waist-high pitch so that the plane of the swing lies at $35°$ from the horizontal.) How fast must the bat be swung in order for the ball to travel 350 feet?

19. (*Cost–Benefit*) Let $f(x)$ be the cost-benefit function from Example 5. If 70% of the pollutant has been removed, what is the added cost to remove another 5%? How does this compare with the cost to remove the final 5% of the pollutant? (See Example 5.)

---

*Robert K. Adair, *The Physics of Baseball* (New York: Harper & Row, 1990).

20. Suppose that the cost (in millions of dollars) to remove $x$ percent of a certain pollutant is given by the cost-benefit function

$$f(x) = \frac{20x}{102 - x} \qquad \text{for } 0 \le x \le 100.$$

   (a) Find the cost to remove 85% of the pollutant.

   (b) Find the cost to remove the final 5% of the pollutant.

*Each of the quadratic functions in Exercises 21–26 has the form $y = ax^2 + bx + c$. Identify $a$, $b$, and $c$.*

21. $y = 3x^2 - 4x$

22. $y = \dfrac{x^2 - 6x + 2}{3}$

23. $y = 3x - 2x^2 + 1$

24. $y = 3 - 2x + 4x^2$

25. $y = 1 - x^2$

26. $y = \frac{1}{2}x^2 + \sqrt{3}\,x - \pi$

*Sketch the graphs of the following functions.*

27. $f(x) = \begin{cases} 3x & \text{for } 0 \le x \le 1 \\ \frac{9}{2} - \frac{3}{2}x & \text{for } x > 1 \end{cases}$

28. $f(x) = \begin{cases} 1 + x & \text{for } x \le 3 \\ 4 & \text{for } x > 3 \end{cases}$

29. $f(x) = \begin{cases} 3 & \text{for } x < 2 \\ 2x + 1 & \text{for } x \ge 2 \end{cases}$

30. $f(x) = \begin{cases} \frac{1}{2}x & \text{for } 0 \le x < 4 \\ 2x - 3 & \text{for } 4 \le x \le 5 \end{cases}$

31. $f(x) = \begin{cases} 4 - x & \text{for } 0 \le x < 2 \\ 2x - 2 & \text{for } 2 \le x < 3 \\ x + 1 & \text{for } x \ge 3 \end{cases}$

32. $f(x) = \begin{cases} 4x & \text{for } 0 \le x < 1 \\ 8 - 4x & \text{for } 1 \le x < 2 \\ 2x - 4 & \text{for } x \ge 2 \end{cases}$

*Evaluate each of the functions in Exercises 33–38 at the given value of $x$.*

33. $f(x) = x^{100}$, $x = -1$

34. $f(x) = x^5$, $x = \frac{1}{2}$

35. $f(x) = |x|$, $x = 10^{-2}$

36. $f(x) = |x|$, $x = \pi$

37. $f(x) = |x|$, $x = -2.5$

38. $f(x) = |x|$, $x = -\frac{2}{3}$

## Technology Exercises

*In Exercises 39–42, give the graphing calculator expression for the function and display its graph.*

39. $f(x) = \begin{cases} 3x - 1 & \text{for } x \le 2 \\ x^2 + 1 & \text{for } x > 2 \end{cases}$

40. $f(x) = \begin{cases} 2x & \text{for } x < 3 \\ \sqrt{x} & \text{for } x \ge 3 \end{cases}$

41. $f(x) = \begin{cases} x^3 + 2x & \text{for } -2 \le x < 2 \\ \sqrt{x - 2} & \text{for } x \ge 2 \end{cases}$

42. $f(x) = \begin{cases} x^4 - 5x & \text{for } x \le 3 \\ 50 - 4x & \text{for } 3 < x \le 8 \end{cases}$

*Graph the functions in Exercises 43–46. Select a window setting for which extraneous nearly vertical lines are avoided.*

43. $f(x) = \dfrac{x}{x - 2}$ for $x \ge 0$

44. $f(x) = \dfrac{.2x^2}{5 - x}$ for $x \ge 0$

45. $f(x) = \dfrac{2x}{50 - x}$ for $0 \le x < 50$

46. $f(x) = \dfrac{20x}{75 - x}$ for $0 \le x < 75$

47. Graph the function $f(x) = |x|$ for $-10 \le x \le 10$.

48. Graph the function $f(x) = |x - 3|$ for $-10 \le x \le 10$.

## SOLUTIONS TO PRACTICE PROBLEMS 0.2

1. If $x$ represents the number of pages copied per month, then the variable cost is $.04x$ dollars. Now [total cost] = [fixed cost] + [variable cost]. If we define

$$f(x) = 2000 + .04x,$$

   then $f(x)$ gives the total cost per month.

2. To find the $y$-intercept, evaluate $f(x)$ at $x = 0$.

$$f(0) = -\tfrac{3}{8}(0) + 6 = 0 + 6 = 6$$

To find the $x$-intercept, set $f(x) = 0$ and solve for $x$.

$$-\tfrac{3}{8}x + 6 = 0$$
$$\tfrac{3}{8}x = 6$$
$$x = \tfrac{8}{3} \cdot 6 = 16$$

Therefore, the $y$-intercept is $(0, 6)$ and the $x$-intercept is $(16, 0)$.

## 0.3 The Algebra of Functions

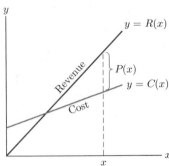

**Figure 1.** Profit equals revenue minus cost.

Many functions we shall encounter later in the text can be viewed as combinations of other functions. For example, let $P(x)$ represent the profit a company makes on the sale of $x$ units of some commodity. If $R(x)$ denotes the revenue received from the sale of $x$ units, and if $C(x)$ is the cost of producing $x$ units, then

$$P(x) = R(x) - C(x)$$

$$[\text{profit}] = [\text{revenue}] - [\text{cost}].$$

Writing the profit function in this way makes it possible to predict the behavior of $P(x)$ from properties of $R(x)$ and $C(x)$. For instance, we can determine when the profit $P(x)$ is positive by observing whether $R(x)$ is greater than $C(x)$. (See Fig. 1.)

The first four examples review the algebraic techniques needed to combine functions by addition, subtraction, multiplication, and division.

**EXAMPLE 1** Let $f(x) = 3x + 4$ and $g(x) = 2x - 6$. Find

$$f(x) + g(x), \quad f(x) - g(x), \quad \frac{f(x)}{g(x)}, \quad \text{and} \quad f(x)g(x).$$

**Solution** For $f(x) + g(x)$ and $f(x) - g(x)$, we add or subtract corresponding terms:

$$f(x) + g(x) = (3x + 4) + (2x - 6) = 3x + 4 + 2x - 6 = 5x - 2$$
$$f(x) - g(x) = (3x + 4) - (2x - 6) = 3x + 4 - 2x + 6 = x + 10.$$

To compute $\dfrac{f(x)}{g(x)}$ and $f(x)g(x)$, we first substitute the formulas for $f(x)$ and $g(x)$:

$$\frac{f(x)}{g(x)} = \frac{3x + 4}{2x - 6}$$
$$f(x)g(x) = (3x + 4)(2x - 6).$$

The expression for $\dfrac{f(x)}{g(x)}$ is already in simplest form. To simplify the expression for $f(x)g(x)$, we carry out the multiplication indicated in $(3x + 4)(2x - 6)$. We must be careful to multiply each term of $3x + 4$ by each term of $2x - 6$. A common order for multiplying such expressions is (1) the First terms, (2) the Outer terms, (3) the Inner terms, and (4) the Last terms. (This procedure may be remembered by the word FOIL.)

$$f(x)g(x) = (3x + 4)(2x - 6) = 6x^2 - 18x + 8x - 24$$
$$= 6x^2 - 10x - 24 \qquad \blacksquare$$

**EXAMPLE 2** Let

$$g(x) = \frac{2}{x} \quad \text{and} \quad h(x) = \frac{3}{x - 1}.$$

Express $g(x) + h(x)$ as a rational function.

**Solution**  First we have

$$g(x) + h(x) = \frac{2}{x} + \frac{3}{x-1}, \qquad x \neq 0, 1.$$

The restriction $x \neq 0, 1$ comes from the fact that $g(x)$ is defined only for $x \neq 0$ and $h(x)$ is defined only for $x \neq 1$. (A rational function is not defined for values of the variable for which the denominator is 0.) In order for us to add two fractions, their denominators must be the same. A common denominator for $\frac{2}{x}$ and $\frac{3}{x-1}$ is $x(x-1)$. If we multiply $\frac{2}{x}$ by $\frac{x-1}{x-1}$, we obtain an equivalent expression whose denominator is $x(x-1)$. Similarly, if we multiply $\frac{3}{x-1}$ by $\frac{x}{x}$, we obtain an equivalent expression whose denominator is $x(x-1)$. Thus

$$\frac{2}{x} + \frac{3}{x-1} = \frac{2}{x} \cdot \frac{x-1}{x-1} + \frac{3}{x-1} \cdot \frac{x}{x}$$

$$= \frac{2(x-1)}{x(x-1)} + \frac{3x}{x(x-1)}$$

$$= \frac{2(x-1) + 3x}{x(x-1)}$$

$$= \frac{5x-2}{x(x-1)}.$$

So

$$g(x) + h(x) = \frac{5x-2}{x(x-1)}. \qquad \blacksquare$$

---

**EXAMPLE 3**  Find $f(t)g(t)$, where

$$f(t) = \frac{t}{t-1} \quad \text{and} \quad g(t) = \frac{t+2}{t+1}.$$

**Solution**  To multiply rational functions, multiply numerator by numerator and denominator by denominator:

$$f(t)g(t) = \frac{t}{t-1} \cdot \frac{t+2}{t+1} = \frac{t(t+2)}{(t-1)(t+1)}.$$

An alternative way of expressing $f(t)g(t)$ is obtained by carrying out the indicated multiplications:

$$f(t)g(t) = \frac{t^2 + 2t}{t^2 + t - t - 1} = \frac{t^2 + 2t}{t^2 - 1}.$$

The choice of which expression to use for $f(t)g(t)$ depends on the particular application. $\blacksquare$

---

**EXAMPLE 4**  Find

$$\frac{f(x)}{g(x)}, \quad \text{where } f(x) = \frac{x}{x-3} \text{ and } g(x) = \frac{x+1}{x-5}.$$

**Solution**  The function $f(x)$ is defined only for $x \neq 3$, and $g(x)$ is defined only for $x \neq 5$. The quotient $f(x)/g(x)$ is therefore not defined for $x = 3, 5$. Moreover, the quotient is not defined for values of $x$ for which $g(x)$ is equal to 0; that is, $x = -1$. Thus, the quotient is defined for $x \neq -1, 3, 5$. To divide $f(x)$ by $g(x)$, we multiply $f(x)$ by the reciprocal of $g(x)$:

$$\frac{f(x)}{g(x)} = \frac{x}{x-3} \cdot \frac{x-5}{x+1} = \frac{x(x-5)}{(x-3)(x+1)} = \frac{x^2 - 5x}{x^2 - 2x - 3}, \quad x \neq -1, 3, 5. \quad \blacksquare$$

**Composition of Functions** Another important way of combining two functions $f(x)$ and $g(x)$ is to substitute the function $g(x)$ for every occurrence of the variable $x$ in $f(x)$. The resulting function is called the *composition* (or *composite*) of $f(x)$ and $g(x)$ and is denoted by $f(g(x))$.

---

**EXAMPLE 5** Let $f(x) = x^2 + 3x + 1$ and $g(x) = x - 5$. What is $f(g(x))$?

**Solution** We substitute $g(x)$ in place of each $x$ in $f(x)$:

$$\begin{aligned} f(g(x)) &= [g(x)]^2 + 3g(x) + 1 \\ &= (x-5)^2 + 3(x-5) + 1 \\ &= (x^2 - 10x + 25) + (3x - 15) + 1 \\ &= x^2 - 7x + 11. \end{aligned} \quad \blacksquare$$

Later in the text we shall need to study expressions of the form $f(x+h)$, where $f(x)$ is a given function and $h$ represents some number. The meaning of $f(x+h)$ is that $x+h$ is to be substituted for each occurrence of $x$ in the formula for $f(x)$. In fact, $f(x+h)$ is just a special case of $f(g(x))$, where $g(x) = x + h$.

---

**EXAMPLE 6** If $f(x) = x^3$, find $f(x+h) - f(x)$.

**Solution**
$$\begin{aligned} f(x+h) &= (x+h)^3 = x^3 + 3x^2 h + 3xh^2 + h^3 \\ f(x+h) - f(x) &= (x^3 + 3x^2 h + 3xh^2 + h^3) - x^3 \\ &= 3x^2 h + 3xh^2 + h^3. \end{aligned} \quad \blacksquare$$

---

**EXAMPLE 7** In a certain lake, the bass feed primarily on minnows, and the minnows feed on plankton. Suppose that the size of the bass population is a function $f(n)$ of the number $n$ of minnows in the lake, and the number of minnows is a function $g(x)$ of the amount $x$ of plankton in the lake. Express the size of the bass population as a function of the amount of plankton, if $f(n) = 50 + \sqrt{n/150}$ and $g(x) = 4x + 3$.

**Solution** We have $n = g(x)$. Substituting $g(x)$ for $n$ in $f(n)$, we find the size of the bass population is given by

$$f(g(x)) = 50 + \sqrt{\frac{g(x)}{150}} = 50 + \sqrt{\frac{4x+3}{150}}. \quad \blacksquare$$

**Incorporating Technology** After expressions have been assigned to $\mathbf{Y_1}$ and $\mathbf{Y_2}$, $\mathbf{Y_3}$ can be set to $\mathbf{Y_1} + \mathbf{Y_2}$, $\mathbf{Y_1} - \mathbf{Y_2}$, $\mathbf{Y_1} * \mathbf{Y_2}$, or $\mathbf{Y_1}/\mathbf{Y_2}$. The composite of the two functions is specified by $\mathbf{Y_1}(\mathbf{Y_2})$ on many, but not all, calculators.

## PRACTICE PROBLEMS 0.3

**1.** Let $f(x) = x^5$, $g(x) = x^3 - 4x^2 + x - 8$.
  **(a)** Find $f(g(x))$.     **(b)** Find $g(f(x))$.

**2.** Let $f(x) = x^2$. Calculate $\dfrac{f(1+h) - f(1)}{h}$ and simplify.

# EXERCISES 0.3

Let $f(x) = x^2 + 1$, $g(x) = 9x$, and $h(x) = 5 - 2x^2$. Calculate the following functions.

**1.** $f(x) + g(x)$

**2.** $f(x) - h(x)$

**3.** $f(x)g(x)$

**4.** $g(x)h(x)$

**5.** $\dfrac{f(t)}{g(t)}$

**6.** $\dfrac{g(t)}{h(t)}$

In Exercises 7–12, express $f(x) + g(x)$ as a rational function. Carry out all multiplications.

**7.** $f(x) = \dfrac{2}{x-3}$,   $g(x) = \dfrac{1}{x+2}$

**8.** $f(x) = \dfrac{3}{x-6}$,   $g(x) = \dfrac{-2}{x-2}$

**9.** $f(x) = \dfrac{x}{x-8}$,   $g(x) = \dfrac{-x}{x-4}$

**10.** $f(x) = \dfrac{-x}{x+3}$,   $g(x) = \dfrac{x}{x+5}$

**11.** $f(x) = \dfrac{x+5}{x-10}$,   $g(x) = \dfrac{x}{x+10}$

**12.** $f(x) = \dfrac{x+6}{x-6}$,   $g(x) = \dfrac{x-6}{x+6}$

Let $f(x) = \dfrac{x}{x-2}$, $g(x) = \dfrac{5-x}{5+x}$, and $h(x) = \dfrac{x+1}{3x-1}$. Express the following as rational functions.

**13.** $f(x) - g(x)$

**14.** $f(t) - h(t)$

**15.** $f(x)g(x)$

**16.** $g(x)h(x)$

**17.** $\dfrac{f(x)}{g(x)}$

**18.** $\dfrac{h(s)}{f(s)}$

**19.** $f(x+1)g(x+1)$

**20.** $f(x+2) + g(x+2)$

**21.** $\dfrac{g(x+5)}{f(x+5)}$

**22.** $f\left(\dfrac{1}{t}\right)$

**23.** $g\left(\dfrac{1}{u}\right)$

**24.** $h\left(\dfrac{1}{x^2}\right)$

Let $f(x) = x^6$, $g(x) = \dfrac{x}{1-x}$, and $h(x) = x^3 - 5x^2 + 1$. Calculate the following functions.

**25.** $f(g(x))$

**26.** $h(f(t))$

**27.** $h(g(x))$

**28.** $g(f(x))$

**29.** $g(h(t))$

**30.** $f(h(x))$

**31.** If $f(x) = x^2$, find $f(x+h) - f(x)$ and simplify.

**32.** If $f(x) = 1/x$, find $f(x+h) - f(x)$ and simplify.

**33.** If $g(t) = 4t - t^2$, find $\dfrac{g(t+h) - g(t)}{h}$ and simplify.

**34.** If $g(t) = t^3 + 5$, find $\dfrac{g(t+h) - g(t)}{h}$ and simplify.

**35.** (*Cost*) After $t$ hours of operation, an assembly line has assembled $A(t) = 20t - \frac{1}{2}t^2$ power lawn mowers, $0 \le t \le 10$. Suppose that the factory's cost of manufacturing $x$ units is $C(x)$ dollars, where $C(x) = 3000 + 80x$.

  **(a)** Express the factory's cost as a (composite) function of the number of hours of operation of the assembly line.

  **(b)** What is the cost of the first 2 hours of operation?

**36.** (*Cost*) During the first $\frac{1}{2}$ hour, the employees of a machine shop prepare the work area for the day's work. After that, they turn out 10 precision machine parts per hour, so that the output after $t$ hours is $f(t)$ machine parts, where $f(t) = 10\left(t - \frac{1}{2}\right) = 10t - 5$, $\frac{1}{2} \le t \le 8$. The total cost of producing $x$ machine parts is $C(x)$ dollars, where $C(x) = .1x^2 + 25x + 200$.

  **(a)** Express the total cost as a (composite) function of $t$.

  **(b)** What is the cost of the first 4 hours of operation?

**37.** (*Conversion Scales*) Table 1 shows a conversion table for men's hat sizes for three countries. The function $g(x) = 8x + 1$ converts from British sizes to French sizes, and the function $f(x) = \frac{1}{8}x$ converts from French sizes to U.S. sizes. Determine the function $h(x) = f(g(x))$ and give its interpretation.

| TABLE 1 | Conversion Table for Men's Hat Sizes | | | | | | | |
|---|---|---|---|---|---|---|---|---|
| Britain | $6\frac{1}{2}$ | $6\frac{5}{8}$ | $6\frac{3}{4}$ | $6\frac{7}{8}$ | $7$ | $7\frac{1}{8}$ | $7\frac{1}{4}$ | $7\frac{3}{8}$ |
| France | $53$ | $54$ | $55$ | $56$ | $57$ | $58$ | $59$ | $60$ |
| United States | $6\frac{5}{8}$ | $6\frac{3}{4}$ | $6\frac{7}{8}$ | $7$ | $7\frac{1}{8}$ | $7\frac{1}{4}$ | $7\frac{3}{8}$ | $7\frac{1}{2}$ |

## Technology Exercises

**38.** Let $f(x) = x^2$. Graph the functions $f(x+1)$, $f(x-1)$, $f(x+2)$, and $f(x-2)$. Make a guess about the relationship between the graph of a general function $f(x)$ and the graph of $f(g(x))$ where $g(x) = x + a$ for some constant $a$. Test your guess on the functions $f(x) = x^3$ and $f(x) = \sqrt{x}$.

**39.** Let $f(x) = x^2$. Graph the functions $f(x)+1$, $f(x)-1$, $f(x)+2$, and $f(x)-2$. Make a guess about the relationship between the graph of a general function $f(x)$ and the graph of $f(x) + c$ for some constant $c$. Test your guess on the functions $f(x) = x^3$ and $f(x) = \sqrt{x}$.

**40.** Based on the results of Exercises 38 and 39, sketch the graph of $f(x) = (x-1)^2 + 2$ without using a graphing

calculator. Check your result with a graphing calculator.

**41.** Based on the results of Exercises 38 and 39, sketch the graph of $f(x) = (x+2)^2 - 1$ without using a graphing calculator. Check your result with a graphing calculator.

**42.** Let $f(x) = x^2 + 3x + 1$ and let $g(x) = x^2 - 3x - 1$. Graph the two functions $f(g(x))$ and $g(f(x))$ together in the window $[-4, 4]$ by $[-10, 10]$ and determine if they are the same function.

**43.** Let $f(x) = \dfrac{x}{x - 1}$ and graph the function $f(f(x))$ in the window $[-15, 15]$ by $[-10, 10]$. Trace to examine the coordinates of several points on the graph and then determine the formula for $f(f(x))$.

---

### SOLUTIONS TO PRACTICE PROBLEMS 0.3

**1. (a)** $f(g(x)) = [g(x)]^5 = (x^3 - 4x^2 + x - 8)^5$

**(b)** $g(f(x)) = [f(x)]^3 - 4[f(x)]^2 + f(x) - 8$

$\qquad = (x^5)^3 - 4(x^5)^2 + x^5 - 8$

$\qquad = x^{15} - 4x^{10} + x^5 - 8$

**2.** $\dfrac{f(1+h) - f(1)}{h} = \dfrac{(1+h)^2 - 1^2}{h}$

$\qquad = \dfrac{1 + 2h + h^2 - 1}{h}$

$\qquad = \dfrac{2h + h^2}{h} = 2 + h$

---

## 0.4  Zeros of Functions—The Quadratic Formula and Factoring

A *zero* of a function $f(x)$ is a value of $x$ for which $f(x) = 0$. For instance, the function $f(x)$ whose graph is shown in Fig. 1 has $x = -3$, $x = 3$, and $x = 7$ as zeros. Throughout this book we shall need to determine zeros of functions or, what amounts to the same thing, to solve the equation $f(x) = 0$.

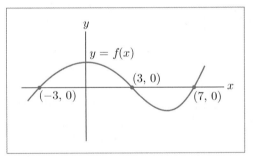

**Figure 1.**  Zeros of a function.

In Section 0.2 we found zeros of linear functions. In this section our emphasis is on zeros of quadratic functions.

**The Quadratic Formula**   Consider the quadratic function $f(x) = ax^2 + bx + c$, $a \neq 0$. The zeros of this function are precisely the solutions of the quadratic equation

$$ax^2 + bx + c = 0.$$

One way of solving such an equation is via the *quadratic formula*.

The solutions of the equation $ax^2 + bx + c = 0$ are

$$x = \frac{-b \pm \sqrt{b^2 - 4ac}}{2a}.$$

The $\pm$ sign tells us to form two expressions, one with $+$ and one with $-$. The quadratic formula implies that a quadratic equation has at most two roots. It will have none if the expression $b^2 - 4ac$ is negative and one if $b^2 - 4ac$ equals 0. The quadratic formula is derived at the end of the section.

**EXAMPLE 1**    Solve the quadratic equation $3x^2 - 6x + 2 = 0$.

**Solution**    Here $a = 3$, $b = -6$, and $c = 2$. Substituting these values into the quadratic formula, we find that

$$\sqrt{b^2 - 4ac} = \sqrt{(-6)^2 - 4(3)(2)} = \sqrt{36 - 24} = \sqrt{12} = \sqrt{4 \cdot 3} = 2\sqrt{3}$$

and

$$x = \frac{-b \pm \sqrt{b^2 - 4ac}}{2a}$$
$$= \frac{-(-6) \pm 2\sqrt{3}}{2(3)}$$
$$= \frac{6 \pm 2\sqrt{3}}{6}$$
$$= 1 \pm \frac{\sqrt{3}}{3}.$$

The solutions of the equation are $1 + \sqrt{3}/3$ and $1 - \sqrt{3}/3$. (See Fig. 2.)    ∎

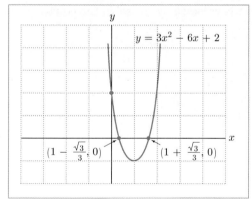

**Figure 2**

**EXAMPLE 2**    Find the zeros of the following quadratic functions.

(a) $f(x) = 4x^2 - 4x + 1$          (b) $f(x) = \frac{1}{2}x^2 - 3x + 5$

**Solution**    (a) We must solve $4x^2 - 4x + 1 = 0$. Here $a = 4$, $b = -4$, and $c = 1$, so that

$$\sqrt{b^2 - 4ac} = \sqrt{(-4)^2 - 4(4)(1)} = \sqrt{0} = 0.$$

Thus there is only one zero, namely,

$$x = \frac{-(-4) \pm 0}{2(4)} = \frac{4}{8} = \frac{1}{2}.$$

The graph of $f(x)$ is sketched in Fig. 3.

(b) We must solve $\frac{1}{2}x^2 - 3x + 5 = 0$. Here $a = \frac{1}{2}$, $b = -3$, and $c = 5$, so that

$$\sqrt{b^2 - 4ac} = \sqrt{(-3)^2 - 4(\tfrac{1}{2})(5)} = \sqrt{9 - 10} = \sqrt{-1}.$$

The square root of a negative number is undefined, so we conclude that $f(x)$ has no zeros. The reason for this is clear from Fig. 4. The graph of $f(x)$ lies entirely above the $x$-axis and has no $x$-intercepts. ∎

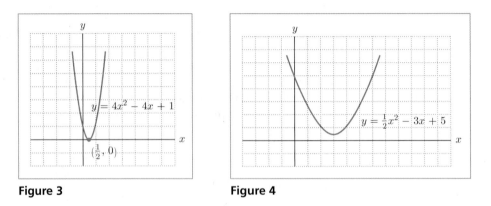

**Figure 3**          **Figure 4**

The common problem of finding where two curves intersect amounts to finding the zeros of a function.

---

**EXAMPLE 3**    Find the points of intersection of the graphs of the functions $y = x^2 + 1$ and $y = 4x$. (See Fig. 5.)

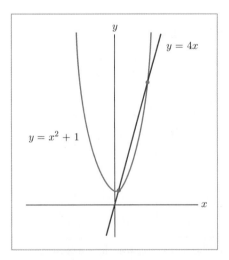

**Figure 5.**   Points of intersection of two graphs.

**Solution**    If a point $(x, y)$ is on both graphs, then its coordinates must satisfy both equations. That is, $x$ and $y$ must satisfy $y = x^2 + 1$ and $y = 4x$. Equating the two expressions for $y$, we have

$$x^2 + 1 = 4x.$$

To use the quadratic formula, we rewrite the equation in the form
$$x^2 - 4x + 1 = 0.$$
From the quadratic formula,
$$x = \frac{4 \pm \sqrt{16 - 4}}{2} = \frac{4 \pm \sqrt{12}}{2} = \frac{4 \pm 2\sqrt{3}}{2} = 2 \pm \sqrt{3}.$$
Thus the $x$-coordinates of the points of intersection are $2 + \sqrt{3}$ and $2 - \sqrt{3}$. To find the $y$-coordinates, we substitute these values of $x$ into either equation, $y = x^2 + 1$ or $y = 4x$. The second equation is simpler. We obtain $y = 4(2 + \sqrt{3}) = 8 + 4\sqrt{3}$ and $y = 4(2 - \sqrt{3}) = 8 - 4\sqrt{3}$. Thus the points of intersection are $(2 + \sqrt{3}, 8 + 4\sqrt{3})$ and $(2 - \sqrt{3}, 8 - 4\sqrt{3})$. ∎

---

**EXAMPLE 4**   **Profit and break-even points**   A cable television company estimates that with $x$ thousand subscribers, its monthly revenue and cost (in thousands of dollars) are
$$R(x) = 32x - .21x^2,$$
$$C(x) = 195 + 12x.$$
Determine the company's *break-even points*; that is, find the number of subscribers at which the revenue equals the cost. See Fig. 6.

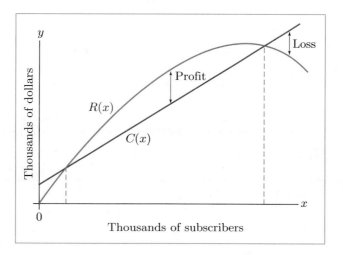

**Figure 6.**   Break-even points.

**Solution**   Let $P(x)$ be the profit function.
$$P(x) = R(x) - C(x)$$
$$= (32x - .21x^2) - (195 + 12x)$$
$$= -.21x^2 + 20x - 195$$
The break-even points occur where the profit is zero. Thus we must solve
$$-.21x^2 + 20x - 195 = 0.$$
From the quadratic formula,
$$x = \frac{-20 \pm \sqrt{20^2 - 4(-.21)(-195)}}{2(-.21)} = \frac{-20 \pm \sqrt{236.2}}{-.42}$$
$$\approx 47.62 \pm 36.59 = 11.03 \text{ and } 84.21.$$
The break-even points occur where the company has 11,030 or 84,210 subscribers. Between those two levels, the company will be profitable. ∎

**Factoring**   If $f(x)$ is a polynomial, we can often write $f(x)$ as a product of linear factors (i.e., factors of the form $ax + b$). If this can be done, then the zeros of $f(x)$ can be determined by setting each of the linear factors equal to zero and solving for $x$. (The reason is that the product of numbers can be zero only when one of the numbers is zero.)

---

**EXAMPLE 5**   Factor the following quadratic polynomials.

(a)  $x^2 + 7x + 12$ ,

(b)  $x^2 - 13x + 12$

(c)  $x^2 - 4x - 12$

(d)  $x^2 + 4x - 12$

**Solution**   Note first that for any numbers $c$ and $d$,

$$(x + c)(x + d) = x^2 + (c + d)x + cd.$$

In the quadratic on the right, the constant term is the product $cd$, whereas the coefficient of $x$ is the sum $c + d$.

(a)  Think of all integers $c$ and $d$ such that $cd = 12$. Then choose the pair that satisfies $c + d = 7$; that is, take $c = 3$, $d = 4$. Thus

$$x^2 + 7x + 12 = (x + 3)(x + 4).$$

(b)  We want $cd = 12$. Since 12 is positive, $c$ and $d$ must be both positive or both negative. We must also have $c + d = -13$. These facts lead us to

$$x^2 - 13x + 12 = (x - 12)(x - 1).$$

(c)  We want $cd = -12$. Since $-12$ is negative, $c$ and $d$ must have opposite signs. Also, they must sum to give $-4$. We find that

$$x^2 - 4x - 12 = (x - 6)(x + 2).$$

(d)  This is almost the same as part (c).

$$x^2 + 4x - 12 = (x + 6)(x - 2).$$   ∎

---

**EXAMPLE 6**   Factor the following polynomials.

(a)  $x^2 - 6x + 9$

(b)  $x^2 - 25$

(c)  $3x^2 - 21x + 30$

(d)  $20 + 8x - x^2$

**Solution**   (a)  We look for $cd = 9$ and $c + d = -6$. The solution is $c = d = -3$, and

$$x^2 - 6x + 9 = (x - 3)(x - 3) = (x - 3)^2.$$

In general,

$$x^2 - 2cx + c^2 = (x - c)(x - c) = (x - c)^2.$$

(b)  We use the identity

$$x^2 - c^2 = (x + c)(x - c).$$

Hence

$$x^2 - 25 = (x + 5)(x - 5).$$

(c)  We first factor out a common factor of 3 and then use the method of Example 5.

$$3x^2 - 21x + 30 = 3(x^2 - 7x + 10)$$
$$= 3(x - 5)(x - 2).$$

(d) We first factor out $-1$ in order to make the coefficient of $x^2$ equal to $+1$.

$$20 + 8x - x^2 = (-1)(x^2 - 8x - 20)$$
$$= (-1)(x - 10)(x + 2). \qquad \blacksquare$$

---

**EXAMPLE 7**  Factor the following polynomials.

(a) $x^2 - 8x$      (b) $x^3 + 3x^2 - 18x$      (c) $x^3 - 10x$

**Solution**  In each case we first factor out a common factor of $x$.

(a) $x^2 - 8x = x(x - 8)$.

(b) $x^3 + 3x^2 - 18x = x(x^2 + 3x - 18) = x(x + 6)(x - 3)$.

(c) $x^3 - 10x = x(x^2 - 10)$. To factor $x^2 - 10$, we use the identity $x^2 - c^2 = (x + c)(x - c)$, where $c^2 = 10$ and $c = \sqrt{10}$. Thus

$$x^3 - 10x = x(x^2 - 10) = x(x + \sqrt{10})(x - \sqrt{10}). \qquad \blacksquare$$

---

**EXAMPLE 8**  Solve the following equations.

(a) $x^2 - 2x - 15 = 0$      (b) $x^2 - 20 = x$      (c) $\dfrac{x^2 + 10x + 25}{x + 1} = 0$

**Solution**  (a) The equation $x^2 - 2x - 15 = 0$ may be written in the form

$$(x - 5)(x + 3) = 0.$$

The product of two numbers is zero if one or the other of the numbers (or both) is zero. Hence

$$x - 5 = 0 \quad \text{or} \quad x + 3 = 0.$$

That is,

$$x = 5 \quad \text{or} \quad x = -3.$$

(b) First we must rewrite the equation $x^2 - 20 = x$ in the form $ax^2 + bx + c = 0$; that is,

$$x^2 - x - 20 = 0$$
$$(x - 5)(x + 4) = 0.$$

We conclude that

$$x - 5 = 0 \quad \text{or} \quad x + 4 = 0;$$

that is,

$$x = 5 \quad \text{or} \quad x = -4.$$

(c) A rational function will be zero only if the numerator is zero. Thus

$$x^2 + 10x + 25 = 0$$
$$(x + 5)^2 = 0$$
$$x + 5 = 0.$$

That is,

$$x = -5.$$

Since the denominator is not 0 at $x = -5$, we conclude that $x = -5$ is the solution. $\blacksquare$

### Derivation of the Quadratic Formula

$$ax^2 + bx + c = 0 \qquad\qquad (a \neq 0)$$

$$ax^2 + bx = -c$$

$$4a^2x^2 + 4abx = -4ac \qquad \text{(both sides multiplied by } 4a)$$

$$4a^2x^2 + 4abx + b^2 = b^2 - 4ac \qquad (b^2 \text{ added to both sides)}$$

Now note that $4a^2x^2 + 4abx + b^2 = (2ax + b)^2$. To check this, simply multiply out the right-hand side. Therefore,

$$(2ax + b)^2 = b^2 - 4ac$$

$$2ax + b = \pm\sqrt{b^2 - 4ac}$$

$$2ax = -b \pm \sqrt{b^2 - 4ac}$$

$$x = \frac{-b \pm \sqrt{b^2 - 4ac}}{2a}.$$

**Incorporating Technology**   Most graphing calculators have a **zero** (or ROOT) routine that computes zeros of functions, and have an **intersect** (or ISECT) routine that computes the intersection points of graphs. See Appendices A–D for the details for TI calculators. Figure 7 shows a zero of the function from Example 1 and Fig. 8 shows an intersection point from Example 3.

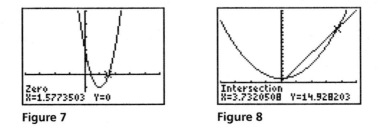

**Figure 7**              **Figure 8**

One of the problems in graphing a function is to find a domain (**Xmin** to **Xmax**) that contains all the zeros. For a polynomial function, there is an easy solution to this problem. Write the polynomial in the form $c(x^n + a_{n-1}x^{n-1} + \cdots + a_0)$ and let $M$ be the number that is one more than the largest magnitude of the coefficients $a_{n-1}, \ldots, a_0$. Then the interval $[-M, M]$ will contain all the zeros of the polynomial. For instance, all the zeros of the polynomial $x^3 - 10x^2 + 9x + 8$ lie in the interval $[-11, 11]$. After you view the polynomial on the domain $[-M, M]$, you can usually find a smaller domain that also contains all the zeros.

## PRACTICE PROBLEMS 0.4

**1.** Solve the equation $x - \dfrac{14}{x} = 5$.

**2.** Use the quadratic formula to solve $7x^2 - 35x + 35 = 0$.

# EXERCISES 0.4

*Use the quadratic formula to find the zeros of the functions in Exercises 1–6.*

**1.** $f(x) = 2x^2 - 7x + 6$      **2.** $f(x) = 3x^2 + 2x - 1$

**3.** $f(t) = 4t^2 - 12t + 9$      **4.** $f(x) = \frac{1}{4}x^2 + x + 1$

**5.** $f(x) = -2x^2 + 3x - 4$      **6.** $f(a) = 11a^2 - 7a + 1$

*Use the quadratic formula to solve the equations in Exercises 7–12.*

**7.** $5x^2 - 4x - 1 = 0$      **8.** $x^2 - 4x + 5 = 0$

**9.** $15x^2 - 135x + 300 = 0$      **10.** $z^2 - \sqrt{2}\,z - \frac{5}{4} = 0$

**11.** $\frac{3}{2}x^2 - 6x + 5 = 0$      **12.** $9x^2 - 12x + 4 = 0$

*Factor the polynomials in Exercises 13–24.*

**13.** $x^2 + 8x + 15$      **14.** $x^2 - 10x + 16$

**15.** $x^2 - 16$      **16.** $x^2 - 1$

**17.** $3x^2 + 12x + 12$      **18.** $2x^2 - 12x + 18$

**19.** $30 - 4x - 2x^2$      **20.** $15 + 12x - 3x^2$

**21.** $3x - x^2$      **22.** $4x^2 - 1$

**23.** $6x - 2x^3$      **24.** $16x + 6x^2 - x^3$

*Find the points of intersection of the pairs of curves in Exercises 25–32.*

**25.** $y = 2x^2 - 5x - 6$, $y = 3x + 4$

**26.** $y = x^2 - 10x + 9$, $y = x - 9$

**27.** $y = x^2 - 4x + 4$, $y = 12 + 2x - x^2$

**28.** $y = 3x^2 + 9$, $y = 2x^2 - 5x + 3$

**29.** $y = x^3 - 3x^2 + x$, $y = x^2 - 3x$

**30.** $y = \frac{1}{2}x^3 - 2x^2$, $y = 2x$

**31.** $y = \frac{1}{2}x^3 + x^2 + 5$, $y = 3x^2 - \frac{1}{2}x + 5$

**32.** $y = 30x^3 - 3x^2$, $y = 16x^3 + 25x^2$

*Solve the equations in Exercises 33–38.*

**33.** $\dfrac{21}{x} - x = 4$      **34.** $x + \dfrac{2}{x - 6} = 3$

**35.** $x + \dfrac{14}{x + 4} = 5$      **36.** $1 = \dfrac{5}{x} + \dfrac{6}{x^2}$

**37.** $\dfrac{x^2 + 14x + 49}{x^2 + 1} = 0$      **38.** $\dfrac{x^2 - 8x + 16}{1 + \sqrt{x}} = 0$

**39.** (*Break-Even Points*) Suppose the cable television company's cost function in Example 4 changes to $C(x) = 275 + 12x$. Determine the new break-even points.

**40.** (*Velocity*) When a car is moving at $x$ miles per hour and the driver decides to slam on the brakes, the car will travel $x + \frac{1}{20}x^2$ feet.[*] If a car travels 175 feet after the driver decides to stop, how fast was the car moving?

## Technology Exercises

*In Exercises 41–44, find the zeros of the function. (Use the specified viewing window.)*

**41.** $f(x) = x^2 - x - 2$; $[-4, 5]$ *by* $[-4, 10]$

**42.** $f(x) = x^3 - 3x + 2$; $[-3, 3]$ *by* $[-10, 10]$

**43.** $f(x) = \sqrt{x + 2} - x + 2$; $[-2, 7]$ *by* $[-2, 4]$

**44.** $f(x) = \dfrac{x}{x + 2} - x^2 + 1$; $[-1.5, 2]$ *by* $[-2, 3]$

*In Exercises 45–48, find the points of intersection of the graphs of the functions. (Use the specified viewing window.)*

**45.** $f(x) = 2x - 1$; $g(x) = x^2 - 2$; $[-4, 4]$ *by* $[-6, 10]$

**46.** $f(x) = -x - 2$; $g(x) = -4x^2 + x + 1$; $[-2, 2]$ *by* $[-5, 2]$

**47.** $f(x) = 3x^4 - 14x^3 + 24x - 3$; $g(x) = 2x - 30$; $[-3, 5]$ *by* $[-80, 30]$

**48.** $f(x) = \dfrac{1}{x}$; $g(x) = \sqrt{x^2 - 1}$; $[0, 4]$ *by* $[-1, 3]$

*In Exercises 49–52, find a good window setting for the graph of the function. The graph should show all the zeros of the polynomial.*

**49.** $f(x) = x^3 - 22x^2 + 17x + 19$

**50.** $f(x) = x^4 - 200x^3 - 100x^2$

**51.** $f(x) = 3x^3 + 52x^2 - 12x - 12$

**52.** $f(x) = 2x^5 - 24x^4 - 24x + 2$

## SOLUTIONS TO PRACTICE PROBLEMS 0.4

**1.** Multiply both sides of the equation by $x$. Then $x^2 - 14 = 5x$. Now, take the term $5x$ to the left side of the equation and solve by factoring.

$$x^2 - 5x - 14 = 0$$
$$(x - 7)(x + 2) = 0$$
$$x = 7 \quad \text{or} \quad x = -2$$

---

[*]The general formula is $f(x) = ax + bx^2$, where the constant $a$ depends on the driver's reaction time and the constant $b$ depends on the weight of the car and the type of tires. This mathematical model was analyzed in D. Burghes, I. Huntley, and J. McDonald, *Applying Mathematics: A Course in Mathematical Modelling* (New York: Halstead Press, 1982), pp. 57–60.

**SOLUTIONS TO PRACTICE PROBLEMS 0.4**

**2.** In this case, each coefficient is a multiple of 7. To simplify the arithmetic, we divide both sides of the equation by 7 before using the quadratic formula.

$$x^2 - 5x + 5 = 0$$

$$\sqrt{b^2 - 4ac} = \sqrt{(-5)^2 - 4(1)(5)} = \sqrt{5}$$

$$x = \frac{-b \pm \sqrt{b^2 - 4ac}}{2a} = \frac{5 \pm \sqrt{5}}{2 \cdot 1} = \frac{5}{2} \pm \frac{1}{2}\sqrt{5}$$

## 0.5  Exponents and Power Functions

In this section we review the operations with exponents that occur frequently throughout the text. We begin with the definition of $b^r$ for various types of numbers $b$ and $r$.

For any nonzero number $b$ and any positive integer $n$, we have by definition that

$$b^n = \underbrace{b \cdot b \cdots \cdots b}_{n \text{ times}},$$

$$b^{-n} = \frac{1}{b^n},$$

and

$$b^0 = 1.$$

For example, $2^4 = 2 \cdot 2 \cdot 2 \cdot 2 = 16$, $2^{-4} = \frac{1}{2^4} = \frac{1}{16}$, and $2^0 = 1$.

Next, we consider numbers of the form $b^{1/n}$, where $n$ is a positive integer. For instance,

$2^{1/2}$ is the positive number whose square is 2:     $2^{1/2} = \sqrt{2}$;

$2^{1/3}$ is the positive number whose cube is 2:     $2^{1/3} = \sqrt[3]{2}$;

$2^{1/4}$ is the positive number whose fourth power is 2:   $2^{1/4} = \sqrt[4]{2}$;

and so on. In general, when $b$ is zero or positive, $b^{1/n}$ is zero or the positive number whose $n$th power is $b$.

If $n$ is even, there is no number whose $n$th power is $b$ if $b$ is negative. Therefore, $b^{1/n}$ is not defined if $n$ is even and $b < 0$. When $n$ is odd, we may permit $b$ to be negative as well as positive. For example, $(-8)^{1/3}$ is the number whose cube is $-8$; that is,

$$(-8)^{1/3} = -2.$$

Thus, when $b$ is negative and $n$ is odd, we again define $b^{1/n}$ to be the number whose $n$th power is $b$.

Finally, let us consider numbers of the form $b^{m/n}$ and $b^{-m/n}$, where $m$ and $n$ are positive integers. We may assume that the fraction $m/n$ is in lowest terms (so that $m$ and $n$ have no common factor). Then we define

$$b^{m/n} = (b^{1/n})^m$$

whenever $b^{1/n}$ is defined, and

$$b^{-m/n} = \frac{1}{b^{m/n}}$$

whenever $b^{m/n}$ is defined and is not zero. For example,

$$8^{5/3} = (8^{1/3})^5 = (2)^5 = 32,$$

$$8^{-5/3} = \frac{1}{8^{5/3}} = \frac{1}{32},$$

$$(-8)^{5/3} = \left[(-8)^{1/3}\right]^5 = (-2)^5 = -32.$$

Exponents may be manipulated algebraically according to the following rules:

---

**Laws of Exponents**

**1.** $b^r b^s = b^{r+s}$                     **4.** $(b^r)^s = b^{rs}$

**2.** $b^{-r} = \dfrac{1}{b^r}$                **5.** $(ab)^r = a^r b^r$

**3.** $\dfrac{b^r}{b^s} = b^r \cdot b^{-s} = b^{r-s}$    **6.** $\left(\dfrac{a}{b}\right)^r = \dfrac{a^r}{b^r}$

---

**EXAMPLE 1**   Use the laws of exponents to calculate the following quantities.

(a) $2^{1/2} 50^{1/2}$          (b) $(2^{1/2} 2^{1/3})^6$          (c) $\dfrac{5^{3/2}}{\sqrt{5}}$

**Solution**   (a) $2^{1/2} 50^{1/2} = (2 \cdot 50)^{1/2}$          (Law 5)

$$= \sqrt{100}$$

$$= 10$$

(b) $\left(2^{1/2} 2^{1/3}\right)^6 = \left(2^{(1/2)+(1/3)}\right)^6$          (Law 1)

$$= \left(2^{5/6}\right)^6$$

$$= 2^{(5/6)6}$$          (Law 4)

$$= 2^5$$

$$= 32$$

(c) $\dfrac{5^{3/2}}{\sqrt{5}} = \dfrac{5^{3/2}}{5^{1/2}}$

$$= 5^{(3/2)-(1/2)}$$          (Law 3)

$$= 5^1$$

$$= 5$$          ◼

Now that we know the definition of $x^r$ for $r$ rational, we can examine some calculations and applications involving power functions.

---

**EXAMPLE 2**   Simplify the following expressions.

(a) $\dfrac{1}{x^{-4}}$          (b) $\dfrac{x^2}{x^5}$          (c) $\sqrt{x}\left(x^{3/2} + 3\sqrt{x}\right)$

**Solution** (a) $\dfrac{1}{x^{-4}} = x^{-(-4)}$        (Law 2 with $r = -4$)

$\qquad\qquad = x^4$

(b) $\dfrac{x^2}{x^5} = x^{2-5}$        (Law 3)

$\qquad\qquad = x^{-3}$

It is also correct to write this answer as $\dfrac{1}{x^3}$.

(c) $\sqrt{x}\left(x^{3/2} + 3\sqrt{x}\right) = x^{1/2}\left(x^{3/2} + 3x^{1/2}\right)$

$\qquad\qquad\qquad = x^{1/2}x^{3/2} + 3x^{1/2}x^{1/2}$

$\qquad\qquad\qquad = x^{(1/2)+(3/2)} + 3x^{(1/2)+(1/2)}$        (Law 1)

$\qquad\qquad\qquad = x^2 + 3x$        ■

A *power function* is a function of the form

$$f(x) = x^r,$$

for some number $r$.

---

**EXAMPLE 3**  Let $f(x)$ and $g(x)$ be the power functions

$$f(x) = x^{-1} \quad \text{and} \quad g(x) = x^{1/2}.$$

Determine the following functions.

(a) $\dfrac{f(x)}{g(x)}$        (b) $f(x)g(x)$        (c) $\dfrac{g(x)}{f(x)}$

**Solution** (a) $\dfrac{f(x)}{g(x)} = \dfrac{x^{-1}}{x^{1/2}} = x^{-1-(1/2)} = x^{-3/2} = \dfrac{1}{x^{3/2}}$

(b) $f(x)g(x) = x^{-1}x^{1/2} = x^{-1+(1/2)} = x^{-1/2} = \dfrac{1}{x^{1/2}} = \dfrac{1}{\sqrt{x}}$

(c) $\dfrac{g(x)}{f(x)} = \dfrac{x^{1/2}}{x^{-1}} = x^{(1/2)-(-1)} = x^{3/2}$        ■

**Compound Interest**  The subject of compound interest provides a significant application of exponents. Let's introduce this topic at this point with a view toward using it as a source of applied problems throughout the book.

When money is deposited in a savings account, interest is paid at stated intervals. If this interest is added to the account and thereafter earns interest itself, then the interest is called *compound interest*. The original amount deposited is called the *principal amount*. The principal amount plus the compound interest is called the *compound amount*. The interval between interest payments is referred to as the *interest period*. In formulas for compound interest, the interest rate is expressed as a decimal rather than a percent. Thus 6% is written as .06.

If $1000 is deposited at 6% annual interest, compounded annually, the compound amount at the end of the first year will be

$$A_1 = \underset{\text{principal}}{1000} + \underset{\text{interest}}{1000(.06)} = 1000(1 + .06).$$

At the end of the second year the compound amount will be

$$A_2 = \underset{\substack{\text{compound} \\ \text{amount}}}{A_1} + \underset{\text{interest}}{A_1(.06)} = A_1(1 + .06)$$

$$= [1000(1 + .06)](1 + .06) = 1000(1 + .06)^2.$$

At the end of 3 years,

$$A_3 = A_2 + A_2(.06) = A_2(1 + .06)$$
$$= \left[1000(1 + .06)^2\right](1 + .06) = 1000(1 + .06)^3.$$

After $n$ years the compound amount will be

$$A = 1000(1 + .06)^n.$$

In this example the interest period was 1 year. The important point to note, however, is that at the end of each interest period the amount on deposit grew by a factor of $(1 + .06)$. In general, if the interest rate is $i$ instead of .06, the compound amount will grow by a factor of $(1 + i)$ at the end of each interest period.

Suppose that a principal amount $P$ is invested at a compound interest rate $i$ per interest period, for a total of $n$ interest periods. Then the compound amount $A$ at the end of the $n$th period will be

$$A = P(1 + i)^n. \tag{1}$$

**EXAMPLE 4** Suppose that $5000 is invested at 8% per year, with interest compounded annually. What is the compound amount after 3 years?

**Solution** Substituting $P = 5000$, $i = .08$, and $n = 3$ into formula (1), we have
$$A = 5000(1 + .08)^3 = 5000(1.08)^3$$
$$= 5000(1.259712) = 6298.56 \text{ dollars.} \qquad \blacksquare$$

It is common practice to state the interest rate as a percentage per year ("per annum"), even though each interest period is often shorter than 1 year. If the annual rate is $r$ and if interest is paid and compounded $m$ times per year, then the interest rate $i$ for each period is given by

$$[\text{rate per period}] = i = \frac{r}{m} = \frac{[\text{annual interest rate}]}{[\text{periods per year}]}.$$

Many banks pay interest quarterly. If the stated annual rate is 5%, then $i = .05/4 = .0125$.

If interest is compounded for $t$ years, with $m$ interest periods each year, there will be a total of $mt$ interest periods. If in formula (1) we replace $n$ by $mt$ and replace $i$ by $r/m$, we obtain the following formula for the compound amount:

$$A = P\left(1 + \frac{r}{m}\right)^{mt},$$

where $P$ = principal amount,
$\quad r$ = interest rate per annum,
$\quad m$ = number of interest periods per year,
$\quad t$ = number of years. $\tag{2}$

**EXAMPLE 5**  Suppose that $1000 is deposited in a savings account that pays 6% per annum, compounded quarterly. If no additional deposits or withdrawals are made, how much will be in the account at the end of 1 year?

**Solution**  We use (2) with $P = 1000$, $r = .06$, $m = 4$, and $t = 1$.

$$A = 1000 \left(1 + \frac{.06}{4}\right)^4 = 1000(1.015)^4$$

$$\approx 1000(1.06136355) \approx 1061.36 \text{ dollars.} \quad \blacksquare$$

Note that the $1000 in Example 5 earned a total of $61.36 in (compound) interest. This is 6.136% of $1000. Savings institutions sometimes advertise this rate as the *effective* annual interest rate. That is, the savings institutions mean that *if* they paid interest only once a year, they would have to pay a rate of 6.136% in order to produce the same earnings as their 6% rate compounded quarterly. The stated rate of 6% is often called the *nominal rate*.

The effective annual rate can be increased by compounding the interest more often. Some savings institutions compound interest monthly or even daily.

**EXAMPLE 6**  Suppose that the interest in Example 5 were compounded monthly. How much would be in the account at the end of 1 year? What about the case when 6% annual interest is compounded daily?

**Solution**  For monthly compounding, $m = 12$. From (2) we have

$$A = 1000 \left(1 + \frac{.06}{12}\right)^{12} = 1000(1.005)^{12} \approx 1061.68 \text{ dollars.}$$

The effective rate in this case is 6.168%.

A "bank year" usually consists of 360 days (in order to simplify calculations). So, for daily compounding, we take $m = 360$. Then

$$A = 1000 \left(1 + \frac{.06}{360}\right)^{360} \approx 1000(1.00016667)^{360}$$

$$\approx 1000(1.06183251) \approx 1061.83 \text{ dollars.}$$

With daily compounding, the effective rate is 6.183%. $\quad \blacksquare$

**EXAMPLE 7**  Suppose that a corporation issues a bond costing $200 and paying interest compounded monthly. The interest is accumulated until the bond reaches maturity. (A security of this sort is called a *zero coupon bond*.) Suppose that after 5 years, the bond is worth $500. What is the annual interest rate?

**Solution**  Let $r$ denote the annual interest rate. The value $A$ of the bond after 5 years $= 60$ months is given by the compound interest formula:

$$A = 200 \left(1 + \frac{r}{12}\right)^{60}.$$

We must find $r$ that satisfies

$$500 = 200 \left(1 + \frac{r}{12}\right)^{60}$$

$$2.5 = \left(1 + \frac{r}{12}\right)^{60}.$$

Raise both sides to the power $\frac{1}{60}$ and apply the laws of exponents to obtain

$$(2.5)^{1/60} = \left[\left(1 + \frac{r}{12}\right)^{60}\right]^{\frac{1}{60}} = \left(1 + \frac{r}{12}\right)^{60 \cdot \frac{1}{60}} = 1 + \frac{r}{12}$$

$$r = 12 \cdot \left((2.5)^{1/60} - 1\right).$$

Using a calculator, we see that $r \approx .18466$. That is, the annual interest rate is 18.466%. (A bond paying a rate of interest this high is generally called a *junk bond*.) ■

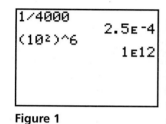

**Figure 1**

**Incorporating Technology** On graphing calculators, very small numbers and very large numbers are automatically displayed in exponential form. In Fig. 1, **2.5E-4** stands for $2.5 \cdot 10^{-4}$ or .00025, and **1E12** stands for $1 \cdot 10^{12}$ or 1,000,000,000,000. (*Note:* Multiplying a number by $10^{-4}$ moves the decimal point four places to the left and multiplying by $10^{12}$ moves the decimal point 12 places to the right.) You may enter numbers in scientific notation using the **EE** (= enter exponent) key found on most calculators.

## PRACTICE PROBLEMS 0.5

**1.** Compute the following.
    **(a)** $-5^2$      **(b)** $16^{.75}$

**2.** Simplify the following.
    **(a)** $(4x^3)^2$      **(b)** $\dfrac{\sqrt[3]{x}}{x^3}$      **(c)** $\dfrac{2 \cdot (x+5)^6}{x^2 + 10x + 25}$

# EXERCISES 0.5

*In Exercises 1–28, compute the numbers.*

**1.** $3^3$      **2.** $(-2)^3$      **3.** $1^{100}$

**4.** $0^{25}$      **5.** $(.1)^4$      **6.** $(100)^4$

**7.** $-4^2$      **8.** $(.01)^3$      **9.** $(16)^{1/2}$

**10.** $(27)^{1/3}$      **11.** $(.000001)^{1/3}$      **12.** $\left(\dfrac{1}{125}\right)^{1/3}$

**13.** $6^{-1}$      **14.** $\left(\dfrac{1}{2}\right)^{-1}$      **15.** $(.01)^{-1}$

**16.** $(-5)^{-1}$      **17.** $8^{4/3}$      **18.** $16^{3/4}$

**19.** $(25)^{3/2}$      **20.** $(27)^{2/3}$      **21.** $(1.8)^0$

**22.** $9^{1.5}$      **23.** $16^{.5}$      **24.** $(81)^{.75}$

**25.** $4^{-1/2}$      **26.** $\left(\dfrac{1}{8}\right)^{-2/3}$      **27.** $(.01)^{-1.5}$

**28.** $1^{-1.2}$

*In Exercises 29–40, use the laws of exponents to compute the numbers.*

**29.** $5^{1/3} \cdot 200^{1/3}$      **30.** $(3^{1/3} \cdot 3^{1/6})^6$      **31.** $6^{1/3} \cdot 6^{2/3}$

**32.** $(9^{4/5})^{5/8}$      **33.** $\dfrac{10^4}{5^4}$      **34.** $\dfrac{3^{5/2}}{3^{1/2}}$

**35.** $(2^{1/3} \cdot 3^{2/3})^3$      **36.** $20^{.5} \cdot 5^{.5}$      **37.** $\left(\dfrac{8}{27}\right)^{2/3}$

**38.** $(125 \cdot 27)^{1/3}$      **39.** $\dfrac{7^{4/3}}{7^{1/3}}$      **40.** $(6^{1/2})^0$

*In Exercises 41–70, use the laws of exponents to simplify the algebraic expressions. Your answer should not involve parentheses or negative exponents.*

**41.** $(xy)^6$      **42.** $(x^{1/3})^6$      **43.** $\dfrac{x^4 \cdot y^5}{xy^2}$

**44.** $\dfrac{1}{x^{-3}}$      **45.** $x^{-1/2}$      **46.** $(x^3 \cdot y^6)^{1/3}$

**47.** $\left(\dfrac{x^4}{y^2}\right)^3$      **48.** $\left(\dfrac{x}{y}\right)^{-2}$      **49.** $(x^3 y^5)^4$

**50.** $\sqrt{1+x}\,(1+x)^{3/2}$      **51.** $x^5 \cdot \left(\dfrac{y^2}{x}\right)^3$

**52.** $x^{-3} \cdot x^7$      **53.** $(2x)^4$      **54.** $\dfrac{-3x}{15x^4}$

**55.** $\dfrac{-x^3 y}{-xy}$      **56.** $\dfrac{x^3}{y^{-2}}$      **57.** $\dfrac{x^{-4}}{x^3}$

**58.** $(-3x)^3$      **59.** $\sqrt[3]{x} \cdot \sqrt[3]{x^2}$      **60.** $(9x)^{-1/2}$

**61.** $\left(\dfrac{3x^2}{2y}\right)^3$      **62.** $\dfrac{x^2}{x^5 y}$      **63.** $\dfrac{2x}{\sqrt{x}}$

**64.** $\dfrac{1}{yx^{-5}}$      **65.** $(16x^8)^{-3/4}$      **66.** $(-8y^9)^{2/3}$

**67.** $\sqrt{x}\left(\dfrac{1}{4x}\right)^{5/2}$

**68.** $\dfrac{(25xy)^{3/2}}{x^2y}$

**69.** $\dfrac{(-27x^5)^{2/3}}{\sqrt[3]{x}}$

**70.** $(-32y^{-5})^{3/5}$

*The expressions in Exercises 71–74 may be factored as shown. Find the missing factors.*

**71.** $\sqrt{x} - \dfrac{1}{\sqrt{x}} = \dfrac{1}{\sqrt{x}}(\quad)$

**72.** $2x^{2/3} - x^{-1/3} = x^{-1/3}(\quad)$

**73.** $x^{-1/4} + 6x^{1/4} = x^{-1/4}(\quad)$

**74.** $\sqrt{\dfrac{x}{y}} - \sqrt{\dfrac{y}{x}} = \sqrt{xy}\,(\quad)$

**75.** Explain why $\sqrt{a} \cdot \sqrt{b} = \sqrt{ab}$.

**76.** Explain why $\sqrt{a}/\sqrt{b} = \sqrt{a/b}$.

*In Exercises 77–84, evaluate $f(4)$.*

**77.** $f(x) = x^2$

**78.** $f(x) = x^3$

**79.** $f(x) = x^{-1}$

**80.** $f(x) = x^{1/2}$

**81.** $f(x) = x^{3/2}$

**82.** $f(x) = x^{-1/2}$

**83.** $f(x) = x^{-5/2}$

**84.** $f(x) = x^0$

*Calculate the compound amount from the given data in Exercises 85–92.*

**85.** principal = \$500, compounded annually, 6 years, annual rate = 6%

**86.** principal = \$700, compounded annually, 8 years, annual rate = 8%

**87.** principal = \$50,000, compounded quarterly, 10 years, annual rate = 9.5%

**88.** principal = \$20,000, compounded quarterly, 3 years, annual rate = 12%

**89.** principal = \$100, compounded monthly, 10 years, annual rate = 5%

**90.** principal = \$500, compounded monthly, 1 year, annual rate = 4.5%

**91.** principal = \$1500, compounded daily, 1 year, annual rate = 6%

**92.** principal = \$1500, compounded daily, 3 years, annual rate = 6%

**93.** (*Annual Compound*) Assume that a couple invests \$1000 upon the birth of their daughter. Assume that the investment earns 6.8% compounded annually. What will the investment be worth on the daughter's 18th birthday?

**94.** (*Annual Compound with Deposits*) Assume that a couple invests \$4000 each year for four years in an investment that earns 8% compounded annually. What will the value of the investment be 8 years after the first amount is invested?

**95.** (*Quarterly Compound*) Assume that a \$500 investment earns interest compounded quarterly. Express the value of the investment after one year as a polynomial in the annual rate of interest $r$.

**96.** (*Semiannual Compound*) Assume that a \$1000 investment earns interest compounded semiannually. Express the value of the investment after two years as a polynomial in the annual rate of interest $r$.

**97.** (*Velocity*) When a car's brakes are slammed on at a speed of $x$ miles per hour, the stopping distance is $\frac{1}{20}x^2$ feet. Show that when the speed is doubled, the stopping distance increases fourfold.

## Technology Exercises
*In Exercises 98–101, convert the numbers from graphing calculator form to standard form (that is, without E).*

**98. 5E-5**

**99. 8.103E-4**

**100. 1.35E13**

**101. 8.23E-6**

## SOLUTIONS TO PRACTICE PROBLEMS 0.5

**1. (a)** $-5^2 = -25$. [Note that $-5^2$ is the same as $-(5^2)$. This number is different from $(-5)^2$, which equals 25. Whenever there are no parentheses, apply the exponent first and then apply the other operations.]

**(b)** Since $.75 = \frac{3}{4}$, $16^{.75} = 16^{3/4} = (\sqrt[4]{16})^3 = 2^3 = 8$.

**2. (a)** Apply Law 5 with $a = 4$ and $b = x^3$. Then use Law 4.

$$(4x^3)^2 = 4^2 \cdot (x^3)^2 = 16 \cdot x^6$$

[A common error is to forget to square the 4. If that had been our intent, we would have asked for $4\left(x^3\right)^2$.]

**(b)** $\dfrac{\sqrt[3]{x}}{x^3} = \dfrac{x^{1/3}}{x^3} = x^{(1/3)-3} = x^{-8/3}$.

[The answer can also be given as $1/x^{8/3}$.] When simplifying expressions involving radicals, it is usually a good idea to first convert the radicals to exponents.

**(c)** $\dfrac{2(x+5)^6}{x^2 + 10x + 25} = \dfrac{2 \cdot (x+5)^6}{(x+5)^2} = 2(x+5)^{6-2}$

$$= 2(x+5)^4.$$

[Here the third law of exponents was applied to $(x+5)$. The laws of exponents apply to any algebraic expression.]

## 0.6  Functions and Graphs in Applications

The key step in solving many applied problems in this text is to construct appropriate functions or equations. Once this is done, the remaining mathematical steps are usually straightforward. This section focuses on representative applied problems and reviews skills needed to set up and analyze functions, equations, and their graphs.

**Geometric Problems**   Many examples and exercises in the text involve dimensions, areas, or volumes of objects similar to those in Fig. 1. When a problem involves a plane figure, such as a rectangle or circle, one must distinguish between the *perimeter* and the *area* of the figure. The perimeter of a figure, or "distance around" the figure, is a *length* or *sum of lengths.* Typical units, if specified, are inches, feet, centimeters, meters, and so on. Area involves the *product of two lengths,* and the units are *square* inches, *square* feet, *square* centimeters, and so on.

---

**EXAMPLE 1**   **Cost**  Suppose the longer side of the rectangle in Fig. 1 has twice the length of the shorter side, and let $x$ denote the length of the shorter side.

(a) Express the perimeter of the rectangle as a function of $x$.

(b) Express the area of the rectangle as a function of $x$.

(c) Suppose the rectangle represents a kitchen countertop to be constructed of a durable material costing \$25 per square foot. Write a function $C(x)$ that expresses the cost of the material as a function of $x$, where lengths are in feet.

**Figure 1.**  Geometric figures.        Rectangle                Rectangular box              Cylinder

**Solution**  (a) The rectangle is shown in Fig. 2. The length of the longer side is $2x$. If the perimeter is denoted by $P$, then $P$ is the sum of the lengths of the four sides of the rectangle, namely, $x + 2x + x + 2x$. That is, $P = 6x$.

**Figure 2**

(b) The area $A$ of the rectangle is the product of the lengths of two adjacent sides. That is, $A = x \cdot 2x = 2x^2$.

(c) Here the area is measured in square feet. The basic principle for this part is

$$\begin{bmatrix} \text{cost of} \\ \text{materials} \end{bmatrix} = \begin{bmatrix} \text{cost per} \\ \text{square foot} \end{bmatrix} \cdot \begin{bmatrix} \text{number of} \\ \text{square feet} \end{bmatrix}$$

$$C(x) \quad = \quad\quad 25 \quad\quad \cdot \quad\quad 2x^2$$

$$= 50x^2 \text{ (dollars)}.\quad\blacksquare$$

When a problem involves a three-dimensional object, such as a box or cylinder, one must distinguish between the *surface area* of the object and the *volume* of the object. Surface area is an area, of course, so it is measured in *square* units. Typically, the surface area is a *sum of areas* (each area is a product of two lengths). The volume of an object is often a *product of three lengths* and is measured in *cubic* units.

**EXAMPLE 2**  **Surface area**  A rectangular box has a square copper base, wooden sides, and a wooden top. The copper costs $21 per square foot and the wood costs $2 per square foot.

(a) Write an expression giving the surface area (that is, the sum of the areas of the bottom, the top, and the four sides of the box) in terms of the dimensions of the box. Also, write an expression giving the volume of the box.

(b) Write an expression giving the total cost of the materials used to make the box in terms of the dimensions.

**Solution** (a) The first step is to assign letters to the dimensions of the box. Denote the length of one (and therefore every) side of the square base by $x$, and denote the height of the box by $h$. See Fig. 3.

The top and bottom each has area $x^2$, and each of the four sides has area $xh$. Therefore, the surface area is $2x^2 + 4xh$. The volume of the box is the product of the length, width, and height. Because the base is square, the volume is $x^2h$.

**Figure 3.** Closed box.

(b) When the various surfaces of the box have different costs per square foot, the cost of each is computed separately:

$$[\text{cost of bottom}] = [\text{cost per sq. ft.}] \cdot [\text{area of bottom}] = 21x^2;$$
$$[\text{cost of top}] = [\text{cost per sq. ft.}] \cdot [\text{area of top}] = 2x^2;$$
$$[\text{cost of one side}] = [\text{cost per sq. ft.}] \cdot [\text{area of one side}] = 2xh.$$

The total cost is

$$C = [\text{cost of bottom}] + [\text{cost of top}] + 4 \cdot [\text{cost of one side}]$$
$$= 21x^2 + 2x^2 + 4 \cdot 2xh = 23x^2 + 8xh.$$  ■

**Business Problems**  Many applications in the text involve cost, revenue, and profit functions.

**EXAMPLE 3**  **Cost**  Suppose a toy manufacturer has fixed costs of $3000 (such as rent, insurance, and business loans) that must be paid no matter how many toys are produced. In addition, there are variable costs of $2 per toy. At a production level of $x$ toys, the variable costs are $2 \cdot x$ (dollars) and the total cost is

$$C(x) = 3000 + 2x \quad \text{(dollars)}.$$

(a) Find the cost of producing 2000 toys.

(b) What additional cost is incurred if the production level is raised from 2000 toys to 2200 toys?

(c) To answer the question "How many toys may be produced at a cost of $5000?" should you compute $C(5000)$ or should you solve the equation $C(x) = 5000$?

**Solution** (a) $C(2000) = 3000 + 2(2000) = 7000$ (dollars).

(b) The total cost when $x = 2200$ is $C(2200) = 3000 + 2(2200) = 7400$ (dollars). So the *increase* in cost when production is raised from 2000 to 2200 toys is

$$C(2200) - C(2000) = 7400 - 7000 = 400 \quad \text{(dollars)}.$$

(c) This is an important type of question. The phrase "how many toys" implies that the quantity $x$ is unknown. Therefore, the answer is found by solving $C(x) = 5000$ for $x$:

$$3000 + 2x = 5000$$
$$2x = 2000$$
$$x = 1000 \quad \text{(toys)}.$$

Another way to analyze this problem is to look at the types of units involved. The input $x$ of the cost function is the *quantity* of toys, and the output of the cost function is the *cost*, measured in dollars. Since the question involves 5000 *dollars*, it is the *output* that is specified. The input $x$ is unknown. ∎

---

**EXAMPLE 4**  **Cost, profit, and revenue**  Suppose the toys in Example 3 sell for $10 apiece. When $x$ toys are sold, the revenue (amount of money received) $R(x)$ is $10x$ dollars. Given the same cost function, $C(x) = 3000 + 2x$, the profit (or loss) $P(x)$ generated by the $x$ toys will be

$$P(x) = R(x) - C(x)$$
$$= 10x - (3000 + 2x) = 8x - 3000.$$

(a) To determine the revenue generated by 8000 toys, should you compute $R(8000)$ or should you solve the equation $R(x) = 8000$?

(b) If the revenue from the production and sale of some toys is $7000, what is the corresponding profit?

**Solution**  (a) The revenue is unknown, but the input to the revenue function is known. So compute $R(8000)$ to find the revenue.

(b) The profit is unknown, so we want to compute the value of $P(x)$. Unfortunately, we don't know the value of $x$. However, the fact that the revenue is $7000 enables us to solve for $x$. Thus the solution has two steps:

(i) Solve $R(x) = 7000$ to find $x$.

$$10x = 7000$$
$$x = 700 \quad \text{(toys)}$$

(ii) Compute $P(x)$ when $x = 700$.

$$P(x) = 8(700) - 3000$$
$$= 2600 \quad \text{(dollars)} \qquad ∎$$

**Functions and Graphs**  When a function arises in an applied problem, the graph of the function provides useful information. Every statement or task involving a function corresponds to a feature or task involving its graph. This "graphical" point of view will broaden your understanding of functions and strengthen your ability to work with them.

Modern graphing calculators and calculus computer software provide excellent tools for thinking geometrically about functions. Most popular graphing calculators and programs provide a *cursor* or *cross hairs* that may be moved to any point on the screen, with the $x$- and $y$-coordinates of the cursor displayed somewhere on the screen. The next example shows how geometric calculations with the graph of a function correspond to the more familiar numerical computations. This example is worth reading even if a computer (or calculator) is unavailable.

**EXAMPLE 5**  To plan for future growth, a company analyzes production costs for one of its products and estimates that the cost (in dollars) of operating at a production level of $x$ units per hour is given by the function

$$C(x) = 150 + 59x - 1.8x^2 + .02x^3.$$

Suppose the graph of this function is available, either displayed on the screen of a graphing utility or perhaps printed on graph paper in a company report. See Fig. 4.

(a) The point $(16, 715.12)$ is on the graph. What does that say about the cost function $C(x)$?

(b) The equation $C(x) = 900$ may be solved graphically by finding a certain point on the graph and reading its $x$- and $y$-coordinates. Describe how to locate the point. How do the coordinates of the point provide the solution to the equation $C(x) = 900$?

(c) The task "Find $C(45)$" may be completed graphically by finding a point on the graph. Describe how to locate the point. How do the coordinates of the point provide the value of $C(45)$?

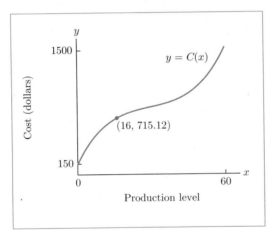

**Figure 4.** Graph of a cost function.

**Solution** (a) The fact that $(16, 715.12)$ is on the graph of $C(x)$ means that $C(16) = 715.12$. That is, if the production level is 16 units per hour, then the cost is $715.12.

(b) To solve $C(x) = 900$ graphically, locate 900 on the $y$-axis and move to the right until you reach the point $(?, 900)$ on the graph of $C(x)$. See Fig. 5. The $x$-coordinate of the point is the solution of $C(x) = 900$. Estimate $x$ graphically. (With a graphing utility, find the coordinates of the point of intersection of the graph with the horizontal line $y = 900$; on graph paper, use a ruler to find $x$ on the $x$-axis. To two decimal places, $x = 39.04$.)

(c) To find $C(45)$ graphically, locate 45 on the $x$-axis and move up until you reach the point $(45, ?)$ on the graph of $C(x)$. See Fig. 6. The $y$-coordinate of the point is the value of $C(45)$. [In fact, $C(45) = 982.50$.]  ■

The two final examples illustrate how to extract information about a function by examining its graph.

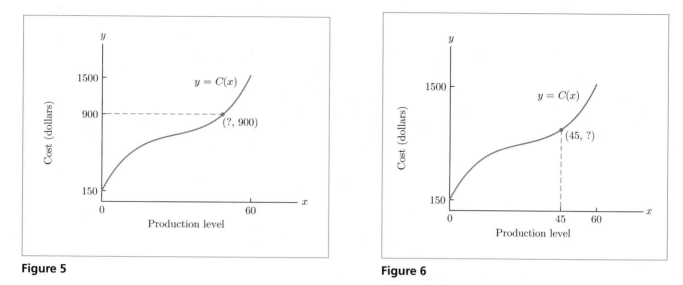

**Figure 5**

**Figure 6**

---

**EXAMPLE 6**  Figure 7 is the graph of the function $R(x)$, the revenue obtained from selling $x$ bicycles.

(a) What is the revenue from the sale of 1000 bicycles?

(b) How many bicycles must be sold to achieve a revenue of $102,000?

(c) What is the revenue from the sale of 1100 bicycles?

(d) What additional revenue is derived from the sale of 100 more bicycles if the current sales level is 1000 bicycles?

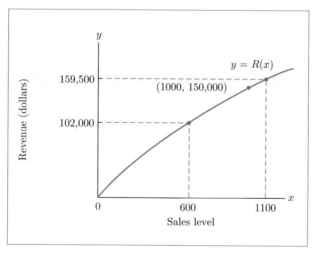

**Figure 7.**  Graph of a revenue function.

**Solution**  (a) Since $(1000, 150,000)$ is on the graph of $R(x)$, the revenue from the sale of 1000 bicycles is $150,000.

(b) The horizontal line at $y = 102,000$ intersects the graph at the point with $x$-coordinate 600. Therefore, the revenue from the sale of 600 bicycles is $102,000.

(c) The vertical line at $x = 1100$ intersects the graph at the point with $y$-coordinate 159,500. Therefore, $R(1100) = 159,500$ and the revenue is $159,500.

(d) When the value of $x$ increases from 1000 to 1100, the revenue increases from 150,000 to 159,500. Therefore, the additional revenue is $9500. ■

**EXAMPLE 7** **Height of a ball** A ball is thrown straight up into the air from the top of a 64-foot tower. The function $h(t)$, the height of the ball (in feet) after $t$ seconds, has the graph shown in Fig. 8. (*Note:* This graph is not a picture of the physical path of the ball; the ball is thrown vertically into the air.)

(a) What is the height of the ball after 1 second?

(b) After how many seconds does the ball reach its greatest height, and what is this height?

(c) After how many seconds does the ball hit the ground?

(d) When is the height 64 feet?

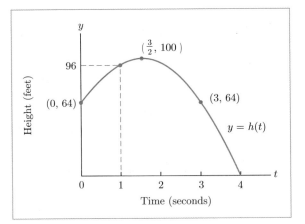

**Figure 8.** Graph of a height function.

**Solution** (a) Since the point $(1, 96)$ is on the graph of $h(t)$, $h(1) = 96$. Therefore, the height of the ball after 1 second is 96 feet.

(b) The highest point on the graph of the function has coordinates $\left(\frac{3}{2}, 100\right)$. Therefore, after $\frac{3}{2}$ seconds the ball achieves its greatest height, 100 feet.

(c) The ball hits the ground when the height is 0. This occurs after 4 seconds.

(d) The height of 64 feet occurs twice, at times $t = 0$ and $t = 3$ seconds. ∎

Table 1 summarizes most of the concepts in Examples 3–7. Although stated here for a profit function, the concepts will arise later for many other types of functions as well. Each statement about the profit is translated into a statement about $f(x)$ and a statement about the graph of $f(x)$. The graph in Fig. 9 illustrates each statement.

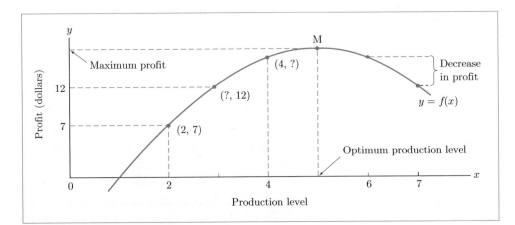

**Figure 9.** Graph of a profit function.

**TABLE 1** Translating an Applied Problem

*Assume that $f(x)$ is the profit in dollars at production level $x$.*

| Applied Problem | Function | Graph |
|---|---|---|
| When production is at 2 units, the profit is $7. | $f(2) = 7$. | The point $(2, 7)$ is on the graph. |
| Determine the number of units that generate a profit of $12. | Solve $f(x) = 12$ for $x$. | Find the $x$-coordinate(s) of the point(s) on the graph whose $y$-coordinate is 12. |
| Determine the profit when the production level is 4 units. | Evaluate $f(4)$. | Find the $y$-coordinate of the point on the graph whose $x$-coordinate is 4. |
| Find the production level that maximizes the profit. | Find $x$ such that $f(x)$ is as large as possible. | Find the $x$-coordinate of the highest point, $M$, on the graph. |
| Determine the maximum profit. | Find the maximum value of $f(x)$. | Find the $y$-coordinate of the highest point on the graph. |
| Determine the change in profit when the production level is changed from 6 to 7 units. | Find $f(7) - f(6)$. | Determine the difference in $y$-coordinates of the points with $x$-coordinates 7 and 6. |
| The profit decreases when the production level is changed from 6 to 7 units. | The function value decreases when $x$ changes from 6 to 7. | The point on the graph with $x$-coordinate 6 is higher than the point with $x$-coordinate 7. |

**Figure 10**

**Incorporating Technology** The function in Fig. 9 has the equation $y = -x^2 + 10x - 9$. Most of the tasks from Table 1 can be carried out with a graphing calculator. In Fig. 10, Task 2 is accomplished by finding the intersection of the graph of the profit function and the graph of $y = 12$. In Fig. 11(a), Task 3 is accomplished by **TRACE** (possibly with some help from **value**), and in Fig. 11(b) Task 3 is accomplished by evaluating the function on the home screen. In Fig. 12, Tasks 4 and 5 are carried out with the **maximum** (or FMAX) routine. The computation for Task 6 can be carried out on the home screen of a TI-82, 83, or 86 as shown in Fig. 13.

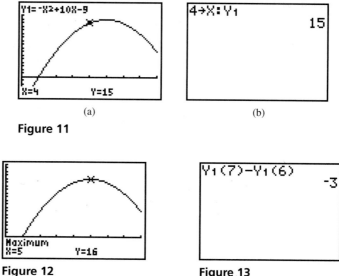

(a)

(b)

**Figure 11**

**Figure 12**

**Figure 13**

## PRACTICE PROBLEMS 0.6

Consider the cylinder shown in Fig. 14.

1. Assign letters to the dimensions of the cylinder.

2. The girth of the cylinder is the circumference of the colored circle in the figure. Express the girth in terms of the dimensions of the cylinder.

3. What is the area of the bottom (or top) of the cylinder?

4. What is the surface area of the side of the cylinder? (*Hint*: Imagine cutting the side of the cylinder and unrolling the cylinder to form a rectangle.)

**Figure 14**

# EXERCISES 0.6

*In Exercises 1–6, assign letters to the dimensions of the geometric object.*

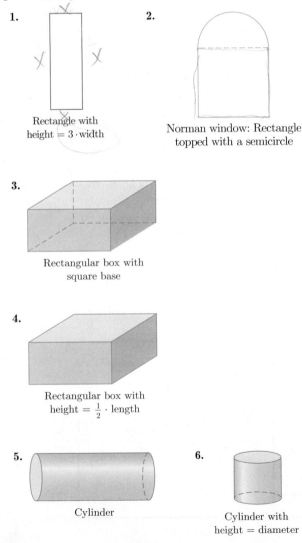

1.

Rectangle with height = 3·width

2.

Norman window: Rectangle topped with a semicircle

3.

Rectangular box with square base

4.

Rectangular box with height = $\frac{1}{2}$ · length

5.

Cylinder

6.

Cylinder with height = diameter

*Exercises 7–14 refer to the letters assigned to the figures in Exercises 1–6.*

7. (*Perimeter, Area*) Consider the rectangle in Exercise 1. Write an expression for the perimeter. Suppose the area is 25 square feet, and write this fact as an equation.

8. (*Perimeter, Area*) Consider the rectangle in Exercise 1. Write an expression for the area. Write an equation expressing the fact that the perimeter is 30 centimeters.

9. (*Area, Circumference*) Consider a circle of radius $r$. Write an expression for the area. Write an equation expressing the fact that the circumference is 15 centimeters.

10. (*Perimeter, Area*) Consider the Norman window of Exercise 2. Write an expression for the perimeter. Write an equation expressing the fact that the area is 2.5 square meters.

11. (*Volume, Surface Area*) Consider the rectangular box in Exercise 3, and suppose that it has no top. Write an expression for the volume. Write an equation expressing the fact that the surface area is 65 square inches.

12. (*Surface Area, Volume*) Consider the closed rectangular box in Exercise 4. Write an expression for the surface area. Write an equation expressing the fact that the volume is 10 cubic feet.

13. (*Volume, Surface Area, and Cost*) Consider the cylinder of Exercise 5. Write an equation expressing the fact that the volume is 100 cubic inches. Suppose the material to construct the left end costs $5 per square inch, the material to construct the right end costs $6 per square inch, and the material to construct the side costs $7 per square inch. Write an expression for the total cost of material for the cylinder.

14. (*Surface Area, Volume*) Consider the cylinder of Exercise 6. Write an equation expressing the fact that the surface area is $30\pi$ square inches. Write an expression for the volume.

**15.** (*Fencing a Rectangular Corral*) Consider a rectangular corral with a partition down the middle, as shown in Fig. 15. Assign letters to the outside dimensions of the corral. Write an equation expressing the fact that 5000 feet of fencing are needed to construct the corral (including the partition). Write an expression for the total area of the corral.

**Figure 15**

**16.** (*Fencing a Rectangular Corral*) Consider a rectangular corral with two partitions, as in Fig. 16. Assign letters to the outside dimensions of the corral. Write an equation expressing the fact that the corral has a total area of 2500 square feet. Write an expression for the amount of fencing needed to construct the corral (including both partitions).

**Figure 16**

**17.** (*Cost of Fencing*) Consider the corral of Exercise 16. Suppose the fencing for the boundary of the corral costs $10 per foot and the fencing for the inner partitions costs $8 per foot. Write an expression for the total cost of the fencing.

**18.** (*Cost of an Open Box*) Consider the rectangular box of Exercise 3. Assume the box has no top, the material needed to construct the base costs $5 per square foot, and the material needed to construct the sides costs $4 per square foot. Write an equation expressing the fact that the total cost of materials is $150. (Use the dimensions assigned in Exercise 3.)

**19.** Suppose the rectangle in Exercise 1 has a perimeter of 40 cm. Find the area of the rectangle.

**20.** Suppose the cylinder in Exercise 6 has a volume of $54\pi$ cubic inches. Find the surface area of the cylinder.

**21.** (*Cost*) A specialty shop prints custom slogans and designs on T-shirts. The shop's total cost at a daily sales level of $x$ T-shirts is $C(x) = 73 + 4x$ dollars.

 **(a)** At what sales level will the cost be $225?

 **(b)** If the sales level is at 40 T-shirts, how much will the cost rise if the sales level changes to 50 T-shirts?

**22.** (*Cost, Revenue, and Profit*) A college student earns income by typing term papers on a computer, which

she leases (along with a printer). The student charges $4 per page for her work, and she estimates that her monthly cost when typing $x$ pages is $C(x) = .10x + 75$ dollars.

 **(a)** What is the student's profit if she types 100 pages in one month?

 **(b)** Determine the change in profit when the typing business rises from 100 to 101 pages per month.

**23.** (*Profit*) A frozen yogurt stand makes a profit of $P(x) = .40x - 80$ dollars when selling $x$ scoops of yogurt per day.

 **(a)** Find the break-even sales level, that is, the level at which $P(x) = 0$.

 **(b)** What sales level generates a daily profit of $30?

 **(c)** How many more scoops of yogurt will have to be sold to raise the daily profit from $30 to $40?

**24.** (*Profit*) A cellular telephone company estimates that if it has $x$ thousand subscribers, then its monthly profit is $P(x)$ thousand dollars, where $P(x) = 12x - 200$.

 **(a)** How many subscribers are needed for a monthly profit of 160 thousand dollars?

 **(b)** How many new subscribers would be needed to raise the monthly profit from 160 to 166 thousand dollars?

**25.** (*Cost, Revenue, and Profit*) An average sale at a small florist shop is $21, so the shop's weekly revenue function is $R(x) = 21x$, where $x$ is the number of sales in 1 week. The corresponding weekly cost is $C(x) = 9x + 800$ dollars.

 **(a)** What is the florist shop's weekly profit function?

 **(b)** How much profit is made when sales are at 120 per week?

 **(c)** If the profit is $1000 for a week, what is the revenue for the week?

**26.** A catering company estimates that if it has $x$ customers in a typical week, then its expenses will be approximately $C(x) = 550x + 6500$ dollars, and its revenue will be approximately $R(x) = 1200x$ dollars.

 **(a)** How much profit will the company earn in a week when it has 12 customers?

 **(b)** How much profit is the company making each week if the weekly costs are running at a level of $14,750?

*Exercises 27–32 refer to the function $f(r)$, which gives the cost (in cents) of constructing a 100-cubic-inch cylinder of radius $r$ inches. The graph of $f(r)$ is shown in Fig. 17.*

**27.** What is the cost of constructing a cylinder of radius 6 inches?

**28.** For what value(s) of $r$ is the cost 330 cents?

**29.** Interpret the fact that the point $(3, 162)$ is on the graph of the function.

**30.** Interpret the fact that the point $(3, 162)$ is the lowest point on the graph of the function. What does this say in terms of cost versus radius?

**31.** What is the additional cost of increasing the radius from 3 inches to 6 inches?

**32.** How much is saved by increasing the radius from 1 inch to 3 inches?

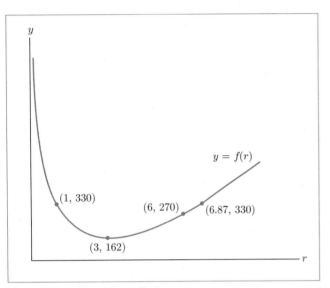

**Figure 17.** Cost of a cylinder.

*Exercises 33–36 refer to the cost and revenue functions in Fig. 18. The cost of producing $x$ units of goods is $C(x)$ dollars and the revenue from selling $x$ units of goods is $R(x)$ dollars.*

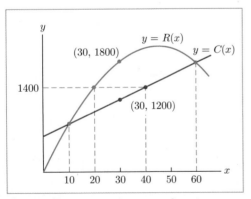

**Figure 18.** Cost and revenue functions.

**33.** What are the revenue and cost from the production and sale of 30 units of goods?

**34.** At what level of production is the revenue $1400?

**35.** At what level of production is the cost $1400?

**36.** What is the profit from the manufacture and sale of 30 units of goods?

*Exercises 37–40 refer to the cost function in Fig. 19.*

**37.** The point $(1000, 4000)$ is on the graph of the function. Restate this fact in terms of the function $C(x)$.

**38.** Translate the task "solve $C(x) = 3500$ for $x$" into a task involving the graph of the function.

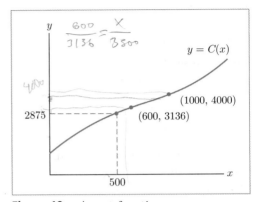

**Figure 19.** A cost function.

**39.** Translate the task "find $C(400)$" into a task involving the graph.

**40.** Suppose 500 units of goods are produced. What is the cost of producing 100 more units of goods?

*Exercises 41–44 refer to the profit function in Fig. 20.*

**41.** The point $(2500, 52{,}500)$ is the highest point on the graph of the function. What does this say in terms of profit versus quantity?

**42.** The point $(1500, 42{,}500)$ is on the graph of the function. Restate this fact in terms of the function $P(x)$.

**43.** Translate the task "solve $P(x) = 30{,}000$" into a task involving the graph of the function.

**44.** Translate the task "find $P(2000)$" into a task involving the graph.

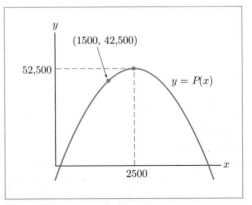

**Figure 20.** A profit function.

*A ball is thrown straight up into the air. The function $h(t)$ gives the height of the ball (in feet) after $t$ seconds. In Exercises 45–50, translate the task into both a statement involving the function and a statement involving the graph of the function.*

**45.** Find the height of the ball after 3 seconds.

**46.** Find the time at which the ball attains its greatest height.

**47.** Find the greatest height attained by the ball.

**48.** Determine when the ball will hit the ground.

**49.** Determine when the height of the ball is 100 feet.

**50.** Find the height of the ball when it is first released.

### Technology Exercises

**51.** A ball thrown straight up into the air has height $-16x^2 + 80x$ feet after $x$ seconds.

(a) Graph the function in the window

$$[0, 6] \ by \ [-30, 120].$$

(b) What is the height of the ball after 3 seconds?

(c) At what times will the height be 64 feet?

(d) At what time will the ball hit the ground?

(e) When will the ball reach its greatest height? What is that height?

**52.** The daily cost (in dollars) of producing $x$ units of a certain product is given by the function $C(x) = 225 + 36.5x - .9x^2 + .01x^3$.

(a) Graph $C(x)$ in the window $[0, 70]$ by $[-400, 2000]$.

(b) What is the cost of producing 50 units of goods?

(c) Consider the situation as in Part (b). What is the additional cost of producing one more unit of goods?

(d) At what production level will the daily cost be $510?

**53.** A store estimates that the total revenue (in dollars) from the sale of $x$ bicycles per year is given by the function $R(x) = 250x - .2x^2$.

(a) Graph $R(x)$ in the window

$$[200, 500] \ by \ [42000, 75000].$$

(b) What sales level produces a revenue of $63,000?

(c) What revenue is received from the sale of 400 bicycles?

(d) Consider the situation of Part (c). If the sales level were to decrease by 50 bicycles, by how much would revenue fall?

(e) The store believes that if it spends $5000 in advertising, it can raise the total sales from 400 to 450 bicycles next year. Should it spend the $5000? Explain your conclusion.

### SOLUTIONS TO PRACTICE PROBLEMS 0.6

**1.** Let $r$ be the radius of circular base and let $h$ be the height of the cylinder.

**2.** Girth $= 2\pi r$ (the circumference of the circle).

**3.** Area of the bottom $= \pi r^2$.

**4.** The cylinder is a rolled-up rectangle of height $h$ and base $2\pi r$ (the circumference of the circle). The area is $2\pi rh$. See Fig. 21.

**Figure 21.** Unrolled side of a cylinder.

# REVIEW OF FUNDAMENTAL CONCEPTS

**1.** Explain the relationships and differences among real numbers, rational numbers, and irrational numbers.

**2.** What are the four types of inequalities and what do they each mean?

**3.** What is the difference between an open interval and a closed interval from $a$ to $b$?

**4.** What is a function?

**5.** What is meant by "the value of a function at $x$"?

**6.** What is meant by the domain and range of a function?

**7.** What is the graph of a function and how is it related to vertical lines?

**8.** What is a linear function? Constant function? Give examples.

**9.** What are the $x$- and $y$-intercepts of a function and how are they found?

**10.** What is a quadratic function? What shape does its graph have?

**11.** Define and give an example of each of the following types of functions.
   **(a)** quadratic function     **(b)** polynomial function
   **(c)** rational function     **(d)** power function

**12.** What is meant by the absolute value of a number?

**13.** What five operations on functions are discussed in this chapter? Give an example of each.

**14.** What is a zero of a function?

**15.** Give two methods for finding the zeros of a quadratic function.

**16.** State the six laws of exponents.

**17.** In the formula $A = P(1 + i)^n$, what do $A$, $P$, $i$, and $n$ represent?

**18.** Explain how to solve $f(x) = b$ geometrically from the graph of $y = f(x)$.

**19.** Explain how to find $f(a)$ geometrically from the graph of $y = f(x)$.

# SUPPLEMENTARY EXERCISES

**1.** Let $f(x) = x^3 + \dfrac{1}{x}$. Evaluate $f(1)$, $f(3)$, $f(-1)$, $f(-\frac{1}{2})$, and $f(\sqrt{2})$.

**2.** Let $f(x) = 2x + 3x^2$. Evaluate $f(0)$, $f(-\frac{1}{4})$, and $f(1/\sqrt{2})$.

**3.** Let $f(x) = x^2 - 2$. Evaluate $f(a - 2)$.

**4.** Let $f(x) = [1/(x+1)] - x^2$. Evaluate $f(a + 1)$.

*Determine the domains of the following functions.*

**5.** $f(x) = \dfrac{1}{x(x+3)}$     **6.** $f(x) = \sqrt{x - 1}$

**7.** $f(x) = \sqrt{x^2 + 1}$     **8.** $f(x) = \dfrac{1}{\sqrt{3x}}$

**9.** Is the point $(\frac{1}{2}, -\frac{3}{5})$ on the graph of the function $h(x) = (x^2 - 1)/(x^2 + 1)$?

**10.** Is the point $(1, -2)$ on the graph of the function $k(x) = x^2 + (2/x)$?

*Factor the polynomials in Exercises 11–14.*

**11.** $5x^3 + 15x^2 - 20x$     **12.** $3x^2 - 3x - 60$

**13.** $18 + 3x - x^2$     **14.** $x^5 - x^4 - 2x^3$

**15.** Find the zeros of the quadratic function $y = 5x^2 - 3x - 2$.

**16.** Find the zeros of the quadratic function $y = -2x^2 - x + 2$.

**17.** Find the points of intersection of the curves $y = 5x^2 - 3x - 2$ and $y = 2x - 1$.

**18.** Find the points of intersection of the curves $y = -x^2 + x + 1$ and $y = x - 5$.

*Let $f(x) = x^2 - 2x$, $g(x) = 3x - 1$, and $h(x) = \sqrt{x}$. Find the following functions.*

**19.** $f(x) + g(x)$     **20.** $f(x) - g(x)$

**21.** $f(x)h(x)$     **22.** $f(x)g(x)$

**23.** $f(x)/h(x)$     **24.** $g(x)h(x)$

*Let $f(x) = x/(x^2 - 1)$, $g(x) = (1 - x)/(1 + x)$, and $h(x) = 2/(3x + 1)$. Express the following as rational functions.*

**25.** $f(x) - g(x)$     **26.** $f(x) - g(x + 1)$

**27.** $g(x) - h(x)$     **28.** $f(x) + h(x)$

**29.** $g(x) - h(x - 3)$     **30.** $f(x) + g(x)$

*Let $f(x) = x^2 - 2x + 4$, $g(x) = 1/x^2$, and $h(x) = 1/(\sqrt{x} - 1)$. Determine the following functions.*

**31.** $f(g(x))$     **32.** $g(f(x))$

**33.** $g(h(x))$     **34.** $h(g(x))$

**35.** $f(h(x))$     **36.** $h(f(x))$

**37.** Simplify $(81)^{3/4}$, $8^{5/3}$, and $(.25)^{-1}$.

**38.** Simplify $(100)^{3/2}$ and $(.001)^{1/3}$.

**39.** The population of a city is estimated to be $750 + 25t + .1t^2$ thousand people $t$ years from the present. Ecologists estimate that the average level of carbon monoxide in the air above the city will be $1 + .4x$ ppm (parts per million) when the population is $x$ thousand people. Express the carbon monoxide level as a function of the time $t$.

**40.** The revenue $R(x)$ (in thousands of dollars) a company receives from the sale of $x$ thousand units is given by $R(x) = 5x - x^2$. The sales level $x$ is in turn a function $f(d)$ of the number $d$ of dollars spent on advertising, where

$$f(d) = 6\left(1 - \frac{200}{d + 200}\right).$$

Express the revenue as a function of the amount spent on advertising.

*In Exercises 41–44, use the laws of exponents to simplify the algebraic expressions.*

**41.** $(\sqrt{x + 1})^4$     **42.** $\dfrac{xy^3}{x^{-5}y^6}$

**43.** $\dfrac{x^{3/2}}{\sqrt{x}}$     **44.** $\sqrt[3]{x}\,(8x^{2/3})$

# The Derivative

T HE derivative is a mathematical tool used to measure rates of change. To illustrate the sort of change we have in mind, consider the following example. Suppose that a colony of yeast cells grows in a culture dish. Its environment (the dish) imposes limits on the space and nutrients available. Based on experiments,[*] the culture can have at most 10,000 yeast cells. At time 0 hours, the number of yeast cells is 385. The experimental data in Table 1 list the number of yeast cells present at 5-hour intervals. Let $N(t)$ denote the number of cells in the colony at time $t$. Then $N(t)$ is a function of $t$, whose values at particular values of $t$ are given by Table 1.

As $t$ changes, the value of $N(t)$ changes. Let's analyze this change. Refer to Table 1. The third column records changes in the value of $t$. The first entry records the time elapsed from $t = 0$ until $t = 5$—that is $5 - 0 = 5$ hours. The second entry records the time elapsed from $t = 5$ until $t = 10$—that is $10 - 5 = 5$. And so forth, down the column. To each change in $t$ (from 0 to 5, from 5 to 10, and so on), there

---

[*]See C. F. Gause, "Experimental Studies on the Struggle of Existence," *J. Exp. Biology*, 9(1932), 389–402.

| TABLE 1 | | | | |
|:---:|:---:|:---:|:---:|:---:|
| $t$ | $N(t)$ | Change in $t$ | Change in $N(t)$ | $\dfrac{\text{Change in } N(t)}{\text{Change in } t}$ |
| 0 | 385 | | | |
| 5 | 619 | 5 | 234 | 46.8 |
| 10 | 981 | 5 | 362 | 72.4 |
| 15 | 1520 | 5 | 539 | 107.8 |
| 20 | 2281 | 5 | 761 | 152.2 |
| 25 | 3276 | 5 | 995 | 199.0 |
| 30 | 4455 | 5 | 1179 | 235.8 |
| 35 | 5698 | 5 | 1243 | 248.6 |
| 40 | 6859 | 5 | 1161 | 232.2 |
| 45 | 7826 | 5 | 967 | 193.4 |
| 50 | 8558 | 5 | 732 | 146.4 |
| 55 | 9073 | 5 | 515 | 103.0 |
| 60 | 9416 | 5 | 343 | 68.6 |
| 65 | 9638 | 5 | 222 | 44.4 |
| 70 | 9777 | 5 | 139 | 27.8 |

is a corresponding change in $N(t)$ (from $N(0)$ to $N(5)$, from $N(5)$ to $N(10)$, and so on). For example, if $t$ changes from 0 to 5, then the value of $N(t)$ changes from 385 to 619, a change of 234. The change in $N(t)$ corresponding to each change in $t$ is recorded in the fourth column of Table 1.

Let's assume that the changes in $N(t)$ are uniform during each time interval. That is, the increase of 234 cells occurs uniformly over the interval from time 0 until time 5. Then the rate of increase in the number of cells can be obtained as the quotient:

$$[\text{Rate of change in number of yeast cells}] = \frac{[\text{Change in number of yeast cells}]}{[\text{Change in time}]}$$

$$\frac{N(5) - N(0)}{5 - 0} = \frac{619 - 385}{5} = \frac{234}{5}$$

$$= 46.8 \text{ cells per hour.}$$

Column 5 of Table 1 records the rate of change in the number of yeast cells per hour, for each time interval described by the table. We see that each rate of change is positive, indicating that the yeast colony is continually increasing in size. However, the rates of change initially grow, then shrink. This indicates that the yeast colony grows at increasingly faster rates, taking advantage of the existing space and nutrients. As the colony becomes more crowded and food becomes more limited (per cell), the rate of increase slows.

We can describe the yeast colony by graphing $N(t)$, as shown in Fig. 1. Note that the colony size increases continually, approaching the maximum colony size of 10,000. Moreover, reading the graph from left to right, we see that the graph initially gets increasingly steep, then decreasingly steep. That is, there appears to be a connection between the rate at which $N(t)$ is changing and the steepness of the graph. This illustrates one of the fundamental ideas of calculus, which may be put roughly as follows: *The rate of change of a function corresponds to the steepness of its graph.*

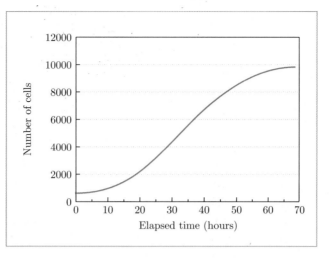

**Figure 1.**  Yeast cells in a culture.

This chapter is devoted to the *derivative*, which provides a numerical measure of the steepness of a curve at a particular point. By studying the derivative, we will be able to deal numerically with rates of change in applied problems.

## 1.1   The Slope of a Straight Line

As we shall see later, the study of straight lines is crucial for the study of the steepness of curves. So this section is devoted to a discussion of the geometric and algebraic properties of straight lines. Since a vertical line is not the graph of a function (it fails the vertical line test), we focus our discussion on nonvertical lines.

> **Equations of Nonvertical Lines**    A nonvertical line $L$ has an equation of the form
>
> $$y = mx + b. \tag{1}$$
>
> The number $m$ is called the *slope* of $L$ and the point $(0, b)$ is called the *y-intercept*. The equation (1) is called the *slope-intercept equation* of $L$.

If we set $x = 0$, we see that $y = b$, so that $(0, b)$ is on the line $L$. Thus the $y$-intercept tells us where the line $L$ crosses the $y$-axis. The slope measures the steepness of the line. In Fig. 1 we give three examples of lines with slope $m = 2$. In Fig. 2 we give three examples of lines with slope $m = -2$.

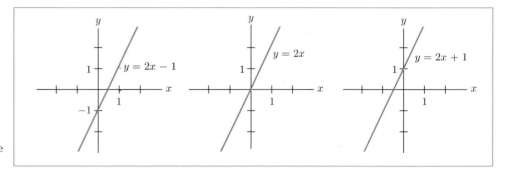

**Figure 1.**   Three lines of slope 2.

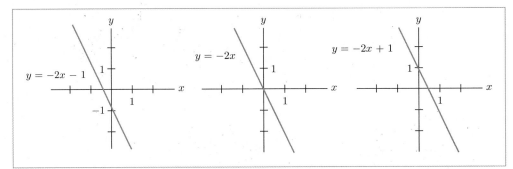

**Figure 2.**   Three lines of slope $-2$.

To conceptualize the meaning of slope, think of walking along a line from left to right. On lines of positive slope we will be walking uphill; the greater the slope, the steeper the ascent. On lines of negative slope we will be walking downhill; the more negative the slope, the steeper the descent. Walking on lines of zero slope corresponds to walking on level ground. In Fig. 3 we have graphed lines with $m = 3, 1, \frac{1}{3}, 0, -\frac{1}{3}, -1, -3$, all having $b = 0$. The reader can readily verify our conceptualization of slope for these lines.

The slope and $y$-intercept of a straight line often have physical interpretations, as the following three examples illustrate.

**Figure 3**

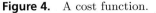

**Figure 4.**   A cost function.

---

**EXAMPLE 1**    A manufacturer finds that the total cost of producing $x$ units of a commodity is $2x + 1000$ dollars. What is the economic significance of the $y$-intercept and the slope of the line $y = 2x + 1000$? (See Fig. 4.)

**Solution**    The $y$-intercept is $(0, 1000)$. In other words, when $x = 0$ (no units produced), the cost is still $y = 1000$ dollars. The number 1000 represents the *fixed costs* of the manufacturer—those overhead costs, such as rent and insurance, that must be paid no matter how many items are produced.

The slope of the line is 2. This number represents the cost of producing each additional unit. To see this, we can calculate some typical costs.

| Quantity Produced | Total Cost |
|:---:|:---:|
| $x = 1500$ | $y = 2(1500) + 1000 = 4000$ |
| $x = 1501$ | $y = 2(1501) + 1000 = 4002$ |
| $x = 1502$ | $y = 2(1502) + 1000 = 4004$ |

Each time $x$ is increased by 1, the value of $y$ increases by 2. The number 2 is called the *marginal cost*. It is equal to the additional cost that is incurred when the production level is increased by one unit, from $x$ to $x + 1$. ∎

**EXAMPLE 2**   An apartment complex has a storage tank to hold its heating oil. The tank was filled on January 1, but no more deliveries of oil will be made until some time in March. Let $t$ denote the number of days after January 1 and let $y$ denote the number of gallons of oil in the tank. Current records of the apartment complex show that $y$ and $t$ are related approximately by the equation

$$y = 30{,}000 - 400t. \qquad (2)$$

What interpretation can be given to the $y$-intercept and slope of this line? (See Fig. 5.)

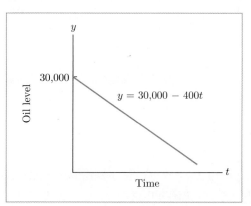

**Figure 5.**   Amount of heating oil in a tank.

**Solution**   The $y$-intercept is $(0, 30{,}000)$. This value of $y$ corresponds to $t = 0$, so there were 30,000 gallons of oil in the tank on January 1. Let us examine how fast the oil is removed from the tank.

| Days After January 1 | Gallons of Oil in the Tank |
|:---:|:---:|
| $t = 0$ | $y = 30{,}000 - 400(0) = 30{,}000$ |
| $t = 1$ | $y = 30{,}000 - 400(1) = 29{,}600$ |
| $t = 2$ | $y = 30{,}000 - 400(2) = 29{,}200$ |
| $t = 3$ | $y = 30{,}000 - 400(3) = 28{,}800$ |
| ⋮ | ⋮ |

The oil level in the tank drops by 400 gallons each day; that is, the oil is being used at the rate of 400 gallons per day. The slope of the line (2) is $-400$. Thus the slope gives the rate at which the level of oil in the tank is changing. The negative sign on the $-400$ indicates that the oil level is decreasing rather than increasing. ∎

**EXAMPLE 3**   For tax purposes, businesses are allowed to regard equipment as decreasing in value (or depreciating) each year. The amount of depreciation may be taken as an income tax deduction.

Suppose that the value $y$ of a piece of equipment $x$ years after its purchase is given by

$$y = 500{,}000 - 50{,}000x.$$

Interpret the $y$-intercept and the slope of the graph.

**Solution** The $y$-intercept is $(0, 500{,}000)$ and corresponds to the value of $y$ when $x = 0$. That is, the $y$-intercept gives the original value, \$500,000, of the equipment. The slope indicates the rate at which the equipment is changing in value. Thus the value of the equipment is decreasing at the rate of 50,000 dollars per year. ■

**Properties of the Slope of a Line** Let us now examine several useful properties of the slope of a straight line. At the end of the section we shall explain why these properties are valid.

**Slope Property 1** Suppose that we start at a point on a line of slope $m$ and move one unit to the right. Then we must move $m$ units in the $y$-direction in order to return to the line.

**Slope Property 2** We can compute the slope of a line by knowing two points on the line. If $(x_1, y_1)$ and $(x_2, y_2)$ are on the line, then the slope of the line is

$$m = \frac{y_2 - y_1}{x_2 - x_1}.$$

**Slope Property 3** The equation of a line can be obtained if we know the slope and one point on the line. If the slope is $m$ and if $(x_1, y_1)$ is on the line, then the equation of the line is

$$y - y_1 = m(x - x_1). \tag{3}$$

This equation is called the *point-slope form* of the equation of the line.

> **Slope Property 4**     Distinct lines of the same slope are parallel. Conversely, if two lines are parallel, they have the same slope.

> **Slope Property 5**     When two lines are perpendicular, the product of their slopes is $-1$.

### Calculations Involving Slope of a Line

**EXAMPLE 4**     Find the slope and the $y$-intercept of the line whose equation is $2x + 3y = 6$.

**Solution**     We put the equation into slope-intercept form by solving for $y$ in terms of $x$.

$$3y = -2x + 6$$
$$y = -\frac{2}{3}x + 2$$

The slope is $-\frac{2}{3}$ and the $y$-intercept is $(0, 2)$.  ∎

**EXAMPLE 5**     Sketch the graph of the line
(a)  passing through $(2, -1)$ with slope 3,
(b)  passing through $(2, 3)$ with slope $-\frac{1}{2}$.

**Solution**     We use Slope Property 1. (See Fig. 6.) In each case, we begin at the given point, move one unit to the right, and then move $m$ units in the $y$-direction (upward for positive $m$, downward for negative $m$). The new point reached will also be on the line. Draw the straight line through these two points.  ∎

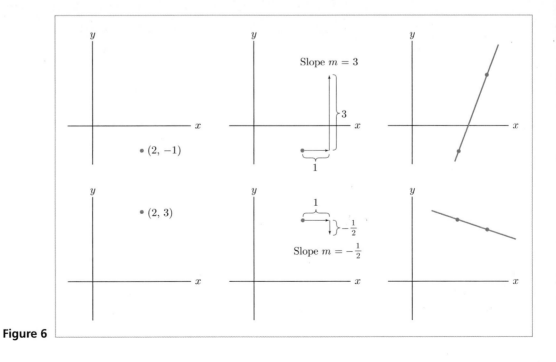

**Figure 6**

**EXAMPLE 6**    Find the slope of the line passing through the points $(6, -2)$ and $(9, 4)$.

**Solution**    We apply Slope Property 2 with $(x_1, y_1) = (6, -2)$ and $(x_2, y_2) = (9, 4)$. Then

$$\frac{y_2 - y_1}{x_2 - x_1} = \frac{4 - (-2)}{9 - 6} = \frac{6}{3} = 2.$$

Thus the slope is 2. [We would have reached the same answer if we had let $(x_1, y_1) = (9, 4)$ and $(x_2, y_2) = (6, -2)$.] The slope is just the difference of the $y$-coordinates divided by the difference of the $x$-coordinates, with each difference formed in the same order.  ■

**EXAMPLE 7**    Find an equation of the line passing through $(-1, 2)$ with slope 3.

**Solution**    We let $(x_1, y_1) = (-1, 2)$ and $m = 3$, and we use Slope Property 3. The equation of the line is

$$y - 2 = 3[x - (-1)]$$

or

$$y - 2 = 3(x + 1).$$

If desired, this equation can be put into the form $y = mx + b$:

$$y - 2 = 3(x + 1) = 3x + 3$$
$$y = 3x + 5.$$

■

**EXAMPLE 8**    Find an equation of the line passing through the points $(1, -2)$ and $(2, -3)$.

**Solution**    By Slope Property 2, the slope of the line is

$$\frac{-3 - (-2)}{2 - 1} = \frac{-3 + 2}{1} = -1.$$

Since $(1, -2)$ is on the line, we can use Property 3 to get the equation of the line:

$$y - (-2) = (-1)(x - 1)$$
$$y + 2 = -x + 1$$
$$y = -x - 1.$$

■

**EXAMPLE 9**    Find an equation of the line passing through $(5, 3)$ parallel to the line $2x + 5y = 7$.

**Solution**    We first find the slope of the line $2x + 5y = 7$.

$$2x + 5y = 7$$
$$5y = 7 - 2x$$
$$y = -\frac{2}{5}x + \frac{7}{5}.$$

The slope of this line is $-\frac{2}{5}$. By Slope Property 4, any line parallel to this line will also have slope $-\frac{2}{5}$. Using the given point $(5, 3)$ and Slope Property 3, we get the desired equation:

$$y - 3 = -\frac{2}{5}(x - 5).$$

This equation can also be written as

$$y = -\frac{2}{5}x + 5.$$

■

## The Slope as a Rate of Change

Let us revisit the notion of rate of change that we discussed in Example 2 and see how it ties in with the notion of slope. Given a linear function $y = L(x) = mx + b$, as $x$ varies from $x_1$ to $x_2$, $y$ varies from $y_1 = L(x_1)$ to $y_2 = L(x_2)$. The *rate of change* of $y = L(x)$ over the interval from $x_1$ to $x_2$ is the change in $y$ divided by the change in $x$:

$$\left[ \begin{array}{c} \text{rate of change of } y = L(x) \\ \text{over the interval from } x_1 \text{ to } x_2 \end{array} \right] = \frac{\text{change in } y}{\text{change in } x}$$

$$= \frac{y_2 - y_1}{x_2 - x_1} = m,$$

by Slope Property 2. Thus the rate of change of a linear function over any interval is constant and equal to the slope $m$. As we did in Example 2, we will refer to the slope as *the* rate of change of the function, without mentioning an underlying interval. The fact that the rate of change of a linear function is constant characterizes linear functions (see Exercise 64) and has many important applications.

---

**EXAMPLE 10**  **Impact of mad cow disease on Canadian beef exports**  The discovery of one case of bovine spongiform encephalopathy, or mad cow disease, in May 2003 in Canada led to an immediate ban on all Canadian beef exports. At the beginning of September 2003, the ban was eased and exports of Canadian beef rose at a steady rate of 42.5 million dollars per month.* Express the value of the monthly exports of Canadian beef as a function of time for the period starting the first day of September 2003. What was the value of the monthly exports at the end of December 2003, when presumably the exports regained their normal level?

**Solution**  Let $y$ denote the value of monthly exports in millions of dollars, and $x$ the number of months that have elapsed since the first day of September 2003. Since the rate of change of $y$ is constant, we conclude that $y$ is a linear function of $x$, whose slope is equal to its rate of change, $m = 42.5$. On the first day of September (when $x = 0$), the value of monthly exports was 0 dollars, since the ban had just ended. So the point $(0, 0)$ is on the graph of $y$. To write $y$ in terms of $x$, we use the point-slope form of the equation of a line and get

$$y - 0 = 42.5(x - 0)$$
$$y = 42.5x.$$

The end of December 2003 corresponds to $x = 4$, at which time the exports had reached the value $y = (42.5) \cdot (4) = 170$ million dollars (Fig. 7).  ■

---

**EXAMPLE 11**  **Temperature in a city**  As a cold front moved through a midwestern town, the temperature dropped at the rate of 3°F per hour between noon and 8 P.M. Express the temperature as a function of time and find the temperature at noon, given that at 1 P.M. the temperature was 47°F.

**Solution**  Let $t$ denote time measured in hours since noon and $T(t)$ the temperature of the town at time $t$. The fact that the temperature was dropping at a (constant) rate of 3°F per hour tells us that the temperature is a linear function whose graph is

---

*Source: International Trade Division Statistics, Canada.

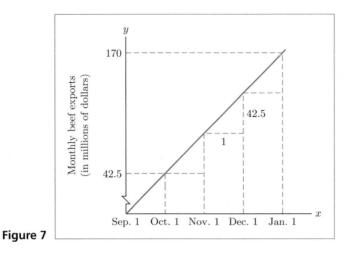

**Figure 7**

falling. Thus its slope is $m = -3$ and so $T(t) = -3t + b$. To determine $b$ we use the fact that at 1 P.M. the temperature was $47°F$ or $T(1) = 47$:

$$(-3)(1) + b = 47$$
$$b = 50.$$

So $T(t) = -3t + 50$. The graph of the temperature is shown in Fig. 8. To determine the temperature at noon, we set $t = 0$ and get $T(0) = 50°F$. ■

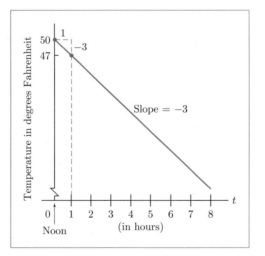

**Figure 8.** Temperature in a city.

Note the importance of the sign of the slope in Examples 10 and 11. An increasing linear function has a positive rate of change and thus a positive slope. A decreasing linear function has a negative rate of change and thus a negative slope.

**Verification of the Properties of Slope** It is convenient to verify the properties of slope in the order 2, 3, 1. Verifications of Properties 4 and 5 are outlined in Exercises 67 and 68.

*Verification of Property 2* Suppose that the equation of the line is $y = mx + b$. Then, since $(x_2, y_2)$ is on the line, we have $y_2 = mx_2 + b$. Similarly, since $(x_1, y_1)$ is on the line, we have $y_1 = mx_1 + b$. Subtracting, we see that

$$y_2 - y_1 = mx_2 - mx_1$$
$$y_2 - y_1 = m(x_2 - x_1),$$

so that

$$\frac{y_2 - y_1}{x_2 - x_1} = m.$$

This is the formula for the slope $m$ stated in Property 2.

***Verification of Property 3*** The equation $y - y_1 = m(x - x_1)$ may be put into the form

$$y = mx + \underbrace{(y_1 - mx_1)}_{b}. \tag{4}$$

This is the equation of a line with slope $m$. Furthermore, the point $(x_1, y_1)$ is on this line because equation (4) remains true when we plug in $x_1$ for $x$ and $y_1$ for $y$. Thus the equation $y - y_1 = m(x - x_1)$ corresponds to the line of slope $m$ passing through $(x_1, y_1)$.

***Verification of Property 1*** Let $P = (x_1, y_1)$ be on a line $l$ and let $y_2$ be the $y$-coordinate of the point $R$ on $l$ obtained by moving $P$ one unit to the right and then moving vertically to return to the line. (See Fig. 9.) By Property 2, the slope $m$ of this line $l$ satisfies

$$m = \frac{y_2 - y_1}{(x_1 + 1) - x_1} = \frac{y_2 - y_1}{1} = y_2 - y_1.$$

Thus the difference of the $y$-coordinates of $R$ and $Q$ is $m$. This is Property 1.

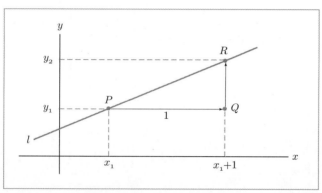

**Figure 9**

**Incorporating Technology**   Here's how to graph nonvertical lines on a graphing calculator.

1. Start from the equation of the line. Write it as a function $y = \dots$ by solving for $y$ in terms of $x$.
2. Enter the function found in Step 1 into the list of functions in the calculator. (Usually obtained by pressing **Y=**).
3. Press **GRAPH** to display the graph of the function.

For example, to plot the graph of $y = 3x - 4$, enter this function into the calculator as **Y1**, as shown in Fig. 10(a). Pressing **GRAPH** yields the graph of the equation, as shown in Fig. 10(b).

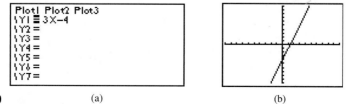

**Figure 10**                    (a)                                    (b)

**PRACTICE PROBLEMS 1.1**

*Find the slopes of the following lines.*

**1.** The line whose equation is $x = 3y - 7$

**2.** The line going through the points $(2, 5)$ and $(2, 8)$

# EXERCISES 1.1*

*Find the slopes and y-intercepts of the following lines.*

**1.** $y = 2 - 5x$

**2.** $y = \dfrac{x+1}{3}$

**3.** $3y - 2x - 1 = 0$

**4.** $y = 5$

**5.** $y = \dfrac{3x - 1}{3}$

**6.** $2x + 7y = -1$

*Find the equations of the following lines.*

**7.** Slope is $-7$; $(5, 0)$ on line.

**8.** Slope is $0$; $(0, 0)$ on line.

**9.** Slope is $4$; $(1, 0)$ on line.

**10.** Slope is $\frac{7}{3}$; $(5, -1)$ on line.

**11.** $\left(\frac{2}{3}, 5\right)$ and $\left(-\frac{5}{6}, -4\right)$ on line.

**12.** $\left(\frac{1}{2}, 1\right)$ and $(1, 2)$ on line.

**13.** $(0, 0)$ and $(1, -2)$ on line.

**14.** $\left(-\frac{1}{2}, 0\right)$ and $(1, 2)$ on line.

**15.** Horizontal through $(3, -1)$.

**16.** $x$-intercept is $2$; $y$-intercept is $-2$.

**17.** $x$-intercept is $-1$; $y$-intercept is $1$.

**18.** Slope is $6$; $x$-intercept is $-1$.

**19.** Slope is $1$; $x$-intercept is $-2$.

**20.** Horizontal through $(-2, \sqrt{2})$.

**21.** Parallel to $y = 2x$; $(2, 0)$ on line.

**22.** Parallel to $x + y = 0$; $(1, 1)$ on line.

**23.** Parallel to $y = -x + 7$; $x$-intercept is $1$.

**24.** Parallel to $y - x = 13$; $y$-intercept is $-\frac{1}{2}$.

**25.** Perpendicular to $y = 2x$; $(2, 0)$ on line.

**26.** Perpendicular to $y = -\frac{1}{2}x + 1$; $(0, 0)$ on line.

*In Exercises 27–30, we specify a line by giving the slope and one point on the line. Start at the given point and use Slope Property 1 to sketch the graph of the line.*

**27.** $m = 1$, $(1, 0)$ on line.

**28.** $m = \frac{1}{2}$, $(-1, 1)$ on line.

**29.** $m = -\frac{1}{3}$, $(1, -1)$ on line.

**30.** $m = 0$, $(0, 2)$ on line.

**31.** Each of the lines (A), (B), (C), and (D) in Fig. 11 is the graph of one of the equations (a), (b), (c), and (d). Match each equation with its graph.

**(a)** $x + y = 1$

**(b)** $x - y = 1$

**(c)** $x + y = -1$

**(d)** $x - y = -1$

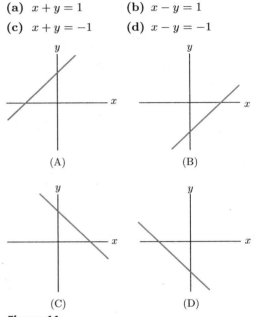

(A)　　　　　(B)

(C)　　　　　(D)

**Figure 11**

**32.** Table 1 gives some points on the line $y = mx + b$. Find $m$ and $b$.

| TABLE 1 | Points on a Line | | | | |
|---|---|---|---|---|---|
| $x$ | 4.8 | 4.9 | 5 | 5.1 | 5.2 |
| $y$ | 3.6 | 4.8 | 6 | 7.2 | 8.4 |

*In Exercises 33–36, refer to a line of slope m. Suppose you begin at a point on the line and move h units in the x-direction. How many units must you move in the y-direction to return to the line?*

**33.** $m = \frac{1}{2}$, $h = 4$

**34.** $m = 2$, $h = \frac{1}{4}$

**35.** $m = -3$, $h = .25$

**36.** $m = .2$, $h = 5$

*In each of Exercises 37–38, we specify a line by giving the slope and one point on the line. We give the first coordinate of some points on the line. Without deriving the equation of the line, find the second coordinate of each of the points.*

---

*A complete solution for every third odd-numbered exercise is given in the *Study Guide* for this text.

**37.** Slope is 2, $(1,3)$ on line; $(2, \ )$; $(3, \ )$; $(0, \ )$.

**38.** Slope is $-3$, $(2,2)$ on line; $(3, \ )$; $(4, \ )$; $(1, \ )$.

**39.** Are the points $A = (1,-1)$, $B = (2,0)$, and $C = (3,1)$ on one line? Answer this question in two ways.

    **(a)** Compare the slope of the line through $A$ and $B$ to the slope of the line through $A$ and $C$.

    **(b)** Plot the points and decide whether they are on the same line.

**40.** Is the line through the points $(3,4)$ and $(-1,2)$ parallel to the line $2x + 3y = 0$? Justify your answer.

*For each pair of lines in the following figures, determine the one with the greater slope.*

**41.**

**42.**

*Find the equation and sketch the graph of the following lines.*

**43.** With slope $-2$ and $y$-intercept $(0,-1)$.

**44.** With slope $\frac{1}{3}$ and $y$-intercept $(0,1)$.

*In Exercises 45–46, two lines intersect the graph of a function $y = f(x)$ as shown in the figure. Find $a$ and $f(a)$.*

**45.**

**46.**

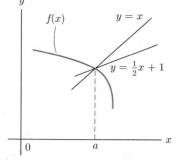

**47.** (*Marginal Cost*) Let $C(x) = 12x + 1100$ denote the total cost (in dollars) of manufacturing $x$ units of a certain commodity per day.

    **(a)** What is the total cost if the production is set at 10 units per day?

    **(b)** What is the marginal cost?

    **(c)** Use (b) to determine the additional cost of raising the daily production level from 10 to 11 units.

**48.** Refer to Exercise 47. Use the formula for $C(x)$ to show directly that $C(x+1) - C(x) = 12$. Interpret your result as it pertains to the marginal cost.

**49.** (*Price of Gasoline*) The price of one gallon of unleaded gasoline at the pump reached \$1.69 on January 1st, 2004, and continued to rise at the rate of 6 cents per month for the next nine months. Express the price of one gallon of unleaded gasoline as a function of time for the period starting the first day of January 2004. What was the price of 15 gallons of gasoline on April 1st, 2004? On September 1st, 2004?

**50.** (*Marginal Cost*) Let $C(x)$ denote the cost (in thousands of dollars) of manufacturing $x$ thousand computer chips. Suppose that $C(x)$ is a linear function. If the marginal cost is 100, what is the additional cost of increasing the current production level by 4 thousand chips?

**51.** (*Depreciation of Assets*) A car rental company purchased 5 cars in the year 2002 at a total cost of \$125,000. The value of these cars will be depreciated to zero at the rate of \$25,000 per year. Let $y(t)$ denote the value of the cars $t$ years after the year 2002. Find an expression of $y(t)$ in terms of $t$. When will the value of the cars reach zero?

**52.** (*Quit Ratio*) In industry, it is important to understand the relationship between wages and the quit ratio of employees, which is the percentage of employees that quit within one year of employment. The quit ratio of a large restaurant chain that paid its employees the minimum hourly wage (\$5.75 per hour) was .2 or 20 employees per 100. When the company raised the hourly wage to \$6, the quit ratio dropped to .18 or 18 employees per 100.

(a) Assuming a linear relationship between the quit ratio $Q(x)$ and the hourly wage $x$, find an expression for $Q(x)$.

(b) What should the hourly wage be in order for the quit ratio to drop to 10 employees per 100?

53. Price affects sales. When the owner of a gas station sets the price of one gallon of unleaded gasoline at \$2.60, he can sell approximately 1500 gallons per day. When he sets the price at \$2.85 per gallon, he can sell approximately 1250 gallons per day. Let $G(x)$ denote the number of gallons of unleaded gasoline sold per day when the price is $x$ dollars. Assume that $G(x)$ is a linear function of $x$. Approximately how many gallons will be sold per day if the price is set at \$2.70 per gallon?

54. Refer to the previous exercise. Where should the owner set the price if he wants to sell 2200 gallons per day?

55. (*Marginal Cost Analysis*) A company manufactures and sells fishing rods. The company has a fixed cost of \$1500 per day and a total cost of \$2200 per day when the production is set at 100 rods per day. Assume that the total cost $C(x)$ is linearly related to the daily production level $x$.

(a) Express the total cost as a function of the daily production level.

(b) What is the marginal cost at production level $x = 100$?

(c) What is the additional cost of raising the daily production level from 100 to 101 rods? Answer this question in two different ways: By using the marginal cost, and by computing $C(101) - C(100)$.

56. A salesperson's weekly pay depends on the volume of sales. If she sells $x$ units of goods, then her pay is $y = 5x + 60$ dollars. Give an interpretation of the slope and the $y$-intercept of this straight line.

57. The demand equation for a monopolist is $y = -.02x + 7$, where $x$ is the number of units produced and $y$ is the price. That is, in order to sell $x$ units of goods, the price must be $y = -.02x + 7$ dollars. Interpret the slope and $y$-intercept of this line.

58. (*Converting Fahrenheit to Celsius*) Temperatures of $32°$F and $212°$F correspond to temperatures of $0°$C and $100°$C. Suppose the linear equation $y = mx + b$ converts Fahrenheit temperatures to Celsius temperatures. Find $m$ and $b$. What is the Celsius equivalent of $98.6°$F?

59. (*Intravenous Injection*) A drug is administered to a patient through an IV (intravenous) injection at the rate of 6 ml per minute. Assuming that the patient's body contained already 1.5 ml of this drug at the beginning of the infusion, find an expression of the amount of the drug in the body $x$ minutes from the start of the infusion.

60. Refer to Exercise 59. Suppose that the patient's body eliminates the drug at the rate of 2 ml per hour. Find

an expression for the amount of the drug in the body $x$ minutes from the start of the infusion.

*In Exercises 61–62, compute the rate of change of the function over the given intervals. In each case verify that your answer equals the slope of the given linear function.*

61. $y = x + 7$ over the intervals $[0, 2]$ and $[-1, 2]$.

62. $y + 2x = 27$ over the intervals $[0, 1]$ and $[0, .5]$.

63. (a) Draw the graph of any function $f(x)$ that passes through the point $(3, 2)$.

(b) Choose a point to the right of $x = 3$ on the $x$-axis and label it $3 + h$.

(c) Draw the straight line through the points $(3, f(3))$ and $(3 + h, f(3 + h))$.

(d) What is the slope of this straight line (in terms of $h$)?

64. Suppose that for some constant $m$,

$$\frac{f(x_2) - f(x_1)}{x_2 - x_1} = m$$

for all $x_1 \neq x_2$. Show that $f(x) = mx + b$, where $b$ is some constant. [*Hint*: Fix $x_1$ and take $x = x_2$; then solve for $f(x)$.]

*In Exercises 65–66, compute the quotient*

$$\frac{L(x + h) - L(x)}{h} \quad (h \neq 0)$$

*and show that it is equal to the slope of the given linear function.*

65. $L(x) = -x + 2$      66. $L(x) = \dfrac{2x + 11}{3}$

67. Prove Slope Property 4 of straight lines. [*Hint*: If $y = mx + b$ and $y = m'x + b'$ are two lines, then they have a point in common if and only if the equation $mx + b = m'x + b'$ has a solution $x$.]

68. Prove Slope Property 5 of straight lines. [*Hint*: Without loss of generality, assume that both lines pass through the origin. Use Slope Property 1 and the Pythagorean theorem. See Fig. 12.]

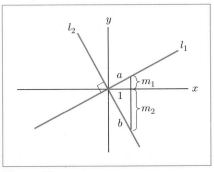

**Figure 12**

### Technology Exercises

**69.** Let $y$ denote the percentage of the world population that is urban $x$ years after 1980. According to recently published data, $y$ has been a linear function of $x$ since 1980. The percent of the world population that is urban was 39.5 in 1980 and 45.2 in 1995.

(a) Determine $y$ as a function of $x$.

(b) Graph this function in the window $[0, 40]$ *by* $[0, 100]$.

(c) Interpret the slope in applied terms.

(d) Determine graphically the percent of the world population that was urban in 1990.

(e) Determine graphically the year in which 50% of the world population will be urban.

(f) By what amount does the percent of the world population that is urban increase every 5 years?

**70.** Let $y$ denote the average amount claimed for itemized deductions on a tax return reporting $x$ dollars of income. According to Internal Revenue Service data, $y$ is a linear function of $x$. Moreover, in a recent year, income tax returns reporting \$20,000 of income averaged \$729 in itemized deductions, while returns reporting \$50,000 averaged \$1380.

(a) Determine $y$ as a function of $x$.

(b) Graph this function in the window $[0, 75000]$ *by* $[0, 2000]$.

(c) Give an interpretation of the slope in applied terms.

(d) Determine graphically the average amount of itemized deductions on a return reporting \$75,000.

(e) Determine graphically the income level at which the average itemized deductions are \$1600.

(f) Suppose that the income level increases by \$15,000. By how much do the average itemized deductions increase?

## SOLUTIONS TO PRACTICE PROBLEMS 1.1

**1.** We solve for $y$ in terms of $x$.

$$y = \frac{1}{3}x + \frac{7}{3}$$

The slope of the line is the coefficient of $x$, that is, $\frac{1}{3}$.

**2.** The line passing through these two points is a vertical line; therefore, its slope is undefined.

## 1.2  The Slope of a Curve at a Point

In order to extend the concept of slope from straight lines to more general curves, we must first discuss the notion of the tangent line to a curve at a point.

We have a clear idea of what is meant by the tangent line to a circle at a point $P$. It is the straight line that touches the circle at just the one point $P$. Let us focus on the region near $P$, designated by the dashed rectangle shown in Fig. 1. The enlarged portion of the circle looks almost straight, and the straight line that it resembles is the tangent line. Further enlargements would make the circle near $P$ look even straighter and have an even closer resemblance to the tangent line. In this sense, the tangent line to the circle at the point $P$ is the straight line through $P$ that best approximates the circle near $P$. In particular, the tangent line at $P$ reflects the steepness of the circle at $P$. Thus it seems reasonable to define the *slope* of the circle at $P$ to be the slope of the tangent line at $P$.

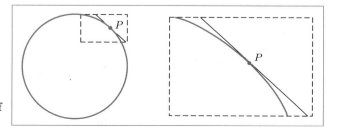

**Figure 1.**  Enlarged portion of a circle.

Similar reasoning leads us to a suitable definition of slope for an arbitrary curve at a point $P$. Consider the three curves drawn in Fig. 2. We have drawn an enlarged version of the dashed box around each point $P$. Notice that the portion of each curve lying in the boxed region looks almost straight. If we further magnify the curve near $P$, it would appear even straighter. Indeed, if we apply higher and higher magnification, the portion of the curve near $P$ would approach a certain straight line more and more exactly. (See Fig. 3.) This straight line is called the *tangent line to the curve at $P$*. This line best approximates the curve near $P$ and leads us to the following definition.

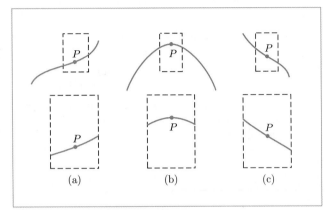

**Figure 2.**   Enlarged portions of curves.

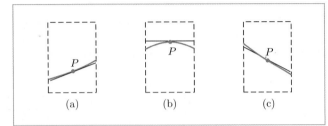

**Figure 3.**   Tangent lines to curves.

**Slope of a Curve**    The *slope of a curve at a point $P$* is defined to be the slope of the tangent line to the curve at $P$.

Let us consider some examples and applications involving slopes of curves and tangent lines. Later, in the next section, we shall describe a process for constructing the tangent line and computing its slope.

**EXAMPLE 1**   **Slope of a graph**   The graph of $f(x) = x^2$ and the tangent line at the point $P = (1, 1)$ are shown in Fig. 4. Find the slope of the graph at $P$.

**Solution**   The slope of the graph at a point is by definition the slope of the tangent line at that point. Figure 4 shows that the tangent line at $P$ rises two units for each unit change in $x$. Thus the slope of the tangent line at $P = (1, 1)$ is

$$\left[\; \text{slope of tangent line at } P \;\right] = \frac{\text{change in } y}{\text{change in } x} = \frac{2}{1} = 2$$

(see Fig. 4), and hence the slope of the graph at $P$ is 2. ■

## Slope of a Graph as a Rate of Change

We learned in the previous section that the slope of a linear function measures its rate of increase or decrease. We have just defined the slope of a curve at a point $P$ to be the slope of the tangent line to the curve at $P$. Now, the portion of the curve near $P$ can be, at least within an approximation, replaced by the tangent line at $P$. So, thinking of the curve as being approximated by the tangent line near $P$, we obtain the following important interpretation of the slope of a curve.

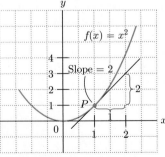

**Figure 4.**   Tangent line and slope at the point $P$.

**Slope of a Curve as a Rate of Change**     The slope of a curve at a point $P$— that is, the slope of the tangent line at $P$—measures the rate of change (the rate of increase or decrease) of the curve as it passes through $P$.

**EXAMPLE 2**     In Example 2 of Section 1.1 the apartment complex used approximately 400 gallons of oil per day. Suppose that we keep a continuous record of the oil level in the storage tank. The graph for a typical 2-day period appears in Fig. 5. What is the physical significance of the slope of the graph at the point $P$?

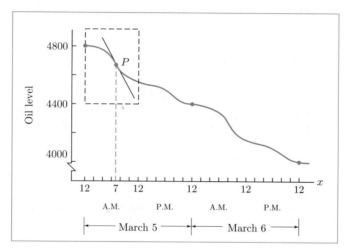

**Figure 5.**   Oil level in a storage tank.

**Solution**   The curve near $P$ is closely approximated by its tangent line. So think of the curve as replaced by its tangent line near $P$. Then the slope at $P$ is just the rate of decrease of the oil level at 7 A.M. on March 5. ■

Notice that during the entire day of March 5, the graph in Fig. 5 seems to be the steepest at 7 A.M. That is, the oil level is falling the fastest at that time. This corresponds to the fact that most people awake around 7 A.M., turn up their thermostats, take showers, and so on. You can estimate the rate of consumption of oil at 7 A.M. by estimating the slope of the tangent line at $P$ in Fig. 5.

**EXAMPLE 3**     **Slope of a graph**   In Fig. 6, we show an enlarged version of a portion of the graph designated by a dashed rectangle in Fig. 5. Estimate the slope of the graph at the point $P$ and give an interpretation of your result.

**Figure 6.**   Oil level in a storage tank around 7 A.M.

**Solution**  The slope of the graph at $\boxed{P}$ is by definition the slope of the tangent line at $P$. From Fig. 6, we see that the tangent line at $P$ falls approximately 250 units for a 4 unit change in $x$. So the slope of the graph at $P$ is

$$\Big[\text{ slope of tangent line at } P\ \Big] = \frac{\text{change in } y}{\text{change in } x} = \frac{-250}{4} = -62.5.$$

Since the slope at $P$ is the rate of increase or decrease of the oil level, we conclude that the oil level at 7 A.M. is decreasing at the rate of 62.5 gallons per hour. ■

Examples 2 and 3 provide typical illustrations of the manner in which slopes can be interpreted as rates of change. We shall return to this important idea in later sections.

## Slope Formulas

We know that the slope of a straight line is constant and does not depend on where we are on the line. This is clearly not the case with other curves. In general, the slope of a curve at a point will depend on the point. We saw in Example 1 that the slope of the tangent line to the graph of $y = x^2$ at the point $(1,1)$ is 2. In fact, we can show that the slope of the tangent line to the graph of $y = x^2$ at the point $(3,9)$ is 6; and the slope at $\left(-\frac{5}{2}, \frac{25}{4}\right)$ is $-5$. The various tangent lines are shown in Fig. 7. As we will demonstrate in the next section, in calculus we can usually compute the slopes by using formulas. For the parabola $y = x^2$, we notice that the slope at each point is two times the $x$-coordinate of the point. This is a general fact for this graph, which is expressed by the following *slope formula*:

$$\Big[\text{ slope of the graph of } y = x^2 \text{ at the point } (x,y)\ \Big] = 2x$$

(see Fig. 8). This simple formula will be derived in the next section. For now we will use it to get further insight into the topics of slopes of curves and tangent lines.

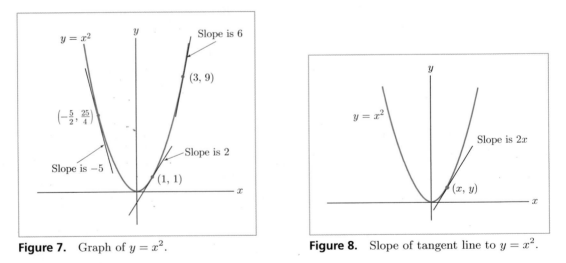

**Figure 7.**  Graph of $y = x^2$.

**Figure 8.**  Slope of tangent line to $y = x^2$.

**EXAMPLE 4**  (a) What is the slope of the graph of $y = x^2$ at the point $\left(\frac{3}{4}, \frac{9}{16}\right)$?

(b) Write the equation of the tangent line to the graph of $y = x^2$ at the point $\left(\frac{3}{4}, \frac{9}{16}\right)$.

**Solution**  (a) The $x$-coordinate of $\left(\frac{3}{4}, \frac{9}{16}\right)$ is $\frac{3}{4}$, so the slope of $y = x^2$ at this point is $2\left(\frac{3}{4}\right) = \frac{3}{2}$.

(b) We shall write the equation of the tangent line in the point-slope form. The point is $\left(\frac{3}{4}, \frac{9}{16}\right)$, and the slope is $\frac{3}{2}$ by part (a). Hence the equation is

$$y - \frac{9}{16} = \frac{3}{2}\left(x - \frac{3}{4}\right).$$

$\blacksquare$

**Figure 9.** Result of TI-83 **Tangent** command.

**Incorporating Technology**   Many graphing calculators have a command that calculates the slope of a curve at a designated point and automatically draws the tangent line through the point. Some calculators display the slope at the bottom of the screen as $dy/dx = slope$, and others give the $x$-coordinate of the point and the equation of the tangent line. Figure 9 shows the tangent line of $y = x^2$ at the point $(1, 1)$.

## PRACTICE PROBLEMS 1.2

1. Refer to Fig. 10.

   (a) What is the slope of the curve at $(3, 4)$?

   (b) What is the equation of the tangent line at the point where $x = 3$?

2. What is the equation of the tangent line to the graph of $y = \frac{1}{2}x + 1$ at the point $(4, 3)$?

**Figure 10**

# EXERCISES 1.2

*Trace the curves in Exercises 1–4 onto another piece of paper and sketch the tangent line in each case at the designated point P.*

1.

2.

**3.**

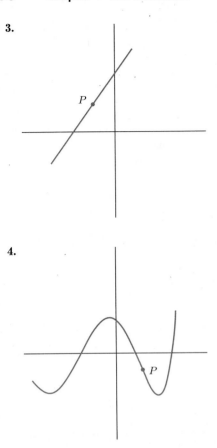

**4.**

*Estimate the slope of each of the following curves at the designated point P.*

**5.**

**6.**

**7.**

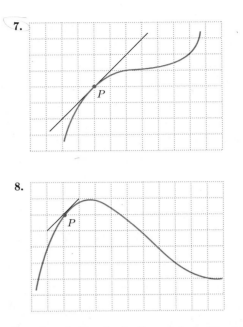

**8.**

*Exercises 9–14 refer to the points in Fig. 11. Assign one of the following descriptors to each point: large positive slope, small positive slope, zero slope, small negative slope, large negative slope.*

**9.** *A*    **10.** *B*    **11.** *C*    **12.** *D*    **13.** *E*    **14.** *F*

**Figure 11**

**15.** (*U.S. Federal Debt*) The federal debt is the total amount of money owed by the U.S. federal government. Figure 12 shows the graph of the federal debt in trillions of dollars for the years 1940 to 2004 (with projections for the year 2006).* Trace the graph onto another sheet of paper and draw the tangent line at the point corresponding to the year 1990. Use the tangent line to approximate the annual rate of increase of the federal debt in 1990.

**16.** Repeat Exercise 15 using the year 2002 instead of 1990.

*The data for Exercises 15 and 17 are based on data from the U.S. Treasury Department.

Federal Debt at the End of Year: 1940–2004
(in Trillions of Dollars)

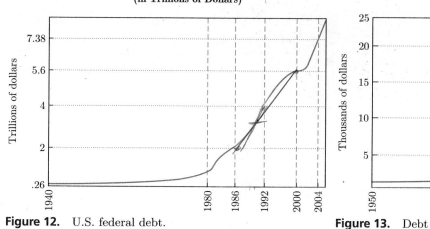

**Figure 12.** U.S. federal debt.

Per Capita Debt: 1950–2004
(in Thousands of Dollars)

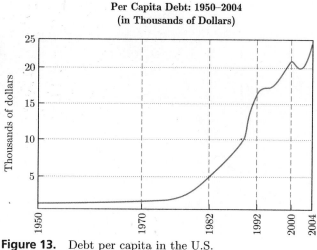

**Figure 13.** Debt per capita in the U.S.

**17.** (*Debt Per Capita*) The debt per capita measures the average amount of money owed by one person in the U.S. Figure 13 shows the graph of the debt per capita in thousands of dollars for the years 1950 to 2004. Use the graph to answer the following questions.

(a) Approximate the debt per capita in 1950, 1990, 2000, and 2004.

(b) Trace the graph onto another sheet of paper and draw the tangent line at the point corresponding to the year 2000. Use the tangent line to approximate the annual rate of increase of the debt per capita in 2000.

**18.** Refer to Fig. 13 to decide whether the following statements about the debt per capita are correct or not. Justify your answers.

(a) The debt per capita rose at a faster rate in 1980 than in 2000.

(b) The debt per capita was almost constant up until the mid-1970s and then rose at an almost constant rate from the mid-1970s to the mid-1980s.

*In Exercises 19–21, find the slope of the tangent line to the graph of $y = x^2$ at the point indicated and then write the corresponding equation of the tangent line.*

**19.** $(-2, 4)$　　　**20.** $(-.4, .16)$　　**21.** $\left(\frac{4}{3}, \frac{16}{9}\right)$

**22.** Find the slope of the tangent line to the graph of $y = x^2$ at the point where $x = -\frac{1}{2}$.

**23.** Write the equation of the tangent line to the graph of $y = x^2$ at the point where $x = 1.5$.

**24.** Write the equation of the tangent line to the graph of $y = x^2$ at the point where $x = .6$.

**25.** Find the point on the graph of $y = x^2$ where the curve has slope $\frac{5}{3}$.

**26.** Find the point on the graph of $y = x^2$ where the curve has slope $-4$.

**27.** Find the point on the graph of $y = x^2$ where the tangent line is parallel to the line $x + 2y = 4$.

**28.** Find the point on the graph of $y = x^2$ where the tangent line is parallel to the line $3x - y = 2$.

*In the next section we shall see that the tangent line to the graph of $y = x^3$ at the point $(x, y)$ has slope $3x^2$. See Fig. 14. Using this result, find the slope of the curve at the points in Exercises 29–31.*

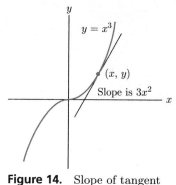

**Figure 14.** Slope of tangent line to $y = x^3$.

**29.** $(2, 8)$　　　**30.** $\left(\frac{3}{2}, \frac{27}{8}\right)$　　**31.** $\left(-\frac{1}{2}, -\frac{1}{8}\right)$

**32.** Write the equation of the line tangent to the graph of $y = x^3$ at the point where $x = -1$.

*In Exercises 33–34, you are shown the tangent line to the graph of $f(x) = x^2$ at the point $(a, f(a))$. Find $a$, $f(a)$, and the slope of the parabola at $(a, f(a))$.*

**33.**

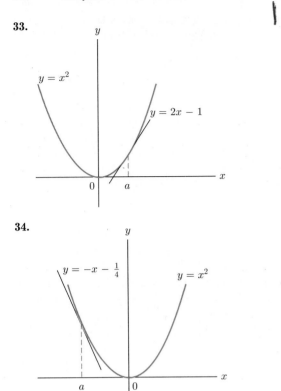

**34.**

*In Exercises 35–36, you are shown the graph of $y = f(x)$, the tangent line through the point $(a, f(a))$, and a line that intersects the graph at $(a, f(a))$. Find $a, f(a)$, and the slope of the curve at $(a, f(a))$.*

**35.**

~~intersects f(a)~~

**36.**

**37.** Let $l$ be the line through the points $P$ and $Q$ in Fig. 15.

   **(a)** Suppose $P = (2, 4)$ and $Q = (5, 13)$. Find the slope of the line $l$ and the length of the line segment $d$.

   **(b)** As the point $Q$ moves toward $P$, does the slope of the line $l$ increase or decrease?

**Figure 15.**   The slope of a secant line.

**38.** In Fig. 16, $h$ represents a positive number, and $3 + h$ is the number $h$ units to the right of 3. Draw line segments on the graph having the following lengths.

   **(a)** $f(3)$          **(b)** $f(3 + h)$

   **(c)** $f(3 + h) - f(3)$     **(d)** $h$

   **(e)** Draw a line of slope $\dfrac{f(3 + h) - f(3)}{h}$.

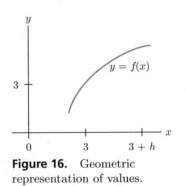

**Figure 16.**   Geometric representation of values.

### Technology Exercises

*In Exercises 39–42, draw the graph of the function with the given window setting, and then use the tangent-drawing feature of your calculator to find the slope of the function at the given value of $x$ and to draw the tangent line. Write the equation of the tangent line in point-slope form.*

**39.** $f(x) = \sqrt{x + 2}$; $[0, 8]$ *by* $[0, 4]$; $x = 4$

**40.** $f(x) = \dfrac{1}{x^2 - 2x + 2}$; $[-1, 3]$ *by* $[-.5, 1.5]$; $x = 1$

**41.** $f(x) = \dfrac{64x + 40}{x + 2}$; $[0, 7]$ *by* $[0, 70]$; $x = 3.5$

**42.** $f(x) = x^3 - 9x^2 + 28x - 28$; $[-.5, 6.5]$ *by* $[-2, 7]$; $x = 3$

## SOLUTIONS TO PRACTICE PROBLEMS 1.2

1. **(a)** The slope of the curve at the point $(3, 4)$ is, by definition, the slope of the tangent line at $(3, 4)$. Note that the point $(4, 6)$ is also on the line. Therefore, the slope is

$$\frac{6 - 4}{4 - 3} = \frac{2}{1} = 2.$$

**(b)** Use the point-slope formula. The equation of the line passing through the point $(3, 4)$ and having

slope 2 is

$$y - 4 = 2(x - 3)$$

or

$$y = 2x - 2.$$

2. The tangent line at $(4, 3)$ is, by definition, the line that best approximates the curve at $(4, 3)$. Since the "curve" in this case is itself a line, the curve and its tangent line at $(4, 3)$ (and at every other point) must be the same. Therefore, the equation is $y = \frac{1}{2}x + 1$.

# 1.3  The Derivative

Suppose that a curve is the graph of a function $f(x)$ and that we have a tangent line at every point on the graph. We saw in the previous section that it is possible to find a formula that gives the slope of the curve $y = f(x)$ at any point. This slope formula is called the *derivative* of $f(x)$ and is written $f'(x)$. For each value of $x$, $f'(x)$ gives the slope of the curve $y = f(x)$ at the point with first coordinate $x$.* (See Fig. 1.)

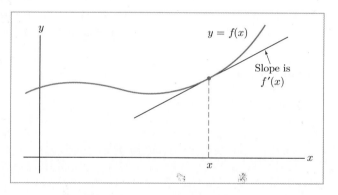

**Figure 1.**  Definition of $f'(x)$.

We considered already some interesting examples of slope formulas in the preceding section. These are our first examples of derivatives. We will list a few more examples and then consider some applications. At the end of this section, we will show how to construct the tangent line at a point of a curve and derive in the process a formula that can be used to compute derivatives and study their properties. The process of computing $f'(x)$ for a given function $f(x)$ is called *differentiation*.

## Examples of Derivatives: The Power Rule

The case of a linear function $f(x) = mx + b$ is particularly simple. The graph of $y = mx + b$ is a straight line $L$ of slope $m$. Recall that, by definition, the tangent line to a curve at a point is the line that best approximates the curve at that point. Since the curve $L$ is itself a line, the tangent line to $L$ at any point is the line $L$ itself. So the slope of the graph is $m$ at every point. (See Fig. 2.) In other words,

---

*As we shall see, there are curves that do not have tangent lines at every point. At values of $x$ corresponding to such points, the derivative $f'(x)$ is not defined. For the sake of the current discussion, which is designed to develop an intuitive feeling for the derivative, let us assume that the graph of $f(x)$ has a tangent line for each $x$ in the domain of $f$.

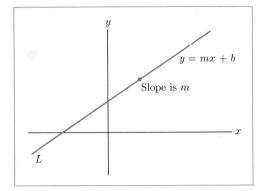

**Figure 2.** Derivative of a linear function.

**Figure 3.** Derivative of a constant function.

the value of the derivative $f'(x)$ is always equal to $m$. We summarize this fact as follows:

If $f(x) = mx + b$, then we have

$$f'(x) = m. \tag{1}$$

Set $m = 0$ in $f(x) = mx + b$. Then the function becomes $f(x) = b$, which has the value $b$ for each value of $x$. The graph is a horizontal line of slope 0, so $f'(x) = 0$ for all $x$. (See Fig. 3.) Thus we have the following fact:

The derivative of a constant function $f(x) = b$ is zero; that is,

$$f'(x) = 0. \tag{2}$$

Next, consider the function $f(x) = x^2$. As we stated in Section 1.2 (and will prove at the end of this section), the slope of the graph of $y = x^2$ at the point $(x, y)$ is equal to $2x$. That is, the value of the derivative $f'(x)$ is $2x$:

If $f(x) = x^2$, then its derivative is the function $2x$. That is,

$$f'(x) = 2x. \tag{3}$$

In Exercises 29–32 of Section 1.2 we made use of the fact that the slope of the graph of $y = x^3$ at the point $(x, y)$ is $3x^2$. This can be restated in terms of derivatives as follows:

If $f(x) = x^3$, then the derivative is $3x^2$. That is,

$$f'(x) = 3x^2. \tag{4}$$

One of the reasons calculus is so useful is that it provides general techniques that can be easily used to determine derivatives. One such general rule, which contains formulas (3) and (4) as special cases, is the so-called power rule.

> **Power Rule**   Let $r$ be any number and let $f(x) = x^r$. Then $f'(x) = rx^{r-1}$.

Indeed, if $r = 2$, then $f(x) = x^2$ and $f'(x) = 2x^{2-1} = 2x$, which is formula (3). If $r = 3$, then $f(x) = x^3$ and $f'(x) = 3x^{3-1} = 3x^2$, which is (4). We shall prove the power rule in Chapter 4. Until then, we shall use it to calculate derivatives.

---

**EXAMPLE 1**   Let $f(x) = \sqrt{x}$. What is $f'(x)$?

**Solution**   Recall that $\sqrt{x} = x^{1/2}$. We may apply the power rule with $r = \frac{1}{2}$.

$$f(x) = x^{1/2}$$
$$f'(x) = \tfrac{1}{2}x^{1/2-1} = \tfrac{1}{2}x^{-1/2}$$
$$= \frac{1}{2} \cdot \frac{1}{x^{1/2}} = \frac{1}{2\sqrt{x}}. \quad \blacksquare$$

Another important special case of the power rule occurs for $r = -1$, corresponding to $f(x) = x^{-1}$. In this case, $f'(x) = (-1)x^{-1-1} = -x^{-2}$. However, since $x^{-1} = 1/x$ and $x^{-2} = 1/x^2$, the power rule for $r = -1$ may also be written as follows:*

> If $f(x) = \dfrac{1}{x}$, then $f'(x) = -\dfrac{1}{x^2}$  $(x \neq 0)$. $\qquad$ (5)

## Geometric Meaning of the Derivative

We should, at this stage at least, keep the geometric meaning of the derivative clearly in mind. Figure 4 shows the graphs of $x^2$ and $x^3$ together with the interpretations of formulas (3) and (4) in terms of slope.

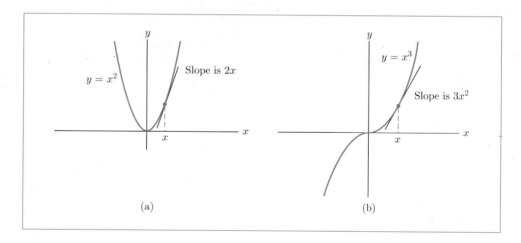

(a) $\qquad\qquad\qquad\qquad\qquad$ (b)

**Figure 4.**   Derivatives of $x^2$ and $x^3$.

---

**EXAMPLE 2**   Find the slope of the curve $y = 1/x$ at $\left(2, \frac{1}{2}\right)$.

---

*The formula gives $f'(x)$ for $x \neq 0$. The derivative of $f(x)$ is not defined at $x = 0$ since $f(x)$ itself is not defined there.

**Solution**  Set $f(x) = 1/x$. The point $\left(2, \frac{1}{2}\right)$ corresponds to $x = 2$, so in order to find the slope at this point, we compute $f'(2)$. From (5) we find that

$$f'(2) = -\frac{1}{2^2} = -\frac{1}{4}.$$

Thus the slope of $y = 1/x$ at the point $\left(2, \frac{1}{2}\right)$ is $-\frac{1}{4}$. (See Fig. 5.)  ■

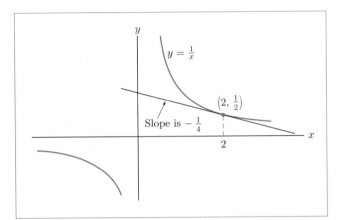

**Figure 5.**   Derivative of $\frac{1}{x}$.

▶ **Warning**  Do not confuse $f'(2)$, the value of the derivative at 2, with $f(2)$, the value of the $y$-coordinate at the point on the graph at which $x = 2$. In Example 2 we have $f'(2) = -\frac{1}{4}$, whereas $f(2) = \frac{1}{2}$. The number $f'(2)$ gives the *slope* of the graph at $x = 2$; the number $f(2)$ gives the *height* of the graph at $x = 2$.  ◀

Suppose that the graph of $y = f(x)$ has a tangent line at the point $(a, f(a))$. Obviously, one point on the tangent line is $(a, f(a))$ and since the slope of the tangent line is $f'(a)$, we obtain the point-slope form of the equation of the tangent line at $(a, f(a))$:

> **Equation of the Tangent Line**
>
> $$y - f(a) = f'(a)(x - a) \tag{6}$$

As an application, let us derive the equation of the tangent line to the graph of $f(x) = 1/x$ at $(2, 1/2)$ (Fig. 5). Take $a = 2$, $f(a) = 1/2$, and $f'(2) = -\frac{1}{4}$, then (6) yields the equation of the tangent line

$$y - \frac{1}{2} = -\frac{1}{4}(x - 2).$$

You do not need to memorize formula (6), but you should be able to derive it in a given situation; in particular, you should be able to distinguish all its different parts. Here is one more illustration.

**EXAMPLE 3**  Find the equation of the tangent line to the graph of $f(x) = \sqrt{x}$ at the point $(1, 1)$.

**Solution**  From Example 1, we have $f'(x) = \frac{1}{2\sqrt{x}}$. Thus the slope of the tangent line at the point $(1, 1)$ is $f'(1) = \frac{1}{2\sqrt{1}} = \frac{1}{2}$. Since the tangent line passes through the point

$(1, 1)$, its equation in point-slope form is

$$y - 1 = \frac{1}{2}(x - 1).$$

Note how this follows from (6) with $a = 1$, $f(a) = 1$, and $f'(a) = f'(1) = 1/2$. Figure 6 shows the graph of $y = \sqrt{x}$ and the tangent line at the point $(1, 1)$. (To plot the tangent line, start at the point $(1, 1)$; move over one unit to the right and then up one-half unit.) ■

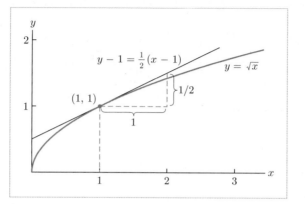

**Figure 6.** Tangent to $f(x) = \sqrt{x}$ at the point $(1, 1)$. Slope of tangent line is $f'(1) = \frac{1}{2\sqrt{1}} = \frac{1}{2}$.

**EXAMPLE 4**   **Tangent line**   The line $y = -\frac{1}{3}x + b$ is tangent to the graph of $y = \frac{1}{x}$ at the point $P = \left(a, \frac{1}{a}\right)$ $(a > 0)$. Find $P$ and determine $b$ (Fig. 7).

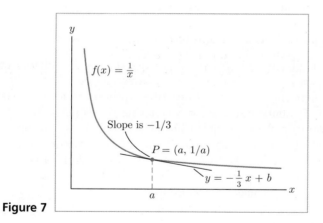

**Figure 7**

**Solution**   Let $f(x) = \frac{1}{x}$; then $f'(x) = -\frac{1}{x^2}$. We know that the slope of the tangent line at any point $P = \left(a, \frac{1}{a}\right)$ is equal to the value of the derivative $f'(x)$ at $x = a$ or $-\frac{1}{a^2}$. Since the slope of the tangent line $y = -\frac{1}{3}x + b$ is $-\frac{1}{3}$, we conclude that

$$-\frac{1}{a^2} = -\frac{1}{3};$$
$$a^2 = 3$$
$$a = \pm\sqrt{3}.$$

We take $a = \sqrt{3}$, since by assumption $a > 0$. Hence $P = \left(\sqrt{3}, \frac{1}{\sqrt{3}}\right)$. To determine $b$, we use the fact that $P$ is also on the tangent line and so its coordinates must satisfy the equation $y = -\frac{1}{3}x + b$. So

$$\frac{1}{\sqrt{3}} = \frac{\sqrt{3}}{3} + b;$$

$$1 = -1 + b\sqrt{3} \quad \text{(multiply both sides by } \sqrt{3})$$

$$\frac{2}{\sqrt{3}} = b.$$

Hence the equation of the tangent line is $y = -\frac{1}{3}x + \frac{2}{\sqrt{3}}$.

**Notation**  The operation of forming a derivative $f'(x)$ from a function $f(x)$ is also indicated by the symbol $\dfrac{d}{dx}$ (read "the derivative with respect to $x$"). Thus

$$\frac{d}{dx}f(x) = f'(x).$$

For example,

$$\frac{d}{dx}(x^6) = 6x^5, \qquad \frac{d}{dx}(x^{5/3}) = \frac{5}{3}x^{2/3}, \qquad \frac{d}{dx}\left(\frac{1}{x}\right) = -\frac{1}{x^2}.$$

When working with an equation of the form $y = f(x)$, we often write $\dfrac{dy}{dx}$ as a symbol for the derivative $f'(x)$. For example, if $y = x^6$, we may write

$$\frac{dy}{dx} = 6x^5.$$

## The Secant-Line Calculation of the Derivative

So far, we have said nothing about how to derive differentiation formulas such as (3), (4), or (5). Let us remedy that omission now. The derivative gives the slope of the tangent line, so we must describe a procedure for calculating that slope.

The fundamental idea for calculating the slope of the tangent line at a point $P$ is to approximate the tangent line very closely by *secant lines*. A secant line at $P$ is a straight line passing through $P$ and a nearby point $Q$ on the curve. (See Fig. 8.) By taking $Q$ very close to $P$, we can make the slope of the secant line approximate the slope of the tangent line to any desired degree of accuracy. Let us see what this amounts to in terms of calculations.

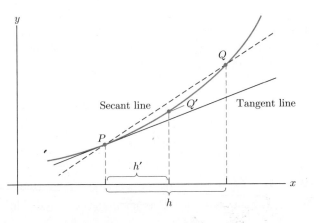

**Figure 8.**  A secant line approximation to a tangent line.

Suppose that the point $P$ is $(x, f(x))$. Suppose also that $Q$ is $h$ horizontal units away from $P$. Then $Q$ has $x$-coordinate $x+h$ and $y$-coordinate $f(x+h)$. The slope of the secant line through the points $P = (x, f(x))$ and $Q = (x+h, f(x+h))$ is simply

$$[\text{slope of secant line}] = \frac{f(x+h) - f(x)}{(x+h) - x} = \frac{f(x+h) - f(x)}{h}.$$

(See Fig. 9.)

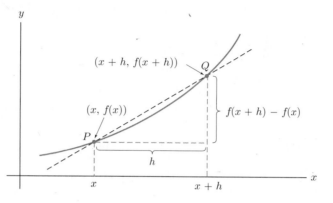

**Figure 9.** Computing the slope of a secant line.

In order to move $Q$ close to $P$ along the curve, we let $h$ approach zero. Then the secant line approaches the tangent line, and so

$$[\text{slope of secant line}] \quad \text{approaches} \quad [\text{slope of tangent line}];$$

that is,

$$\frac{f(x+h) - f(x)}{h} \quad \text{approaches} \quad f'(x).$$

Since we can make the secant line as close to the tangent line as we wish by taking $h$ sufficiently small, the quantity $[f(x+h) - f(x)]/h$ can be made to approximate $f'(x)$ to any desired degree of accuracy. Thus we arrive at the following method to compute the derivative $f'(x)$.

To calculate $f'(x)$,

1. First calculate the quotient $\dfrac{f(x+h) - f(x)}{h}$ for $h \neq 0$.

2. Then let $h$ approach zero.

3. The quantity $\dfrac{f(x+h) - f(x)}{h}$ will approach $f'(x)$.

The quotient in Step 1 turns out to be so important we give it a name: the *difference quotient.*

As an illustration, we apply our three-step method to compute the derivative of $f(x) = x^2$ at $x = 2$. (Since $f'(x) = 2x$, we expect the method to yield the result $f'(2) = 4$.)

***Step 1***    Take $x = 2$ and calculate the difference quotient:

$$\frac{f(2+h) - f(2)}{h} = \frac{(2+h)^2 - 2^2}{h}. \tag{7}$$

| **TABLE 1** | **Values of a Difference Quotient as $h$ Approaches 0** | | |
|---|---|---|---|
| $h$ | $\dfrac{(2+h)^2 - 2^2}{h}$ | $h$ | $\dfrac{(2+h)^2 - 2^2}{h}$ |
| 1 | $\dfrac{(2+1)^2 - 2^2}{1} = 5$ | $-1$ | $\dfrac{(2+(-1))^2 - 2^2}{-1} = 3$ |
| .1 | $\dfrac{(2+.1)^2 - 2^2}{.1} = 4.1$ | $-.1$ | $\dfrac{(2+(-.1))^2 - 2^2}{-.1} = 3.9$ |
| .01 | $\dfrac{(2+.01)^2 - 2^2}{.01} = 4.01$ | $-.01$ | $\dfrac{(2+(-.01))^2 - 2^2}{-.01} = 3.99$ |
| .001 | $\dfrac{(2+.001)^2 - 2^2}{.001} = 4.001$ | $-.001$ | $\dfrac{(2+(-.001))^2 - 2^2}{-.001} = 3.999$ |
| .0001 | $\dfrac{(2+.0001)^2 - 2^2}{.0001} = 4.0001$ | $-.0001$ | $\dfrac{(2+(-.0001))^2 - 2^2}{-.0001} = 3.9999$ |

*Step 2*     In this step, we let $h$ approach 0. Something very special happens if we replace $h$ by 0 in the expression on the right side of (7): Both the numerator and denominator become 0. So it is not clear what will become of the right side of (7) as $h$ approaches 0. Table 1 gives some typical values of the difference quotient for progressively smaller values of $h$, both positive and negative.

*Step 3*     It is clear from Table 1 that the values of the difference quotient

$$\frac{(2+h)^2 - 2^2}{h}$$

are approaching 4 as $h$ approaches 0. Since the values of the difference quotient approach the derivative $f'(2)$, we conclude that $f'(2) = 4$.

Using a terminology and notation that will be discussed in the next section, we express our result by saying that 4 is the *limit* of

$$\frac{(2+h)^2 - 2^2}{h}$$

as $h$ approaches 0, and write

$$\lim_{h \to 0} \frac{(2+h)^2 - 2^2}{h} = 4. \tag{8}$$

To compute $f'(x)$ for an arbitrary $x$, we must repeat the above three steps with an arbitrary value of $x$. Since $x$ is arbitrary, this will require figuring out the limit of the difference quotient without recourse to a table of numerical values. Is this possible? Let us take a closer look at the difference quotient on the right side in (7) and simplify it by expanding the square in the numerator: for $h \neq 0$,

$$\frac{(2+h)^2 - 2^2}{h} = \frac{\cancel{2^2} + 4h + h^2 - \cancel{2^2}}{h}$$

$$= \frac{(4+h)\cancel{h}}{\cancel{h}} = 4 + h.$$

$h < 0$     $h > 0$

$4 + h \dashrightarrow 4 \dashleftarrow 4 + h$

**Figure 10.** The distance from 4 to $4 + h$ is $|h|$. As $h$ approaches 0, this distance shrinks to 0, and $4 + h$ approaches 4.

(We were able to simplify by dividing both the numerator and denominator by $h$ because $h$ is never 0.) Thus the difference quotient is equal to $4+h$ and, since $4+h$ approaches 4 as $h$ approaches 0 (Fig. 10), we conclude that $f'(2) = 4$.

As we now show, we can repeat the last computation with minor modifications to evaluate the derivative of $f(x) = x^2$, for any $x$.

**EXAMPLE 5**  **Verification of the power rule for $r = 2$**  Apply the three-step method to confirm formula (3); that is, show that the derivative of $f(x) = x^2$ is $f'(x) = 2x$.

**Solution**  **Step 1**  We calculate the difference quotient and simplify as much as possible:

$$\frac{f(x+h) - f(x)}{h} = \frac{(x+h)^2 - x^2}{h} \qquad (h \neq 0)$$

$$= \frac{\cancel{x^2} + 2xh + h^2 - \cancel{x^2}}{h} = \frac{(2x+h)\cancel{h}}{\cancel{h}}$$

$$= 2x + h.$$

**Step 2**  As $h$ approaches 0, the expression $2x + h$ approaches $2x$, and hence the difference quotient approaches $2x$.

**Step 3**  Since the difference quotient approaches the derivative of $f(x) = x^2$ as $h$ approaches 0, we conclude that $f'(x) = 2x$. ∎

## The Limit Definition of the Derivative

Our discussion of the derivative has been based on an intuitive geometric concept of the tangent line. This geometric approach led us to the three-step method for computing the derivative as a limit of a difference quotient. This process may be considered independently of its geometric interpretation and used to *define* $f'(x)$. In fact, we say that $f$ is *differentiable* at $x$ if

$$\frac{f(x+h) - f(x)}{h}$$

approaches some number as $h$ approaches 0, and we denote this limiting number by $f'(x)$; in symbols, using the limit notation,

---

LIMIT DEFINITION OF THE DERIVATIVE

$$f'(x) = \lim_{h \to 0} \frac{f(x+h) - f(x)}{h} \tag{9}$$

---

If the difference quotient

$$\frac{f(x+h) - f(x)}{h}$$

does not approach any specific number as $h$ approaches 0, we say that $f$ is *non-differentiable* at $x$. Essentially all the functions in this text are differentiable at all points in their domain. A few exceptions are described in Section 1.5.

Formula (9) expresses analytically the geometric process of computing the slope of the tangent line as a limit of slopes of secant lines. It is of great importance in the theoretical development of calculus and can be used to compute derivatives, as we saw in Example 5. Do you need to use formula (9) every time you want to compute a derivative? The answer is no. The beauty and power of calculus is that it offers you general formulas, such as the power rule, to compute derivatives without going through the process of limits. These formulas, as well as many properties of the derivative, are usually established by using formula (9).

Let us end our discussion with two more applications of formula (9).

---

**EXAMPLE 6**  **Limit calculation of the derivative**  Use limits to find the derivative of $f(x) = x^2 + 2$.

**Solution** We shall follow our three-step method.

**Step 1**
$$\frac{f(x+h) - f(x)}{h} = \frac{\overbrace{(x+h)^2 + 2}^{f(x+h)} - \overbrace{(x^2+2)}^{f(x)}}{h} \quad (h \neq 0)$$
$$= \frac{x^2 + 2xh + h^2 + 2 - x^2 - 2}{h} = \frac{(2x+h)h}{h} = 2x + h.$$

**Step 2** As $h$ approaches 0, the expression $2x + h$ approaches $2x$.

**Step 3** Since the difference quotient approaches $f'(x)$, we conclude that $f'(x) = 2x$. ∎

It is interesting to note that $x^2 + 2$ and $x^2$ have the same derivative $2x$. This is not a coincidence and will follow easily from properties of the derivative of a sum in Section 1.6.

---

**EXAMPLE 7** **Derivative of a linear function** Confirm formula (1) by using the limit definition of the derivative to show that if $f(x) = mx + b$, then $f'(x) = m$.

**Solution** **Step 1** Simplify the difference quotient as much as possible.

$$\frac{f(x+h) - f(x)}{h} = \frac{\overbrace{m(x+h) + b}^{f(x+h)} - \overbrace{(mx+b)}^{f(x)}}{h}$$
$$= \frac{mx + mh + b - mx - b}{h} = \frac{mh}{h} = m$$

**Step 2** As $h$ approaches 0, the constant $m$ does not change and hence its limit is $m$.

**Step 3** Since the difference quotient approaches $f'(x)$, we conclude that $f'(x) = m$. ∎

Successful applications of formula (9) will clearly depend on our ability to evaluate limits. For this purpose, we present additional properties and tricks with limits in the following section. These will be used occasionally in proofs.

**Incorporating Technology** Graphing calculators have a numerical derivative function (sometimes called **nDeriv** or **nDer**) that gives approximate derivatives by essentially using a small value of $h$. Some calculators (such as the TI-85 and TI-86) have a function (called **der1**) that computes exact derivatives. In Fig. 11 the value of $f'(2)$ is first approximated and then calculated exactly for $f(x) = 1/x$. In Fig. 12, **Y₂** is specified as the derivative of $f(x)$. In Fig. 13, **Y₂** is graphed and **TRACE** is used to obtain the value of $f'(2)$.

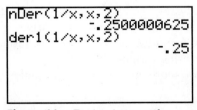

**Figure 11.** Derivatives on the TI-86.

**Figure 12.** Derivative function on the TI-83.

**Figure 13.** Graph of the function **Y₂** from Fig. 9.

## PRACTICE PROBLEMS 1.3

1. Consider the curve $y = f(x)$ in Fig. 14.

   (a) Find $f(5)$.

   (b) Find $f'(5)$.

2. Let $f(x) = 1/x^4$.

   (a) Find its derivative.

   (b) Find $f'(2)$.

**Figure 14**

# EXERCISES 1.3

*Use (1), (2), and the power rule to find the derivatives of the following functions.*

1. $f(x) = 2x - 5$

2. $f(x) = -3x + 7$

3. $f(x) = x^5$

4. $f(x) = x^{19}$

5. $f(x) = x^{\frac{2}{5}}$

6. $f(x) = \dfrac{1}{\sqrt{x}}$

7. $f(x) = \sqrt[3]{x}$

8. $f(x) = x^{4/3}$

9. $f(x) = \dfrac{1}{x^2}$

10. $f(x) = 3^2$

11. $f(x) = x^{-5/7}$

12. $f(x) = \dfrac{1}{x^{-5}}$

13. $f(x) = \frac{3}{4}$

14. $f(x) = \dfrac{1}{\sqrt[7]{x}}$

15. $f(x) = \dfrac{1}{\sqrt[3]{x}}$

16. $f(x) = \sqrt[5]{x}$

*In Exercises 17–24, find the derivative of $f(x)$ at the designated value of $x$.*

17. $f(x) = x^3$ at $x = 1$

18. $f(x) = x^5$ at $x = \frac{2}{3}$

19. $f(x) = \dfrac{1}{x}$ at $x = 3$

20. $f(x) = 3^2$ at $x = 1$

21. $f(x) = 4x + 11$ at $x = -1$

22. $f(x) = x^{3/2}$ at $x = 4$

23. $f(x) = \sqrt{x}$ at $x = 16$

24. $f(x) = \dfrac{1}{x^6}$ at $x = 2$

25. Find the slope of the curve $y = x^4$ at $x = 3$.

26. Find the slope of the curve $y = x^5$ at $x = -2$.

27. If $f(x) = x^2$, compute $f(-5)$ and $f'(-5)$.

28. If $f(x) = x + 6$, compute $f(3)$ and $f'(3)$.

29. If $f(x) = x^{4/3}$, compute $f(8)$ and $f'(8)$.

30. If $f(x) = 1/x^2$, compute $f(5)$ and $f'(5)$.

31. If $f(x) = 1/x^5$, compute $f(2)$ and $f'(2)$.

32. If $f(x) = x^{3/2}$, compute $f(16)$ and $f'(16)$.

*In Exercises 33–40, find the equation of the tangent line to the graph of $y = f(x)$ at the given $x$. Do not apply (6), but proceed as we did in Example 3.*

33. $f(x) = x^2$, $x = -\frac{1}{2}$

34. $f(x) = x^3$, $x = -2$

35. $f(x) = 3x + 1$, $x = 4$

36. $f(x) = 5$, $x = -2$

37. $f(x) = \sqrt{x}$, $x = 9$

38. $f(x) = \dfrac{1}{x}$, $x = .01$

39. $f(x) = \dfrac{1}{\sqrt{x}}$, $x = 4$

40. $f(x) = \dfrac{1}{x^2}$, $x = 1$

41. The point-slope form of the equation of the tangent line to the graph of $y = x^4$ at the point $(1, 1)$ is $y - 1 = 4(x - 1)$. Explain how this equation follows from formula (6).

42. The tangent to the graph of $y = \frac{1}{x}$ at the point $P = \left(a, \frac{1}{a}\right)$ is perpendicular to the line $y = 4x + 1$. Find $P$.

43. The line $y = 2x + b$ is tangent to the graph of $y = \sqrt{x}$ at the point $P = (a, \sqrt{a})$. Find $P$ and determine $b$.

44. The line $y = ax + b$ is tangent to the graph of $y = x^3$ at the point $P = (-3, -27)$. Find $a$ and $b$.

45. (a) Find the point on the curve $y = \sqrt{x}$ where the tangent line is parallel to the line $y = \frac{x}{8}$.

    (b) On the same axes, plot the curve $y = \sqrt{x}$, the line $y = \frac{x}{8}$, and the tangent line to $y = \sqrt{x}$ that is parallel to $y = \frac{x}{8}$.

46. Is there any point on the graph of $y = \frac{1}{x}$ where the slope of the graph is positive? What does your answer imply about the graph of $y = \frac{1}{x}$?

47. (a) Draw the graphs of $f(x) = x^2$ and $g(x) = x^3$ on the same axes.

    (b) Compute $f'(1)$ and $g'(1)$. Which is larger? Interpret your answer as a fact about the graphs. [*Hint*: The derivative is a slope formula.]

48. (a) Draw the graphs of $f(x) = x$ and $g(x) = \sqrt{x}$ on the same axes.

**(b)** Compute $f(1)$, $g(1)$, $f'(1)$ and $g'(1)$. Interpret your answer as a fact about the graphs.

*In Exercises 49–56, find the indicated derivative.*

**49.** $\dfrac{d}{dx}\left(x^8\right)$

**50.** $\dfrac{d}{dx}\left(x^{-3}\right)$

**51.** $\dfrac{d}{dx}\left(x^{3/4}\right)$

**52.** $\dfrac{d}{dx}\left(x^{-1/3}\right)$

**53.** $\dfrac{dy}{dx}$ if $y = 1$

**54.** $\dfrac{dy}{dx}$ if $y = x^{-4}$

**55.** $\dfrac{dy}{dx}$ if $y = x^{1/5}$

**56.** $\dfrac{dy}{dx}$ if $y = \dfrac{x-1}{3}$

**57.** Consider the curve $y = f(x)$ in Fig. 15. Find $f(6)$ and $f'(6)$.

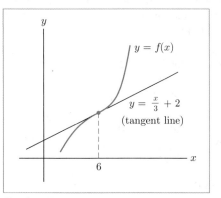

**Figure 15**

**58.** Consider the curve $y = f(x)$ in Fig. 16. Find $f(1)$ and $f'(1)$.

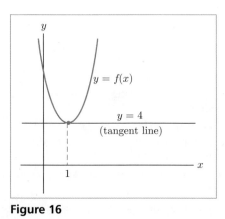

**Figure 16**

**59.** In Fig. 17 the straight line $y = \frac{1}{4}x + b$ is tangent to the graph of $f(x) = \sqrt{x}$. Find the values of $a$ and $b$.

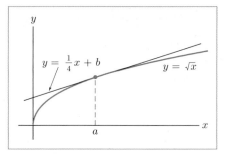

**Figure 17**

**60.** In Fig. 18 the straight line is tangent to the graph of $f(x) = 1/x$. Find the value of $a$.

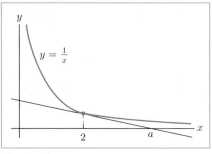

**Figure 18**

**61.** Consider the curve $y = f(x)$ in Fig. 19. Find $a$ and $f(a)$. Estimate $f'(a)$.

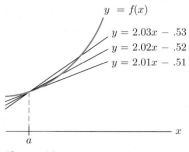

**Figure 19**

**62.** Consider the curve $y = f(x)$ in Fig. 20. Estimate $f'(1)$.

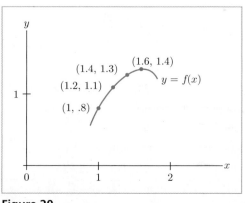

**Figure 20**

*In Exercises 63–70, compute the difference quotient*

$$\frac{f(x+h)-f(x)}{h}.$$

*Simplify your answer as much as possible.*

**63.** $f(x) = x^2 - 2x + 3$     **64.** $f(x) = -2x^2 + x + 1$

**65.** $f(x) = -x^2 - x - 1$     **66.** $f(x) = \frac{1}{2}x^2 + \frac{1}{2}$

**67.** $f(x) = x^3 + x + 2$     **68.** $f(x) = x^3 - 2x^2 + \sqrt{5}$

**69.** $f(x) = \dfrac{2}{x+2}$     **70.** $f(x) = \dfrac{1}{x^2+1}$

*In Exercises 71–74, apply the three-step method to compute the following derivatives. Follow the steps that we used before Example 5. In particular, you should simplify the difference quotient as much as possible before taking limits.*

**71.** $f'(1)$, where $f(x) = x^2 + 1$

**72.** $f'(-1)$, where $f(x) = -x^2 + 2$

**73.** $f'(2)$, where $f(x) = x^3$ [*Hint:* $(a+b)^3 = a^3 + 3a^2b + 3ab^2 + b^3$.]

**74.** $f'(-1)$, where $f(x) = -3x^3 + 1$

*In Exercises 75–82, apply the three-step method to compute the derivative of the given function using limits. (See Examples 5 and 6.)*

**75.** $f(x) = x^2 + 4$     **76.** $f(x) = 2x^2 + x$

**77.** $f(x) = -2x^2 - 1$     **78.** $f(x) = -x^2 + \frac{1}{2}x - 2$

**79.** $f(x) = x^3$     **80.** $f(x) = 2x^3 + 1$

**81.** $f(x) = 2x - 1$     **82.** $f(x) = 5$

**83.** Use the limit definition of the derivative to verify formula (2).

**84.** Use the limit definition to show that the derivative of $f(x) = ax^2$ is $f'(x) = 2ax$, where $a$ is an arbitrary constant.

**85.** In Fig. 21, find the equation of the tangent line to $f(x)$ at the point $A$.

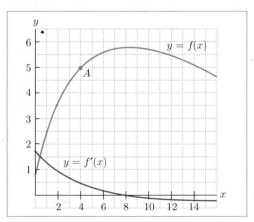

**Figure 21**

**86.** In Fig. 22, find the equation of the tangent line to $f(x)$ at the point $P$.

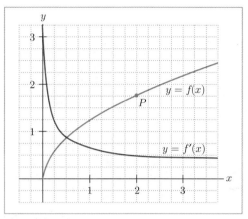

**Figure 22**

**87.** (a) Draw two graphs (of your choice) that represent a function $y = f(x)$ and its vertical shift $y = f(x) + 3$.

(b) Pick two points on these graphs with the same first coordinates: $(x, f(x))$ and $(x, f(x)+3)$. Draw the tangent lines to the curves at these points and describe what you observe about the tangent lines.

(c) Based on your observation in part (b), explain why

$$\frac{d}{dx}f(x) = \frac{d}{dx}(f(x) + 3).$$

**88.** Use the approach of Exercise 87 to show that

$$\frac{d}{dx}f(x) = \frac{d}{dx}(f(x) + c),$$

for any constant $c$. [*Hint:* Compare the tangent lines to the graphs of $f(x)$ and $f(x) + c$.]

*Technology Exercises*

*In Exercises 89–94, use a derivative routine to obtain the value of the derivative. Give the value to 5 decimal places.*

**89.** $f'(0)$, where $f(x) = 2^x$

**90.** $f'(1)$, where $f(x) = \dfrac{1}{1+x^2}$

**91.** $f'(1)$, where $f(x) = \sqrt{1+x^2}$

**92.** $f'(3)$, where $f(x) = \sqrt{25 - x^2}$

**93.** $f'(2)$, where $f(x) = \dfrac{x}{1+x}$

**94.** $f'(0)$, where $f(x) = 10^{1+x}$

*In Exercises 95 and 96, let $Y_1$ be the specified function and use a derivative routine to set $Y_2$ as its derivative. For instance, you might use $Y_2 = \mathbf{nDeriv}(Y_1, X, X)$ or $y2 = \mathbf{der1}(y1, x, x)$. Then graph $Y_2$ in the specified window and use TRACE to obtain the value of the derivative of $Y_1$ at $x = 2$.*

**95.** $f(x) = 3x^2 - 5$, $[0, 4]$ by $[-5, 40]$
**96.** $f(x) = \sqrt{2x}$, $[0, 4]$ by $[-.5, 3]$

*In Exercises 97–100, use paper and pencil to find the equation of the tangent line to the graph of the function at the designated point. Then graph both the function and the line to confirm it is indeed the sought-after tangent line.*

**97.** $f(x) = \sqrt{x}$, $(9, 3)$       **98.** $f(x) = \dfrac{1}{x}$, $(.5, 2)$

**99.** $f(x) = \dfrac{1}{x^2}$, $(.5, 4)$       **100.** $f(x) = x^3$, $(1, 1)$

*In Exercises 101–104, $g(x)$ is the tangent line to the graph of $f(x)$ at $x = a$. Graph $f(x)$ and $g(x)$ and determine the value of $a$.*

**101.** $f(x) = \dfrac{5}{x}$, $g(x) = 5 - 1.25x$

**102.** $f(x) = \sqrt{8x}$, $g(x) = x + 2$

**103.** $f(x) = x^3 - 12x^2 + 46x - 50$, $g(x) = 14 - 2x$

**104.** $f(x) = (x - 3)^3 + 4$, $g(x) = 4$

## SOLUTIONS TO PRACTICE PROBLEMS 1.3

**1.** **(a)** The number $f(5)$ is the $y$-coordinate of the point $P$. Since the tangent line passes through $P$, the coordinates of $P$ satisfy the equation $y = -x + 8$. Since its $x$-coordinate is 5, its $y$-coordinate is $-5 + 8 = 3$. Therefore, $f(5) = 3$.

**(b)** The number $f'(5)$ is the slope of the tangent line at $P$, which is readily seen to be $-1$.

**2.** **(a)** The function $1/x^4$ can be written as the power function $x^{-4}$. Here $r = -4$. Therefore,

$$f'(x) = (-4)x^{(-4)-1} = -4x^{-5} = \frac{-4}{x^5}.$$

**(b)** $f'(2) = -4/2^5 = -4/32 = -\frac{1}{8}$.

## 1.4   Limits and the Derivative

The notion of a limit is one of the fundamental ideas of calculus. Indeed, any "theoretical" development of calculus rests on an extensive use of the theory of limits. Even in this book, where we have adopted an intuitive viewpoint, limit arguments are used occasionally (although in an informal way). In the previous section, limits were used to define the derivative. They arose naturally from our geometric construction of the tangent line as a *limit* of secant lines. In this section we continue our brief introduction to limits and state some additional properties that are required in our development of the derivative and several other areas of calculus. We start with a definition.

> **DEFINITION OF A LIMIT**   Let $g(x)$ be a function, $a$ a number. We say that the number $L$ is the *limit of $g(x)$ as $x$ approaches $a$* provided that $g(x)$ can be made arbitrarily close to $L$ by taking $x$ sufficiently close (but not equal) to $a$. In this case we write
> $$\lim_{x \to a} g(x) = L.$$

If, as $x$ approaches $a$, the values $g(x)$ do *not* approach a specific number, then we say that the limit of $g(x)$ as $x$ approaches $a$ *does not exist*. We saw several examples of limits in the previous section. Let us give some further basic examples before turning to the properties of limits.

**EXAMPLE 1**    Determine $\lim_{x \to 2}(3x - 5)$.

**Solution**    Let us make a table of values of $x$ approaching 2 and the corresponding values of $3x - 5$:

| $x$ | $3x - 5$ | $x$ | $3x - 5$ |
|------|---------|--------|---------|
| 2.1 | 1.3 | 1.9 | .7 |
| 2.01 | 1.03 | 1.99 | .97 |
| 2.001 | 1.003 | 1.999 | .997 |
| 2.0001 | 1.0003 | 1.9999 | .9997 |

As $x$ approaches 2, we see that $3x - 5$ approaches 1. In terms of our notation,

$$\lim_{x \to 2}(3x - 5) = 1.$$

**EXAMPLE 2**    For each of the functions in Fig. 1, determine if $\lim_{x \to 2} g(x)$ exists. (The circles drawn on the graphs are meant to represent breaks in the graph, indicating that the functions under consideration are not defined at $x = 2$.)

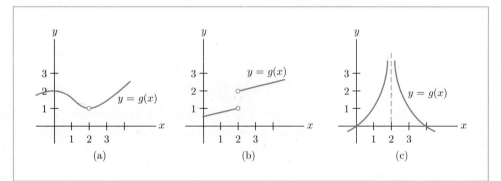

**Figure 1.**    Functions with an undefined point.

**Solution**    (a) $\lim_{x \to 2} g(x) = 1$. We can see that as $x$ gets closer and closer to 2, the values of $g(x)$ get closer and closer to 1. This is true for values of $x$ to both the right and the left of 2.

(b) $\lim_{x \to 2} g(x)$ does not exist. As $x$ approaches 2 from the right, $g(x)$ approaches 2. However, as $x$ approaches 2 from the left, $g(x)$ approaches 1. In order for a limit to exist, the function must approach the *same* value from each direction.

(c) $\lim_{x \to 2} g(x)$ does not exist. As $x$ approaches 2, the values of $g(x)$ become larger and larger and do not approach a fixed number.

The following limit theorems, which we cite without proof, allow us to reduce the computation of limits for combinations of functions to computations of limits involving the constituent functions.

**Limit Theorems**    Suppose that $\lim\limits_{x \to a} f(x)$ and $\lim\limits_{x \to a} g(x)$ both exist. Then we have the following results.

(I)  If $k$ is a constant, then $\lim\limits_{x \to a} k \cdot f(x) = k \cdot \lim\limits_{x \to a} f(x)$.

(II)  If $r$ is a positive constant, then $\lim\limits_{x \to a} [f(x)]^r = \left[\lim\limits_{x \to a} f(x)\right]^r$.

(III)  $\lim\limits_{x \to a} [f(x) + g(x)] = \lim\limits_{x \to a} f(x) + \lim\limits_{x \to a} g(x)$.

(IV)  $\lim\limits_{x \to a} [f(x) - g(x)] = \lim\limits_{x \to a} f(x) - \lim\limits_{x \to a} g(x)$.

(V)  $\lim\limits_{x \to a} [f(x) \cdot g(x)] = \left[\lim\limits_{x \to a} f(x)\right] \cdot \left[\lim\limits_{x \to a} g(x)\right]$.

(VI)  If $\lim\limits_{x \to a} g(x) \neq 0$, then $\lim\limits_{x \to a} \dfrac{f(x)}{g(x)} = \dfrac{\lim\limits_{x \to a} f(x)}{\lim\limits_{x \to a} g(x)}$.

---

**EXAMPLE 3**    Use the limit theorems to compute the following limits.

(a)  $\lim\limits_{x \to 2} x^3$            (b)  $\lim\limits_{x \to 2} 5x^3$            (c)  $\lim\limits_{x \to 2} (5x^3 - 15)$

(d)  $\lim\limits_{x \to 2} \sqrt{5x^3 - 15}$        (e)  $\lim\limits_{x \to 2} \left(\sqrt{5x^3 - 15}/x^5\right)$

**Solution**  (a) Since $\lim\limits_{x \to 2} x = 2$, we have by Limit Theorem II that

$$\lim_{x \to 2} x^3 = \left(\lim_{x \to 2} x\right)^3 = 2^3 = 8.$$

(b)  $\lim\limits_{x \to 2} 5x^3 = 5 \lim\limits_{x \to 2} x^3$    (Limit Theorem I with $k = 5$)

$\qquad\qquad = 5 \cdot 8$        [by part (a)]

$\qquad\qquad = 40.$

(c)  $\lim\limits_{x \to 2} (5x^3 - 15) = \lim\limits_{x \to 2} 5x^3 - \lim\limits_{x \to 2} 15$    (Limit Theorem IV)

Note that $\lim\limits_{x \to 2} 15 = 15$. This is because the constant function $g(x) = 15$ always has the value 15, and so its limit as $x$ approaches *any* number is 15. By part (b), $\lim\limits_{x \to 2} 5x^3 = 40$. Thus

$$\lim_{x \to 2} (5x^3 - 15) = 40 - 15 = 25.$$

(d)  $\lim\limits_{x \to 2} \sqrt{5x^3 - 15} = \lim\limits_{x \to 2} (5x^3 - 15)^{1/2}$

$$= \left[\lim_{x \to 2} (5x^3 - 15)\right]^{1/2} \quad \begin{array}{l}\text{[Limit Theorem II with}\\ r = \tfrac{1}{2},\ f(x) = 5x^3 - 15]\end{array}$$

$\qquad\qquad = 25^{1/2}$        [by part (c)]

$\qquad\qquad = 5$

(e)  The limit of the denominator is $\lim\limits_{x \to 2} x^5$, which is $2^5 = 32$, a nonzero number. So by Limit Theorem VI we have

$$\lim_{x \to 2} \frac{\sqrt{5x^3 - 15}}{x^5} = \frac{\lim\limits_{x \to 2} \sqrt{5x^3 - 15}}{\lim\limits_{x \to 2} x^5}$$

$$= \frac{5}{32} \quad \text{[by part (d)]}.\qquad\blacksquare$$

The following facts, which may be deduced by repeated applications of the various limit theorems, are extremely handy in evaluating limits.

---

**Limit Theorems (Continued)**

(VII) **Limit of a Polynomial Function** Let $p(x)$ be a polynomial function, $a$ any number. Then
$$\lim_{x \to a} p(x) = p(a).$$

(VIII) **Limit of a Rational Function** Let $r(x) = p(x)/q(x)$ be a rational function, where $p(x)$ and $q(x)$ are polynomials. Let $a$ be a number such that $q(a) \neq 0$. Then
$$\lim_{x \to a} r(x) = r(a).$$

---

In other words, to determine a limit of a polynomial or a rational function, simply evaluate the function at $x = a$, provided, of course, that the function is defined at $x = a$. For instance, we can rework the solution to Example 3(c) as follows:

$$\lim_{x \to 2} (5x^3 - 15) = 5(2)^3 - 15 = 25.$$

Many situations require algebraic simplifications before the limit theorems can be applied.

---

**EXAMPLE 4**     Compute the following limits.

(a) $\lim\limits_{x \to 3} \dfrac{x^2 - 9}{x - 3}$

(b) $\lim\limits_{x \to 0} \dfrac{\sqrt{x + 4} - 2}{x}$

**Solution** (a) The function $\dfrac{x^2 - 9}{x - 3}$ is not defined when $x = 3$, since

$$\frac{3^2 - 9}{3 - 3} = \frac{0}{0},$$

which is undefined. That causes no difficulty, since the limit as $x$ approaches 3 depends only on the values of $x$ *near* 3 and excludes considerations of the value at $x = 3$ itself. To evaluate the limit, note that $x^2 - 9 = (x - 3)(x + 3)$. So for $x \neq 3$,

$$\frac{x^2 - 9}{x - 3} = \frac{(x - 3)(x + 3)}{x - 3} = x + 3.$$

As $x$ approaches 3, $x + 3$ approaches 6. Therefore,

$$\lim_{x \to 3} \frac{x^2 - 9}{x - 3} = 6.$$

(b) Since the denominator approaches zero when taking the limit, we may not apply Limit Theorem VI directly. However, if we first apply an algebraic "rationalization" trick, the limit may be evaluated. Multiply numerator and denominator by $\sqrt{x+4}+2$.

$$\frac{\sqrt{x+4}-2}{x} \cdot \frac{\sqrt{x+4}+2}{\sqrt{x+4}+2} = \frac{(x+4)-4}{x(\sqrt{x+4}+2)}$$

$$= \frac{x}{x(\sqrt{x+4}+2)}$$

$$= \frac{1}{\sqrt{x+4}+2}$$

Thus

$$\lim_{x \to 0} \frac{\sqrt{x+4}-2}{x} = \lim_{x \to 0} \frac{1}{\sqrt{x+4}+2}$$

$$= \frac{\lim_{x \to 0} 1}{\lim_{x \to 0} (\sqrt{x+4}+2)} \qquad \text{(Limit Theorem VI)}$$

$$= \frac{1}{4}.$$

## Limits and the Derivative

Our geometric definition of the derivative as the slope of the tangent line led us naturally to the concept of a limit. Indeed, we defined the derivative of a function $f$ at $x$ by

$$f'(x) = \lim_{h \to 0} \frac{f(x+h) - f(x)}{h}, \tag{1}$$

provided the limit exists. We recall that the difference quotient $\frac{f(x+h)-f(x)}{h}$ represents the slope of the secant line through the points $(x, f(x))$ and $(x+h, f(x+h))$ on the graph of $f$, and the limit in (1) expresses the fact that the slope of the tangent line (the derivative) is equal to the limit of the slopes of the secant lines. You may wish at this point to review this topic in the previous section (in particular, the discussion leading to (9)). To evaluate the limit in (1), proceed in three basic steps:

---

**Using Limits to Calculate $f'(a)$:**

***Step 1***    Write the difference quotient $\dfrac{f(a+h) - f(a)}{h}$.

***Step 2***    Simplify the difference quotient.

***Step 3***    Find the limit as $h \to 0$. This limit is $f'(a)$.

---

**EXAMPLE 5**    Use limits to compute the derivative $f'(5)$ for the following functions.

(a) $f(x) = 15 - x^2$          (b) $f(x) = \dfrac{1}{2x - 3}$

**Solution**   In each case, we must calculate $\lim\limits_{h \to 0} \dfrac{f(5+h) - f(5)}{h}$.

(a) $\dfrac{f(5+h) - f(5)}{h} = \dfrac{\left[15 - (5+h)^2\right] - \left(15 - 5^2\right)}{h}$       **(step 1)**

$$= \frac{15 - \left(25 + 10h + h^2\right) - \left(15 - 25\right)}{h} \qquad \text{(step 2)}$$

$$= \frac{-10h - h^2}{h} = -10 - h.$$

Therefore, $f'(5) = \lim_{h \to 0} (-10 - h) = -10.$ \qquad (step 3)

(b) $\dfrac{f(5 + h) - f(5)}{h} = \dfrac{\dfrac{1}{2(5 + h) - 3} - \dfrac{1}{2(5) - 3}}{h}$ \qquad (step 1)

$$= \frac{\dfrac{1}{7 + 2h} - \dfrac{1}{7}}{h} = \frac{\dfrac{7 - (7 + 2h)}{(7 + 2h)7}}{h} \qquad \text{(step 2)}$$

$$= \frac{-2h}{(7 + 2h)7 \cdot h} = \frac{-2}{(7 + 2h)7} = \frac{-2}{49 + 14h}.$$

Therefore,

$$f'(5) = \lim_{h \to 0} \frac{-2}{49 + 14h} = -\frac{2}{49} \qquad \text{(step 3)}$$

▪

▶ *Remark*  When computing the limits in Example 5, we considered only values of $h$ near zero (and not $h = 0$ itself). Therefore, we were freely able to divide both numerator and denominator by $h$. ◀

It is important to distinguish between the derivative at a specific point (as in Example 5) and a derivative formula. A derivative at a specific point is a number, while a derivative formula is a function. The three-step method that we outlined can be used to find derivative formulas as we now illustrate.

**EXAMPLE 6**  Find the derivative of $f(x) = x^2 + 2x + 2$.

**Solution**  We have

$$\frac{f(x + h) - f(x)}{h} = \frac{(x + h)^2 + 2(x + h) + 2 - (x^2 + 2x + 2)}{h} \qquad \text{(step 1)}$$

$$= \frac{\cancel{x^2} + 2xh + h^2 + \cancel{2x} + 2h + \cancel{2} - \cancel{x^2} - \cancel{2x} - \cancel{2}}{h} \qquad \text{(step 2)}$$

$$= \frac{2xh + h^2 + 2h}{h} = \frac{(2x + h + 2)\cancel{h}}{\cancel{h}}$$

$$= 2x + 2 + h.$$

Therefore,

$$f'(x) = \lim_{h \to 0} (2x + 2 + h) = 2x + 2. \qquad \text{(step 3)}$$

▪

Let us go back to the power rule in the previous section and verify it in two special cases: $r = -1$ and $r = \frac{1}{2}$.

**EXAMPLE 7**  Find the derivative of $f(x) = \frac{1}{x}$.

**Solution** We have

$$\frac{f(x+h) - f(x)}{h} = \frac{\frac{1}{x+h} - \frac{1}{x}}{h} \qquad \text{(step 1)}$$

$$= \frac{1}{h} \left[ \frac{1}{x+h} - \frac{1}{x} \right] \qquad \text{(step 2)}$$

$$= \frac{1}{h} \left[ \frac{x - (x+h)}{(x+h)x} \right] = \frac{1}{\not{h}} \left[ \frac{-\not{h}}{(x+h)x} \right]$$

$$= \frac{-1}{(x+h)x}$$

**Step 3**   Using Limit Theorem VIII, we find that as $h$ approaches 0, $\frac{-1}{(x+h)x}$ approaches $\frac{-1}{x^2}$. Hence

$$f'(x) = \frac{-1}{x^2}. \qquad \blacksquare$$

In evaluating the next limit, we use the rationalization trick of Example 4(b).

---

**EXAMPLE 8**   Find the derivative of $f(x) = \sqrt{x}$.

**Solution** In simplifying the difference quotient, we use the algebraic identity $(a+b)(a-b) = a^2 - b^2$. So

$$(\sqrt{x+h} - \sqrt{x})(\sqrt{x+h} + \sqrt{x}) = (\sqrt{x+h})^2 - (\sqrt{x})^2 = x + h - x = h.$$

We have

$$\frac{f(x+h) - f(x)}{h} = \frac{\sqrt{x+h} - \sqrt{x}}{h} \qquad \text{(step 1)}$$

$$= \frac{\sqrt{x+h} - \sqrt{x}}{h} \cdot \frac{\sqrt{x+h} + \sqrt{x}}{\sqrt{x+h} + \sqrt{x}} \qquad \text{(step 2)}$$

$$= \frac{x + h - x}{h(\sqrt{x+h} + \sqrt{x})} = \frac{\not{h}}{\not{h}(\sqrt{x+h} + \sqrt{x})}$$

$$= \frac{1}{\sqrt{x+h} + \sqrt{x}}.$$

Therefore,

$$f'(x) = \lim_{h \to 0} \frac{1}{\sqrt{x+h} + \sqrt{x}} = \frac{1}{\sqrt{x} + \sqrt{x}} = \frac{1}{2\sqrt{x}}. \qquad \text{(step 3)}$$

$$\blacksquare$$

Sometimes we can use our knowledge of differentiation to evaluate some difficult limits. Here is one illustration.

---

**EXAMPLE 9**   Match the limit

$$\lim_{x \to 0} \frac{(1+x)^{1/3} - 1}{x}$$

with a derivative and find its value.

Solution The idea here is to identify the given limit as a derivative given by (1) for a specific choice of $f$ and $x$. Toward this end, let us replace $x$ by $h$ in the given limit. This does not change the limit, since we are only renaming the variables. The limit becomes

$$\lim_{h \to 0} \frac{(1+h)^{1/3} - 1}{h}.$$

Now go back to (1). Take $f(x) = x^{1/3}$ and evaluate $f'(1)$ according to (1):

$$f'(1) = \lim_{h \to 0} \frac{(1+h)^{1/3} - 1}{h}.$$

On the right side we have the desired limit; while on the left side, $f'(1)$, can be computed using the power rule:

$$f'(x) = \frac{1}{3} x^{-2/3},$$

$$f'(1) = \frac{1}{3}(1)^{-2/3} = \frac{1}{3}.$$

Hence

$$\lim_{h \to 0} \frac{(1+h)^{1/3} - 1}{h} = f'(1) = \frac{1}{3}. \qquad \blacksquare$$

**Infinity and Limits** Consider the function $f(x)$ whose graph is sketched in Fig. 2. As $x$ grows large the value of $f(x)$ approaches 2. In this circumstance, we

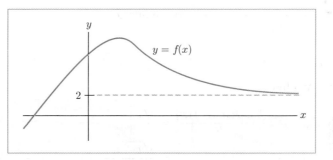

**Figure 2.** Function with a limit as $x$ approaches infinity.

say that 2 is the *limit of $f(x)$ as $x$ approaches infinity.* Infinity is denoted by the symbol $\infty$. The preceding limit statement is expressed in the following notation:

$$\lim_{x \to \infty} f(x) = 2.$$

In a similar vein, consider the function whose graph is sketched in Fig. 3. As $x$ grows large in the negative direction, the value of $f(x)$ approaches 0. In this circumstance, we say that 0 is the *limit of $f(x)$ as $x$ approaches minus infinity.* In symbols,

$$\lim_{x \to -\infty} f(x) = 0.$$

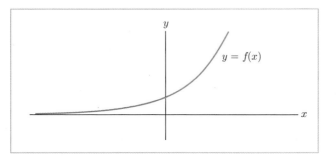

**Figure 3.** Function with a limit as $x$ approaches minus infinity.

**EXAMPLE 10**  Calculate the following limits.

(a) $\lim_{x \to \infty} \dfrac{1}{x^2 + 1}$    (b) $\lim_{x \to \infty} \dfrac{6x - 1}{2x + 1}$

**Solution**  (a) As $x$ increases without bound, so does $x^2 + 1$. Therefore, $1/(x^2 + 1)$ approaches zero as $x$ approaches $\infty$.

(b) Both $6x - 1$ and $2x + 1$ increase without bound as $x$ does. To determine the limit of their quotient, we employ an algebraic trick. Divide both numerator and denominator by $x$ to obtain

$$\lim_{x \to \infty} \frac{6x - 1}{2x + 1} = \lim_{x \to \infty} \frac{6 - \dfrac{1}{x}}{2 + \dfrac{1}{x}}.$$

As $x$ increases without bound, $1/x$ approaches zero, so $6 - (1/x)$ approaches 6 and $2 + (1/x)$ approaches 2. Thus the desired limit is $6/2 = 3$.  ∎

**Incorporating Technology**   Figure 4 shows the graph of the function

$$y = \frac{x^2 - 9}{x - 3}$$

from Example 4(a) with the window setting $[0, 6]$ *by* $[-2, 9]$. (Notice that 3 is in the middle of the $x$-range.) If you look closely, you will see a little blank space at the point where $x = 3$. However, if you examine function values near $x = 3$, you will be convinced that

$$\lim_{x \to 3} \frac{x^2 - 9}{x - 3} = 6.$$

**Figure 4**          **Figure 5**

Figure 5(a) shows the graph of the function

$$y = \frac{6x - 1}{2x + 1}$$

from Example 10(b) with the window $[0, 41]$ *by* $[-1, 4]$, and Fig. 5(b) shows the function values for increasing values of $x$. These figures provide convincing evidence that

$$\lim_{x \to \infty} \frac{6x - 1}{2x + 1} = 3.$$

**PRACTICE PROBLEMS 1.4**

*Determine which of the following limits exist. Compute the limits that exist.*

**1.** $\lim_{x \to 6} \dfrac{x^2 - 4x - 12}{x - 6}$    **2.** $\lim_{x \to 6} \dfrac{4x + 12}{x - 6}$

# EXERCISES 1.4

*For each of the following functions $g(x)$, determine whether or not $\lim_{x \to 3} g(x)$ exists. If so, give the limit.*

**1.**

**2.**

**3.**

**4.**

**5.**

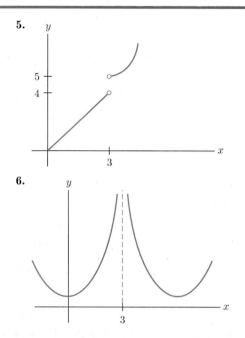

**6.**

*Determine which of the following limits exist. Compute the limits that exist.*

**7.** $\lim_{x \to 1} (1 - 6x)$

**8.** $\lim_{x \to 2} \dfrac{x}{x - 2}$

**9.** $\lim_{x \to 3} \sqrt{x^2 + 16}$

**10.** $\lim_{x \to 4} (x^3 - 7)$

**11.** $\lim_{x \to 5} \dfrac{x^2 + 1}{5 - x}$

**12.** $\lim_{x \to 6} \left( \sqrt{6x} + 3x - \dfrac{1}{x} \right) (x^2 - 4)$

**13.** $\lim_{x \to 7} (x + \sqrt{x - 6})(x^2 - 2x + 1)$

**14.** $\lim_{x \to 8} \dfrac{\sqrt{5x - 4} - 1}{3x^2 + 2}$

**15.** $\lim_{x \to 9} \dfrac{\sqrt{x^2 - 5x - 36}}{8 - 3x}$

**16.** $\lim_{x \to 10} \left( 2x^2 - 15x - 50 \right)^{20}$

**17.** $\lim_{x \to 0} \dfrac{x^2 + 3x}{x}$

**18.** $\lim_{x \to 1} \dfrac{x^2 - 1}{x - 1}$

**19.** $\lim_{x \to 2} \dfrac{-2x^2 + 4x}{x - 2}$

**20.** $\lim_{x \to 3} \dfrac{x^2 - x - 6}{x - 3}$

**21.** $\lim_{x \to 4} \dfrac{x^2 - 16}{4 - x}$

**22.** $\lim_{x \to 5} \dfrac{2x - 10}{x^2 - 25}$

**23.** $\lim_{x \to 6} \dfrac{x^2 - 6x}{x^2 - 5x - 6}$

**24.** $\lim_{x \to 7} \dfrac{x^3 - 2x^2 + 3x}{x^2}$

**25.** $\lim_{x \to 8} \dfrac{x^2 + 64}{x - 8}$

**26.** $\lim_{x \to 9} \dfrac{1}{(x - 9)^2}$

*In Exercises 27–28, compute the limits that exist, given that*

$$\lim_{x \to 0} f(x) = -\frac{1}{2} \quad and \quad \lim_{x \to 0} g(x) = \frac{1}{2}.$$

**27. (a)** $\lim\limits_{x \to 0} (f(x) + g(x))$  **(b)** $\lim\limits_{x \to 0} (f(x) - 2g(x))$

**(c)** $\lim\limits_{x \to 0} f(x) \cdot g(x)$  **(d)** $\lim\limits_{x \to 0} \dfrac{f(x)}{g(x)}$

**28. (a)** $\lim\limits_{x \to 0} x[f(x)]^2$  **(b)** $\lim\limits_{x \to 0} (f(x) + 1)$

**(c)** $\lim\limits_{x \to 0} \dfrac{\sqrt{2}}{\sqrt{g(x)}}$  **(d)** $\lim\limits_{x \to 0} \dfrac{1}{f(x) + g(x)}$

*Use limits to compute the following derivatives.*

**29.** $f'(3)$ where $f(x) = x^2 + 1$

**30.** $f'(2)$ where $f(x) = x^3$

**31.** $f'(0)$ where $f(x) = x^3 + 3x + 1$

**32.** $f'(0)$ where $f(x) = x^2 + 2x + 2$

**33.** $f'(3)$ where $f(x) = \dfrac{1}{2x + 5}$

**34.** $f'(4)$ where $f(x) = \sqrt{2x - 1}$

**35.** $f'(2)$ where $f(x) = \sqrt{5 - x}$

**36.** $f'(3)$ where $f(x) = \dfrac{1}{7 - 2x}$

*In Exercises 37–48, use limits to compute $f'(x)$. [Hint: In Exercises 45–48, use the rationalization trick of Example 8.]*

**37.** $f(x) = 3x + 1$  **38.** $f(x) = -x + 11$

**39.** $f(x) = x + \dfrac{1}{x}$  **40.** $f(x) = \dfrac{1}{x^2}$

**41.** $f(x) = \dfrac{x}{x + 1}$  **42.** $f(x) = -1 + \dfrac{2}{x - 2}$

**43.** $f(x) = \dfrac{1}{x^2 + 1}$  **44.** $f(x) = \dfrac{x}{x + 2}$

**45.** $f(x) = \sqrt{x + 2}$  **46.** $f(x) = \sqrt{x^2 + 1}$

**47.** $f(x) = \dfrac{1}{\sqrt{x}}$  **48.** $f(x) = x\sqrt{x}$

*Each limit in Exercises 49–54 is a definition of $f'(a)$. Determine the function $f(x)$ and the value of $a$.*

**49.** $\lim\limits_{h \to 0} \dfrac{(1 + h)^2 - 1}{h}$  **50.** $\lim\limits_{h \to 0} \dfrac{(2 + h)^3 - 8}{h}$

**51.** $\lim\limits_{h \to 0} \dfrac{\dfrac{1}{10 + h} - .1}{h}$  **52.** $\lim\limits_{h \to 0} \dfrac{(64 + h)^{1/3} - 4}{h}$

**53.** $\lim\limits_{h \to 0} \dfrac{\sqrt{9 + h} - 3}{h}$  **54.** $\lim\limits_{h \to 0} \dfrac{(1 + h)^{-1/2} - 1}{h}$

*In Exercises 55–60, match the given limit with a derivative then find the limit by computing the derivative.*

**55.** $\lim\limits_{h \to 0} \dfrac{(h + 2)^2 - 4}{h}$  **56.** $\lim\limits_{h \to 0} \dfrac{(h - 1)^3 + 1}{h}$

**57.** $\lim\limits_{h \to 0} \dfrac{\sqrt{h + 2} - \sqrt{2}}{h}$  **58.** $\lim\limits_{h \to 0} \dfrac{\sqrt{h + 4} - 2}{h}$

**59.** $\lim\limits_{h \to 0} \dfrac{(8 + h)^{1/3} - 2}{h}$  **60.** $\lim\limits_{h \to 0} \dfrac{1}{h} \left[ \dfrac{1}{h + 1} - 1 \right]$

*Compute the following limits.*

**61.** $\lim\limits_{x \to \infty} \dfrac{1}{x^2}$  **62.** $\lim\limits_{x \to -\infty} \dfrac{1}{x^2}$

**63.** $\lim\limits_{x \to \infty} \dfrac{1}{x - 8}$  **64.** $\lim\limits_{x \to \infty} \dfrac{5x + 3}{3x - 2}$

**65.** $\lim\limits_{x \to \infty} \dfrac{10x + 100}{x^2 - 30}$  **66.** $\lim\limits_{x \to \infty} \dfrac{x^2 + x}{x^2 - 1}$

*In Exercises 67–72, refer to Figure 6 to find the given limit.*

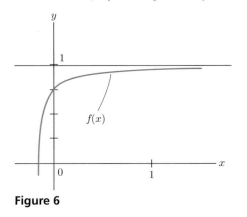

**Figure 6**

**67.** $\lim\limits_{x \to 0} f(x)$  **68.** $\lim\limits_{x \to \infty} f(x)$  **69.** $\lim\limits_{x \to 0} x f(x)$

**70.** $\lim\limits_{x \to \infty} (1 + 2f(x))$  **71.** $\lim\limits_{x \to \infty} (1 - f(x))$  **72.** $\lim\limits_{x \to 0} [f(x)]^2$

## Technology Exercises

*Examine the graph of the function and evaluate the function at large values of $x$ to guess the value of the limit.*

**73.** $\lim\limits_{x \to \infty} \sqrt{25 + x} - \sqrt{x}$  **74.** $\lim\limits_{x \to \infty} \dfrac{x^2}{2^x}$

**75.** $\lim\limits_{x \to \infty} \dfrac{x^2 - 2x + 3}{2x^2 + 1}$  **76.** $\lim\limits_{x \to \infty} \dfrac{-8x^2 + 1}{x^2 + 1}$

## SOLUTIONS TO PRACTICE PROBLEMS 1.4

1. The function under consideration is a rational function. Since the denominator has value 0 at $x = 6$, we cannot immediately determine the limit by just evaluating the function at $x = 6$. Also,

$$\lim_{x \to 6} (x - 6) = 0.$$

Since the function in the denominator has the limit zero, we cannot apply Limit Theorem VI. However, since the definition of limit considers only values of $x$ different from 6, the quotient can be simplified by factoring and canceling.

$$\frac{x^2 - 4x - 12}{x - 6} = \frac{(x + 2)(x - 6)}{(x - 6)} = x + 2 \qquad \text{for } x \neq 6.$$

Now $\lim_{x \to 6} (x + 2) = 8$. Therefore,

$$\lim_{x \to 6} \frac{x^2 - 4x - 12}{x - 6} = 8.$$

2. No limit exists. It is easily seen that $\lim_{x \to 6} (4x + 12) = 36$ and $\lim_{x \to 6} (x - 6) = 0$. As $x$ approaches 6, the denominator gets very small and the numerator approaches 36. For example, if $x = 6.00001$, then the numerator is 36.00004 and the denominator is .00001. The quotient is 3,600,004. As $x$ approaches 6 even more closely, the quotient gets arbitrarily large and cannot possibly approach a limit.

## 1.5  Differentiability and Continuity

In the preceding section we defined differentiability of $f(x)$ at $x = a$ in terms of a limit. If this limit does not exist, then we say that $f(x)$ is *nondifferentiable* at $x = a$. Geometrically, the nondifferentiability of $f(x)$ at $x = a$ can manifest itself in several different ways. First of all, the graph of $f(x)$ could have no tangent line at $x = a$. Second, the graph could have a vertical tangent line at $x = a$. (Recall that slope is not defined for vertical lines.) Some of the various geometric possibilities are illustrated in Fig. 1.

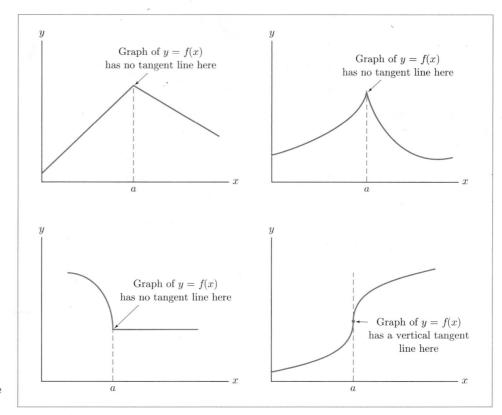

**Figure 1.**  Functions that are nondifferentiable at $x = a$.

The following example illustrates how nondifferentiable functions can arise in practice.

---

**EXAMPLE 1**    A railroad company charges $10 per mile to haul a boxcar up to 200 miles and $8 per mile for each mile exceeding 200. In addition, the railroad charges a $1000 handling charge per boxcar. Graph the cost of sending a boxcar $x$ miles.

**Solution**    If $x$ is at most 200 miles, then the cost $C(x)$ is given by $C(x) = 1000 + 10x$ dollars. The cost for 200 miles is $C(200) = 1000 + 2000 = 3000$ dollars. If $x$ exceeds 200 miles, then the total cost will be

$$C(x) = \underbrace{3000}_{\substack{\text{cost of first} \\ \text{200 miles}}} + \underbrace{8(x - 200)}_{\substack{\text{cost of miles in} \\ \text{excess of 200}}} = 1400 + 8x.$$

Thus

$$C(x) = \begin{cases} 1000 + 10x, & \text{for } 0 < x \le 200, \\ 1400 + 8x, & \text{for } x > 200. \end{cases}$$

The graph of $C(x)$ is sketched in Fig. 2. Note that $C(x)$ is not differentiable at $x = 200$. ∎

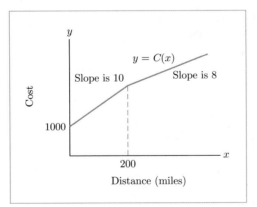

**Figure 2.**    Cost of hauling a boxcar.

Closely related to the concept of differentiability is that of continuity. We say that a function $f(x)$ is *continuous* at $x = a$ provided that, roughly speaking, its graph has no breaks (or gaps) as it passes through the point $(a, f(a))$. That is, $f(x)$ is continuous at $x = a$ provided that we can draw the graph through $(a, f(a))$ without lifting our pencil from the paper. The functions whose graphs are drawn in Figs. 1 and 2 are continuous for all values of $x$. By contrast, however, the function whose graph is drawn in Fig. 3(a) is not continuous (we say it is *discontinuous*) at $x = 1$ and $x = 2$, since the graph has breaks there. Similarly, the function whose graph is drawn in Fig. 3(b) is discontinuous at $x = 2$.

Discontinuous functions can occur in applications, as the following example shows.

---

**EXAMPLE 2**    Suppose that a manufacturing plant is capable of producing 15,000 units in one shift of 8 hours. For each shift worked, there is a fixed cost of $2000 (for light, heat, etc.). Suppose that the variable cost (the cost of labor and raw materials) is $2 per unit. Graph the cost $C(x)$ of manufacturing $x$ units.

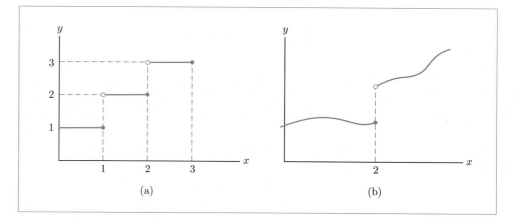

**Figure 3.** Functions with discontinuities.

**Solution** If $x \leq 15{,}000$, a single shift will suffice, so that

$$C(x) = 2000 + 2x, \qquad 0 \leq x \leq 15{,}000.$$

If $x$ is between 15,000 and 30,000, one extra shift will be required, and

$$C(x) = 4000 + 2x, \qquad 15{,}000 < x \leq 30{,}000.$$

If $x$ is between 30,000 and 45,000, the plant will need to work three shifts, and

$$C(x) = 6000 + 2x, \qquad 30{,}000 < x \leq 45{,}000.$$

The graph of $C(x)$ for $0 \leq x \leq 45{,}000$ is drawn in Fig. 4. Note that the graph has breaks at two points. ■

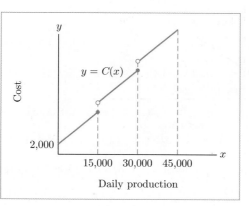

**Figure 4.** Cost function of a manufacturing plant.

The relationship between differentiability and continuity is this:

**Theorem I**    If $f(x)$ is differentiable at $x = a$, then $f(x)$ is continuous at $x = a$.

Note, however, that the converse statement is definitely false: A function may be continuous at $x = a$ but still not be differentiable there. The functions whose graphs are drawn in Fig. 1 provide examples of this phenomenon. We prove Theorem I at the end of the section.

Just as with differentiability, the notion of continuity can be phrased in terms of limits. In order for $f(x)$ to be continuous at $x = a$, the values of $f(x)$ for all $x$ near $a$ must be close to $f(a)$ (otherwise, the graph would have a break at $x = a$).

In fact, the closer $x$ is to $a$, the closer $f(x)$ must be to $f(a)$ (again, in order to avoid a break in the graph). In terms of limits, we must therefore have

$$\lim_{x \to a} f(x) = f(a).$$

Conversely, an intuitive argument shows that if the preceding limit relation holds, then the graph of $y = f(x)$ has no break at $x = a$.

---

**LIMIT DEFINITION OF CONTINUITY**  A function $f(x)$ is continuous at $x = a$ provided the following limit relation holds:

$$\lim_{x \to a} f(x) = f(a). \tag{1}$$

---

In order for (1) to hold, three conditions must be fulfilled.

**1.** $f(x)$ must be defined at $x = a$.

**2.** $\lim_{x \to a} f(x)$ must exist.

**3.** The limit $\lim_{x \to a} f(x)$ must have the value $f(a)$.

A function will fail to be continuous at $x = a$ when any one of these conditions fails to hold. The various possibilities are illustrated in the next example.

---

**EXAMPLE 3**  Determine whether the functions whose graphs are drawn in Fig. 5 are continuous at $x = 3$. Use the limit definition.

**Solution**  (a) Here $\lim_{x \to 3} f(x) = 2$. However, $f(3) = 4$. So

$$\lim_{x \to 3} f(x) \neq f(3)$$

and $f(x)$ is not continuous at $x = 3$. (Geometrically, this is clear. The graph has a break at $x = 3$.)

(b) $\lim_{x \to 3} g(x)$ does not exist, so $g(x)$ is not continuous at $x = 3$.

(c) $\lim_{x \to 3} h(x)$ does not exist, so $h(x)$ is not continuous at $x = 3$.

(d) $f(x)$ is not defined at $x = 3$, so $f(x)$ is not continuous at $x = 3$.  ■

Using our result on the limit of a polynomial function (Section 1.4), we see that

$$p(x) = a_0 + a_1 x + \cdots + a_n x^n, \qquad a_0, \ldots, a_n \text{ constants,}$$

is continuous for all $x$. Similarly, a rational function

$$\frac{p(x)}{q(x)}, \qquad p(x), q(x) \text{ polynomials,}$$

is continuous at all $x$ for which $q(x) \neq 0$.

**Proof of Theorem I**  Suppose that $f$ is differentiable at $x = a$ and let us show that $f$ is continuous at $x = a$. It is enough to show that

$$\lim_{x \to a} (f(x) - f(a)) = 0. \tag{2}$$

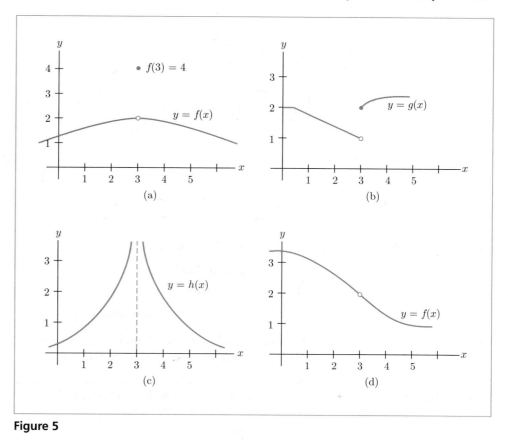

**Figure 5**

For $x \neq a$, write $x = a + h$, where $h \neq 0$, and note that $x$ approaches $a$ if and only if $h$ approaches 0. Then

$$\lim_{x \to a} (f(x) - f(a)) = \lim_{h \to 0} \left[ \overbrace{\frac{f(a+h)}{h} - f(a)}^{f(x)} \cdot h \right]$$

$$= \lim_{h \to 0} \left[ \frac{f(a+h) - f(a)}{h} \right] \cdot \lim_{h \to 0} h,$$

by Limit Theorem V of the previous section. To evaluate the product of these limits, note that $\lim_{h \to 0} h = 0$ and, since $f$ is differentiable at $a$,

$$\lim_{h \to 0} \frac{f(a+h) - f(a)}{h} = f'(a).$$

So the product of the limits is $f'(a) \cdot 0 = 0$ and hence (2) holds.

## PRACTICE PROBLEMS 1.5

Let $f(x) = \begin{cases} \dfrac{x^2 - x - 6}{x - 3} & \text{for } x \neq 3 \\ 4 & \text{for } x = 3. \end{cases}$

**1.** Is $f(x)$ continuous at $x = 3$?

**2.** Is $f(x)$ differentiable at $x = 3$?

# EXERCISES 1.5

*Is the function whose graph is drawn in Fig. 6 continuous at the following values of x?*

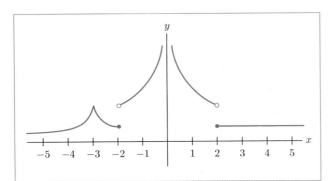

**Figure 6**

1. $x = 0$    2. $x = -3$    3. $x = 3$
4. $x = .001$    5. $x = -2$    6. $x = 2$

*Is the function whose graph is drawn in Fig. 6 differentiable at the following values of x?*

7. $x = 0$    8. $x = -3$    9. $x = 3$
10. $x = .001$    11. $x = -2$    12. $x = 2$

*Determine whether each of the following functions is continuous and/or differentiable at $x = 1$.*

13. $f(x) = x^2$

14. $f(x) = \dfrac{1}{x}$

15. $f(x) = \begin{cases} x + 2 & \text{for } -1 \le x \le 1 \\ 3x & \text{for } 1 < x \le 5 \end{cases}$

16. $f(x) = \begin{cases} x^3 & \text{for } 0 \le x < 1 \\ x & \text{for } 1 \le x \le 2 \end{cases}$

17. $f(x) = \begin{cases} 2x - 1 & \text{for } 0 \le x \le 1 \\ 1 & \text{for } 1 < x \end{cases}$

18. $f(x) = \begin{cases} x & \text{for } x \ne 1 \\ 2 & \text{for } x = 1 \end{cases}$

19. $f(x) = \begin{cases} \dfrac{1}{x - 1} & \text{for } x \ne 1 \\ 0 & \text{for } x = 1 \end{cases}$

20. $f(x) = \begin{cases} x - 1 & \text{for } 0 \le x < 1 \\ 1 & \text{for } x = 1 \\ 2x - 2 & \text{for } x > 1 \end{cases}$

*The functions in Exercises 21–26 are defined for all x except for one value of x. If possible, define $f(x)$ at the exceptional point in a way that makes $f(x)$ continuous for all x.*

21. $f(x) = \dfrac{x^2 - 7x + 10}{x - 5}, \ x \ne 5$

22. $f(x) = \dfrac{x^2 + x - 12}{x + 4}, \ x \ne -4$

23. $f(x) = \dfrac{x^3 - 5x^2 + 4}{x^2}, \ x \ne 0$

24. $f(x) = \dfrac{x^2 + 25}{x - 5}, \ x \ne 5$

25. $f(x) = \dfrac{(6 + x)^2 - 36}{x}, \ x \ne 0$

26. $f(x) = \dfrac{\sqrt{9 + x} - \sqrt{9}}{x}, \ x \ne 0$

27. The tax that you pay to the federal government is a percentage of your *taxable income*, which is what remains of your gross income after you subtract your allowed deductions. In the year 2001, there were five rates or brackets for a single taxpayer, as shown in Table 1.

| TABLE 1 | Single taxpayer rates for year 2001 | |
| --- | --- | --- |
| **Amount over** | **But Not Over** | **Tax Rate** |
| $0 | $27,050 | 15% |
| $27,050 | $65,550 | 27.5% |
| $65,550 | $136,750 | 30.5% |
| $136,750 | $297,350 | 35.5% |
| $297,350 | . . . | 39.1% |

So if you are single and your taxable income was less than $27,050 in 2001, then your tax is your taxable income times 15% (.15). The maximum amount of tax that you will pay on your income in this first bracket is 15% of $27,050 or (.15) × 27,050 = 4057.50 dollars. If your taxable income is more than $27,050 but less than $65,550, then your tax is $4057.50 plus 27.5% of the amount in excess of $27,050. So, for example, if your taxable income is $50,000, then your tax is $4057.5 + .275(50,000 − 27,050) = 4057.5 + .275 × 22,950 = 10,368.75 dollars. Let x denote your taxable income in 2001 and $T(x)$ your tax for that year.

(a) Find a formula for $T(x)$ if x is not over $136,750.

(b) Plot the graph of $T(x)$ for $0 \le x \le 136,750$.

(c) Find the maximum amount of tax that you will pay on the portion of your income in the second tax bracket. Express this amount as a difference between two values of $T(x)$.

28. Refer to Exercise 27.

(a) Find a formula for $T(x)$ for all taxable income x.

(b) Plot $T(x)$.

(c) Determine the maximum amount of tax that you will pay on the portion of your income in the fourth tax bracket.

**29.** The owner of a photocopy store charges 7 cents per copy for the first 100 copies and 4 cents per copy for each copy exceeding 100. In addition, there is a setup fee of \$2.50 for each photocopying job.

    **(a)** Determine $R(x)$, the revenue from selling $x$ copies.

    **(b)** If it costs the store owner 3 cents per copy, what is the profit from selling $x$ copies? (Recall that profit is revenue minus cost.)

**30.** Do Exercise 29 if it costs 10 cents per copy for the first 50 copies and 5 cents per copy for each copy exceeding 50, and there is no setup fee.

**31.** The graph in Fig. 7 shows the total sales in thousands of dollars in a department store during a typical 24-hour period.

    **(a)** Estimate the rate of sales during the period between 8 A.M. and 10 A.M.

    **(b)** Which 2-hour interval in the day sees the highest rate of sales and what is this rate?

**32.** Refer to Exercise 31.

    **(a)** Which 2-hour time intervals have the same rate of sales and what is this rate?

    **(b)** What is the total amount of sales between midnight and 8 A.M.? Compare this amount to the total sales in the period between 8 A.M. and 10 A.M.

*In Exercises 33–34, determine the value of $a$ that makes the function $f(x)$ continuous at $x = 0$.*

**33.** $f(x) = \begin{cases} 1 & \text{for } x \geq 0 \\ x + a & \text{for } x < 0 \end{cases}$

**34.** $f(x) = \begin{cases} 2(x - a) & \text{for } x \geq 0 \\ x^2 + 1 & \text{for } x < 0 \end{cases}$

**Figure 7**

---

## SOLUTIONS TO PRACTICE PROBLEMS 1.5

**1.** The function $f(x)$ is defined at $x = 3$, namely, $f(3) = 4$. When computing $\lim\limits_{x \to 3} f(x)$, we exclude consideration of $x = 3$; therefore, we can simplify the expression for $f(x)$ as follows:

$$f(x) = \frac{x^2 - x - 6}{x - 3} = \frac{(x - 3)(x + 2)}{x - 3} = x + 2$$

$$\lim_{x \to 3} f(x) = \lim_{x \to 3} (x + 2) = 5.$$

Since $\lim\limits_{x \to 3} f(x) = 5 \neq 4 = f(3)$, $f(x)$ is not continuous at $x = 3$.

**2.** There is no need to compute any limits in order to answer this question. By Theorem 1, since $f(x)$ is not continuous at $x = 3$, it cannot possibly be differentiable there.

---

# 1.6 Some Rules for Differentiation

Three additional rules of differentiation greatly extend the number of functions that we can differentiate.

**1.** Constant-multiple rule

$$\frac{d}{dx}[k \cdot f(x)] = k \cdot \frac{d}{dx}[f(x)], \quad k \text{ a constant.}$$

**2.** Sum rule

$$\frac{d}{dx}[f(x) + g(x)] = \frac{d}{dx}[f(x)] + \frac{d}{dx}[g(x)].$$

**3.  General power rule**

$$\frac{d}{dx}\left([g(x)]^r\right) = r \cdot [g(x)]^{r-1} \cdot \frac{d}{dx}[g(x)].$$

We shall discuss these rules, present some applications, and then prove the first two rules at the end of the section.

**The Constant-Multiple Rule**   Starting with a function $f(x)$, we can multiply it by a constant number $k$ in order to obtain a new function $k \cdot f(x)$. For instance, if $f(x) = x^2 - 4x + 1$ and $k = 2$, then

$$2f(x) = 2(x^2 - 4x + 1) = 2x^2 - 8x + 2.$$

The constant-multiple rule says that the derivative of the new function $k \cdot f(x)$ is just $k$ times the derivative of the original function.*  In other words, when faced with the differentiation of a constant times a function, simply carry along the constant and differentiate the function.

**EXAMPLE 1**   Calculate.

(a)  $\dfrac{d}{dx}\left(2x^5\right)$      (b)  $\dfrac{d}{dx}\left(\dfrac{x^3}{4}\right)$      (c)  $\dfrac{d}{dx}\left(-\dfrac{3}{x}\right)$      (d)  $\dfrac{d}{dx}\left(5\sqrt{x}\right)$

**Solution**   (a)  With $k = 2$ and $f(x) = x^5$, we have

$$\frac{d}{dx}\left(2 \cdot x^5\right) = 2 \cdot \frac{d}{dx}\left(x^5\right) = 2\left(5x^4\right) = 10x^4.$$

(b)  Write $\dfrac{x^3}{4}$ in the form $\dfrac{1}{4} \cdot x^3$. Then

$$\frac{d}{dx}\left(\frac{x^3}{4}\right) = \frac{1}{4} \cdot \frac{d}{dx}\left(x^3\right) = \frac{1}{4}\left(3x^2\right) = \frac{3}{4}x^2.$$

(c)  Write $-\dfrac{3}{x}$ in the form $(-3) \cdot \dfrac{1}{x}$. Then

$$\frac{d}{dx}\left(-\frac{3}{x}\right) = (-3) \cdot \frac{d}{dx}\left(\frac{1}{x}\right) = (-3) \cdot \frac{-1}{x^2} = \frac{3}{x^2}.$$

(d)  $\dfrac{d}{dx}\left(5\sqrt{x}\right) = 5\dfrac{d}{dx}\left(\sqrt{x}\right) = 5\dfrac{d}{dx}\left(x^{1/2}\right) = \dfrac{5}{2}x^{-1/2}.$

This answer may also be written in the form $\dfrac{5}{2\sqrt{x}}$.  ∎

**The Sum Rule**   To differentiate a sum of functions, differentiate each function individually and add the derivatives together.†  Another way of saying this is "the derivative of a sum of functions is the sum of the derivatives."

---

*More precisely, the constant-multiple rule asserts that if $f(x)$ is differentiable at $x = a$, then so is the function $k \cdot f(x)$, and the derivative of $k \cdot f(x)$ at $x = a$ may be computed using the given formula.

†More precisely, the sum rule asserts that if both $f(x)$ and $g(x)$ are differentiable at $x = a$, then so is $f(x) + g(x)$, and the derivative (at $x = a$) of the sum is then the sum of the derivatives (at $x = a$).

**EXAMPLE 2**  Find each of the following.

(a) $\dfrac{d}{dx}\left(x^3 + 5x\right)$     (b) $\dfrac{d}{dx}\left(x^4 - \dfrac{3}{x^2}\right)$     (c) $\dfrac{d}{dx}\left(2x^7 - x^5 + 8\right)$

**Solution**  (a) Let $f(x) = x^3$ and $g(x) = 5x$. Then

$$\frac{d}{dx}(x^3 + 5x) = \frac{d}{dx}(x^3) + \frac{d}{dx}(5x) = 3x^2 + 5.$$

(b) The sum rule applies to differences as well as sums (see Exercise 46). Indeed, by the sum rule,

$$\frac{d}{dx}\left(x^4 - \frac{3}{x^2}\right) = \frac{d}{dx}(x^4) + \frac{d}{dx}\left(-\frac{3}{x^2}\right) \qquad \text{(sum rule)}$$

$$= \frac{d}{dx}(x^4) - 3\frac{d}{dx}(x^{-2}) \qquad \text{(constant-multiple rule)}$$

$$= 4x^3 - 3(-2x^{-3})$$

$$= 4x^3 + 6x^{-3}.$$

After some practice, one usually omits most or all of the intermediate steps and simply writes

$$\frac{d}{dx}\left(x^4 - \frac{3}{x^2}\right) = 4x^3 + 6x^{-3}.$$

(c) We apply the sum rule repeatedly and use the fact that the derivative of a constant function is 0:

$$\frac{d}{dx}(2x^7 - x^5 + 8) = \frac{d}{dx}(2x^7) - \frac{d}{dx}(x^5) + \frac{d}{dx}(8)$$

$$= 2(7x^6) - 5x^4 + 0$$

$$= 14x^6 - 5x^4. \qquad \blacksquare$$

▶ *Remark*  The differentiation of a function *plus* a constant is different from the differentiation of a constant *times* a function. Figure 1 shows the graphs of $f(x)$, $f(x) + 2$, and $2 \cdot f(x)$, where $f(x) = x^3 - \frac{3}{2}x^2$. For each $x$, the graphs of $f(x)$ and $f(x) + 2$ have the same slope. In contrast, for each $x$, the slope of the graph of $2 \cdot f(x)$ is twice the slope of the graph of $f(x)$. Upon differentiation, an added constant disappears, whereas a constant that multiplies a function is carried along.  ◀

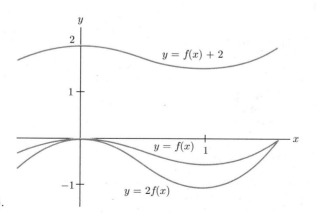

**Figure 1.**  Two effects of a constant on the graph of $f(x)$.

**The General Power Rule** Frequently, we will encounter expressions of the form $[g(x)]^r$—for instance, $(x^3 + 5)^2$, where $g(x) = x^3 + 5$ and $r = 2$. The general power rule says that, to differentiate $[g(x)]^r$, we must first treat $g(x)$ as if it were simply an $x$, form $r[g(x)]^{r-1}$, and then multiply it by a "correction factor" $g'(x)$.* Thus

$$\frac{d}{dx}(x^3 + 5)^2 = 2(x^3 + 5)^1 \cdot \frac{d}{dx}(x^3 + 5)$$

$$= 2(x^3 + 5) \cdot (3x^2)$$

$$= 6x^2(x^3 + 5).$$

In this special case it is easy to verify that the general power rule gives the correct answer. We first expand $(x^3 + 5)^2$ and then differentiate.

$$(x^3 + 5)^2 = (x^3 + 5)(x^3 + 5) = x^6 + 10x^3 + 25$$

From the constant-multiple rule and the sum rule, we have

$$\frac{d}{dx}(x^3 + 5)^2 = \frac{d}{dx}(x^6 + 10x^3 + 25)$$

$$= 6x^5 + 30x^2 + 0$$

$$= 6x^2(x^3 + 5).$$

The two methods give the same answer.

Note that if we set $g(x) = x$ in the general power rule, we recover the power rule. So the general power rule contains the power rule as a special case.

---

**EXAMPLE 3** Differentiate $\sqrt{1 - x^2}$.

**Solution**
$$\frac{d}{dx}(\sqrt{1 - x^2}) = \frac{d}{dx}[(1 - x^2)^{1/2}] = \frac{1}{2}(1 - x^2)^{-1/2} \cdot \frac{d}{dx}(1 - x^2)$$

$$= \frac{1}{2}(1 - x^2)^{-1/2} \cdot (-2x)$$

$$= \frac{-x}{(1 - x^2)^{1/2}} = \frac{-x}{\sqrt{1 - x^2}} \qquad \blacksquare$$

---

**EXAMPLE 4** Differentiate
$$y = \frac{1}{x^3 + 4x}.$$

**Solution**
$$y = \frac{1}{x^3 + 4x} = (x^3 + 4x)^{-1}$$

$$\frac{dy}{dx} = (-1)(x^3 + 4x)^{-2} \cdot \frac{d}{dx}(x^3 + 4x)$$

$$= \frac{-1}{(x^3 + 4x)^2}(3x^2 + 4)$$

$$= -\frac{3x^2 + 4}{(x^3 + 4x)^2} \qquad \blacksquare$$

---

*More precisely, the general power rule asserts that if $g(x)$ is differentiable at $x = a$ and if $[g(x)]^{r-1}$ is defined at $x = a$, then $[g(x)]^r$ is also differentiable at $x = a$ and its derivative is given by the formula stated.

Some derivatives require more than one differentiation rule.

---

**EXAMPLE 5**    Differentiate $5\sqrt[3]{1 + x^3}$.

**Solution**    $\dfrac{d}{dx}(5\sqrt[3]{1 + x^3}) = 5\dfrac{d}{dx}\left[\sqrt[3]{1 + x^3}\right]$ $\qquad\qquad$ (constant-multiple rule)

$\qquad\qquad\qquad = 5\dfrac{d}{dx}\left[(1 + x^3)^{1/3}\right]$

$\qquad\qquad\qquad = 5\left(\dfrac{1}{3}\right)(1 + x^3)^{\frac{1}{3} - 1}\dfrac{d}{dx}\left[(1 + x^3)\right]$ $\qquad$ (general power rule)

$\qquad\qquad\qquad = \dfrac{5}{3}(1 + x^3)^{-\frac{2}{3}}(3x^2) = 5x^2(1 + x^3)^{-\frac{2}{3}}$ $\qquad\qquad$ ■

Example 5 illustrates how the differentiation rules can be combined together. For example, as a byproduct of the constant multiple rule and the general power rule, we have

$$\frac{d}{dx}(k \cdot [g(x)]^r) = kr \cdot [g(x)]^{r-1}g'(x).$$

## The Derivative as a Rate of Change

Suppose that $y = f(x)$ is a function and $P = (a, f(a))$ a point on its graph (Fig. 2). Recall two important properties of the tangent line to the graph at $P$.

- The tangent line at $P$ has slope $f'(a)$.
- The tangent line at $P$ is the line that best approximates the graph near $P$.

The second property allows us to think of the portion of the graph near $P$ as being a straight line with slope $f'(a)$. Now we know from Section 1.1 that the rate of change of a linear function is equal to its slope. So let us think of the graph of $f$ near $P$ as being a line with slope $f'(a)$, then it follows that the rate of change of the graph of $f$, as it goes through $P$, is equal to the slope $f'(a)$. This gives us the following important interpretation of the derivative:

$$f'(a) = \text{rate of change of } f(x) \text{ at } x = a.$$

Thus the derivative can be used to measure the rate of change of quantities. We will see plenty of applications of this observation in this book.

---

**EXAMPLE 6**    **Declining sales**    At the end of the holiday season in January, the sales at a department store are expected to fall (Fig. 3). It is estimated that for the $x$ day of January, the sales will be

$$S(x) = 3 + \frac{9}{(x + 1)^2} \text{ thousand dollars.}$$

Compute $S(2)$ and $S'(2)$ and interpret your results.

**Solution**    We have

$$S(2) = 3 + \frac{9}{(2 + 1)^2} = 4 \text{ thousand dollars.}$$

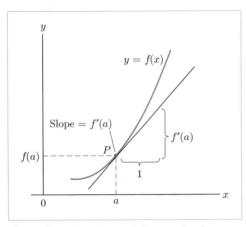

**Figure 2.** The slope of the graph of $y = f(x)$ at $P$ is the slope of the tangent line at $P$. This slope measures the rate of change of $f$ as its graph goes through $P$.

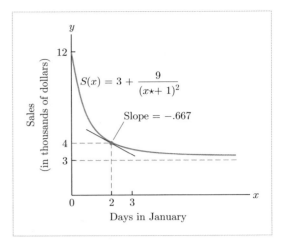

**Figure 3.** The sales decrease starting the first day of January.

To compute $S'(2)$, write $\frac{9}{(x+1)^2}$ as $9 \cdot (x + 1)^{-2}$; then, using all three rules of differentiation, we find

$$S'(x) = \frac{d}{dx}[S(x)] = \overbrace{\frac{d}{dx}[3]}^{=0} + \frac{d}{dx}(9 \cdot (x + 1)^{-2})$$

$$= 9 \cdot \frac{d}{dx}((x + 1)^{-2})$$

$$= 9(-2)(x + 1)^{-3}\frac{d}{dx}(x + 1) = -18(x + 1)^{-3}(1) = \frac{-18}{(x + 1)^3};$$

$$S'(2) = \frac{-18}{(2 + 1)^3} = -\frac{2}{3} \approx -.667.$$

The equations $S(2) = 4$ and $S'(2) = -.667$ tell us that on January 2nd the sales are \$4000 and are falling at the rate of .667 thousand dollars per day or \$667 per day. ∎

---

**EXAMPLE 7** **Predicting sales** Refer to Example 6. Use the values $S(2) = 4$ and $S'(2) = -.667$ to estimate the sales on January 3. Compare your result with the exact value of sales on January 3 by computing $S(3)$.

**Solution** On January 2 the sales are \$4000 and are falling at the rate of 667 dollars per day. Thus we expect the sales on January 3 to be \$667 less than the sales on January 2. Thus our estimate of the sales on January 3 is $4000 - 667 = \$3333$. To compare with the exact value of sales on January 3, we compute $S(3)$ from the formula in Example 6:

$$S(3) = 3 + \frac{9}{(3 + 1)^2} = 3 + \frac{9}{16} = \frac{57}{16} = 3.5625 \text{ thousand dollars}$$

or \$3562.5. This is close to our estimate of \$3333. ∎

*Proofs of the Constant-Multiple and Sum Rules*    Let us verify both rules when $x$ has the value $a$. Recall that if $f(x)$ is differentiable at $x = a$, then its derivative is the limit

$$\lim_{h \to 0} \frac{f(a+h) - f(a)}{h}.$$

**Constant-Multiple Rule**    We assume that $f(x)$ is differentiable at $x = a$. We must prove that $k \cdot f(x)$ is differentiable at $x = a$ and that its derivative is $k \cdot f'(x)$. This amounts to showing that the limit

$$\lim_{h \to 0} \frac{k \cdot f(a+h) - k \cdot f(a)}{h}$$

exists and has the value $k \cdot f'(a)$. However,

$$\lim_{h \to 0} \frac{k \cdot f(a+h) - k \cdot f(a)}{h}$$

$$= \lim_{h \to 0} k \left[ \frac{f(a+h) - f(a)}{h} \right]$$

$$= k \cdot \lim_{h \to 0} \frac{f(a+h) - f(a)}{h} \qquad \text{(by Limit Theorem I)}$$

$$= k \cdot \lim_{h \to 0} \frac{f(a+h) - f(a)}{h} = k f'(a)$$

$$\text{[since } f(x) \text{ is differentiable at } x = a\text{]},$$

which is what we desired to show.

**Sum Rule**    We assume that both $f(x)$ and $g(x)$ are differentiable at $x = a$. We must prove that $f(x) + g(x)$ is differentiable at $x = a$ and that its derivative is $f'(a) + g'(a)$. That is, we must show that the limit

$$\lim_{h \to 0} \frac{[f(a+h) + g(a+h)] - [f(a) + g(a)]}{h}$$

exists and equals $f'(a) + g'(a)$. Using Limit Theorem III and the fact that $f(x)$ and $g(x)$ are differentiable at $x = a$, we have

$$\lim_{h \to 0} \frac{[f(a+h) + g(a+h)] - [f(a) + g(a)]}{h}$$

$$= \lim_{h \to 0} \left[ \frac{f(a+h) - f(a)}{h} + \frac{g(a+h) - g(a)}{h} \right]$$

$$= \lim_{h \to 0} \frac{f(a+h) - f(a)}{h} + \lim_{h \to 0} \frac{g(a+h) - g(a)}{h}$$

$$= f'(a) + g'(a).$$

The general power rule will be proven as a special case of the chain rule in Chapter 3.

## PRACTICE PROBLEMS 1.6

**1.** Find the derivative $\dfrac{d}{dx}(x)$.

**2.** Differentiate the function $y = \dfrac{x + (x^5 + 1)^{10}}{3}$.

# EXERCISES 1.6

*Differentiate.*

**1.** $y = x^3 + x^2$

**2.** $y = 3x^4$

**3.** $y = 2\sqrt{x}$

**4.** $y = \dfrac{1}{3x^3}$

**5.** $y = \dfrac{x}{2} - \dfrac{2}{x}$

**6.** $f(x) = 12 + \dfrac{1}{7^3}$

**7.** $f(x) = x^4 + x^3 + x$

**8.** $y = 4x^3 - 2x^2 + x + 1$

**9.** $y = (2x + 4)^3$

**10.** $y = (x^2 - 1)^3$

**11.** $y = (x^3 + x^2 + 1)^7$

**12.** $y = (x^2 + x)^{-2}$

**13.** $y = \dfrac{4}{x^2}$

**14.** $y = 4(x^2 - 6)^{-3}$

**15.** $y = 3\sqrt[3]{2x^2 + 1}$

**16.** $y = 2\sqrt{x + 1}$

**17.** $y = 2x + (x + 2)^3$

**18.** $y = (x - 1)^3 + (x + 2)^4$

**19.** $y = \dfrac{1}{5x^5}$

**20.** $y = (x^2 + 1)^2 + 3(x^2 - 1)^2$

**21.** $y = \dfrac{1}{x^3 + 1}$

**22.** $y = \dfrac{2}{x + 1}$

**23.** $y = x + \dfrac{1}{x + 1}$

**24.** $y = 2\sqrt[4]{x^2 + 1}$

**25.** $f(x) = 5\sqrt{3x^3 + x}$

**26.** $y = \dfrac{1}{x^3 + x + 1}$

**27.** $y = 3x + \pi^3$

**28.** $y = \sqrt{1 + x^2}$

**29.** $y = \sqrt{1 + x + x^2}$

**30.** $y = \dfrac{1}{2x + 5}$

**31.** $y = \dfrac{2}{1 - 5x}$

**32.** $y = \dfrac{7}{\sqrt{1 + x}}$

**33.** $y = \dfrac{45}{1 + x + \sqrt{x}}$

**34.** $y = (1 + x + x^2)^{11}$

**35.** $y = x + 1 + \sqrt{x + 1}$

**36.** $y = \pi^2 x$

**37.** $f(x) = \left(\dfrac{\sqrt{x}}{2} + 1\right)^{3/2}$

**38.** $y = \left(x - \dfrac{1}{x}\right)^{-1}$

*In Exercises 39 and 40, find the slope of the graph of $y = f(x)$ at the designated point.*

**39.** $f(x) = 3x^2 - 2x + 1$, $(1, 2)$

**40.** $f(x) = x^{10} + 1 + \sqrt{1 - x}$, $(0, 2)$

**41.** Find the slope of the tangent line to the curve $y = x^3 + 3x - 8$ at $(2, 6)$.

**42.** Write the equation of the tangent line to the curve $y = x^3 + 3x - 8$ at $(2, 6)$.

**43.** Find the slope of the tangent line to the curve $y = (x^2 - 15)^6$ at $x = 4$. Then write the equation of this tangent line.

**44.** Find the equation of the tangent line to the curve $y = \dfrac{8}{x^2 + x + 2}$ at $x = 2$.

**45.** Differentiate the function $f(x) = (3x^2 + x - 2)^2$ in two ways.

   **(a)** Use the general power rule.

**(b)** Multiply $3x^2 + x - 2$ by itself and then differentiate the resulting polynomial.

**46.** Using the sum rule and the constant-multiple rule, show that for any functions $f(x)$ and $g(x)$

$$\frac{d}{dx}[f(x) - g(x)] = \frac{d}{dx}f(x) - \frac{d}{dx}g(x).$$

**47.** Figure 4 contains the curves $y = f(x)$ and $y = g(x)$ and the tangent line to $y = f(x)$ at $x = 1$, with $g(x) = 3 \cdot f(x)$. Find $g(1)$ and $g'(1)$.

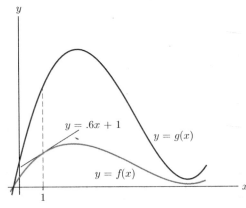

**Figure 4.** Graphs of $f(x)$ and $g(x) = 3f(x)$.

**48.** Figure 5 contains the curves $y = f(x)$, $y = g(x)$, and $y = h(x)$ and the tangent lines to $y = f(x)$ and $y = g(x)$ at $x = 1$, with $h(x) = f(x) + g(x)$. Find $h(1)$ and $h'(1)$.

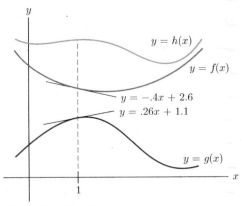

**Figure 5.** Graphs of $f(x)$, $g(x)$, and $h(x) = f(x) + g(x)$.

**49.** Suppose $f(5) = 2$, $f'(5) = 3$, $g(5) = 4$, and $g'(5) = 1$. Find $h(5)$ and $h'(5)$, where $h(x) = 3f(x) + 2g(x)$.

**50.** Suppose $g(3) = 2$ and $g'(3) = 4$. Find $f(3)$ and $f'(3)$, where $f(x) = 2 \cdot [g(x)]^3$.

**51.** Suppose $g(1) = 4$ and $g'(1) = 3$. Find $f(1)$ and $f'(1)$, where $f(x) = 5 \cdot \sqrt{g(x)}$.

**52.** Suppose that $h(x) = [f(x)]^2 + \sqrt{g(x)}$. Determine $h(1)$ and $h'(1)$ given that $f(1) = 1$, $g(1) = 4$, $f'(1) = -1$, and $g'(1) = 4$.

**53.** The tangent line to the curve $y = \frac{1}{3}x^3 - 4x^2 + 18x + 22$ is parallel to the line $6x - 2y = 1$ at two points on the curve. Find the two points.

**54.** The tangent line to the curve $y = x^3 - 6x^2 - 34x - 9$ has slope 2 at two points on the curve. Find the two points.

**55.** In Fig. 6 the straight line is tangent to the graph of $f(x)$. Find $f(4)$ and $f'(4)$.

**Figure 6**

**56.** In Fig. 7 the straight line is tangent to the parabola. Find the value of $b$.

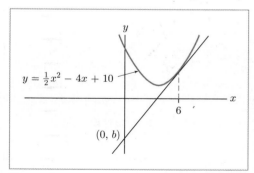

**Figure 7**

**57.** Let $S(x)$ represent the total sales (in thousands of dollars) for the month $x$ in the year 2005 at a certain department store. Represent each statement below by an equation involving $S$ or $S'$.

**(a)** The sales at the end of January reached $120,560 and were rising at the rate of $1500 per month.

**(b)** At the end of March, the sales for this month dropped to $80,000 and were falling by about $200 a day. (Use 1 month = 30 days.)

**58.** (*Comparing Rates of Change*)

**(a)** In Example 6, find the total sales for January 10 and determine the rate at which sales are falling on that day.

**(b)** Compare the rate of change of sales on January 2 (Example 6) to the rate on January 10. What can you infer about the rate of change of sales?

**59.** (*Predicting Sales*) Refer to Example 6.

**(a)** Compute $S(10)$ and $S'(10)$.

**(b)** Use the data in (a) to estimate the total sales on January 11. Compare your estimate to the actual value given by $S(11)$.

**60.** (*Correcting a Prediction*) The financial analysts at the store in Example 6 corrected their projections and are now expecting the total sales for the $x$ day of January to be

$$T(x) = \frac{24}{5} + \frac{36}{5(3x + 1)^2} \text{ thousand dollars.}$$

**(a)** Let $S(x)$ be as in Example 6. Compute $T(1)$, $T'(1)$, $S(1)$, and $S'(1)$.

**(b)** Compare and interpret the data in part (a), as it pertains to the sales on January 1.

**61. (a)** Let $A(x)$ denote the number (in hundreds) of computers sold when $x$ thousand dollars are spent on advertising. Represent the following statement by equations involving $A$ or $A'$: When eight thousand dollars were spent on advertising the number of computers sold was 1200 and it was rising at the rate of 50 computers for each 1000 dollars spent on advertising.

**(b)** Estimate the number of computers that will be sold if $9000 are spent on advertising.

**62.** A toy company introduces a new video game on the market. Let $S(x)$ denote the number of videos sold on the day $x$ since the item was introduced. Let $n$ be a positive integer. Interpret $S(n)$, $S'(n)$, and $S(n) + S'(n)$.

**63.** (*U.S. Federal Debt*) Based on data from the U.S. Treasury Department, the federal debt (in trillions of dollars) for the years 1995 to 2004 was given approximately by the formula

$$D(x) = 4.95 + .402x - .1067x^2 + .0124x^3 - .00024x^4,$$

where $x$ is the number of years elapsed since the end of 1995. Estimate the federal debt at the end of 1999 ($x = 4$) and the rate at which it was increasing at that time.

**64.** Refer to the previous exercise.

**(a)** Is it true that by the end of 2003 the federal debt was twice as large as it was at the end of 2001?

**(b)** Is it true that by the end of 2003 the federal debt was increasing at a rate that is more than twice the rate at the end of 2001?

## SOLUTIONS TO PRACTICE PROBLEMS 1.6

1. The problem asks for the derivative of the function $y = x$, a straight line of slope 1. Therefore,

$$\frac{d}{dx}(x) = 1.$$

The result can also be obtained from the power rule with $r = 1$. If $f(x) = x^1$, then

$$\frac{d}{dx}(f(x)) = 1 \cdot x^{1-1} = x^0 = 1.$$

See Fig. 8.

**Figure 8**

2. All three rules are required to differentiate this function.

$$\frac{dy}{dx} = \frac{d}{dx}\frac{1}{3} \cdot [x + (x^5 + 1)^{10}]$$

$$= \frac{1}{3}\frac{d}{dx}[x + (x^5 + 1)^{10}] \qquad \text{(constant-multiple rule)}$$

$$= \frac{1}{3}\left[\frac{d}{dx}(x) + \frac{d}{dx}(x^5 + 1)^{10}\right] \qquad \text{(sum rule)}$$

$$= \frac{1}{3}[1 + 10(x^5 + 1)^9 \cdot (5x^4)] \qquad \text{(general power rule)}$$

$$= \frac{1}{3}[1 + 50x^4(x^5 + 1)^9]$$

## 1.7 More About Derivatives

In many applications it is convenient to use variables other than $x$ and $y$. One might, for instance, study the function $f(t) = t^2$ instead of writing $f(x) = x^2$. In this case, the notation for the derivative involves $t$ rather than $x$, but the concept of the derivative as a slope formula is unaffected. (See Fig. 1.) When the independent variable is $t$ instead of $x$, we write $\frac{d}{dt}$ in place of $\frac{d}{dx}$. For instance,

$$\frac{d}{dt}(t^3) = 3t^2, \qquad \frac{d}{dt}(2t^2 + 3t) = 4t + 3.$$

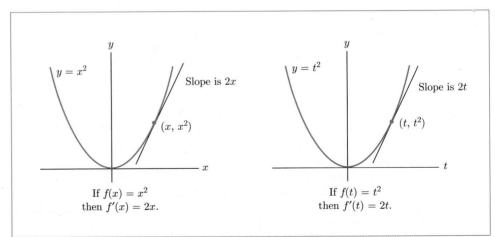

**Figure 1.** The same function but different variables.

Recall that if $y$ is a function of $x$, say $y = f(x)$, then we may write $\dfrac{dy}{dx}$ in place of $f'(x)$. We sometimes call $\dfrac{dy}{dx}$ "the derivative of $y$ with respect to $x$." Similarly, if $v$ is a function of $t$, then the derivative of $v$ with respect to $t$ is written as $\dfrac{dv}{dt}$. For example, if $v = 4t^2$, then $\dfrac{dv}{dt} = 8t$.

Of course, other letters can be used to denote variables. The formulas

$$\frac{d}{dP}(P^3) = 3P^2, \qquad \frac{d}{ds}(s^3) = 3s^2, \qquad \frac{d}{dz}(z^3) = 3z^2$$

all express the same basic fact that the slope formula for the cubic curve $y = x^3$ is given by $3x^2$.

---

**EXAMPLE 1**   Compute.

(a) $\dfrac{ds}{dp}$ if $s = 3(p^2 + 5p + 1)^{10}$    (b) $\dfrac{d}{dt}(at^2 + St^{-1} + S^2)$

**Solution** (a) $\dfrac{d}{dp}\,3(p^2 + 5p + 1)^{10} = 30(p^2 + 5p + 1)^9 \cdot \dfrac{d}{dp}(p^2 + 5p + 1)$

$$= 30(p^2 + 5p + 1)^9(2p + 5)$$

(b) Although the expression $at^2 + St^{-1} + S^2$ contains several letters, the notation $\dfrac{d}{dt}$ indicates that for purposes of calculating the derivative with respect to $t$, all letters except $t$ are to be considered as constants. Hence

$$\frac{d}{dt}(at^2 + St^{-1} + S^2) = \frac{d}{dt}(at^2) + \frac{d}{dt}(St^{-1}) + \frac{d}{dt}(S^2)$$

$$= a \cdot \frac{d}{dt}(t^2) + S \cdot \frac{d}{dt}(t^{-1}) + 0$$

$$= 2at - St^{-2}.$$

$$\left[\text{The derivative } \frac{d}{dt}(S^2) \text{ is zero because } S^2 \text{ is a constant.}\right] \qquad ■$$

**The Second Derivative**   When we differentiate a function $f(x)$, we obtain a new function $f'(x)$ that is a formula for the slope of the curve $y = f(x)$. If we differentiate the function $f'(x)$, we obtain what is called the *second derivative* of $f(x)$, denoted by $f''(x)$. That is,

$$\frac{d}{dx}f'(x) = f''(x).$$

---

**EXAMPLE 2**   Find the second derivatives of the following functions.

(a) $f(x) = x^3 + (1/x)$    (b) $f(x) = 2x + 1$    (c) $f(t) = t^{1/2} + t^{-1/2}$

**Solution** (a)  $f(x) = x^3 + (1/x) = x^3 + x^{-1}$

$f'(x) = 3x^2 - x^{-2}$

$f''(x) = 6x + 2x^{-3}$

(b) $f(x) = 2x + 1$

$f'(x) = 2$ (a constant function whose value is 2)

$f''(x) = 0$ (The derivative of a constant function is zero.)

(c) $f(t) = t^{1/2} + t^{-1/2}$

$f'(t) = \frac{1}{2}t^{-1/2} - \frac{1}{2}t^{-3/2}$

$f''(t) = -\frac{1}{4}t^{-3/2} + \frac{3}{4}t^{-5/2}$ ∎

The first derivative of a function $f(x)$ gives the slope of the graph of $f(x)$ at any point. The second derivative of $f(x)$ gives important additional information about the shape of the curve near any point. We shall examine this subject carefully in the next chapter.

**Other Notation for Derivatives** Unfortunately, the process of differentiation does not have a standardized notation. Consequently, it is important to become familiar with alternative terminology.

If $y$ is a function of $x$, say $y = f(x)$, then we may denote the first and second derivatives of this function in several ways.

| Prime Notation | $\dfrac{d}{dx}$ Notation |
|:---:|:---:|
| $f'(x)$ | $\dfrac{d}{dx}f(x)$ |
| $y'$ | $\dfrac{dy}{dx}$ |
| $f''(x)$ | $\dfrac{d^2}{dx^2}f(x)$ |
| $y''$ | $\dfrac{d^2y}{dx^2}$ |

The notation $\dfrac{d^2}{dx^2}$ is purely symbolic. It reminds us that the second derivative is obtained by differentiating $\dfrac{d}{dx}f(x)$; that is,

$$f'(x) = \frac{d}{dx}f(x)$$

$$f''(x) = \frac{d}{dx}\left[\frac{d}{dx}f(x)\right].$$

If we evaluate the derivative $f'(x)$ at a specific value of $x$, say $x = a$, we get a number $f'(a)$ that gives the slope of the curve $y = f(x)$ at the point $(a, f(a))$. Another way of writing $f'(a)$ is

$$\left.\frac{dy}{dx}\right|_{x=a}.$$

If we have a second derivative $f''(x)$, then its value when $x = a$ is written

$$f''(a) \quad \text{or} \quad \left.\frac{d^2y}{dx^2}\right|_{x=a}.$$

**EXAMPLE 3**  If $y = x^4 - 5x^3 + 7$, find

$$\left.\frac{d^2y}{dx^2}\right|_{x=3}.$$

Solution

$$\frac{dy}{dx} = \frac{d}{dx}(x^4 - 5x^3 + 7) = 4x^3 - 15x^2$$

$$\frac{d^2y}{dx^2} = \frac{d}{dx}(4x^3 - 15x^2) = 12x^2 - 30x$$

$$\left.\frac{d^2y}{dx^2}\right|_{x=3} = 12(3)^2 - 30(3) = 108 - 90 = 18 \qquad ■$$

**EXAMPLE 4**  If $s = t^3 - 2t^2 + 3t$, find

$$\left.\frac{ds}{dt}\right|_{t=-2} \quad \text{and} \quad \left.\frac{d^2s}{dt^2}\right|_{t=-2}.$$

Solution

$$\frac{ds}{dt} = \frac{d}{dt}(t^3 - 2t^2 + 3t) = 3t^2 - 4t + 3$$

$$\left.\frac{ds}{dt}\right|_{t=-2} = 3(-2)^2 - 4(-2) + 3 = 12 + 8 + 3 = 23$$

To find the value of the second derivative at $t = -2$, we must first differentiate $\dfrac{ds}{dt}$.

$$\frac{d^2s}{dt^2} = \frac{d}{dt}(3t^2 - 4t + 3) = 6t - 4$$

$$\left.\frac{d^2s}{dt^2}\right|_{t=-2} = 6(-2) - 4 = -12 - 4 = -16 \qquad ■$$

## The Marginal Concept in Economics

In the previous section, we talked about the derivative as a rate of change. Let us revisit this topic in a setting that arises in economics. For the sake of this discussion, suppose that $C(x)$ is a cost function (the cost of producing $x$ units of a commodity), measured in thousands of dollars.

A problem of interest to economists is to approximate the quantity $C(a + 1) - C(a)$, which is the additional cost that is incurred if the production level is increased by one unit from $x = a$ to $x = a + 1$. Note that $C(a + 1) - C(a)$ is also the cost of producing the $(a + 1)$ unit.

To be more specific, suppose you are given $C(2) = 4$ and $C'(2) = .5$. Can you approximate the additional cost $C(3) - C(2)$?

To tackle this problem, we turn to the equation $C'(2) = .5$, which says that the slope of the graph of $C(x)$ is .5 when $x = 2$ (Fig. 2). That is, as the graph of $y = C(x)$ passes through the point $P = (2, 4)$, it is rising at the rate of .5 units for each unit increase in $x$. If $C(x)$ were a linear function (equal to its tangent line at $P$), then the additional cost would be exactly .5 thousand dollars or 500 dollars (Fig. 2). In general, the tangent line approximates the graph and all we can say is that the additional cost is *approximately* .5 thousand dollars. In symbols,

$$\text{additional cost} = C(3) - C(2) \approx .5 \text{ thousand dollars.}$$

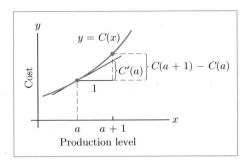

**Figure 2.** Approximating the additional cost by using the derivative.

**Figure 3.** Approximating the change in a cost function.

Recall that $C'(2) = .5$, and so

$$\text{additional cost} = C(3) - C(2) \approx C'(2).$$

The argument that we used at production level $x = 2$ works at any level $x = a$ and shows that

$$\text{additional cost} = C(a + 1) - C(a) \approx C'(a), \tag{1}$$

where the symbol $\approx$ is used to indicate that, in general, we have an approximation and not an equality (Fig. 3). Economists refer to the derivative $C'(a)$ as the *marginal cost at production level a* or the *marginal cost of producing a units* of a commodity. In summary, we have the following important result.

**Marginal Cost** If $C(x)$ is a cost function, then the *marginal cost function* is $C'(x)$. The marginal cost of producing $a$ units, $C'(a)$, is approximately equal to $C(a+1) - C(a)$, which is the additional cost that is incurred when the production level is increased by one unit from $a$ to $a + 1$.

Before we give an example, let us note that if $C(x)$ is measured in dollars, where $x$ is the number of items, then $C'(x)$, being a rate of change, is measured in dollars per item.

**EXAMPLE 5** **Marginal cost** Suppose the cost of producing $x$ items is $C(x) = .005x^3 - .5x^2 + 28x + 300$ dollars, and daily production is 50 items.

(a) What is the extra cost of increasing daily production from 50 to 51 items?

(b) What is the marginal cost when $x = 50$?

**Solution** (a) The change in cost when daily production is raised from 50 to 51 items is $C(51) - C(50)$, which equals

$$[.005(51)^3 - .5(51)^2 + 28(51) + 300] - [.005(50)^3 - .5(50)^2 + 28(50) + 300]$$
$$= 1090.755 - 1075$$
$$= 15.755.$$

(b) The marginal cost at production level 50 is $C'(50)$.

$$C'(x) = .015x^2 - x + 28$$
$$C'(50) = 15.5$$

Notice that 15.5 is close to the actual cost in (a) of increasing production by one item. ∎

Our discussion of cost and marginal cost applies as well to other economic quantities such as profit and revenue. In fact, in economics the derivatives are often described by the adjective "marginal." Here are two more definitions of marginal functions and their interpretations.

> **Marginal Revenue and Marginal Profit**   If $R(x)$ is the revenue generated from the production of $x$ units of a certain commodity and $P(x)$ is the corresponding profit, then the *marginal revenue function* is $R'(x)$ and the *marginal profit function* is $P'(x)$.

The marginal revenue of producing $a$ units, $R'(a)$, is an approximation of the additional revenue that results from increasing the production level by one unit, from $a$ to $a + 1$:

$$R(a + 1) - R(a) \approx R'(a) \tag{2}$$

Similarly, for the marginal profit, we have

$$P(a + 1) - P(a) \approx P'(a) \tag{3}$$

The following example illustrates how marginal functions help in the decision-making process in economics.

---

**EXAMPLE 6**   **Predicting profits**   Let $R(x)$ denote the revenue (in thousands of dollars) generated from the production of $x$ units of a certain commodity.

(a) Given that $R(4) = 7$ and $R'(4) = -.5$, estimate the additional revenue that results from increasing the production level by one unit from $x = 4$ to $x = 5$.

(b) Estimate the revenue generated from the production of 5 units.

(c) Is it profitable to raise the production to 5 units if the cost (in thousands of dollars) to produce $x$ units is given by $C(x) = x + \frac{4}{x+1}$?

**Solution**   (a) Using (2) with $a = 4$, we find

$$R(5) - R(4) \approx R'(4) = -.5 \text{ thousand dollars.}$$

Thus, if the production level is raised to 5 units, the revenue will drop by about $500 dollars.

(b) The revenue that results from the production of 4 units is $R(4) = 7$ or $7000. By part (a), this revenue will drop by about $500 if we raise the production from 4 to 5 units. Thus at production level $x = 5$, the revenue will be approximately $7000 - 500 = \$6500$.

(c) Let $P(x)$ denote the profit (in thousands of dollars) that results from the production of $x$ units. We have $P(x) = R(x) - C(x)$. At production level $x = 5$, we have

$$C(5) = 5 + \frac{4}{5+1} = \frac{17}{3} \approx 5.667 \text{ thousand dollars or } \$5667;$$

also, from (b), the revenue is $R(5) \approx \$6500$. So the profit at production level $x = 5$ is $P(5) \approx 6500 - 5667 = \$833$. So it is still profitable to raise the production level to $x = 5$, even though the cost will rise and the revenue will fall. ∎

**Figure 4.** Graph of $x^2$ and its first two derivatives.

**Incorporating Technology** Although functions can be specified (and differentiated) in graphing calculators with letters other than X, only functions in X can be graphed. Therefore, we will always use X as the variable.

Figure 4 contains the graphs of $f(x) = x^2$ and its first and second derivatives. These three graphs can be obtained with any of the function editor settings in Figs. 5(a), 5(b), and 5(c).

(a) TI-83         (b) TI-83         (c) TI-86

**Figure 5**

## PRACTICE PROBLEMS 1.7

**1.** Let $f(t) = t + (1/t)$. Find $f''(2)$.

**2.** Differentiate $g(r) = 2\pi rh$.

## EXERCISES 1.7

*Find the first derivatives.*

**1.** $f(t) = (t^2 + 1)^5$

**2.** $f(P) = P^3 + 3P^2 - 7P + 2$

**3.** $v(t) = 4t^2 + 11\sqrt{t} + 1$

**4.** $g(y) = y^2 - 2y + 4$

**5.** $y = T^5 - 4T^4 + 3T^2 - T - 1$

**6.** $x = 16t^2 + 45t + 10$

**7.** Find $\dfrac{d}{dP}(3P^2 - \frac{1}{2}P + 1)$.   **8.** Find $\dfrac{d}{ds}\sqrt{s^2 + 1}$.

**9.** Find $\dfrac{d}{dt}(a^2t^2 + b^2t + c^2)$.   **10.** Find $\dfrac{d}{dP}(T^2 + 3P)^3$.

*Find the first and second derivatives.*

**11.** $y = x + 1$      **12.** $y = (x + 12)^3$

**13.** $y = \sqrt{x}$

**14.** $y = 100$

**15.** $y = \sqrt{x + 1}$

**16.** $v = 2t^2 + 3t + 11$

**17.** $f(r) = \pi r^2$

**18.** $y = \pi^2 + 3x^2$

**19.** $f(P) = (3P + 1)^5$

**20.** $T = (1 + 2t)^2 + t^3$

*Compute the following.*

**21.** $\dfrac{d}{dx}(2x + 7)^2 \Big|_{x=1}$

**22.** $\dfrac{d}{dt}\left(t^2 + \dfrac{1}{t+1}\right)\Big|_{t=0}$

**23.** $\dfrac{d}{dz}(z^2 + 2z + 1)^7 \Big|_{z=-1}$

**24.** $\dfrac{d^2}{dx^2}(3x^4 + 4x^2)\Big|_{x=2}$

**25.** $\dfrac{d^2}{dx^2}(3x^3 - x^2 + 7x - 1)\Big|_{x=2}$

**26.** $\dfrac{d}{dx}\left(\dfrac{dy}{dx}\right)\Big|_{x=1}$, where $y = x^3 + 2x - 11$

**27.** $f'(1)$ and $f''(1)$, when $f(t) = \dfrac{1}{2+t}$

**28.** $g'(0)$ and $g''(0)$, when $g(T) = (T+2)^3$

**29.** $\dfrac{d}{dt}\left(\dfrac{dv}{dt}\right)\Big|_{t=32}$, where $v = 20t + 12$

**30.** $\dfrac{d}{dt}\left(\dfrac{dv}{dt}\right)$, where $v = 2t^2 + \dfrac{1}{t+1}$

**31.** Suppose a company finds that the revenue $R$ generated by spending $x$ dollars on advertising is given by $R = 1000 + 80x - .02x^2$, for $0 \le x \le 2000$. Find $\dfrac{dR}{dx}\Big|_{x=1500}$.

**32.** A supermarket finds that its average daily volume of business $V$ (in thousands of dollars) and the number of hours $t$ the store is open for business each day are approximately related by the formula

$$V = 20\left(1 - \frac{100}{100 + t^2}\right), \qquad 0 \le t \le 24.$$

Find $\dfrac{dV}{dt}\Big|_{t=10}$.

**33.** The *third derivative* of a function $f(x)$ is the derivative of the second derivative $f''(x)$ and is denoted by $f'''(x)$. Compute $f'''(x)$ for the following functions.
  **(a)** $f(x) = x^5 - x^4 + 3x$     **(b)** $f(x) = 4x^{5/2}$

**34.** Compute the third derivatives of the following functions.
  **(a)** $f(t) = t^{10}$     **(b)** $f(z) = \dfrac{1}{z+5}$

**35.** Suppose that $s = Tx^2 + 3xP + T^2$. Find
  **(a)** $\dfrac{ds}{dx}$     **(b)** $\dfrac{ds}{dP}$     **(c)** $\dfrac{ds}{dT}$

**36.** Suppose that $s = 7x^2 y\sqrt{z}$. Find
  **(a)** $\dfrac{d^2 s}{dx^2}$     **(b)** $\dfrac{d^2 s}{dy^2}$     **(c)** $\dfrac{ds}{dz}$

**37.** Let $C(x)$ be the cost (in dollars) of manufacturing $x$ bicycles per day in a certain factory. Interpret $C(50) = 5000$ and $C'(50) = 45$.

**38.** Estimate the cost of manufacturing 51 bicycles per day in the previous exercise.

**39.** Suppose the revenue from producing (and selling) $x$ units of a product is given by $R(x) = 3x - .01x^2$ dollars.
  **(a)** Find the marginal revenue at a production level of 20.
  **(b)** Find the production levels where the revenue is $200.

**40.** Let $P(x)$ be the profit from producing (and selling) $x$ units of goods. Match each question with the proper solution.

*Questions:*
  **A.** What is the profit from producing 1000 units of goods?
  **B.** At what level of production will the marginal profit be 1000 dollars?
  **C.** What is the marginal profit from producing 1000 units of goods?
  **D.** For what level of production will the profit be 1000 dollars?

*Solutions:*
  **a.** Compute $P'(1000)$.
  **b.** Find a value of $a$ for which $P'(a) = 1000$.
  **c.** Set $P(x) = 1000$ and solve for $x$.
  **d.** Compute $P(1000)$.

**41.** Let $R(x)$ denote the revenue (in thousands of dollars) generated from the production of $x$ units of computer chips per day, where each unit consists of 100 chips.
  **(a)** Represent the following statement by equations involving $R$ or $R'$: "When 1200 chips are produced per day, the revenue is $22,000 and the marginal revenue is $.75 per chip."
  **(b)** If the marginal cost of producing 1200 chips is $1.5 per chip, what is the marginal profit at this production level?

**42.** Refer to the previous exercise. Is it profitable to produce 1300 chips per day if the cost of producing 1200 chips per day is $14,000?

## Technology Exercises

**43.** For the given function, simultaneously graph the functions $f(x)$, $f'(x)$, and $f''(x)$ with the specified window setting. *Note:* Since we have not yet learned how to differentiate the given function, you must use your graphing utility's differentiation command to define the derivatives.

$$f(x) = \frac{x}{1+x^2}, \ [-4, 4] \ by \ [-2, 2]$$

**44.** Consider the cost function of Example 5.
  **(a)** Graph $C(x)$ in the window $[0, 60]$ *by* $[-300, 1260]$.
  **(b)** For what level of production will the cost be $535?
  **(c)** For what level of production will the marginal cost be $14?

## SOLUTIONS TO PRACTICE PROBLEMS 1.7

**1.** $f(t) = t + t^{-1}$

$f'(t) = 1 + (-1)t^{(-1)-1} = 1 - t^{-2}$

$f''(t) = -(-2)t^{(-2)-1} = 2t^{-3} = \dfrac{2}{t^3}$

Therefore,

$$f''(2) = \frac{2}{2^3} = \frac{1}{4}.$$

[*Note*: It is essential first to compute the function $f''(t)$ and *then* to evaluate the function at $t = 2$.]

**2.** The expression $2\pi rh$ contains two numbers, 2 and $\pi$, and two letters, $r$ and $h$. The notation $g(r)$ tells us that the expression $2\pi rh$ is to be regarded as a function of $r$. Therefore, $h$—and hence $2\pi h$—is to be treated as a constant, and differentiation is done with respect to the variable $r$. That is,

$$g(r) = (2\pi h)r$$
$$g'(r) = 2\pi h.$$

## 1.8  The Derivative as a Rate of Change

As we saw in previous sections, an important interpretation of the slope of a function at a point is as a rate of change. In this section, we reexamine this interpretation and discuss some additional applications in which it proves useful. The first step is to understand what is meant by "average rate of change" of a function $f(x)$.

Consider a function $y = f(x)$ defined on the interval $a \leq x \leq b$. The average rate of change of $f(x)$ over this interval is the change in $f(x)$ divided by the length of the interval. That is,

$$\left[ \begin{array}{c} \text{average rate of change of } f(x) \\ \text{over the interval } a \leq x \leq b \end{array} \right] = \frac{f(b) - f(a)}{b - a}.$$

**EXAMPLE 1**    Suppose that $f(x) = x^2$. Calculate the average rate of change of $f(x)$ over the following intervals:

(a) $1 \leq x \leq 2$ \qquad (b) $1 \leq x \leq 1.1$ \qquad (c) $1 \leq x \leq 1.01$

**Solution**    (a) The average rate of change over the interval $1 \leq x \leq 2$ is

$$\frac{2^2 - 1^2}{2 - 1} = \frac{3}{1} = 3.$$

(b) The average rate of change over the interval $1 \leq x \leq 1.1$ is

$$\frac{1.1^2 - 1^2}{1.1 - 1} = \frac{.21}{.1} = 2.1.$$

(c) The average rate of change over the interval $1 \leq x \leq 1.01$ is

$$\frac{1.01^2 - 1^2}{1.01 - 1} = \frac{.0201}{.01} = 2.01. \qquad \blacksquare$$

In the special case where $b$ is $a + h$, the value of $b - a$ is $(a + h) - a$ or $h$, and the average rate of change of the function over the interval is the familiar difference quotient

$$\frac{f(a + h) - f(a)}{h}.$$

Geometrically, this quotient is the slope of the secant line in Fig. 1. Recall that as $h$ approaches 0, the slope of the secant line approaches the slope of the tangent line. Thus the average rate of change approaches $f'(a)$. For this reason, $f'(a)$ is called the (*instantaneous*) *rate of change* of $f(x)$ exactly at the point where $x = a$. From

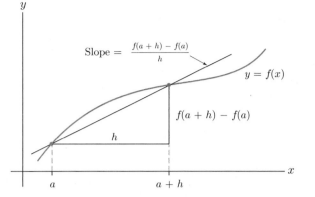

**Figure 1.** Average rate of change is the slope of the secant line.

now on, unless we explicitly use the word "average" when we refer to the rate of change of a function, we mean the "instantaneous" rate of change.

> The derivative $f'(a)$ measures the rate of change of $f(x)$ at $x = a$.

**EXAMPLE 2**    Consider the function $f(x) = x^2$ of Example 1. Calculate the rate of change of $f(x)$ at $x = 1$.

**Solution**    The rate of change of $f(x)$ at $x = 1$ is equal to $f'(1)$. We have

$$f'(x) = 2x$$
$$f'(1) = 2 \cdot 1 = 2.$$

That is, the rate of change is 2 units per unit change in $x$. Notice how the average rates of change in Example 1 approach the (instantaneous) rate of change as the intervals beginning at $x = 1$ shrink. ∎

**EXAMPLE 3**    The function $f(t)$ in Fig. 2 gives the population of the United States $t$ years after the beginning of 1800. The figure also shows the tangent line through the point $(40, 17)$.

(a) What was the average rate of growth of the United States population from 1840 to 1870?

(b) How fast was the population growing in 1840?

(c) Was the population growing faster in 1810 or 1880?

**Solution**    (a) Since 1870 is 70 years after 1800 and 1840 is 40 years after 1800, the question asks for the average rate of change of $f(t)$ over the interval $40 \le t \le 70$. This value is

$$\frac{f(70) - f(40)}{70 - 40} = \frac{40 - 17}{30} = \frac{23}{30} \approx .77.$$

Therefore, from 1840 to 1870, the population grew at the average rate of about .77 million or 770,000 people per year.

(b) The rate of growth of $f(t)$ at $t = 40$ is $f'(40)$; that is, the slope of the tangent line at $t = 40$. Since $(40, 17)$ and $(90, 42)$ are two points on the tangent line, the slope of the tangent line is

$$\frac{42 - 17}{90 - 40} = \frac{25}{50} = .5.$$

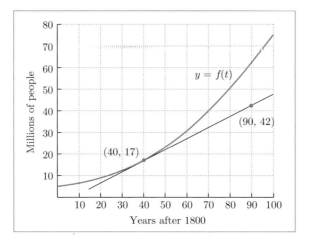

**Figure 2.**   United States population from 1800 to 1900.

Therefore, in 1840 the population was growing at the rate of .5 million or 500,000 people per year.

(c) The graph is clearly steeper in 1880 than in 1810. Therefore, the population was growing faster in 1880 than in 1810. ∎

**Velocity and Acceleration**    An everyday illustration of rate of change is given by the velocity of a moving object. Suppose that we are driving a car along a straight road and at each time $t$ we let $s(t)$ be our position on the road, measured from some convenient reference point. See Fig. 3, where distances are positive to the right of the car. For the moment we shall assume that we are proceeding in only the positive direction along the road.

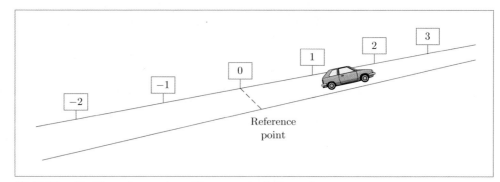

**Figure 3.**   Position of a car traveling on a straight road.

At any instant, the car's speedometer tells us how fast we are moving—that is, how fast our position $s(t)$ is changing. To show how the speedometer reading is related to our calculus concept of a derivative, let us examine what is happening at a specific time, say $t = 2$. Consider a short time interval of duration $h$ from $t = 2$ to $t = 2 + h$. Our car will move from position $s(2)$ to position $s(2 + h)$, a distance of $s(2 + h) - s(2)$. Thus the *average velocity from $t = 2$ to $t = 2 + h$* is

$$\frac{[\text{distance traveled}]}{[\text{time elapsed}]} = \frac{s(2 + h) - s(2)}{h}. \tag{1}$$

If the car is traveling at a steady speed during this time period, then the speedometer reading will equal the average velocity in (1).

From our discussion in Section 1.3 the ratio (1) approaches the derivative $s'(2)$ as $h$ approaches zero. For this reason we call $s'(2)$ *the* (instantaneous) *velocity at* $t = 2$. This number will agree with the speedometer reading at $t = 2$ because when $h$ is very small, the car's speed will be nearly steady over the time interval from $t = 2$ to $t = 2 + h$, and so the average velocity over this time interval will be nearly the same as the speedometer reading at $t = 2$.

The reasoning used for $t = 2$ holds for an arbitrary $t$ as well. Thus the following definition makes sense:

> If $s(t)$ denotes the position function of an object moving in a straight line, then the velocity $v(t)$ of the object at time $t$ is given by
>
> $$v(t) = s'(t).$$

In our discussion we assumed that the car moved in the positive direction. If the car moves in the opposite direction, the ratio (1) and the limiting value $s'(2)$ will be negative. So we interpret negative velocity as movement in the negative direction along the road.

The derivative of the velocity function $v(t)$ is called the acceleration function and is often written as $a(t)$:

> $$a(t) = v'(t).$$

Because $v'(t)$ measures the rate of change of the velocity $v(t)$, this use of the word "acceleration" agrees with our common usage in connection with automobiles. Note that since $v(t) = s'(t)$, the acceleration is actually the second derivative of the position function $s(t)$,

$$a(t) = s''(t).$$

---

**EXAMPLE 4** When a ball is thrown straight up into the air, its position may be measured as the vertical distance from the ground. Regard "up" as the positive direction, and let $s(t)$ be the height of the ball in feet after $t$ seconds. Suppose that $s(t) = -16t^2 + 128t + 5$. Figure 4 contains a graph of the function $s(t)$. *Note:* The graph does *not* show the path of the ball. The ball is rising *straight* up and then falling *straight* down.

(a) What is the velocity after 2 seconds?

(b) What is the acceleration after 2 seconds?

(c) At what time is the velocity $-32$ feet per second? (The negative sign indicates that the ball's height is decreasing; that is, the ball is falling.)

(d) When is the ball at a height of 117 feet?

**Solution** (a) The velocity is the rate of change of the height function, so

$$v(t) = s'(t) = -32t + 128.$$

The velocity when $t = 2$ is $v(2) = -32(2) + 128 = 64$ feet per second.

(b) $a(t) = v'(t) = -32$. The acceleration is $-32$ feet per second per second for all $t$. This constant acceleration is due to the downward (and therefore negative) force of gravity.

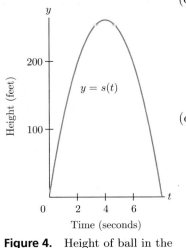

**Figure 4.** Height of ball in the air.

(c) Since the velocity is given and the time is unknown, we set $v(t) = -32$ and solve for $t$:

$$-32t + 128 = -32$$
$$-32t = -160$$
$$t = 5.$$

The velocity is $-32$ feet per second when $t$ is 5 seconds.

(d) The question here involves the height function, not the velocity. Since the height is given and the time is unknown, we set $s(t) = 117$ and solve for $t$:

$$-16t^2 + 128t + 5 = 117$$
$$-16(t^2 - 8t + 7) = 0$$
$$-16(t - 1)(t - 7) = 0.$$

The ball is at a height of 117 feet once on the way up (when $t = 1$ second) and once on the way down (when $t = 7$ seconds). ∎

**Approximating the Change in a Function**   Consider the function $f(x)$ near $x = a$. As we just saw, the average rate of change of $f$ over a small interval of length $h$ is approximately equal to the instantaneous change at an endpoint of the interval. That is,

$$\frac{f(a + h) - f(a)}{h} \approx f'(a).$$

Multiplying both sides of this approximation by $h$, we have

$$f(a + h) - f(a) \approx f'(a) \cdot h. \tag{2}$$

If $x$ changes from $a$ to $a + h$, then the change in the value of the function $f(x)$ is approximately $f'(a)$ times the change $h$ in the value of $x$. This result holds for both positive and negative values of $h$. In applications, the right side of (2) is calculated and used to estimate the left side.

Figure 5 contains a geometric interpretation of (2). Given the small change $h$ in $x$, the quantity $f'(a) \cdot h$ gives the corresponding change in $y$ *along the tangent line* at $(a, f(a))$. In contrast, the quantity $f(a + h) - f(a)$ gives the change in $y$ *along the curve* $y = f(x)$. When $h$ is small, $f'(a) \cdot h$ is a good approximation to the change in $f(x)$.

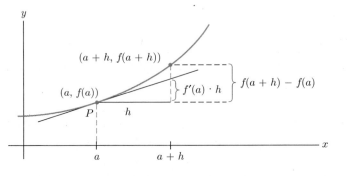

**Figure 5.** Change in $y$ along the tangent line and along the graph of $y = f(x)$.

Before we consider an application, it is instructive to present an alternative derivation of (2). From Section 1.3, the equation of the tangent line at $P$ is

$$y - f(a) = f'(a)(x - a). \tag{3}$$

Now the tangent line approximates the graph of $f$ near $P$, so if in (3) we replace $y$ (the value of the tangent line) by $f(x)$, we get

$$f(x) - f(a) \approx f'(a)(x - a), \qquad (4)$$

and (2) follows from (4) upon setting $h = x - a$ and $x = a + h$.

---

**EXAMPLE 5**    Let the production function $p(x)$ give the number of units of goods produced when employing $x$ units of labor. Suppose 5000 units of labor are currently employed, $p(5000) = 300$, and $p'(5000) = 2$.

(a) Interpret $p(5000) = 300$.

(b) Interpret $p'(5000) = 2$.

(c) Estimate the number of *additional* units of goods produced when $x$ is increased from 5000 to $5000\frac{1}{2}$ units of labor.

(d) Estimate the *change* in the number of units of goods produced when $x$ is decreased from 5000 to 4999 units of labor.

**Solution**    (a) When 5000 units of labor are employed, 300 units of goods will be produced.

(b) If 5000 units of labor are currently employed and we consider adding more labor, productivity will increase at approximately the rate of 2 units of goods for each additional unit of labor.

(c) Here $h = \frac{1}{2}$. By (2), the change in $p(x)$ will be approximately

$$p'(5000) \cdot \tfrac{1}{2} = 2 \cdot \tfrac{1}{2} = 1.$$

About one additional unit will be produced. Therefore, about 301 units of goods will be produced when $5000\frac{1}{2}$ units of labor are employed.

(d) Here $h = -1$, since the amount of labor is reduced. The change in $p(x)$ will be approximately

$$p'(5000) \cdot (-1) = 2 \cdot (-1) = -2.$$

About two fewer (that is, 298 units) of goods will be produced.    ■

To simplify the use of formula (2) in marginal cost analysis, let us rewrite it with a cost function, as follows:

$$C(a + h) - C(a) \approx C'(a) \cdot h. \qquad (5)$$

Taking $h = 1$, we obtain the familiar formula (1) of the previous section,

$$C(a + 1) - C(a) \approx C'(a). \qquad (6)$$

Thus for a whole unit increase in production ($h = 1$), the change in cost is approximately equal to the marginal cost, but for an increase by $h$ units, we must modify the marginal cost proportionally.

---

**EXAMPLE 6**    **Marginal cost**    The total cost in thousands of dollars of producing $x$ units of a certain commodity is $C(x) = 6x^2 + 2x + 10$.

(a) Find the marginal cost function.

(b) Find the cost and marginal cost when 10 units are produced.

(c) Use marginal cost to approximate the cost of producing the 11th unit.

(d) Use marginal cost to estimate the additional cost incurred if the production level is increased from 10 to 10.5 units.

**Solution** (a) The marginal cost function is the derivative of the cost function:

$$C'(x) = 12x + 2 \text{ (thousand dollars per unit)}.$$

(b) When 10 units are produced, the cost is

$$C(10) = 6(10^2) + 2(10) + 10 = 630 \text{ thousand dollars;}$$

and the marginal cost is $C'(10) = 12(10) + 2 = 122$ thousand dollars per unit.

(c) The cost of producing the 11th unit is the difference in cost as $x$ varies from 10 to 11 or $C(11) - C(10)$. By (6), this difference can be approximated by the marginal cost $C'(10) = 122$. Hence to produce the 11th unit it will cost approximately 122 thousand dollars.

(d) The additional cost when we raise the production level by half a unit from 10 to 10.5 is $C(10.5) - C(10)$. According to (5), with $a = 10$ and $h = .5$,

$$C(10.5) - C(10) \approx C'(10) \cdot (.5) = (122)(.5) = 61 \text{ thousand dollars.} \quad \blacksquare$$

**Units for Rate of Change**    Table 1 shows the units appearing in several examples from this section. In general,

$$[\text{unit of measure for } f'(x)]$$
$$= [\text{unit of measure for } f(x)] \text{ per } [\text{unit of measure for } x].$$

**TABLE 1    Units of Measure**

| Example | Unit for $f(t)$ or $f(x)$ | Unit for $t$ or $x$ | Unit for $f'(t)$ or $f'(x)$ |
|---|---|---|---|
| U.S. population | millions of people | year | millions of people per year |
| Ball in air | feet | second | feet per second |
| Cost function | dollars | item | dollars per item |

## PRACTICE PROBLEMS 1.8

*Let $f(t)$ be the temperature (in degrees Celsius) of a liquid at time $t$ (in hours). The rate of temperature change at time $a$ has the value $f'(a)$. Listed next are typical questions about $f(t)$ and its slope at various points. Match each question with the proper method of solution.*

**Questions:**

1. What is the temperature of the liquid after 6 hours (that is, when $t = 6$)?

2. When is the temperature rising at the rate of 6 degrees per hour?

3. By how many degrees did the temperature rise during the first 6 hours?

4. When is the liquid's temperature only 6 degrees?

5. How fast is the temperature of the liquid changing after 6 hours?

6. What is the average rate of increase in the temperature during the first 6 hours?

**Methods of Solution:**

(a) Compute $f(6)$.

(b) Set $f(t) = 6$ and solve for $t$.

(c) Compute $[f(6) - f(0)]/6$.

(d) Compute $f'(6)$.

(e) Find a value of $a$ for which $f'(a) = 6$.

(f) Compute $f(6) - f(0)$.

# EXERCISES 1.8

**1.** Suppose that $f(x) = 4x^2$.

  **(a)** What is the average rate of change of $f(x)$ over each of the intervals 1 to 2, 1 to 1.5, and 1 to 1.1?

  **(b)** What is the (instantaneous) rate of change of $f(x)$ when $x = 1$?

**2.** Suppose that $f(x) = -6/x$.

  **(a)** What is the average rate of change of $f(x)$ over each of the intervals 1 to 2, 1 to 1.5, and 1 to 1.2?

  **(b)** What is the (instantaneous) rate of change of $f(x)$ when $x = 1$?

**3.** Suppose that $f(t) = t^2 + 3t - 7$.

  **(a)** What is the average rate of change of $f(t)$ over the interval 5 to 6?

  **(b)** What is the (instantaneous) rate of change of $f(t)$ when $t = 5$?

**4.** Suppose that $f(t) = 3t + 2 - \dfrac{12}{t}$.

  **(a)** What is the average rate of change of $f(t)$ over the interval 2 to 3?

  **(b)** What is the (instantaneous) rate of change of $f(t)$ when $t = 2$?

**5.** Refer to Fig. 6, where $f(t)$ is the percentage yield (interest rate) on a 3-month T-bill (U.S. Treasury bill) $t$ years after January 1, 1980.

  **(a)** What was the average rate of change in the yield from January 1, 1981, to January 1, 1985?

  **(b)** How fast was the percentage yield rising on January 1, 1989?

  **(c)** Was the percentage yield rising faster on January 1, 1980, or January 1, 1989?

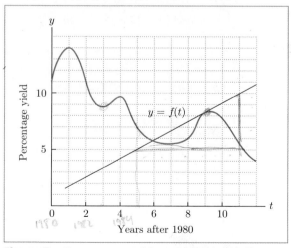

**Figure 6.**  Percentage yield on 3-month T-bill (1980–1992).

**6.** Refer to Fig. 7, where $f(t)$ is the mean (average) size of a farm (in acres) $t$ years after January 1, 1900.

  **(a)** What was the average rate of growth in the mean size of farms from 1920 to 1960?

  **(b)** How fast was the mean farm size rising on January 1, 1950?

  **(c)** Was the mean farm size rising faster on January 1, 1960, or January 1, 1980?

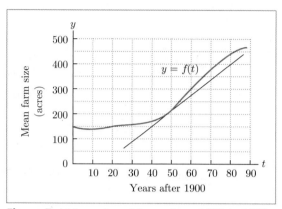

**Figure 7.**  Mean size of farms from 1900 to 1990.

**7.** Suppose that the weight of a cancerous tumor after $t$ weeks is $W(t) = .1t^2$ grams. What is the average rate of growth of the tumor during the fifth week, that is, from time $t = 4$ to $t = 5$? What is the (instantaneous) rate of growth of the tumor at time $t = 4$?

**8.** After an advertising campaign, the sales of a product often increase and then decrease. Suppose that $t$ days after the end of the advertising, the daily sales are $f(t) = -3t^2 + 32t + 100$ units. What is the average rate of growth in sales during the fourth day, that is, from time $t = 3$ to $t = 4$? At what (instantaneous) rate are the sales changing when $t = 2$?

**9.** An analysis of the daily output of a factory assembly line shows that about $60t + t^2 - \frac{1}{12}t^3$ units are produced after $t$ hours of work, $0 \le t \le 8$. What is the rate of production (in units per hour) when $t = 2$?

**10.** Liquid is pouring into a large vat. After $t$ hours, there are $5t - \sqrt{t}$ gallons in the vat. At what rate is the liquid flowing into the vat (in gallons per hour) when $t = 4$?

**11.** An object moving in a straight line travels $s(t)$ kilometers in $t$ hours, where $s(t) = 2t^2 + 4t$.

  **(a)** What is the object's velocity when $t = 6$?

  **(b)** How far has the object traveled in 6 hours?

  **(c)** When is the object traveling at the rate of 6 kilometers per hour?

**12.** Suppose that the position of a car at time $t$ (in hours) is given by $s(t) = 50t - 7/(t + 1)$, where the position is measured in kilometers. Find the velocity and acceleration of the car at $t = 0$.

**13.** A toy rocket fired straight up into the air has height $s(t) = 160t - 16t^2$ feet after $t$ seconds.

(a) What is the rocket's initial velocity (when $t = 0$)?

(b) What is the velocity after 2 seconds?

(c) What is the acceleration when $t = 3$?

(d) At what time will the rocket hit the ground?

(e) At what velocity will the rocket be traveling just as it smashes into the ground?

**14.** A helicopter is rising straight up in the air. Its distance from the ground $t$ seconds after takeoff is $s(t)$ feet, where $s(t) = t^2 + t$.

(a) How long will it take for the helicopter to rise 20 feet?

(b) Find the velocity and the acceleration of the helicopter when it is 20 feet above the ground.

**15.** Let $s(t)$ be the height (in feet) after $t$ seconds of a ball thrown straight up into the air. Match each question with the proper solution.

*Questions:*

**A.** What will be the velocity of the ball after 3 seconds?

**B.** When will the velocity be 3 feet per second?

**C.** What is the average velocity during the first 3 seconds?

**D.** When will the ball be 3 feet above the ground?

**E.** When will the ball hit the ground?

**F.** How high will the ball be after 3 seconds?

**G.** How far did the ball travel during the first 3 seconds?

*Solutions:*

**a.** Set $s(t) = 0$ and solve for $t$.

**b.** Compute $s'(3)$.

**c.** Compute $s(3)$.

**d.** Set $s'(t) = 3$ and solve for $t$.

**e.** Find a value of $a$ for which $s(a) = 3$.

**f.** Compute $[s(3) - s(0)]/3$.

**g.** Compute $s(3) - s(0)$.

**16.** Table 2 gives a car's trip-meter reading (in miles) at 1 hour into a trip and at several nearby times. What is the average speed during the time interval from 1 to 1.05 hours? Estimate the speed at time 1 hour into the trip.

| TABLE 2 | Trip-Meter Readings at Several Times |
|---|---|
| **Time** | **Trip Meter** |
| .96 | 43.2 |
| .97 | 43.7 |
| .98 | 44.2 |
| .99 | 44.6 |
| 1.00 | 45.0 |
| 1.01 | 45.4 |
| 1.02 | 45.8 |
| 1.03 | 46.3 |
| 1.04 | 46.8 |
| 1.05 | 47.4 |

**17.** A particle is moving in a straight line in such a way that its position at time $t$ (in seconds) is $s(t) = t^2 + 3t + 2$ feet to the right of a reference point, for $t \geq 0$.

(a) What is the velocity of the object when the time is 6 seconds?

(b) Is the object moving toward the reference point when $t = 6$? Explain your answer.

(c) What is the object's velocity when the object is 6 feet from the reference point?

**18.** A car is traveling from New York to Boston and is partway between the two cities. Let $s(t)$ be the distance from New York during the next minute. Match each behavior with the corresponding graph of $s(t)$ in Fig. 8.

(a) The car travels at a steady speed.

(b) The car is stopped.

(c) The car is backing up.

(d) The car is accelerating.

(e) The car is decelerating.

**19.** Suppose $f(100) = 5000$ and $f'(100) = 10$. Estimate each of the following.

(a) $f(101)$    (b) $f(100.5)$    (c) $f(99)$

(d) $f(98)$    (e) $f(99.75)$

**20.** Suppose $f(25) = 10$ and $f'(25) = -2$. Estimate each of the following.

(a) $f(27)$    (b) $f(26)$    (c) $f(25.25)$

(d) $f(24)$    (e) $f(23.5)$

**21.** Let $f(t)$ be the temperature of a cup of coffee, $t$ minutes after it has been poured. Interpret $f(4) = 120$ and $f'(4) = -5$. Estimate the temperature of the coffee after 4 minutes and 6 seconds, that is, after 4.1 minutes.

**22.** Suppose 5 mg of a drug is injected into the bloodstream. Let $f(t)$ be the amount present in the bloodstream after $t$ hours. Interpret $f(3) = 2$ and $f'(3) = -.5$. Estimate the number of milligrams of the drug in the bloodstream after $3\frac{1}{2}$ hours.

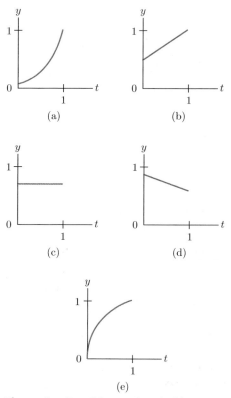

**Figure 8.** Possible graphs of $s(t)$.

**23.** Price affects sales. Let $f(p)$ be the number of cars sold when the price is $p$ dollars per car. Interpret the statements $f(10{,}000) = 200{,}000$ and $f'(10{,}000) = -3$.

**24.** Advertising affects sales. Let $f(x)$ be the number of toys sold when $x$ dollars are spent on advertising. Interpret the statements $f(100{,}000) = 3{,}000{,}000$ and $f'(100{,}000) = 30$.

**25.** Let $f(x)$ be the number (in thousands) of computers sold when the price is $x$ hundred dollars per computer. Interpret the statements $f(12) = 60$ and $f'(12) = -2$. Then estimate the number of computers sold if the price is set at \$1250 per computer.

**26.** Let $C(x)$ be the cost (in dollars) of manufacturing $x$ radios. Interpret the statements $C(2000) = 50{,}000$ and $C'(2000) = 10$. Estimate the cost of manufacturing 1998 radios.

**27.** Let $P(x)$ be the profit (in dollars) from manufacturing and selling $x$ luxury cars. Interpret $P(100) = 90{,}000$ and $P'(100) = 1200$. Estimate the profit from manufacturing and selling 99 cars.

**28.** Let $f(x)$ be the value in dollars of one share of a company $x$ days since the company went public.

**(a)** Interpret the statements $f(100) = 16$ and $f'(100) = .25$.

**(b)** Estimate the value of one share on the 101st day since the company went public.

**29.** (*Marginal Cost Analysis*) Consider the cost function $C(x) = 6x^2 + 14x + 18$ (thousand dollars).

**(a)** What is the marginal cost at production level $x = 5$?

**(b)** Estimate the cost of raising the production level from $x = 5$ to $x = 5.25$.

**(c)** Let $R(x) = -x^2 + 37x + 38$ denote the revenue in thousands of dollars generated from the production of $x$ units. What is the break-even point? (Recall that the break-even point is when revenue is equal to cost.)

**(d)** Compute and compare the marginal revenue and marginal cost at the break-even point. Should the company increase production beyond the break-even point? Justify your answer using marginals.

**30.** Estimate how much the function

$$f(x) = \frac{1}{1 + x^2}$$

will change if $x$ decreases from 1 to .9.

**31.** National health expenditures (in billions of dollars) from 1980 to 1998 are given by the function $f(t)$ in Fig. 9.

**(a)** How much money was spent in 1987?

**(b)** Approximately how fast were expenditures rising in 1987?

**(c)** When did expenditures reach one trillion dollars?

**(d)** When were expenditures rising at the rate of \$100 billion per year?

**32.** In an 8-second test run, a vehicle accelerates for several seconds and then decelerates. The function $s(t)$ gives the number of feet traveled after $t$ seconds and is graphed in Fig. 10.

**(a)** How far has the vehicle traveled after 3.5 seconds?

**(b)** What is the velocity after 2 seconds?

**(c)** What is the acceleration after 1 second?

**(d)** When will the vehicle have traveled 120 feet?

**(e)** When, during the second part of the test run, will the vehicle be traveling at the rate of 20 feet per second?

**(f)** What is the greatest velocity? At what time is this greatest velocity reached? How far has the vehicle traveled at this time?

## Technology Exercises

**33.** In a psychology experiment,[†] people improved their ability to recognize common verbal and semantic information with practice. Their judgment time after $t$ days of practice was $f(t) = .36 + .77(t-.5)^{-.36}$ seconds.

---

[†]John R. Anderson, "Automaticity and the ACT Theory," *American Journal of Psychology*, 105:2 (Summer 1992), 165–180.

**Figure 9**

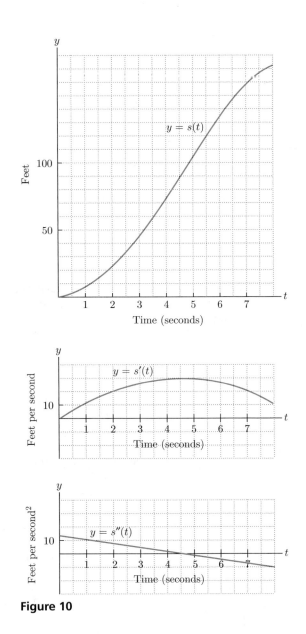

**Figure 10**

(a) Display the graphs of $f(t)$ and $f'(t)$ in the window $[.5, 6]$ by $[-3, 3]$. Use these graphs to answer the following questions.

(b) What was the judgment time after 4 days of practice?

(c) After how many days of practice was the judgment time about .8 seconds?

(d) After 4 days of practice, at what rate was judgment time changing with respect to days of practice?

(e) After how many days was judgment time changing at the rate of $-.08$ seconds per day of practice?

**34.** A ball thrown straight up into the air has height $s(t) = 102t - 16t^2$ feet after $t$ seconds.

(a) Display the graphs of $s(t)$ and $s'(t)$ in the window $[0, 7]$ by $[-100, 200]$. Use these graphs to answer the remaining questions.

(b) How high is the ball after 2 seconds?

(c) When, during descent, is the height 110 feet?

(d) What is the velocity after 6 seconds?

(e) When is the velocity 70 feet per second?

(f) How fast is the ball traveling when it hits the ground?

## SOLUTIONS TO PRACTICE PROBLEMS 1.8

1. Method (a). The question involves $f(t)$, the temperature at time $t$. Since the time is given, compute $f(6)$.

2. Method (e). The question involves $f'(t)$, the rate of change of temperature. The interrogative "when" indicates that the time is unknown. Set $f'(t) = 6$ and solve for $t$.

3. Method (f). The question asks for the change in the value of the function from time 0 to time 6, $f(6) - f(0)$.

4. Method (b). The question involves $f(t)$, and the time is unknown. Set $f(t) = 6$ and solve for $t$.

5. Method (d). The question involves $f'(t)$, and the time is given. Compute $f'(6)$.

6. Method (c). The question asks for the average rate of change of the function during the time interval 0 to 6, $[f(6) - f(0)]/6$.

## REVIEW OF FUNDAMENTAL CONCEPTS

1. Define the slope of a line and give a physical description.

2. What is the point-slope form of the equation of a line?

3. Describe how to find an equation for a line when you know the coordinates of two points on the graph of a line.

4. What can you say about the slopes of parallel lines? Perpendicular lines?

5. Give a physical description of what is meant by the slope of $f(x)$ at the point $(2, f(2))$.

6. What does $f'(2)$ represent?

7. Explain why the derivative of a constant function is 0.

8. State the power rule, constant-multiple rule, and sum rule, and give an example of each.

9. Explain how to calculate $f'(2)$ as the limit of slopes of secant lines through the point $(2, f(2))$.

10. In your own words, explain the meaning of $\lim_{x \to 2} f(x) = 3$. Give an example of a function with this property.

11. Give the limit definition of $f'(2)$, that is, the slope of $f(x)$ at the point $(2, f(2))$.

12. In your own words, explain the meaning of $\lim_{x \to \infty} f(x) = 3$. Give an example of such a function $f(x)$. Do the same for $\lim_{x \to -\infty} f(x) = 3$.

13. In your own words, explain the meaning of "$f(x)$ is continuous at $x = 2$." Give an example of a function $f(x)$ that is not continuous at $x = 2$.

14. In your own words, explain the meaning of "$f(x)$ is differentiable at $x = 2$." Give an example of a function $f(x)$ that is not differentiable at $x = 2$.

15. State the general power rule and give an example.

16. Give two different notations for the first derivative of $f(x)$ at $x = 2$. Do the same for the second derivative.

17. What is meant by the average rate of change of a function over an interval?

18. How is an (instantaneous) rate of change related to average rates of change?

19. Explain the relationship between derivatives and velocity and acceleration.

20. What expression involving a derivative gives an approximation to $f(a + h) - f(a)$?

21. Describe marginal cost in your own words.

22. How do you determine the proper units for a rate of change? Give an example.

## SUPPLEMENTARY EXERCISES

*Find the equation and sketch the graph of the following lines.*

1. With slope $-2$, $y$-intercept $(0, 3)$.

2. With slope $\frac{3}{4}$, $y$-intercept $(0, -1)$.

3. Through $(2, 0)$, with slope 5.

4. Through $(1, 4)$, with slope $-\frac{1}{3}$.

5. Parallel to $y = -2x$, passing through $(3, 5)$.

6. Parallel to $-2x + 3y = 6$, passing through $(0, 1)$.

7. Through $(-1, 4)$ and $(3, 7)$.

8. Through $(2, 1)$ and $(5, 1)$.

9. Perpendicular to $y = 3x + 4$ and passing through $(1, 2)$.

10. Perpendicular to $3x + 4y = 5$ and passing through $(6, 7)$.

11. Horizontal with height 3 units above the $x$-axis.

12. Vertical and 4 units to the right of the $y$-axis.

13. The $y$-axis.

14. The $x$-axis.

*Differentiate.*

15. $y = x^7 + x^3$        16. $y = 5x^8$

**17.** $y = 6\sqrt{x}$

**18.** $y = x^7 + 3x^5 + 1$

**19.** $y = \dfrac{3}{x}$

**20.** $y = x^4 - \dfrac{4}{x}$

**21.** $y = (3x^2 - 1)^8$

**22.** $y = \frac{3}{4}x^{4/3} + \frac{4}{3}x^{3/4}$

**23.** $y = \dfrac{1}{5x - 1}$

**24.** $y = (x^3 + x^2 + 1)^5$

**25.** $y = \sqrt{x^2 + 1}$

**26.** $y = \dfrac{5}{7x^2 + 1}$

**27.** $f(x) = 1/\sqrt[4]{x}$

**28.** $f(x) = (2x + 1)^3$

**29.** $f(x) = 5$

**30.** $f(x) = \dfrac{5x}{2} - \dfrac{2}{5x}$

**31.** $f(x) = \left[x^5 - (x - 1)^5\right]^{10}$

**32.** $f(t) = t^{10} - 10t^9$

**33.** $g(t) = 3\sqrt{t} - \dfrac{3}{\sqrt{t}}$

**34.** $h(t) = 3\sqrt{2}$

**35.** $f(t) = \dfrac{2}{t - 3t^3}$

**36.** $g(P) = 4P^{.7}$

**37.** $h(x) = \frac{3}{2}x^{3/2} - 6x^{2/3}$

**38.** $f(x) = \sqrt{x + \sqrt{x}}$

**39.** If $f(t) = 3t^3 - 2t^2$, find $f'(2)$.

**40.** If $V(r) = 15\pi r^2$, find $V'\left(\frac{1}{3}\right)$.

**41.** If $g(u) = 3u - 1$, find $g(5)$ and $g'(5)$.

**42.** If $h(x) = -\frac{1}{2}$, find $h(-2)$ and $h'(-2)$.

**43.** If $f(x) = x^{5/2}$, what is $f''(4)$?

**44.** If $g(t) = \frac{1}{4}(2t - 7)^4$, what is $g''(3)$?

**45.** Find the slope of the graph of $y = (3x - 1)^3 - 4(3x - 1)^2$ at $x = 0$.

**46.** Find the slope of the graph of $y = (4 - x)^5$ at $x = 5$.

*Compute.*

**47.** $\dfrac{d}{dx}(x^4 - 2x^2)$

**48.** $\dfrac{d}{dt}(t^{5/2} + 2t^{3/2} - t^{1/2})$

**49.** $\dfrac{d}{dP}(\sqrt{1 - 3P})$

**50.** $\dfrac{d}{dn}(n^{-5})$

**51.** $\dfrac{d}{dz}(z^3 - 4z^2 + z - 3)\Big|_{z=-2}$

**52.** $\dfrac{d}{dx}(4x - 10)^5\Big|_{x=3}$

**53.** $\dfrac{d^2}{dx^2}(5x + 1)^4$

**54.** $\dfrac{d^2}{dt^2}(2\sqrt{t})$

**55.** $\dfrac{d^2}{dt^2}(t^3 + 2t^2 - t)\Big|_{t=-1}$

**56.** $\dfrac{d^2}{dP^2}(3P + 2)\Big|_{P=4}$

**57.** $\dfrac{d^2y}{dx^2}$, where $y = 4x^{3/2}$

**58.** $\dfrac{d}{dt}\left(\dfrac{dy}{dt}\right)$, where $y = \dfrac{1}{3t}$

**59.** What is the slope of the graph of $f(x) = x^3 - 4x^2 + 6$ at $x = 2$? Write the equation of the line tangent to the graph of $f(x)$ at $x = 2$.

**60.** What is the slope of the curve $y = 1/(3x - 5)$ at $x = 1$? Write the equation of the line tangent to this curve at $x = 1$.

**61.** Find the equation of the tangent line to the curve $y = x^2$ at the point $\left(\frac{3}{2}, \frac{9}{4}\right)$. Sketch the graph of $y = x^2$ and sketch the tangent line at $\left(\frac{9}{2}, \frac{9}{4}\right)$.

**62.** Find the equation of the tangent line to the curve $y = x^2$ at the point $(-2, 4)$. Sketch the graph of $y = x^2$ and sketch the tangent line at $(-2, 4)$.

**63.** Determine the equation of the tangent line to the curve $y = 3x^3 - 5x^2 + x + 3$ at $x = 1$.

**64.** Determine the equation of the tangent line to the curve $y = (2x^2 - 3x)^3$ at $x = 2$.

**65.** In Fig. 1 the straight line has slope $-1$ and is tangent to the graph of $f(x)$. Find $f(2)$ and $f'(2)$.

**Figure 1**

**66.** In Fig. 2 the straight line is tangent to the graph of $f(x) = x^3$. Find the value of $a$.

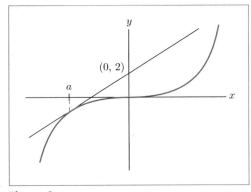

**Figure 2**

**67.** A helicopter is rising at a rate of 32 feet per second. At a height of 128 feet the pilot drops a pair of binoculars. After $t$ seconds, the binoculars have height $s(t) = -16t^2 + 32t + 128$ feet from the ground. How fast will they be falling when they hit the ground?

**68.** Each day, the total output of a coal mine after $t$ hours of operation is approximately $40t + t^2 - \frac{1}{15}t^3$ tons, $0 \leq t \leq 12$. What is the rate of output (in tons of coal per hour) at $t = 5$ hours?

*Exercises 69–72 refer to Fig. 3, where s(t) is the number of feet traveled by a person after t seconds of walking along a straight path.*

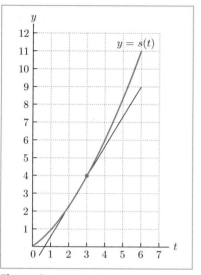

**Figure 3.** Walker's progress.

**69.** How far has the person traveled after 6 seconds?

**70.** What is the person's average velocity from time $t = 1$ to $t = 4$?

**71.** What is the person's velocity at time $t = 3$?

**72.** Without calculating velocities, determine whether the person is traveling faster at $t = 5$ or at $t = 6$.

**73.** A manufacturer estimates that the hourly cost of producing $x$ units of a product on an assembly line is $C(x) = .1x^3 - 6x^2 + 136x + 200$ dollars.

**(a)** Compute $C(21) - C(20)$, the extra cost of raising the production from 20 to 21 units.

**(b)** Find the marginal cost when the production level is 20 units.

**74.** The number of people riding the subway daily from Silver Spring to Metro Center is a function $f(x)$ of the fare, $x$ cents. Suppose $f(235) = 4600$ and $f'(235) = -100$. Approximate the daily number of riders for each of the following costs:

**(a)** 237 cents      **(b)** 234 cents

**(c)** 240 cents      **(d)** 232 cents

**75.** Let $h(t)$ be a boy's height (in inches) after $t$ years. If $h'(12) = 1.5$, how much will his height increase (approximately) between ages 12 and $12\frac{1}{2}$?

**76.** If you deposit \$100 into a savings account at the end of each month for 2 years, the balance will be a function $f(r)$ of the interest rate, $r\%$. At 7% interest (compounded monthly), $f(7) = 2568.10$ and $f'(7) = 25.06$. Approximately how much additional money would you earn if the bank paid $7\frac{1}{2}\%$ interest?

*Determine whether the following limits exist. If so, compute the limit.*

**77.** $\displaystyle\lim_{x \to 2} \frac{x^2 - 4}{x - 2}$

**78.** $\displaystyle\lim_{x \to 3} \frac{1}{x^2 - 4x + 3}$

**79.** $\displaystyle\lim_{x \to 4} \frac{x - 4}{x^2 - 8x + 16}$

**80.** $\displaystyle\lim_{x \to 5} \frac{x - 5}{x^2 - 7x + 2}$

*Use limits to compute the following derivatives.*

**81.** $f'(5)$, where $f(x) = 1/(2x)$.

**82.** $f'(3)$, where $f(x) = x^2 - 2x + 1$.

**83.** What geometric interpretation may be given to

$$\frac{(3 + h)^2 - 3^2}{h}$$

in connection with the graph of $f(x) = x^2$?

**84.** As $h$ approaches 0, what value is approached by

$$\frac{\dfrac{1}{2 + h} - \dfrac{1}{2}}{h}?$$

# Applications of the Derivative

C ALCULUS techniques can be applied to a wide variety of problems in real life, such as the description of the growth of a yeast culture in Chapter 1. We consider many examples in this chapter. In each case we construct a function as a "mathematical model" of some problem and then analyze the function and its derivatives in order to gain information about the original problem. Our principal method for analyzing a function will be to sketch its graph. For this reason we devote the first part of the chapter to curve sketching.

## 2.1 Describing Graphs of Functions

Let's examine the graph of a typical function, such as the one shown in Fig. 1, and introduce some terminology to describe its behavior. First observe that the graph is either rising or falling, depending on whether we look at it from left to right or from right to left. To avoid confusion, we shall always follow the accepted practice of reading a graph from left to right.

Let's now examine the behavior of a function $f(x)$ in an interval throughout which it is defined.

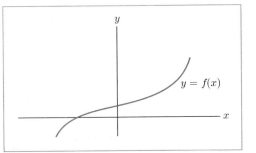

**Figure 1.** An increasing function.

**Increasing Functions** We say that a function $f(x)$ is *increasing in the interval* if the graph continuously rises as $x$ goes from from left to right through the interval. That is, whenever $x_1$ and $x_2$ are in the interval with $x_1 < x_2$, then we have $f(x_1) < f(x_2)$. We say that $f(x)$ is *increasing at* $x = c$ provided that $f(x)$ is increasing in some open interval on the $x$-axis that contains the point $c$.

**Decreasing Functions** We say that a function $f(x)$ is *decreasing in an interval* provided that the graph continuously falls as $x$ goes from left to right through the interval. That is, whenever $x_1$ and $x_2$ are in the interval with $x_1 < x_2$, then we have $f(x_1) > f(x_2)$. We say that $f(x)$ is *decreasing at* $x = c$ provided that $f(x)$ is decreasing in some open interval that contains the point $c$.

Figure 2 shows graphs that are increasing and decreasing at $x = c$. Observe in Fig. 2(d) that when $f(c)$ is negative and $f(x)$ is decreasing, the values of $f(x)$ become *more* negative. When $f(c)$ is negative and $f(x)$ is increasing, as in Fig. 2(e), the values of $f(x)$ become *less* negative.

**Extreme Points** A *relative extreme point* or an *extremum* of a function is a point at which its graph changes from increasing to decreasing, or vice versa. We distinguish the two possibilities in an obvious way.

**Relative Maximum and Minimum** A *relative maximum point* is a point at which the graph changes from increasing to decreasing; a *relative minimum point* is a point at which the graph changes from decreasing to increasing. (See Fig. 3.)

The adjective "relative" in these definitions indicates that a point is maximal or minimal only relative to nearby points on the graph. The adjective "local" is also used in place of "relative."

**Absolute Maximum and Minimum** The *maximum value* (or *absolute maximum value*) of a function is the largest value that the function assumes on its domain. The *minimum value* (or *absolute minimum value*) of a function is the smallest value that the function assumes on its domain.

Functions may or may not have maximum or minimum values. (See Fig. 4.) However, it can be shown that a continuous function whose domain is an interval of the form $a \leq x \leq b$ has both a maximum and a minimum value.

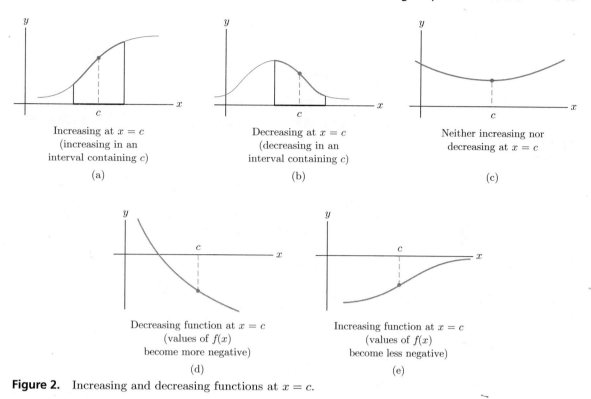

Figure 2. Increasing and decreasing functions at $x = c$.

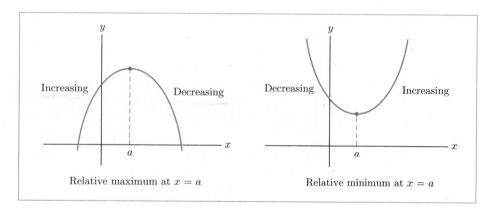

**Figure 3.** Relative extreme points.

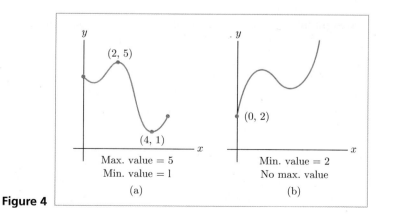

**Figure 4**

Maximum values and minimum values of functions usually occur at relative maximum points and relative minimum points, as in Fig. 4(a). However, they can occur at endpoints of the domain, as in Fig. 4(b). If so, we say that the function has an *endpoint extreme value* (or *endpoint extremum*).

Relative maximum points and endpoint maximum points are higher than any nearby points. The maximum value of a function is the $y$-coordinate of the highest point on its graph. (The highest point is called the *absolute maximum point*.) Similar considerations apply to minima.

---

**EXAMPLE 1**  When a drug is injected intramuscularly (into a muscle), the concentration of the drug in the veins has the time-concentration curve shown in Fig. 5. Describe this graph, using the terms introduced above.

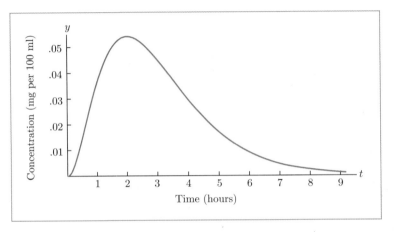

**Figure 5.** A drug time-concentration curve.

**Solution**  Initially (when $t = 0$), there is no drug in the veins. When the drug is injected into the muscle, it begins to diffuse into the bloodstream. The concentration in the veins increases until it reaches its maximum value at $t = 2$. After this time the concentration begins to decrease, as the body's metabolic processes remove the drug from the blood. Eventually the drug concentration decreases to a level so small that, for all practical purposes, it is zero.  ■

**Changing Slope**  An important but subtle feature of a graph is the way the graph's slope *changes* (as we look from left to right). The graphs in Fig. 6 are both increasing, but there is a fundamental difference in the way they are increasing. Graph I, which describes the U.S. gross federal debt per person, is steeper for 1990 than for 1960. That is, the *slope* of graph I *increases* as we move from left to right. A newspaper description of graph I might read,

> U.S. gross federal debt per capita rose at an increasing rate during the years from 1960 to 1990.

In contrast, the *slope* of graph II *decreases* as we move from left to right from 1960 to 1990. Although the U.S. population is rising each year, the rate of increase declines throughout the years from 1960 to 1990. That is, the slope becomes less positive. The media might say,

> During the 1960s, '70s and '80s, U.S. population rose at a decreasing rate.

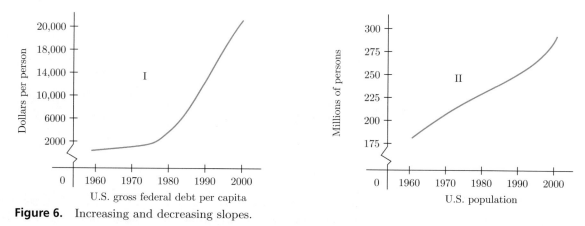

**Figure 6.** Increasing and decreasing slopes.

---

**EXAMPLE 2**   The daily number of hours of sunlight in Washington, D.C., increased from 9.45 hours on December 21 to 12 hours on March 21 and then increased to 14.9 hours on June 21. From December 22 to March 21, the daily increase was greater than the previous daily increase and from March 22 to June 21, the daily increase was less than the previous daily increase. Draw a possible graph of the number of hours of daylight as a function of time.

**Solution**   Let $f(t)$ be the number of hours of daylight $t$ months after December 21. See Fig. 7. The first part of the graph, December 21 to March 21, is increasing at an increasing rate. The second part of the graph, March 21 to June 21, is increasing at a decreasing rate.   ■

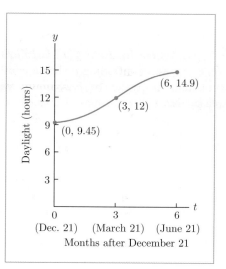

**Figure 7.** Hours of daylight in Washington, D.C.

▶ *Warning*   Recall that when a negative quantity decreases, it becomes more negative. (Think about the temperature outside when it is below zero and the temperature is falling.) So if the slope of a graph is negative and the slope is decreasing, then the slope is becoming more negative, as in Fig. 8. This technical use of the term "decreasing" runs counter to our intuition, because in popular discourse, decrease often means to become smaller in size.   ◀

It is true that the curve in Fig. 9 is becoming "less steep" in a nontechnical sense (since steepness, if it were defined, would probably refer to the magnitude

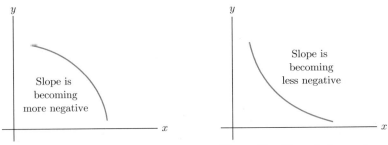

**Figure 8.** Slope is decreasing.      **Figure 9.** Slope is increasing.

of the slope). However, the slope of the curve in Fig. 9 is increasing because it is becoming less negative. The popular press would probably describe the curve in Fig. 9 as decreasing at a decreasing rate, since the rate of fall tends to taper off. Since this terminology is potentially confusing, we shall not use it.

**Concavity** The U.S. debt and population graphs in Fig. 6 may also be described in geometric terms: Between 1960 and 1990, graph I opens up and lies above its tangent line at each point, whereas graph II opens down and lies below its tangent line at each point between 1960 and 1990 (Fig. 10).

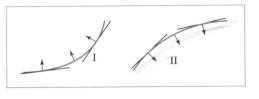

**Figure 10.** Relationship between concavity and tangent lines.

> **Concavity** We say that a function $f(x)$ is *concave up* at $x = a$ if there is an open interval on the $x$-axis containing $a$ throughout which the graph of $f(x)$ lies above its tangent line. Equivalently, $f(x)$ is concave up at $x = a$ if the slope of the graph increases as we move from left to right through $(a, f(a))$.

The graph in Fig. 10(I) is an example of a function that is concave up at each point.

Similarly, we say that a function $f(x)$ is *concave down* at $x = a$ if there is an open interval on the $x$-axis containing $a$ throughout which the graph of $f(x)$ lies below its tangent line. Equivalently, $f(x)$ is concave down at $x = a$ if the slope of the graph decreases as we move from left to right through $(a, f(a))$. The graph in Fig. 10(II) is concave down at each point.

> **Inflection Point** An *inflection point* is a point on the graph of a function at which the function is continuous and at which the graph changes from concave up to concave down, or vice versa.

At such a point, the graph crosses its tangent line (Fig. 11). (The continuity condition means that the graph cannot break at an inflection point.)

---

  **EXAMPLE 3**    Use the terms defined earlier to describe the graph shown in Fig. 12.

**Figure 11.** Inflection points.

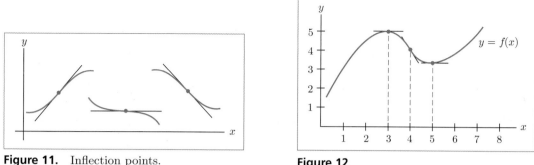

**Figure 12**

**Solution** (a) For $x < 3$, $f(x)$ is increasing and concave down.

(b) Relative maximum point at $x = 3$.

(c) For $3 < x < 4$, $f(x)$ is decreasing and concave down.

(d) Inflection point at $x = 4$.

(e) For $4 < x < 5$, $f(x)$ is decreasing and concave up.

(f) Relative minimum point at $x = 5$.

(g) For $x > 5$, $f(x)$ is increasing and concave up.

**Intercepts, Undefined Points, and Asymptotes** A point at which a graph crosses the $y$-axis is called a *y-intercept*, and a point at which it crosses the $x$-axis is called an *x-intercept*. A function can have at most one $y$-intercept. Otherwise its graph would violate the vertical line test for a function. Note, however, that a function can have any number of $x$-intercepts (possibly none). The $x$-coordinate of an $x$-intercept is sometimes called a "zero" of the function, since the function has the value zero there. (See Fig. 13.)

Some functions are not defined for all values of $x$. For instance, $f(x) = 1/x$ is not defined for $x = 0$, and $f(x) = \sqrt{x}$ is not defined for $x < 0$. (See Fig. 14.) Many functions that arise in applications are defined only for $x \geq 0$. A properly drawn graph should leave no doubt as to the values of $x$ for which the function is defined.

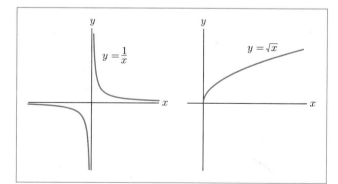

**Figure 13.** Intercepts of a graph.

**Figure 14.** Graphs with undefined points.

Graphs in applied problems sometimes straighten out and approach some straight line as $x$ gets large (Fig. 15). Such a straight line is called an *asymptote* of the curve. The most common asymptotes are horizontal as in (a) and (b) of Fig. 15. In Example 1, the $t$-axis is an asymptote of the drug time-concentration curve.

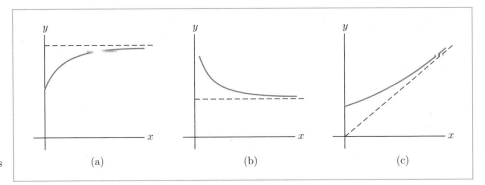

**Figure 15.** Graphs that approach asymptotes as $x$ gets large.

The horizontal asymptotes of a graph may be determined by calculating the limits

$$\lim_{x \to \infty} f(x) \quad \text{and} \quad \lim_{x \to -\infty} f(x).$$

If either limit exists, then the value of the limit determines a horizontal asymptote.

Occasionally, a graph will approach a vertical line as $x$ approaches some fixed value, as in Fig. 16. Such a line is a *vertical asymptote*. Most often, we expect a vertical asymptote at a value $x$ that would result in division by zero in the definition of $f(x)$. For example, $f(x) = 1/(x-3)$ has a vertical asymptote $x = 3$.

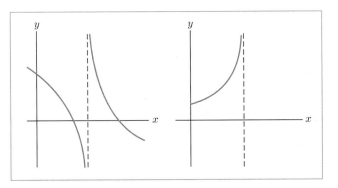

**Figure 16.** Examples of vertical asymptotes.

We now have six categories for describing the graph of a function.

---

### Describing a Graph

**1.** Intervals in which the function is increasing (resp. decreasing), relative maximum points, relative minimum points

**2.** Maximum value, minimum value

**3.** Intervals in which the function is concave up (resp. concave down), inflection points

**4.** $x$-intercepts, $y$-intercept

**5.** Undefined points

**6.** Asymptotes

---

For us, the first three categories will be the most important. However, the last three categories should not be forgotten.

## PRACTICE PROBLEMS 2.1

**1.** Does the slope of the curve in Fig. 17 increase or decrease as $x$ increases?

**Figure 17**

**2.** At what value of $x$ is the slope of the curve in Fig. 18 minimized?

**Figure 18**

# EXERCISES 2.1

*Exercises 1–4 refer to graphs (a)–(f) in Fig. 19.*

**1.** Which functions are increasing for all $x$?

**2.** Which functions are decreasing for all $x$?

**3.** Which functions have the property that the slope always increases as $x$ increases?

**4.** Which functions have the property that the slope always decreases as $x$ increases?

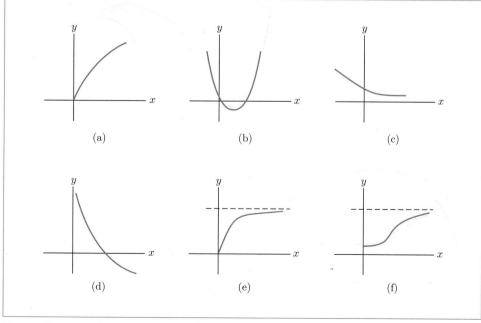

**Figure 19**

*Describe each of the following graphs. Your description should include each of the six categories mentioned previously.*

**5.**

**6.**

**7.**

**8.**

**9.**

**10.**

**11.**

**12.**

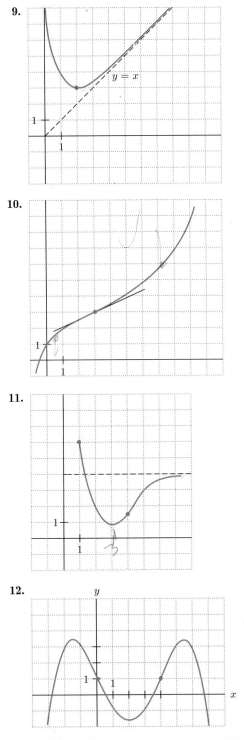

**13.** Describe the way the *slope* changes as you move along the graph (from left to right) in Exercise 5.

**14.** Describe the way the *slope* changes on the graph in Exercise 6.

**15.** Describe the way the *slope* changes on the graph in Exercise 8.

**16.** Describe the way the *slope* changes on the graph in Exercise 10.

*Exercises 17 and 18 refer to the graph in Fig. 20.*

**17.** **(a)** At which labeled points is the function increasing?

**(b)** At which labeled points is the graph concave up?

**(c)** Which labeled point has the most positive slope?

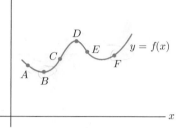

**Figure 20**

**18.** **(a)** At which labeled points is the function decreasing?

**(b)** At which labeled points is the graph concave down?

**(c)** Which labeled point has the most negative slope (that is, negative and with the greatest magnitude)?

*In Exercises 19–22, draw the graph of a function $y = f(x)$ with the stated properties.*

**19.** Both the function and the slope increase as $x$ increases.

**20.** The function increases and the slope decreases as $x$ increases.

**21.** The function decreases and the slope increases as $x$ increases. [*Note:* The slope is negative but becomes less negative.]

**22.** Both the function and the slope decrease as $x$ increases. [*Note:* The slope is negative and becomes more negative.]

**23.** The annual world consumption of oil rises each year. Furthermore, the amount of the annual *increase* in oil consumption is also rising each year. Sketch a graph that could represent the annual world consumption of oil.

**24.** In certain professions the average annual income has been rising at an increasing rate. Let $f(T)$ denote the average annual income at year $T$ for persons in one of these professions and sketch a graph that could represent $f(T)$.

**25.** At noon a child's temperature is $101°$ and is rising at an increasing rate. At 1 P.M. the child is given medicine. After 2 P.M. the temperature is still increasing but at a decreasing rate. The temperature reaches a peak of $103°$ at 3 P.M. and decreases to $100°$ by 5 P.M. Draw a possible graph of the function $T(t)$, the child's temperature at time $t$.

**26.** The number of parking tickets given out each year in the District of Columbia increased from 114,000 in 1950 to 1,500,000 in 1990. Also, each year's increase was greater than the increase for the previous year. Draw a possible graph of the annual number of parking tickets as a function of time $t$. [*Note:* If $f(t)$ is the yearly rate $t$ years after 1950, then $f(0) = 114$ and $f(40) = 1500$ thousand tickets.]

**27.** Let $C(x)$ denote the total cost of manufacturing $x$ units of some product. Then $C(x)$ is an increasing function for all $x$. For small values of $x$, the rate of increase of $C(x)$ decreases. (This is because of the savings that are possible with "mass production.") Eventually, however, for large values of $x$, the cost $C(x)$ increases at an increasing rate. (This happens when production facilities are strained and become less efficient.) Sketch a graph that could represent $C(x)$.

**28.** One method of determining the level of blood flow through the brain requires the person to inhale air containing a fixed concentration of $N_2O$, nitrous oxide. During the first minute, the concentration of $N_2O$ in the jugular vein grows at an increasing rate to a level of .25%. Thereafter it grows at a decreasing rate and reaches a concentration of about 4% after 10 minutes. Draw a possible graph of the concentration of $N_2O$ in the vein as a function of time.

**29.** Suppose that some organic waste products are dumped into a lake at time $t = 0$, and suppose that the oxygen content of the lake at time $t$ is given by the graph in Fig. 21. Describe the graph in physical terms. Indicate the significance of the inflection point at $t = b$.

**30.** Figure 22 gives the U.S. electrical energy production in trillion kilowatt-hours from 1935 ($t = 0$) to 1995 ($t = 60$) with projections. In what year was the level of production growing at the greatest rate?

**31.** Figure 23 gives the number of U.S. farms in millions from 1920 ($t = 20$) to 2000 ($t = 100$). In what year was the number of farms decreasing most rapidly?

**32.** Figure 24 shows the graph of the consumer price index for the years 1983 ($t = 0$) through 2002 ($t = 19$). This index measures how much a basket of commodities that cost \$100 in the beginning of 1983 would cost at any given time. In what year was the rate of increase of the index greatest? The least?

**33.** Let $s(t)$ be the distance (in feet) traveled by a parachutist after $t$ seconds from the time of opening the chute, and suppose that $s(t)$ has the line $y = -15t + 10$ as an asymptote. What does this imply about the velocity of the parachutist? [*Note:* Distance traveled downward is given a negative value.]

**34.** Let $P(t)$ be the population of a bacteria culture after $t$ days and suppose that $P(t)$ has the line $y = 25,000,000$ as an asymptote. What does this imply about the size of the population?

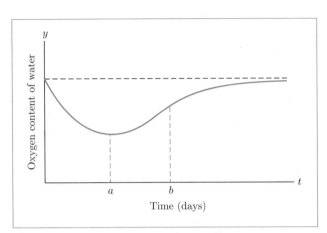

**Figure 21.** A lake's recovery from pollution.

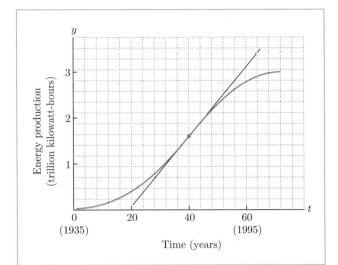

**Figure 22.** U.S. electrical energy production.

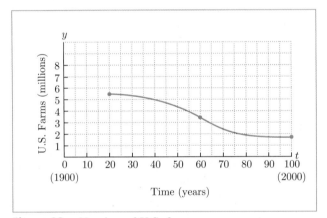

**Figure 23.** Number of U.S. farms.

**Figure 24.** Consumer price index.

*In Exercises 35–38, sketch the graph of a function having the given properties.*

**35.** Defined for $0 \leq x \leq 10$; relative maximum point at $x = 3$; absolute maximum value at $x = 10$.

**36.** Relative maximum points at $x = 1$ and $x = 5$; relative minimum point at $x = 3$; inflection points at $x = 2$ and $x = 4$.

**37.** Defined and increasing for all $x \geq 0$; inflection point at $x = 5$; asymptotic to the line $y = (3/4)x + 5$.

**38.** Defined for $x \geq 0$; absolute minimum value at $x = 0$; relative maximum point at $x = 4$; asymptotic to the line $y = (x/2) + 1$.

**39.** Consider a smooth curve with no undefined points.

    **(a)** If it has two relative maximum points, must it have a relative minimum point?

    **(b)** If it has two relative extreme points, must it have an inflection point?

**40.** Suppose the function $f(x)$ has a relative minimum at $x = a$ and a relative maximum at $x = b$. Must $f(a)$ be less than $f(b)$?

*The difference between the pressure inside the lungs and the pressure surrounding the lungs is called the* transmural pres- *sure (or transmural pressure gradient). Figure 25 shows for three different persons how the volume of the lungs is related to the transmural pressure, based on static measurements taken while there is no air flowing through the mouth. (The functional reserve capacity mentioned on the vertical axis is the volume of air in the lungs at the end of a normal expiration.) The rate of change in lung volume with respect to transmural pressure is called the* lung compliance.

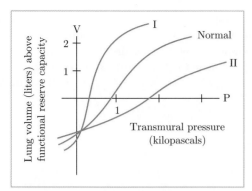

**Figure 25.** Lung pressure-volume curves.

**41.** If the lungs are less flexible than normal, an increase in pressure will cause a smaller change in lung volume than in a normal lung. In this case, is the compliance relatively high or low?

**42.** Most lung diseases cause a decrease in lung compliance. However, the compliance of a person with emphysema is higher than normal. Which curve (I or II) in the figure could correspond to a person with emphysema?

*Technology Exercises*

**43.** Graph the function

$$f(x) = \frac{1}{x^3 - 2x^2 + x - 2}$$

in the window $[0, 4]$ *by* $[-15, 15]$. For what value of $x$ does $f(x)$ have a vertical asymptote?

**44.** The graph of the function

$$f(x) = \frac{2x^2 - 1}{.5x^2 + 6}$$

has a horizontal asymptote of the form $y = c$. Estimate the value of $c$ by graphing $f(x)$ in the window $[0, 50]$ *by* $[-1, 6]$.

**45.** Simultaneously graph the functions

$$y = \frac{1}{x} + x \quad \text{and} \quad y = x$$

in the window $[-6, 6]$ *by* $[-6, 6]$. Convince yourself that the straight line is an asymptote of the first curve. How far apart are the two graphs when $x = 6$?

**46.** Simultaneously graph the functions

$$y = \frac{1}{x} + x + 2 \quad \text{and} \quad y = x + 2$$

in the window $[-6, 6]$ *by* $[-4, 8]$. Convince yourself that the straight line is an asymptote of the first curve. How far apart are the two graphs when $x = 6$?

## SOLUTIONS TO PRACTICE PROBLEMS 2.1

**1.** The curve is concave up, so the slope increases. Even though the curve itself is decreasing, the slope becomes less negative as we move from left to right.

**2.** At $x = 3$. We have drawn in tangent lines at various points (Fig. 26). Note that as we move from left to right, the slopes decrease steadily until the point $(3, 2)$, at which time they start to increase. This is consistent with the fact that the graph is concave down (hence, slopes are decreasing) to the left of $(3, 2)$ and concave up (hence, slopes are increasing) to the right of $(3, 2)$. Extreme values of slopes always occur at inflection points.

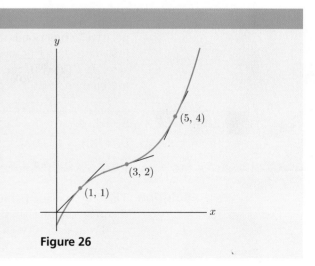

**Figure 26**

## 2.2 The First and Second Derivative Rules

We shall now show how properties of the graph of a function $f(x)$ are determined by properties of the derivatives, $f'(x)$ and $f''(x)$. These relationships will provide the key to the curve sketching and optimization discussed in the rest of the chapter.*

---

*Throughout this chapter, we shall assume that we are dealing with functions that are not "too badly behaved." More precisely, it suffices to assume that all of our functions have continuous first and second derivatives in the interval(s) (in $x$) where we are considering their graphs.

We begin with a discussion of the first derivative of a function $f(x)$. Suppose that for some value of $x$, say $x = a$, the derivative $f'(a)$ is positive. Then the tangent line at $(a, f(a))$ has positive slope and is a rising line (moving from left to right, of course). Since the graph of $f(x)$ near $(a, f(a))$ resembles its tangent line, the function must be increasing at $x = a$. Similarly, when $f'(a) < 0$, the function is decreasing at $x = a$. (See Fig. 1.)

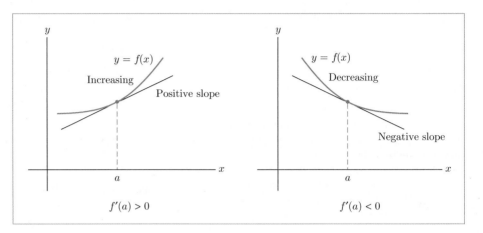

**Figure 1.** Illustration of the first derivative rule.

Thus we have the following useful result.

**First Derivative Rule** If $f'(a) > 0$, then $f(x)$ is increasing at $x = a$. If $f'(a) < 0$, then $f(x)$ is decreasing at $x = a$.

In other words, a function is increasing whenever the value of its derivative is positive; a function is decreasing whenever the value of its derivative is negative. The first derivative rule says nothing about the case when the derivative of a function is zero. If $f'(a) = 0$, the function might be increasing or decreasing or have a relative extreme point at $x = a$.

**EXAMPLE 1** Sketch the graph of a function $f(x)$ that has all the following properties.
(a) $f(3) = 4$.
(b) $f'(x) > 0$ for $x < 3$, $f'(3) = 0$, and $f'(x) < 0$ for $x > 3$.

**Solution** The only specific point on the graph is $(3, 4)$ [property (a)]. We plot this point and then use the fact that $f'(3) = 0$ to sketch the tangent line at $x = 3$ (Fig. 2).

From property (b) and the first derivative rule, we know that $f(x)$ must be increasing for $x$ less than 3 and decreasing for $x$ greater than 3. A graph with these properties might look like the curve in Fig. 3. ∎

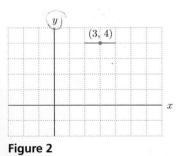

**Figure 2**

The second derivative of a function $f(x)$ gives useful information about the concavity of the graph of $f(x)$. Suppose that $f''(a)$ is negative. Then since $f''(x)$ is the derivative of $f'(x)$, we conclude that $f'(x)$ has a negative derivative at $x = a$. In this case, $f'(x)$ must be a decreasing function at $x = a$; that is, the slope of the graph of $f(x)$ is decreasing as we move from left to right on the graph near $(a, f(a))$. (See Fig. 4.) This means that the graph of $f(x)$ is concave down at $x = a$. A similar analysis shows that if $f''(a)$ is positive, then $f(x)$ is concave up at $x = a$. Thus we have the following rule.

**Figure 3**

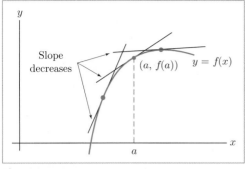

**Figure 4.**  Illustration of the second derivative rule.

> **Second Derivative Rule**  If $f''(a) > 0$, then $f(x)$ is concave up at $x = a$. If $f''(a) < 0$, then $f(x)$ is concave down at $x = a$.

When $f''(a) = 0$, the second derivative rule gives no information. In this case, the function might be concave up, concave down, or neither at $x = a$.

The following chart shows how a graph may combine the properties of increasing, decreasing, concave up, and concave down.

| Conditions on the Derivatives | Description of $f(x)$ at $x = a$ | Graph of $y = f(x)$ Near $x = a$ |
|---|---|---|
| 1. $f'(a)$ positive<br>   $f''(a)$ positive | $f(x)$ increasing<br>$f(x)$ concave up | |
| 2. $f'(a)$ positive<br>   $f''(a)$ negative | $f(x)$ increasing<br>$f(x)$ concave down | |
| 3. $f'(a)$ negative<br>   $f''(a)$ positive | $f(x)$ decreasing<br>$f(x)$ concave up | |
| 4. $f'(a)$ negative<br>   $f''(a)$ negative | $f(x)$ decreasing<br>$f(x)$ concave down | |

**EXAMPLE 2**  Sketch the graph of a function $f(x)$ with all the following properties.

(a)  $(2, 3)$, $(4, 5)$, and $(6, 7)$ are on the graph.

(b)  $f'(6) = 0$ and $f'(2) = 0$.

(c)  $f''(x) > 0$ for $x < 4$, $f''(4) = 0$, and $f''(x) < 0$ for $x > 4$.

**Solution**  First we plot the three points from property (a) and then sketch two tangent lines, using the information from property (b).  (See Fig. 5.)  From property

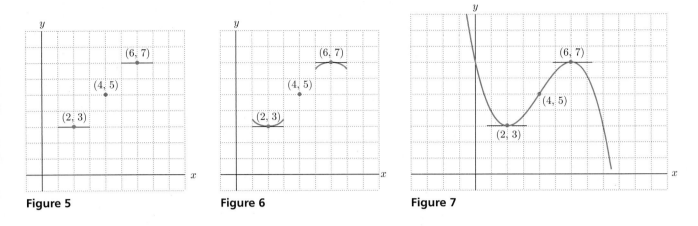

**Figure 5**          **Figure 6**          **Figure 7**

(c) and the second derivative rule, we know that $f(x)$ is concave up for $x < 4$. In particular, $f(x)$ is concave up at $(2,3)$. Also $f(x)$ is concave down for $x > 4$, in particular, at $(6,7)$. Note that $f(x)$ must have an inflection point at $x = 4$ because the concavity changes there. We now sketch small portions of the curve near $(2,3)$ and $(6,7)$. (See Fig. 6.) We can now complete the sketch (Fig. 7), taking care to make the curve concave up for $x < 4$ and concave down for $x > 4$. ■

**Connections Between the Graphs of $f(x)$ and $f'(x)$**   Think of the derivative of $f(x)$ as a "slope function" for $f(x)$. The "$y$-values" on the graph of $y = f'(x)$ are the *slopes* of the corresponding points on the original graph $y = f(x)$. This important connection is illustrated in the next three examples.

---

**EXAMPLE 3**   The function $f(x) = 8x - x^2$ is graphed in Fig. 8 along with the slope at several points. How is the slope changing on the graph? Compare the slopes on the graph with the $y$-coordinates of the points on the graph of $f'(x)$ in Fig. 9.

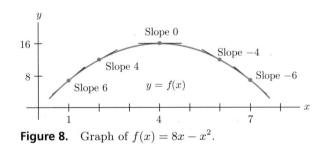

**Figure 8.**  Graph of $f(x) = 8x - x^2$.

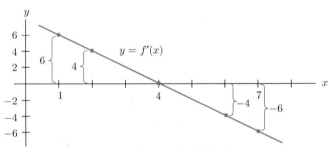

**Figure 9.**  Graph of the derivative of the function in Fig. 8.

**Solution**   The slopes are decreasing (as we move from left to right). That is, $f'(x)$ is a decreasing function. Observe that the $y$-values of $f'(x)$ decrease to zero at $x = 4$ and then continue to decrease for $x$-values greater than 4. ■

The graph in Fig. 8 is the shape of a typical revenue curve for a manufacturer. In this case, the graph of $f'(x)$ in Fig. 9 would be the *marginal revenue curve*. The graph in the next example has the shape of a typical cost curve. Its derivative produces a *marginal cost curve*.

**EXAMPLE 4**  The function $f(x) = \frac{1}{3}x^3 - 4x^2 + 18x + 10$ is graphed in Fig. 10. The slope decreases at first and then increases. Use the graph of $f'(x)$ to verify that the slope in Fig. 10 is minimum at the inflection point, where $x = 4$.

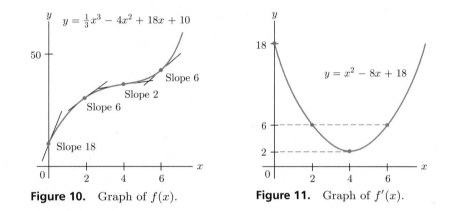

**Figure 10.** Graph of $f(x)$.    **Figure 11.** Graph of $f'(x)$.

**Solution**  Several slopes are marked on the graph of $f(x)$. These values are $y$-coordinates of points on the graph of the derivative, $f'(x) = x^2 - 8x + 18$. Observe in Fig. 11 that the $y$-values on the graph of $f'(x)$ decrease at first and then begin to increase. The minimum value of $f'(x)$ occurs at $x = 4$. ∎

**EXAMPLE 5**  Figure 12 shows the graph of $y = f'(x)$, the derivative of a function $f(x)$.

    (a) What is the slope of the graph of $f(x)$ when $x = 1$?

    (b) Describe how the values of $f'(x)$ change on the interval $1 \le x \le 2$.

    (c) Describe the shape of the graph of $f(x)$ on the interval $1 \le x \le 2$.

    (d) For what values of $x$ does the graph of $f(x)$ have a horizontal tangent line?

    (e) Explain why $f(x)$ has a relative maximum at $x = 3$.

**Solution**  (a) Since $f'(1)$ is 2, $f(x)$ has slope 2 when $x = 1$.

    (b) The values of $f'(x)$ are positive and decreasing as $x$ increases from 1 to 2.

    (c) On the interval $1 \le x \le 2$, the slope of the graph of $f(x)$ is positive and is decreasing as $x$ increases. Therefore, the graph of $f(x)$ is increasing and concave down.

    (d) The graph of $f(x)$ has a horizontal tangent line when the slope is 0—that is, when $f'(x)$ is 0. This occurs at $x = 3$.

    (e) Since $f'(x)$ is positive to the left of $x = 3$ and negative to the right of $x = 3$, the graph of $f(x)$ changes from increasing to decreasing at $x = 3$. Therefore, $f(x)$ has a relative maximum at $x = 3$. ∎

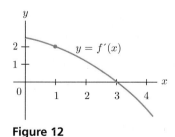

**Figure 12**

Figure 13 shows the graph of a function $f(x)$ whose derivative has the shape shown in Fig. 12. Reread Example 5 and its solution while referring to Fig. 13.

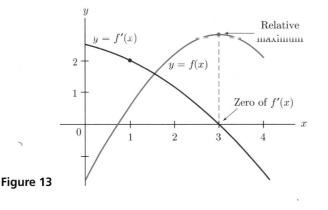

**Figure 13**

## PRACTICE PROBLEMS 2.2

1. Make a good sketch of the function $f(x)$ near the point where $x = 2$, given that $f(2) = 5$, $f'(2) = 1$, and $f''(2) = -3$.

2. The graph of $f(x) = x^3$ is shown in Fig. 14.

**Figure 14**

(a) Is the function increasing at $x = 0$?

(b) Compute $f'(0)$.

(c) Reconcile your answers to parts (a) and (b) with the first derivative rule.

3. The graph of $y = f'(x)$ is shown in Fig. 15. Explain why $f(x)$ must have a relative minimum point at $x = 3$.

**Figure 15**

## EXERCISES 2.2

*Exercises 1–4 refer to the functions whose graphs are given in Fig. 16.*

1. Which functions have a positive first derivative for all $x$?

2. Which functions have a negative first derivative for all $x$?

3. Which functions have a positive second derivative for all $x$?

4. Which functions have a negative second derivative for all $x$?

5. Which one of the graphs in Fig. 17 could represent a function $f(x)$ for which $f(a) > 0$, $f'(a) = 0$, and $f''(a) < 0$?

6. Which one of the graphs in Fig. 17 could represent a function $f(x)$ for which $f(a) = 0$, $f'(a) < 0$, and $f''(a) > 0$?

*In Exercises 7–12, sketch the graph of a function that has the properties described.*

7. $f(2) = 1$; $f'(2) = 0$; concave up for all $x$.

8. $f(-1) = 0$; $f'(x) < 0$ for $x < -1$, $f'(-1) = 0$ and $f'(x) > 0$ for $x > -1$.

9. $f(3) = 5$; $f'(x) > 0$ for $x < 3$, $f'(3) = 0$ and $f'(x) > 0$ for $x > 3$.

10. $(-2, -1)$ and $(2, 5)$ are on the graph; $f'(-2) = 0$ and $f'(2) = 0$; $f''(x) > 0$ for $x < 0$, $f''(0) = 0$, $f''(x) < 0$ for $x > 0$.

11. $(0, 6)$, $(2, 3)$, and $(4, 0)$ are on the graph; $f'(0) = 0$ and $f'(4) = 0$; $f''(x) < 0$ for $x < 2$, $f''(2) = 0$, $f''(x) > 0$ for $x > 2$.

12. $f(x)$ defined only for $x \geq 0$; $(0, 0)$ and $(5, 6)$ are on the graph; $f'(x) > 0$ for $x \geq 0$; $f''(x) < 0$ for $x < 5$, $f''(5) = 0$, $f''(x) > 0$ for $x > 5$.

Figure 16

Figure 17

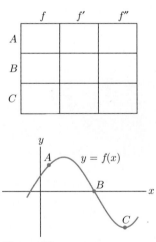

Figure 18

*In Exercises 13–18, use the given information to make a good sketch of the function $f(x)$ near $x = 3$.*

**13.** $f(3) = 4$, $f'(3) = -\frac{1}{2}$, $f''(3) = 5$

**14.** $f(3) = -2$, $f'(3) = 0$, $f''(3) = 1$

**15.** $f(3) = 1$, $f'(3) = 0$, inflection point at $x = 3$, $f'(x) > 0$ for $x > 3$

**16.** $f(3) = 4$, $f'(3) = -\frac{3}{2}$, $f''(3) = -2$

**17.** $f(3) = -2$, $f'(3) = 2$, $f''(3) = 3$

**18.** $f(3) = 3$, $f'(3) = 1$, inflection point at $x = 3$, $f''(x) < 0$ for $x > 3$

**19.** Refer to the graph in Fig. 18. Fill in each entry of the grid with POS, NEG, or 0.

**20.** The first and second derivatives of the function $f(x)$ have the values given in Table 1.

   **(a)** Find the $x$-coordinates of all relative extreme points.

   **(b)** Find the $x$-coordinates of all inflection points.

**21.** Suppose that Fig. 19 contains the graph of $y = s(t)$, the distance traveled by a car after $t$ hours. Is the car going faster at $t = 1$ or $t = 2$?

**22.** Suppose that Fig. 19 contains the graph of $y = v(t)$, the velocity of a car after $t$ hours. Is the car going faster at $t = 1$ or $t = 2$?

| TABLE 1 | Values of the First Two Derivatives of a Function | |
| --- | --- | --- |
| $x$ | $f'(x)$ | $f''(x)$ |
| $0 \le x < 2$ | Positive | Negative |
| 2 | 0 | Negative |
| $2 < x < 3$ | Negative | Negative |
| 3 | Negative | 0 |
| $3 < x < 4$ | Negative | Positive |
| 4 | 0 | 0 |
| $4 < x \le 6$ | Negative | Negative |

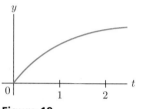

**Figure 19**

**23.** Refer to Fig. 20.

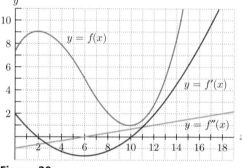

**Figure 20**

**(a)** Looking at the graph of $f'(x)$, determine whether $f(x)$ is increasing or decreasing at $x = 9$. Look at the graph of $f(x)$ to confirm your answer.

**(b)** Looking at the values of $f'(x)$ for $1 \le x < 2$ and $2 < x \le 3$, explain why the graph of $f(x)$ must have a relative maximum at $x = 2$. What are the coordinates of the relative maximum point?

**(c)** Looking at the values of $f'(x)$ for $x$ close to 10, explain why the graph of $f(x)$ has a relative minimum at $x = 10$.

**(d)** Looking at the graph of $f''(x)$, determine whether $f(x)$ is concave up or concave down at $x = 2$. Look at the graph of $f(x)$ to confirm your answer.

**(e)** Looking at the graph of $f''(x)$, determine where $f(x)$ has an inflection point. Look at the graph of $f(x)$ to confirm your answer. What are the coordinates of the inflection point?

**(f)** Find the $x$-coordinate of the point on the graph of $f(x)$ at which $f(x)$ is increasing at the rate of 6 units per unit change in $x$.

**24.** In Fig. 21, the $t$-axis represents time in minutes.

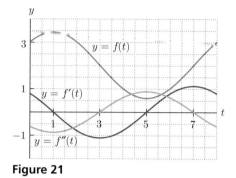

**Figure 21**

**(a)** What is $f(2)$? **(b)** Solve $f(t) = 1$.

**(c)** When does $f(t)$ attain its greatest value?

**(d)** When does $f(t)$ attain its least value?

**(e)** What is the rate of change of $f(t)$ at $t = 7.5$?

**(f)** When is $f(t)$ decreasing at the rate of 1 unit per minute? That is, when is the rate of change equal to $-1$?

**(g)** When is $f(t)$ decreasing at the greatest rate?

**(h)** When is $f(t)$ increasing at the greatest rate?

*Exercises 25–36 refer to Fig. 22, which contains the graph of $f'(x)$, the derivative of the function $f(x)$.*

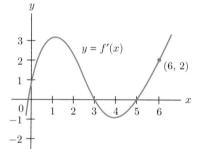

**Figure 22**

**25.** Explain why $f(x)$ must be increasing at $x = 6$.

**26.** Explain why $f(x)$ must be decreasing at $x = 4$.

**27.** Explain why $f(x)$ has a relative maximum at $x = 3$.

**28.** Explain why $f(x)$ has a relative minimum at $x = 5$.

**29.** Explain why $f(x)$ must be concave up at $x = 0$.

**30.** Explain why $f(x)$ must be concave down at $x = 2$.

**31.** Explain why $f(x)$ has an inflection point at $x = 1$.

**32.** Explain why $f(x)$ has an inflection point at $x = 4$.

**33.** If $f(6) = 3$, what is the equation of the tangent line to the graph of $y = f(x)$ at $x = 6$?

**34.** If $f(6) = 8$, what is an approximate value of $f(6.5)$?

**35.** If $f(0) = 3$, what is an approximate value of $f(.25)$?

**36.** If $f(0) = 3$, what is the equation of the tangent line to the graph of $y = f(x)$ at $x = 0$?

**37.** Suppose melting snow causes a river to overflow its banks and $h(t)$ is the number of inches of water on Main Street $t$ hours after the melting begins.

(a) If $h'(100) = \frac{1}{3}$, by approximately how much will the water level change during the next half-hour?

(b) Which of the following two conditions are the better news?

(i) $h(100) = 3$, $h'(100) = 2$, $h''(100) = -5$
(ii) $h(100) = 3$, $h'(100) = -2$, $h''(100) = 5$

**38.** Suppose $T(t)$ is the temperature on a hot summer day at time $t$ hours.

(a) If $T'(10) = 4$, by approximately how much will the temperature rise from 10:00 to 10:45?

(b) Which of the following two conditions are the better news?

(i) $T(10) = 95$, $T'(10) = 4$, $T''(10) = -3$
(ii) $T(10) = 95$, $T'(10) = -4$, $T''(10) = 3$

**39.** By looking at the first derivative, decide which of the curves in Fig. 23 could *not* be the graph of $f(x) = (3x^2 + 1)^4$ for $x \geq 0$.

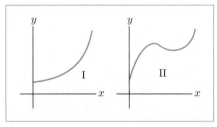

**Figure 23**

**40.** By looking at the first derivative, decide which of the curves in Fig. 23 could *not* be the graph of $f(x) = x^3 - 9x^2 + 24x + 1$ for $x \geq 0$. [*Hint:* Factor the formula for $f'(x)$.]

**41.** By looking at the second derivative, decide which of the curves in Fig. 24 could be the graph of $f(x) = x^{5/2}$.

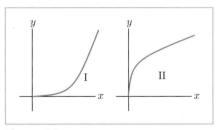

**Figure 24**

**42.** Match each observation (a)–(e) with a conclusion (A)–(E).

*Observations:*

(a) The point $(3, 4)$ is on the graph of $f'(x)$.

(b) The point $(3, 4)$ is on the graph of $f(x)$.

(c) The point $(3, 4)$ is on the graph of $f''(x)$.

(d) The point $(3, 0)$ is on the graph of $f'(x)$ and the point $(3, 4)$ is on the graph of $f''(x)$.

(e) The point $(3, 0)$ is on the graph of $f'(x)$ and the point $(3, -4)$ is on the graph of $f''(x)$.

*Conclusions:*

(A) $f(x)$ has a relative minimum point at $x = 3$.

(B) When $x = 3$, the graph of $f(x)$ is concave up.

(C) When $x = 3$, the tangent line to the graph of $y = f(x)$ has slope 4.

(D) When $x = 3$, the value of $f(x)$ is 4.

(E) $f(x)$ has a relative maximum point at $x = 3$.

**43.** The number of farms in the United States $t$ years after 1925 is $f(t)$ million, where $f$ is the function graphed in Fig. 25(a). [The graphs of $f'(t)$ and $f''(t)$ are shown in Fig. 25(b).]

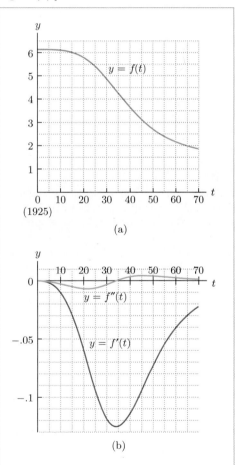

**Figure 25**

(a) Approximately how many farms were there in 1990?

(b) At what rate was the number of farms declining in 1990?

**(c)** In what year were there about 6 million farms?

**(d)** When was the number of farms declining at the rate of 60,000 farms per year?

**(e)** When was the number of farms declining fastest?

**44.** After a drug is taken orally, the amount of the drug in the bloodstream after $t$ hours is $f(t)$ units. Figure 26 shows partial graphs of $f'(t)$ and $f''(t)$.

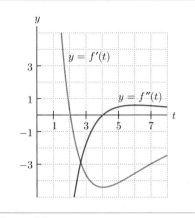

**Figure 26**

**(a)** Is the amount of the drug in the bloodstream increasing or decreasing at $t = 5$ hours?

**(b)** Is the graph of $f(t)$ concave up or concave down at $t = 5$ hours?

**(c)** When is the level of the drug in the bloodstream decreasing the fastest?

**(d)** At what time is the greatest level of drug in the bloodstream reached?

**(e)** When is the level of the drug in the bloodstream decreasing at the rate of 3 units per hour?

### Technology Exercises

*In Exercises 45 and 46, display the graph of the derivative of $f(x)$ in the specified window. Then use the graph of $f'(x)$ to determine the approximate values of $x$ at which the graph of $f(x)$ has relative extreme points and inflection points. Then check your conclusions by displaying the graph of $f(x)$.*

**45.** $3x^5 - 20x^3 - 120x$; $[-4, 4]$ *by* $[-325, 325]$

**46.** $x^4 - x^2$; $[-1.5, 1.5]$ *by* $[-.75, 1]$

## SOLUTIONS TO PRACTICE PROBLEMS 2.2

**1.** Since $f(2) = 5$, the point $(2, 5)$ is on the graph [Fig. 27(a)]. Since $f'(2) = 1$, the tangent line at the point $(2, 5)$ has slope 1. Draw in the tangent line [Fig. 27(b)]. Near the point $(2, 5)$ the graph looks approximately like the tangent line. Since $f''(2) = -3$, a negative number, the graph is concave down at the point $(2, 5)$. Now we are ready to sketch the graph [Fig. 27(c)].

**2. (a)** Yes. The graph is steadily increasing as we pass through the point $(0, 0)$.

**(b)** Since $f'(x) = 3x^2$, $f'(0) = 3 \cdot 0^2 = 0$.

**(c)** There is no contradiction here. The first deriva-

tive rule says that if the derivative is positive, the function is increasing. However, it does not say that this is the only condition under which a function is increasing. As we have just seen, sometimes we can have the first derivative zero and the function still increasing.

**3.** Since $f'(x)$, the derivative of $f(x)$, is negative to the left of $x = 3$ and positive to the right of $x = 3$, $f(x)$ is decreasing to the left of $x = 3$ and increasing to the right of $x = 3$. Therefore, by the definition of a relative minimum point, $f(x)$ has a relative minimum point at $x = 3$.

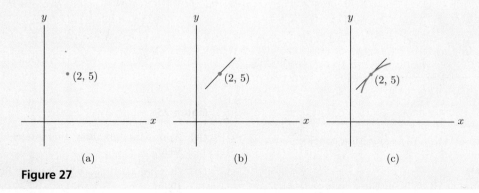

**Figure 27**

## 2.3   The First and Second Derivative Tests and Curve Sketching

In this section and the next we develop our ability to sketch the graphs of functions. There are two important reasons for doing so. First, a geometric "picture" of a function is often easier to comprehend than its abstract formula. Second, the material in this section will provide a foundation for the applications in Sections 2.5 through 2.7.

A "sketch" of the graph of a function $f(x)$ should convey the general shape of the graph—it should show where $f(x)$ is defined and where it is increasing and decreasing and it should indicate, insofar as possible, where $f(x)$ is concave up and concave down. In addition, one or more key points should be accurately located on the graph. These points usually include extreme points, inflection points, and $x$- and $y$-intercepts. Other features of a graph may be important, too, but we shall discuss them as they arise in examples and applications.

Our general approach to curve sketching will involve four main steps:

**1.** Starting with $f(x)$, we compute $f'(x)$ and $f''(x)$.
**2.** Next, we locate all relative maximum and relative minimum points and make a partial sketch.
**3.** We study the concavity of $f(x)$ and locate all inflection points.
**4.** We consider other properties of the graph, such as the intercepts, and complete the sketch.

The first step was the main subject of the preceding chapter. We discuss the second and third steps in this section and then present several completely worked examples that combine all four steps in the next section.

**Locating Relative Extreme Points**   The tangent line at a relative maximum or a relative minimum point of a function $f(x)$ has zero slope; that is, the derivative is zero there. Thus we may state the following useful rule.

> Look for possible relative extreme points of $f(x)$ by setting $f'(x) = 0$ and solving for $x$. $\qquad(1)$

A number $a$ in the domain of $f$ such that $f'(a) = 0$ is called a **critical number** or **critical value** of $f$. (We also call a value $a$ in the domain of $f$ a critical value if $f'(a)$ does not exist.) If $a$ is a critical value, then the point $(a, f(a))$ is called a **critical point**. Our rule states that if $f$ has a relative extreme value at $x = a$, then $a$ must be a critical value of $f$. But not every critical value of $f$ yields a relative extreme point. The condition $f'(a) = 0$ tells us only that the tangent line is horizontal at $a$. In Fig. 1, we show four cases that can occur when $f'(a) = 0$. We can see from the figure that we have an extreme point when the first derivative, $f'(x)$, changes sign at $x = a$. But if the sign of the first derivative does not change, then we do not have an extreme point. Based on these observations and the first derivative rule of the previous section, we obtain the following useful test for extreme points.

> ### The First Derivative Test (for local extreme points)
>
> Suppose that $f'(a) = 0$.
> (a) If $f'$ changes from positive to negative at $x = a$, then $f$ has a local maximum at $a$ (Fig. 1(a)).

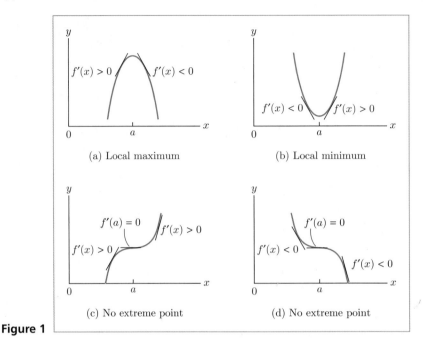

Figure 1

(a) Local maximum

(b) Local minimum

(c) No extreme point

(d) No extreme point

### The First Derivative Test (for local extreme points) (Continued)

(b) If $f'$ changes from negative to positive at $x = a$, then $f$ has a local minimum at $a$ (Fig. 1(b)).

(c) If $f'$ does not change sign at $a$ (that is, $f'$ is either positive on both sides of $a$ (Fig. 1(c)), or negative on both sides of $a$ (Fig. 1(d)), then $f$ has no local extremum at $a$.

---

**EXAMPLE 1** **Applying the first derivative test** Find the local maximum and minimum points of $f(x) = \frac{1}{3}x^3 - 2x^2 + 3x + 1$.

**Solution** First we find the critical values and critical points of $f$:

$$f'(x) = \frac{1}{3}(3)x^2 - 2(2)x + 3 = x^2 - 4x + 3$$

$$= (x - 1)(x - 3).$$

The first derivative $f'(x) = 0$ if $x - 1 = 0$ or $x - 3 = 0$. Thus the critical values are

$$x = 1 \quad \text{and} \quad x = 3.$$

Substituting the critical values into the expression of $f$:

$$f(1) = \frac{1}{3}(1)^3 - 2(1)^2 + 3(1) + 1 = \frac{1}{3} + 2 = \frac{7}{3};$$

$$f(3) = \frac{1}{3}(3)^3 - 2(3)^2 + 3(3) + 1 = 1.$$

Thus the critical points are $(1, \frac{7}{3})$ and $(3, 1)$. To determine whether we have a relative maximum, minimum, or neither at a critical point, we shall apply the first derivative test. This requires a careful study of the sign of $f'(x)$, which can be facilitated with the help of a chart. Here is how you can set up the chart.

- Divide the real line into intervals with the critical values as endpoints.
- Since the sign of $f'$ depends on the signs of its two factors $x - 1$ and $x - 3$, determine the signs of the factors of $f'$ over each interval. Usually this is done by testing the sign of a factor at points selected from each interval.
- In each interval, use a plus sign if the factor is positive and a minus sign if it is negative. Then determine the sign of $f'$ over each interval by multiplying the signs of the factors and using

$$(+) \cdot (+) = +; \quad (+) \cdot (-) = -; \quad (-) \cdot (+) = -; \quad (-) \cdot (-) = +.$$

- A plus sign of $f'$ corresponds to an increasing portion of the graph of $f$ and a minus sign to a decreasing portion. Denote an increasing portion with an upward arrow and a decreasing portion with a downward arrow. The sequence of arrows should convey the general shape of the graph and, in particular, tell you whether or not your critical values correspond to extreme points. Here is the chart:

| Critical Values | | 1 | | 3 | |
|---|---|---|---|---|---|
| | $x < 1$ | | $1 < x < 3$ | | $3 < x$ |
| $x - 1$ | $-$ | 0 | $+$ | | $+$ |
| $x - 3$ | $-$ | | $-$ | 0 | $+$ |
| $f'(x)$ | $+$ | 0 | $-$ | 0 | $+$ |
| $f(x)$ | Increasing on $(-\infty, 1)$ | 7/3 | Decreasing on $(1, 3)$ | 1 | Increasing on $(3, \infty)$ |
| | | Local maximum $(1, 7/3)$ | | Local minimum $(3, 1)$ | |

You can see from the chart that the sign of $f'$ varies from positive to negative at $x = 1$. Thus according to the first derivative test, $f$ has a local maximum at $x = 1$. Also, the sign of $f'$ varies from negative to positive at $x = 3$; and so $f$ has a local minimum at $x = 3$. You can confirm these assertions by following the direction of the arrows in the last row of the chart. In conclusion, $f$ has a local maximum at $(1, \frac{7}{3})$ and a local minimum at $(3, 1)$ (Fig. 2). ■

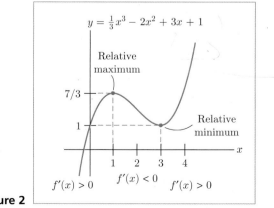

$$y = \tfrac{1}{3}x^3 - 2x^2 + 3x + 1$$

Relative maximum

7/3

1

Relative minimum

$x$

1  2  3  4

$f'(x) > 0$   $f'(x) < 0$   $f'(x) > 0$

**Figure 2**

The previous example illustrates cases (a) and (b) in Fig. 1, which correspond to local extreme points on the graph at each critical value. The following example illustrates cases (c) and (d) in Fig. 1.

**EXAMPLE 2**    **Applying the first derivative test**   Find the local maximum and minimum points of $f(x) = (3x - 1)^3$.

**Solution**   We begin by finding the critical values and critical points. Using the general power rule, we have

$$f'(x) = 3(3x - 1)^2(3) = 9(3x - 1)^2$$
$$f'(x) = 0 \Rightarrow 3x - 1 = 0$$
$$\Rightarrow x = \tfrac{1}{3}.$$

Plugging this value into the original expression for $f$, we find

$$f\left(\tfrac{1}{3}\right) = \left(3 \cdot \tfrac{1}{3} - 1\right)^3 = 0.$$

Thus $(\tfrac{1}{3}, 0)$ is the only critical point. To determine whether it is a local maximum, minimum, or neither, we use the first derivative test. Look at $f'(x) = 9(3x - 1)^2$. This expression is always nonnegative, because a square is always $\geq 0$ (there is no need for a sign chart in this case). Consequently, $f'$ is positive on both sides of $x = \tfrac{1}{3}$ and so, since $f'$ does not change sign at $x = \tfrac{1}{3}$, we conclude from part (c) of the first derivative test that there is no relative extreme point. Indeed, since $f'(x) \geq 0$ for all $x$, the graph is always increasing (see Fig. 3).   ■

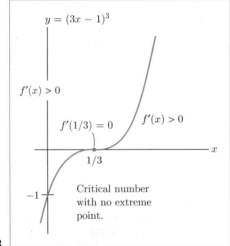

**Figure 3**

**Using Concavity to Determine Extreme Points**   Studying the variation of the graph of $f$ and determining the extreme points can be tedious tasks. Any shortcut that helps us avoid excessive computations is always appreciated. Let us look back at Fig. 1 and give an alternative description of the extreme points using concavity. At $x = a$ the concavity is downward in Fig. 1(a), upward in Fig. 1(b), and neither in Fig. 1(c) and (d). These observations together with the second derivative rule from the previous section are the basis for the following important test.

> **The Second Derivative Test (for local extreme points)**
>
> (a) If $f'(a) = 0$ and $f''(a) < 0$, then $f$ has a local maximum at $a$ (Fig. 1(a)).
>
> (b) If $f'(a) = 0$ and $f''(a) > 0$, then $f$ has a local minimum at $a$ (Fig. 1(b)).

**EXAMPLE 3**  The graph of the quadratic function $f(x) = \frac{1}{4}x^2 - x + 2$ is a parabola and so has one relative extreme point. Find it and sketch the graph.

**Solution**  We begin by computing the first and second derivatives of $f(x)$.

$$f(x) = \tfrac{1}{4}x^2 - x + 2$$
$$f'(x) = \tfrac{1}{2}x - 1$$
$$f''(x) = \tfrac{1}{2}$$

Setting $f'(x) = 0$, we have $\frac{1}{2}x - 1 = 0$, so that $x = 2$ is the only critical value. Thus, $f'(2) = 0$. Geometrically, this means that the graph of $f(x)$ will have a horizontal tangent line at the point where $x = 2$. To plot this point, we substitute the value 2 for $x$ in the original expression for $f(x)$.

$$f(2) = \tfrac{1}{4}(2)^2 - (2) + 2 = 1$$

Figure 4 shows the point $(2, 1)$ together with the horizontal tangent line. Is $(2, 1)$ a relative extreme point? In order to decide, we look at $f''(x)$. Since $f''(x) = \frac{1}{2}$, which is positive, the graph of $f(x)$ is concave up at $x = 2$, and so $(2, 1)$ is a local minimum by the second derivative test. A partial sketch of the graph near $(2, 1)$ should look something like Fig. 5.

**Figure 4**

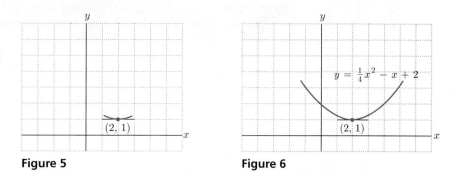

**Figure 5**          **Figure 6**

We see that $(2, 1)$ is a relative minimum point. In fact, it is the only relative extreme point, for there is no other place where the tangent line is horizontal. Since the graph has no other "turning points," it must be decreasing before it gets to $(2, 1)$ and then increasing to the right of $(2, 1)$. Note that since $f''(x)$ is positive (and equal to $\frac{1}{2}$) for all $x$, the graph is concave up at each point. A completed sketch is given in Fig. 6. ∎

**EXAMPLE 4**  **Applying the second derivative test**  Locate all possible relative extreme points on the graph of the function $f(x) = x^3 - 3x^2 + 5$. Check the concavity at these points and use this information to sketch the graph of $f(x)$.

**Solution** We have

$$f(x) = x^3 - 3x^2 + 5$$
$$f'(x) = 3x^2 - 6x$$
$$f''(x) = 6x - 6.$$

The easiest way to find the critical values is to factor the expression for $f'(x)$:

$$3x^2 - 6x = 3x(x - 2).$$

From this factorization it is clear that $f'(x)$ will be zero if and only if $x = 0$ or $x = 2$. In other words, the graph will have horizontal tangent lines when $x = 0$ and $x = 2$, and nowhere else.

To plot the points on the graph where $x = 0$ and $x = 2$, we substitute these values back into the original expression for $f(x)$. That is, we compute

$$f(0) = (0)^3 - 3(0)^2 + 5 = 5$$
$$f(2) = (2)^3 - 3(2)^2 + 5 = 1.$$

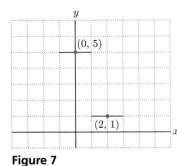

**Figure 7**

Figure 7 shows the points $(0, 5)$ and $(2, 1)$, along with the corresponding tangent lines.

Next, we check the sign of $f''(x)$ at $x = 0$ and $x = 2$ and apply the second derivative test:

$$f''(0) = 6(0) - 6 = -6 < 0 \quad \text{(local maximum)}$$
$$f''(2) = 6(2) - 6 = 6 > 0 \quad \text{(local minimum)}.$$

Since $f''(0)$ is negative, the graph is concave down at $x = 0$; since $f''(2)$ is positive, the graph is concave up at $x = 2$. A partial sketch of the graph is given in Fig. 8.

It is clear from Fig. 8 that $(0, 5)$ is a relative maximum point and $(2, 1)$ is a relative minimum point. Since they are the only turning points, the graph must be increasing before it gets to $(0, 5)$, decreasing from $(0, 5)$ to $(2, 1)$, and then increasing again to the right of $(2, 1)$. A sketch incorporating these properties appears in Fig. 9. ∎

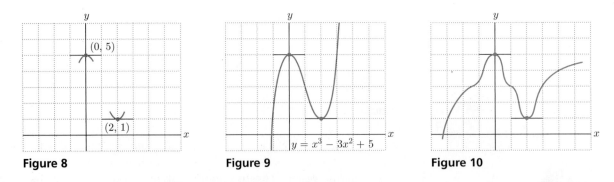

**Figure 8**          **Figure 9**          **Figure 10**

The facts that we used to sketch Fig. 9 could equally well be used to produce the graph in Fig. 10. Which graph really corresponds to $f(x) = x^3 - 3x^2 + 5$? The answer will be clear when we find the inflection points on the graph of $f(x)$.

**Locating Inflection Points** An inflection point of a function $f(x)$ can occur only at a value of $x$ for which $f''(x)$ is zero, because the curve is concave up where $f''(x)$ is positive and concave down where $f''(x)$ is negative. Thus we have the following rule.

> Look for possible inflection points by setting
> $$f''(x) = 0 \text{ and solving for } x.$$
> (2)

Once we have a value of $x$ where the second derivative is zero, say at $x = b$, we must check the concavity of $f(x)$ at nearby points to see if the concavity really changes at $x = b$.

**EXAMPLE 5**   Find the inflection points of the function $f(x) = x^3 - 3x^2 + 5$ and explain why the graph in Fig. 9 has the correct shape.

**Solution**   From Example 4 we have $f''(x) = 6x - 6 = 6(x - 1)$. Clearly, $f''(x) = 0$ if and only if $x = 1$. We will want to plot the corresponding point on the graph, so we compute

$$f(1) = (1)^3 - 3(1)^2 + 5 = 3.$$

**Figure 11**

Therefore, the only possible inflection point is $(1, 3)$.

Now look back at Fig. 8, where we indicated the concavity of the graph at the relative extreme points. Since $f(x)$ is concave down at $(0, 5)$ and concave up at $(2, 1)$, the concavity must reverse somewhere between these points. Hence $(1, 3)$ must be an inflection point. Furthermore, since the concavity of $f(x)$ reverses nowhere else, the concavity at all points to the left of $(1, 3)$ must be the same (i.e., concave down). Similarly, the concavity of all points to the right of $(1, 3)$ must be the same (i.e., concave up). Thus the graph in Fig. 9 has the correct shape. The graph in Fig. 10 has too many "wiggles," caused by frequent changes in concavity; that is, there are too many inflection points. A correct sketch showing the one inflection point at $(1, 3)$ is given in Fig. 11.  ■

**EXAMPLE 6**   Sketch the graph of $y = -\frac{1}{3}x^3 + 3x^2 - 5x$.

**Solution**   Let

$$f(x) = -\tfrac{1}{3}x^3 + 3x^2 - 5x.$$

Then

$$f'(x) = -x^2 + 6x - 5$$
$$f''(x) = -2x + 6.$$

***Step 1: Finding the critical points***   We set $f'(x) = 0$ and solve for $x$:

$$-(x^2 - 6x + 5) = 0$$
$$-(x - 1)(x - 5) = 0$$
$$x = 1 \quad \text{or} \quad x = 5 \quad \text{(critical values)}.$$

Substituting these values of $x$ back into $f(x)$, we find that

$$f(1) = -\tfrac{1}{3}(1)^3 + 3(1)^2 - 5(1) = -\tfrac{7}{3}$$
$$f(5) = -\tfrac{1}{3}(5)^3 + 3(5)^2 - 5(5) = \tfrac{25}{3}.$$

***Step 2: Determining the extreme points***   The information we have so far is given in Fig. 12(a). The sketch in Fig. 12(b) is obtained by computing

$$f''(1) = -2(1) + 6 = 4 > 0 \quad \text{(local minimum)}$$
$$f''(5) = -2(5) + 6 = -4 < 0 \quad \text{(local maximum)}.$$

The curve is concave up at $x = 1$ because $f''(1)$ is positive (local minimum), and the curve is concave down at $x = 5$ because $f''(5)$ is negative (local maximum).

**Step 3: Concavity and inflection points** Since the concavity reverses somewhere between $x = 0$ and $x = 5$, there must be at least one inflection point. If we set $f''(x) = 0$, we find that

$$-2x + 6 = 0$$
$$x = 3.$$

So the inflection point must occur at $x = 3$. In order to plot the inflection point, we compute

$$f(3) = -\tfrac{1}{3}(3)^3 + 3(3)^2 - 5(3) = 3.$$

The final sketch of the graph is given in Fig. 13. ∎

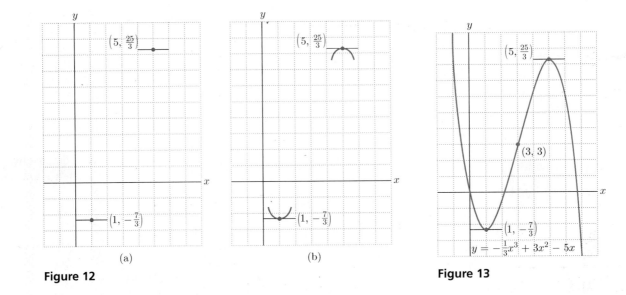

(a)      (b)

**Figure 12**

**Figure 13**

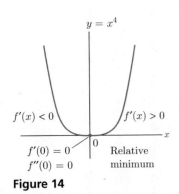

**Figure 14**

The argument in Example 6 that there must be an inflection point because concavity reverses is valid whenever $f(x)$ is a polynomial. However, it does not always apply to a function whose graph has a break in it. For example, the function $f(x) = 1/x$ is concave down at $x = -1$ and concave up at $x = 1$, but there is no inflection point in between.

In Examples 1 and 2 we used the first derivative test in order to locate the extreme points; but in the remaining examples we used the second derivative test. In general, which test should we use? If $f''(x)$ is straightforward to compute (e.g., when $f(x)$ is a polynomial), you should try the second derivative test first. If the second derivative is tedious to compute or if $f''(a) = 0$, use the first derivative test. Remember that when $f''(a) = 0$ the second derivative test is inconclusive. Keep the following examples in mind. The function $f(x) = x^3$ is always increasing and has no local maximum or minimum; even though $f'(0) = 0$ and $f''(0) = 0$. The function $f(x) = x^4$ is shown in Fig. 14. We have $f'(x) = 4x^3$ and $f''(x) = 12x^2$. So $f'(0) = 0$ and $f''(0) = 0$, but, as you can see from Fig. 14, you have a local minimum at $x = 0$. The reason for this follows from the first derivative test, since $f'$ changes sign at $x = 0$ and goes from negative to positive.

A summary of curve-sketching techniques appears at the end of the next section. You may find steps 1, 2, and 3 helpful when working the exercises.

## PRACTICE PROBLEMS 2.3

1. Which of the curves in Fig. 15 could possibly be the graph of a function of the form $f(x) = ax^2 + bx + c$, where $a \neq 0$?

2. Which of the curves in Fig. 16 could possibly be the graph of a function of the form $f(x) = ax^3 + bx^2 + cx + d$, where $a \neq 0$?

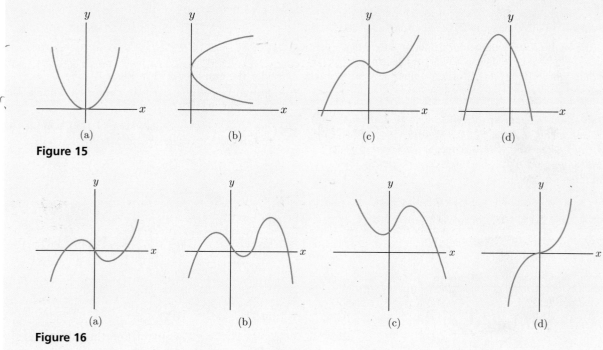

(a)   (b)

**Figure 15**

(a)   (b)

**Figure 16**

# EXERCISES 2.3

*Each of the graphs of the functions in Exercises 1–8 has one relative maximum and one relative minimum point. Find these points using the first derivative test. Use a variation chart as in Example 1.*

1. $f(x) = x^3 - 27x$

2. $f(x) = x^3 - 6x^2 + 1$

3. $f(x) = -x^3 + 6x^2 - 9x + 1$

4. $f(x) = -6x^3 - \frac{3}{2}x^2 + 3x - 3$

5. $f(x) = \frac{1}{3}x^3 - x^2 + 1$

6. $f(x) = \frac{4}{3}x^3 - x + 2$

7. $f(x) = -x^3 - 12x^2 - 2$

8. $f(x) = 2x^3 + 3x^2 - 3$

*Each of the graphs of the functions in Exercises 9–16 has one relative extreme point. Plot this point and check the concavity there. Using only this information, sketch the graph. [As you work the problems, observe that if $f(x) = ax^2 + bx + c$, then $f(x)$ has a relative minimum point when $a > 0$ and a relative maximum point when $a < 0$.]*

9. $f(x) = 2x^2 - 8$

10. $f(x) = x^2$

11. $f(x) = \frac{1}{2}x^2 + x - 4$

12. $f(x) = -3x^2 + 12x + 2$

13. $f(x) = 1 + 6x - x^2$

14. $f(x) = \frac{1}{2}x^2 + \frac{1}{2}$

15. $f(x) = -x^2 - 8x - 10$

16. $f(x) = -x^2 + 2x - 5$

*Each of the graphs of the functions in Exercises 17–24 has one relative maximum and one relative minimum point. Plot these two points and check the concavity there. Using only this information, sketch the graph.*

17. $f(x) = x^3 + 6x^2 + 9x$

18. $f(x) = \frac{1}{9}x^3 - x^2$

19. $f(x) = x^3 - 12x$

20. $f(x) = -\frac{1}{3}x^3 + 9x - 2$

21. $f(x) = -\frac{1}{9}x^3 + x^2 + 9x$

22. $f(x) = 2x^3 - 15x^2 + 36x - 24$

23. $f(x) = -\frac{1}{3}x^3 + 2x^2 - 12$

24. $f(x) = \frac{1}{3}x^3 + 2x^2 - 5x + \frac{8}{3}$

*Sketch the following curves, indicating all relative extreme points and inflection points.*

25. $y = x^3 - 3x + 2$

26. $y = x^3 - 6x^2 + 9x + 3$

27. $y = 1 + 3x^2 - x^3$

28. $y = -x^3 + 12x - 4$

29. $y = \frac{1}{3}x^3 - x^2 - 3x + 5$

30. $y = x^4 + \frac{1}{3}x^3 - 2x^2 - x + 1$ [*Hint:* $4x^3 + x^2 - 4x + 1 = (x^2 - 1)(4x - 1)$]

31. $y = 2x^3 - 3x^2 - 36x + 20$

32. $y = x^4 - \frac{4}{3}x^3$

33. Let $a$, $b$, $c$ be fixed numbers with $a \neq 0$ and let $f(x) = ax^2 + bx + c$. Is it possible for the graph of $f(x)$ to have an inflection point? Explain your answer.

34. Let $a$, $b$, $c$, $d$ be fixed numbers with $a \neq 0$ and let $f(x) = ax^3 + bx^2 + cx + d$. Is it possible for the graph of $f(x)$ to have more than one inflection point? Explain your answer.

*The graph of each function in Exercises 35–40 has one relative extreme point. Find it (giving both x- and y-coordinates) and determine if it is a relative maximum or a relative minimum point. Do not include a sketch of the graph of the function.*

35. $f(x) = \frac{1}{4}x^2 - 2x + 7$

36. $f(x) = 5 - 12x - 2x^2$

37. $g(x) = 3 + 4x - 2x^2$

38. $g(x) = x^2 + 10x + 10$

39. $f(x) = 5x^2 + x - 3$

40. $f(x) = 30x^2 - 1800x + 29{,}000$

*In Exercises 41 and 42, determine which function is the derivative of the other.*

41.

42.

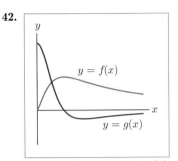

43. Consider the graph of $g(x)$ in Fig. 17.

   (a) If $g(x)$ is the first derivative of $f(x)$, what is the nature of $f(x)$ when $x = 2$?

   (b) If $g(x)$ is the second derivative of $f(x)$, what is the nature of $f(x)$ when $x = 2$?

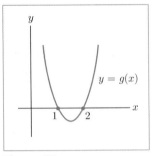

**Figure 17**

44. (*U.S. Population*) The population (in millions) of the United States (excluding Alaska and Hawaii) $t$ years after 1800 is given by the function $f(t)$ in Fig. 18(a). The graphs of $f'(t)$ and $f''(t)$ are shown in Figs. 18(b) and 18(c).

   (a) What was the population in 1925?

   (b) Approximately when was the population 25 million?

   (c) How fast was the population growing in 1950?

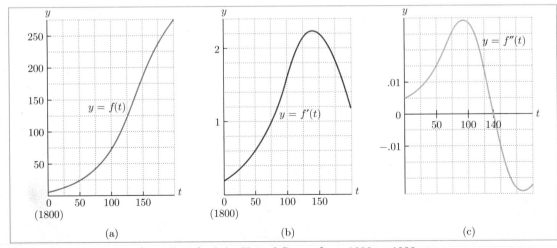

**Figure 18.** Population (in millions) of the United States from 1800 to 1998.

**(d)** When during the last 50 years was the population growing at the rate of 1.8 million people per year?

**(e)** In what year was the population growing at the greatest rate?

**45.** (*Index-Fund Fees*) When a mutual fund company charges a fee of 0.47% on its index funds, its assets in the fund are $41 billion. And when it charges a fee of 0.18%, its assets in the fund are $300 billion.*

**(a)** Let $x$ denote the fee that the company charges as a percentage of the index fund and $A(x)$ its assets in the fund. Express $A(x)$ as a linear function of $x$.

**(b)** In September 2004, Fidelity Mutual lowered its fees on various index funds from an average of 0.3% to 0.10%. Let $R(x)$ denote the revenue of the company from fees when its index-fund fee is $x$%. Compare the revenue of the company before and after lowering the fees. [*Hint:* Revenue is $x$% of assets.]

**(c)** Find the fee that maximizes the revenue of the company and determine the maximum revenue.

**46.** (*Index-Fund Fees*) Suppose that the cost function in Exercise 45 is $C(x) = -2.5x + 1$, where $x$ is the index-fund fee. (The company has a fixed cost of $1 billion dollars, and the cost decreases as a function of the index-fund fee.) Find the value of $x$ that maximizes profit. How well did Fidelity Mutual do before and after lowering the index-fund fees?

**47.** The function $f(t)$ shown in Fig. 19 gives the percentage of homes (with TV sets) that also had a VCR in year $t$, where $t = 0$ corresponds to 1980.

**(a)** In what year did 10% of the homes have a VCR?

**(b)** What percentage of homes had a VCR in 1990?

**(c)** When was the percentage of homes with TVs that also had VCRs growing at the greatest rate? Approximately what percentage of homes with TVs also had VCRs at that time?

## Technology Exercises

**48.** Draw the graph of $f(x) = \frac{1}{6}x^3 - x^2 + 3x + 3$ in the window $[-2, 6]$ *by* $[-10, 20]$. It has an inflection point at $x = 2$, but no relative extreme points. Enlarge the window a few times to convince yourself that there are

no relative extreme points anywhere. What does this tell you about $f'(x)$?

**49.** Draw the graph of $f(x) = \frac{1}{6}x^3 - \frac{5}{2}x^2 + 13x - 20$ in the window $[0, 10]$ *by* $[-20, 30]$. Algebraically determine the coordinates of the inflection point. Zoom in and zoom out to convince yourself that there are no relative extreme points anywhere.

**50.** Draw the graph of

$$f(x) = 2x + \frac{18}{x} - 10$$

in the window $[0, 16]$ *by* $[0, 16]$. In what ways is this graph like the graph of a parabola that opens upward? In what ways is it different?

**51.** Draw the graph of

$$f(x) = 3x + \frac{75}{x} - 25$$

in the window $[0, 25]$ *by* $[0, 50]$. Use the trace feature of the calculator or computer to estimate the coordinates of the relative minimum point. Then determine the coordinates algebraically. Convince yourself both graphically (with the calculator or computer) and algebraically that this function has no inflection points.

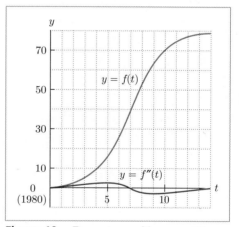

**Figure 19.** Percentage of homes with TVs that also had VCRs, from 1980 to 1994.

---

*Based on the article "Fidelity slashes index-fund fees," *Boston Globe*, September 1, 2004.

# 2.4 Curve Sketching (Conclusion)

In Section 2.3 we discussed the main techniques for curve sketching. Here we add a few finishing touches and examine some slightly more complicated curves.

The more points we plot on a graph, the more accurate the graph becomes. This statement is true even for the simple quadratic and cubic curves in Section 2.3. Of course, the most important points on a curve are the relative extreme points and the inflection points. In addition, the $x$- and $y$-intercepts often have some intrinsic interest in an applied problem. The $y$-intercept is $(0, f(0))$. To find the $x$-intercepts on the graph of $f(x)$, we must find those values of $x$ for which $f(x) = 0$. Since this can be a difficult (or impossible) problem, we shall find $x$-intercepts only when they are easy to find or when a problem specifically requires us to find them.

When $f(x)$ is a quadratic function, as in Example 1, we can easily compute the $x$-intercepts (if they exist) either by factoring the expression for $f(x)$ or by using the quadratic formula, which we now recall from Section 0.4:

---

**The Quadratic Formula**

$$x = \frac{-b \pm \sqrt{b^2 - 4ac}}{2a}$$

---

The sign $\pm$ tells us to form two expressions, one with $+$ and one with $-$. The equation has two distinct roots, if $b^2 - 4ac > 0$; one double root, if $b^2 - 4ac = 0$; and no (real) roots if $b^2 - 4ac < 0$.

---

**EXAMPLE 1** | **Applying the second derivative test**  Sketch the graph of $y = \frac{1}{2}x^2 - 4x + 7$.

**Solution** Let

$$f(x) = \tfrac{1}{2}x^2 - 4x + 7.$$

Then

$$f'(x) = x - 4$$
$$f''(x) = 1.$$

Since $f'(x) = 0$ only when $x = 4$ and since $f''(4)$ is positive, $f(x)$ must have a relative minimum point at $x = 4$ (the second derivative test). The relative minimum point is $(4, f(4)) = (4, -1)$.

The $y$-intercept is $(0, f(0)) = (0, 7)$. To find the $x$-intercepts, we set $f(x) = 0$ and solve for $x$:

$$\tfrac{1}{2}x^2 - 4x + 7 = 0.$$

The expression for $f(x)$ is not easily factored, so we use the quadratic formula to solve the equation:

$$x = \frac{-(-4) \pm \sqrt{(-4)^2 - 4(\frac{1}{2})(7)}}{2(\frac{1}{2})} = 4 \pm \sqrt{2}.$$

The $x$-intercepts are $(4 - \sqrt{2}, 0)$ and $(4 + \sqrt{2}, 0)$. To plot these points we use the approximation $\sqrt{2} \approx 1.4$. (See Fig. 1.) ∎

The next example is interesting. The function that we consider has no critical points.

**Figure 1**

---

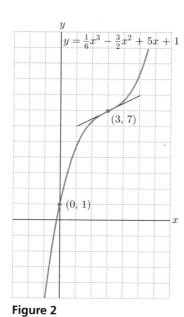

**Figure 2**

**EXAMPLE 2**  **A function with no critical points**  Sketch the graph of $f(x) = \frac{1}{6}x^3 - \frac{3}{2}x^2 + 5x + 1$.

**Solution**

$$f(x) = \tfrac{1}{6}x^3 - \tfrac{3}{2}x^2 + 5x + 1$$

$$f'(x) = \tfrac{1}{2}x^2 - 3x + 5$$

$$f''(x) = x - 3$$

Searching for critical points, we set $f'(x) = 0$ and try to solve for $x$:

$$\tfrac{1}{2}x^2 - 3x + 5 = 0. \tag{1}$$

If we apply the quadratic formula with $a = \frac{1}{2}$, $b = -3$, and $c = 5$, we see that $b^2 - 4ac$ is negative, and so there is no solution to (1). In other words, $f'(x)$ is never zero. Thus the graph cannot have relative extreme points. If we evaluate $f'(x)$ at some $x$, say $x = 0$, we see that the first derivative is positive, and so $f(x)$ is increasing there. Since the graph of $f(x)$ is a smooth curve with no relative extreme points and no breaks, $f(x)$ must be increasing for all $x$. (If a function were increasing at $x = a$ and decreasing at $x = b$, then it would have a relative extreme point between $a$ and $b$.)

Now let us check the concavity.

| | $f''(x) = x - 3$ | Graph of $f(x)$ |
|---|---|---|
| $x < 3$ | Negative | Concave down |
| $x = 3$ | Zero | Concavity reverses |
| $x > 3$ | Positive | Concave up |

Since $f''(x)$ changes sign at $x = 3$, we have an inflection point at $(3, 7)$. The $y$-intercept is $(0, f(0)) = (0, 1)$. We omit the $x$-intercept because it is difficult to solve the cubic equation $\frac{1}{6}x^3 - \frac{3}{2}x^2 + 5x + 1 = 0$.

The quality of our sketch of the curve will be improved if we first sketch the tangent line at the inflection point. To do this, we need to know the slope of the graph at $(3, 7)$:

$$f'(3) = \tfrac{1}{2}(3)^2 - 3(3) + 5 = \tfrac{1}{2}.$$

We draw a line through $(3, 7)$ with slope $\frac{1}{2}$ and then complete the sketch as shown in Fig. 2. ■

**EXAMPLE 3** **Using the first derivative test** Sketch the graph of $f(x) = (x-2)^4 - 1$.

**Solution**

$$f(x) = (x-2)^4 - 1$$
$$f'(x) = 4(x-2)^3$$
$$f''(x) = 12(x-2)^2.$$

Clearly, $f'(x) = 0$ only if $x = 2$. So the curve has a horizontal tangent at $(2, f(2)) = (2, -1)$. Since $f''(2) = 0$, the second derivative test is inconclusive. We shall apply the first derivative test. Note that

$$f'(x) = 4(x-2)^3 \quad \begin{cases} \text{negative} & \text{if } x < 2 \\ \text{positive} & \text{if } x > 2 \end{cases}$$

since the cube of a negative number is negative and the cube of a positive number is positive. Therefore, as $x$ goes from left to right in the vicinity of 2, the first derivative changes sign and goes from negative to positive. By the first derivative test, the point $(2, -1)$ is a relative minimum.

The $y$-intercept is $(0, f(0)) = (0, 15)$. To find the $x$-intercepts, we set $f(x) = 0$ and solve for $x$:

$$(x-2)^4 - 1 = 0$$
$$(x-2)^4 = 1$$
$$x - 2 = 1 \quad \text{or} \quad x - 2 = -1$$
$$x = 3 \quad \text{or} \quad x = 1.$$

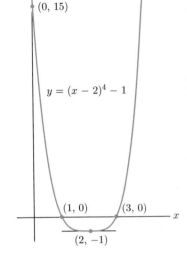

$(0, 15)$

$y = (x-2)^4 - 1$

$(1, 0)$ $(3, 0)$

$(2, -1)$

**Figure 3**

(See Fig. 3.)

**A Graph with Asymptotes** Graphs similar to the one in the next example will arise in several applications later in this chapter.

**EXAMPLE 4** **Asymptotes** Sketch the graph of $f(x) = x + (1/x)$, for $x > 0$.

**Solution**

$$f(x) = x + \frac{1}{x}$$
$$f'(x) = 1 - \frac{1}{x^2}$$
$$f''(x) = \frac{2}{x^3}$$

We set $f'(x) = 0$ and solve for $x$:

$$1 - \frac{1}{x^2} = 0$$
$$1 = \frac{1}{x^2}$$
$$x^2 = 1$$
$$x = 1.$$

(We exclude the case $x = -1$ because we are only considering positive values of $x$.) The graph has a horizontal tangent at $(1, f(1)) = (1, 2)$. Since $f''(1) = 2 > 0$, the graph is concave up at $x = 1$ and, by the second derivative test, $(1, 2)$ is a relative minimum point. In fact, $f''(x) = (2/x^3) > 0$ for all positive $x$, and therefore the graph is concave up at all points.

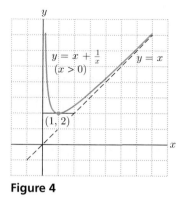

$y = x + \frac{1}{x}$
$(x > 0)$

$y = x$

$(1, 2)$

**Figure 4**

Before sketching the graph, notice that as $x$ approaches zero [a point at which $f(x)$ is not defined], the term $1/x$ in the formula for $f(x)$ is dominant. That is, this term becomes arbitrarily large, whereas the term $x$ contributes a diminishing proportion to the function value as $x$ approaches 0. Thus $f(x)$ has the $y$-axis as an asymptote. For large values of $x$, the term $x$ is dominant. The value of $f(x)$ is only slightly larger than $x$ since the term $1/x$ has decreasing significance as $x$ becomes arbitrarily large; that is, the graph of $f(x)$ is slightly above the graph of $y = x$. As $x$ increases, the graph of $f(x)$ has the line $y = x$ as an asymptote. (See Fig. 4.) ■

## Summary of Curve-Sketching Techniques

1. Compute $f'(x)$ and $f''(x)$.

2. Find all relative extreme points.

   (a) Find the critical values and critical points: Set $f'(x) = 0$ and solve for $x$. Suppose that $x = a$ is a solution (a critical value). Substitute $x = a$ into $f(x)$ to find $f(a)$, plot the critical point $(a, f(a))$, and draw a small horizontal tangent line through the point. Compute $f''(a)$.

   *The second derivative test*

   (i) If $f''(a) > 0$, draw a small concave up arc with $(a, f(a))$ as its lowest point. The curve has a relative minimum at $x = a$.

   (ii) If $f''(a) < 0$, draw a small concave down arc with $(a, f(a))$ as its peak. The curve has a relative maximum at $x = a$.

   *The first derivative test*

   (iii) If $f''(a) = 0$, examine $f'(x)$ to the left and right of $x = a$ in order to determine if the function changes from increasing to decreasing, or vice versa. If a relative extreme point is indicated, draw an appropriate arc as in parts (i) and (ii).

   (b) Repeat the preceding steps for each of the solutions to $f'(x) = 0$.

3. Find all the inflection points of $f(x)$.

   (a) Set $f''(x) = 0$ and solve for $x$. Suppose that $x = b$ is a solution. Compute $f(b)$ and plot the point $(b, f(b))$.

   (b) Test the concavity of $f(x)$ to the right and left of $b$. If the concavity changes at $x = b$, then $(b, f(b))$ is an inflection point.

4. Consider other properties of the function and complete the sketch.

   (a) If $f(x)$ is defined at $x = 0$, the $y$-intercept is $(0, f(0))$.

   (b) Does the partial sketch suggest that there are $x$-intercepts? If so, they are found by setting $f(x) = 0$ and solving for $x$. (Solve only in easy cases or when a problem essentially requires you to calculate the $x$-intercepts.)

   (c) Observe where $f(x)$ is defined. Sometimes the function is given only for restricted values of $x$. Sometimes the formula for $f(x)$ is meaningless for certain values of $x$.

   (d) Look for possible asymptotes.

   (i) Examine the formula for $f(x)$. If some terms become insignificant as $x$ gets large and if the rest of the formula gives the equation of a straight line, then that straight line is an asymptote.

   (ii) Suppose that there is some point $a$ such that $f(x)$ is defined for $x$ near $a$ but not at $a$ (e.g., $1/x$ at $x = 0$). If $f(x)$ gets arbitrarily large (in the positive or negative sense) as $x$ approaches $a$, then the vertical line $x = a$ is an asymptote for the graph.

   (e) Complete the sketch.

## PRACTICE PROBLEMS 2.4

*Determine whether each of the following functions has an asymptote as x gets large. If so, give the equation of the straight line which is the asymptote.*

**1.** $f(x) = \dfrac{3}{x} - 2x + 1$     **2.** $f(x) = \sqrt{x} + x$

**3.** $f(x) = \dfrac{1}{2x}$

## EXERCISES 2.4

*Find the x-intercepts of the given function.*

**1.** $y = x^2 - 3x + 1$     **2.** $y = x^2 + 5x + 5$

**3.** $y = 2x^2 + 5x + 2$     **4.** $y = 4 - 2x - x^2$

**5.** $y = 4x - 4x^2 - 1$     **6.** $y = 3x^2 + 10x + 3$

**7.** Show that the function $f(x) = \frac{1}{3}x^3 - 2x^2 + 5x$ has no relative extreme points.

**8.** Show that the function $f(x) = -x^3 + 2x^2 - 6x + 3$ is always decreasing.

*Sketch the graphs of the following functions.*

**9.** $f(x) = x^3 - 6x^2 + 12x - 6$

**10.** $f(x) = -x^3$     **11.** $f(x) = x^3 + 3x + 1$

**12.** $f(x) = x^3 + 2x^2 + 4x$

**13.** $f(x) = 5 - 13x + 6x^2 - x^3$

**14.** $f(x) = 2x^3 + x - 2$     **15.** $f(x) = \frac{4}{3}x^3 - 2x^2 + x$

**16.** $f(x) = -3x^3 - 6x^2 - 9x - 6$

**17.** $f(x) = 1 - 3x + 3x^2 - x^3$

**18.** $f(x) = x^3 - 6x^2 + 12x - 5$

**19.** $f(x) = x^4 - 6x^2$     **20.** $f(x) = 3x^4 - 6x^2 + 3$

**21.** $f(x) = (x - 3)^4$     **22.** $f(x) = (x + 2)^4 - 1$

*Sketch the graphs of the following functions for x > 0.*

**23.** $y = \dfrac{1}{x} + \dfrac{1}{4}x$     **24.** $y = \dfrac{2}{x}$

**25.** $y = \dfrac{9}{x} + x + 1$     **26.** $y = \dfrac{12}{x} + 3x + 1$

**27.** $y = \dfrac{2}{x} + \dfrac{x}{2} + 2$

**28.** $y = \dfrac{1}{x^2} + \dfrac{x}{4} - \dfrac{5}{4}$  [*Hint:* $(1,0)$ is an x-intercept.]

**29.** $y = 6\sqrt{x} - x$     **30.** $y = \dfrac{1}{\sqrt{x}} + \dfrac{x}{2}$

*In Exercises 31 and 32, determine which function is the derivative of the other.*

**31.**

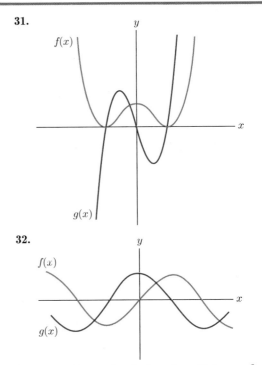

**32.**

**33.** Find the quadratic function $f(x) = ax^2 + bx + c$ that goes through $(2,0)$ and has a local maximum at $(0,1)$.

**34.** Find the quadratic function $f(x) = ax^2 + bx + c$ that goes through $(0,1)$ and has a local minimum at $(1,-1)$.

**35.** Suppose that $f'(a) = 0$ and that $f'(x)$ is increasing at $x = a$. Explain why $f(x)$ must have a local minimum at $x = a$. [*Hint:* Use the first derivative test.]

**36.** Suppose that $f'(a) = 0$ and that $f'(x)$ is decreasing at $x = a$. Explain why $f(x)$ must have a local maximum at $x = a$.

### Technology Exercises

**37.** In a medical experiment,* the body weight of a baby rat in the control group after $t$ days was $f(t) = 4.96 + .48t + .17t^2 - .0048t^3$ grams.

  **(a)** Graph $f(t)$ in the window $[0, 20]$ by $[-12, 50]$.

  **(b)** Approximately how much did the rat weigh after 7 days?

---

*Johnson, Wogenrich, Hsi, Skipper, and Greenberg, "Growth Retardation During the Sucking Period in Expanded Litters of Rats; Observations of Growth Patterns and Protein Turnover," *Growth, Development and Aging*, 55(1991), 263–273.

(c) Approximately when did the rat's weight reach 27 grams?

(d) Approximately how fast was the rat gaining weight after 4 days?

(e) Approximately when was the rat gaining weight at the rate of 2 grams per day?

(f) Approximately when was the rat gaining weight at the fastest rate?

**38.** The canopy height (in meters) of the tropical bunchgrass elephantmillet[†] $t$ days after mowing (for $t \geq 32$) is $f(t) = -3.14 + .142t - .0016t^2 + .0000079t^3 - .0000000133t^4$.

(a) Graph $f(t)$ in the window $[32, 250]$ $by$ $[-1.2, 4.5]$.

(b) How tall was the canopy after 100 days?

(c) When was the canopy 2 meters high?

(d) How fast was the canopy growing after 80 days?

(e) When was the canopy growing at the rate of .02 meters per day?

(f) Approximately when was the canopy growing slowest?

(g) Approximately when was the canopy growing fastest?

## SOLUTIONS TO PRACTICE PROBLEMS 2.4

*Functions with asymptotes as $x$ gets large have the form $f(x) = g(x) + mx + b$, where $g(x)$ approaches zero as $x$ gets large. The function $g(x)$ often looks like $c/x$ or $c/(ax + d)$. The asymptote will be the straight line $y = mx + b$.*

**1.** Here $g(x)$ is $3/x$ and the asymptote is $y = -2x + 1$.

**2.** This function has no asymptote as $x$ gets large. Of

course, it can be written as $g(x) + mx + b$, where $m = 1$ and $b = 0$. However, $g(x) = \sqrt{x}$ does not approach zero as $x$ gets large.

**3.** Here $g(x)$ is $\dfrac{1}{2x}$ and the asymptote is $y = 0$. That is, the function has the $x$-axis as an asymptote.

# 2.5 Optimization Problems

One of the most important applications of the derivative concept is to "optimization" problems, in which some quantity must be maximized or minimized. Examples of such problems abound in many areas of life. An airline must decide how many daily flights to schedule between two cities in order to maximize its profits. A doctor wants to find the minimum amount of a drug that will produce a desired response in one of her patients. A manufacturer needs to determine how often to replace certain equipment in order to minimize maintenance and replacement costs.

Our purpose in this section is to illustrate how calculus can be used to solve optimization problems. In each example we will find or construct a function that provides a "mathematical model" for the problem. Then, by sketching the graph of this function, we will be able to determine the answer to the original optimization problem by locating the highest or lowest point on the graph. The $y$-coordinate of this point will be the maximum value or minimum value of the function.

The first two examples are quite simple because the functions to be studied are given explicitly.

**EXAMPLE 1** Find the minimum value of the function $f(x) = 2x^3 - 15x^2 + 24x + 19$ for $x \geq 0$.

**Solution** Using the curve-sketching techniques from Section 2.3, we obtain the graph in Fig. 1. As part of that process, we compute the derivatives:

$$f'(x) = 6x^2 - 30x + 24$$
$$f''(x) = 12x - 30.$$

---

[†]Woodward and Prine, "Crop Quality and Utilization," *Crop Science*, 33(1993), 818–824.

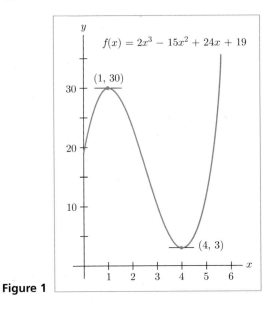

$$f(x) = 2x^3 - 15x^2 + 24x + 19$$

**Figure 1**

The $x$-coordinate of the minimum point satisfies the equation

$$f'(x) = 0$$
$$6x^2 - 30x + 24 = 0$$
$$6(x - 4)(x - 1) = 0$$
$$x = 1, \ 4 \quad \text{(critical values)}.$$

The corresponding critical points on the curve are

$$(1, f(1)) = (1, 30)$$
$$(4, f(4)) = (4, 3).$$

Applying the second derivative test, we see that

$$f''(1) = 12 \cdot 1 - 30 = -18 < 0 \quad \text{(maximum)}$$
$$f''(4) = 12 \cdot 4 - 30 = 18 > 0 \quad \text{(minimum)}.$$

That is, the point $(1, 30)$ is a relative maximum and the point $(4, 3)$ is a relative minimum. The lowest point on the graph is $(4, 3)$. The minimum *value* of the function $f(x)$ is the $y$-coordinate of this point—namely, 3. ∎

---

**EXAMPLE 2**     Suppose that a ball is thrown straight up into the air and its height after $t$ seconds is $4 + 48t - 16t^2$ feet. Determine how long it will take for the ball to reach its maximum height and determine the maximum height.

**Solution**     Consider the function $f(t) = 4 + 48t - 16t^2$. For each value of $t$, $f(t)$ is the height of the ball at time $t$. We want to find the value of $t$ for which $f(t)$ is the greatest. Using the techniques of Section 2.3, we sketch the graph of $f(t)$. (See Fig. 2.) Note that we may neglect the portions of the graph corresponding to points for which either $t < 0$ or $f(t) < 0$. [A negative value of $f(t)$ would correspond to the ball being underneath the ground.] The $t$-coordinate giving the maximum height is the solution of the equation:

$$f'(t) = 48 - 32t = 0$$
$$t = \tfrac{3}{2}.$$

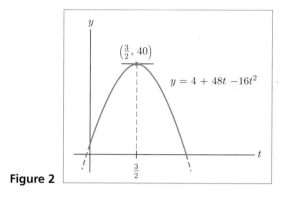

**Figure 2**

Since

$$f''(t) = -32$$

$$f''(\tfrac{3}{2}) = -32 < 0 \quad \text{(maximum)},$$

we see from the second derivative test that $t = \tfrac{3}{2}$ is the location of a relative maximum. So $f(t)$ is greatest when $t = \tfrac{3}{2}$. At this value of $t$, the ball attains a height of 40 feet. Thus the ball reaches its maximum height of 40 feet in 1.5 seconds. [Note that the curve in Fig. 2 is the graph of $f(t)$, *not* a picture of the physical path of the ball.] ■

---

**EXAMPLE 3**   **Maximizing an area**   You want to plant a rectangular garden along one side of a house, with a picket fence on the other three sides of the garden. Find the dimensions of the largest garden that can be enclosed using 40 feet of fencing.

**Solution**   Let us think about this problem before embarking on the solution. With 40 feet of fencing material, you can enclose a rectangular garden along your house in many different ways. Here are three illustrations: In Fig. 3(a), the enclosed area is $10 \times 15 = 150$ square feet; in Fig. 3(b), it is $16 \times 12 = 192$ square feet; and in Fig. 3(c), it is $32 \times 4 = 128$ square feet. Clearly, the enclosed area varies with your choice of dimensions. Which dimensions, that total 40 feet, yield the largest area? We next show how we can use calculus and optimization techniques to solve this problem.

(a) Area = $10 \times 15 = 150$ sq. ft.      (b) Area = $16 \times 12 = 192$ sq. ft.      (c) Area = $32 \times 4 = 128$ sq. ft.

**Figure 3.**   Rectangular gardens enclosed with 40 ft of fencing.

Since we do not know the dimensions, the first step is to make a simple diagram that represents the general case, and assign letters to the quantities that may vary. Let us denote the dimensions of the rectangular garden by $w$ and $x$ (Fig. 4). The phrase "largest garden" indicates that we must maximize the area, $A$, of the garden. In terms of the variables $w$ and $x$,

$$A = wx. \tag{1}$$

The fencing on three sides must total 40 running feet; that is,

$$2x + w = 40. \tag{2}$$

We now solve equation (2) for $w$ in terms of $x$:

$$w = 40 - 2x. \tag{3}$$

Substituting this expression for $w$ into equation (1), we have

$$A = (40 - 2x)x = 40x - 2x^2. \tag{4}$$

We now have a formula for the area $A$ that depends on just one variable. From the statement of the problem, the value of $2x$ can be at most 40, so that the domain of the function consists of $x$ in the interval $(0, 20)$. So the function that we wish to maximize is

$$A(x) = 40x - 2x^2 \quad 0 \le x \le 20.$$

Its graph is a parabola looking downward (Fig. 5). To find the maximum point, we compute

$$A'(x) = 0$$
$$40 - 4x = 0$$
$$x = 10.$$

House

Garden

$x$         $x$

$w$

Area $= w \cdot x$ sq. ft.
Fencing required $= 2x + w$ ft.

**Figure 4.**   General case.

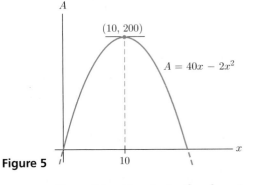

$A$

$(10, 200)$

$A = 40x - 2x^2$

$x$

$10$

**Figure 5**

Since this value is in the domain of our function, we conclude that the absolute maximum of $A(x)$ occurs at $x = 10$. [Alternatively, since $A''(x) = -4 < 0$, the concavity of the curve is always downward, and so the local maximum at $x = 10$ must be an absolute maximum.] (The maximum area is $A(10) = 200$ square feet, but this fact is not needed for the problem.) From equation (3) we find that when $x = 10$,

$$w = 40 - 2(10) = 20.$$

In conclusion, the rectangular garden with the largest area that can be enclosed using 40 feet of fencing has dimensions $x = 10$ feet and $w = 20$ feet.  ■

Equation (1) in Example 3 is called an *objective equation*. It expresses the quantity to be optimized (the area of the garden) in terms of the variables $w$ and $x$. Equation (2) is called a *constraint equation* because it places a limit or constraint on the way $x$ and $w$ may vary.

**EXAMPLE 4**   **Minimizing cost**   The manager of a department store wants to build a 600-square-foot rectangular enclosure on the store's parking lot in order to display some equipment. Three sides of the enclosure will be built of redwood fencing, at a cost of \$14 per running foot. The fourth side will be built of cement blocks, at a cost of \$28 per running foot. Find the dimensions of the enclosure that will minimize the total cost of the building materials.

**Figure 6.**   Rectangular enclosure.          **Figure 7**

**Solution**   Let $x$ be the length of the side built out of cement blocks and let $y$ be the length of an adjacent side, as shown in Fig. 6. The phrase "minimize the total cost" tells us that the objective equation should be a formula giving the total cost of the building materials.

$$[\text{cost of redwood}] = [\text{length of redwood fencing}] \times [\text{cost per foot}]$$
$$= (x + 2y) \cdot 14 = 14x + 28y$$
$$[\text{cost of cement blocks}] = [\text{length of cement wall}] \times [\text{cost per foot}]$$
$$= x \cdot 28$$

If $C$ denotes the total cost of the materials, then

$$C = (14x + 28y) + 28x$$
$$C = 42x + 28y \quad \text{(objective equation)}. \tag{5}$$

Since the area of the enclosure must be 600 square feet, the constraint equation is

$$xy = 600. \tag{6}$$

We simplify the objective equation by solving (6) for one of the variables, say $y$, and substituting into (5). Since $y = 600/x$,

$$C = 42x + 28\left(\frac{600}{x}\right) = 42x + \frac{16{,}800}{x}.$$

We now have $C$ as a function of the single variable $x$. From the context, we must have $x > 0$, since a length must be positive. However, to any positive value for $x$, there is a corresponding value for $C$. So the domain of $C$ consists of all $x > 0$. We may now sketch the graph of $C$ (Fig. 7). (A similar curve was sketched in Example 4 of Section 2.4.) The $x$-coordinate of the minimum point is a solution of

$$C'(x) = 0$$
$$42 - \frac{16{,}800}{x^2} = 0$$
$$42x^2 = 16{,}800$$
$$x^2 = 400$$
$$x = 20.$$

The corresponding value of $C$ is

$$C(20) = 42(20) + \frac{16{,}800}{20} = \$1680.$$

That is, the minimum total cost of $1680 occurs where $x = 20$. From equation (6) we find that the corresponding value of $y$ is $\frac{600}{20} = 30$. To minimize the total cost of building the 600-square-foot rectangular enclosure, the manager must take the dimensions to be $x = 20$ feet and $y = 30$ feet. ■

---

**EXAMPLE 5** **Maximizing volume** U.S. parcel post regulations state that packages must have length plus girth of no more than 84 inches. Find the dimensions of the cylindrical package of greatest volume that is mailable by parcel post.

**Solution** Let $l$ be the length of the package and let $r$ be the radius of the circular end. (See Fig. 8.) The phrase "greatest volume" tells us that the objective equation should express the volume of the package in terms of the dimensions $l$ and $r$. Let $V$ denote the volume. Then

$$V = [\text{area of base}] \cdot [\text{length}]$$
$$V = \pi r^2 l \quad (\text{objective equation}). \tag{7}$$

The girth equals the circumference of the end—that is, $2\pi r$. Since we want the package to be as large as possible, we must use the entire 84 inches allowable:

$$\text{length} + \text{girth} = 84$$
$$l + 2\pi r = 84 \quad (\text{constraint equation}). \tag{8}$$

We now solve equation (8) for one of the variables, say $l = 84 - 2\pi r$. Substituting this expression into (7), we obtain

$$V = \pi r^2 (84 - 2\pi r) = 84\pi r^2 - 2\pi^2 r^3. \tag{9}$$

Let $f(r) = 84\pi r^2 - 2\pi^2 r^3$. Then, for each value of $r$, $f(r)$ is the volume of the parcel with end radius $r$ that meets the postal regulations. We want to find that value of $r$ for which $f(r)$ is as large as possible.

Using curve-sketching techniques, we obtain the graph of $f(r)$ in Fig. 9. The domain excludes values of $r$ that are negative and values of $r$ for which the volume $f(r)$ is negative. Points corresponding to values of $r$ not in the domain are shown with a dashed curve. We see that the volume is greatest when $r = 28/\pi$.

**Figure 8.** Cylindrical mailing package.

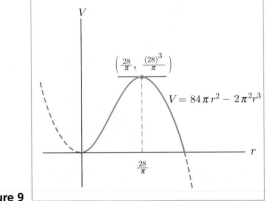

**Figure 9**

From (8) we find that the corresponding value of $l$ is

$$l = 84 - 2\pi r = 84 - 2\pi \left( \frac{28}{\pi} \right) = 84 - 56 = 28.$$

The girth when $r = 28/\pi$ is

$$2\pi r = 2\pi \left( \frac{28}{\pi} \right) = 56.$$

Thus the dimensions of the cylindrical package of greatest volume are $l = 28$ inches and $r = 28/\pi$ inches. ◼

### Suggestions for Solving an Optimization Problem

1. Draw a picture, if possible.
2. Decide what quantity $Q$ is to be maximized or minimized.
3. Assign letters to other quantities that may vary.
4. Determine the "objective equation" that expresses $Q$ as a function of the variables assigned in step 3.
5. Find the "constraint equation" that relates the variables to each other and to any constants that are given in the problem.
6. Use the constraint equation to simplify the objective equation in such a way that $Q$ becomes a function of only one variable. Determine the domain of this function.
7. Sketch the graph of the function obtained in step 6 and use this graph to solve the optimization problem. Alternatively, you can use the second derivative test.

▶ *Note* Optimization problems often involve geometric formulas. The most common formulas are shown in Fig. 10.

Area $= xy$  Volume $= xyz$  Area $= \pi r^2$  Volume $= \pi r^2 h$
**Figure 10** Perimeter $= 2x + 2y$  Circumference $= 2\pi r$  ◀

## PRACTICE PROBLEMS 2.5

**1.** *(Volume)* A canvas wind shelter for the beach has a back, two square sides, and a top (Fig. 11). Suppose that 96 square feet of canvas are to be used. Find the dimensions of the shelter for which the space inside the shelter (i.e., the volume) will be maximized.

**2.** In Practice Problem 1, what are the objective equation and the constraint equation?

**Figure 11.** Wind shelter.

## EXERCISES 2.5

**1.** For what $x$ does the function $g(x) = 10 + 40x - x^2$ have its maximum value?

**2.** Find the maximum value of the function $f(x) = 12x - x^2$ and give the value of $x$ where this maximum occurs.

**3.** Find the minimum value of $f(t) = t^3 - 6t^2 + 40$, $t \geq 0$, and give the value of $t$ where this minimum occurs.

**4.** For what $t$ does the function $f(t) = t^2 - 24t$ have its minimum value?

**5.** Find the maximum of $Q = xy$ if $x + y = 2$.

**6.** Find two positive numbers $x$ and $y$ that maximize $Q = x^2y$ if $x + y = 2$.

**7.** Find the minimum of $Q = x^2 + y^2$ if $x + y = 6$.

**8.** In Exercise 7, can there be a maximum for $Q = x^2 + y^2$ if $x + y = 6$? Justify your answer.

**9.** Find the positive values of $x$ and $y$ that minimize $S = x + y$ if $xy = 36$, and find this minimum value.

**10.** Find the positive values of $x$, $y$, and $z$ that maximize $Q = xyz$, if $x + y = 1$ and $y + z = 2$. What is this maximum value?

**11.** *(Area)* Three hundred twenty dollars are available to fence in a rectangular garden. The fencing for the side of the garden facing the road costs $6 per foot and the fencing for the other three sides costs $2 per foot. [See Fig. 12(a).] Consider the problem of finding the dimensions of the largest possible garden.

   **(a)** Determine the objective and constraint equations.

   **(b)** Express the quantity to be maximized as a function of $x$.

   **(c)** Find the optimal values of $x$ and $y$.

**12.** *(Volume)* Figure 12(b) shows an open rectangular box with a square base. Consider the problem of finding the values of $x$ and $h$ for which the volume is 32 cubic feet and the total surface area of the box is minimal. (The surface area is the sum of the areas of the five faces of the box.)

   **(a)** Determine the objective and constraint equations.

   **(b)** Express the quantity to be minimized as a function of $x$.

   **(c)** Find the optimal values of $x$ and $h$.

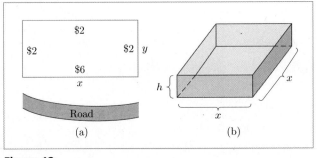

**Figure 12**

13. (*Volume*) Postal requirements specify that parcels must have length plus girth at most 84 inches. Consider the problem of finding the dimensions of the square-ended rectangular package of greatest volume that is mailable.

    (a) Draw a square-ended rectangular box. Label each edge of the square end with the letter $x$ and label the remaining dimension of the box with the letter $h$.

    (b) Express the length plus the girth in terms of $x$ and $h$.

    (c) Determine the objective and constraint equations.

    (d) Express the quantity to be maximized as a function of $x$.

    (e) Find the optimal values of $x$ and $h$.

14. (*Perimeter*) Consider the problem of finding the dimensions of the rectangular garden of area 100 square meters for which the amount of fencing needed to surround the garden is as small as possible.

    (a) Draw a picture of a rectangle and select appropriate letters for the dimensions.

    (b) Determine the objective and constraint equations.

    (c) Find the optimal values for the dimensions.

15. (*Cost*) A rectangular garden of area 75 square feet is to be surrounded on three sides by a brick wall costing $10 per foot and on one side by a fence costing $5 per foot. Find the dimensions of the garden such that the cost of materials is minimized.

16. (*Cost*) A closed rectangular box with square base and a volume of 12 cubic feet is to be constructed using two different types of materials. The top is made of a metal costing $2 per square foot and the remainder of wood costing $1 per square foot. Find the dimension of the box for which the cost of materials is minimized.

17. (*Surface Area*) Find the dimensions of the closed rectangular box with square base and volume 8000 cubic centimeters that can be constructed with the least amount of material.

18. (*Volume*) A canvas wind shelter for the beach has a back, two square sides, and a top. Find the dimensions for which the volume will be 250 cubic feet and that requires the least possible amount of canvas.

19. (*Area*) A farmer has $1500 available to build an E-shaped fence along a straight river so as to create two identical rectangular pastures. (See Fig. 13.) The materials for the side parallel to the river cost $6 per foot and the materials for the three sections perpendicular to the river cost $5 per foot. Find the dimensions for which the total area is as large as possible.

20. (*Area*) Find the dimensions of the rectangular garden of greatest area that can be fenced off (all four sides) with 300 meters of fencing.

21. Find two positive numbers, $x$ and $y$, whose sum is 100 and whose product is as large as possible.

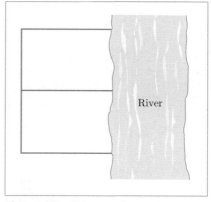

**Figure 13.** Rectangular pastures along a river.

22. Find two positive numbers, $x$ and $y$, whose product is 100 and whose sum is as small as possible.

23. (*Area*) Figure 14(a) shows a Norman window, which consists of a rectangle capped by a semicircular region. Find the value of $x$ such that the perimeter of the window will be 14 feet and the area of the window will be as large as possible.

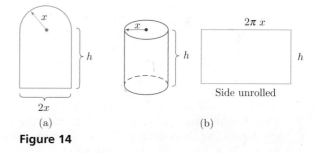

**Figure 14**

24. (*Surface Area*) A large soup can is to be designed so that the can will hold $16\pi$ cubic inches (about 28 ounces) of soup. [See Fig. 14(b).] Find the values of $x$ and $h$ for which the amount of metal needed is as small as possible.

25. In Example 3 one can solve the constraint equation (2) for $x$ instead of $w$ to get $x = 20 - \frac{1}{2}w$. Substituting this for $x$ in (1), one has

$$A = xw = \left(20 - \tfrac{1}{2}w\right)w.$$

    Sketch the graph of the equation

$$A = 20w - \tfrac{1}{2}w^2$$

    and show that the maximum occurs when $w = 20$ and $x = 10$.

26. (*Cost*) A ship uses $5x^2$ dollars of fuel per hour when traveling at a speed of $x$ miles per hour. The other expenses of operating the ship amount to $2000 per hour. What speed minimizes the cost of a 500-mile trip? [*Hint*: Express cost in terms of speed and time. The constraint equation is *distance* = *speed* × *time*.]

**Figure 15**

**Figure 16**

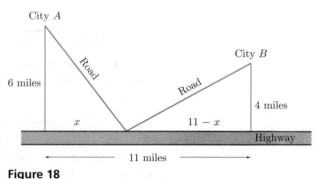

**Figure 17.** Shortest distance from a point to a curve.

**27.** (*Cost*) A cable is to be installed from one corner $C$ of a rectangular factory to a machine $M$ on the floor. The cable will run along one edge of the floor from $C$ to a point $P$ at a cost of \$6 per meter, and then from $P$ to $M$ in a straight line buried under the floor at a cost of \$10 per meter (see Fig. 15). Let $x$ denote the distance from $C$ to $P$. If $M$ is 24 meters from the nearest point $A$ on the edge of the floor on which $P$ lies, and $A$ is 20 meters from $C$, find the value of $x$ that minimizes the cost of installing the cable and determine the minimum cost.

**28.** (*Area*) A rectangular page is to contain 50 square inches of print. The page has to have a 1-inch margin on top and at the bottom and a 1/2-inch margin on each side (see Fig. 16). Find the dimensions of the page that minimize the amount of paper used.

**29.** (*Distance*) Find the point on the graph of $y = \sqrt{x}$ that is closest to the point $(2,0)$. See Fig. 17. [*Hint:* $\sqrt{(x-2)^2 + y^2}$ has its smallest value when $(x-2)^2 + y^2$ does. Therefore, just minimize the second expression.]

**30.** (*Distance*) Developers from two towns, $A$ and $B$, want to connect their towns to a major highway and plan to build rest stops and gas stations at the highway entrance. In order to minimize the cost of road construction, the developers must find the location for the highway entrance that minimizes the total distance, $d_1 + d_2$, from the towns $A$ and $B$ to the highway. Let $x$ be as in Fig. 18. Find the value of $x$ that solves the developers' problem, and find the minimum total distance.

**31.** (*Distance*) Find the point on the line $y = -2x + 5$ that is closest to the origin.

**Figure 18**

## *Technology Exercises*

**32.** Find the value of $x$ for which the rectangle inscribed in the semicircle of radius 3 in Fig. 19 has the greatest area. (*Note:* The equation of the semicircle is $y = \sqrt{9 - x^2}$.)

**Figure 19**

## SOLUTIONS TO PRACTICE PROBLEMS 2.5

**1.** Since the sides of the wind shelter are square, we may let $x$ represent the length of each side of the square. The remaining dimension of the wind shelter can be denoted by the letter $h$. (See Fig. 20.)

**Figure 20**

The volume of the shelter is $x^2 h$, and this is to be maximized. Since we have learned to maximize only functions of a single variable, we must express $h$ in terms of $x$. We must use the information that 96 feet of canvas are used—that is, $2x^2 + 2xh = 96$. (*Note:* The roof and the back each have an area $xh$, and each end has an area $x^2$.) We now solve this equation for $h$.

$$2x^2 + 2xh = 96$$
$$2xh = 96 - 2x^2$$
$$h = \frac{96}{2x} - \frac{2x^2}{2x} = \frac{48}{x} - x.$$

The volume $V$ is

$$x^2 h = x^2 \left(\frac{48}{x} - x\right) = 48x - x^3.$$

By sketching the graph of $V = 48x - x^3$, we see that $V$ has a maximum value when $x = 4$. Then, $h = \frac{48}{4} - 4 = 12 - 4 = 8$. So each end of the shelter should be a 4-foot by 4-foot square and the top should be 8 feet long.

**2.** The objective equation is $V = x^2 h$, since it expresses the volume (the quantity to be maximized) in terms of the variables. The constraint equation is $2x^2 + 2xh = 96$, for it relates the variables to each other; that is, it can be used to express one of the variables in terms of the other.

## 2.6  Further Optimization Problems

In this section we apply the optimization techniques developed in the preceding section to some practical situations.

**Inventory Control**   When a firm regularly orders and stores supplies for later use or resale, it must decide on the size of each order. If it orders enough supplies to last an entire year, the business will incur heavy *carrying costs*. Such costs include insurance, storage costs, and cost of capital that is tied up in inventory. To reduce these carrying costs, the firm could order small quantities of the supplies at frequent intervals. However, such a policy increases the *ordering costs*. These might consist of minimum freight charges, the clerical costs of preparing the orders, and the costs of receiving and checking the orders when they arrive. Clearly, the firm must find an inventory ordering policy that lies between these two extremes.

To appreciate the inventory control problem, let us first consider a simple problem that does not involve optimization.

**EXAMPLE 1**   A supermarket manager anticipates that 1200 cases of frozen orange juice will be sold at a steady rate during the next year. The manager plans on ordering these cases by placing equally spaced orders of the same size throughout the year. Given that it costs \$8 to carry one case of orange juice in inventory for one year, find the carrying cost if:

(a) The manager places only one order during the year.

(b) The manager places two orders during the year.

(c) The manager places four orders during the year.

Carrying costs should be computed on the average inventory during the order–reorder period.

**Solution** (a) This plan calls for ordering 1200 cases all at once. The inventory of frozen orange juice in this case is shown in Fig. 1(a). It starts at 1200 and decreases steadily to 0 during this one order–reorder period. At any given time during the year, the inventory is between 1200 and 0. So there are $\frac{1200}{2}$, or 600 cases on average in inventory. Since carrying costs are computed on the average inventory, the carrying cost in this case is $C = 600 \times 8 = 4800$ dollars.

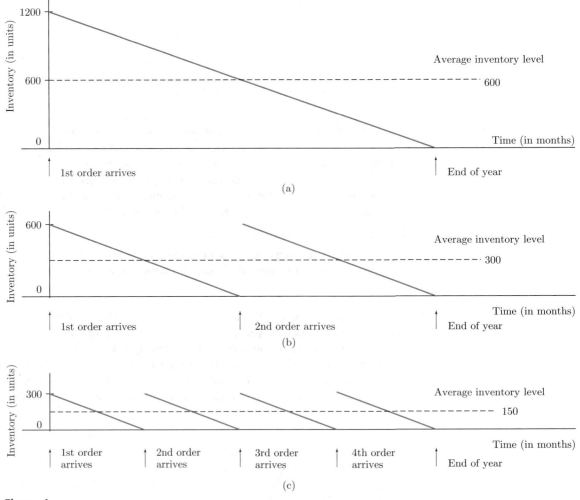

**Figure 1**

(b) This plan calls for ordering 1200 cases by placing two equally spaced orders of the same size. Thus the size of each order is equal to $\frac{1200}{2}$ or 600 cases. The inventory of frozen orange juice in this case is shown in Fig. 1(b). At the beginning of each order–reorder period, the inventory starts at 600 and decreases steadily to 0 at the end of the order–reorder period. At any given time during the order–reorder period, the inventory is between 600 and 0. So there are $\frac{600}{2}$, or 300 cases on average in inventory. Since the carrying costs are computed on the average inventory, the carrying cost in this case is $C = 300 \times 8 = 2400$ dollars.

(c) This plan calls for ordering 1200 cases by placing four equally spaced orders of the same size. The size of each order is $\frac{1200}{4} = 300$ cases. The inventory in this case is shown in Fig. 1(c). The average inventory during an order–reorder period is $\frac{300}{2}$, or 150 cases, and so the carrying cost in this case is $C = 150 \times 8 = 1200$ dollars.  ∎

By increasing the number of orders in Example 1, the manager was able to reduce the carrying cost from \$4800 down to \$2400 down to \$1200. Clearly, the manager can reduce the carrying cost even further by ordering more frequently. In reality, things are not so simple. There is a cost attached to each order, and by increasing the number of orders, the manager will increase the annual inventory cost, where

$$[\text{inventory cost}] = [\text{ordering cost}] + [\text{carrying cost}].$$

The inventory control problem consists of finding the *economic order quantity*, commonly referred to in business as the EOQ,* which minimizes the inventory cost. We now show how we can use calculus to solve this problem.

---

**EXAMPLE 2**  **Inventory control**  Suppose that the manager in Example 1 wants to establish an optimal inventory policy for frozen orange juice. Again, it is estimated that a total of 1200 cases will be sold at a steady rate during the next year. The manager plans to place several orders of the same size spaced equally throughout the year. Use the following data to determine the economic order quantity; that is, the order size that minimizes the total ordering and carrying cost.

1.  The ordering cost for each delivery is \$75.

2.  It costs \$8 to carry one case of orange juice in inventory for one year. (Carrying costs should be computed on the average inventory during the order–reorder period.)

**Solution**  Since we are not given the size of each order or the number of orders that will be placed throughout the year, we let $x$ be the order quantity and $r$ the number of orders placed during the year. As in Example 1, the number of cases of orange juice in inventory declines steadily from $x$ cases (each time a new order is filled) to 0 cases at the end of each order–reorder period. Figure 2 shows that the average number of cases in storage during the year is $x/2$. Since the carrying cost for one

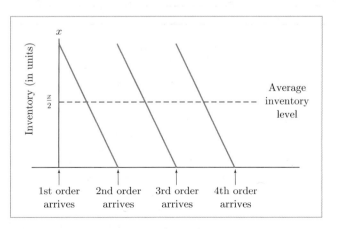

**Figure 2.**  Average inventory level.

---

*James C. Van Horne, *Financial Management and Policy*, 6th ed. (Englewood Cliffs, N.J.: Prentice-Hall, 1983), pp. 416–420.

case is \$8 per year, the cost for $x/2$ cases is $8 \cdot (x/2)$ dollars. Now

$$[\text{inventory cost}] = [\text{ordering cost}] + [\text{carrying cost}]$$
$$= 75r + 8 \cdot \frac{x}{2}$$
$$= 75r + 4x.$$

If $C$ denotes the inventory cost, then the objective equation is

$$C = 75r + 4x.$$

Since there are $r$ orders of $x$ cases each, the total number of cases ordered during the year is $r \cdot x$. Therefore, the constraint equation is

$$r \cdot x = 1200.$$

The constraint equation says that $r = 1200/x$. Substitution into the objective equation yields

$$C = \frac{90,000}{x} + 4x.$$

Figure 3 is the graph of $C$ as a function of $x$, for $x > 0$. The minimum point of the graph occurs where the first derivative equals zero. We find this point using the calculation

$$C'(x) = -\frac{90,000}{x^2} + 4 = 0$$
$$4x^2 = 90,000$$
$$x^2 = 22,500$$
$$x = 150 \quad (\text{ignore the negative root since } x > 0).$$

The total cost is at a minimum when $x = 150$. Therefore, the optimum inventory policy is to order 150 cases at a time and to place $1200/150 = 8$ orders during the year. ◼

**Figure 3.** Cost function for inventory problem.

---

**EXAMPLE 3** What should the inventory policy of Example 2 be if sales of frozen orange juice increase fourfold (i.e., 4800 cases are sold each year), but all other conditions are the same?

**Solution**  The only change in our previous solution is in the constraint equation, which now becomes

$$r \cdot x = 4800.$$

The objective equation is, as before,

$$C = 75r + 4x.$$

Since $r = 4800/x$,

$$C = 75 \cdot \frac{4800}{x} + 4x = \frac{360{,}000}{x} + 4x.$$

Now

$$C' = -\frac{360{,}000}{x^2} + 4.$$

Setting $C' = 0$ yields

$$\frac{360{,}000}{x^2} = 4$$
$$90{,}000 = x^2$$
$$x = 300.$$

Therefore, the economic order quantity is 300 cases.

Notice that although the sales increased by a factor of 4, the economic order quantity increased by only a factor of 2 ($= \sqrt{4}$). In general, a store's inventory of an item should be proportional to the square root of the expected sales. (See Exercise 9 for a derivation of this result.) Many stores tend to keep their average inventories at a fixed percentage of sales. For example, each order may contain enough goods to last for 4 or 5 weeks. This policy is likely to create excessive inventories of high-volume items and uncomfortably low inventories of slower-moving items.

Manufacturers have an inventory-control problem similar to that of retailers. They have the carrying costs of storing finished products and the startup costs of setting up each production run. The size of the production run that minimizes the sum of these two costs is called the *economic lot size*. See Exercises 6 and 7.

---

**EXAMPLE 4**  **Lungs and trachea clearing**  Doctors studying coughing problems need to know the most effective condition for clearing the lungs and trachea (windpipe); that is, what radius of the trachea gives the greatest air velocity. Using the following principles of fluid flow, determine this optimum value. (See Fig. 4.) Let

$$r_0 = \text{normal radius of the trachea,}$$
$$r = \text{radius during a cough,}$$
$$P = \text{increase in air pressure in the trachea during cough,}$$
$$v = \text{velocity of air through trachea during cough.}$$

(1) $r_0 - r = aP$, for some positive constant $a$. (Experiment has shown that, during coughing, the decrease in the radius of the trachea is nearly proportional to the increase in the air pressure.)

(2) $v = b \cdot P \cdot \pi r^2$, for some positive constant $b$. (The theory of fluid flow requires that the velocity of the air forced through the trachea is proportional to the product of the increase in the air pressure and the area of a cross section of the trachea.)

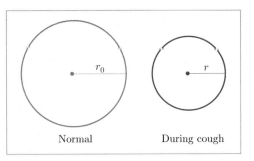

**Figure 4.** Contraction of the windpipe during coughing.

Solution In this problem the constraint equation (1) and the objective equation (2) are given directly. Solving equation (1) for $P$ and substituting this result into equation (2), we have

$$v = b\left(\frac{r_0 - r}{a}\right)\pi r^2 = k(r_0 - r)r^2,$$

where $k = b\pi/a$. To find the radius at which the velocity $v$ is a maximum, we first compute the derivatives with respect to $r$:

$$v = k(r_0 r^2 - r^3)$$

$$\frac{dv}{dr} = k(2r_0 r - 3r^2) = kr(2r_0 - 3r)$$

$$\frac{d^2v}{dr^2} = k(2r_0 - 6r).$$

We see that $\frac{dv}{dr} = 0$ when $r = 0$ or when $2r_0 - 3r = 0$; that is, when $r = \frac{2}{3}r_0$. It is easy to see that $\frac{d^2v}{dr^2}$ is positive at $r = 0$ and is negative at $r = \frac{2}{3}r_0$. The graph of $v$ as a function of $r$ is drawn in Fig. 5. The air velocity is maximized at $r = \frac{2}{3}r_0$. ∎

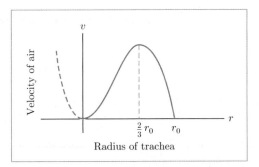

**Figure 5.** Graph for the windpipe problem.

When solving optimization problems, we look for the maximum or minimum point on a graph. From our discussion of curve sketching, we have seen that this point occurs either at a relative extreme point or at an endpoint of the domain of definition. In all our optimization problems so far, the maximum or minimum points were at relative extreme points. In the next example, the optimum point is an endpoint. This occurs often in applied problems.

**EXAMPLE 5** **Area** A rancher has 204 meters of fencing from which to build two corrals: one square and the other rectangular with length that is twice the width. Find the dimensions that result in the greatest combined area.

**Solution** Let $x$ be the width of the rectangular corral and $h$ be the length of each side of the square corral. (See Fig. 6.) Let $A$ be the combined area. Then

$$A = [\text{area of square}] + [\text{area of rectangle}] = h^2 + 2x^2.$$

The constraint equation is

$$204 = [\text{perimeter of square}] + [\text{perimeter of rectangle}] = 4h + 6x.$$

Since the perimeter of the rectangle cannot exceed 204, we must have $0 \le 6x \le 204$, or $0 \le x \le 34$. Solving the constraint equation for $h$ and substituting into the objective equation leads to the function graphed in Fig. 7. The graph reveals that the area is minimized when $x = 18$. However, the problem asks for the *maximum* possible area. From Fig. 7 we see that this occurs at the endpoint where $x = 0$. Therefore, the rancher should build only the square corral, with $h = 204/4 = 51$ meters. In this example, the objective function has an endpoint extremum; namely, the maximum value occurs at the endpoint where $x = 0$. ∎

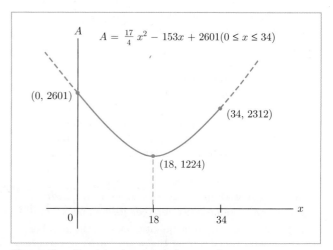

$A = \frac{17}{4} x^2 - 153x + 2601 (0 \le x \le 34)$

(0, 2601)

(34, 2312)

(18, 1224)

**Figure 6.** Two corrals.    **Figure 7.** Combined area of the corrals.

## PRACTICE PROBLEMS 2.6

**1.** In the inventory problem of Example 2, suppose that the ordering cost is the same but that it costs \$9 to carry one case of orange juice in inventory for one year, and carrying costs are computed on the *maximum* inventory during the order–reorder period. What is the new economic order quantity?

**2.** In the inventory problem of Example 2, suppose that the sales of frozen orange juice increase ninefold; that is, 10,800 cases are sold each year. What is the new economic order quantity?

## EXERCISES 2.6

**1.** Figure 8 shows the inventory levels of dried Rainier cherries at a natural food store in Seattle and the order–reorder periods over one year. Refer to the figure to answer the following questions.

   **(a)** What is the average amount of cherries in inventory during one order–reorder period?

   **(b)** What is the maximum amount of cherries in inventory during one order–reorder period?

   **(c)** How many orders were placed during the year?

   **(d)** How many pounds of dried cherries were sold during one year?

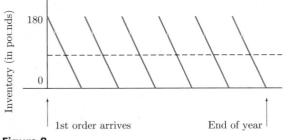

**Figure 8**

**2.** Refer to Fig. 8. Suppose that

(i) The ordering cost for each delivery of dried cherries is $50.

(ii) It costs $7 to carry one pound of dried cherries in inventory for one year.

(a) What is the inventory cost (carrying cost plus ordering cost) if carrying costs are computed on the average inventory during the order–reorder period?

(b) What is the inventory cost if carrying costs are computed on the maximum inventory during the order–reorder period?

**3.** (*Inventory Control*) A pharmacist wants to establish an optimal inventory policy for a new antibiotic that requires refrigeration in storage. The pharmacist expects to sell 800 packages of this antibiotic at a steady rate during the next year. She plans to place several orders of the same size spaced equally throughout the year. The ordering cost for each delivery is $16, and carrying costs, based on the average number of packages in inventory, amount to $4 per year for one package.

(a) Let $x$ be the order quantity and $r$ the number of orders placed during the year. Find the inventory cost (ordering cost plus carrying cost) in terms of $x$ and $r$.

(b) Find the constraint function.

(c) Determine the economic order quantity that minimizes the inventory cost and then find the minimum inventory cost.

**4.** (*Inventory Control*) A furniture store expects to sell 640 sofas at a steady rate next year. The manager of the store plans to order these sofas from the manufacturer by placing several orders of the same size spaced equally throughout the year. The ordering cost for each delivery is $160, and carrying costs, based on the average number of sofas in inventory, amount to $32 per year for one sofa.

(a) Let $x$ be the order quantity and $r$ the number of orders placed during the year. Find the inventory cost in terms of $x$ and $r$.

(b) Find the constraint function.

(c) Determine the economic order quantity that minimizes the inventory cost and then find the minimum inventory cost.

**5.** (*Inventory Control*) A California distributor of sporting equipment expects to sell 10,000 cases of tennis balls during the coming year at a steady rate. Yearly carrying costs (to be computed on the average number of cases in stock during the year) are $10 per case, and the cost of placing an order with the manufacturer is $80.

(a) Find the inventory cost incurred if the distributor orders 500 cases at a time during the year.

(b) Determine the economic order quantity, that is, the order quantity that minimizes the inventory cost.

**6.** (*Economic Lot Size*) The Great American Tire Co. expects to sell 600,000 tires of a particular size and grade during the next year. Sales tend to be roughly the same from month to month. Setting up each production run costs the company $15,000. Carrying costs, based on the average number of tires in storage, amount to $5 per year for one tire.

(a) Determine the costs incurred if there are 10 production runs during the year.

(b) Find the economic lot size (i.e., the production run size that minimizes the overall cost of producing the tires).

**7.** (*Economic Lot Size*) Foggy Optics, Inc. makes laboratory microscopes. Setting up each production run costs $2500. Insurance costs, based on the average number of microscopes in the warehouse, amount to $20 per microscope per year. Storage costs, based on the maximum number of microscopes in the warehouse, amount to $15 per microscope per year. Suppose that the company expects to sell 1600 microscopes at a fairly uniform rate throughout the year. Determine the number of production runs that will minimize the company's overall expenses.

**8.** (*Inventory Control*) A bookstore is attempting to determine the economic order quantity for a popular book. The store sells 8000 copies of this book a year. The store figures that it costs $40 to process each new order for books. The carrying cost (due primarily to interest payments) is $2 per book, to be figured on the maximum inventory during an order–reorder period. How many times a year should orders be placed?

**9.** (*Inventory Control*) A store manager wants to establish an optimal inventory policy for an item. Sales are expected to be at a steady rate and should total $Q$ items sold during the year. Each time an order is placed, a cost of $h$ dollars is incurred. Carrying costs for the year will be $s$ dollars per item, to be figured on the average number of items in storage during the year. Show that the total inventory cost is minimized when each order calls for $\sqrt{2hQ/s}$ items.

**10.** Refer to the inventory problem of Example 2. Suppose that the distributor offers a discount of $1 per case for orders of 600 or more cases. Should the manager change the quantity ordered?

**11.** (*Area*) Starting with a 100-foot-long stone wall, a farmer would like to construct a rectangular enclosure by adding 400 feet of fencing as shown in Fig. 9(a). Find the values of $x$ and $w$ that result in the greatest possible area.

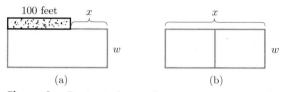

(a)                (b)

**Figure 9.** Rectangular enclosures.

**12.** Rework Exercise 11 for the case where only 200 feet of fencing is added to the stone wall.

**13.** (*Length*) A rectangular corral of 54 square meters is to be fenced off and then divided by a fence into two sections, as shown in Fig. 9(b). Find the dimensions of the corral so that the amount of fencing required is minimized.

**14.** Referring to Exercise 13, suppose that the cost of the fencing for the boundary is $5 per meter and the dividing fence costs $2 per meter. Find the dimensions of the corral that minimize the cost of the fencing.

**15.** (*Revenue*) Shakespear's Pizza sells 1000 large Vegi Pizzas per week for $18 a pizza. When the owner offers a $5 discount, the weekly sales increase to 1500.

   **(a)** Assume a linear relation between the weekly sales $A(x)$ and the discount $x$. Find $A(x)$.

   **(b)** Find the value of $x$ that maximizes the weekly revenue. [*Hint*: Revenue = $A(x) \cdot$ (Price).]

   **(c)** Answer parts (a) and (b) if the price of one pizza is $9 and all other data are unchanged.

**16.** (*Surface Area*) Design an open rectangular box with square ends, having volume 36 cubic inches, that minimizes the amount of material required for construction.

**17.** (*Cost*) A storage shed is to be built in the shape of a box with a square base. It is to have a volume of 150 cubic feet. The concrete for the base costs $4 per square foot, the material for the roof costs $2 per square foot, and the material for the sides costs $2.50 per square foot. Find the dimensions of the most economical shed.

**18.** (*Cost*) A supermarket is to be designed as a rectangular building with a floor area of 12,000 square feet. The front of the building will be mostly glass and will cost $70 per running foot for materials. The other three walls will be constructed of brick and cement block, at a cost of $50 per running foot. Ignore all other costs (labor, cost of foundation and roof, etc.) and find the dimensions of the base of the building that will minimize the cost of the materials for the four walls of the building.

**19.** (*Volume*) A certain airline requires that rectangular packages carried on an airplane by passengers be such that the sum of the three dimensions is at most 120 centimeters. Find the dimensions of the square-ended rectangular package of greatest volume that meets this requirement.

**20.** (*Area*) An athletic field [Fig. 10(a)] consists of a rectangular region with a semicircular region at each end. The perimeter will be used for a 440-yard track. Find the value of $x$ for which the area of the rectangular region is as large as possible.

**21.** (*Volume*) An open rectangular box is to be constructed by cutting square corners out of a 16- by 16-inch piece of cardboard and folding up the flaps. [See Fig. 10(b).] Find the value of $x$ for which the volume of the box will be as large as possible.

**22.** (*Volume*) A closed rectangular box is to be constructed with a base that is twice as long as it is wide. Suppose that the total surface area must be 27 square feet. Find the dimensions of the box that will maximize the volume.

**23.** Let $f(t)$ be the amount of oxygen (in suitable units) in a lake $t$ days after sewage is dumped into the lake, and suppose that $f(t)$ is given approximately by

$$f(t) = 1 - \frac{10}{t + 10} + \frac{100}{(t + 10)^2}.$$

At what time is the oxygen content increasing the fastest?

**24.** The daily output of a coal mine after $t$ hours of operation is approximately $40t + t^2 - \frac{1}{15}t^3$ tons, $0 \le t \le 12$. Find the maximum rate of output (in tons of coal per hour).

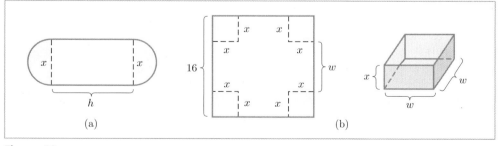

(a)                         (b)

**Figure 10**

**25.** (*Area*) Consider a parabolic arch whose shape may be represented by the graph of $y = 9 - x^2$, where the base of the arch lies on the x-axis from $x = -3$ to $x = 3$. Find the dimensions of the rectangular window of maximum area that can be constructed inside the arch.

**26.** Advertising for a certain product is terminated, and $t$ weeks later the weekly sales are $f(t)$ cases, where

$$f(t) = 1000(t + 8)^{-1} - 4000(t + 8)^{-2}.$$

At what time is the weekly sales amount falling the fastest?

**27.** (*Surface Area*) An open rectangular box of volume 400 cubic inches has a square base and a partition down the middle. See Fig. 11. Find the dimensions of the box for which the amount of material needed to construct the box is as small as possible.

**Figure 11.**   Open rectangular box with dividing partition.

**28.** Suppose $f(x)$ is defined on the interval $0 \le x \le 5$ and $f'(x)$ is negative for all $x$. For what value of $x$ will $f(x)$ have its greatest value?

### Technology Exercises

**29.** (*Volume*) A pizza box is formed from a 20 cm by 40 cm rectangular piece of cardboard by cutting out six squares of equal size, three from each long side of the rectangle, and then folding the cardboard in the obvious manner to create a box. See Fig. 12. Let $x$ be the

length of each side of the six squares. For what value of $x$ will the box have greatest volume?

**Figure 12**

**30.** Coffee consumption in the United States is greater on a per capita basis than anywhere else in the world. However, due to price fluctuations of coffee beans and worries over health effects of caffeine, coffee consumption has varied considerably over the years. According to data published in the *Wall Street Journal*, the number of cups $f(x)$ consumed daily per adult in year $x$ (with 1955 corresponding to $x = 0$) is given by the mathematical model

$$f(x) = 2.77 + 0.0848x - 0.00832x^2 + 0.000144x^3.$$

**(a)** Graph $y = f(x)$ to show daily coffee consumption from 1955 through 1994.

**(b)** Use $f'(x)$ to determine the year in which coffee consumption was least during this period. What was the daily coffee consumption at that time?

**(c)** Use $f'(x)$ to determine the year in which coffee consumption was greatest during this period. What was the daily coffee consumption at that time?

**(d)** Use $f''(x)$ to determine the year in which coffee consumption was decreasing at the greatest rate.

### SOLUTIONS TO PRACTICE PROBLEMS 2.6

**1.** Let $x$ denote the order quantity and $r$ the number of orders placed during the year. It is clear from Fig. 2 that the maximum inventory at any time during the order–reorder period is $x$ cases. This is the number of cases in inventory at the moment an order arrives. Since the carrying cost for one case is $9, the carrying cost for $x$ cases is $9 \cdot x$ dollars. This is the new carrying cost. The ordering cost is the same as in Example 2. So the objective function becomes

$$C = [\text{ordering cost}] + [\text{carrying cost}] = 75r + 9x.$$

The constraint equation is also the same as in Example 2, because the total number of cases ordered

during the year has not changed. Thus

$$r \cdot x = 1200.$$

From this point, the solution proceeds as in Example 2. We have

$$r = \frac{1200}{x}$$

$$C = 75 \cdot \frac{1200}{x} + 9x = \frac{90,000}{x} + 9x$$

$$C'(x) = -\frac{90,000}{x^2} + 9.$$

Solving $C'(x) = 0$, we find

$$-\frac{90,000}{x^2} + 9 = 0$$

$$9x^2 = 90,000$$

$$x^2 = 10,000$$

$$x = \sqrt{10,000} = 100 \quad \text{(ignore the negative root since } x > 0\text{)}.$$

The total cost is at a minimum when $x = 100$. The optimum inventory policy is to order 100 cases at a time and to place $\frac{1200}{100}$ or 12 orders during the year.

2. This problem can be solved in the same manner that Example 3 was solved. However, the comment made at the end of Example 3 indicates that the economic order quantity should increase by a factor of 3, since $3 = \sqrt{9}$. Therefore, the economic order quantity is $3 \cdot 150 = 450$ cases.

## 2.7  Applications of Derivatives to Business and Economics

In recent years, economic decision making has become more and more mathematically oriented. Faced with huge masses of statistical data, depending on hundreds or even thousands of different variables, business analysts and economists have increasingly turned to mathematical methods to help them describe what is happening, predict the effects of various policy alternatives, and choose reasonable courses of action from the myriad possibilities. Among the mathematical methods employed is calculus. In this section we illustrate just a few of the many applications of calculus to business and economics. All our applications will center around what economists call *the theory of the firm*. In other words, we study the activity of a business (or possibly a whole industry) and restrict our analysis to a time period during which background conditions (such as supplies of raw materials, wage rates, taxes) are fairly constant. We then show how derivatives can help the management of such a firm make vital production decisions.

Management, whether or not it knows calculus, utilizes many functions of the sort we have been considering. Examples of such functions are

$C(x) = $ cost of producing $x$ units of the product,

$R(x) = $ revenue generated by selling $x$ units of the product,

$P(x) = R(x) - C(x) = $ the profit (or loss) generated by producing and selling $x$ units of the product.

Note that the functions $C(x)$, $R(x)$, $P(x)$ often are defined only for nonnegative integers—that is, for $x = 0, 1, 2, 3, \ldots$. The reason is that it does not make sense to speak about the cost of producing $-1$ cars or the revenue generated by selling 3.62 refrigerators. Thus each of the functions may give rise to a set of discrete points on a graph, as in Fig. 1(a). In studying these functions, however, economists usually draw a smooth curve through the points and assume that $C(x)$ is actually defined for all positive $x$. Of course, we must often interpret answers to problems in light of the fact that $x$ is, in most cases, a nonnegative integer.

**Cost Functions**  If we assume that a cost function $C(x)$ has a smooth graph as in Fig. 1(b), we can use the tools of calculus to study it. A typical cost function is analyzed in Example 1.

---

**EXAMPLE 1**  **Marginal cost analysis**  Suppose that the cost function for a manufacturer is given by $C(x) = (10^{-6})x^3 - .003x^2 + 5x + 1000$ dollars.

(a) Describe the behavior of the marginal cost.

(b) Sketch the graph of $C(x)$.

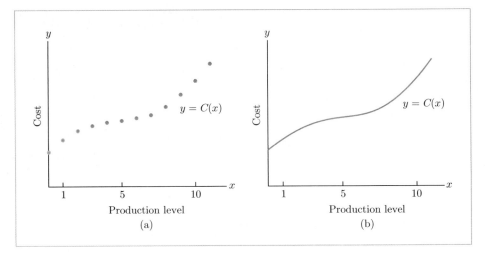

**Figure 1.** A cost function.

**Solution** The first two derivatives of $C(x)$ are given by

$$C'(x) = (3 \cdot 10^{-6})x^2 - .006x + 5$$
$$C''(x) = (6 \cdot 10^{-6})x - .006.$$

Let us sketch the marginal cost $C'(x)$ first. From the behavior of $C'(x)$, we will be able to graph $C(x)$. The marginal cost function $y = (3 \cdot 10^{-6})x^2 - .006x + 5$ has as its graph a parabola that opens upward. Since $y' = C''(x) = .000006(x - 1000)$, we see that the parabola has a horizontal tangent at $x = 1000$. So the minimum value of $C'(x)$ occurs at $x = 1000$. The corresponding $y$-coordinate is

$$(3 \cdot 10^{-6})(1000)^2 - .006 \cdot (1000) + 5 = 3 - 6 + 5 = 2.$$

The graph of $y = C'(x)$ is shown in Fig. 2. Consequently, at first the marginal cost decreases. It reaches a minimum of 2 at production level 1000 and increases thereafter. This answers part (a). Let us now graph $C(x)$. Since the graph shown in Fig. 2 is the graph of the derivative of $C(x)$, we see that $C'(x)$ is never zero, so there are no relative extreme points. Since $C'(x)$ is always positive, $C(x)$ is always increasing (as any cost curve should be). Moreover, since $C'(x)$ decreases for $x$ less than 1000 and increases for $x$ greater than 1000, we see that $C(x)$ is concave down for $x$ less than 1000, is concave up for $x$ greater than 1000, and has an inflection point at $x = 1000$. The graph of $C(x)$ is drawn in Fig. 3. Note that the inflection point of $C(x)$ occurs at the value of $x$ for which marginal cost is a minimum. ∎

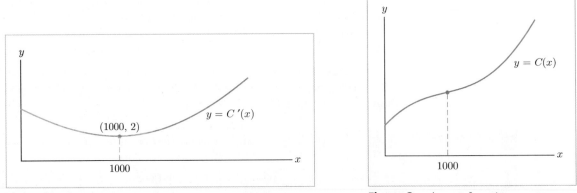

**Figure 2.** A marginal cost function.

**Figure 3.** A cost function.

Actually, most marginal cost functions have the same general shape as the marginal cost curve of Example 1. For when $x$ is small, production of additional units is subject to economies of production, which lowers unit costs. Thus, for $x$ small, marginal cost decreases. However, increased production eventually leads to overtime, use of less efficient, older plants, and competition for scarce raw materials. As a result, the cost of additional units will increase for very large $x$. So we see that $C'(x)$ initially decreases and then increases.

**Revenue Functions**   In general, a business is concerned not only with its costs, but also with its revenues. Recall that if $R(x)$ is the revenue received from the sale of $x$ units of some commodity, then the derivative $R'(x)$ is called the *marginal revenue*. Economists use this to measure the rate of increase in revenue per unit increase in sales.

If $x$ units of a product are sold at a price $p$ per unit, then the total revenue $R(x)$ is given by

$$R(x) = x \cdot p.$$

If a firm is small and is in competition with many other companies, its sales have little effect on the market price. Then since the price is constant as far as the one firm is concerned, the marginal revenue $R'(x)$ equals the price $p$ [that is, $R'(x)$ is the amount that the firm receives from the sale of one additional unit]. In this case, the revenue function will have a graph as in Fig. 4.

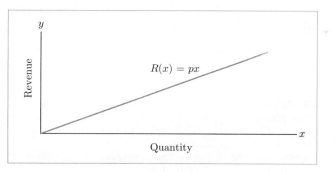

**Figure 4.**   A revenue curve.

An interesting problem arises when a single firm is the only supplier of a certain product or service—that is, when the firm has a monopoly. Consumers will buy large amounts of the commodity if the price per unit is low and less if the price is raised. For each quantity $x$, let $f(x)$ be the highest price per unit that can be set in order to sell all $x$ units to customers. Since selling greater quantities requires a lowering of the price, $f(x)$ will be a decreasing function. Figure 5 shows a typical "demand curve" that relates the quantity demanded, $x$, to the price $p = f(x)$.

The *demand equation* $p = f(x)$ determines the total revenue function. If the firm wants to sell $x$ units, the highest price it can set is $f(x)$ dollars per unit, and so the total revenue from the sale of $x$ units is

$$R(x) = x \cdot p = x \cdot f(x). \tag{1}$$

The concept of a demand curve applies to an entire industry (with many producers) as well as to a single monopolistic firm. In this case, many producers offer the same product for sale. If $x$ denotes the total output of the industry, then $f(x)$ is the market price per unit of output and $x \cdot f(x)$ is the total revenue earned from the sale of the $x$ units.

**EXAMPLE 2**   **Maximizing revenue**   The demand equation for a certain product is $p = 6 - \frac{1}{2}x$ dollars. Find the level of production that results in maximum revenue.

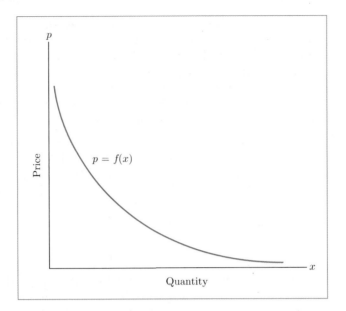

**Figure 5.** A demand curve.

**Solution** In this case, the revenue function $R(x)$ is

$$R(x) = x \cdot p = x \left(6 - \tfrac{1}{2}x\right) = 6x - \tfrac{1}{2}x^2$$

dollars. The marginal revenue is given by

$$R'(x) = 6 - x.$$

The graph of $R(x)$ is a parabola that opens downward (Fig. 6). It has a horizontal tangent precisely at those $x$ for which $R'(x) = 0$—that is, for those $x$ at which marginal revenue is 0. The only such $x$ is $x = 6$. The corresponding value of revenue is

$$R(6) = 6 \cdot 6 - \tfrac{1}{2}(6)^2 = 18 \text{ dollars.}$$

Thus the rate of production resulting in maximum revenue is $x = 6$, which results in total revenue of 18 dollars. ■

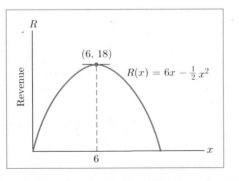

**Figure 6.** Maximizing revenue.

---

**EXAMPLE 3** **Setting up a demand equation** The WMA Bus Lines offers sightseeing tours of Washington, D.C. One of the tours, priced at \$7 per person, had an average demand of about 1000 customers per week. When the price was lowered to \$6, the weekly demand jumped to about 1200 customers. Assuming that the demand equation is linear, find the tour price that should be charged per person in order to maximize the total revenue each week.

**Solution** First we must find the demand equation. Let $x$ be the number of customers per week and let $p$ be the price of a tour ticket. Then $(x, p) = (1000, 7)$ and $(x, p) = (1200, 6)$ are on the demand curve (Fig. 7). Using the point-slope formula for the line through these two points, we have

$$p - 7 = \frac{7 - 6}{1000 - 1200} \cdot (x - 1000) = -\frac{1}{200}(x - 1000) = -\frac{1}{200}x + 5,$$

so

$$p = 12 - \frac{1}{200}x. \tag{2}$$

From equation (1) we obtain the revenue function

$$R(x) = x\left(12 - \frac{1}{200}x\right) = 12x - \frac{1}{200}x^2.$$

The marginal revenue function is

$$R'(x) = 12 - \frac{1}{100}x = -\frac{1}{100}(x - 1200).$$

Using $R(x)$ and $R'(x)$, we can sketch the graph of $R(x)$ (Fig. 8). The maximum revenue occurs when the marginal revenue is zero—that is, when $x = 1200$. The price corresponding to this number of customers is found from the demand equation (2),

$$p = 12 - \frac{1}{200}(1200) = 6 \text{ dollars}.$$

Thus the price of \$6 is most likely to bring the greatest revenue per week. ■

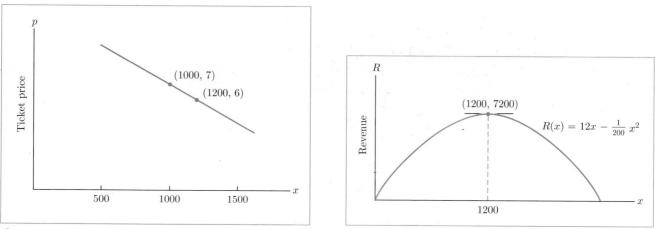

**Figure 7.**  A demand curve.

**Figure 8.**  Maximizing revenue.

**Profit Functions**  Once we know the cost function $C(x)$ and the revenue function $R(x)$, we can compute the profit function $P(x)$ from

$$P(x) = R(x) - C(x).$$

---

**EXAMPLE 4**  **Maximizing profits**  Suppose that the demand equation for a monopolist is $p = 100 - .01x$ and the cost function is $C(x) = 50x + 10{,}000$. Find the value of $x$ that maximizes the profit and determine the corresponding price and total profit for this level of production. (See Fig. 9.)

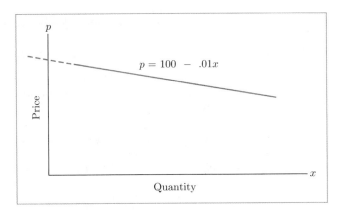

**Figure 9.** A demand curve.

**Solution** The total revenue function is

$$R(x) = x \cdot p = x(100 - .01x) = 100x - .01x^2.$$

Hence the profit function is

$$\begin{aligned} P(x) &= R(x) - C(x) \\ &= 100x - .01x^2 - (50x + 10,000) \\ &= -.01x^2 + 50x - 10,000. \end{aligned}$$

The graph of this function is a parabola that opens downward. (See Fig. 10.) Its highest point will be where the curve has zero slope—that is, where the marginal

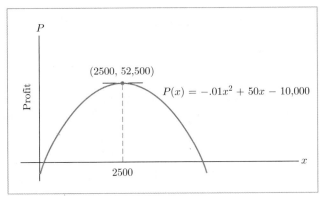

**Figure 10.** Maximizing profit.

profit $P'(x)$ is zero. Now

$$P'(x) = -.02x + 50 = -.02(x - 2500).$$

So $P'(x) = 0$ when $x = 2500$. The profit for this level of production is

$$P(2500) = -.01(2500)^2 + 50(2500) - 10,000 = 52,500 \text{ dollars}.$$

Finally, we return to the demand equation to find the highest price that can be charged per unit to sell all 2500 units:

$$p = 100 - .01(2500) = 100 - 25 = 75 \text{ dollars}.$$

Thus, to maximize the profit, produce 2500 units and sell them at \$75 per unit. The profit will be \$52,500. ∎

**EXAMPLE 5** Rework Example 4 under the condition that the government imposes an excise tax of $10 per unit.

**Solution** For each unit sold, the manufacturer will have to pay $10 to the government. In other words, $10x$ dollars are added to the cost of producing and selling $x$ units. The cost function is now

$$C(x) = (50x + 10{,}000) + 10x = 60x + 10{,}000.$$

The demand equation is unchanged by this tax, so the revenue is still

$$R(x) = 100x - .01x^2.$$

Proceeding as before, we have

$$\begin{aligned} P(x) &= R(x) - C(x) \\ &= 100x - .01x^2 - (60x + 10{,}000) \\ &= -.01x^2 + 40x - 10{,}000. \end{aligned}$$
$$P'(x) = -.02x + 40 = -.02(x - 2000).$$

The graph of $P(x)$ is still a parabola that opens downward, and the highest point is where $P'(x) = 0$—that is, where $x = 2000$. (See Fig. 11.) The corresponding profit is

$$P(2000) = -.01(2000)^2 + 40(2000) - 10{,}000 = 30{,}000 \text{ dollars.}$$

From the demand equation, $p = 100 - .01x$, we find the price that corresponds to $x = 2000$:

$$p = 100 - .01(2000) = 80 \text{ dollars.}$$

To maximize profit, produce 2000 units and sell them at $80 per unit. The profit will be $30,000. ■

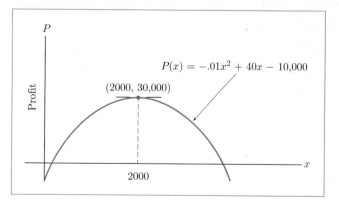

**Figure 11.** Profit after an excise tax.

Notice in Example 5 that the optimal price is raised from $75 to $80. If the monopolist wishes to maximize profits, he or she should pass only half the $10 tax on to the customer. The monopolist cannot avoid the fact that profits will be substantially lowered by the imposition of the tax. This is one reason why industries lobby against taxation.

**Setting Production Levels** Suppose that a firm has cost function $C(x)$ and revenue function $R(x)$. In a free-enterprise economy the firm will set production $x$ in such a way as to maximize the profit function

$$P(x) = R(x) - C(x).$$

We have seen that if $P(x)$ has a maximum at $x = a$, then $P'(a) = 0$. In other words, since $P'(x) = R'(x) - C'(x)$,

$$R'(a) - C'(a) = 0$$
$$R'(a) = C'(a).$$

Thus profit is maximized at a production level for which marginal revenue equals marginal cost. (See Fig. 12.)

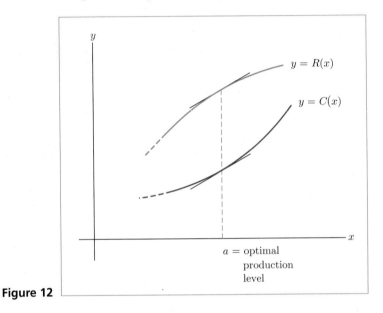

**Figure 12**

## PRACTICE PROBLEMS 2.7

1. Rework Example 4 by finding the production level at which marginal revenue equals marginal cost.

2. Rework Example 4 under the condition that the fixed cost is increased from $10,000 to $15,000.

3. Suppose that, on a certain route, a regional airline carries 8000 passengers per month, each paying $50.

The airline wants to increase the fare. However, the market research department estimates that for each $1 increase in fare, the airline will lose 100 passengers. Determine the price that maximizes the airline's revenue.

## EXERCISES 2.7

1. Given the cost function $C(x) = x^3 - 6x^2 + 13x + 15$, find the minimum marginal cost.

2. Suppose that a total cost function is $C(x) = .0001x^3 - .06x^2 + 12x + 100$. Is the marginal cost increasing, decreasing, or not changing at $x = 100$? Find the minimum marginal cost.

3. The revenue function for a one-product firm is

$$R(x) = 200 - \frac{1600}{x + 8} - x.$$

Find the value of $x$ that results in maximum revenue.

4. The revenue function for a particular product is $R(x) = x(4 - .0001x)$. Find the largest possible revenue.

5. (*Cost and profit*) A one-product firm estimates that its daily total cost function (in suitable units) is $C(x) = x^3 - 6x^2 + 13x + 15$ and its total revenue function is $R(x) = 28x$. Find the value of $x$ that maximizes the daily profit.

6. A small tie shop sells ties for $3.50 each. The daily cost function is estimated to be $C(x)$ dollars, where $x$ is the number of ties sold on a typical day and $C(x) = .0006x^3 - .03x^2 + 2x + 20$. Find the value of $x$ that will maximize the store's daily profit.

7. (*Demand and Revenue*) The demand equation for a certain commodity is

$$p = \tfrac{1}{12}x^2 - 10x + 300,$$

$0 \le x \le 60$. Find the value of $x$ and the corresponding price $p$ that maximize the revenue.

8. The demand equation for a product is $p = 2 - .001x$. Find the value of $x$ and the corresponding price $p$ that maximize the revenue.

9. (*Profit*) Some years ago it was estimated that the demand for steel approximately satisfied the equation $p = 256 - 50x$, and the total cost of producing $x$ units of steel was $C(x) = 182 + 56x$. (The quantity $x$ was measured in millions of tons and the price and total cost were measured in millions of dollars.) Determine the level of production and the corresponding price that maximize the profits.

10. Consider a rectangle in the $xy$-plane, with corners at $(0,0)$, $(a,0)$, $(0,b)$, and $(a,b)$. Suppose that $(a,b)$ lies on the graph of the equation $y = 30 - x$. Find $a$ and $b$ such that the area of the rectangle is maximized. What economic interpretations can be given to your answer if the equation $y = 30 - x$ represents a demand curve and $y$ is the price corresponding to the demand $x$?

11. (*Demand, Revenue, and Profit*) Until recently hamburgers at the city sports arena cost $2 each. The food concessionaire sold an average of 10,000 hamburgers on a game night. When the price was raised to $2.40, hamburger sales dropped off to an average of 8000 per night.

   (a) Assuming a linear demand curve, find the price of a hamburger that will maximize the nightly hamburger revenue.

   (b) Suppose that the concessionaire has fixed costs of $1000 per night and the variable cost is $.60 per hamburger. Find the price of a hamburger that will maximize the nightly hamburger profit.

12. (*Demand and Revenue*) The average ticket price for a concert at the opera house was $50. The average attendance was 4000. When the ticket price was raised to $52, attendance declined to an average of 3800 persons per performance. What should the ticket price be in order to maximize the revenue for the opera house? (Assume a linear demand curve.)

13. (*Demand and Revenue*) An artist is planning to sell signed prints of her latest work. If 50 prints are offered for sale, she can charge $400 each. However, if she makes more than 50 prints, she must lower the price of all the prints by $5 for each print in excess of the 50. How many prints should the artist make in order to maximize her revenue?

14. (*Demand and Revenue*) A swimming club offers memberships at the rate of $200, provided that a minimum of 100 people join. For each member in excess of 100, the membership fee will be reduced $1 per person (for each member). At most 160 memberships will be sold. How many memberships should the club try to sell in order to maximize its revenue?

15. (*Profit*) In the planning of a sidewalk cafe, it is estimated that if there are 12 tables, the daily profit will be $10 per table. Because of overcrowding, for each additional table the profit per table (for every table in the cafe) will be reduced by $.50. How many tables should be provided to maximize the profit from the cafe?

16. (*Demand and Revenue*) A certain toll road averages 36,000 cars per day when charging $1 per car. A survey concludes that increasing the toll will result in 300 fewer cars for each cent of increase. What toll should be charged in order to maximize the revenue?

17. (*Price Setting*) The monthly demand equation for an electric utility company is estimated to be

$$p = 60 - (10^{-5})x,$$

where $p$ is measured in dollars and $x$ is measured in thousands of kilowatt-hours. The utility has fixed costs of $7,000,000 per month and variable costs of $30 per 1000 kilowatt-hours of electricity generated, so that the cost function is

$$C(x) = 7 \cdot 10^6 + 30x.$$

   (a) Find the value of $x$ and the corresponding price for 1000 kilowatt-hours that maximize the utility's profit.

   (b) Suppose that rising fuel costs increase the utility's variable costs from $30 to $40 so that its new cost function is

$$C_1(x) = 7 \cdot 10^6 + 40x.$$

   Should the utility pass all this increase of $10 per thousand kilowatt-hours on to consumers? Explain your answer.

18. (*Taxes, Profit, and Revenue*) The demand equation for a monopolist is $p = 200 - 3x$, and the cost function is

$$C(x) = 75 + 80x - x^2, \qquad 0 \le x \le 40.$$

   (a) Determine the value of $x$ and the corresponding price that maximize the profit.

**(b)** Suppose that the government imposes a tax on the monopolist of $4 per unit quantity produced. Determine the new price that maximizes the profit.

**(c)** Suppose that the government imposes a tax of $T$ dollars per unit quantity produced so that the new cost function is

$$C(x) = 75 + (80 + T)x - x^2, \qquad 0 \le x \le 40.$$

Determine the new value of $x$ that maximizes the monopolist's profit as a function of $T$. Assuming that the monopolist cuts back production to this level, express the tax revenues received by the government as a function of $T$. Finally, determine the value of $T$ that will maximize the tax revenue received by the government.

**19.** (*Interest Rate*) A savings and loan association estimates that the amount of money on deposit will be 1,000,000 times the percentage rate of interest. For instance, a 4% interest rate will generate $4,000,000 in deposits. Suppose the savings and loan association can loan all the money it takes in at 10% interest. What interest rate on deposits generates the greatest profit?

**20.** Let $P(x)$ be the annual profit for a certain product, where $x$ is the amount of money spent on advertising. See Fig. 13.

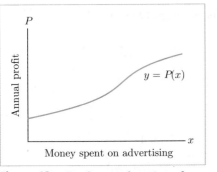

**Figure 13.** Profit as a function of advertising.

**(a)** Interpret $P(0)$.

**(b)** Describe how the marginal profit changes as the amount of money spent on advertising increases.

**(c)** Explain the economic significance of the inflection point.

**21.** The revenue for a manufacturer is $R(x)$ thousand dollars, where $x$ is the number of units of goods produced (and sold) and $R$ and $R'$ are the functions given in Figs. 14(a) and 14(b).

**(a)** What is the revenue from producing 40 units of goods?

**(b)** What is the marginal revenue when 17.5 units of goods are produced?

**(c)** At what level of production is the revenue $45,000?

**(d)** At what level(s) of production is the marginal revenue $800?

**(e)** At what level of production is the revenue greatest?

**22.** The cost function for a manufacturer is $C(x)$ dollars, where $x$ is the number of units of goods produced and $C$, $C'$, and $C''$ are the functions given in Fig. 15.

**(a)** What is the cost of manufacturing 60 units of goods?

**(b)** What is the marginal cost when 40 units of goods are manufactured?

**(c)** At what level of production is the cost $1200?

**(d)** At what level(s) of production is the marginal cost $22.50?

**(e)** At what level of production does the marginal cost have the least value? What is the marginal cost at this level of production?

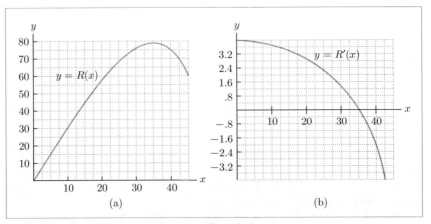

(a)               (b)

**Figure 14.** Revenue function and its first derivative.

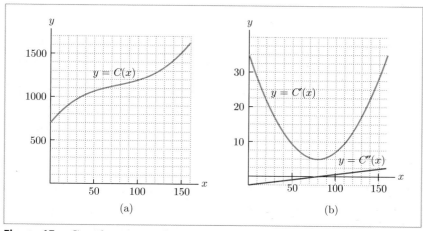

**Figure 15.** Cost function and its derivatives.

## SOLUTIONS TO PRACTICE PROBLEMS 2.7

**1.** The revenue function is $R(x) = 100x - .01x^2$, so the marginal revenue function is $R'(x) = 100 - .02x$. The cost function is $C(x) = 50x + 10,000$, so the marginal cost function is $C'(x) = 50$. Let us now equate the two marginal functions and solve for $x$.

$$R'(x) = C'(x)$$
$$100 - .02x = 50$$
$$-.02x = -50$$
$$x = \frac{-50}{-.02} = \frac{5000}{2} = 2500.$$

Of course, we obtain the same level of production as before.

**2.** If the fixed cost is increased from \$10,000 to \$15,000, the new cost function will be $C(x) = 50x + 15,000$ but the marginal cost function will still be $C'(x) = 50$. Therefore, the solution will be the same: 2500 units should be produced and sold at \$75 per unit. (Increases in fixed costs should not necessarily be passed on to the consumer if the objective is to maximize the profit.)

**3.** Let $x$ denote the number of passengers per month and $p$ the price per ticket. The number of passengers lost due to a fare increase is obtained by multiplying the number of dollars of fare increase, $p - 50$, by the number of passengers lost for each dollar of fare increase. So

$$x = 8000 - (p - 50)100 = -100p + 13,000.$$

Solving for $p$, we get the demand equation

$$p = -\frac{1}{100}x + 130.$$

From (1), the revenue function is

$$R(x) = x \cdot p = x\left(-\frac{1}{100}x + 130\right).$$

The graph is a parabola that opens downward, with $x$-intercepts at $x = 0$ and $x = 13,000$ (Fig. 16). Its maximum is located at the midpoint of the $x$-intercepts, or $x = 6500$. The price corresponding to this number of passengers is $p = -\frac{1}{100}(6500) + 130 = 65$ dollars. Thus the price of \$65 per ticket will bring the highest revenue to the airline company per month.

**Figure 16**

# REVIEW OF FUNDAMENTAL CONCEPTS

1. State as many terms used to describe graphs of functions as you can recall.

2. What is the difference between having a relative maximum at $x = 2$ and having an absolute maximum at $x = 2$?

3. Give three characterizations of what it means for the graph of $f(x)$ to be concave up at $x = 2$. Concave down.

4. What does it mean to say that the graph of $f(x)$ has an inflection point at $x = 2$?

5. What is the difference between an $x$-intercept and a zero of a function?

6. How do you determine the $y$-intercept of a function?

7. What is an asymptote? Give an example.

8. State the first derivative rule. Second derivative rule.

9. Give two connections between the graphs of $f(x)$ and $f'(x)$.

10. Outline a method for locating the relative extreme points of a function.

11. Outline a method for locating the inflection points of a function.

12. Outline a procedure for sketching the graph of a function.

13. What is an objective equation?

14. What is a constraint equation?

15. Outline the procedure for solving an optimization problem.

16. How are the cost, revenue, and profit functions related?

# SUPPLEMENTARY EXERCISES

1. Figure 1 contains the graph of $f'(x)$, the derivative of $f(x)$. Use the graph to answer the following questions about the graph of $f(x)$.

   (a) For what values of $x$ is the graph of $f(x)$ increasing? Decreasing?

   (b) For what values of $x$ is the graph of $f(x)$ concave up? Concave down?

2. Figure 2 shows the graph of the function $f(x)$ and its tangent line at $x = 3$. Find $f(3)$, $f'(3)$, and $f''(3)$.

*In Exercises 3–6, draw the graph of a function $f(x)$ for which the function and its first derivative have the stated property for all $x$.*

3. $f(x)$ and $f'(x)$ increasing.

4. $f(x)$ and $f'(x)$ decreasing.

5. $f(x)$ increasing and $f'(x)$ decreasing.

6. $f(x)$ decreasing and $f'(x)$ increasing.

*Exercises 7–12 refer to the graph in Fig. 3. List the labeled values of $x$ at which the derivative has the stated property.*

7. $f'(x)$ is positive.      8. $f'(x)$ is negative.

9. $f''(x)$ is positive.      10. $f''(x)$ is negative.

11. $f'(x)$ is maximized.      12. $f'(x)$ is minimized.

*Properties of various functions are described next. In each case draw some conclusion about the graph of the function.*

13. $f(1) = 2$, $f'(1) > 0$      14. $g(1) = 5$, $g'(1) = -1$

15. $h'(3) = 4$, $h''(3) = 1$      16. $F'(2) = -1$, $F''(2) < 0$

**Figure 1**

**Figure 2**

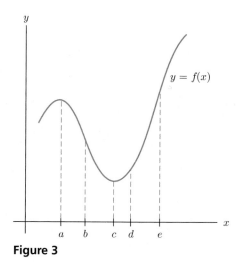

**Figure 3**

**17.** $G(10) = 2$, $G'(10) = 0$, $G''(10) > 0$

**18.** $f(4) = -2$, $f'(4) > 0$, $f''(4) = -1$

**19.** $g(5) = -1$, $g'(5) = -2$, $g''(5) = 0$

**20.** $H(0) = 0$, $H'(0) = 0$, $H''(0) = 1$

**21.** In Figs. 4(a) and 4(b), the $t$-axis represents time in hours.

   **(a)** When is $f(t) = 1$?

   **(b)** Find $f(5)$.

   **(c)** When is $f(t)$ changing at the rate of $-.08$ units per hour?

   **(d)** How fast is $f(t)$ changing after 8 hours?

**22.** U.S. electrical energy production (in trillions of kilowatt-hours) in year $t$ (with 1900 corresponding to $t = 0$) is given by $f(t)$, where $f$ and its derivatives are graphed in Figs. 5(a) and 5(b).

   **(a)** How much electrical energy was produced in 1950?

   **(b)** How fast was energy production rising in 1950?

   **(c)** When did energy production reach 3000 trillion kilowatt-hours?

   **(d)** When was the level of energy production rising at the rate of 10 trillion kilowatt-hours per year?

   **(e)** When was energy production growing at the greatest rate? What was the level of production at that time?

*Sketch the following parabolas. Include their x- and y-intercepts.*

**23.** $y = 3 - x^2$                    **24.** $y = 7 + 6x - x^2$

**25.** $y = x^2 + 3x - 10$          **26.** $y = 4 + 3x - x^2$

**27.** $y = -2x^2 + 10x - 10$     **28.** $y = x^2 - 9x + 19$

**29.** $y = x^2 + 3x + 2$            **30.** $y = -x^2 + 8x - 13$

**31.** $y = -x^2 + 20x - 90$       **32.** $y = 2x^2 + x - 1$

*Sketch the following curves.*

**33.** $y = 2x^3 + 3x^2 + 1$       **34.** $y = x^3 - \frac{3}{2}x^2 - 6x$

**35.** $y = x^3 - 3x^2 + 3x - 2$

**36.** $y = 100 + 36x - 6x^2 - x^3$

**37.** $y = \frac{11}{3} + 3x - x^2 - \frac{1}{3}x^3$

**38.** $y = x^3 - 3x^2 - 9x + 7$

**39.** $y = -\frac{1}{3}x^3 - 2x^2 - 5x$

**40.** $y = x^3 - 6x^2 - 15x + 50$

**41.** $y = x^4 - 2x^2$

**42.** $y = x^4 - 4x^3$

**43.** $y = \frac{x}{5} + \frac{20}{x} + 3 \quad (x > 0)$

**44.** $y = \frac{1}{2x} + 2x + 1 \quad (x > 0)$

(a)                                        (b)

**Figure 4**

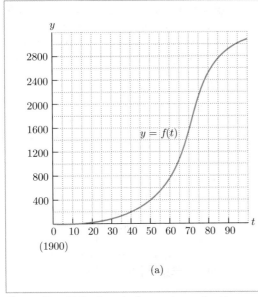

**Figure 5.**   U.S. electrical energy production.

**45.** Let $f(x) = (x^2 + 2)^{3/2}$. Show that the graph of $f(x)$ has a possible relative extreme point at $x = 0$.

**46.** Show that the function $f(x) = (2x^2+3)^{3/2}$ is decreasing for $x < 0$ and increasing for $x > 0$.

**47.** Let $f(x)$ be a function whose *derivative* is

$$f'(x) = \frac{1}{1 + x^2}.$$

Note that $f'(x)$ is always positive. Show that the graph of $f(x)$ definitely has an inflection point at $x = 0$.

**48.** Let $f(x)$ be a function whose *derivative* is

$$f'(x) = \sqrt{5x^2 + 1}.$$

Show that the graph of $f(x)$ definitely has an inflection point at $x = 0$.

**49.** Suppose a car is traveling on a straight road and $s(t)$ is the distance traveled after $t$ hours. Match each set of information about $s(t)$ and its derivatives with the corresponding description of the car's motion.

*Information:*

  **A.** $s(t)$ is a constant function.

  **B.** $s'(t)$ is a positive constant function.

  **C.** $s'(t)$ is positive at $t = a$.

  **D.** $s'(t)$ is negative at $t = a$.

  **E.** $s'(t)$ and $s''(t)$ are positive at $t = a$.

  **F.** $s'(t)$ is positive and $s''(t)$ is negative at $t = a$.

*Descriptions:*

  **a.** The car is moving forward and speeding up at time $a$.

  **b.** The car is backing up at time $a$.

  **c.** The car is standing still.

  **d.** The car is moving forward but slowing down at time $a$.

  **e.** The car is moving forward at a steady rate.

  **f.** The car is moving forward at time $a$.

**50.** The water level in a reservoir varies during the year. Let $h(t)$ be the depth (in feet) of the water at time $t$ days, where $t = 0$ at the beginning of the year. Match each set of information about $h(t)$ and its derivatives with the corresponding description of the reservoir's activity.

*Information:*

  **A.** $h(t)$ has the value 50 for $1 \le t \le 2$.

  **B.** $h'(t)$ has the value .5 for $1 \le t \le 2$.

  **C.** $h'(t)$ is positive at $t = a$.

  **D.** $h'(t)$ is negative at $t = a$.

  **E.** $h'(t)$ and $h''(t)$ are positive at $t = a$.

  **F.** $h'(t)$ is positive and $h''(t)$ is negative at $t = a$.

*Descriptions:*

  **a.** The water level is rising at an increasing rate at time $a$.

  **b.** The water level is receding at time $a$.

  **c.** The water level is constant at 50 feet on January 2.

  **d.** At time $a$ the water level is rising, but the rate of increase is slowing down.

  **e.** On January 2 the water rose steadily at a rate of .5 feet per day.

  **f.** The water level is rising at time $a$.

**51.** Let $f(x)$ be the number of people living within $x$ miles of the center of New York City.

**(a)** What does $f(10+h) - f(10)$ represent?

**(b)** Explain why $f'(10)$ cannot be negative.

**52.** For what $x$ does the function $f(x) = \frac{1}{4}x^2 - x + 2$, $0 \le x \le 8$, have its maximum value?

**53.** Find the maximum value of the function $f(x) = 2 - 6x - x^2$, $0 \le x \le 5$, and give the value of $x$ where this maximum occurs.

**54.** Find the minimum value of the function $g(t) = t^2 - 6t + 9$, $1 \le t \le 6$.

**55.** (*Surface Area*) An open rectangular box is to be 4 feet long and have a volume of 200 cubic feet. Find the dimensions for which the amount of material needed to construct the box is as small as possible.

**56.** (*Volume*) A closed rectangular box with a square base is to be constructed using two different types of wood. The top is made of wood costing $3 per square foot and the remainder is made of wood costing $1 per square foot. Suppose that $48 is available to spend. Find the dimensions of the box of greatest volume that can be constructed.

**57.** (*Volume*) A long rectangular sheet of metal 30 inches wide is to be made into a gutter by turning up strips vertically along the two sides. How many inches should be turned up on each side in order to maximize the amount of water that the gutter can carry?

**58.** (*Maximizing the Total Yield*) A small orchard yields 25 bushels of fruit per tree when planted with 40 trees. Because of overcrowding, the yield per tree (for each tree in the orchard) is reduced by $\frac{1}{2}$ bushel for each additional tree that is planted. How many trees should be planted in order to maximize the total yield of the orchard?

**59.** (*Inventory Control*) A publishing company sells 400,000 copies of a certain book each year. Ordering the entire amount printed at the beginning of the year ties up valuable storage space and capital. However, running off the copies in several partial runs throughout the year results in added costs for setting up each printing run. Setting up each production run costs $1000. The carrying costs, figured on the average number of books in storage, are 50¢ per book. Find the economic lot size, that is, the production run size that minimizes the total setting up and carrying costs.

**60.** (*Area*) A poster is to have an area of 125 square inches. The printed material is to be surrounded by a margin

of 3 inches at the top and margins of 2 inches at the bottom and sides. Find the dimensions of the poster that maximize the area of the printed material.

**61.** (*Profit*) Suppose that the demand equation for a monopolist is $p = 150 - .02x$ and the cost function is $C(x) = 10x + 300$. Find the value of $x$ that maximizes the profit.

**62.** (*Minimizing Time*) Jane wants to drive her tractor from point $A$ on one side of her 5-mile-wide field to a point $B$ on the opposite side of the field, as shown in Fig. 6. Jane could drive her tractor directly across the field to point $C$ and then drive 15 miles down a road to $B$, or she could drive to some point $P$ between $B$ and $C$ and then drive to $B$. If Jane drives her tractor 8 miles per hour across the field and 17 miles per hour on the road, find the point $P$ where she should cut across the field in order to minimize the time it takes to reach $B$. [*Hint*: time $= \frac{\text{distance}}{\text{rate}}$.]

**Figure 6**

**63.** A travel agency offers a boat tour of several Caribbean islands for 3 days and 2 nights. For a group of 12 people, the cost per person is $800. For each additional person above the 12-person minimum, the cost per person is reduced by $20 for each person in the group. The maximum tour group size is 25. What tour group size produces the greatest revenue for the travel agency?

## CHAPTER 3

# Techniques of Differentiation

**3.1** The Product and Quotient Rules
**3.2** The Chain Rule and the General Power Rule
**3.3** Implicit Differentiation and Related Rates

W E have seen that the derivative is useful in many applications. However, our ability to differentiate functions is somewhat limited. For example, we cannot yet readily differentiate

$$(x^2 - 1)^4(x^2 + 1)^5, \qquad \frac{x^3}{(x^2 + 1)^4}.$$

In this chapter we develop differentiation techniques that apply to functions like those just given. Two new rules are the *product rule* and the *quotient rule*. In Section 3.2 we extend the general power rule into a powerful formula called the *chain rule*.

## 3.1 The Product and Quotient Rules

We observed in our discussion of the sum rule for derivatives that the derivative of the sum of two differentiable functions is the sum of the derivatives. Unfortunately, however, the derivative of the product $f(x)g(x)$ is *not* the product of the derivatives. Rather, the derivative of a product is determined from the following rule.

**Product Rule**

$$\frac{d}{dx}[f(x)g(x)] = f(x)g'(x) + g(x)f'(x)$$

*The derivative of the product of two functions is the first function times the derivative of the second plus the second function times the derivative of the first.* At the end of the section we show why this statement is true.

---

**EXAMPLE 1** **Verifying the product rule** Show that the product rule works for the case $f(x) = x^2$, $g(x) = x^3$.

**Solution** Since $x^2 \cdot x^3 = x^5$, we know that

$$\frac{d}{dx}(x^2 \cdot x^3) = \frac{d}{dx}(x^5) = 5x^4.$$

On the other hand, using the product rule,

$$\frac{d}{dx}(x^2 \cdot x^3) = x^2 \cdot \frac{d}{dx}(x^3) + x^3 \cdot \frac{d}{dx}(x^2)$$

$$= x^2(3x^2) + x^3(2x)$$

$$= 3x^4 + 2x^4 = 5x^4.$$

Thus the product rule gives the correct answer. ∎

---

**EXAMPLE 2** **Finding the derivative of a product** Differentiate $y = (2x^3 - 5x)(3x + 1)$.

**Solution** Let $f(x) = 2x^3 - 5x$ and $g(x) = 3x + 1$. Then

$$\frac{d}{dx}\left[(2x^3 - 5x)(3x + 1)\right] = (2x^3 - 5x) \cdot \frac{d}{dx}(3x + 1) + (3x + 1) \cdot \frac{d}{dx}(2x^3 - 5x)$$

$$= (2x^3 - 5x)(3) + (3x + 1)(6x^2 - 5)$$

$$= 6x^3 - 15x + 18x^3 - 15x + 6x^2 - 5$$

$$= 24x^3 + 6x^2 - 30x - 5.$$ ∎

---

**EXAMPLE 3** Apply the product rule to $y = g(x) \cdot g(x)$.

**Solution**
$$\frac{d}{dx}[g(x) \cdot g(x)] = g(x) \cdot g'(x) + g(x) \cdot g'(x)$$
$$= 2g(x)g'(x)$$

This answer is the same as that given by the general power rule:

$$\frac{d}{dx}[g(x) \cdot g(x)] = \frac{d}{dx}[g(x)]^2 = 2g(x)g'(x).$$ ∎

---

**EXAMPLE 4** Find

$$\frac{dy}{dx}, \quad \text{where } y = (x^2 - 1)^4(x^2 + 1)^5.$$

**Solution** Let $f(x) = (x^2 - 1)^4$, $g(x) = (x^2 + 1)^5$, and use the product rule. The general power rule is needed to compute $f'(x)$ and $g'(x)$.

$$\frac{dy}{dx} = (x^2 - 1)^4 \cdot \frac{d}{dx}(x^2 + 1)^5 + (x^2 + 1)^5 \cdot \frac{d}{dx}(x^2 - 1)^4$$

$$= (x^2 - 1)^4 \cdot 5(x^2 + 1)^4(2x) + (x^2 + 1)^5 \cdot 4(x^2 - 1)^3(2x) \qquad (1)$$

This form of $\dfrac{dy}{dx}$ is suitable for some purposes. For example, if we need to compute $\dfrac{dy}{dx}\Big|_{x=2}$, it is easier just to substitute 2 for $x$ than to simplify and then substitute. However, it is often helpful to simplify the formula for $\dfrac{dy}{dx}$, such as in the case when we need to find $x$ such that $\dfrac{dy}{dx} = 0$.

To simplify the answer in (1), we shall write $\dfrac{dy}{dx}$ as a single product rather than as a sum of two products. The first step is to identify the common factors.

$$\frac{dy}{dx} = \overbrace{(x^2 - 1)^4 \cdot 5(x^2 + 1)^4}\underbrace{(2x)} + \overbrace{(x^2 + 1)^5 \cdot 4(x^2 - 1)^3}\underbrace{(2x)}.$$

Both terms contain $2x$ and powers of $x^2 - 1$ and $x^2 + 1$. The most we can factor out of each term is $2x(x^2 - 1)^3(x^2 + 1)^4$. We obtain

$$\frac{dy}{dx} = 2x(x^2 - 1)^3(x^2 + 1)^4[5(x^2 - 1) + 4(x^2 + 1)].$$

Simplifying the right-most factor in this product, we have

$$\frac{dy}{dx} = 2x(x^2 - 1)^3(x^2 + 1)^4[9x^2 - 1]. \tag{2}$$

(See Fig. 3 and Exercise 34 for a related application.) ■

The answers to the exercises for this section appear in two forms, similar to those in (1) and (2). The unsimplified answers will allow you to check if you have mastered the differentiation rules. In each case you should make an effort to transform your original answer into the simplified version. Examples 4 and 6 show how to do this.

**The Quotient Rule**  Another useful formula for differentiating functions is the quotient rule.

> **Quotient Rule**
> $$\frac{d}{dx}\left[\frac{f(x)}{g(x)}\right] = \frac{g(x)f'(x) - f(x)g'(x)}{[g(x)]^2}$$

One must be careful to remember the order of the terms in this formula because of the minus sign in the numerator.

**EXAMPLE 5**  Differentiate

$$y = \frac{x}{2x + 3}.$$

**Solution** Let $f(x) = x$ and $g(x) = 2x + 3$.

$$\frac{d}{dx}\left(\frac{x}{2x+3}\right) = \frac{(2x+3)\cdot\frac{d}{dx}(x) - (x)\cdot\frac{d}{dx}(2x+3)}{(2x+3)^2}$$

$$= \frac{(2x+3)\cdot 1 - x\cdot 2}{(2x+3)^2} = \frac{3}{(2x+3)^2}.$$ ∎

---

**EXAMPLE 6** **Simplifying after the quotient rule** Find

$$\frac{dy}{dx}, \quad \text{where } y = \frac{x^3}{(x^2+1)^4}.$$

**Solution** Let $f(x) = x^3$ and $g(x) = (x^2+1)^4$.

$$\frac{d}{dx}\left(\frac{x^3}{(x^2+1)^4}\right) = \frac{(x^2+1)^4\cdot\frac{d}{dx}(x^3) - (x^3)\cdot\frac{d}{dx}(x^2+1)^4}{[(x^2+1)^4]^2}$$

$$= \frac{(x^2+1)^4\cdot 3x^2 - x^3\cdot 4(x^2+1)^3(2x)}{(x^2+1)^8}$$

If a simplified form of $\dfrac{dy}{dx}$ is desired, we can divide the numerator and the denominator by the common factor $(x^2+1)^3$.

$$\frac{dy}{dx} = \frac{(x^2+1)\cdot 3x^2 - x^3\cdot 4(2x)}{(x^2+1)^5}$$

$$= \frac{3x^4 + 3x^2 - 8x^4}{(x^2+1)^5}$$

$$= \frac{3x^2 - 5x^4}{(x^2+1)^5} = \frac{x^2(3 - 5x^2)}{(x^2+1)^5}$$ ∎

The derivative in Example 6 is a rational function. To simplify it, you can cancel common factors of the entire numerator and denominator, but terms in fractions should never be canceled. Here is another illustration of these rules.

---

**EXAMPLE 7** Differentiate

$$y = \frac{x}{x + (x+1)^3}.$$

**Solution** Let $f(x) = x$ and $g(x) = x + (x+1)^3$. Then $f'(x) = 1$, $g'(x) = 1 + 3(x+1)^2$, and so

$$\frac{d}{dx}\left(\frac{x}{x+(x+1)^3}\right) = \frac{(x+(x+1)^3) - x\cdot(1+3(x+1)^2)}{(x+(x+1)^3)^2}.$$

Although there are several common terms in the denominator and numerator, none of them can be canceled because none of them are common factors of the entire numerator and denominator. You can continue simplifying as follows:

$$\frac{dy}{dx} = \frac{\cancel{x} + (x+1)^3 - \cancel{x} - 3x(x+1)^2}{(x+(x+1)^3)^2}$$

$$= \frac{\overbrace{(x+1)^3} - 3x\overbrace{(x+1)^2}}{(x+(x+1)^3)^2}$$

$$= \frac{(x+1)^2[(x+1) - 3x]}{(x+(x+1)^3)^2} = \frac{(x+1)^2(-2x+1)}{(x+(x+1)^3)^2}. \qquad \blacksquare$$

---

**EXAMPLE 8**    **Minimizing the average cost**    Suppose that the total cost of manufacturing $x$ units of a certain product is given by the function $C(x)$. Then the *average cost per unit*, $AC$, is defined by

$$AC = \frac{C(x)}{x}.$$

Recall that the *marginal cost*, $MC$, is defined by

$$MC = C'(x).$$

Show that at the level of production where the average cost is at a minimum, the average cost equals the marginal cost.

**Solution**    In practice, the marginal cost and average cost curves will have the general shapes shown in Fig. 1. So the minimum point on the average cost curve will occur when $\frac{d}{dx}(AC) = 0$. To compute the derivative, we need the quotient rule,

$$\frac{d}{dx}(AC) = \frac{d}{dx}\left(\frac{C(x)}{x}\right) = \frac{x \cdot C'(x) - C(x)}{x^2}.$$

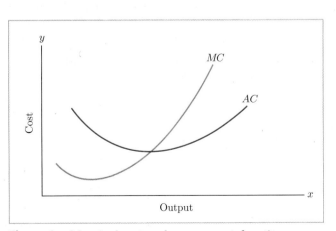

**Figure 1.**    Marginal cost and average cost functions.

Setting the derivative equal to zero and multiplying by $x^2$, we obtain

$$0 = x \cdot C'(x) - C(x)$$

$$C(x) = x \cdot C'(x)$$

$$\frac{C(x)}{x} = C'(x)$$

$$AC = MC.$$

Thus when the output $x$ is chosen so that the average cost is minimized, the average cost equals the marginal cost. ∎

## Verification of the Product and Quotient Rules

*Verification of the Product Rule* From our discussion of limits we compute the derivative of $f(x)g(x)$ at $x = a$ as the limit

$$\frac{d}{dx}[f(x)g(x)]\Big|_{x=a} = \lim_{h \to 0} \frac{f(a+h)g(a+h) - f(a)g(a)}{h}.$$

Let us add and subtract the quantity $f(a)g(a+h)$ in the numerator. After factoring and applying Limit Theorem III, Section 1.4, we obtain

$$\lim_{h \to 0} \frac{[f(a+h)g(a+h) - f(a)g(a+h)] + [f(a)g(a+h) - f(a)g(a)]}{h}$$

$$= \lim_{h \to 0} g(a+h) \cdot \frac{f(a+h) - f(a)}{h} + \lim_{h \to 0} f(a) \cdot \frac{g(a+h) - g(a)}{h}.$$

And this expression may be rewritten by Limit Theorem V, Section 1.4, as

$$\lim_{h \to 0} g(a+h) \cdot \lim_{h \to 0} \frac{f(a+h) - f(a)}{h} + \lim_{h \to 0} f(a) \cdot \lim_{h \to 0} \frac{g(a+h) - g(a)}{h}.$$

Note, however, that since $g(x)$ is differentiable at $x = a$, it is continuous there, so that $\lim_{h \to 0} g(a+h) = g(a)$. Therefore, the preceding expression equals

$$g(a)f'(a) + f(a)g'(a).$$

That is, we have proved that

$$\frac{d}{dx}[f(x)g(x)]\Big|_{x=a} = g(a)f'(a) + f(a)g'(a),$$

which is the product rule. An alternative verification of the product rule not involving limit arguments is outlined in Exercise 66.

*Verification of the Quotient Rule* From the general power rule, we know that

$$\frac{d}{dx}\left[\frac{1}{g(x)}\right] = \frac{d}{dx}[g(x)]^{-1} = (-1)[g(x)]^{-2} \cdot g'(x).$$

We can now derive the quotient rule from the product rule.

$$\frac{d}{dx}\left[\frac{f(x)}{g(x)}\right] = \frac{d}{dx}\left[\frac{1}{g(x)} \cdot f(x)\right]$$

$$= \frac{1}{g(x)} \cdot f'(x) + f(x) \cdot \frac{d}{dx}\left[\frac{1}{g(x)}\right]$$

$$= \frac{g(x)f'(x)}{[g(x)]^2} + f(x) \cdot (-1)[g(x)]^{-2} \cdot g'(x)$$

$$= \frac{g(x)f'(x) - f(x)g'(x)}{[g(x)]^2}.$$

**Incorporating Technology**  A quotient of continuous functions typically has a vertical asymptote at each zero of the denominator. Graphing calculators do not do well drawing graphs with vertical asymptotes. For example, the graph of

$$y = \frac{1}{x-1}$$

has a vertical asymptote at $x = 1$, which is shown on the calculator screen as a spike. The problem is that the standard graphing mode on most calculators (**Connected** mode on TI calculators) connects consecutive dots along the graph. There is no point over $x = 1$ since the function is not defined there. When the calculator connects the dots on either side of the asymptote, a spike results. To display a graph more accurately with an asymptote, most calculators have a **Dot** mode in which consecutive dots are displayed, but not connected with line segments. On the TI-83, the **Dot** mode is selected from the **Mode** screen. The **Dot** mode setting is retained until you change it.

## PRACTICE PROBLEMS 3.1

**1.** Consider the function $y = (\sqrt{x} + 1)x$.

   **(a)** Differentiate $y$ by the product rule.

   **(b)** First multiply out the expression for $y$ and then differentiate.

**2.** Differentiate $y = \dfrac{5}{x^4 - x^3 + 1}$.

## EXERCISES 3.1

*Differentiate the functions in Exercises 1–28.*

**1.** $y = (x+1)(x^3 + 5x + 2)$  **2.** $y = (x^3 + 3x + 11)(x-1)^5$

**3.** $y = (2x^4 - x + 1)(-x^5 + 1)$  **4.** $y = (x^2 + 3)(2x - 1)$

**5.** $y = x(x^2 + 1)^4$  **6.** $y = x^2(7x - 1)^2$

**7.** $y = (x^2 + 3)(x^2 - 3)^{10}$  **8.** $y = [(x-1)(7x+2)]^4$

**9.** $y = (5x+1)(x^2 - 1) + \dfrac{2x+1}{3}$

**10.** $y = x^7(3x^4 + 12x - 1)^2$

**11.** $y = \dfrac{x-1}{x+1}$  **12.** $y = \dfrac{x^2 + 2x - 1}{x^2 + 2x - 2}$

**13.** $y = \dfrac{x^2 - 1}{x^2 + 1}$  **14.** $y = \dfrac{1}{x+1} + \dfrac{1}{x-1}$

**15.** $y = \dfrac{x+3}{(2x+1)^2}$  **16.** $y = \dfrac{2}{7}\dfrac{2x-11}{2x+1}$

**17.** $y = \dfrac{1}{\pi} + \dfrac{2}{x^2 + 1}$  **18.** $y = \dfrac{7}{9} + \dfrac{x^2 + x + 1}{x^5 + 1}$

**19.** $y = \dfrac{3x^2 + 5x + 1}{3 - x^2}$  **20.** $y = \dfrac{x^2}{(x^2 + 1)^2}$

**21.** $y = [(3x^2 + 2x + 2)(x - 2)]^2$  **22.** $y = \dfrac{1}{\sqrt{x} - 2}$

**23.** $y = \dfrac{1}{\sqrt{x} + 1}$  **24.** $y = \dfrac{3}{\sqrt[3]{x} + 1}$

**25.** $y = \dfrac{x^4 - 4x^2 + 3}{x}$  **26.** $y = \dfrac{x+2}{(x+3)(x+4)}$

**27.** $y = \sqrt{x+2}(2x+1)^2$  **28.** $y = \dfrac{\sqrt{3x-1}}{x}$

**29.** Find the equation of the tangent line to the curve $y = (x-2)^5(x+1)^2$ at the point $(3, 16)$.

**30.** Find the equation of the tangent line to the curve $y = (x+1)/(x-1)$ at the point $(2, 3)$.

**31.** Find all $x$-coordinates of points $(x, y)$ on the curve $y = (x-2)^5/(x-4)^3$ where the tangent line is horizontal.

**32.** Find the inflection points on the graph of $y = \dfrac{1}{x^2 + 1}$. (See Fig. 2.)

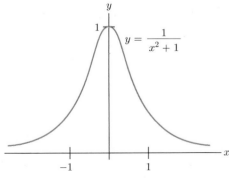

**Figure 2**

**33.** Find all $x$ such that $\dfrac{dy}{dx} = 0$, where

$$y = (x^2 - 4)^3(2x^2 + 5)^5.$$

**34.** The graph of $y = (x^2 - 1)^4(x^2 + 1)^5$ is shown in Fig. 3. Find the coordinates of the local maxima and minima. [*Hint*: : Use Example 4.]

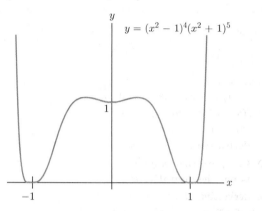

**Figure 3**

**35.** Find the point(s) on the graph of $y = (x^2 + 3x - 1)/x$ where the slope is 5.

**36.** Find the point(s) on the graph of $y = (2x^4 + 1)(x - 5)$ where the slope is 1.

*Find* $\dfrac{d^2y}{dx^2}$.

**37.** $y = (x^2 + 1)^4$           **38.** $y = \sqrt{x^2 + 1}$

**39.** $y = x\sqrt{x + 1}$           **40.** $y = \dfrac{2}{2 + x^2}$

*In Exercises 41–44, a function $h(x)$ is defined in terms of a differentiable $f(x)$. Find an expression for $h'(x)$.*

**41.** $h(x) = xf(x)$           **42.** $h(x) = (x^2 + 2x - 1)f(x)$

**43.** $h(x) = \dfrac{f(x)}{x^2 + 1}$           **44.** $h(x) = \left(\dfrac{f(x)}{x}\right)^2$

**45.** (*Volume*) An open rectangular box is 3 feet long and has a surface area of 16 square feet. Find the dimensions of the box for which the volume is as large as possible.

**46.** (*Volume*) A closed rectangular box is to be constructed with one side 1 meter long. The material for the top costs $20 per square meter, and the material for the sides and bottom costs $10 per square meter. Find the dimensions of the box with the largest possible volume that can be built at a cost of $240 for materials.

**47.** (*Average Cost*) A sugar refinery can produce $x$ tons of sugar per week at a weekly cost of $.1x^2 + 5x + 2250$ dollars. Find the level of production for which the average cost is at a minimum and show that the average cost equals the marginal cost at that level of production.

**48.** (*Average Cost*) A cigar manufacturer produces $x$ cases of cigars per day at a daily cost of $50x(x+200)/(x+100)$ dollars. Show that his cost increases and his average cost decreases as the output $x$ increases.

**49.** (*Average Revenue*) Let $R(x)$ be the revenue received from the sale of $x$ units of a product. The *average revenue per unit* is defined by $AR = R(x)/x$. Show that at the level of production where the average revenue is maximized, the average revenue equals the marginal revenue.

**50.** (*Average Velocity*) Let $s(t)$ be the number of miles a car travels in $t$ hours. Then the average velocity during the first $t$ hours is $\bar{v}(t) = s(t)/t$ miles per hour. Suppose that the average velocity is maximized at time $t_0$. Show that at this time the average velocity $\bar{v}(t_0)$ equals the instantaneous velocity $s'(t_0)$. [*Hint*: Compute the derivative of $\bar{v}(t)$.]

**51.** (*Rate of Change*) The width of a rectangle is increasing at a rate of 3 inches per second and its length is increasing at the rate of 4 inches per second. At what rate is the area of the rectangle increasing when its width is 5 inches and its length is 6 inches? [*Hint*: Let $W(t)$ and $L(t)$ be the widths and lengths, respectively, at time $t$.]

**52.** (*Cost-benefit of Emission Control*) A manufacturer plans to decrease the amount of sulfur dioxide escaping from its smokestacks. The estimated cost-benefit function is

$$f(x) = \frac{3x}{105 - x}, \qquad 0 \le x \le 100,$$

where $f(x)$ is the cost in millions of dollars for eliminating $x$ percent of the total sulfur dioxide. (See Fig. 4.) Find the value of $x$ at which the rate of increase of the cost-benefit function is 1.4 million dollars per unit. (Each unit is 1 percentage point increase in pollutant removed.)

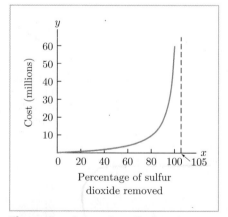

**Figure 4.**   A cost-benefit function.

*In Exercises 53 and 54, use the fact that at the beginning of 1998 the population of the United States was 268,924,000 people and growing at the rate of 1,856,000 people per year.*

**53.** At the beginning of 1998, the annual per capita consumption of gasoline in the United States was 52.3 gallons and growing at the rate of .2 gallons per year. At what rate was the total annual consumption of gasoline in the United States increasing at that time? [*Hint:* [total annual consumption] = [population]·[annual per capita consumption].]

**54.** At the beginning of 1998, the annual consumption of ice cream in the United States was 12,582 million pints and growing at the rate of 212 million pints per year. At what rate was the annual per capita consumption of ice cream increasing at that time? [*Hint:* [annual per capita consumption] = $\frac{\text{annual consumption}}{\text{population size}}$.]

**55.** Figure 5 shows the graph of

$$y = \frac{10x}{1 + .25x^2}$$

for $x \geq 0$. Find the coordinates of the maximum point.

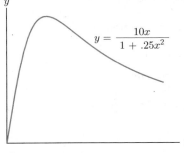

$$y = \frac{10x}{1 + .25x^2}$$

**Figure 5**

**56.** Figure 6 shows the graph of

$$y = \frac{1}{2} + \frac{x^2 - 2x + 1}{x^2 - 2x + 2}$$

for $0 \leq x \leq 2$. Find the coordinates of the minimum point.

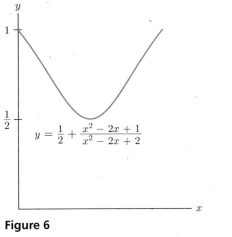

$$y = \frac{1}{2} + \frac{x^2 - 2x + 1}{x^2 - 2x + 2}$$

**Figure 6**

**57.** Show that the derivative of $\dfrac{x^4}{x^2 + 1}$ is not $\dfrac{4x^3}{2x}$.

**58.** Show that the derivative of $(5x^3)(2x^4)$ is not $(15x^2)(8x^3)$.

**59.** Suppose that $f(x)$ is a function whose derivative is $f'(x) = \dfrac{1}{1 + x^2}$. Find the derivative of $\dfrac{f(x)}{1 + x^2}$.

**60.** Suppose that $f(x)$ and $g(x)$ are differentiable functions such that $f(1) = 2$, $f'(1) = 3$, $g(1) = 4$, and $g'(1) = 5$. Find $\dfrac{d}{dx}[f(x)g(x)]\Big|_{x=1}$.

**61.** Consider the functions of Exercise 60. Find $\dfrac{d}{dx}\left[\dfrac{f(x)}{g(x)}\right]\Big|_{x=1}$.

**62.** Suppose that $f(x)$ is a function whose derivative is $f'(x) = 1/x$. Find the derivative of $xf(x) - x$.

**63.** Let $f(x) = 1/x$ and $g(x) = x^3$.

(a) Show that the product rule yields the correct derivative of $(1/x)x^3 = x^2$.

(b) Compute the product $f'(x)g'(x)$ and note that it is *not* the derivative of $f(x)g(x)$.

**64.** The derivative of $(x^3 - 4x)/x$ is obviously $2x$ for $x \neq 0$, because $(x^3 - 4x)/x = x^2 - 4$ for $x \neq 0$. Verify that the quotient rule gives the same derivative.

**65.** Let $f(x)$, $g(x)$, and $h(x)$ be differentiable functions. Find a formula for the derivative of $f(x)g(x)h(x)$. [*Hint:* First differentiate $f(x)[g(x)h(x)]$.]

**66.** (*Alternative Verification of the Product Rule*) Apply the special case of the general power rule

$$\frac{d}{dx}[h(x)]^2 = 2h(x)h'(x)$$

and the identity

$$fg = \tfrac{1}{4}[(f + g)^2 - (f - g)^2]$$

to prove the product rule.

**67.** (*Body Mass Index*) The body mass index, or BMI, is a ratio of a person's weight divided by the square of his or her height. Let $b(t)$ denote the BMI; then

$$b(t) = \frac{w(t)}{[h(t)]^2},$$

where $t$ is the age of the person, $w(t)$ the weight in kilograms, and $h(t)$ the height in meters. Find an expression for $b'(t)$.

**68.** The BMI is usually used as a guideline to determine whether a person is overweight or underweight. For example, according to the Center for Disease Control, a 12-year-old boy is at risk of being overweight if his BMI is between 21 and 24, and is overweight if his BMI is above 24. A 13-year-old boy is at risk of being overweight if his BMI is between 22 and 25, and is overweight if his BMI is above 25.

(a) At age 12, a boy's weight was 50 kilograms and his height was 1.55 meters. Find his BMI and determine whether he is overweight or at risk of being overweight.

**(b)** Suppose that the boy's weight was increasing at the rate of 7 kilograms per year and his height was increasing at the rate of 5 centimeters per year. Find the rate of change of his BMI at age 12 or $b'(12)$.

**(c)** Use the values $b(12)$ and $b'(12)$ to estimate $b(13)$, the BMI of the child at age 13. Is the child at risk of being overweight when he turns 13? [*Hint*: Use formula (2), Section 1.8, to approximate $b(13)$.]

### *Technology Exercises*

**69.** (*Area of the Pupil*) The relationship between the area of the pupil of the eye and the intensity of light was analyzed by B. H. Crawford.* Crawford concluded that

the area of the pupil is

$$f(x) = \frac{160x^{-.4} + 94.8}{4x^{-.4} + 15.8} \qquad (0 \le x \le 37)$$

square millimeters when $x$ units of light are entering the eye per unit time.

**(a)** Graph $f(x)$ and $f'(x)$ in the window $[0, 6]$ by $[-5, 20]$.

**(b)** How large is the pupil when 3 units of light are entering the eye per unit time?

**(c)** For what light intensity is the pupil size 11 square millimeters?

**(d)** When 3 units of light are entering the eye per unit time, what is the rate of change of pupil size with respect to a unit change in light intensity?

## SOLUTIONS TO PRACTICE PROBLEMS 3.1

**1.** **(a)** Apply the product rule to $y = (\sqrt{x} + 1)x$ with

$$f(x) = \sqrt{x} + 1 = x^{1/2} + 1$$
$$g(x) = x.$$

$$\frac{dy}{dx} = (x^{1/2} + 1) \cdot 1 + x \cdot \tfrac{1}{2} x^{-1/2}$$
$$= x^{1/2} + 1 + \tfrac{1}{2} x^{1/2}$$
$$= \tfrac{3}{2} \sqrt{x} + 1.$$

**(b)** $y = (\sqrt{x} + 1)x = (x^{1/2} + 1)x = x^{3/2} + x$

$$\frac{dy}{dx} = \tfrac{3}{2} x^{1/2} + 1.$$

Comparing parts (a) and (b), we note that it is helpful to simplify the function before differentiating.

**2.** We may apply the quotient rule to $y = \dfrac{5}{x^4 - x^3 + 1}$.

$$\frac{dy}{dx} = \frac{(x^4 - x^3 + 1) \cdot 0 - 5 \cdot (4x^3 - 3x^2)}{(x^4 - x^3 + 1)^2}$$
$$= \frac{-5x^2(4x - 3)}{(x^4 - x^3 + 1)^2}.$$

However, it is slightly faster to use the general power rule, since $y = 5(x^4 - x^3 + 1)^{-1}$. Thus

$$\frac{dy}{dx} = -5(x^4 - x^3 + 1)^{-2}(4x^3 - 3x^2)$$
$$= -5x^2(x^4 - x^3 + 1)^{-2}(4x - 3).$$

## 3.2 The Chain Rule and the General Power Rule

In this section we show that the general power rule is a special case of a powerful differentiation technique called the chain rule. Applications of the chain rule appear throughout the text.

A useful way of combining functions $f(x)$ and $g(x)$ is to replace each occurrence of the variable $x$ in $f(x)$ by the function $g(x)$. The resulting function is called the *composition* (or *composite*) of $f(x)$ and $g(x)$ and is denoted by $f(g(x))$.[†]

---

**EXAMPLE 1** **Composition of functions**  Let

$$f(x) = \frac{x - 1}{x + 1}, \quad g(x) = x^3.$$

What is $f(g(x))$?

---

*The dependence of pupil size upon the external light stimulus under static and variable conditions; *Proceedings of the Royal Society Series B*, 121: 376–395, 1937.

[†]See Section 0.3 for additional information on function composition.

**Solution** Replace each occurrence of $x$ in $f(x)$ by $g(x)$ to obtain

$$f(g(x)) = \frac{g(x) - 1}{g(x) + 1} = \frac{x^3 - 1}{x^3 + 1}.$$

Given a composite function $f(g(x))$, we may think of $f(x)$ as the "outside" function that acts on the values of the "inside" function $g(x)$. This point of view often helps us to recognize a composite function.

---

**EXAMPLE 2** **Decomposing composite functions** Write the following functions as composites of simpler functions.

(a) $h(x) = (x^5 + 9x + 3)^8$      (b) $k(x) = \sqrt{4x^2 + 1}$

**Solution** (a) $h(x) = f(g(x))$, where the outside function is the power function, $(\ldots)^8$, that is $f(x) = x^8$. Inside this power function is $g(x) = x^5 + 9x + 3$.

(b) $k(x) = f(g(x))$, where the outside function is the square root function, $f(x) = \sqrt{x}$, and the inside function is $g(x) = 4x^2 + 1$.

A function of the form $[g(x)]^r$ is a composite $f(g(x))$, where the outside function is $f(x) = x^r$. We have already given a rule for differentiating this function—namely,

$$\frac{d}{dx}[g(x)]^r = r[g(x)]^{r-1}g'(x).$$

The *chain rule* has the same form, except that the outside function $f(x)$ can be *any* differentiable function.

> **The Chain Rule** To differentiate $f(g(x))$, first differentiate the outside function $f(x)$ and substitute $g(x)$ for $x$ in the result. Then multiply by the derivative of the inside function $g(x)$. Symbolically,
>
> $$\frac{d}{dx}f(g(x)) = f'(g(x))g'(x).$$

---

**EXAMPLE 3** **The general power rule as a chain rule** Use the chain rule to compute the derivative of $f(g(x))$, where $f(x) = x^8$ and $g(x) = x^5 + 9x + 3$.

**Solution** $$f'(x) = 8x^7, \qquad g'(x) = 5x^4 + 9$$
$$f'(g(x)) = 8(x^5 + 9x + 3)^7.$$

Finally, by the chain rule,

$$\frac{d}{dx}f(g(x)) = f'(g(x))g'(x) = 8(x^5 + 9x + 3)^7(5x^4 + 9).$$

Since in this example the outside function is a power function, the calculations are the same as when the general power rule is used to compute the derivative of $y = (x^5 + 9x + 3)^8$. However, the organization of the solution here emphasizes the notation of the chain rule.

---

**EXAMPLE 4** **Derivative of a function that is not specified** Suppose that $h(x) = f(\sqrt{x})$, where $f$ is a differentiable function that is not specified. Find $h'(x)$ in terms of $f'$.

**Solution** Let $g(x) = \sqrt{x}$; then $h$ is the composition of $f(x)$ and $g(x)$: $h(x) = f(g(x))$. We have $g'(x) = \frac{1}{2\sqrt{x}}$ and, by the chain rule,

$$h'(x) = \frac{d}{dx}(f(g(x)) = f'(g(x))g'(x)$$

$$= f'(\sqrt{x}) \cdot \frac{1}{2\sqrt{x}} = \frac{f'(\sqrt{x})}{2\sqrt{x}}. \qquad \blacksquare$$

There is another way to write the chain rule. Given the function $y = f(g(x))$, set $u = g(x)$ so that $y = f(u)$. Then $y$ may be regarded either as a function of $u$ or, indirectly through $u$, as a function of $x$. With this notation, we have $\dfrac{du}{dx} = g'(x)$ and $\dfrac{dy}{du} = f'(u) = f'(g(x))$. Thus the chain rule says that

$$\frac{dy}{dx} = \frac{dy}{du}\frac{du}{dx}. \qquad (1)$$

Although the derivative symbols in (1) are not really fractions, the apparent cancellation of the "$du$" symbols provides a mnemonic device for remembering this form of the chain rule.

Here is an illustration that makes (1) a plausible formula. Suppose that $y$, $u$, and $x$ are three varying quantities, with $y$ varying three times as fast as $u$ and $u$ varying two times as fast as $x$. It seems reasonable that $y$ should vary six times as fast as $x$. That is, $\dfrac{dy}{dx} = \dfrac{dy}{du}\dfrac{du}{dx} = 3 \cdot 2 = 6$.

---

**EXAMPLE 5** **The chain rule** Find

$$\frac{dy}{dx} \quad \text{if } y = u^5 - 2u^3 + 8 \text{ and } u = x^2 + 1.$$

**Solution** Since $y$ is not given directly as a function of $x$, we cannot compute $\frac{dy}{dx}$ by differentiating $y$ directly with respect to $x$. We can, however, differentiate with respect to $u$ the relation $y = u^5 - 2u^3 + 8$, and get

$$\frac{dy}{du} = 5u^4 - 6u^2.$$

Similarly, we can differentiate with respect to $x$ the relation $u = x^2 + 1$ and get

$$\frac{du}{dx} = 2x.$$

Applying the chain rule, as prescribed by (1), we obtain

$$\frac{dy}{dx} = \frac{dy}{du}\frac{du}{dx} = (5u^4 - 6u^2) \cdot (2x).$$

It is usually desirable to express $\dfrac{dy}{dx}$ as a function of $x$ alone, so we substitute $x^2 + 1$ for $u$ to obtain

$$\frac{dy}{dx} = [5(x^2 + 1)^4 - 6(x^2 + 1)^2] \cdot 2x. \qquad \blacksquare$$

You may have noticed another entirely mechanical way to work Example 5. Namely, first substitute $u = x^2 + 1$ into the original formula for $y$ and obtain $y = (x^2 + 1)^5 - 2(x^2 + 1)^3 + 8$. Then $\dfrac{dy}{dx}$ is easily calculated by the sum rule, the general power rule, and the constant-multiple rule. The solution in Example 5, however, lays the foundation for applications of the chain rule in Section 3.3 and elsewhere.

In many situations involving composition of functions, the basic variable is time, $t$. It may happen that $x$ is a function of $t$, say $x = g(t)$, and some other variable such as $R$ is a function of $x$, say $R = f(x)$. Then $R = f(g(t))$, and the chain rule says that

$$\frac{dR}{dt} = \frac{dR}{dx}\frac{dx}{dt}.$$

---

**EXAMPLE 6**  **Marginal revenue and time rate of change**   A store sells ties for $12 apiece. Let $x$ be the number of ties sold in one day and let $R$ be the revenue received from the sale of $x$ ties, so that $R = 12x$. Suppose that daily sales are rising at the rate of four ties per day. How fast is the revenue rising?

**Solution**  Clearly, revenue is rising at the rate of $48 per day, because each of the additional four ties brings in $12. This intuitive conclusion also follows from the chain rule,

$$\frac{dR}{dt} \qquad = \qquad \frac{dR}{dx} \qquad \cdot \qquad \frac{dx}{dt}$$

$$\begin{bmatrix} \text{rate of change} \\ \text{of revenue with} \\ \text{respect to time} \end{bmatrix} = \begin{bmatrix} \text{rate of change} \\ \text{of revenue with} \\ \text{respect to sales} \end{bmatrix} \cdot \begin{bmatrix} \text{rate of change} \\ \text{of sales with} \\ \text{respect to time} \end{bmatrix}$$

$$\begin{bmatrix} \$48 \text{ increase} \\ \text{per day} \end{bmatrix} = \begin{bmatrix} \$12 \text{ increase} \\ \text{per add'l tie} \end{bmatrix} \cdot \begin{bmatrix} \text{four additional} \\ \text{ties per day} \end{bmatrix}.$$

Notice that $\dfrac{dR}{dx}$ is actually the marginal revenue, studied earlier. This example shows that the time rate of change of revenue, $\dfrac{dR}{dt}$, is the marginal revenue multiplied by the time rate of change of sales.   ■

---

**EXAMPLE 7**  **Time rate of change of revenue**   The demand equation for a certain brand of graphing calculators is $p = 86 - .002x$, where $p$ is the price of one calculator (in dollars) and $x$ is the number of calculators produced and sold. Determine the (time) rate of change of the total revenue if the company increases production by 200 calculators per day, when the production level is at 6000 calculators.

**Solution**  The total revenue received from the production and sale of $x$ calculators is $R(x) = x \cdot p = x(86 - .002x) = -.002x^2 + 86x$. We are told that, when $x = 6000$, the production will increase at the rate of 200 calculators per day; that is, $\frac{dx}{dt} = 200$, when $x = 6000$. We are asked to find the time rate of change of revenue, $\frac{dR}{dt}$, when $x = 6000$. Since $R$ is not given directly as a function of $t$, to differentiate $R$ with respect to $t$, we appeal to the chain rule. We have

$$\frac{dR}{dt} = \frac{dR}{dx} \cdot \frac{dx}{dt}.$$

(As in the previous example, the time rate of change of revenue, $\frac{dR}{dt}$, is the marginal revenue, $\frac{dR}{dx}$, multiplied by the time rate of change of sales, $\frac{dx}{dt}$.) Now

$$\frac{dR}{dx} = \frac{d}{dx}\overbrace{(-.002x^2 + 86x)}^{R(x)} = -.004x + 86.$$

So

$$\frac{dR}{dt} = \overbrace{(-.004x + 86)}^{dR/dx} \cdot \frac{dx}{dt}.$$

Substituting $x = 6000$ and $\frac{dx}{dt} = 200$, we find

$$\frac{dR}{dt} = (-.004(6000) + 86)(200) = 12{,}400 \text{ dollars per day.}$$

Thus if at production level $x = 6000$ the company increases production by 200 calculators per day, its total revenue will increase at the rate of \$12,400 per day.  ■

**Verification of the Chain Rule**  Suppose that $f(x)$ and $g(x)$ are differentiable, and let $x = a$ be a number in the domain of $f(g(x))$. Since every differentiable function is continuous, we have

$$\lim_{h \to 0} g(a + h) = g(a),$$

which implies that

$$\lim_{h \to 0} [g(a + h) - g(a)] = 0. \tag{2}$$

Now $g(a)$ is a number in the domain of $f$, and the limit definition of the derivative gives us

$$f'(g(a)) = \lim_{k \to 0} \frac{f(g(a) + k) - f(g(a))}{k}. \tag{3}$$

Let $k = g(a + h) - g(a)$. By equation (2), $k$ approaches zero as $h$ approaches zero. Also, $g(a + h) = g(a) + k$. Therefore, (3) may be rewritten in the form

$$f'(g(a)) = \lim_{h \to 0} \frac{f(g(a + h)) - f(g(a))}{g(a + h) - g(a)}. \tag{4}$$

[Strictly speaking, we must assume that the denominator in (4) is never zero. This assumption may be avoided by a somewhat different and more technical argument that we omit.] Finally, we show that the function $f(g(x))$ has a derivative at $x = a$. We use the limit definition of the derivative, Limit Theorem V, and equation (4).

$$\left. \frac{d}{dx} f(g(x)) \right|_{x=a} = \lim_{h \to 0} \frac{f(g(a + h)) - f(g(a))}{h}$$

$$= \lim_{h \to 0} \left[ \frac{f(g(a + h)) - f(g(a))}{g(a + h) - g(a)} \cdot \frac{g(a + h) - g(a)}{h} \right]$$

$$= \lim_{h \to 0} \frac{f(g(a + h)) - f(g(a))}{g(a + h) - g(a)} \cdot \lim_{h \to 0} \frac{g(a + h) - g(a)}{h}$$

$$= f'(g(a)) \cdot g'(a)$$

## PRACTICE PROBLEMS 3.2

Consider the function $h(x) = (2x^3 - 5)^5 + (2x^3 - 5)^4$.

1. Write $h(x)$ as a composite function, $f(g(x))$.

2. Compute $f'(x)$ and $f'(g(x))$.

3. Use the chain rule to differentiate $h(x)$.

# EXERCISES 3.2

Compute $f(g(x))$, where $f(x)$ and $g(x)$ are the following.

**1.** $f(x) = \dfrac{x}{x+1}$, $g(x) = x^3$

**2.** $f(x) = x - 1$, $g(x) = \dfrac{1}{x+1}$

**3.** $f(x) = x(x^2 + 1)$, $g(x) = \sqrt{x}$

**4.** $f(x) = \dfrac{x+1}{x-3}$, $g(x) = x + 3$

Each of the following functions may be viewed as a composite function $h(x) = f(g(x))$. Find $f(x)$ and $g(x)$.

**5.** $h(x) = (x^3 + 8x - 2)^5$ **6.** $h(x) = (9x^2 + 2x - 5)^7$

**7.** $h(x) = \sqrt{4 - x^2}$ **8.** $h(x) = (5x^2 + 1)^{-1/2}$

**9.** $h(x) = \dfrac{1}{x^3 - 5x^2 + 1}$ **10.** $h(x) = (4x - 3)^3 + \dfrac{1}{4x - 3}$

Differentiate the functions in Exercises 11–20 using one or more of the differentiation rules discussed thus far.

**11.** $y = (x^2 + 5)^{15}$ **12.** $y = (x^4 + x^2)^{10}$

**13.** $y = 6x^2(x - 1)^3$ **14.** $y = 5x^3(2 - x)^4$

**15.** $y = 2(x^3 - 1)(3x^2 + 1)^4$

**16.** $y = 2(2x - 1)^{5/4}(2x + 1)^{3/4}$

**17.** $y = \left(\dfrac{4}{1-x}\right)^3$ **18.** $y = \dfrac{4x^2 + x}{\sqrt{x}}$

**19.** $y = \left(\dfrac{4x - 1}{3x + 1}\right)^3$ **20.** $y = \left(\dfrac{x}{x^2 + 1}\right)^2$

In Exercises 21–26, a function $h(x)$ is defined in terms of a differentiable $f(x)$. Find an expression for $h'(x)$.

**21.** $h(x) = f(x^2)$ **22.** $h(x) = 2f(2x + 1)$

**23.** $h(x) = -f(-x)$ **24.** $h(x) = f(f(x))$

**25.** $h(x) = \dfrac{f(x^2)}{x}$ **26.** $h(x) = \sqrt{f(x^2)}$

**27.** Sketch the graph of $y = 4x/(x + 1)^2$, $x > -1$.

**28.** Sketch the graph of $y = 2/(1 + x^2)$.

Compute $\dfrac{d}{dx} f(g(x))$, where $f(x)$ and $g(x)$ are the following.

**29.** $f(x) = x^5$, $g(x) = 6x - 1$

**30.** $f(x) = \sqrt{x}$, $g(x) = x^2 + 1$

**31.** $f(x) = \dfrac{1}{x}$, $g(x) = 1 - x^2$

**32.** $f(x) = \dfrac{1}{1 + \sqrt{x}}$, $g(x) = \dfrac{1}{x}$

**33.** $f(x) = x^4 - x^2$, $g(x) = x^2 - 4$

**34.** $f(x) = \dfrac{4}{x} + x^2$, $g(x) = 1 - x^4$

**35.** $f(x) = (x^3 + 1)^2$, $g(x) = x^2 + 5$

**36.** $f(x) = x(x - 2)^4$, $g(x) = x^3$

Compute $\dfrac{dy}{dx}$ using the chain rule in formula (1).

**37.** $y = u^{3/2}$, $u = 4x + 1$

**38.** $y = \sqrt{u + 1}$, $u = 2x^2$

**39.** $y = \dfrac{u}{2} + \dfrac{2}{u}$, $u = x - x^2$

**40.** $y = \dfrac{u^2 + 2u}{u + 1}$, $u = x(x + 1)$

Compute $\dfrac{dy}{dt}\bigg|_{t=t_0}$.

**41.** $y = x^2 - 3x$, $x = t^2 + 3$, $t_0 = 0$.

**42.** $y = (x^2 - 2x + 4)^2$, $x = \dfrac{1}{t+1}$, $t_0 = 1$.

**43.** $y = \dfrac{x+1}{x-1}$, $x = \dfrac{t^2}{4}$, $t_0 = 3$.

**44.** $y = \sqrt{x + 1}$, $x = \sqrt{t + 1}$, $t_0 = 0$.

**45.** Find the equation of the line tangent to the graph of $y = 2x(x - 4)^6$ at the point $(5, 10)$.

**46.** Find the equation of the line tangent to the graph of $y = \dfrac{x}{\sqrt{2 - x^2}}$ at the point $(1, 1)$.

**47.** Find the $x$-coordinates of all points on the curve $y = (-x^2 + 4x - 3)^3$ with a horizontal tangent line.

**48.** The function $f(x) = \sqrt{x^2 - 6x + 10}$ has one relative minimum point for $x \geq 0$. Find it.

**49.** The length, $x$, of the edge of a cube is increasing.

(a) Write the chain rule for $\dfrac{dV}{dt}$, the time rate of change of the volume of the cube.

(b) For what value of $x$ is $\dfrac{dV}{dt}$ equal to 12 times the rate of increase of $x$?

50. *(Allometric Equation)* Many relations in biology are expressed by power functions, known as **allometric equations**, of the form $y = kx^a$, where $k$ and $a$ are constants. For example, the weight of a male hognose snake is approximately $446x^3$ grams, where $x$ is its length in meters.* Suppose a snake has length .4 meters and is growing at the rate of .2 meters per year. At what rate is the snake gaining weight?

51. Suppose that $P$, $y$, and $t$ are variables, where $P$ is a function of $y$ and $y$ is a function of $t$.

   (a) Write the derivative symbols for the following quantities: the rate of change of $y$ with respect to $t$, the rate of change of $P$ with respect to $y$, and the rate of change of $P$ with respect to $t$. Select your answers from the following:

   $$\frac{dP}{dy}, \quad \frac{dy}{dP}, \quad \frac{dy}{dt}, \quad \frac{dP}{dt}, \quad \frac{dt}{dP}, \quad \text{and} \quad \frac{dt}{dy}.$$

   (b) Write the chain rule for $\dfrac{dP}{dt}$.

52. Suppose that $Q$, $x$, and $y$ are variables, where $Q$ is a function of $x$ and $x$ is a function of $y$. (Read this carefully.)

   (a) Write the derivative symbols for the following quantities: the rate of change of $x$ with respect to $y$, the rate of change of $Q$ with respect to $y$, and the rate of change of $Q$ with respect to $x$. Select your answers from the following:

   $$\frac{dy}{dx}, \quad \frac{dx}{dy}, \quad \frac{dQ}{dx}, \quad \frac{dx}{dQ}, \quad \frac{dQ}{dy}, \quad \text{and} \quad \frac{dy}{dQ}.$$

   (b) Write the chain rule for $\dfrac{dQ}{dy}$.

53. *(Marginal Profit and Time Rate of Change)* When a company produces and sells $x$ thousand units per week its total weekly profit is $P$ thousand dollars, where

   $$P = \frac{200x}{100 + x^2}.$$

   The production level at $t$ weeks from the present is $x = 4 + 2t$.

   (a) Find the marginal profit, $\dfrac{dP}{dx}$.

   (b) Find the time rate of change of profit, $\dfrac{dP}{dt}$.

   (c) How fast (with respect to time) are profits changing when $t = 8$?

54. *(Marginal Cost and Time Rate of Change)* The cost of manufacturing $x$ cases of cereal is $C$ dollars, where $C = 3x + 4\sqrt{x} + 2$. Weekly production at $t$ weeks from the present is estimated to be $x = 6200 + 100t$ cases.

   (a) Find the marginal cost, $\dfrac{dC}{dx}$.

   (b) Find the time rate of change of cost, $\dfrac{dC}{dt}$.

   (c) How fast (with respect to time) are costs rising when $t = 2$?

55. *(A Model for Carbon Monoxide Levels)* Ecologists estimate that when the population of a certain city is $x$ thousand persons, the average level $L$ of carbon monoxide in the air above the city will be $L$ ppm (parts per million), where $L = 10 + .4x + .0001x^2$. The population of the city is estimated to be $x = 752 + 23t + .5t^2$ thousand persons $t$ years from the present.

   (a) Find the rate of change of carbon monoxide with respect to the population of the city.

   (b) Find the time rate of change of the population when $t = 2$.

   (c) How fast (with respect to time) is the carbon monoxide level changing at time $t = 2$?

56. *(Profit)* A manufacturer of microcomputers estimates that $t$ months from now it will sell $x$ thousand units of its main line of microcomputers per month, where $x = .05t^2 + 2t + 5$. Because of economies of scale, the profit $P$ from manufacturing and selling $x$ thousand units is estimated to be $P = .001x^2 + .1x - .25$ million dollars. Calculate the rate at which the profit will be increasing 5 months from now.

57. Suppose that $f(x)$ and $g(x)$ are differentiable functions. Find $g(x)$ if you know that

   $$\frac{d}{dx}f(g(x)) = 3x^2 \cdot f'(x^3 + 1).$$

58. Suppose that $f(x)$ and $g(x)$ are differentiable functions. Find $g(x)$ if you know that $f'(x) = 1/x$ and

   $$\frac{d}{dx}f(g(x)) = \frac{2x + 5}{x^2 + 5x - 4}.$$

59. Suppose that $f(x)$ and $g(x)$ are differentiable functions such that $f(1) = 2$, $f'(1) = 3$, $f'(5) = 4$, $g(1) = 5$, $g'(1) = 6$, $g'(2) = 7$, and $g'(5) = 8$. Find $\left.\dfrac{d}{dx}f(g(x))\right|_{x=1}$.

60. Consider the functions of Exercise 59. Find $\left.\dfrac{d}{dx}g(f(x))\right|_{x=1}$.

---

*D. R. Platt, "Natural History of the Hognose Snakes *Heterodon Platyrhinos* and *Heterodon Nasicus*," University of Kansas Publications, *Museum of Natural History* 18:4 (1969), 253–420.

**61.** *(Effect of Stocks on Total Assets of a Company)* After a computer software company went public, the price of one of its shares on the stock market fluctuated according to the graph in Fig. 1(a). The total worth of the company depended on the value of one of its shares and was estimated to be

$$W(x) = 10\frac{12 + 8x}{3 + x},$$

where $x$ is the value of one share (in dollars) and $W(x)$ is the total value of the company in millions of dollars [see Fig. 1(b)].

**(a)** Find the total value of the company when $t = 1.5$ and when $t = 3.5$.

**(b)** Find $\left.\dfrac{dx}{dt}\right|_{t=1.5}$ and $\left.\dfrac{dx}{dt}\right|_{t=3.5}$. Give an interpretation for these values.

**62.** Refer to Exercise 61. Use the chain rule to find $\left.\dfrac{dW}{dt}\right|_{t=1.5}$ and $\left.\dfrac{dW}{dt}\right|_{t=3.5}$. Give an interpretation for these values.

**63.** Refer to Exercise 61.

**(a)** Find $\left.\dfrac{dx}{dt}\right|_{t=2.5}$ and $\left.\dfrac{dx}{dt}\right|_{t=4}$. Give an interpretation for these values.

**(b)** Use the chain rule to find $\left.\dfrac{dW}{dt}\right|_{t=2.5}$ and $\left.\dfrac{dW}{dt}\right|_{t=4}$. Give an interpretation for these values.

**64.** Refer to Exercise 61.

**(a)** What was the maximum value of the company during the first six months since it went public and when was it attained?

**(b)** Assuming that the value of one share will continue to increase at the rate that it did during the period following the sixth month, what is the limit of the total value of the company as $t$ increases?

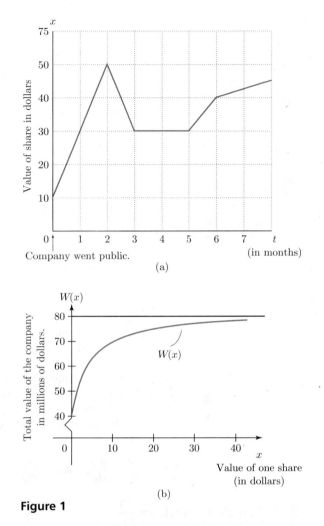

**Figure 1**

**65.** In an expression of the form $f(g(x))$, $f(x)$ is called the *outer* function and $g(x)$ is called the *inner* function. Give a written description of the chain rule using the words "inner" and "outer."

## SOLUTIONS TO PRACTICE PROBLEMS 3.2

**1.** Let $f(x) = x^5 + x^4$ and $g(x) = 2x^3 - 5$.

**2.** $f'(x) = 5x^4 + 4x^3$, $f'(g(x)) = 5(2x^3 - 5)^4 + 4(2x^3 - 5)^3$.

**3.** We have $g'(x) = 6x^2$. Then, from the chain rule and

the result of Problem 2, we have

$$h'(x) = f'(g(x))g'(x) = [5(2x^3 - 5)^4 + 4(2x^3 - 5)^3](6x^2).$$

# 3.3 Implicit Differentiation and Related Rates

This section presents two different applications of the chain rule. In each case we will need to differentiate one or more composite functions where the "inside" functions are not known explicitly.

## Implicit Differentiation

In some applications, the variables are related by an equation rather than a function. In these cases, we can still determine the rate of change of one variable with respect to the other by the technique of implicit differentiation. As an illustration, consider the equation

$$x^2 + y^2 = 4. \tag{1}$$

The graph of this equation is the circle in Fig. 1. This graph obviously is not the graph of a function since, for instance, there are two points on the graph whose $x$-coordinate is 1. (Functions must satisfy the vertical line test. See Section 0.1.)

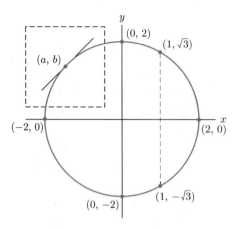

**Figure 1.** Graph of $x^2 + y^2 = 4$.

We denote the slope of the curve at the point $(1, \sqrt{3})$ by

$$\left. \frac{dy}{dx} \right|_{\substack{x=1 \\ y=\sqrt{3}}}.$$

In general, the slope at the point $(a, b)$ is denoted by

$$\left. \frac{dy}{dx} \right|_{\substack{x=a \\ y=b}}.$$

In a small vicinity of the point $(a, b)$, the curve looks like the graph of a function. [That is, on this part of the curve, $y = g(x)$ for some function $g(x)$.] We say that this function is defined *implicitly* by the equation.* We obtain a formula for $\frac{dy}{dx}$ by differentiating both sides of the equation with respect to $x$ while treating $y$ as a function of $x$.

---

**EXAMPLE 1**  Consider the graph of the equation $x^2 + y^2 = 4$.

(a) Use implicit differentiation to compute $\dfrac{dy}{dx}$.

(b) Find the slope of the graph at the points $(1, \sqrt{3})$ and $(1, -\sqrt{3})$.

**Solution** (a) The first term $x^2$ has derivative $2x$, as usual. We think of the second term $y^2$ as having the form $[g(x)]^2$. To differentiate we use the chain rule (specifically, the general power rule):

$$\frac{d}{dx}[g(x)]^2 = 2[g(x)]g'(x)$$

---

*Of course, the tangent line at the point $(a, b)$ must not be vertical. In this section, we assume that the given equations implicitly determine differentiable functions.

or, equivalently,

$$\frac{d}{dx}y^2 = 2y\frac{dy}{dx}.$$

On the right side of the original equation, the derivative of the constant function 4 is zero. Thus implicit differentiation of $x^2 + y^2 = 4$ yields

$$2x + 2y\frac{dy}{dx} = 0.$$

Solving for $\frac{dy}{dx}$, we have

$$2y\frac{dy}{dx} = -2x.$$

If $y \neq 0$, then

$$\frac{dy}{dx} = \frac{-2x}{2y} = -\frac{x}{y}.$$

Notice that this slope formula involves $y$ as well as $x$. This reflects the fact that the slope of the circle at a point depends on the $y$-coordinate of the point as well as the $x$-coordinate.

(b) At the point $(1, \sqrt{3})$ the slope is

$$\frac{dy}{dx}\bigg|_{\substack{x=1 \\ y=\sqrt{3}}} = -\frac{x}{y}\bigg|_{\substack{x=1 \\ y=\sqrt{3}}} = -\frac{1}{\sqrt{3}}.$$

At the point $(1, -\sqrt{3})$ the slope is

$$\frac{dy}{dx}\bigg|_{\substack{x=1 \\ y=-\sqrt{3}}} = -\frac{x}{y}\bigg|_{\substack{x=1 \\ y=-\sqrt{3}}} = -\frac{1}{-\sqrt{3}} = \frac{1}{\sqrt{3}}.$$

(See Fig. 2.) The formula for $\frac{dy}{dx}$ gives the slope at every point on the graph of $x^2 + y^2 = 4$ except $(-2, 0)$ and $(2, 0)$. At these two points the tangent line is vertical and the slope of the curve is undefined. ∎

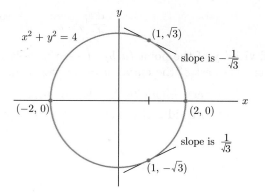

**Figure 2.** Slope of the tangent line.

The difficult step in Example 1(a) was to differentiate $y^2$ correctly. The derivative of $y^2$ *with respect to $y$* would be $2y$, by the ordinary power rule. But the derivative of $y^2$ *with respect to $x$* must be computed by the general power rule. To understand this distinction, think of $y$ as a function of $x$, say, $y = f(x)$. Then

$$\frac{dy}{dx} = f'(x);$$
$$y^2 = [f(x)]^2;$$
$$\frac{d}{dx}y^2 = \frac{d}{dx}([f(x)]^2) = 2f(x)f'(x),$$

by the general power rule. Substituting $y = f(x)$ and $f'(x) = \frac{dy}{dx}$, we obtain

$$\frac{d}{dx}y^2 = 2y\frac{dy}{dx}.$$

Similarly, if $a$ is a constant, then

$$\frac{d}{dx}(ay) = a\frac{dy}{dx};$$

$$\frac{d}{dx}(ay^2) = 2ay\frac{dy}{dx}.$$

Also, if $r$ is a constant, the general power rule yields

$$\frac{d}{dx}y^r = ry^{r-1}\frac{dy}{dx}. \tag{2}$$

This rule is used to compute slope formulas in the next two examples.

---

**EXAMPLE 2** Use implicit differentiation to calculate

$$\frac{dy}{dx}$$

for the equation $x^2y^6 = 1$.

**Solution** Differentiate each side of the equation $x^2y^6 = 1$ with respect to $x$. On the left side of the equation use the product rule and treat $y$ as a function of $x$.

$$x^2\frac{d}{dx}(y^6) + y^6\frac{d}{dx}(x^2) = \frac{d}{dx}(1)$$

$$x^2 \cdot 6y^5\frac{dy}{dx} + y^6 \cdot 2x = 0.$$

Solve for $\frac{dy}{dx}$ by moving the term not involving $\frac{dy}{dx}$ to the right side and dividing by the factor that multiplies $\frac{dy}{dx}$.

$$6x^2y^5\frac{dy}{dx} = -2xy^6$$

$$\frac{dy}{dx} = \frac{-2xy^6}{6x^2y^5} = -\frac{y}{3x}. \qquad \blacksquare$$

---

**EXAMPLE 3** Use implicit differentiation to calculate

$$\frac{dy}{dx}$$

when $y$ is related to $x$ by the equation $x^2y + xy^3 - 3x = 5$.

**Solution** Differentiate the equation term by term, taking care to differentiate $x^2y$ and $xy^3$ by the product rule.

$$x^2\frac{d}{dx}(y) + y\frac{d}{dx}(x^2) + x\frac{d}{dx}(y^3) + y^3\frac{d}{dx}(x) - 3 = 0$$

$$x^2\frac{dy}{dx} + y \cdot 2x + x \cdot 3y^2\frac{dy}{dx} + y^3 \cdot 1 - 3 = 0.$$

Solve for $\frac{dy}{dx}$ in terms of $x$ and $y$ following these steps:

**Step 1** Move all terms involving $\frac{dy}{dx}$ to the left side of the equation and move the other terms to the right side.

$$x^2\frac{dy}{dx} + 3xy^2\frac{dy}{dx} = 3 - y^3 - 2xy$$

**Step 2** Factor out $\frac{dy}{dx}$ on the left side of the equation.

$$(x^2 + 3xy^2)\frac{dy}{dx} = 3 - y^3 - 2xy$$

**Step 3** Divide both sides of the equation by the factor that multiplies $\frac{dy}{dx}$.

$$\frac{dy}{dx} = \frac{3 - y^3 - 2xy}{x^2 + 3xy^2}$$  ∎

▶ *Reminder* When a power of $y$ is differentiated with respect to $x$, the result must include the factor $\frac{dy}{dx}$. When a power of $x$ is differentiated, there is no factor $\frac{dy}{dx}$. ◀

Here is the general procedure for implicit differentiation.

---

### Finding $\frac{dy}{dx}$ by Implicit Differentiation

Differentiate each term of the equation *with respect to* $x$, treating $y$ as a function of $x$.

Move all terms involving $\frac{dy}{dx}$ to the left side of the equation and move the other terms to the right side.

Factor out $\frac{dy}{dx}$ on the left side of the equation.

Divide both sides of the equation by the factor that multiplies $\frac{dy}{dx}$.

---

Equations that implicitly define functions arise frequently in economic models. The economic background for the equation in the next example is discussed in Section 7.1.

---

**EXAMPLE 4**  **Isoquants and marginal rate of substitution**  Suppose that $x$ and $y$ represent the amounts of two basic inputs into a production process and the equation

$$60x^{3/4}y^{1/4} = 3240$$

describes all input amounts $(x, y)$ for which the output of the process is 3240 units. (The graph of this equation is called a *production isoquant* or *constant product curve*.) See Fig. 3. Use implicit differentiation to calculate the slope of the graph at the point on the curve where $x = 81$, $y = 16$.

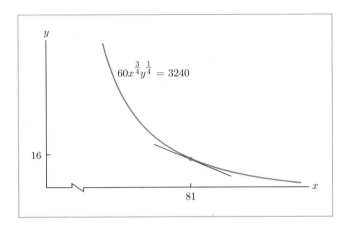

**Figure 3.** A production isoquant.

**Solution** We use the product rule and treat $y$ as a function of $x$.

$$60x^{3/4}\frac{d}{dx}(y^{1/4}) + y^{1/4}\frac{d}{dx}(60x^{3/4}) = 0$$

$$60x^{3/4}\cdot\left(\frac{1}{4}\right)y^{-3/4}\frac{dy}{dx} + y^{1/4}\cdot 60\left(\frac{3}{4}\right)x^{-1/4} = 0$$

$$15x^{3/4}y^{-3/4}\frac{dy}{dx} = -45x^{-1/4}y^{1/4}$$

$$\frac{dy}{dx} = \frac{-45x^{-1/4}y^{1/4}}{15x^{3/4}y^{-3/4}} = \frac{-3y}{x}.$$

When $x = 81$ and $y = 16$, we have

$$\left.\frac{dy}{dx}\right|_{\substack{x=81\\y=16}} = \frac{-3(16)}{81} = -\frac{16}{27}.$$

The number $-\frac{16}{27}$ is the slope of the production isoquant at the point $(81, 16)$. If the first input (corresponding to $x$) is increased by one small unit, then the second input (corresponding to $y$) must decrease by approximately $\frac{16}{27}$ unit in order to keep the production output unchanged [i.e., to keep $(x, y)$ on the curve]. In economic terminology, the absolute value of $\frac{dy}{dx}$ is called the *marginal rate of substitution* of the first input for the second input. ■

**Related Rates** In implicit differentiation we differentiate an equation involving $x$ and $y$, with $y$ treated as a function of $x$. However, in some applications where $x$ and $y$ are related by an equation, both variables are functions of a third variable $t$ (which may represent time). Often the formulas for $x$ and $y$ as functions of $t$ are not known. When we differentiate such an equation with respect to $t$, we derive a relationship between the rates of change $\frac{dy}{dt}$ and $\frac{dx}{dt}$. We say that these derivatives are *related rates*. The equation relating the rates may be used to find one of the rates when the other is known.

**EXAMPLE 5** **A point moving along a graph** Suppose that $x$ and $y$ are both differentiable functions of $t$ and are related by the equation

$$x^2 + 5y^2 = 36. \tag{3}$$

(a) Differentiate each term in the equation with respect to $t$ and solve the resulting equation for $\dfrac{dy}{dt}$.

(b) Calculate $\dfrac{dy}{dt}$ at a time when $x = 4$, $y = 2$, and $\dfrac{dx}{dt} = 5$.

**Solution**  (a) Since $x$ is a function of $t$, the general power rule gives

$$\frac{d}{dt}(x^2) = 2x\frac{dx}{dt}.$$

A similar formula holds for the derivative of $y^2$. Differentiating each term in (3) with respect to $t$, we obtain

$$\frac{d}{dt}(x^2) + \frac{d}{dt}(5y^2) = \frac{d}{dt}(36)$$

$$2x\frac{dx}{dt} + 5 \cdot 2y\frac{dy}{dt} = 0$$

$$10y\frac{dy}{dt} = -2x\frac{dx}{dt}$$

$$\frac{dy}{dt} = -\frac{x}{5y}\frac{dx}{dt}.$$

(b) When $x = 4$, $y = 2$, and $\dfrac{dx}{dt} = 5$,

$$\frac{dy}{dt} = -\frac{4}{5(2)} \cdot (5) = -2. \qquad \blacksquare$$

**Figure 4.**  Graph of $x^2 + 5y^2 = 36$.

There is a helpful graphical interpretation of the calculations in Example 5. Imagine a point that is moving clockwise along the graph of the equation $x^2 + 5y^2 = 36$, which is an ellipse. (See Fig. 4.) Suppose that when the point is at $(4, 2)$, the $x$-coordinate of the point is changing at the rate of 5 units per minute, so that $\dfrac{dx}{dt} = 5$. In Example 5(b) we found that $\dfrac{dy}{dt} = -2$. This means that the $y$-coordinate of the point is decreasing at the rate of 2 units per minute when the moving point reaches $(4, 2)$.

---

**EXAMPLE 6**  **Relating weekly sales to price**  Suppose that $x$ thousand units of a commodity can be sold weekly when the price is $p$ dollars per unit, and suppose that $x$ and $p$ satisfy the demand equation

$$p + 2x + xp = 38.$$

See Fig. 5. How fast are weekly sales changing at a time when $x = 4$, $p = 6$, and the price is falling at the rate of \$.40 per week?

**Solution**  Assume that $p$ and $x$ are differentiable functions of $t$, and differentiate the demand equation with respect to $t$.

$$\frac{d}{dt}(p) + \frac{d}{dt}(2x) + \frac{d}{dt}(xp) = \frac{d}{dt}(38)$$

$$\frac{dp}{dt} + 2\frac{dx}{dt} + x\frac{dp}{dt} + p\frac{dx}{dt} = 0. \qquad (4)$$

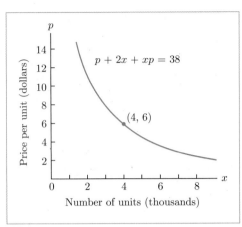

**Figure 5.** A demand curve.

We want to know $\dfrac{dx}{dt}$ at a time when $x = 4$, $p = 6$, and $\dfrac{dp}{dt} = -.40$. (The derivative $\dfrac{dp}{dt}$ is negative because the price is decreasing.) We could solve (4) for $\dfrac{dx}{dt}$ and then substitute the given values, but since we do not need a general formula for $\dfrac{dx}{dt}$, it is easier to substitute first and then solve.

$$-.40 + 2\frac{dx}{dt} + 4(-.40) + 6\frac{dx}{dt} = 0$$

$$8\frac{dx}{dt} = 2$$

$$\frac{dx}{dt} = .25.$$

Thus sales are rising at the rate of .25 thousand units (i.e., 250 units) per week. ■

### Suggestions for Solving Related Rates Problems

1.  Draw a picture, if possible.
2.  Assign letters to quantities that vary, and identify one variable, say $t$, on which the other variables depend.
3.  Find an equation that relates the variables to each other.
4.  Differentiate the equation with respect to the independent variable $t$. Use the chain rule whenever appropriate.
5.  Substitute all specified values for the variables and their derivatives.
6.  Solve for the derivative that gives the unknown rate.

**Incorporating Technology**   The graph of an equation in $x$ and $y$ can be easily obtained when $y$ can be expressed as one or more functions of $x$. For instance, the graph of $x^2 + y^2 = 4$ can be plotted by simultaneously graphing $y = \sqrt{4 - x^2}$ and $y = -\sqrt{4 - x^2}$.

## PRACTICE PROBLEMS 3.3

*Suppose that $x$ and $y$ are related by the equation $3y^2 - 3x^2 + y = 1$.*

**1.** Use implicit differentiation to find a formula for the slope of the graph of the equation.

**2.** Suppose that the $x$ and $y$ in the preceding equation are both functions of $t$. Differentiate both sides of the equation with respect to $t$ and find a formula for $\dfrac{dy}{dt}$ in terms of $x$, $y$, and $\dfrac{dx}{dt}$.

# EXERCISES 3.3

*In Exercises 1–18, suppose that $x$ and $y$ are related by the given equation and use implicit differentiation to determine $\dfrac{dy}{dx}$.*

**1.** $x^2 - y^2 = 1$

**2.** $x^3 + y^3 - 6 = 0$

**3.** $y^5 - 3x^2 = x$

**4.** $x^4 + (y+3)^4 = x^2$

**5.** $y^4 - x^4 = y^2 - x^2$

**6.** $x^3 + y^3 = x^2 + y^2$

**7.** $2x^3 + y = 2y^3 + x$

**8.** $x^4 + 4y = x - 4y^3$

**9.** $xy = 5$

**10.** $xy^3 = 2$

**11.** $x(y+2)^5 = 8$

**12.** $x^2 y^3 = 6$

**13.** $x^3 y^2 - 4x^2 = 1$

**14.** $(x+1)^2(y-1)^2 = 1$

**15.** $x^3 + y^3 = x^3 y^3$

**16.** $x^2 + 4xy + 4y = 1$

**17.** $x^2 y + y^2 x = 3$

**18.** $x^3 y + xy^3 = 4$

*Use implicit differentiation of the equations in Exercises 19–24 to determine the slope of the graph at the given point.*

**19.** $4y^3 - x^2 = -5;\ x = 3,\ y = 1$

**20.** $y^2 = x^3 + 1;\ x = 2,\ y = -3$

**21.** $xy^3 = 2;\ x = -\frac{1}{4},\ y = -2$

**22.** $\sqrt{x} + \sqrt{y} = 7;\ x = 9,\ y = 16$

**23.** $xy + y^3 = 14;\ x = 3,\ y = 2$

**24.** $y^2 = 3xy - 5;\ x = 2,\ y = 1$

**25.** Find the equation of the tangent line to the graph of $x^2 y^4 = 1$ at the point $(4, \frac{1}{2})$ and at the point $(4, -\frac{1}{2})$.

**26.** Find the equation of the tangent line to the graph of $x^4 y^2 = 144$ at the point $(2, 3)$ and at the point $(2, -3)$.

**27.** The graph of $x^4 + 2x^2 y^2 + y^4 = 4x^2 - 4y^2$ is the "lemniscate" in Fig. 6.

(a) Find $\dfrac{dy}{dx}$ by implicit differentiation.

(b) Find the slope of the tangent line to the lemniscate at $(\sqrt{6}/2, \sqrt{2}/2)$.

**28.** The graph of $x^4 + 2x^2 y^2 + y^4 = 9x^2 - 9y^2$ is a lemniscate similar to that in Fig. 6.

(a) Find $\dfrac{dy}{dx}$ by implicit differentiation.

(b) Find the slope of the tangent line to the lemniscate at $(\sqrt{5}, -1)$.

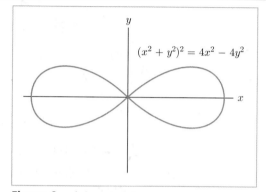

$(x^2 + y^2)^2 = 4x^2 - 4y^2$

**Figure 6.** A lemniscate.

**29.** (*Marginal Rate of Substitution*) Suppose that $x$ and $y$ represent the amounts of two basic inputs for a production process and suppose that the equation

$$30x^{1/3} y^{2/3} = 1080$$

describes all input amounts where the output of the process is 1080 units.

(a) Find $\dfrac{dy}{dx}$.

(b) What is the marginal rate of substitution of $x$ for $y$, when $x = 16$ and $y = 54$?

**30.** Suppose that $x$ and $y$ represent the amounts of two basic inputs for a production process and

$$10x^{1/2} y^{1/2} = 600.$$

Find $\dfrac{dy}{dx}$ when $x = 50$, $y = 72$.

*In Exercises 31–36, suppose that $x$ and $y$ are both differentiable functions of $t$ and are related by the given equation. Use implicit differentiation with respect to $t$ to determine $\dfrac{dy}{dt}$ in terms of $x$, $y$, and $\dfrac{dx}{dt}$.*

**31.** $x^4 + y^4 = 1$

**32.** $y^4 - x^2 = 1$

**33.** $3xy - 3x^2 = 4$

**34.** $y^2 = 8 + xy$

**35.** $x^2 + 2xy = y^3$

**36.** $x^2 y^2 = 2y^3 + 1$

**37.** (*Point on a Curve*) A point is moving along the graph of $x^2 - 4y^2 = 9$. When the point is at $(5, -2)$, its $x$-coordinate is increasing at the rate of 3 units per second. How fast is the $y$-coordinate changing at that moment?

**38.** (*Point on a Curve*) A point is moving along the graph of $x^3 y^2 = 200$. When the point is at $(2, 5)$, its $x$-coordinate is changing at the rate of $-4$ units per minute. How fast is the $y$-coordinate changing at that moment?

**39.** (*Demand Equation*) Suppose that the price $p$ (in dollars) and the weekly sales $x$ (in thousands of units) of a certain commodity satisfy the demand equation

$$2p^3 + x^2 = 4500.$$

Determine the rate at which sales are changing at a time when $x = 50$, $p = 10$, and the price is falling at the rate of $\$.50$ per week.

**40.** Suppose that the price $p$ (in dollars) and the demand $x$ (in thousands of units) of a commodity satisfy the demand equation

$$6p + x + xp = 94.$$

How fast is the demand changing at a time when $x = 4$, $p = 9$, and the price is rising at the rate of $\$2$ per week?

**41.** (*Advertising Affects Revenue*) The monthly advertising revenue $A$ and the monthly circulation $x$ of a magazine are related approximately by the equation

$$A = 6\sqrt{x^2 - 400}, \qquad x \geq 20,$$

where $A$ is given in thousands of dollars and $x$ is measured in thousands of copies sold. At what rate is the advertising revenue changing if the current circulation is $x = 25$ thousand copies and the circulation is changing at the rate of 2 thousand copies per month?
$\left[ Hint: \text{ Use the chain rule } \dfrac{dA}{dt} = \dfrac{dA}{dx} \dfrac{dx}{dt}. \right]$

**42.** Suppose that in Boston the wholesale price $p$ of oranges (in dollars per crate) and the daily supply $x$ (in thousands of crates) are related by the equation $px + 7x + 8p = 328$. If there are 4 thousand crates available today at a price of $\$25$ per crate, and if the supply is changing at the rate of $-.3$ thousand crates per day, at what rate is the price changing?

**43.** (*Volume and Pressure of Gas*) Under certain conditions (called adiabatic expansion) the pressure $P$ and volume $V$ of a gas such as oxygen satisfy the equation $P^5 V^7 = k$, where $k$ is a constant. Suppose that at some moment the volume of the gas is 4 liters, the pressure is 200 units, and the pressure is increasing at the rate of 5 units per second. Find the (time) rate at which the volume is changing.

**44.** (*Spherical Volume*) The volume $V$ of a spherical cancer tumor is given by $V = \pi x^3 / 6$, where $x$ is the diameter of the tumor. A physician estimates that the diameter is growing at the rate of .4 millimeters per day, at a time when the diameter is already 10 millimeters. How fast is the volume of the tumor changing at that time?

**45.** (*Related Rates*) Figure 7 shows a 10-foot ladder leaning against a wall.

(a) Use the Pythagorean theorem to find an equation relating $x$ and $y$.

(b) Suppose that the foot of the ladder is being pulled along the ground at the rate of 3 feet per second. How fast is the top end of the ladder sliding down the wall at the time when the foot of the ladder is 8 feet from the wall? That is, what is $\dfrac{dy}{dt}$ at the time when $\dfrac{dx}{dt} = 3$ and $x = 8$?

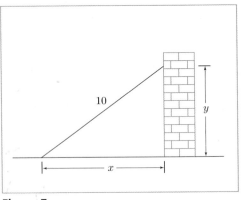

**Figure 7**

**46.** (*Related Rates*) An airplane flying 390 feet per second at an altitude of 5000 feet flew directly over an observer. Figure 8 shows the relationship of the airplane to the observer at a later time.

(a) Find an equation relating $x$ and $y$.

(b) Find the value of $x$ when $y$ is 13,000.

(c) How fast is the distance from the observer to the airplane changing at the time when the airplane is 13,000 feet from the observer? That is, what is $\dfrac{dy}{dt}$ at the time when $\dfrac{dx}{dt} = 390$ and $y = 13,000$?

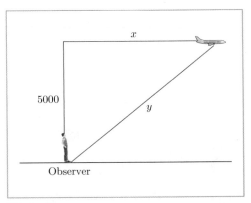

**Figure 8**

**47.** (*Related Rates*) A baseball diamond is a 90-foot by 90-foot square. See Fig. 9. A player runs from first to second base at the speed of 22 feet per second. How fast is the player's distance from third base changing when he is halfway between first and second base? [*Hint*: If $x$ is the distance from the player to second base and $y$ is his distance from third base, then $x^2 + 90^2 = y^2$.]

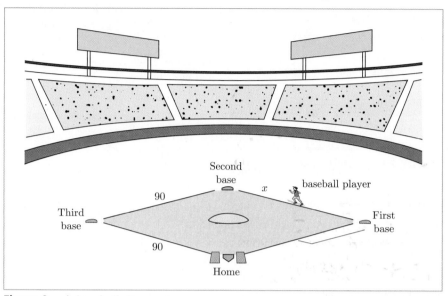

**Figure 9.** A baseball diamond.

## SOLUTIONS TO PRACTICE PROBLEMS 3.3

**1.**
$$\frac{d}{dx}(3y^2) - \frac{d}{dx}(3x^2) + \frac{d}{dx}(y) = \frac{d}{dx}(1)$$

$$6y\frac{dy}{dx} - 6x + \frac{dy}{dx} = 0$$

$$(6y+1)\frac{dy}{dx} = 6x$$

$$\frac{dy}{dx} = \frac{6x}{6y+1}$$

**2.** First, here is the solution without reference to Practice Problem 1.

$$\frac{d}{dt}(3y^2) - \frac{d}{dt}(3x^2) + \frac{d}{dt}(y) = \frac{d}{dt}(1)$$

$$6y\frac{dy}{dt} - 6x\frac{dx}{dt} + \frac{dy}{dt} = 0$$

$$(6y+1)\frac{dy}{dt} = 6x\frac{dx}{dt}$$

$$\frac{dy}{dt} = \frac{6x}{6y+1}\frac{dx}{dt}$$

Alternatively, we can use the chain rule and the formula for $\frac{dy}{dx}$ from Practice Problem 1.

$$\frac{dy}{dt} = \frac{dy}{dx}\frac{dx}{dt} = \frac{6x}{6y+1}\frac{dx}{dt}$$

## REVIEW OF FUNDAMENTAL CONCEPTS

**1.** State the product rule and quotient rule.

**2.** State the chain rule. Give an example.

**3.** What is the relationship between the chain rule and the general power rule?

**4.** What does it mean for a function to be defined implicitly by an equation?

**5.** State the formula for $\frac{d}{dx}y^r$, where $y$ is defined implicitly as a function of $x$.

**6.** Outline the procedure for solving a related rates problem.

## Supplementary Exercises

*Differentiate the following functions.*

**1.** $y = (4x - 1)(3x + 1)^4$

**2.** $y = 2(5 - x)^3(6x - 1)$

**3.** $y = x(x^5 - 1)^3$

**4.** $y = (2x + 1)^{5/2}(4x - 1)^{3/2}$

**5.** $y = 5(\sqrt{x} - 1)^4(\sqrt{x} - 2)^2$

**6.** $y = \dfrac{\sqrt{x}}{\sqrt{x} + 4}$

**7.** $y = 3(x^2 - 1)^3(x^2 + 1)^5$

**8.** $y = \dfrac{1}{(x^2 + 5x + 1)^6}$

**9.** $y = \dfrac{x^2 - 6x}{x - 2}$

**10.** $y = \dfrac{2x}{2 - 3x}$

**11.** $y = \left(\dfrac{3 - x^2}{x^3}\right)^2$

**12.** $y = \dfrac{x^3 + x}{x^2 - x}$

**13.** Let $f(x) = (3x + 1)^4(3 - x)^5$. Find all $x$ such that $f'(x) = 0$.

**14.** Let $f(x) = (x^2 + 1)/(x^2 + 5)$. Find all $x$ such that $f'(x) = 0$.

**15.** Find the equation of the line tangent to the graph of $y = (x^3 - 1)(x^2 + 1)^4$ at the point where $x = -1$.

**16.** Find the equation of the line tangent to the graph of $y = \dfrac{x - 3}{\sqrt{4 + x^2}}$ at the point where $x = 0$.

**17.** A botanical display is to be constructed as a rectangular region with a river as one side and a sidewalk 2 meters wide along the inside edges of the other three sides. (See Fig. 1.) The area for the plants must be 800 square meters. Find the outside dimensions of the region that minimizes the area of the sidewalk (and hence minimizes the amount of concrete needed for the sidewalk).

**18.** Repeat Exercise 17, with the sidewalk on the inside of all four sides. In this case, the 800-square-meter planted region has dimensions $x - 4$ meters by $y - 4$ meters.

**19.** A store estimates that its cost when selling $x$ lamps per day is $C$ dollars, where $C = 40x + 30$ (i.e., the marginal cost per lamp is \$40). Suppose that daily sales are rising at the rate of three lamps per day. How fast are the costs rising? Explain your answer using the chain rule.

**20.** A company pays $y$ dollars in taxes when its annual profit is $P$ dollars. Suppose that $y$ is some (differentiable) function of $P$ and $P$ is some function of time $t$. Give a chain rule formula for the time rate of change of taxes $\dfrac{dy}{dt}$.

*Exercises 21–26 refer to the graphs of the functions $f(x)$ and $g(x)$ in Fig. 2.*

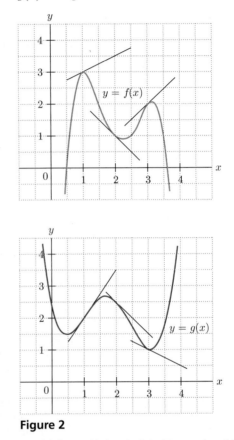

**Figure 2**

**21.** Let $h(x) = 2f(x) - 3g(x)$. Determine $h(1)$ and $h'(1)$.

**22.** Let $h(x) = f(x) \cdot g(x)$. Determine $h(1)$ and $h'(1)$.

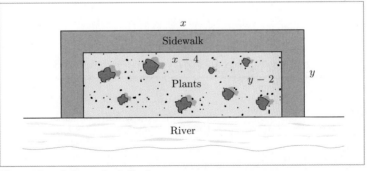

**Figure 1.** A botanical display.

**23.** Let $h(x) = \dfrac{f(x)}{g(x)}$. Determine $h(1)$ and $h'(1)$.

**24.** Let $h(x) = [f(x)]^2$. Determine $h(1)$ and $h'(1)$.

**25.** Let $h(x) = f(g(x))$. Determine $h(1)$ and $h'(1)$.

**26.** Let $h(x) = g(f(x))$. Determine $h(1)$ and $h'(1)$.

*In Exercises 27–29, find a formula for $\dfrac{d}{dx}f(g(x))$, where $f(x)$ is a function such that $f'(x) = 1/(x^2 + 1)$.*

**27.** $g(x) = x^3$  **28.** $g(x) = \dfrac{1}{x}$  **29.** $g(x) = x^2 + 1$

*In Exercises 30–32, find a formula for $\dfrac{d}{dx}f(g(x))$, where $f(x)$ is a function such that $f'(x) = x\sqrt{1 - x^2}$.*

**30.** $g(x) = x^2$  **31.** $g(x) = \sqrt{x}$  **32.** $g(x) = x^{3/2}$

*In Exercises 33–35, find $\dfrac{dy}{dx}$, where $y$ is a function of $u$ such that $\dfrac{dy}{du} = \dfrac{u}{u^2 + 1}$.*

**33.** $u = x^{3/2}$  **34.** $u = x^2 + 1$  **35.** $u = \dfrac{5}{x}$

*In Exercises 36–38, find $\dfrac{dy}{dx}$, where $y$ is a function of $u$ such that $\dfrac{dy}{du} = \dfrac{u}{\sqrt{1 + u^4}}$.*

**36.** $u = x^2$  **37.** $u = \sqrt{x}$  **38.** $u = \dfrac{2}{x}$

**39.** The revenue $R$ a company receives is a function of the weekly sales $x$. Also, the sales level $x$ is a function of the weekly advertising expenditures $A$, and $A$ in turn is a varying function of time.

**(a)** Write the derivative symbols for the following quantities: rate of change of revenue with respect to advertising expenditures, time rate of change of advertising expenditures, marginal revenue, and rate of change of sales with respect to advertising expenditures. Select your answers from the following:

$$\frac{dR}{dx}, \quad \frac{dR}{dt}, \quad \frac{dA}{dt}, \quad \frac{dA}{dR}, \quad \frac{dA}{dx}, \quad \frac{dx}{dA}, \quad \text{and} \quad \frac{dR}{dA}.$$

**(b)** Write a type of chain rule that expresses the time rate of change of revenue, $\dfrac{dR}{dt}$, in terms of three of the derivatives described in part (a).

**40.** The amount $A$ of anesthetics that a certain hospital uses each week is a function of the number $S$ of surgical operations performed each week. Also, $S$ in turn is a function of the population $P$ of the area served by the hospital, while $P$ is a function of time $t$.

**(a)** Write the derivative symbols for the following quantities: population growth rate, rate of change of anesthetic usage with respect to the population size, rate of change of surgical operations with respect to the population size, and rate of change of anesthetic usage with respect to the number of surgical operations. Select your answers from the following:

$$\frac{dS}{dP}, \quad \frac{dS}{dt}, \quad \frac{dP}{dS}, \quad \frac{dP}{dt}, \quad \frac{dA}{dS}, \quad \frac{dA}{dP}, \quad \text{and} \quad \frac{dS}{dA}.$$

**(b)** Write a type of chain rule that expresses the time rate of change of anesthetic usage, $\dfrac{dA}{dt}$, in terms of three of the derivatives described in part (a).

**41.** The graph of $x^{2/3} + y^{2/3} = 8$ is the *astroid* in Fig. 3.

**(a)** Find $\dfrac{dy}{dx}$ by implicit differentiation.

**(b)** Find the slope of the tangent line at $(8, -8)$.

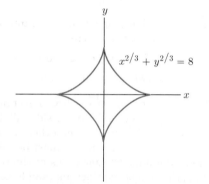

**Figure 3.** Astroid.

**42.** The graph of $x^3 + y^3 = 9xy$ is the folium of Descartes, shown in Fig. 4.

**(a)** Find $\dfrac{dy}{dx}$ by implicit differentiation.

**(b)** Find the slope of the curve at $(2, 4)$.

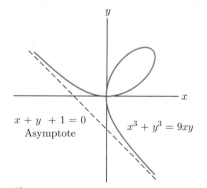

**Figure 4.** Folium of Descartes.

*In Exercises 43–46, $x$ and $y$ are related by the given equation. Use implicit differentiation to calculate the value of $\dfrac{dy}{dx}$ for the given values of $x$ and $y$.*

**43.** $x^2y^2 = 9$; $x = 1$, $y = 3$

**44.** $xy^4 = 48$; $x = 3$, $y = 2$

**45.** $x^2 - xy^3 = 20$; $x = 5$, $y = 1$

**46.** $xy^2 - x^3 = 10$; $x = 2$, $y = 3$

**47.** A factory's weekly production costs $y$ and its weekly production quantity $x$ are related by the equation $y^2 - 5x^3 = 4$, where $y$ is in thousands of dollars and $x$ is in thousands of units of output.

(a) Use implicit differentiation to find a formula for $\dfrac{dy}{dx}$, the marginal cost of production.

(b) Find the marginal cost of production when $x = 4$ and $y = 18$.

(c) Suppose that the factory begins to vary its weekly production level. Assuming that $x$ and $y$ are differentiable functions of time $t$, use the method of related rates to find a formula for $\dfrac{dy}{dt}$, the time rate of change of production costs.

(d) Compute $\dfrac{dy}{dt}$ when $x = 4$, $y = 18$, and the production level is rising at the rate of .3 thousand units per week (i.e., when $\dfrac{dx}{dt} = .3$).

**48.** A town library estimates that when the population is $x$ thousand persons, approximately $y$ thousand books will be checked out of the library during one year, where $x$ and $y$ are related by the equation $y^3 - 8000x^2 = 0$.

(a) Use implicit differentiation to find a formula for $\dfrac{dy}{dx}$, the rate of change of library circulation with respect to population size.

(b) Find the value of $\dfrac{dy}{dx}$ when $x = 27$ thousand persons and $y = 180$ thousand books per year.

(c) Assume that $x$ and $y$ are both differentiable functions of time $t$, and use the method of related rates to find a formula for $\dfrac{dy}{dt}$, the time rate of change of library circulation.

(d) Compute $\dfrac{dy}{dt}$ when $x = 27$, $y = 180$, and the population is rising at the rate of 1.8 thousand persons per year (i.e., $\dfrac{dx}{dt} = 1.8$). Either use part (c), or use part (b) and the chain rule.

**49.** Suppose that the price $p$ and quantity $x$ of a certain commodity satisfy the demand equation $6p + 5x + xp = 50$, and suppose that $p$ and $x$ are functions of time, $t$. Determine the rate at which the quantity $x$ is changing when $x = 4$, $p = 3$, and $\dfrac{dp}{dt} = -2$.

**50.** An offshore oil well is leaking oil onto the ocean surface, forming a circular oil slick about .005 meter thick. If the radius of the slick is $r$ meters, then the volume of oil spilled is $V = .005\pi r^2$ cubic meters. Suppose that the oil is leaking at a constant rate of 20 cubic meters per hour, so that $\dfrac{dV}{dt} = 20$. Find the rate at which the radius of the oil slick is increasing, at a time when the radius is 50 meters. $\left[ \textit{Hint:} \text{ Find a relation between } \dfrac{dV}{dt} \text{ and } \dfrac{dr}{dt}. \right]$

**51.** Animal physiologists have determined experimentally that the weight $W$ (in kilograms) and the surface area $S$ (in square meters) of a typical horse are related by the empirical equation $S = 0.1W^{2/3}$. How fast is the surface area of a horse increasing at a time when the horse weighs 350 kg and is gaining weight at the rate of 200 kg per year? [*Hint:* Use the chain rule.]

**52.** Suppose that a kitchen appliance company's monthly sales and advertising expenses are approximately related by the equation $xy - 6x + 20y = 0$, where $x$ is thousands of dollars spent on advertising and $y$ is thousands of dishwashers sold. Currently, the company is spending 10 thousand dollars on advertising and is selling 2 thousand dishwashers each month. If the company plans to increase monthly advertising expenditures at the rate of \$1.5 thousand per month, how fast will sales rise? Use implicit differentiation to answer the question.

## CHAPTER 4

# The Exponential and Natural Logarithm Functions

WHEN an investment grows steadily at 15% per year, the rate of growth of the investment at any time is proportional to the value of the investment at that time. When a bacteria culture grows in a laboratory dish, the rate of growth of the culture at any moment is proportional to the total number of bacteria in the dish at that moment. These situations are examples of what is called _exponential growth_. A pile of radioactive uranium $^{235}$U decays at a rate that at each moment is proportional to the amount of $^{235}$U present. This decay of uranium (and of radioactive elements in general) is called _exponential decay_. Both exponential growth and exponential decay can be described and studied in terms of exponential functions and the natural logarithm function. The properties of these functions are investigated in this chapter. Subsequently, we shall explore a wide range of applications, in fields such as business, biology, archeology, public health, and psychology.

## 4.1 Exponential Functions

Throughout this section, $b$ will denote a positive number. The function

$$f(x) = b^x$$

is called an *exponential function*, because the variable $x$ is in the exponent. The number $b$ is called the *base* of the exponential function. In Section 0.5 we reviewed the definition of $b^x$ for various values of $b$ and $x$ (although we used the letter $r$ there instead of $x$). For instance, if $f(x)$ is the exponential function with base 2,

$$f(x) = 2^x,$$

then

$$f(0) = 2^0 = 1, \qquad f(1) = 2^1 = 2, \qquad f(4) = 2^4 = 2 \cdot 2 \cdot 2 \cdot 2 = 16,$$

and

$$f(-1) = 2^{-1} = \tfrac{1}{2}, \qquad f(\tfrac{1}{2}) = 2^{1/2} = \sqrt{2}, \qquad f(\tfrac{3}{5}) = (2^{1/5})^3 = (\sqrt[5]{2})^3.$$

Actually, in Section 0.5, we defined $b^x$ only for rational (i.e., integer or fractional) values of $x$. For other values of $x$ (such as $\sqrt{3}$ or $\pi$), it is possible to define $b^x$ by first approximating $x$ with rational numbers and then applying a limiting process. We shall omit the details and simply assume henceforth that $b^x$ can be defined for all numbers $x$ in such a way that the usual laws of exponents remain valid.

Let us state the laws of exponents for reference.

(i) $b^x \cdot b^y = b^{x+y}$          (ii) $b^{-x} = \dfrac{1}{b^x}$

(iii) $\dfrac{b^x}{b^y} = b^x \cdot b^{-y} = b^{x-y}$          (iv) $(b^y)^x = b^{xy}$

(v) $a^x b^x = (ab)^x$          (vi) $\dfrac{a^x}{b^x} = \left(\dfrac{a}{b}\right)^x$

Property (iv) may be used to change the appearance of an exponential function. For instance, the function $f(x) = 8^x$ may also be written as $f(x) = (2^3)^x = 2^{3x}$, and $g(x) = (\tfrac{1}{9})^x$ may be written as $g(x) = (1/3^2)^x = (3^{-2})^x = 3^{-2x}$.

---

**EXAMPLE 1**    Use properties of exponents to write the following expressions in the form $2^{kx}$ for a suitable constant $k$.

(a) $4^{5x/2}$    (b) $(2^{4x} \cdot 2^{-x})^{1/2}$    (c) $8^{x/3} \cdot 16^{3x/4}$    (d) $\dfrac{10^x}{5^x}$

**Solution**    (a) First express the base 4 as a power of 2, and then use Property (iv):

$$4^{5x/2} = (2^2)^{5x/2} = 2^{2(5x/2)} = 2^{5x}.$$

(b) First use Property (i) to simplify the quantity inside the parentheses, and then use Property (iv):

$$(2^{4x} \cdot 2^{-x})^{1/2} = (2^{4x-x})^{1/2} = (2^{3x})^{1/2} = 2^{(3/2)x}.$$

(c) First express the bases 8 and 16 as powers of 2, and then use (iv) and (i):

$$8^{x/3} \cdot 16^{3x/4} = (2^3)^{x/3} \cdot (2^4)^{3x/4} = 2^x \cdot 2^{3x} = 2^{4x}.$$

(d) Use (v) to change the numerator $10^x$, and then cancel the common term $5^x$:

$$\frac{10^x}{5^x} = \frac{(2\cdot 5)^x}{5^x} = \frac{2^x\cdot 5^x}{5^x} = 2^x.$$

An alternative method is to use Property (vi):

$$\frac{10^x}{5^x} = \left(\frac{10}{5}\right)^x = 2^x. \qquad\blacksquare$$

Let us now study the graph of the exponential function $y = b^x$ for various values of $b$. We begin with the special case $b = 2$.

We have tabulated the values of $2^x$ for $x = 0, \pm1, \pm2, \pm3$ and plotted these values in Fig. 1. Other intermediate values of $2^x$ for $x = \pm.1, \pm.2, \pm.3, \ldots$, may be obtained from a table or from a graphing calculator. [See Fig. 2(a).] By passing a smooth curve through these points, we obtain the graph of $y = 2^x$, Fig. 2(b).

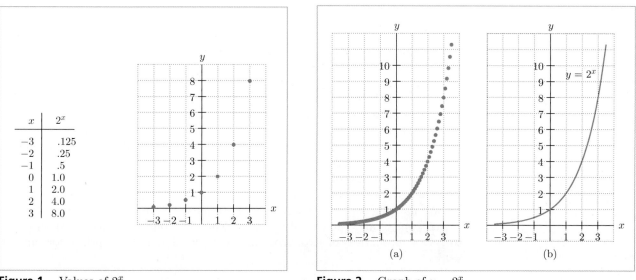

**Figure 1.**  Values of $2^x$.

**Figure 2.**  Graph of $y = 2^x$.

In the same manner, we have sketched the graph of $y = 3^x$ (Fig. 3). The graphs of $y = 2^x$ and $y = 3^x$ have the same basic shape. Also note that they both pass through the point $(0, 1)$ (because $2^0 = 1$, $3^0 = 1$).

In Fig. 4 we have sketched the graphs of several more exponential functions. Notice that the graph of $y = 5^x$ has a large slope at $x = 0$, since the graph at $x = 0$ is quite steep; however, the graph of $y = (1.1)^x$ is nearly horizontal at $x = 0$, and hence the slope is close to zero.

There is an important property of the function $3^x$ that is readily apparent from its graph. Since the graph is always increasing, the function $3^x$ never assumes the same $y$-value twice. That is, the only way $3^r$ can equal $3^s$ is to have $r = s$. This fact is useful when solving certain equations involving exponentials.

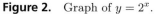

**EXAMPLE 2**  Let $f(x) = 3^{5x}$. Determine all $x$ for which $f(x) = 27$.

**Solution**  Since $27 = 3^3$, we must determine all $x$ for which

$$3^{5x} = 3^3.$$

**Figure 3**

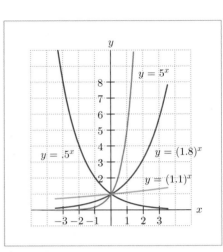

**Figure 4**

Equating exponents, we have

$$5x = 3$$

$$x = \tfrac{3}{5}.$$

In general, for $b > 1$, the equation $b^r = b^s$ implies that $r = s$. This is because the graph of $y = b^x$ has the same basic shape as $y = 2^x$ and $y = 3^x$. Similarly, when $0 < b < 1$, the equation $b^r = b^s$ implies that $r = s$, because the graph of $y = b^x$ resembles the graph of $y = .5^x$ and is always decreasing.

There is no need at this point to become familiar with the graphs of the function $b^x$. We have shown a few graphs merely to make the reader more comfortable with the concept of an exponential function. The main purpose of this section has been to review properties of exponents in a context that is appropriate for our future work.

## PRACTICE PROBLEMS 4.1

**1.** Can a function such as $f(x) = 5^{3x}$ be written in the form $f(x) = b^x$? If so, what is $b$?

**2.** Solve the equation $7 \cdot 2^{6-3x} = 28$.

## EXERCISES 4.1

*Write each expression in Exercises 1–14 in the form $2^{kx}$ or $3^{kx}$, for a suitable constant $k$.*

**1.** $4^x$, $(\sqrt{3})^x$, $\left(\tfrac{1}{9}\right)^x$

**2.** $27^x$, $(\sqrt[3]{2})^x$, $\left(\tfrac{1}{8}\right)^x$

**3.** $8^{2x/3}$, $9^{3x/2}$, $16^{-3x/4}$

**4.** $9^{-x/2}$, $8^{4x/3}$, $27^{-2x/3}$

**5.** $\left(\tfrac{1}{4}\right)^{2x}$, $\left(\tfrac{1}{8}\right)^{-3x}$, $\left(\tfrac{1}{81}\right)^{x/2}$

**6.** $\left(\tfrac{1}{9}\right)^{2x}$, $\left(\tfrac{1}{27}\right)^{x/3}$, $\left(\tfrac{1}{16}\right)^{-x/2}$

**7.** $6^x \cdot 3^{-x}$, $\dfrac{15^x}{5^x}$, $\dfrac{12^x}{2^{2x}}$

**8.** $7^{-x} \cdot 14^x$, $\dfrac{2^x}{6^x}$, $\dfrac{3^{2x}}{18^x}$

**9.** $\dfrac{3^{4x}}{3^{2x}}$, $\dfrac{2^{5x+1}}{2 \cdot 2^{-x}}$, $\dfrac{9^{-x}}{27^{-x/3}}$

**10.** $\dfrac{2^x}{6^x}$, $\dfrac{3^{-5x}}{3^{-2x}}$, $\dfrac{16^x}{8^{-x}}$

**11.** $2^{3x} \cdot 2^{-5x/2}$, $3^{2x} \cdot \left(\tfrac{1}{3}\right)^{2x/3}$

**12.** $2^{5x/4} \cdot \left(\tfrac{1}{2}\right)^x$, $3^{-2x} \cdot 3^{5x/2}$

**13.** $(2^{-3x} \cdot 2^{-2x})^{2/5}$, $(9^{1/2} \cdot 9^4)^{x/9}$

**14.** $(3^{-x} \cdot 3^{x/5})^5$, $(16^{1/4} \cdot 16^{-3/4})^{3x}$

**15.** Find a number $b$ such that the function $f(x) = 3^{-2x}$ can be written in the form $f(x) = b^x$.

**16.** Find $b$ so that $8^{-x/3} = b^x$ for all $x$.

*Solve the following equations for x.*

**17.** $5^{2x} = 5^2$

**18.** $10^{-x} = 10^2$

**19.** $(2.5)^{2x+1} = (2.5)^5$

**20.** $(3.2)^{x-3} = (3.2)^5$

**21.** $10^{1-x} = 100$

**22.** $2^{4-x} = 8$

**23.** $3(2.7)^{5x} = 8.1$

**24.** $4(2.7)^{2x-1} = 10.8$

**25.** $(2^{x+1} \cdot 2^{-3})^2 = 2$

**26.** $(3^{2x} \cdot 3^2)^4 = 3$

**27.** $2^{3x} = 4 \cdot 2^{5x}$

**28.** $3^{5x} \cdot 3^x - 3 = 0$

**29.** $(1+x)2^{-x} - 5 \cdot 2^{-x} = 0$

**30.** $(2 - 3x)5^x + 4 \cdot 5^x = 0$

**31.** $2^x - \dfrac{8}{2^{2x}} = 0$

**32.** $2^x - \dfrac{1}{2^x} = 0$

[*Hint*: In Exercises 33–36, let $X = 2^x$ or $X = 3^x$.]

**33.** $2^{2x} - 6 \cdot 2^x + 8 = 0$

**34.** $2^{2x+2} - 17 \cdot 2^x + 4 = 0$

**35.** $3^{2x} - 12 \cdot 3^x + 27 = 0$

**36.** $2^{2x} - 4 \cdot 2^x - 32 = 0$

*The expressions in Exercises 37–42 may be factored as shown. Find the missing factors.*

**37.** $2^{3+h} = 2^3(\quad)$

**38.** $5^{2+h} = 25(\quad)$

**39.** $2^{x+h} - 2^x = 2^x(\quad)$

**40.** $5^{x+h} + 5^x = 5^x(\quad)$

**41.** $3^{x/2} + 3^{-x/2} = 3^{-x/2}(\quad)$

**42.** $5^{7x/2} - 5^{x/2} = \sqrt{5^x}(\quad)$

## Technology Exercises

**43.** Graph the function $f(x) = 2^x$ in the window $[-1, 2]$ *by* $[-1, 4]$, and estimate the slope of the graph at $x = 0$.

**44.** Graph the function $f(x) = 3^x$ in the window $[-1, 2]$ *by* $[-1, 8]$, and estimate the slope of the graph at $x = 0$.

**45.** By trial and error, find a number of the form $b = 2.\square$ (just one decimal place) with the property that the slope of the graph of $y = b^x$ at $x = 0$ is as close to 1 as possible.

## SOLUTIONS TO PRACTICE PROBLEMS 4.1

**1.** If $5^{3x} = b^x$, then when $x = 1$, $5^{3(1)} = b^1$, which says that $b = 125$. This value of $b$ certainly works, because
$$5^{3x} = (5^3)^x = 125^x.$$

**2.** Divide both sides of the equation by 7. We then obtain
$$2^{6-3x} = 4.$$
Now 4 can be written as $2^2$. So we have
$$2^{6-3x} = 2^2.$$

Equate exponents to obtain
$$6 - 3x = 2$$
$$4 = 3x$$
$$x = \tfrac{4}{3}.$$

## 4.2  The Exponential Function $e^x$

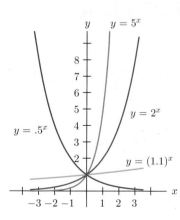

**Figure 1.**  Several exponential functions.

Let us begin by examining the graphs of the exponential functions shown in Fig. 1. They all pass through $(0, 1)$, but with different slopes there. Notice that the graph of $y = 5^x$ is quite steep at $x = 0$, while the graph of $y = (1.1)^x$ is nearly horizontal at $x = 0$. It turns out that at $x = 0$, the graph of $y = 2^x$ has a slope of approximately .693, while the graph of $3^x$ has a slope of approximately 1.1.

Evidently, there is a particular value of the base $b$, between 2 and 3, where the graph of $y = b^x$ has slope *exactly* 1 at $x = 0$. We denote this special value of $b$ by the letter $e$, and we call
$$f(x) = e^x$$
*the* exponential function. The number $e$ is an important constant of nature that has been calculated to thousands of decimal places. To 10 significant digits, we have $e = 2.718281828$. For our purposes, it is usually sufficient to think of $e$ as "approximately 2.7."

Our goal in this section is to find a formula for the derivative of $y = e^x$. It turns out that the calculations for $e^x$ and $2^x$ are very similar. Since many people are more comfortable working with $2^x$ rather than $e^x$, we shall first analyze the graph of $y = 2^x$. Then we shall draw the appropriate conclusions about the graph of $y = e^x$.

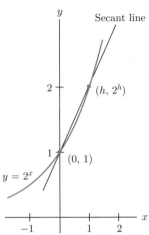

**Figure 2.** A secant line to the graph of $y = 2^x$.

Before computing the slope of $y = 2^x$ at an arbitrary $x$, let us consider the special case $x = 0$. Denote the slope at $x = 0$ by $m$. We shall use the secant-line approximation of the derivative to approximate $m$. We proceed by constructing the secant line in Fig. 2. The slope of the secant line through $(0, 1)$ and $(h, 2^h)$ is $\dfrac{2^h - 1}{h}$. As $h$ approaches zero, the slope of the secant line approaches the slope of $y = 2^x$ at $x = 0$. That is,

$$m = \lim_{h \to 0} \frac{2^h - 1}{h}. \tag{1}$$

We can estimate the value of $m$ by taking $h$ smaller and smaller. Table 1 shows the values of the expressions for $h = .1, .01, \ldots, .0000001$. From the table it is

**TABLE 1**   $Y_1 = (2^{\wedge}X - 1)/X$

| X | Y1 |
|------|--------|
| .1 | .71773 |
| .01 | .69556 |
| .001 | .69339 |
| 1E-4 | .69317 |
| 1E-5 | .69315 |
| 1E-6 | .69315 |
| 1E-7 | .69315 |

$Y_1 = .693147$

reasonable to conclude that $m \approx .693$. Since $m$ equals the slope of $y = 2^x$ at $x = 0$, we have

$$m = \frac{d}{dx}(2^x)\bigg|_{x=0} \approx .693. \tag{2}$$

Now that we have estimated the slope of $y = 2^x$ at $x = 0$, let us compute the slope for an arbitrary value of $x$. We construct a secant line through $(x, 2^x)$ and a nearby point $(x + h, 2^{x+h})$ on the graph. The slope of the secant line is

$$\frac{2^{x+h} - 2^x}{h}. \tag{3}$$

By a law of exponents, we have $2^{x+h} - 2^x = 2^x(2^h - 1)$ so that, by (1), we see that

$$\lim_{h \to 0} \frac{2^{x+h} - 2^x}{h} = \lim_{h \to 0} 2^x \frac{2^h - 1}{h} = 2^x \lim_{h \to 0} \frac{2^h - 1}{h} = m\, 2^x. \tag{4}$$

However, the slope of the secant (3) approaches the derivative of $2^x$ as $h$ approaches zero. Consequently, we have

$$\frac{d}{dx}(2^x) = m\, 2^x, \quad \text{where } m = \frac{d}{dx}(2^x)\bigg|_{x=0}. \tag{5}$$

**EXAMPLE 1**   Calculate

(a) $\dfrac{d}{dx}(2^x)\bigg|_{x=3}$     and     (b) $\dfrac{d}{dx}(2^x)\bigg|_{x=-1}$.

**Solution**  (a) $\dfrac{d}{dx}(2^x)\bigg|_{x=3} = m \cdot 2^3 = 8m \approx 8(.693) = 5.544.$

(b) $\left.\dfrac{d}{dx}(2^x)\right|_{x=-1} = m \cdot 2^{-1} = .5m \approx .5(.693) = .3465.$   ■

The calculations just carried out for $y = 2^x$ can be carried out for $y = b^x$, where $b$ is any positive number. Equation (5) will read exactly the same except that 2 will be replaced by $b$. Thus we have the following formula for the derivative of the function $f(x) = b^x$.

$$\frac{d}{dx}(b^x) = m\,b^x, \quad \text{where } m = \left.\frac{d}{dx}(b^x)\right|_{x=0}. \tag{6}$$

Our calculations showed that if $b = 2$, then $m \approx .693$. If $b = 3$, then it turns out that $m \approx 1.1$. (See Exercise 1.) The derivative formula in (6) is simple when $m = 1$, that is, when the graph of $y = b^x$ has slope 1 at $x = 0$. As we said earlier, this special value of $b$ is denoted by the letter $e$. Thus the number $e$ has the property that

$$\left.\frac{d}{dx}(e^x)\right|_{x=0} = 1 \tag{7}$$

and

$$\frac{d}{dx}(e^x) = 1 \cdot e^x = e^x. \tag{8}$$

The graphical interpretation of (7) is that the curve $y = e^x$ has slope 1 at $x = 0$. The graphical interpretation of (8) is that the slope of the curve $y = e^x$ at an arbitrary value of $x$ is exactly equal to the value of the function $e^x$ at that point. (See Fig. 3.)

The function $e^x$ is the same type of function as $2^x$ and $3^x$ except that differentiating $e^x$ is much easier. In fact, the next section shows that $2^x$ can be written as $e^{kx}$ for a suitable constant $k$. The same is true for $3^x$. For this reason, functions of the form $e^{kx}$ are used in almost all applications that require an exponential type of function to describe a physical, economic, or biological phenomenon.

---

**EXAMPLE 2**   Sketch the graph of $y = e^{-x}$.

**Solution**   Note that, since $e > 1$, we have $0 < 1/e < 1$. In fact, $1/e \approx .37$. Now

$$y = e^{-x} = \frac{1}{e^x} = \left(\frac{1}{e}\right)^x;$$

and so $f(x) = e^{-x}$ is an exponential function with base $b = \frac{1}{e}$, which is less than one. Hence the graph of $y = e^{-x}$ is always decreasing and looks like the graph of the exponential function $y = .5^x$ (see Fig. 1). With the help of a calculator, we form a table of values of $y = e^{-x}$ for selected values of $x$:

| $x$ | $-3$ | $-2$ | $-1$ | 0 | 1 | 2 | 3 |
|---|---|---|---|---|---|---|---|
| $e^{-x}$ | 20.09 | 7.39 | 2.72 | 1 | .37 | .14 | .05 |

Using these values, we sketch the graph of $y = e^{-x}$ in Fig. 4.   ■

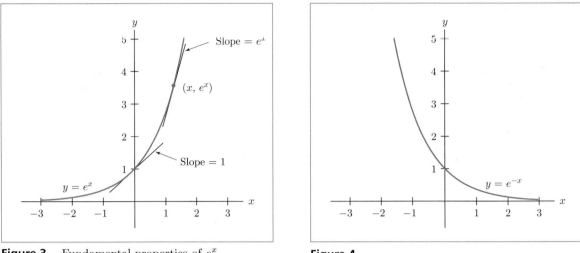

**Figure 3.** Fundamental properties of $e^x$.  **Figure 4**

In the following section, we will compute the derivative of $y = e^{g(x)}$ for any differentiable function $g$, and, in particular, $y = e^{kx}$. However, as we now illustrate, by relating $e^{kx}$ to $e^x$, we can also differentiate $y = e^{kx}$ by using the generalized power rule for differentiation.

---

**EXAMPLE 3**  Find the derivative of $y = e^{-x}$.

**Solution**  Let $f(x) = e^x$, then

$$e^{-x} = (e^x)^{-1} = (f(x))^{-1};$$

and so

$$\frac{d}{dx}[e^{-x}] = \frac{d}{dx}[(f(x))^{-1}]$$

$$= (-1)(f(x))^{-2}\frac{d}{dx}[f(x)] \quad \text{(the general power rule)}$$

$$= (-1)(e^x)^{-2}\frac{d}{dx}[e^x] \quad (f(x) = e^x)$$

$$= (-1)e^{-2x}e^x \quad \text{(by (8))}$$

$$= -e^{-x}$$

■

**Incorporating Technology**  Calculators usually have $e^x$ as the secondary function of the LN key. Figures 5(a) and 5(b) show three computations with $e^x$.

Parentheses are often necessary when evaluating $e^x$. In Fig. 6, **e^2/3** is calculated as $e^2/3$ instead of $e^{2/3}$.

(a) TI-83  (b) TI-86

**Figure 5**  **Figure 6.** TI-86.

## PRACTICE PROBLEMS 4.2

*In the following problems use the number 20 as the (approximate) value of $e^3$.*

**1.** Find the equation of the tangent line to the graph of $y = e^x$ at $x = 3$.

**2.** Solve the following equation for $x$:

$$4e^{6x} = 80.$$

**3.** Differentiate $y = 3(x^2 + 1)e^x$.

# EXERCISES 4.2

**1.** Show that

$$\frac{d}{dx}(3^x)\Big|_{x=0} \approx 1.1$$

by calculating the slope $\dfrac{3^h - 1}{h}$ of the secant line passing through the points $(0, 1)$ and $(h, 3^h)$. Take $h = .1$, .01, and .001.

**2.** Show that

$$\frac{d}{dx}(2.7^x)\Big|_{x=0} \approx .99$$

by calculating the slope $\dfrac{2.7^h - 1}{h}$ of the secant line passing through the points $(0, 1)$ and $(h, 2.7^h)$. Take $h = .1$, .01, and .001.

*In Exercises 3–6, compute the given derivatives with the help of formulas (2), (5), and (6)–(8).*

**3. (a)** $\dfrac{d}{dx}(2^x)\Big|_{x=1}$   **(b)** $\dfrac{d}{dx}(2^x)\Big|_{x=-2}$

**4. (a)** $\dfrac{d}{dx}(2^x)\Big|_{x=1/2}$   **(b)** $\dfrac{d}{dx}(2^x)\Big|_{x=2}$

**5. (a)** $\dfrac{d}{dx}(e^x)\Big|_{x=1}$   **(b)** $\dfrac{d}{dx}(e^x)\Big|_{x=-1}$

**6. (a)** $\dfrac{d}{dx}(e^x)\Big|_{x=e}$   **(b)** $\dfrac{d}{dx}(e^x)\Big|_{x=1/e}$

**7.** Estimate the slope of $e^x$ at $x = 0$ by calculating the slope $\dfrac{e^h - 1}{h}$ of the secant line passing through the points $(0, 1)$ and $(h, e^h)$. Take $h = .01$, .001, and .0001.

**8.** Use (8) and a familiar rule for differentiation to find

$$\frac{d}{dx}(5e^x).$$

**9.** Use (8) and a familiar rule for differentiation to find

$$\frac{d}{dx}(e^x)^{10}.$$

**10.** Use the fact that $e^{2+x} = e^2 \cdot e^x$ to find

$$\frac{d}{dx}(e^{2+x}).$$

[Remember that $e^2$ is just a constant—approximately $(2.7)^2$.]

**11. (a)** Use the fact that $e^{4x} = (e^x)^4$ to find $\dfrac{d}{dx}(e^{4x})$. Simplify the derivative as much as possible.

**(b)** Let $k$ represent any constant. Generalize the calculations in part (a) to find a simple formula for $\dfrac{d}{dx}(e^{kx})$.

**12.** Find $\dfrac{d}{dx}(e^x + x^2)$.

*Write each expression in the form $e^{kx}$ for a suitable constant $k$.*

**13.** $(e^2)^x$, $\left(\dfrac{1}{e}\right)^x$   **14.** $(e^3)^{x/5}$, $\left(\dfrac{1}{e^2}\right)^x$

**15.** $\left(\dfrac{1}{e^3}\right)^{2x}$, $e^{1-x} \cdot e^{3x-1}$   **16.** $\left(\dfrac{e^5}{e^3}\right)^x$, $e^{4x+2} \cdot e^{x-2}$

**17.** $(e^{4x} \cdot e^{6x})^{3/5}$, $\dfrac{1}{e^{-2x}}$   **18.** $\sqrt{e^{-x} \cdot e^{7x}}$, $\dfrac{e^{-3x}}{e^{-4x}}$

*Solve each equation for $x$.*

**19.** $e^{5x} = e^{20}$   **20.** $e^{1-x} = e^2$

**21.** $e^{x^2-2x} = e^8$   **22.** $e^{-x} = 1$

**23.** $e^x(x^2 - 1) = 0$   **24.** $4e^x(x^2 + 1) = 0$

*Differentiate the following functions.*

**25.** $y = \dfrac{e^x}{2}$   **26.** $y = 7e^x - 2x^2$

**27.** $y = (1 + 2e^x)^2$   **28.** $y = \dfrac{1}{1 + e^x}$

**29.** $y = \dfrac{1}{e^x}$   **30.** $y = \dfrac{x}{e^x}$

**31.** $y = xe^x$   **32.** $y = \dfrac{e^x}{x^2}$

**33.** $y = \dfrac{e^x}{1 + e^x}$   **34.** $y = (1 + x^2)e^x$

**35.** $y = (1 + 5e^x)^4$   **36.** $y = (xe^x - 1)^{-3}$

**37.** Find the point on the graph of $y = (1 + x^2)e^x$ where the tangent line is horizontal.

**38.** Show that the tangent line to the graph of $y = e^x$ at the point $(a, e^a)$ is perpendicular to the tangent line to the graph of $y = e^{-x}$ at the point $(a, e^{-a})$.

**39.** Find the slope of the tangent line to the curve $y = xe^x$ at $(0,0)$.

**40.** Find the slope of the tangent line to the curve $y = xe^x$ at $(1, e)$.

**41.** Find the equation of the tangent line to the curve $y = \dfrac{e^x}{1 + 2e^x}$ at $(0, \frac{1}{3})$.

**42.** Find the equation of the tangent line to the curve $y = \dfrac{e^x}{x + e^x}$ at $(0, 1)$.

*Find the first and second derivatives.*

**43.** $f(x) = e^x(1 + x)^2$     **44.** $f(x) = \dfrac{e^x}{x}$

**45.** The atmospheric pressure at an altitude of $x$ kilometers is $f(x)$ g/cm$^2$ (grams per square centimeter), where $f(x) = 1035e^{-.12x}$. Give approximate answers to the following questions using the graphs of $f(x)$ and $f'(x)$ shown in Fig. 7.

   **(a)** What is the pressure at an altitude of 2 kilometers?

   **(b)** At what altitude is the pressure 200 g/cm$^2$?

   **(c)** At an altitude of 8 kilometers, at what rate is the atmospheric pressure changing (with respect to change in altitude)?

   **(d)** At what altitude is atmospheric pressure falling at the rate of 100 g/cm$^2$ per kilometer?

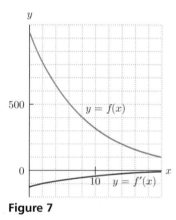

**Figure 7**

**46.** (*Health Expenditures*) The national health expenditures (in billions of dollars) from 1960 to 1980 are given approximately by $f(t) = 27e^{.106t}$, with time in years measured from 1960. Give approximate answers to the following questions using the graphs of $f(t)$ and $f'(t)$ shown in Fig. 8.

   **(a)** How much money was spent in 1978?

   **(b)** How fast were expenditures rising in 1972?

   **(c)** When did expenditures reach 120 billion dollars?

   **(d)** When were expenditures rising at the rate of 20 billion dollars per year?

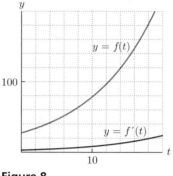

**Figure 8**

## Technology Exercises

**47.** Find the equation of the tangent line to the graph of $y = e^x$ at $x = 0$. Then graph the function and the tangent line together to confirm that your answer is correct.

**48.** **(a)** Graph $y = e^x$.

   **(b)** Zoom in on the region near $x = 0$ until the curve appears as a straight line and estimate the slope of the line. This number is an estimate of $\dfrac{d}{dx}e^x$ at $x = 0$. Compare your answer with the actual slope, 1.

   **(c)** Repeat parts (a) and (b) for $y = 2^x$. Observe that the slope at $x = 0$ is not 1.

**49.** Set $Y_1 = e^x$ and use your calculator's derivative command to specify $Y_2$ as the derivative of $Y_1$. Graph the two functions simultaneously in the window $[-1, 3]$ *by* $[-3, 20]$ and observe that the graphs overlap.

**50.** Demonstrate that $e^x$ grows faster than any power function. For instance, graph $y = \dfrac{x^n}{e^x}$ for $n = 3$ with the window $[0, 16]$ *by* $[0, 25]$ and observe that the function approaches zero as $x$ gets large. Repeat for $n = 4$ and $n = 5$.

**51.** Calculate values of $\dfrac{10^x - 1}{x}$ for small values of $x$ and use them to estimate $\dfrac{d}{dx}(10^x)\Big|_{x=0}$. What is the formula for $\dfrac{d}{dx}(10^x)$?

**52.** The graph of $y = e^{x-2}$ may be obtained by translating the graph of $y = e^x$ two units to the right. Find a constant $k$ such that the graph of $y = ke^x$ is the same as the graph of $y = e^{x-2}$. Verify your result by graphing both functions.

## SOLUTIONS TO PRACTICE PROBLEMS 4.2

**1.** When $x = 3$, $y = e^3 \approx 20$. So the point $(3, 20)$ is on the tangent line. Since $\dfrac{d}{dx}(e^x) = e^x$, the slope of the tangent line is $e^3$ or 20. Therefore, the equation of the tangent line in point-slope form is $y - 20 = 20(x - 3)$.

**2.** This problem is similar to Practice Problem 2 of Section 4.1. First divide both sides of the equation by 4.

$$e^{6x} = 20.$$

The idea is to express 20 as a power of $e$ and then equate exponents.

$$e^{6x} = e^3$$
$$6x = 3$$
$$x = \tfrac{1}{2}.$$

**3.** $\dfrac{d}{dx}[3(x^2 + 1)e^x] = 3\dfrac{d}{dx}[(x^2 + 1)e^x]$

(constant-multiple rule)

$$= 3(x^2 + 1)\dfrac{d}{dx}[e^x] + 3e^x \dfrac{d}{dx}[(x^2 + 1)]$$

(product rule)

$$= 3(x^2 + 1)e^x + 3e^x(2x)$$

(by (8))

$$= 3e^x(x^2 + 2x + 1) = 3e^x(x + 1)^2$$

## 4.3 Differentiation of Exponential Functions

We have shown that $\dfrac{d}{dx}(e^x) = e^x$. Using this fact and the chain rule, we can differentiate functions of the form $y = e^{g(x)}$, where $g(x)$ is any differentiable function. This is because $e^{g(x)}$ is the composite of two functions. Indeed, if $f(x) = e^x$, then

$$e^{g(x)} = f(g(x)).$$

Thus, by the chain rule, we have

$$\dfrac{d}{dx}(e^{g(x)}) = f'(g(x))g'(x)$$
$$= f(g(x))g'(x) \qquad [\text{since } f'(x) = f(x)]$$
$$= e^{g(x)}g'(x).$$

So we have the following result:

**Chain Rule for Exponential Functions**    Let $g(x)$ be any differentiable function. Then

$$\dfrac{d}{dx}(e^{g(x)}) = e^{g(x)}g'(x). \tag{1}$$

If we write $u = g(x)$, then (1) can be written in the form

$$\dfrac{d}{dx}(e^u) = e^u \dfrac{du}{dx}. \tag{1a}$$

**EXAMPLE 1**    Differentiate $y = e^{x^2 + 1}$.

**Solution**    Here $g(x) = x^2 + 1$, $g'(x) = 2x$, so

$$\dfrac{d}{dx}(e^{x^2 + 1}) = e^{x^2 + 1} \cdot 2x = 2xe^{x^2 + 1}.$$

∎

**EXAMPLE 2** Differentiate $y = e^{3x^2 - (1/x)}$.

**Solution** $\dfrac{d}{dx}\left(e^{3x^2 - (1/x)}\right) = e^{3x^2 - (1/x)} \cdot \dfrac{d}{dx}\left(3x^2 - \dfrac{1}{x}\right) = e^{3x^2 - (1/x)}\left(6x + \dfrac{1}{x^2}\right)$.  ■

**EXAMPLE 3** Differentiate $y = e^{5x}$.

**Solution** $\dfrac{d}{dx}(e^{5x}) = e^{5x} \cdot \dfrac{d}{dx}(5x) = e^{5x} \cdot 5 = 5e^{5x}$.  ■

Using a computation similar to that used in Example 3, we may differentiate $y = e^{kx}$ for any constant $k$. (In Example 3 we have $k = 5$.) The result is the following useful formula.

$$\frac{d}{dx}(e^{kx}) = ke^{kx}. \tag{2}$$

Many applications involve exponential functions of the form $y = Ce^{kx}$, where $C$ and $k$ are constants. In the next example we differentiate such functions.

**EXAMPLE 4** Differentiate the following exponential functions.
(a) $y = 3e^{5x}$
(b) $y = 3e^{kx}$, where $k$ is a constant
(c) $y = Ce^{kx}$, where $C$ and $k$ are constants

**Solution** (a) $\dfrac{d}{dx}(3e^{5x}) = 3\dfrac{d}{dx}(e^{5x}) = 3 \cdot 5e^{5x} = 15e^{5x}$.

(b) $\dfrac{d}{dx}(3e^{kx}) = 3\dfrac{d}{dx}(e^{kx})$
$= 3 \cdot ke^{kx} \qquad \text{[by (2)]}$
$= 3ke^{kx}$.

(c) $\dfrac{d}{dx}(Ce^{kx}) = C\dfrac{d}{dx}(e^{kx}) = Cke^{kx}$.  ■

The result of part (c) may be summarized in an extremely useful fashion as follows: Suppose that we let $y = Ce^{kx}$. By part (c) we have

$$y' = Cke^{kx} = k \cdot (Ce^{kx}) = ky.$$

In other words, the derivative of the function $Ce^{kx}$ is $k$ times the function itself. Let us record this fact.

Let $C$, $k$ be any constants and let $y = Ce^{kx}$. Then $y$ satisfies the equation
$$y' = ky.$$

The equation $y' = ky$ expresses a relationship between the function $y$ and its derivative $y'$. Any equation expressing a relationship between a function $y$ and one or more of its derivatives is called a *differential equation*.

Very often an applied problem will involve a function $y = f(x)$ that satisfies the differential equation $y' = ky$. It can be shown that $y$ must then necessarily be an exponential function of the form $y = Ce^{kx}$. That is, we have the following result.

Suppose that $y = f(x)$ satisfies the differential equation

$$y' = ky. \qquad (3)$$

Then $y$ is an exponential function of the form

$$y = Ce^{kx}, \qquad C \text{ a constant.}$$

We leave the verification of this result to Exercise 45.

**EXAMPLE 5**  Determine all functions $y = f(x)$ such that $y' = -.2y$.

**Solution**  The equation $y' = -.2y$ has the form $y' = ky$ with $k = -.2$. Therefore, any solution of the equation has the form

$$y = Ce^{-.2x},$$

where $C$ is a constant.  ■

**EXAMPLE 6**  Determine all functions $y = f(x)$ such that $y' = y/2$ and $f(0) = 4$.

**Solution**  The equation $y' = y/2$ has the form $y' = ky$ with $k = \frac{1}{2}$. Therefore,

$$f(x) = Ce^{(1/2)x}$$

for some constant $C$. We also require that $f(0) = 4$. That is,

$$4 = f(0) = Ce^{(1/2)0} = Ce^0 = C.$$

So $C = 4$ and

$$f(x) = 4e^{(1/2)x}.$$  ■

**The Functions $f(x) = e^{kx}$**  Exponential functions of the form $f(x) = e^{kx}$ occur in many applications. Figure 1 shows the graphs of several functions of this type when $k$ is a positive number. These curves $y = e^{kx}$, $k$ positive, have several properties in common:

**1.** $(0, 1)$ is on the graph.

**2.** The graph lies strictly above the $x$-axis ($e^{kx}$ is never zero).

**3.** The $x$-axis is an asymptote as $x$ becomes large negatively.

**4.** The graph is always increasing and concave up.

When $k$ is negative, the graph of $y = e^{kx}$ is decreasing. (See Fig. 2.) Note the following properties of the curves $y = e^{kx}$, $k$ negative:

**1.** $(0, 1)$ is on the graph.

**2.** The graph lies strictly above the $x$-axis.

**3.** The $x$-axis is an asymptote as $x$ becomes large positively.

**4.** The graph is always decreasing and concave up.

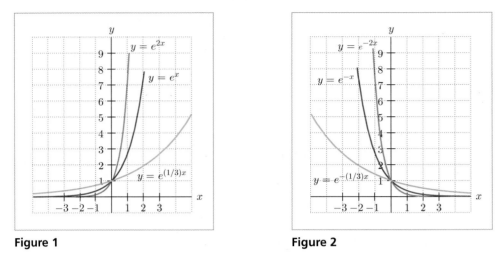

**Figure 1**

**Figure 2**

**The Functions $f(x) = b^x$**   If $b$ is a positive number, then the function $f(x) = b^x$ may be written in the form $f(x) = e^{kx}$ for some $k$. For example, take $b = 2$. From Fig. 3 of the preceding section it is clear that there is some value of $x$ such that $e^x = 2$. Call this value $k$, so that $e^k = 2$. Then

$$2^x = (e^k)^x = e^{kx}$$

for all $x$. In general, if $b$ is any positive number, there is a value of $x$, say $x = k$, such that $e^k = b$. In this case, $b^x = (e^k)^x = e^{kx}$. Thus all the curves $y = b^x$ discussed in Section 4.1 can be written in the form $y = e^{kx}$. This is one reason why we have focused on exponential functions with base $e$ instead of studying $y = 2^x$, $y = 3^x$, and so on.

```
3→x:e^2  x
        22.1671682968
3→x:e^(2x)
       403.428793493
3→x:(e^2)x
        22.1671682968
```

**Figure 3.**   TI-86.

**Incorporating Technology**   Parentheses are often necessary when evaluating exponential functions on graphing calculators. With some calculators, **e^2x** is evaluated as $e^2x$ instead of the intended $e^{2x}$. See Fig. 3.

## PRACTICE PROBLEMS 4.3

**1.** Differentiate $[e^{-3x}(1 + e^{6x})]^{12}$.

**2.** Determine all functions $y = f(x)$ such that $y' = -y/20$, $f(0) = 2$.

## EXERCISES 4.3

*Differentiate the following functions.*

**1.** $f(x) = 4e^{2x}$

**2.** $f(x) = 3e^7$

**3.** $f(x) = 1 + 4x + e^{-2x}$

**4.** $f(x) = \dfrac{e^{3x}}{3}$

**5.** $f(t) = (1 + t)e^{t^2}$

**6.** $f(t) = \dfrac{1}{e^{-t}}$

**7.** $y = (e^x + e^{-x})^3$

**8.** $f(x) = 2e^{\sqrt{x}}$

**9.** $y = \dfrac{e^{\frac{t}{2}+1}}{4}$

**10.** $g(x) = (e^{-2x} - 2x)^3$

**11.** $g(t) = e^{3/t}$

**12.** $y = \dfrac{e^{-2x^2+2x+1}}{10}$

**13.** $y = e^{x^2-5x+4}$

**14.** $g(t) = t^2 e^{1/t}$

**15.** $y = \dfrac{x + e^{2x}}{e^x}$

**16.** $f(t) = (e^{2t})^{-4}$

[*Hint*: In Exercises 16–18, simplify $f(t)$ before differentiating.]

**17.** $f(t) = \dfrac{e^{3t} + e^{-3t}}{e^t}$

**18.** $f(t) = e^t(e^{2t} - e^{-2t})$

**19.** $f(x) = (x + 1)e^{-x+2}$

**20.** $f(x) = \sqrt{e^{x/2} + 1}$

**21.** $f(x) = x^3 e^{2x}$

**22.** $f(x) = e^{e^x}$

**23.** $f(x) = e^{4x}/(4 + x)$

**24.** $f(x) = e^{e^x}$

**25.** $f(x) = \left(\dfrac{1}{x} + 3\right)e^{2x}$

**26.** $f(x) = \dfrac{4x^2}{x^2 + e^{2x}}$

In Exercises 27–32, find the values of $x$ at which the function has a possible relative maximum or minimum point. (Recall that $e^x$ is positive for all $x$.) Use the second derivative to determine the nature of the function at these points.

**27.** $f(x) = (1 + x)e^{-3x}$

**28.** $f(x) = (1 - x)e^{2x}$

**29.** $f(x) = \dfrac{3 - 4x}{e^{2x}}$

**30.** $f(x) = \dfrac{4x - 1}{e^{x/2}}$

**31.** $f(x) = (5x - 2)e^{1-2x}$

**32.** $f(x) = (2x - 5)e^{3x-1}$

**33.** (*Appreciation of Assets*) A painting purchased in 1998 for \$100,000 is estimated to be worth $v(t) = 100{,}000e^{t/5}$ dollars after $t$ years. At what rate will the painting be appreciating in 2003?

**34.** (*Depreciation of Assets*) The value of a computer $t$ years after purchase is $v(t) = 2000e^{-.35t}$ dollars. At what rate is the computer's value falling after 3 years?

**35.** (*Velocity*) The velocity of a parachutist during free fall is

$$f(t) = 60(1 - e^{-.17t})$$

meters per second. Answer the following questions by reading the graph in Fig. 4. (Recall that acceleration is the derivative of velocity.)

**(a)** What is the velocity when $t = 8$ seconds?

**(b)** What is the acceleration when $t = 0$?

**(c)** When is the parachutist's velocity 30 m/sec?

**(d)** When is the acceleration 5 m/sec$^2$?

**36.** Suppose that the velocity of a parachutist is

$$v(t) = 65(1 - e^{-.16t})$$

meters per second. The graph of $v(t)$ is similar to that in Fig. 4. Calculate the parachutist's velocity and acceleration when $t = 9$ seconds.

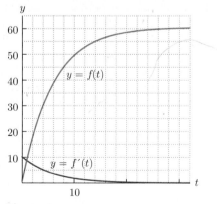

**Figure 4**

**37.** (*Height of a Plant*) The height of a certain plant after $t$ weeks is

$$f(t) = \frac{1}{.05 + e^{-.4t}}$$

inches. The graph of $f(t)$ resembles the graph in Fig. 5. Calculate the rate of growth of the plant after 7 weeks.

**38.** (*Height of a Plant*) The length of a certain weed after $t$ weeks is

$$f(t) = \frac{6}{.2 + 5e^{-.5t}}$$

centimeters. Answer the following questions by reading the graph in Fig. 5.

**(a)** How fast is the weed growing after 10 weeks?

**(b)** When is the weed 10 centimeters long?

**(c)** When is the weed growing at the rate of 2 cm/week?

**(d)** What is the maximum rate of growth?

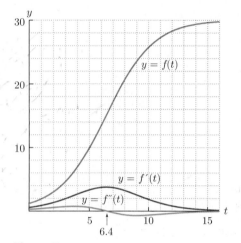

**Figure 5**

**39.** (*Gompertz Growth Curve*) Let $a$ and $b$ be positive numbers. A curve whose equation is $y = e^{-ae^{-bx}}$ is called a *Gompertz growth curve*. These curves are used in biology to describe certain types of population growth. Compute the derivative of $y = e^{-2e^{-.01x}}$.

**40.** Find $\dfrac{dy}{dx}$ if $y = e^{-(1/10)e^{-x/2}}$.

**41.** Determine all solutions of the differential equation $y' = -4y$.

**42.** Determine all solutions of the differential equation $y' = \frac{1}{3}y$.

**43.** Determine all functions $y = f(x)$ such that $y' = -.5y$ and $f(0) = 1$.

**44.** Determine all functions $y = f(x)$ such that $y' = 3y$ and $f(0) = \frac{1}{2}$.

**45.** Verify the result (3). [*Hint:* Let $g(x) = f(x)e^{-kx}$. Show that $g'(x) = 0$.] You may assume that only a constant function has a zero derivative.

**46.** Let $f(x)$ be a function with the property that $f'(x) = 1/x$. Let $g(x) = f(e^x)$, and compute $g'(x)$.

**47.** As $h$ approaches 0, what value is approached by the difference quotient $\dfrac{e^h - 1}{h}$? [*Hint:* $1 = e^0$.]

**48.** As $h$ approaches $0$, what value is approached by $\dfrac{e^{2h} - 1}{h}$? [*Hint:* $1 = e^0$.]

### *Technology Exercises*

**49.** (*Size of a Tumor*) In a study, a cancerous tumor was found to have a volume of

$$f(t) = 1.825^3(1 - 1.6e^{-.4196t})^3$$

milliliters after $t$ weeks, with $t > 1$.*

**(a)** Sketch the graphs of $f(t)$ and $f'(t)$ for $1 \le t \le 15$. What do you notice about the tumor's volume?

**(b)** How large is the tumor after 5 weeks?

**(c)** When will the tumor have a volume of 5 milliliters?

**(d)** How fast is the tumor growing after 5 weeks?

**(e)** When is the tumor growing at the fastest rate?

**(f)** What is the fastest rate of growth of the tumor?

**50.** (*Height of a Plant*) Let $f(t)$ be the function from Exercise 37 that gives the height (inches) of a plant at time $t$ (weeks).

**(a)** When is the plant 11 inches tall?

**(b)** When is the plant growing at the rate of 1 inch per week?

**(c)** What is the fastest rate of growth of the plant, and when does this occur?

---

## SOLUTIONS TO PRACTICE PROBLEMS 4.3

**1.** We must use the general power rule. However, this is most easily done if we first use the laws of exponents to simplify the function inside the brackets.

$$e^{-3x}(1 + e^{6x}) = e^{-3x} + e^{-3x} \cdot e^{6x} = e^{-3x} + e^{3x}.$$

Now

$$\frac{d}{dx}[e^{-3x} + e^{3x}]^{12} = 12 \cdot [e^{-3x} + e^{3x}]^{11} \cdot (-3e^{-3x} + 3e^{3x})$$

$$= 36 \cdot [e^{-3x} + e^{3x}]^{11} \cdot (-e^{-3x} + e^{3x}).$$

**2.** The differential equation $y' = -y/20$ is of the type $y' = ky$, where $k = -\frac{1}{20}$. Therefore, any solution has the form $f(x) = Ce^{-(1/20)x}$. Now

$$f(0) = Ce^{-(1/20)0} = Ce^0 = C,$$

so that $C = 2$ when $f(0) = 2$. Therefore, the desired function is $f(x) = 2e^{-(1/20)x}$.

---

## 4.4   The Natural Logarithm Function

As a preparation for the definition of the natural logarithm, we shall make a geometric digression. In Fig. 1 we have plotted several pairs of points. Observe how they are related to the line $y = x$.

The points $(5, 7)$ and $(7, 5)$, for example, are the same distance from the line $y = x$. If we were to plot the point $(5, 7)$ with wet ink and then fold the page along the line $y = x$, the ink blot would produce a second blot at the point $(7, 5)$. If we think of the line $y = x$ as a mirror, then $(7, 5)$ is the mirror image of $(5, 7)$. We say that $(7, 5)$ is the *reflection* of $(5, 7)$ through the line $y = x$. Similarly, $(5, 7)$ is the reflection of $(7, 5)$ through the line $y = x$.

Now let us consider all points lying on the graph of the exponential function $y = e^x$ [see Fig. 2(a)]. If we reflect each such point through the line $y = x$, we obtain a new graph [see Fig. 2(b)]. For each positive $x$, there is exactly one value of $y$ such that $(x, y)$ is on the new graph. We call this value of $y$ the *natural logarithm of $x$*, denoted $\ln x$. Thus the reflection of the graph of $y = e^x$ through the line $y = x$ is the graph of the natural logarithm function $y = \ln x$.

We may deduce some properties of the natural logarithm function from an inspection of its graph.

---

*Baker, Goddard, Clark, and Whimster, "Proportion of Necrosis in Transplanted Murine Adeno-carcinomas and Its Relationship to Tumor Growth," *Growth, Development and Aging*, 54 (1990), 85–93.

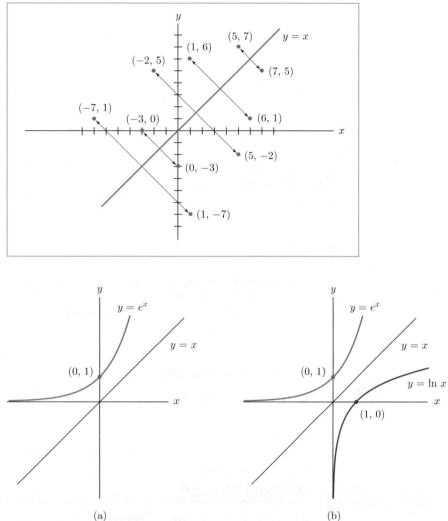

**Figure 1.** Reflections of points through the line $y = x$.

**Figure 2.** Obtaining the graph of $y = \ln x$ as a reflection of $y = e^x$.

(a)                    (b)

**1.** The point $(1, 0)$ is on the graph of $y = \ln x$ [because $(0, 1)$ is on the graph of $y = e^x$]. In other words,

$$\ln 1 = 0. \tag{1}$$

**2.** $\ln x$ is defined only for positive values of $x$.

**3.** $\ln x$ is negative for $x$ between 0 and 1.

**4.** $\ln x$ is positive for $x$ greater than 1.

**5.** $y = \ln x$ is an increasing function and concave down.

Let us study the relationship between the natural logarithm and exponential functions more closely. From the way in which the graph of $y = \ln x$ was obtained we know that $(a, b)$ is on the graph of $y = \ln x$ if and only if $(b, a)$ is on the graph of $y = e^x$. However, a typical point on the graph of $y = \ln x$ is of the form $(a, \ln a)$, $a > 0$. So for any positive value of $a$, the point $(\ln a, a)$ is on the graph of $y = e^x$.

That is,

$$e^{\ln a} = a.$$

Since $a$ was an arbitrary positive number, we have the following important relationship between the natural logarithm and exponential functions.

$$e^{\ln x} = x \quad \text{for } x > 0. \tag{2}$$

Equation (2) can be put into verbal form.

For each positive number $x$, $\ln x$ is that exponent to which we must raise $e$ in order to get $x$.

If $b$ is any number, then $e^b$ is positive and hence $\ln(e^b)$ makes sense. What is $\ln(e^b)$? Since $(b, e^b)$ is on the graph of $e^x$, we know that $(e^b, b)$ must be on the graph of $y = \ln x$. That is, $\ln(e^b) = b$. Thus we have shown that

$$\ln(e^x) = x \quad \text{for any } x. \tag{3}$$

The identities (2) and (3) express the fact that the natural logarithm function is the *inverse* of the exponential function (for $x > 0$). For instance, if we take a number $x$ and compute $e^x$, then, by (3), we can undo the effect of the exponential by taking the natural logarithm; that is, the logarithm of $e^x$ equals the original number $x$. Similarly, if we take a positive number $x$ and compute $\ln x$, then, by (2), we can undo the effect of the logarithm by raising $e$ to the $\ln x$ power; that is, $e^{\ln x}$ equals the original number $x$.

**EXAMPLE 1** Simplify $e^{\ln 4 + 2 \ln 3}$.

**Solution** Using properties of the exponential function, we have

$$e^{\ln 4 + 2 \ln 3} = e^{\ln 4} \cdot e^{2 \ln 3}$$

and

$$e^{2 \ln 3} = e^{(\ln 3) \cdot 2} = \left(e^{\ln 3}\right)^2.$$

By (2), we have $e^{\ln 4} = 4$ and $\left(e^{\ln 3}\right)^2 = 3^2$. So

$$e^{\ln 4 + 2 \ln 3} = e^{\ln 4} \cdot \left(e^{\ln 3}\right)^2 = 4 \cdot 3^2 = 36.$$  ■

Scientific calculators have an "LN" key that will compute the natural logarithm of a number to as many as ten significant figures. For instance, pressing the LN key and entering the number 2 into the calculator, one obtains $\ln 2 = .6931471806$ (to 10 significant figures).

The relationships (2) and (3) between $e^x$ and $\ln x$ may be used to solve equations, as the next examples show.

**EXAMPLE 2** Solve the equation $5e^{x-3} = 4$ for $x$.

**Solution**  First divide each side by 5,

$$e^{x-3} = .8.$$

Taking the logarithm of each side and using (3), we have

$$\ln(e^{x-3}) = \ln .8$$
$$x - 3 = \ln .8$$
$$x = 3 + \ln .8.$$

[If desired, the numerical value of $x$ can be obtained by using a scientific calculator, namely, $x = 3 - .22314 = 2.77686$ (to five decimal places).] ■

---

**EXAMPLE 3**  Solve the equation $2 \ln x + 7 = 0$ for $x$.

**Solution**
$$2 \ln x = -7$$
$$\ln x = -3.5$$
$$e^{\ln x} = e^{-3.5}$$
$$x = e^{-3.5} \qquad [\text{by } (2)]. \qquad ■$$

**Other Exponential and Logarithm Functions**  In our discussion of the exponential function, we mentioned that all exponential functions of the form $y = b^x$, where $b$ is a fixed positive number, can be expressed in terms of *the* exponential function $y = e^x$. Now we can be quite explicit. For since $b = e^{\ln b}$, we see that

$$b^x = (e^{\ln b})^x = e^{(\ln b)x}.$$

Hence we have shown that

$$b^x = e^{kx}, \quad \text{where } k = \ln b.$$

The natural logarithm function is sometimes called the *logarithm to the base e*, for it is the inverse of the exponential function $y = e^x$. If we reflect the graph of the function $y = 2^x$ through the line $y = x$, we obtain the graph of a function called the *logarithm to the base* 2, denoted by $\log_2 x$. Similarly, if we reflect the graph of $y = 10^x$ through the line $y = x$, we obtain the graph of a function called the *logarithm to the base* 10, denoted by $\log_{10} x$. (See Fig. 3.)

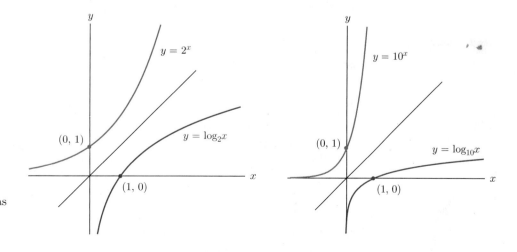

**Figure 3.**  Graphs of $y = \log_2 x$ and $y = \log_{10} x$ as reflections of $y = 2^x$ and $y = 10^x$.

Logarithms to the base 10 are sometimes called *common* logarithms. Common logarithms are usually introduced into algebra courses for the purpose of simplifying certain arithmetic calculations. However, with the advent of the modern computer and the widespread availability of graphing calculators, the need for common logarithms has diminished considerably. It can be shown that

$$\log_{10} x = \frac{1}{\ln 10} \cdot \ln x,$$

so that $\log_{10} x$ is simply a constant multiple of $\ln x$. However, we shall not need this fact.

The natural logarithm function is used in calculus because differentiation and integration formulas are simpler than for $\log_{10} x$ or $\log_2 x$, and so on. (Recall that we prefer the function $e^x$ over the functions $10^x$ and $2^x$ for the same reason.) Also, $\ln x$ arises "naturally" in the process of solving certain differential equations that describe various growth processes.

## PRACTICE PROBLEMS 4.4

**1.** Find $\ln e$.

**2.** Solve $e^{-3x} = 2$ using the natural logarithm function.

## EXERCISES 4.4

**1.** Find $\ln(\sqrt{e})$.

**2.** Find $\ln\left(\frac{1}{e^2}\right)$.

**3.** If $e^x = 5$, write $x$ in terms of the natural logarithm.

**4.** If $e^{-x} = 3.2$, write $x$ in terms of the natural logarithm.

**5.** If $\ln x = -1$, write $x$ using the exponential function.

**6.** If $\ln x = 4.5$, write $x$ using the exponential function.

*Simplify the following expressions.*

**7.** $\ln e^{-3}$

**8.** $e^{\ln 4.1}$

**9.** $e^{e^{\ln 1}}$

**10.** $\ln(e^{-2\ln e})$

**11.** $\ln(\ln e)$

**12.** $e^{4\ln 1}$

**13.** $e^{2\ln x}$

**14.** $e^{x\ln 2}$

**15.** $e^{-2\ln 7}$

**16.** $\ln(e^{-2}e^4)$

**17.** $e^{\ln x + \ln 2}$

**18.** $e^{\ln 3 - 2\ln x}$

*Solve the following equations for $x$.*

**19.** $e^{2x} = 5$

**20.** $e^{1-3x} = 4$

**21.** $\ln(4 - x) = \frac{1}{2}$

**22.** $\ln 3x = 2$

**23.** $\ln x^2 = 9$

**24.** $e^{x^2} = 25$

**25.** $6e^{-.00012x} = 3$

**26.** $4 - \ln x = 0$

**27.** $\ln 3x = \ln 5$

**28.** $\ln(x^2 - 5) = 0$

**29.** $\ln(\ln 3x) = 0$

**30.** $2\ln x = 7$

**31.** $2e^{x/3} - 9 = 0$

**32.** $e^{\sqrt{x}} = \sqrt{e^x}$

**33.** $5\ln 2x = 8$

**34.** $750e^{-.4x} = 375$

**35.** $(e^2)^x \cdot e^{\ln 1} = 4$

**36.** $e^{5x} \cdot e^{\ln 5} = 2$

**37.** $4e^x \cdot e^{-2x} = 6$

**38.** $(e^x)^2 \cdot e^{2-3x} = 4$

**39.** The graph of $f(x) = -5x + e^x$ is shown in Fig. 4. Find the coordinates of the minimum point.

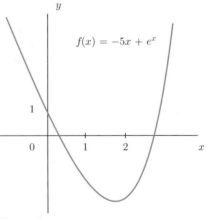

**Figure 4**

**40.** The graph of $f(x) = -1 + (x - 1)^2 e^x$ is shown in Fig. 5. Find the coordinates of the maximum and minimum points.

**41. (a)** Find the point on the graph of $y = e^{-x}$ where the tangent line has slope $-2$.

   **(b)** Plot the graphs of $y = e^{-x}$ and the tangent line in part (a).

**42.** Find the $x$-intercepts of $y = (x - 1)^2 \ln(x + 1)$, $x > -1$.

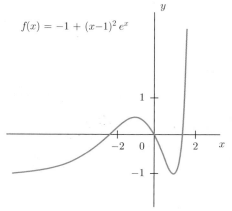

$$f(x) = -1 + (x-1)^2 e^x$$

**Figure 5**

*In Exercises 43–44, find the coordinates of each relative extreme point of the given function and determine if the point is a relative maximum point or a relative minimum point.*

**43.** $f(x) = e^{-x} + 3x$         **44.** $f(x) = 5x - 2e^x$

**45.** (*Concentration of a Drug in the Blood*) When a drug or vitamin is administered intramuscularly (into a muscle), the concentration in the blood at time $t$ after injection can be approximated by a function of the form $f(t) = c(e^{-k_1 t} - e^{-k_2 t})$. The graph of $f(t) = 5(e^{-.01t} - e^{-.51t})$, for $t \geq 0$, is shown in Fig. 6. Find the value of $t$ at which this function reaches its maximum value.

**46.** (*Wind Velocity*) Under certain geographic conditions, the wind velocity $v$ at a height $x$ centimeters above the ground is given by $v = K \ln(x/x_0)$, where $K$ is a positive constant (depending on the air density, average wind velocity, etc.), and $x_0$ is a roughness parameter (depending on the roughness of the vegetation on the ground).* Suppose that $x_0 = .7$ centimeter (a value that applies to lawn grass 3 centimeters high) and $K = 300$ centimeters per second.

**(a)** At what height above the ground is the wind velocity zero?

**(b)** At what height is the wind velocity 1200 centimeters per second?

**47.** Find $k$ such that $2^x = e^{kx}$ for all $x$.

**48.** Find $k$ such that $2^{-x/5} = e^{kx}$ for all $x$.

### Technology Exercises

**49.** Graph the function $y = \ln(e^x)$ and use TRACE to convince yourself that it is the same as the function $y = x$. What do you observe about the graph of $y = e^{\ln x}$?

**50.** Graph $y = e^{2x}$ and $y = 5$ together and determine the $x$-coordinate of their point of intersection (to four decimal places). Express this number in terms of a logarithm.

**51.** Graph $y = \ln 5x$ and $y = 2$ together and determine the $x$-coordinate of their point of intersection (to four decimal places). Express this number in terms of a power of $e$.

$$f(t) = 5(e^{-.01t} - e^{-.51t})$$

**Figure 6**

**SOLUTIONS TO PRACTICE PROBLEMS 4.4**

**1.** Answer: 1. The number $\ln e$ is that exponent to which $e$ must be raised in order to obtain $e$.

**2.** Take the logarithm of each side and use (3) to simplify the left side:

$$\ln e^{-3x} = \ln 2$$
$$-3x = \ln 2$$
$$x = -\frac{\ln 2}{3}.$$

*G. Cox, B. Collier, A. Johnson, and P. Miller, *Dynamic Ecology* (Englewood Cliffs, N.J.: Prentice-Hall, 1973), pp. 113–115.

## 4.5 The Derivative of ln x

Let us now compute the derivative of $y = \ln x$ for $x > 0$. Since $e^{\ln x} = x$, we have

$$\frac{d}{dx}(e^{\ln x}) = \frac{d}{dx}(x) = 1. \tag{1}$$

On the other hand, if we differentiate $y = e^{\ln x}$ by the chain rule, we find that

$$\frac{d}{dx}(e^{\ln x}) = e^{\ln x} \cdot \frac{d}{dx}(\ln x) = x \cdot \frac{d}{dx}(\ln x), \tag{2}$$

where the last equality used the fact that $e^{\ln x} = x$. By combining equations (1) and (2) we obtain

$$x \cdot \frac{d}{dx}(\ln x) = 1.$$

In other words,

$$\frac{d}{dx}(\ln x) = \frac{1}{x}, \qquad x > 0. \tag{3}$$

By combining this differentiation formula with the chain rule, product rule, and quotient rule, we can differentiate many functions involving $\ln x$.

---

**EXAMPLE 1**  Differentiate.

(a) $y = (\ln x)^5$        (b) $y = x \ln x$        (c) $y = \ln(x^3 + 5x^2 + 8)$

**Solution**  (a) By the general power rule,

$$\frac{d}{dx}(\ln x)^5 = 5(\ln x)^4 \cdot \frac{d}{dx}(\ln x) = 5(\ln x)^4 \cdot \frac{1}{x} = \frac{5(\ln x)^4}{x}.$$

(b) By the product rule,

$$\frac{d}{dx}(x \ln x) = x \cdot \frac{d}{dx}(\ln x) + (\ln x) \cdot 1 = x \cdot \frac{1}{x} + \ln x = 1 + \ln x.$$

(c) By the chain rule,

$$\frac{d}{dx}\ln(x^3 + 5x^2 + 8) = \frac{1}{x^3 + 5x^2 + 8} \cdot \frac{d}{dx}(x^3 + 5x^2 + 8) = \frac{3x^2 + 10x}{x^3 + 5x^2 + 8}. \quad \blacksquare$$

Let $g(x)$ be any differentiable function. For any value of $x$ for which $g(x)$ is positive, the function $y = \ln(g(x))$ is defined. For such a value of $x$, the derivative is given by the chain rule as

$$\frac{d}{dx}[\ln g(x)] = \frac{1}{g(x)} \cdot \frac{d}{dx}g(x) = \frac{g'(x)}{g(x)} \tag{4}$$

If $u = g(x)$, equation (4) can be written in the form

$$\frac{d}{dx}[\ln u] = \frac{1}{u}\frac{du}{dx}. \tag{4a}$$

Example 1(c) illustrates a special case of this formula.

**EXAMPLE 2**   The function $f(x) = (\ln x)/x$ has a relative extreme point for some $x > 0$. Find the point and determine whether it is a relative maximum or a relative minimum point.

**Solution**   By the quotient rule,

$$f'(x) = \frac{x \cdot \dfrac{1}{x} - (\ln x) \cdot 1}{x^2} = \frac{1 - \ln x}{x^2}$$

$$f''(x) = \frac{x^2 \cdot \left(-\dfrac{1}{x}\right) - (1 - \ln x)(2x)}{x^4} = \frac{2 \ln x - 3}{x^3}.$$

If we set $f'(x) = 0$, then

$$1 - \ln x = 0$$
$$\ln x = 1$$
$$e^{\ln x} = e^1 = e$$
$$x = e.$$

Therefore, the only possible relative extreme point is at $x = e$. When $x = e$, $f(e) = (\ln e)/e = 1/e$. Furthermore,

$$f''(e) = \frac{2 \ln e - 3}{e^3} = -\frac{1}{e^3} < 0,$$

which implies that the graph of $f(x)$ is concave down at $x = e$. Therefore, $(e, 1/e)$ is a relative maximum point of the graph of $f(x)$ (see Fig. 1). ∎

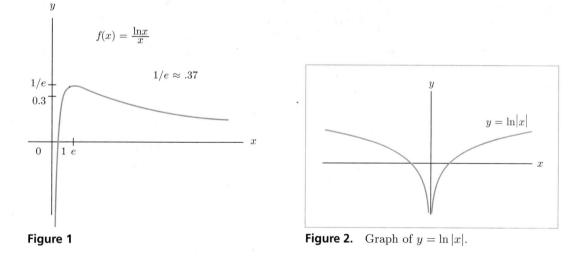

**Figure 1**                           **Figure 2.**   Graph of $y = \ln |x|$.

The next example introduces a function that will be needed later when we study integration.

**EXAMPLE 3**   The function $y = \ln |x|$ is defined for all nonzero values of $x$. Its graph is sketched in Fig. 2. Compute the derivative of $y = \ln |x|$.

**Solution** If $x$ is positive, then $|x| = x$, so

$$\frac{d}{dx} \ln |x| = \frac{d}{dx} \ln x = \frac{1}{x}.$$

If $x$ is negative, then $|x| = -x$; and, by the chain rule,

$$\frac{d}{dx} \ln |x| = \frac{d}{dx} \ln(-x) = \frac{1}{-x} \cdot \frac{d}{dx}(-x) = \frac{1}{-x} \cdot (-1) = \frac{1}{x}.$$

Therefore, we have established the following useful fact.

$$\frac{d}{dx} \ln |x| = \frac{1}{x}, \qquad x \neq 0$$

## PRACTICE PROBLEMS 4.5

*Differentiate.*

**1.** $f(x) = \dfrac{1}{\ln(x^4 + 5)}$

**2.** $f(x) = \ln(\ln x)$

## EXERCISES 4.5

*Differentiate the following functions.*

**1.** $y = \ln 2x$

**2.** $y = \dfrac{\ln x}{2}$

**3.** $y = \ln(x + 3) + \ln 5$

**4.** $y = \ln(6x^2 - 3x + 1)$

**5.** $y = \ln(e^x + e^{-x})$

**6.** $y = \ln \left[ e^{x^2 + 2} \right]$

**7.** $y = e^{\ln x + x}$

**8.** $y = (1 + \ln x)^3$

**9.** $y = \ln x \cdot \ln 2x$

**10.** $y = \ln 2 \cdot \ln x$

**11.** $y = (\ln x)^2 + \ln(x^2)$

**12.** $y = \dfrac{\ln x}{\ln 2x}$

**13.** $y = \ln(kx)$, $k$ constant

**14.** $y = \ln \dfrac{1}{x}$

**15.** $y = (x^2 + 1) \ln(x^2 + 1)$

**16.** $y = \dfrac{1}{\ln x}$

**17.** $y = \ln \left( \dfrac{x - 1}{x + 1} \right)$

**18.** $y = [\ln(e^{2x} + 1)]^2$

**19.** $y = \ln(e^{5x} + 1)$

**20.** $y = \sqrt{\ln 2x}$

*Find.*

**21.** $\dfrac{d}{dt}(t^2 \ln 4)$

**22.** $\dfrac{d^2}{dx^2} \ln(x + 1)$

**23.** $\dfrac{d^2}{dt^2}(\ln t)^3$

**24.** Find the equations of the tangent lines to the graph of $y = \ln |x|$ at $x = 1$ and $x = -1$.

**25.** Write the equation of the tangent line to the graph of $y = \ln(x^2 + e)$ at $x = 0$.

**26.** The function $f(x) = (\ln x + 1)/x$ has a relative extreme point for $x > 0$. Find the coordinates of the point. Is it a relative maximum point?

**27.** The graph of $f(x) = (\ln x)/\sqrt{x}$ is shown in Fig. 3. Find the coordinates of the maximum point.

**Figure 3**

**28.** The graph of $f(x) = x/(\ln x + x)$ is shown in Fig. 4. Find the coordinates of the minimum point.

**Figure 4**

**29.** Find the coordinates of the relative extreme point of $y = x^2 \ln x$, $x > 0$. Then use the second derivative test to decide if the point is a relative maximum point or a relative minimum point.

**30.** Repeat the previous exercise with $y = \sqrt{x}\ln x$.

**31.** If a cost function is $C(x) = (100\ln x)/(40 - 3x)$, find the marginal cost when $x = 10$.

**32.** If a cost function is $C(x) = (1000\ln x)/\sqrt{100 - 3x}$, find the marginal cost when $x = 25$.

**33.** Suppose that the demand equation for a certain commodity is $p = 45/(\ln x)$. Determine the marginal revenue function for this commodity, and compute the marginal revenue when $x = 20$.

**34.** Suppose that the total revenue function for a manufacturer is $R(x) = 300\ln(x + 1)$, so that the sale of $x$ units of a product brings in about $R(x)$ dollars. Suppose also that the total cost of producing $x$ units is $C(x)$ dollars, where $C(x) = 2x$. Find the value of $x$ at which the profit function $R(x) - C(x)$ will be maximized. Show that the profit function has a relative maximum and not a relative minimum point at this value of $x$.

**35.** Evaluate $\displaystyle\lim_{h \to 0} \frac{\ln(7 + h) - \ln 7}{h}$.

**36.** (*An Area Problem*) Find the maximum area of a rectangle in the first quadrant with one corner at the origin, two sides on the coordinate axes, and one corner on the graph of $y = -\ln x$.

*Technology Exercises*

**37.** Graph the function $f(x) = \ln|x|$ in the window $[-5, 5]$ *by* $[-2, 2]$, and verify that the derivative formula gives the correct values of $f'(x)$ at $x = \pm 1$, $\pm 2$, and $\pm 4$.

**38.** (*Analysis of the Effectiveness of an Insect Repellent*) Human hands covered with cotton fabrics that had been impregnated with the insect repellent DEPA were inserted for five minutes into a test chamber containing 200 female mosquitoes.[*] The function $f(x) = 26.48 - 14.09\ln x$ gives the number of mosquito bites received when the concentration was $x$ percent. [*Note:* The answers to parts (b)–(e) can be obtained either algebraically or from the graphs. You might consider trying both methods.]

(a) Graph $f(x)$ and $f'(x)$ for $0 \le x \le 6$.

(b) How many bites were received when the concentration was 3.25%?

(c) What concentration resulted in 15 bites?

(d) At what rate is the number of bites changing with respect to concentration of DEPA when $x = 2.75$?

(e) For what concentration does the rate of change of bites with respect to concentration equal $-10$ bites per percentage increase in concentration?

---

## SOLUTIONS TO PRACTICE PROBLEMS 4.5

**1.** Here $f(x) = [\ln(x^4 + 5)]^{-1}$. By the chain rule,

$$f'(x) = (-1) \cdot [\ln(x^4 + 5)]^{-2} \cdot \frac{d}{dx}\ln(x^4 + 5)$$

$$= -[\ln(x^4 + 5)]^{-2} \cdot \frac{4x^3}{x^4 + 5}.$$

**2.** $f'(x) = \dfrac{d}{dx}\ln(\ln x) = \dfrac{1}{\ln x}\cdot\dfrac{d}{dx}\ln x = \dfrac{1}{\ln x}\cdot\dfrac{1}{x} = \dfrac{1}{x\ln x}.$

---

# 4.6  Properties of the Natural Logarithm Function

The natural logarithm function $\ln x$ possesses many of the familiar properties of logarithms to base 10 (or common logarithms) that are encountered in algebra.

Let $x$ and $y$ be positive numbers, $b$ any number.

LI  $\ln(xy) = \ln x + \ln y$

LIII  $\ln\left(\dfrac{x}{y}\right) = \ln x - \ln y$

LII  $\ln\left(\dfrac{1}{x}\right) = -\ln x$

LIV  $\ln(x^b) = b\ln x$

---

[*]Rao, K. M., Prakash, S., Kumar, S., Suryanarayana, M. V. S., Bhagwat, M. M., Gharia, M. M., and Bhavsar, R. B., "N-diethylphenylacetamide in Treated Fabrics, as a Repellent Against *Aedes aegypti* and *Culex quinquefasclatus* (Deptera: Culiciae)," *Journal of Medical Entomology*, 28 (January 1991), 1.

***Verification of LI*** By equation (2) of Section 4.4 we have $e^{\ln(xy)} = xy$, $e^{\ln x} = x$, and $e^{\ln y} = y$. Therefore,

$$e^{\ln(xy)} = xy = e^{\ln x} \cdot e^{\ln y} = e^{\ln x + \ln y}.$$

By equating exponents, we get LI.

***Verification of LII*** Since $e^{\ln(1/x)} = 1/x$, we have

$$e^{\ln(1/x)} = \frac{1}{x} = \frac{1}{e^{\ln x}} = e^{-\ln x}.$$

By equating exponents, we get LII.

***Verification of LIII*** By LI and LII, we have

$$\ln\left(\frac{x}{y}\right) = \ln\left(x \cdot \frac{1}{y}\right) = \ln x + \ln\left(\frac{1}{y}\right) = \ln x - \ln y.$$

***Verification of LIV*** Since $e^{\ln(x^b)} = x^b$, we have

$$e^{\ln(x^b)} = x^b = \left(e^{\ln x}\right)^b = e^{b \ln x}.$$

Equating exponents, we get LIV.

These properties of the natural logarithm should be learned thoroughly. You will find them useful in many calculations involving $\ln x$ and the exponential function.

---

**EXAMPLE 1** Write $\ln 5 + 2 \ln 3$ as a single logarithm.

**Solution**
$$\ln 5 + 2 \ln 3 = \ln 5 + \ln 3^2 \qquad \text{(LIV)}$$
$$= \ln 5 + \ln 9$$
$$= \ln(5 \cdot 9) \qquad \text{(LI)}$$
$$= \ln 45. \qquad \blacksquare$$

---

**EXAMPLE 2** Write $\frac{1}{2} \ln(4t) - \ln(t^2 + 1)$ as a single logarithm.

**Solution**
$$\frac{1}{2} \ln(4t) - \ln(t^2 + 1) = \ln\left[(4t)^{1/2}\right] - \ln(t^2 + 1) \qquad \text{(LIV)}$$
$$= \ln(2\sqrt{t}) - \ln(t^2 + 1)$$
$$= \ln\left(\frac{2\sqrt{t}}{t^2 + 1}\right). \qquad \text{(LIII)}$$
$$\blacksquare$$

---

**EXAMPLE 3** Simplify $\ln x + \ln 3 + \ln y - \ln 5$.

**Solution** Use (LI) twice and (LIII) once.
$$(\ln x + \ln 3) + \ln y - \ln 5 = \ln 3x + \ln y - \ln 5$$
$$= \ln 3xy - \ln 5$$
$$= \ln\left(\frac{3xy}{5}\right). \qquad \blacksquare$$

**EXAMPLE 4**   Differentiate $f(x) = \ln[x(x+1)(x+2)]$.

**Solution**   First rewrite $f(x)$, using (LI).

$$f(x) = \ln[x(x+1)(x+2)] = \ln x + \ln(x+1) + \ln(x+2).$$

Then $f'(x)$ is easily calculated:

$$f'(x) = \frac{1}{x} + \frac{1}{x+1} + \frac{1}{x+2}. \qquad \blacksquare$$

The natural logarithm function can be used to simplify the task of differentiating products. Suppose, for example, that we wish to differentiate the function

$$g(x) = x(x+1)(x+2).$$

As we showed in Example 4,

$$\frac{d}{dx} \ln g(x) = \frac{1}{x} + \frac{1}{x+1} + \frac{1}{x+2}.$$

However,

$$\frac{d}{dx} \ln g(x) = \frac{g'(x)}{g(x)}.$$

Therefore, equating the two expressions for $\dfrac{d}{dx} \ln g(x)$, we have

$$\frac{g'(x)}{g(x)} = \frac{1}{x} + \frac{1}{x+1} + \frac{1}{x+2}.$$

Finally, we solve for $g'(x)$:

$$g'(x) = g(x) \cdot \left( \frac{1}{x} + \frac{1}{x+1} + \frac{1}{x+2} \right)$$

$$= x(x+1)(x+2) \left( \frac{1}{x} + \frac{1}{x+1} + \frac{1}{x+2} \right).$$

In a similar way, we differentiate the product of any number of factors by first taking natural logarithms, then differentiating, and finally solving for the desired derivative. This procedure is called *logarithmic differentiation*.

**EXAMPLE 5**   Differentiate the function $g(x) = (x^2 + 1)(x^3 - 3)(2x + 5)$ using logarithmic differentiation.

**Solution**   *Step 1*   Take the natural logarithm of both sides of the given equation:

$$\ln g(x) = \ln[(x^2 + 1)(x^3 - 3)(2x + 5)]$$
$$= \ln(x^2 + 1) + \ln(x^3 - 3) + \ln(2x + 5).$$

*Step 2*   Differentiate:

$$\frac{d}{dx} (\ln g(x)) = \frac{g'(x)}{g(x)} = \frac{2x}{x^2 + 1} + \frac{3x^2}{x^3 - 3} + \frac{2}{2x + 5}$$

***Step 3***    Solve for $g'(x)$:

$$g'(x) = g(x) \left( \frac{2x}{x^2 + 1} + \frac{3x^2}{x^3 - 3} + \frac{2}{2x + 5} \right)$$

$$= (x^2 + 1)(x^3 - 3)(2x + 5) \left( \frac{2x}{x^2 + 1} + \frac{3x^2}{x^3 - 3} + \frac{2}{2x + 5} \right). \quad \blacksquare$$

Let us now use logarithmic differentiation to finally establish the power rule:

$$\frac{d}{dx}(x^r) = rx^{r-1}.$$

***Verification of***   Let $f(x) = x^r$. Then
***the Power Rule***

$$\ln f(x) = \ln x^r = r \ln x.$$

Differentiation of this equation yields

$$\frac{f'(x)}{f(x)} = r \cdot \frac{1}{x}$$

$$f'(x) = r \cdot \frac{1}{x} \cdot f(x) = r \cdot \frac{1}{x} \cdot x^r = rx^{r-1}.$$

## PRACTICE PROBLEMS 4.6

1. Differentiate $f(x) = \ln\left[ \dfrac{e^x \sqrt{x}}{(x+1)^6} \right]$.

2. Use logarithmic differentiation to differentiate

$$f(x) = (x+1)^7 (x+2)^8 (x+3)^9.$$

## EXERCISES 4.6

*Simplify the following expressions.*

1. $\ln 5 + \ln x$

2. $\ln x^5 - \ln x^3$

3. $\frac{1}{2} \ln 9$

4. $3 \ln \frac{1}{2} + \ln 16$

5. $\ln 4 + \ln 6 - \ln 12$

6. $\ln 2 - \ln x + \ln 3$

7. $e^{2 \ln x}$

8. $\frac{3}{2} \ln 4 - 5 \ln 2$

9. $5 \ln x - \frac{1}{2} \ln y + 3 \ln z$

10. $e^{\ln x^2 + 3 \ln y}$

11. $\ln x - \ln x^2 + \ln x^4$

12. $\frac{1}{2} \ln xy + \frac{3}{2} \ln \frac{x}{y}$

13. Which is larger, $2 \ln 5$ or $3 \ln 3$?

14. Which is larger, $\frac{1}{2} \ln 16$ or $\frac{1}{3} \ln 27$?

*Evaluate the given expression. Use $\ln 2 = .69$ and $\ln 3 = 1.1$.*

15. **(a)** $\ln 4$     **(b)** $\ln 6$     **(c)** $\ln 54$

16. **(a)** $\ln 12$    **(b)** $\ln 16$    **(c)** $\ln(9 \cdot 2^4)$

17. **(a)** $\ln \frac{1}{6}$    **(b)** $\ln \frac{2}{9}$    **(c)** $\ln \frac{1}{\sqrt{2}}$

18. **(a)** $\ln 100 - 2 \ln 5$   **(b)** $\ln 10 + \ln \frac{1}{5}$   **(c)** $\ln \sqrt{108}$

19. Which of the following is the same as $4 \ln 2x$?

**(a)** $\ln 8x$        **(b)** $8 \ln x$

**(c)** $\ln 8 + \ln x$    **(d)** $\ln 16x^4$

20. Which of the following is the same as $\ln(9x) - \ln(3x)$?

**(a)** $\ln 6x$       **(b)** $\ln(9x)/\ln(3x)$

**(c)** $6 \cdot \ln(x)$    **(d)** $\ln 3$

21. Which of the following is the same as $\dfrac{\ln 8x^2}{\ln 2x}$?

**(a)** $\ln 4x$        **(b)** $4x$

**(c)** $\ln 8x^2 - \ln 2x$    **(d)** none of these

22. Which of the following is the same as $\ln 9x^2$?

**(a)** $2 \cdot \ln 9x$      **(b)** $3x \cdot \ln 3x$

**(c)** $2 \cdot \ln 3x$      **(d)** none of these

*Solve the given equation for $x$.*

23. $\ln x - \ln x^2 + \ln 3 = 0$    24. $\ln \sqrt{x} - 2 \ln 3 = 0$

25. $\ln x^4 - 2 \ln x = 1$        26. $\ln x^2 - \ln 2x + 1 = 0$

27. $\ln \sqrt{x} = \sqrt{\ln x}$         28. $3 \ln x - \ln 3x = 0$

29. $(\ln x)^2 - 1 = 0$          30. $2(\ln x)^2 + \ln x - 1 = 0$

31. $\ln(x+1) - \ln(x-2) = 1$

**32.** $\ln\left[(x-3)(x+2)\right] - \ln(x+2)^2 - \ln 7 = 0$

*Differentiate.*

**33.** $y = \ln[(x+5)(2x-1)(4-x)]$

**34.** $y = \ln[(x+1)(2x+1)(3x+1)]$

**35.** $y = \ln\left[(1+x)^2(2+x)^3(3+x)^4\right]$

**36.** $y = \ln\left[e^{2x}(x^3+1)(x^4+5x)\right]$

**37.** $y = \ln\left[\dfrac{e^{5x}(x+4)(3x-2)}{x+1}\right]$

**38.** $y = \ln\left[\dfrac{\sqrt{x}\,(x+1)^2(x+2)^3}{4x+1}\right]$

**39.** $y = \ln\left[\dfrac{(5x+1)(4x+1)(\ln x)}{\sqrt{2x+1}}\right]$

**40.** $y = \ln\left[\dfrac{x^5 e^{4x}\sqrt{3x+1}}{1-x^2}\right]$

**41.** $y = \ln(3x+1)\ln(5x+1)$    **42.**  $y = (\ln 4x)(\ln 2x)$

*Use logarithmic differentiation to differentiate the following functions.*

**43.** $f(x) = (x+1)^4(4x-1)^2$    **44.**  $f(x) = e^x(3x-4)^8$

**45.** $f(x) = \dfrac{(x+1)(2x+1)(3x+1)}{\sqrt{4x+1}}$

**46.** $f(x) = \dfrac{(x-2)^3(x-3)^4}{(x+4)^5}$    **47.**  $f(x) = 2^x$

**48.** $f(x) = 10^x$    **49.**  $f(x) = x^x$    **50.**  $f(x) = x^{1/x}$

**51.** (*Allometric Equation*) There are substantial empirical data to show that if $x$ and $y$ measure the sizes of two organs of a particular animal, then $x$ and $y$ are related by an *allometric equation* of the form

$$\ln y - k \ln x = \ln c,$$

where $k$ and $c$ are positive constants that depend only on the type of parts or organs that are measured, and are constant among animals belonging to the same species.[*]  Solve this equation for $y$ in terms of $x$, $k$, and $c$.

**52.** (*Epidemics Model*) In the study of epidemics, one finds the equation

$$\ln(1-y) - \ln y = C - rt,$$

where $y$ is the fraction of the population that has a specific disease at time $t$. Solve the equation for $y$ in terms of $t$ and the constants $C$ and $r$.

**53.** Determine the values of $h$ and $k$ for which the graph of $y = he^{kx}$ passes through the points $(1,6)$ and $(4,48)$.

**54.** Find values of $k$ and $r$ for which the graph of $y = kx^r$ passes through the points $(2,3)$ and $(4,15)$.

## SOLUTIONS TO PRACTICE PROBLEMS 4.6

**1.** Use the properties of the natural logarithm to express $f(x)$ as a sum of simple functions before differentiating.

$$f(x) = \ln\left[\frac{e^x\sqrt{x}}{(x+1)^6}\right]$$

$$f(x) = \ln e^x + \ln\sqrt{x} - \ln(x+1)^6$$
$$= x + \tfrac{1}{2}\ln x - 6\ln(x+1)$$

$$f'(x) = 1 + \frac{1}{2x} - \frac{6}{x+1}.$$

**2.**   $f(x) = (x+1)^7(x+2)^8(x+3)^9$

$\ln f(x) = 7\ln(x+1) + 8\ln(x+2) + 9\ln(x+3)$.

Now we differentiate both sides of the equation.

$$\frac{f'(x)}{f(x)} = \frac{7}{x+1} + \frac{8}{x+2} + \frac{9}{x+3}$$

$$f'(x) = f(x)\left(\frac{7}{x+1} + \frac{8}{x+2} + \frac{9}{x+3}\right)$$
$$= (x+1)^7(x+2)^8(x+3)^9\cdot$$
$$\left(\frac{7}{x+1} + \frac{8}{x+2} + \frac{9}{x+3}\right).$$

# REVIEW OF FUNDAMENTAL CONCEPTS

**1.** State as many laws of exponents as you can recall.

**2.** What is $e$?

**3.** Write the differential equation satisfied by $y = Ce^{kt}$.

**4.** State the properties that graphs of the form $y = e^{kx}$

have in common when $k$ is positive. When $k$ is negative.

**5.** What are the coordinates of the reflection of the point $(a,b)$ in the line $y = x$?

[*]E. Batschelet, *Introduction to Mathematics for Life Scientists* (New York: Springer-Verlag, 1971), pp. 305–307.

6. What is a logarithm?
7. What is the $x$-intercept of the graph of the natural logarithm function?
8. State the main features of the graph of $y = \ln x$.
9. State the two key equations giving the relationships between $e^x$ and $\ln x$. (*Hint*: The right side of each equation is just $x$.)
10. What is the difference between a natural logarithm and a common logarithm?
11. Give the formula that converts a function of the form $b^x$ to an exponential function with base $e$.

12. State the differentiation formula for each of the following functions:
    (a) $f(x) = e^{kx}$     (b) $f(x) = e^{g(x)}$
    (c) $f(x) = \ln g(x)$

13. State the four algebraic properties of the natural logarithm function.

14. Give an example of the use of logarithmic differentiation.

# SUPPLEMENTARY EXERCISES

*Calculate the following.*

1. $27^{4/3}$     2. $4^{1.5}$     3. $5^{-2}$     4. $16^{-.25}$
5. $(2^{5/7})^{14/5}$     6. $8^{1/2} \cdot 2^{1/2}$     7. $\dfrac{9^{5/2}}{9^{3/2}}$     8. $4^{.2} \cdot 4^{.3}$

*Simplify the following.*

9. $\left(e^{x^2}\right)^3$     10. $e^{5x} \cdot e^{2x}$
11. $\dfrac{e^{3x}}{e^x}$     12. $2^x \cdot 3^x$
13. $(e^{8x} + 7e^{-2x})e^{3x}$     14. $\dfrac{e^{5x/2} - e^{3x}}{\sqrt{e^x}}$

*Solve the following equations for $x$.*

15. $e^{-3x} = e^{-12}$     16. $e^{x^2 - x} = e^2$
17. $(e^x \cdot e^2)^3 = e^{-9}$     18. $e^{-5x} \cdot e^4 = e$

*Differentiate the following functions.*

19. $y = 10e^{7x}$     20. $y = e^{\sqrt{x}}$
21. $y = xe^{x^2}$     22. $y = \dfrac{e^x + 1}{x - 1}$
23. $y = e^{e^x}$     24. $y = (\sqrt{x} + 1)e^{-2x}$
25. $y = \dfrac{x^2 - x + 5}{e^{3x} + 3}$     26. $y = x^e$

27. Determine all solutions of the differential equation $y' = -y$.

28. Determine all functions $y = f(x)$ such that $y' = -1.5y$ and $f(0) = 2000$.

29. Determine all solutions of the differential equation $y' = 1.5y$ and $f(0) = 2$.

30. Determine all solutions of the differential equation $y' = \frac{1}{3}y$.

*Graph the following functions.*

31. $y = e^{-x} + x$     32. $y = e^x - x$     33. $y = e^{-(1/2)x^2}$
34. $y = 100(x - 2)e^{-x}$ for $x \geq 2$

35. $y = e^{2x} + 2e^{-x}$     36. $y = \dfrac{1}{e^x + e^{-x}}$

37. Find the equation of the tangent line to the graph of $y = \dfrac{e^x}{1 + e^x}$ at $(0, .5)$.

38. Show that the tangent lines to the graph of $y = \dfrac{e^x - e^{-x}}{e^x + e^{-x}}$ at $x = 1$ and $x = -1$ are parallel.

*Simplify the following expressions.*

39. $e^{(\ln 5)/2}$     40. $e^{\ln(x^2)}$     41. $\dfrac{\ln x^2}{\ln x^3}$
42. $e^{2 \ln 2}$     43. $e^{-5 \ln 1}$     44. $\left[e^{\ln x}\right]^2$

*Solve the following equations for $t$.*

45. $t^{\ln t} = e$     46. $\ln(\ln 3t) = 0$
47. $3e^{2t} = 15$     48. $3e^{t/2} - 12 = 0$
49. $2 \ln t = 5$     50. $2e^{-.3t} = 1$

*Differentiate the following functions.*

51. $y = \ln(x^6 + 3x^4 + 1)$     52. $y = \dfrac{x}{\ln x}$
53. $y = \ln(5x - 7)$     54. $y = \ln(9x)$
55. $y = (\ln x)^2$     56. $y = (x \ln x)^3$
57. $y = \ln\left(\dfrac{xe^x}{\sqrt{1 + x}}\right)$
58. $y = \ln\left[e^{6x}(x^2 + 3)^5(x^3 + 1)^{-4}\right]$
59. $y = x \ln x - x$     60. $y = e^{2 \ln(x+1)}$
61. $y = \ln(\ln \sqrt{x})$     62. $y = \dfrac{1}{\ln x}$
63. $y = e^x \ln x$     64. $y = \ln(x^2 + e^x)$
65. $y = \ln\sqrt{\dfrac{x^2 + 1}{2x + 3}}$     66. $y = \ln|-2x + 1|$
67. $y = \ln\left(\dfrac{e^{x^2}}{x}\right)$     68. $y = \ln\sqrt[3]{x^3 + 3x - 2}$
69. $y = \ln(2^x)$     70. $y = \ln(3^{x+1}) - \ln 3$
71. $y = \ln|x - 1|$     72. $y = e^{2 \ln(2x+1)}$
73. $y = \ln\left(\dfrac{1}{e^{\sqrt{x}}}\right)$     74. $y = \ln(e^x + 3e^{-x})$

*Use logarithmic differentiation to differentiate the following functions.*

**75.** $f(x) = \sqrt[5]{\dfrac{x^5 + 1}{x^5 + 5x + 1}}$     **76.** $f(x) = 2^x$

**77.** $f(x) = x^{\sqrt{x}}$     **78.** $f(x) = b^x$, where $b > 0$

**79.** $f(x) = (x^2 + 5)^6(x^3 + 7)^8(x^4 + 9)^{10}$

**80.** $f(x) = x^{1+x}$

**81.** $f(x) = 10^x$     **82.** $f(x) = \sqrt{x^2 + 5}\, e^{x^2}$

**83.** $f(x) = \sqrt{\dfrac{xe^x}{x^3 + 3}}$

**84.** $f(x) = \dfrac{e^x\sqrt{x + 1}(x^2 + 2x + 3)^2}{4x^2}$

**85.** $f(x) = e^{x+1}(x^2 + 1)x$     **86.** $f(x) = e^x x^2 2^x$

*Graph the following functions.*

**87.** $y = \ln(1 + e^x)$     **88.** $y = \ln(1 + e^{-x})$

**89.** $y = \ln(e^x + e^{-x})$     **90.** $y = \ln(e^x + .5e^{-x})$

**91.** $y = (\ln x)^2$     **92.** $y = \ln(x^2 + 1)$

**93.** $y = x - \ln x$     **94.** $y = (\ln x)^3$

**95.** Find $\dfrac{dy}{dx}$ if $e^y - e^x = 3$.

**96.** Find $\dfrac{dy}{dt}$ if $e^{t+y} = t$.

**97.** Find $\dfrac{dy}{dx}$ at the point $(1, 0)$ if $e^{xy} = x$.

**98.** Find $\dfrac{dy}{dx}$ if $e^{2x} + e^{2y} = 4$.

## CHAPTER 5

# Applications of the Exponential and Natural Logarithm Functions

ARLIER, we introduced the exponential function $y = e^x$ and the natural logarithm function $y = \ln x$ and studied their most important properties. From the way we introduced these functions, it is by no means clear that they have any substantial connection with the physical world. However, as this chapter will demonstrate, the exponential and natural logarithm functions are involved in the study of many physical problems, often in a very curious and unexpected way.

Here the most significant fact that we require is that the exponential function is uniquely characterized by its differential equation. In other words, we will constantly make use of the following fact, stated previously.

The function* $y = Ce^{kt}$ satisfies the differential equation

$$y' = ky.$$

Conversely, if $y = f(t)$ satisfies the differential equation, then $y = Ce^{kt}$ for some constant $C$.

---

*Note that we use the independent variable $t$ instead of $x$ throughout this chapter. The reason is that, in most applications, the variable of our exponential function is time.

If $f(t) = Ce^{kt}$, then, by setting $t = 0$, we have

$$f(0) = Ce^0 = C.$$

Therefore, $C$ is the value of $f(t)$ at $t = 0$.

## 5.1 Exponential Growth and Decay

In biology, chemistry, and economics it is often necessary to study the behavior of a quantity that is increasing as time passes. If, at every instant, the rate of increase of the quantity is proportional to the quantity at that instant, then we say that the quantity is *growing exponentially* or is *exhibiting exponential growth*. A simple example of exponential growth is exhibited by the growth of bacteria in a culture. Under ideal laboratory conditions a bacteria culture grows at a rate proportional to the number of bacteria present. It does so because the growth of the culture is accounted for by the division of the bacteria. The more bacteria there are at a given instant, the greater the possibilities for division and hence the greater the rate of growth.

Let us study the growth of a bacteria culture as a typical example of exponential growth. Suppose that $P(t)$ denotes the number of bacteria in a certain culture at time $t$. The rate of growth of the culture at time $t$ is $P'(t)$. We assume that this rate of growth is proportional to the size of the culture at time $t$, so that

$$P'(t) = kP(t), \tag{1}$$

where $k$ is a positive constant of proportionality. If we let $y = P(t)$, then (1) can be written as

$$y' = ky.$$

Therefore, from our discussion at the beginning of this chapter we see that

$$y = P(t) = P_0 e^{kt}, \tag{2}$$

where $P_0$ is the number of bacteria in the culture at time $t = 0$. The number $k$ is called the *growth constant*.

---

**EXAMPLE 1** **Exponential growth** Suppose that a certain bacteria culture grows at a rate proportional to its size. At time $t = 0$, approximately 20,000 bacteria are present. In 5 hours there are 400,000 bacteria. Determine a function that expresses the size of the culture as a function of time, measured in hours.

**Solution** Let $P(t)$ be the number of bacteria present at time $t$. By assumption, $P(t)$ satisfies a differential equation of the form $y' = ky$, so $P(t)$ has the form

$$P(t) = P_0 e^{kt},$$

where the constants $P_0$ and $k$ must be determined. The value of $P_0$ and $k$ can be obtained from the data that give the population size at two different times. We are told that

$$P(0) = 20{,}000, \qquad P(5) = 400{,}000. \tag{3}$$

The first condition immediately implies that $P_0 = 20{,}000$, so

$$P(t) = 20{,}000 e^{kt}.$$

Using the second condition in (3), we have

$$20{,}000e^{k(5)} = P(5) = 400{,}000$$
$$e^{5k} = 20$$
$$5k = \ln 20$$
$$k = \frac{\ln 20}{5} \approx .60. \tag{4}$$

So we may take

$$P(t) = 20{,}000e^{.6t}.$$

This function is a mathematical model of the growth of the bacteria culture. (See Fig. 1.) ∎

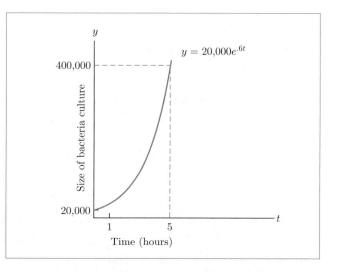

**Figure 1.** A model for bacteria growth.

---

| EXAMPLE 2 | **Determining the growth constant** Suppose that a colony of fruit flies is growing according to the exponential law $P(t) = P_0 e^{kt}$, and suppose that the size of the colony doubles in 9 days. Determine the growth constant $k$. |

Solution   We do not know the initial size of the population at $t = 0$. However, we are told that $P(9) = 2P(0)$; that is,

$$P_0 e^{k(9)} = 2P_0$$
$$e^{9k} = 2$$
$$9k = \ln 2$$
$$k = \frac{\ln 2}{9} \approx .077. \quad ∎$$

Notice that the initial size $P_0$ of the population was not given in Example 2. We were able to determine the growth constant because we were told the amount of time required for the colony to double in size. Thus the growth constant does not depend on the initial size of the population. This property is characteristic of exponential growth.

**EXAMPLE 3** Suppose that the initial size of the colony in Example 2 was 100.
(a) How large will the colony be after 41 days?
(b) How fast will the colony be growing at that time?
(c) At what time will the colony contain 800 fruit flies?

**Solution** (a) From Example 2 we have $P(t) = P_0 e^{.077t}$. Since $P(0) = 100$, we conclude that

$$P(t) = 100 e^{.077t}.$$

Therefore, after 41 days the size of the colony is

$$P(41) = 100 e^{.077(41)} = 100 e^{3.157} \approx 2350.$$

(b) Since the function satisfies the differential equation $y' = .077y$,

$$P'(t) = .077 P(t).$$

In particular, when $t = 41$,

$$P'(41) = .077 P(41) = .077 \cdot 2350 \approx 181.$$

Therefore, after 41 days the colony is growing at the rate of about 181 fruit flies per day.

(c) Use the formula for $P(t)$ and set $P(t) = 800$.

$$100 e^{.077t} = 800$$
$$e^{.077t} = 8$$
$$.077t = \ln 8$$
$$t = \frac{\ln 8}{.077} \approx 27 \text{ days.} \qquad \blacksquare$$

**Exponential Decay** An example of negative exponential growth, or *exponential decay*, is given by the disintegration of a radioactive element such as uranium 235. It is known that, at any instant, the rate at which a radioactive substance is decaying is proportional to the amount of the substance that has not yet disintegrated. If $P(t)$ is the quantity present at time $t$, then $P'(t)$ is the rate of decay. Of course, $P'(t)$ must be negative, since $P(t)$ is decreasing. Thus we may write $P'(t) = kP(t)$ for some negative constant $k$. To emphasize the fact that the constant is negative, $k$ is often replaced by $-\lambda$, where $\lambda$ is a positive constant.* Then $P(t)$ satisfies the differential equation

$$P'(t) = -\lambda P(t). \tag{5}$$

The general solution of (5) has the form

$$P(t) = P_0 e^{-\lambda t}$$

for some positive number $P_0$. We call such a function an *exponential decay function*. The constant $\lambda$ is called the *decay constant*.

**EXAMPLE 4** **Exponential decay** The decay constant for strontium 90 is $\lambda = .0244$, where time is measured in years. How long will it take for a quantity $P_0$ of strontium 90 to decay to one-half its original mass?

---
*$\lambda$ is the Greek lowercase letter lambda.

**Solution**  We have

$$P(t) = P_0 e^{-.0244t}.$$

Next, set $P(t)$ equal to $\frac{1}{2}P_0$ and solve for $t$:

$$P_0 e^{-.0244t} = \tfrac{1}{2}P_0$$
$$e^{-.0244t} = \tfrac{1}{2} = .5$$
$$-.0244t = \ln .5$$
$$t = \frac{\ln .5}{-.0244} \approx 28 \text{ years.} \qquad \blacksquare$$

The *half-life* of a radioactive element is the length of time required for a given quantity of that element to decay to one-half its original mass. Thus strontium 90 has a half-life of about 28 years. This means that it takes 28 years for it to decay to half its original mass and another 28 years for it to decay to $\frac{1}{4}$ its original mass, another 28 years to decay to $\frac{1}{8}$, and so forth. (See Fig. 2.) Notice from Example 4 that the half-life does not depend on the initial amount $P_0$.

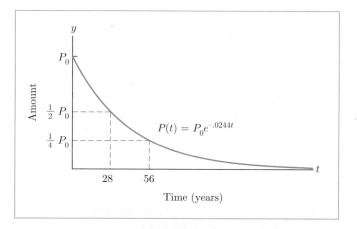

**Figure 2.**  Half-life of radioactive strontium 90.

---

**EXAMPLE 5**  **Half-life and decay constant**  Radioactive carbon 14 has a half-life of about 5730 years. Find its decay constant.

**Solution**  If $P_0$ denotes the initial amount of carbon 14, then the amount after $t$ years will be

$$P(t) = P_0 e^{-\lambda t}.$$

After 5730 years, $P(t)$ will equal $\frac{1}{2}P_0$. That is,

$$P_0 e^{-\lambda(5730)} = P(5730) = \tfrac{1}{2}P_0 = .5P_0.$$

Solving for $\lambda$ gives

$$e^{-5730\lambda} = .5$$
$$-5730\lambda = \ln .5$$
$$\lambda = \frac{\ln .5}{-5730} \approx .00012. \qquad \blacksquare$$

One of the problems connected with aboveground nuclear explosions is the radioactive debris ("fallout") that falls on plants and grass, thereby contaminating the food supply of animals. Strontium 90 is one of the most dangerous components of radioactive debris because it has a relatively long half-life and because it is chemically similar to calcium and is absorbed into the bone structure of animals (including humans) who eat contaminated food. Iodine 131 is also produced by nuclear explosions, but it presents less of a hazard because it has a half-life of 8 days.

---

**EXAMPLE 6**    **Iodine level in dairy products**    If dairy cows eat hay containing too much iodine 131, their milk will be unfit to drink. Suppose that some hay contains 10 times the maximum allowable level of iodine 131. How many days should the hay be stored before it is fed to dairy cows?

**Solution**    Let $P_0$ be the amount of iodine 131 present in the hay. Then the amount at time $t$ is $P(t) = P_0 e^{-\lambda t}$ ($t$ in days). The half-life of iodine 131 is 8 days, so

$$P_0 e^{-\lambda(8)} = .5P_0$$

$$e^{-8\lambda} = .5$$

$$-8\lambda = \ln .5$$

$$\lambda = \frac{\ln .5}{-8} \approx .087$$

and

$$P(t) = P_0 e^{-.087t}.$$

Now that we have the formula for $P(t)$, we want to find $t$ such that $P(t) = \frac{1}{10}P_0$. We have

$$P_0 e^{-.087t} = .1P_0,$$

so

$$e^{-.087t} = .1$$

$$-.087t = \ln .1$$

$$t = \frac{\ln .1}{-.087} \approx 26 \text{ days.} \qquad \blacksquare$$

**Radiocarbon Dating**    Knowledge about radioactive decay is valuable to archaeologists and anthropologists who want to estimate the age of objects belonging to ancient civilizations. Several different substances are useful for radioactive-dating techniques; the most common is radiocarbon, $^{14}C$. Carbon 14 is produced in the upper atmosphere when cosmic rays react with atmospheric nitrogen. Because the $^{14}C$ eventually decays, the concentration of $^{14}C$ cannot rise above certain levels. An equilibrium is reached where $^{14}C$ is produced at the same rate as it decays. Scientists usually assume that the total amount of $^{14}C$ in the biosphere has remained constant over the past 50,000 years. Consequently, it is assumed that the *ratio* of $^{14}C$ to ordinary nonradioactive carbon 12, $^{12}C$, has been constant during this same period. (The ratio is about one part $^{14}C$ to $10^{12}$ parts of $^{12}C$.) Both $^{14}C$ and $^{12}C$ are in the atmosphere as constituents of carbon dioxide. All living vegetation and most forms of animal life contain $^{14}C$ and $^{12}C$ in the same proportion as the atmosphere. The reason is that plants absorb carbon dioxide through photosynthesis. The $^{14}C$ and $^{12}C$ in plants are distributed through the food chain to almost all animal life.

When an organism dies, it stops replacing its carbon, and therefore the amount of $^{14}C$ begins to decrease through radioactive decay. (The $^{12}C$ in the dead organism remains constant.) The ratio of $^{14}C$ to $^{12}C$ can be later measured in order to determine when the organism died. (See Fig. 3.)

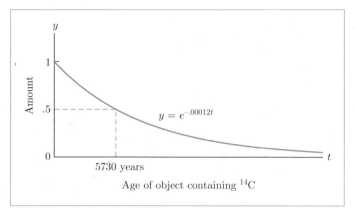

**Figure 3.** $^{14}\text{C}$ - $^{12}\text{C}$ ratio compared to the ratio in living plants.

---

**EXAMPLE 7**    **Carbon dating**  A parchment fragment was discovered that had about 80% of the $^{14}\text{C}$ level found today in living matter. Estimate the age of the parchment.

Solution    We assume that the original $^{14}\text{C}$ level in the parchment was the same as the level in living organisms today. Consequently, about eight-tenths of the original $^{14}\text{C}$ remains. From Example 5 we obtain the formula for the amount of $^{14}\text{C}$ present $t$ years after the parchment was made from an animal skin:

$$P(t) = P_0 e^{-.00012t},$$

where $P_0 =$ initial amount. We want to find $t$ such that $P(t) = .8P_0$.

$$P_0 e^{-.00012t} = .8P_0$$
$$e^{-.00012t} = .8$$
$$-.00012t = \ln .8$$
$$t = \frac{\ln .8}{-.00012} \approx 1860 \text{ years old.} \qquad \blacksquare$$

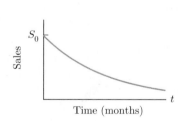

**Figure 4.** Exponential decay of sales.

**A Sales Decay Curve**  Marketing studies* have demonstrated that if advertising and other promotions of a particular product are stopped and if other market conditions remain fairly constant, then, at any time $t$, the sales of that product will be declining at a rate proportional to the amount of current sales at time $t$. (See Fig. 4.) If $S_0$ is the number of sales in the last month during which advertising occurred and if $S(t)$ is the monthly level of sales $t$ months after the cessation of promotional effort, then a good mathematical model for $S(t)$ is

$$S(t) = S_0 e^{-\lambda t},$$

where $\lambda$ is a positive number called the *sales decay constant*. The value of $\lambda$ depends on many factors, such as the type of product, the number of years of prior advertising, the number of competing products, and other characteristics of the market.

**The Time Constant**  Consider an exponential decay function $y = Ce^{-\lambda t}$. In Fig. 5, we have drawn the tangent line to the decay curve when $t = 0$. The slope there is the initial rate of decay. If the decay process were to continue at this rate, the decay curve would follow the tangent line and $y$ would be zero at some time

---

*M. Vidale and H. Wolfe, "An Operations-Research Study of Sales Response to Advertising," *Operations Research*, 5 (1957), 370–381. Reprinted in F. Bass et al., *Mathematical Models and Methods in Marketing* (Homewood, Ill.: Richard D. Irwin, Inc., 1961).

$T$. This time is called the *time constant* of the decay curve. It can be shown (see Exercise 30) that $T = 1/\lambda$ for the curve $y = Ce^{-\lambda t}$. Thus $\lambda = 1/T$, and the decay curve can be written in the form

$$y = Ce^{-t/T}.$$

If one has experimental data that tend to lie along an exponential decay curve, then the numerical constants for the curve may be obtained from Fig. 5. First, sketch the curve and estimate the $y$-intercept, $C$. Then sketch an approximate tangent line and from this estimate the time constant, $T$. This procedure is sometimes used in biology and medicine.

**Incorporating Technology**   Exponential equations also can be solved graphically with a graphing calculator. Figure 6 shows the solution to the equation $e^{-.0244x} = .5$ from Example 4.

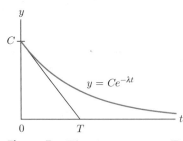

**Figure 5.**   The time constant $T$ in exponential decay: $T = 1/\lambda$.

**Figure 6.**   $Y_1 = .5$ $Y_2 = e^{\hat{}}(-.0244X)$.

## PRACTICE PROBLEMS 5.1

1. **(a)** Solve the differential equation $P'(t) = -.6P(t)$, $P(0) = 50$.

   **(b)** Solve the differential equation $P'(t) = kP(t)$, $P(0) = 4000$, where $k$ is some constant.

   **(c)** Find the value of $k$ in part (b) for which $P(2) = 100P(0)$.

2. Under ideal conditions a colony of *Escherichia coli* bacteria can grow by a factor of 100 every 2 hours. If initially 4000 bacteria are present, how long will it take before there are 1,000,000 bacteria?

## EXERCISES 5.1

1. (*Population Growth*) Let $P(t)$ be the population (in millions) of a certain city $t$ years after 1990, and suppose that $P(t)$ satisfies the differential equation

   $$P'(t) = .02P(t), \qquad P(0) = 3.$$

   **(a)** Find the formula for $P(t)$.

   **(b)** What was the initial population, that is, the population in 1990?

   **(c)** What is the growth constant?

   **(d)** What was the population in 1998?

   **(e)** Use the differential equation to determine how fast the population is growing when it reaches 4 million people.

   **(f)** How large is the population when it is growing at the rate of 70,000 people per year?

2. (*Growth of Bacteria*) Approximately 10,000 bacteria are placed in a culture. Let $P(t)$ be the number of bacteria present in the culture after $t$ hours, and suppose that $P(t)$ satisfies the differential equation

   $$P'(t) = .55P(t).$$

   **(a)** Find the formula for $P(t)$.

   **(b)** What is $P(0)$?

   **(c)** How many bacteria are there after 5 hours?

**(d)** What is the growth constant?

**(e)** Use the differential equation to determine how fast the bacteria culture is growing when it reaches 100,000.

**(f)** What is the size of the bacteria culture when it is growing at the rate of 34,000 bacteria per hour?

**3.** (*Growth of Cells*) Suppose that after $t$ hours there are $P(t)$ cells present in a culture, where $P(t) = 5000e^{.2t}$.

**(a)** How many cells were present initially?

**(b)** Give a differential equation satisfied by $P(t)$.

**(c)** When will the population double?

**(d)** When will 20,000 cells be present?

**4.** (*Insect Population*) The size of a certain insect population is given by $P(t) = 300e^{.01t}$, where $t$ is measured in days.

**(a)** How many insects were present initially?

**(b)** Give a differential equation satisfied by $P(t)$.

**(c)** At what time will the population double?

**(d)** At what time will the population equal 1200?

**5.** Determine the growth constant of a population that is growing at a rate proportional to its size, where the population doubles in size every 40 days.

**6.** Determine the growth constant of a population that is growing at a rate proportional to its size, where the population triples in size every 10 years.

**7.** Suppose a population is growing exponentially with growth constant .05. In how many years will the current population triple?

**8.** Suppose a population is growing exponentially with growth constant .04. In how many years will the current population double?

**9.** The rate of growth of a certain cell culture is proportional to its size. In 10 hours a population of 1 million cells grew to 9 million. How large will the cell culture be after 15 hours?

**10.** (*World's Population*) The world's population was 5.51 billion on January 1, 1993 and 5.88 billion on January 1, 1998. Assume that at any time the population grows at a rate proportional to the population at that time. In what year will the world's population reach 7 billion?

**11.** (*Population of Mexico City*) At the beginning of 1990, 20.2 million people lived in the metropolitan area of Mexico City, and the population was growing exponentially. The 1995 population was 23 million. (Part of the growth is due to immigration.) If this trend continues, how large will the population be in the year 2010?

**12.** The population (in millions) of a state $t$ years after 1970 is given by the graph of the exponential function $y = P(t)$ with growth constant .025 in Fig. 7. [In parts (c) and (d) use the differential equation satisfied by $P(t)$.]

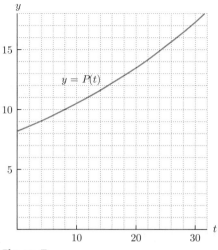

**Figure 7**

**(a)** What was the population in 1974?

**(b)** When was the population 10 million?

**(c)** How fast was the population growing in 1974?

**(d)** When is the population growing at the rate of 275,000 people per year?

**13.** (*Radioactive Decay*) A sample of 8 grams of radioactive material is placed in a vault. Let $P(t)$ be the amount remaining after $t$ years, and suppose that $P(t)$ satisfies the differential equation $P'(t) = -.021P(t)$.

**(a)** Find the formula for $P(t)$.

**(b)** What is $P(0)$?

**(c)** What is the decay constant?

**(d)** How much of the material will remain after 10 years?

**(e)** Use the differential equation to determine how fast the sample is disintegrating when just one gram remains.

**(f)** What amount of radioactive material remains when it is disintegrating at the rate of .105 grams per year?

**(g)** The radioactive material has a half-life of 33 years. How much will remain after 33 years? 66 years? 99 years?

**14.** (*Radioactive Decay*) Radium-226 is used in cancer radiotherapy, as a neutron source for some research purposes, and as a constituent of luminescent paints. Let $P(t)$ be the number of grams of radium-226 in a sample remaining after $t$ years, and suppose that $P(t)$ satisfies the differential equation

$$P'(t) = -.00043P(t), \qquad P(0) = 12.$$

**(a)** Find the formula for $P(t)$.

**(b)** What was the initial amount?

**(c)** What is the decay constant?

**(d)** Approximately how much of the radium will remain after 943 years?

**(e)** How fast is the sample disintegrating when just one gram remains? Use the differential equation.

**(f)** What is the weight of the sample when it is disintegrating at the rate of .004 grams per year?

**(g)** The radioactive material has a half-life of about 1612 years. How much will remain after 1612 years? 3224 years? 4836 years?

**15.** (*Decay of Penicillin in the Bloodstream*) Suppose a person is given an injection of 300 milligrams of penicillin at time $t = 0$, and let $f(t)$ be the amount (in milligrams) of penicillin present in the person's bloodstream $t$ hours after the injection. Then the amount of penicillin decays exponentially and a typical formula is $f(t) = 300e^{-.6t}$.

**(a)** Give the differential equation satisfied by $f(t)$.

**(b)** How much will remain at time $t = 5$ hours?

**(c)** What is the biological half-life of the penicillin (that is, the time required for half of a given amount to decompose) in this case?

**16.** (*Radioactive Decay*) Ten grams of a radioactive substance with decay constant .04 is stored in a vault. Assume that time is measured in days and let $P(t)$ be the amount remaining at time $t$.

**(a)** Give the formula for $P(t)$.

**(b)** Give the differential equation satisfied by $P(t)$.

**(c)** How much will remain after 5 days?

**(d)** What is the half-life of this radioactive substance?

**17.** The decay constant for the radioactive element cesium-137 is .023 when time is measured in years. Find the half-life of cesium-137.

**18.** Radioactive cobalt-60 has a half-life of 5.3 years. Find the decay constant of cobalt-60.

**19.** A sample of radioactive material disintegrates from 5 grams to 2 grams in 100 days. After how many days will just 1 gram remain?

**20.** Ten grams of a radioactive material disintegrates to 3 grams in 5 years. What is the half-life of the radioactive material?

**21.** (*Decay of Sulfate in the Bloodstream*) In an animal hospital, 8 units of a sulfate were injected into a dog. After 50 minutes only 4 units remained in the dog. Let $f(t)$ be the amount of sulfate present after $t$ minutes. At any time, the rate of change of $f(t)$ is proportional to the value of $f(t)$. Find the formula for $f(t)$.

**22.** Five grams of a certain radioactive material decays to 3 grams in 1 year. After how many years will just 1 gram remain?

**23.** (*Radioactive Decay*) A sample of radioactive material decays over time (measured in hours) with decay constant .2. The graph of the exponential function $y =$

$P(t)$ in Fig. 8 gives the number of grams remaining after $t$ hours. [*Hint*: In parts (c) and (d) use the differential equation satisfied by $P(t)$.]

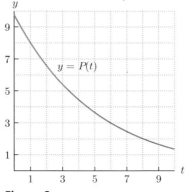

**Figure 8**

**(a)** How much was remaining after 1 hour?

**(b)** Approximate the half-life of the material.

**(c)** How fast was the sample decaying after 6 hours?

**(d)** When was the sample decaying at the rate of .4 grams per hour?

**24.** A sample of radioactive material has decay constant .25, where time is measured in hours. How fast will the sample be disintegrating when the sample size is 8 grams? For what sample size will the sample size be decreasing at the rate of 2 grams per day?

**25.** (*Carbon Dating*) In 1947, a cave with beautiful prehistoric wall paintings was discovered in Lascaux, France. Some charcoal found in the cave contained 20% of the $^{14}C$ expected in living trees. How old are the Lascaux cave paintings? (Recall that the decay constant for $^{14}C$ is .00012.)

**26.** (*King Arthur's Round Table*) According to legend, in the fifth century King Arthur and his knights sat at a huge round table. A round table alleged to have belonged to King Arthur was found at Winchester Castle in England. In 1976 carbon dating revealed the amount of radiocarbon in the table to be 91% of the radiocarbon present in living wood. Could the table possibly have belonged to King Arthur? Why? (Recall that the decay constant for $^{14}C$ is .00012.)

**27.** A 4500-year-old wooden chest was found in the tomb of the twenty-fifth century B.C. Chaldean king Meskalumdug of Ur. What percentage of the original $^{14}C$ would you expect to find in the wooden chest?

**28.** (*Population of the Pacific Northwest*) In 1938, sandals woven from strands of tree bark were found in Fort Rock Creek Cave in Oregon. The bark contained 34% of the level of $^{14}C$ found in living bark. Approximately how old were the sandals? (*Note*: This discovery by University of Oregon anthropologist Luther Cressman forced scientists to double their estimate of how long ago people came to the Pacific Northwest.)

**29.** (*Time of the Fourth Ice Age*) Many scientists believe there have been four ice ages in the past 1 million years. Before the technique of carbon dating was known, geologists erroneously believed that the retreat of the Fourth Ice Age began about 25,000 years ago. In 1950, logs from ancient spruce trees were found under glacial debris near Two Creeks, Wisconsin. Geologists determined that these trees had been crushed by the advance of ice during the Fourth Ice Age. Wood from the spruce trees contained 27% of the level of $^{14}$C found in living trees. Approximately how long ago did the Fourth Ice Age actually occur?

**30.** Let $T$ be the time constant of the curve $y = Ce^{-\lambda t}$ as defined in Fig. 5. Show that $T = 1/\lambda$. (*Hint:* Express the slope of the tangent line in Fig. 5 in terms of $C$ and $T$. Then set this slope equal to the slope of the curve $y = Ce^{-\lambda t}$ at $t = 0$.)

**31.** The amount (in grams) of a certain radioactive material present after $t$ years is given by the function $P(t)$. Match each of the following answers with its corresponding question.

*Answers:*

**a.** Solve $P(t) = .5P(0)$ for $t$.
**b.** Solve $P(t) = .5$ for $t$.
**c.** $P(.5)$
**d.** $P'(.5)$
**e.** $P(0)$
**f.** Solve $P'(t) = -.5$ for $t$.
**g.** $y' = ky$
**h.** $P_0 e^{kt}$, $k < 0$

*Questions:*

**A.** Give a differential equation satisfied by $P(t)$.
**B.** How fast will the radioactive material be disintegrating in $\frac{1}{2}$ year?

**C.** Give the general form of the function $P(t)$.
**D.** Find the half-life of the radioactive material.
**E.** How many grams of the material will remain after $\frac{1}{2}$ year?
**F.** When will the radioactive material be disintegrating at the rate of $\frac{1}{2}$ gram per year?
**G.** When will there be $\frac{1}{2}$ gram remaining?
**H.** How much radioactive material was present initially?

### Technology Exercises

**32.** Figure 9 shows the point of intersection of a horizontal line and the decay curve for one gram of plutonium 237. Make up a question that can be answered by Fig. 9.

**Figure 9**

**33.** Figure 10 shows the point of intersection of a horizontal line and the growth curve for a population that initially has 100 individuals. Make up a question that can be answered by Fig. 10.

**Figure 10**

## SOLUTIONS TO PRACTICE PROBLEMS 5.1

**1. (a)** Answer: $P(t) = 50e^{-.6t}$. Differential equations of the type $y' = ky$ have as their solution $P(t) = Ce^{kt}$, where $C$ is $P(0)$.

**(b)** Answer: $P(t) = 4000e^{kt}$. This problem is like the previous one except that the constant is not specified. Additional information is needed if one wants to determine a specific value for $k$.

**(c)** Answer: $P(t) = 4000e^{2.3t}$. From the solution to part (b) we know that $P(t) = 4000e^{kt}$. We

are given that $P(2) = 100P(0) = 100(4000) = 400{,}000$. So

$$P(2) = 4000e^{k(2)} = 400{,}000$$
$$e^{2k} = 100$$
$$2k = \ln 100$$
$$k = \frac{\ln 100}{2} \approx 2.3.$$

**2.** Let $P(t)$ be the number of bacteria present after $t$ hours. We must first find an expression for $P(t)$ and then determine the value of $t$ for which $P(t) = 1,000,000$. From the discussion at the beginning of the section we know that $P'(t) = k \cdot P(t)$. Also, we are given that $P(2)$ (the population after 2 hours) is $100P(0)$ (100 times the initial population). From Problem 1(c) we have an expression for $P(t)$:

$$P(t) = 4000e^{2.3t}.$$

Now we must solve $P(t) = 1,000,000$ for $t$.

$$4000e^{2.3t} = 1,000,000$$

$$e^{2.3t} = 250$$

$$2.3t = \ln 250$$

$$t = \frac{\ln 250}{2.3} \approx 2.4.$$

Therefore, after 2.4 hours there will be 1,000,000 bacteria.

## 5.2 Compound Interest

In Section 0.5, we introduced the notion of compound interest. Recall the fundamental formula developed there. If a principal amount $P$ is compounded $m$ times per year at an annual rate of interest $r$ for $t$ years, then the compound amount $A$, the balance at the end of time $t$, is given by the formula

$$A = P\left(1 + \frac{r}{m}\right)^{mt}.$$

For instance, suppose $1000 is invested at 6% interest for one year. In the formula this corresponds to $P = \$1000$, $m = 1$, $r = .06$, and $t = 1$ year. If interest is compounded annually, the amount in the account at the end of the year is

$$A = P(1 + r)^1 = 1000(1 + .06) = \$1060.00.$$

If interest is compounded quarterly, the amount in the account at the end of the year is

$$A = P\left(1 + \frac{r}{4}\right)^{4t} = 1000\left(1 + \frac{.06}{4}\right)^4 \approx \$1061.36.$$

Table 1 contains these results along with those for monthly and daily compounding for one year.

**TABLE 1** Effect of Increased Compounding Periods

| Frequency of compounding | Annually | Quarterly | Monthly | Daily |
|---|---|---|---|---|
| $m$ | 1 | 4 | 12 | 360 |
| Balance after one year ($) | 1060.00 | 1061.36 | 1061.68 | 1061.83 |

Would the account balance in Table 1 be much more if the interest were compounded every hour? Every minute? It turns out that no matter how frequently interest is compounded, the balance will not exceed 1061.84 (to the nearest cent). To understand why, we connect the concept of compound interest to the exponential function $e^{rt}$. Start from the formula for the compound amount and write it in the form

$$A = P\left(1 + \frac{r}{m}\right)^{mt} = P\left(1 + \frac{r}{m}\right)^{(m/r) \cdot rt}.$$

If we set $h = r/m$, then $1/h = m/r$, and

$$A = P(1 + h)^{(1/h) \cdot rt}.$$

As the frequency of compounding is increased, $m$ gets large and $h = r/m$ approaches 0. To determine what happens to the compound amount, we must therefore examine the limit

$$\lim_{h \to 0} P(1 + h)^{(1/h) \cdot rt}.$$

The following remarkable fact is proved in the appendix at the end of this section:

$$\lim_{h \to 0} (1 + h)^{1/h} = e.$$

Using this fact together with two limit theorems, we have

$$\lim_{h \to 0} P(1 + h)^{(1/h)rt} = P\left[\lim_{h \to 0} (1 + h)^{1/h}\right]^{rt} = Pe^{rt}.$$

These calculations show that the compound amount calculated from the formula

$$P\left(1 + \frac{r}{m}\right)^{mt}$$

gets closer to $Pe^{rt}$ as the number $m$ of interest periods per year is increased. When the formula

$$A = Pe^{rt} \tag{1}$$

is used to calculate the compound amount, we say that the interest is *compounded continuously*.

When \$1000 is invested for one year at 6% interest compounded continuously, $P = 1000$, $r = .06$, and $t = 1$. The compound amount is

$$1000e^{.06(1)} \approx \$1061.84.$$

This is only one cent more than the result of the daily compounding shown in Table 1. Consequently, frequent compounding (such as every hour or every second) will produce at most one cent more.

In many computations, it is simpler to use the formula for interest compounded continuously than the formula for ordinary compound interest. In these instances, it is commonplace to use interest compounded continuously as an approximation to ordinary compound interest.

When interest is compounded continuously, the compound amount $A(t)$ is an exponential function of the number of years $t$ that interest is earned, $A(t) = Pe^{rt}$. Hence $A(t)$ satisfies the differential equation

$$\frac{dA}{dt} = rA.$$

The rate of growth of the compound amount is proportional to the amount of money present. Since the growth comes from the interest, we conclude that under continuous compounding, interest is earned continuously at a rate of growth proportional to the amount of money present.

The formula $A = Pe^{rt}$ contains four variables. (Remember that the letter $e$ represents a specific constant, $e = 2.718\ldots$) In a typical problem, we are given values for three of these variables and must solve for the remaining variable.

---

**EXAMPLE 1**  **Compound interest**  One thousand dollars is invested at 5% interest compounded continuously.

(a) Give the formula for $A(t)$, the compound amount after $t$ years.

(b) How much will be in the account after 6 years?

(c) After 6 years, at what rate will $A(t)$ be growing?

(d) How long is required for the initial investment to double?

**Solution**  (a) $P = 1000$ and $r = .05$. By formula (1), $A(t) = 1000e^{.05t}$.

(b) $A(6) = 1000e^{.05(6)} = 1000e^{.3} \approx \$1349.86$.

(c) Rate of growth is different from interest rate. Interest rate is fixed at 5% and does not change with time. On the other hand, the rate of growth $A'(t)$ is always changing. Since $A(t) = 1000e^{.05t}$, $A'(t) = 1000 \cdot .05e^{.05t} = 50e^{.05t}$.

$$A'(6) = 50e^{.05(6)} = 50e^{.3} \approx \$67.49.$$

After 6 years the investment is growing at the rate of \$67.49 per year.

There is an easier way to answer part (c). Since $A(t)$ satisfies the differential equation $A'(t) = rA(t)$,

$$A'(6) = .05A(6) = .05 \cdot 1349.86 \approx \$67.49.$$

(d) We must find $t$ such that $A(t) = \$2000$. So we set $1000e^{.05t} = 2000$ and solve for $t$.

$$1000e^{.05t} = 2000$$

$$e^{.05t} = 2$$

$$\ln e^{.05t} = \ln 2$$

$$.05t = \ln 2$$

$$t = \frac{\ln 2}{.05} \approx 13.86 \text{ years} \qquad \blacksquare$$

▶ *Remark*  The calculations in Example 1(d) would be essentially unchanged after the first step if the initial amount of the investment were changed from \$1000 to any arbitrary amount $P$. When this investment doubles, the compound amount will be $2P$. So one sets $2P = Pe^{.05t}$ and solves for $t$ as we did previously, to conclude that, at 5% interest compounded continuously, any amount doubles in about 13.86 years.  ◀

---

**EXAMPLE 2**  **Appreciation of a painting**  Pablo Picasso's "The Dream" was purchased in 1941 for a war-distressed price of \$7000. The painting was sold in 1997 for \$48.4 million, the second highest price ever paid for a Picasso painting at auction. What rate of interest compounded continuously did this investment earn?

**Solution**  Let $Pe^{rt}$ be the value (in millions) of the painting $t$ years after 1941. Since the initial value is .007 million, $P = .007$. Since the value after 56 years is 48.4 million dollars, $.007e^{r(56)} = 48.4$. Divide both sides of the equation by .007, take the logarithms of both sides, and then solve for $r$.

$$e^{r(56)} = \frac{48.4}{.007} \approx 6914.29$$

$$r(56) = \ln 6914.29$$

$$r = \frac{\ln 6914.29}{56} \approx .158$$

Therefore, as an investment, the painting earned an interest rate of 15.8%. ◼

If $P$ dollars are invested today, the formula $A = Pe^{rt}$ gives the value of this investment after $t$ years (assuming continuously compounded interest). We say that $P$ is the *present value* of the amount $A$ to be received in $t$ years. If we solve for $P$ in terms of $A$, we obtain

$$P = Ae^{-rt}. \tag{2}$$

The concept of the present value of money is an important theoretical tool in business and economics. Problems involving depreciation of equipment, for example, may be analyzed by calculus techniques when the present value of money is computed from (2) using continuously compounded interest.

**EXAMPLE 3** Find the present value of $5000 to be received in 2 years if money can be invested at 12% compounded continuously.

**Solution** Use (2) with $A = 5000$, $r = .12$, and $t = 2$.

$$P = 5000e^{-(.12)(2)} = 5000e^{-.24}$$

$$\approx \$3933.14. \qquad ◼$$

**APPENDIX A Limit Formula for e**

For $h \neq 0$, we have

$$\ln(1 + h)^{1/h} = (1/h)\ln(1 + h).$$

Taking the exponential of both sides, we find that

$$(1 + h)^{1/h} = e^{(1/h)\ln(1+h)}.$$

Since the exponential function is continuous,

$$\lim_{h \to 0}(1 + h)^{1/h} = e^{\left[\lim_{h \to 0}(1/h)\ln(1 + h)\right]}. \tag{3}$$

To examine the limit inside the exponential function, we note that $\ln 1 = 0$, and hence

$$\lim_{h \to 0}\left(\frac{1}{h}\right)\ln(1 + h) = \lim_{h \to 0}\frac{\ln(1 + h) - \ln 1}{h}.$$

The limit on the right is a difference quotient of the type used to compute a derivative. In fact,

$$\lim_{h \to 0} \frac{\ln(1+h) - \ln 1}{h} = \frac{d}{dx} \ln x \Big|_{x=1} = \frac{1}{x} \Big|_{x=1} = 1.$$

Thus the limit inside the exponential function in (3) is 1. That is,

$$\lim_{h \to 0} (1+h)^{1/h} = e^{[1]} = e.$$

## PRACTICE PROBLEMS 5.2

**1.** One thousand dollars is to be invested in a bank for 4 years. Would 8% interest compounded semiannually be better than $7\frac{3}{4}\%$ interest compounded continuously?

**2.** A building was bought for $150,000 and sold 10 years later for $400,000. What interest rate (compounded continuously) was earned on the investment?

## EXERCISES 5.2

**1.** (*Savings Account*) Let $A(t) = 5000e^{.04t}$ be the balance in a savings account after $t$ years.

  **(a)** How much money was originally deposited?

  **(b)** What is the interest rate?

  **(c)** How much money will be in the account after 10 years?

  **(d)** What differential equation is satisfied by $y = A(t)$?

  **(e)** Use the results of (c) and (d) to determine how fast the balance is growing after 10 years.

  **(f)** How large will the balance be when it is growing at the rate of $280 per year?

**2.** (*Savings Account*) Let $A(t)$ be the balance in a savings account after $t$ years, and suppose that $A(t)$ satisfies the differential equation

$$A'(t) = .045A(t), \qquad A(0) = 3000.$$

  **(a)** How much money was originally deposited into the account?

  **(b)** What interest rate is being earned?

  **(c)** Find the formula for $A(t)$.

  **(d)** What is the balance after 5 years?

  **(e)** Use part (d) and the differential equation to determine how fast the balance is growing after 5 years.

  **(f)** How large will the balance be when it is growing at the rate of $270 per year?

**3.** (*Savings Account*) Four thousand dollars is deposited into a savings account at 3.5% interest compounded continuously.

  **(a)** What is the formula for $A(t)$, the balance after $t$ years?

  **(b)** What differential equation is satisfied by $A(t)$, the balance after $t$ years?

  **(c)** How much money will be in the account after 2 years?

  **(d)** When will the balance reach $5000?

  **(e)** How fast is the balance growing when it reaches $5000?

**4.** (*Savings Account*) Ten thousand dollars is deposited into a savings account at 4.6% interest compounded continuously.

  **(a)** What differential equation is satisfied by $A(t)$, the balance after $t$ years?

  **(b)** What is the formula for $A(t)$, the balance after $t$ years?

  **(c)** How much money will be in the account after 3 years?

  **(d)** When will the balance triple?

  **(e)** How fast is the balance growing when it triples?

**5.** (*Investment Analysis*) An investment earns 4.2% interest compounded continuously. How fast is the investment growing when its value is $9000?

**6.** (*Investment Analysis*) An investment earns 5.1% interest compounded continuously and is currently growing at the rate of $765 per year. What is the current value of the investment?

**7.** (*Continuous Compound*) One thousand dollars is deposited in a savings account at 6% interest compounded continuously. How many years are required for the balance in the account to reach $2500?

**8.** (*Continuous Compound*) Ten thousand dollars is invested at 6.5% interest compounded continuously. When will the investment be worth $41,787?

9. (*Technology Stock*) One hundred shares of a technology stock were purchased on January 2, 1990 for $1200 and sold on January 2, 1998 for $12,500. What rate of interest compounded continuously did this investment earn?

10. (*Appreciation of Art Work*) Pablo Picasso's "Angel Fernandez de Soto" was acquired in 1946 for a postwar splurge of $22,220. The painting was sold in 1995 for $29.1 million. What rate of interest compounded continuously did this investment earn?

11. (*Investment Analysis*) How many years are required for an investment to double in value if it is appreciating at the rate of 4% compounded continuously?

12. What interest rate (compounded continuously) is earned by an investment that doubles in 10 years?

13. If an investment triples in 15 years, what interest rate (compounded continuously) does the investment earn?

14. (*Real Estate Investment*) If real estate in a certain city appreciates at the rate of 15% compounded continuously, when will a building purchased in 1998 triple in value?

15. (*Real Estate Investment*) In a certain town, property values tripled from 1980 to 1995. If this trend continues, when will property values be at five times their 1980 level? (Use an exponential model for the property value at time $t$.)

16. (*Real Estate Investment*) A lot purchased in 1980 for $5000 was appraised at $60,000 in 1998. If the lot continues to appreciate at the same rate, when will it be worth $100,000?

17. (*Real Estate Investment*) A farm purchased in 1985 for $1,000,000 was valued at $3,000,000 in 1995. If the farm continues to appreciate at the same rate (with continuous compounding), when will it be worth $10,000,000?

18. (*Real Estate Investment*) A parcel of land bought in 1990 for $10,000 was worth $16,000 in 1995. If the land continues to appreciate at this rate, in what year will it be worth $45,000?

19. (*Present Value*) Find the present value of $1000 payable at the end of 3 years, if money may be invested at 8% with interest compounded continuously.

20. (*Present Value*) Find the present value of $2000 to be received in 10 years, if money may be invested at 8% with interest compounded continuously.

21. (*Present Value*) How much money must you invest now at 4.5% interest compounded continuously in order to have $10,000 at the end of 5 years?

22. (*Present Value*) Suppose that the present value of $1000 to be received in 5 years is $559.90. What rate of interest, compounded continuously, was used to compute this present value?

23. Investment A is currently worth $70,200 and is growing at the rate of 13% per year compounded continuously.

Investment B is currently worth $60,000 and is growing at the rate of 14% per year compounded continuously. After how many years will the two investments have the same value?

24. Ten thousand dollars is deposited into a money market fund paying 8% interest compounded continuously. How much interest will be earned during the second year of the investment?

25. A small amount of money is deposited in a savings account with interest compounded continuously. Let $A(t)$ be the balance in the account after $t$ years. Match each of the following answers with its corresponding question.

*Answers:*

**a.** $Pe^{rt}$    **b.** $A(3)$    **c.** $A(0)$    **d.** $A'(3)$

**e.** Solve $A'(t) = 3$ for $t$.

**f.** Solve $A(t) = 3$ for $t$.

**g.** $y' = ry$

**h.** Solve $A(t) = 3A(0)$ for $t$.

*Questions:*

**A.** How fast will the balance be growing in 3 years?

**B.** Give the general form of the function $A(t)$.

**C.** How long will it take for the initial deposit to triple?

**D.** Find the balance after 3 years.

**E.** When will the balance be 3 dollars?

**F.** When will the balance be growing at the rate of 3 dollars per year?

**G.** What was the principal amount?

**H.** Give a differential equation satisfied by $A(t)$.

26. The curve in Fig. 1 shows the growth of money in a savings account with interest compounded continuously.

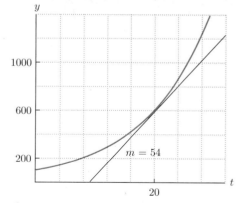

**Figure 1.** Growth of money in a savings account.

**(a)** What is the balance after 20 years?

**(b)** At what rate is the money growing after 20 years?

**(c)** Use the answers to (a) and (b) to determine the interest rate.

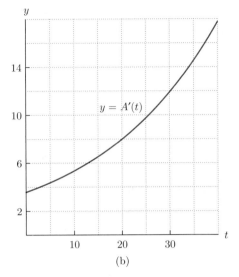

(a)

(b)

**Figure 2**

27. The function $A(t)$ in Fig. 2(a) gives the balance in a savings account after $t$ years with interest compounded continuously. Fig. 2(b) shows the derivative of $A(t)$.

   (a) What is the balance after 20 years?

   (b) How fast is the balance increasing after 20 years?

   (c) Use the answers to (a) and (b) to determine the interest rate.

   (d) When is the balance $300?

   (e) When is the balance increasing at the rate of $12 per year?

   (f) Why do the graphs of $A(t)$ and $A'(t)$ look the same?

28. When $1000 is invested at $r\%$ interest (compounded continuously) for 10 years, the balance is $f(r)$ dollars, where $f$ is the function shown in Fig. 3.

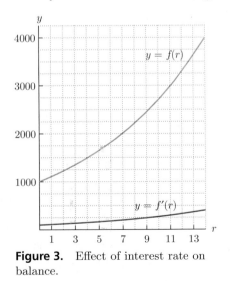

**Figure 3.** Effect of interest rate on balance.

   (a) What will the balance be at 7% interest?

   (b) For what interest rate will the balance be $3000?

   (c) If the interest rate is 9%, what is the growth rate of the balance with respect to a unit increase in interest?

### Technology Exercises

29. Convince yourself that $\lim_{h \to 0}(1 + h)^{1/h} = e$ by graphing $f(x) = (1 + x)^{1/x}$ near $x = 0$ and examining values of $f(x)$ for $x$ near 0. For instance, use a window such as $[-.5, .5]$ by $[-.5, 4]$.

30. Convince yourself that daily compounding is nearly the same as continuous compounding by graphing $y = 100(1+(.05/360))^{360x}$ together with $y = 100e^{.05x}$ in the window $[0, 64]$ by $[250, 2500]$. The two graphs should appear the same on the screen. Approximately how far apart are they when $x = 32$? When $x = 64$?

31. (*Internal Rate of Return*) Suppose that an investment of $2000 yields payments of $1200 in 3 years, $800 in 4 years, and $500 in 5 years. Thereafter, the investment is worthless. What constant rate of return $r$ would the investment need to produce in order to yield the payments specified? The number $r$ is called the *internal rate of return* on the investment. We can consider the investment as consisting of three parts, each part yielding one payment. The sum of the present values of the three parts must total $2000. This yields the equation

$$2000 = 1200e^{-3r} + 800e^{-4r} + 500e^{-5r}.$$

Solve this equation to find the value of $r$.

## SOLUTIONS TO PRACTICE PROBLEMS 5.2

1. Let us compute the balance after 4 years for each type of interest.

   *8% compounded semiannually:* Use the formula at the beginning of this section. Here $P = 1000$, $r = .08$, $m = 2$ (semiannually means there are two interest periods per year), and $t = 4$. Therefore,

   $$A = 1000 \left(1 + \frac{.08}{2}\right)^{2 \cdot 4} = 1000(1.04)^8 \approx \$1368.57.$$

   $7\frac{3}{4}\%$ *compounded continuously:* Use the formula $A = Pe^{rt}$, where $P = 1000$, $r = .0775$, and $t = 4$. Then

   $$A = 1000e^{(.0775) \cdot 4} = 1000e^{.31} \approx \$1363.43.$$

   Therefore, 8% compounded semiannually is better.

2. If the \$150,000 had been compounded continuously for 10 years at interest rate $r$, the balance would be $150{,}000e^{r \cdot 10}$. The question asks: For what value of $r$ will the balance be 400,000? We need just to solve an equation for $r$.

   $$150{,}000e^{r \cdot 10} = 400{,}000$$
   $$e^{r \cdot 10} \approx 2.67$$
   $$r \cdot 10 = \ln 2.67$$
   $$r = \frac{\ln 2.67}{10} \approx .098.$$

   Therefore, the investment earned 9.8% interest per year.

## 5.3 Applications of the Natural Logarithm Function to Economics

In this section we consider two applications of the natural logarithm to the field of economics. Our first application is concerned with relative rates of change and the second with elasticity of demand.

**Relative Rates of Change**   The *logarithmic derivative* of a function $f(t)$ is defined by the equation

$$\frac{d}{dt} \ln f(t) = \frac{f'(t)}{f(t)}. \tag{1}$$

The quantity on either side of equation (1) is often called the *relative rate of change of $f(t)$ per unit change of $t$.* Indeed, this quantity compares the rate of change of $f(t)$ [namely, $f'(t)$] with $f(t)$ itself. The *percentage rate of change* is the relative rate of change of $f(t)$ expressed as a percentage.

A simple example will illustrate these concepts. Suppose that $f(t)$ denotes the average price per pound of sirloin steak at time $t$ and $g(t)$ denotes the average price of a new car (of a given make and model) at time $t$, where $f(t)$ and $g(t)$ are given in dollars and time is measured in years. Then the ordinary derivatives $f'(t)$ and $g'(t)$ may be interpreted as the rate of change of the price of a pound of sirloin steak and of a new car, respectively, where both are measured in dollars per year. Suppose that, at a given time $t_0$, we have $f(t_0) = \$5.25$ and $g(t_0) = \$12{,}000$. Moreover, suppose that $f'(t_0) = \$.75$ and $g'(t_0) = \$1500$. Then at time $t_0$ the price per pound of steak is increasing at a rate of \$.75 per year, while the price of a new car is increasing at a rate of \$1500 per year. Which price is increasing more quickly? It is not meaningful to say that the car price is increasing faster simply because \$1500 is larger than \$.75. We must take into account the vast difference between the actual cost of a car and the cost of steak. The usual basis of comparison of price increases is the percentage rate of increase. In other words, at $t = t_0$, the price of sirloin steak is increasing at the percentage rate

$$\frac{f'(t_0)}{f(t_0)} = \frac{.75}{5.25} \approx .143 = 14.3\%$$

per year, but at the same time the price of a new car is increasing at the percentage rate

$$\frac{g'(t_0)}{g(t_0)} = \frac{1500}{12{,}000} = .125 = 12.5\%$$

per year. Thus the price of sirloin steak is increasing at a faster percentage rate than the price of a new car.

Economists often use relative rates of change (or percentage rates of change) when discussing the growth of various economic quantities, such as national income or national debt, because such rates of change can be meaningfully compared.

---

**EXAMPLE 1** **Gross domestic product** Suppose that a certain school of economists modeled the Gross Domestic Product of the United States at time $t$ (measured in years from January 1, 1990) by the formula

$$f(t) = 3.4 + .04t + .13e^{-t},$$

where the Gross Domestic Product is measured in trillions of dollars (see Fig. 1). What was the predicted percentage rate of growth (or decline) of the economy at $t = 0$ and $t = 1$?

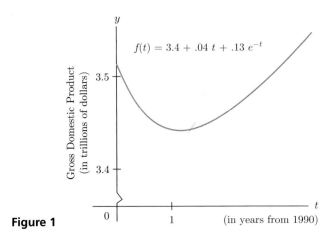

**Figure 1**

**Solution** Since

$$f'(t) = .04 - .13e^{-t},$$

we see that

$$\frac{f'(0)}{f(0)} = \frac{.04 - .13}{3.4 + .13} = -\frac{.09}{3.53} \approx -2.5\%.$$

$$\frac{f'(1)}{f(1)} = \frac{.04 - .13e^{-1}}{3.4 + .04 + .13e^{-1}} \approx -\frac{.00782}{3.4878} \approx -.2\%.$$

So on January 1, 1990, the economy is predicted to contract or decline at a relative rate of 2.5% per year; on January 1, 1991, the economy is predicted to be still contracting but only at a relative rate of .2% per year. ∎

---

**EXAMPLE 2** **Value of an investment** Suppose that the value in dollars of a certain business investment at time $t$ may be approximated empirically by the function $f(t) = 750{,}000e^{.6\sqrt{t}}$. Use a logarithmic derivative to describe how fast the value of the investment is increasing when $t = 5$ years.

**Solution**  We have

$$\frac{f'(t)}{f(t)} = \frac{d}{dt} \ln f(t) = \frac{d}{dt} \left( \ln 750{,}000 + \ln e^{.6\sqrt{t}} \right)$$

$$= \frac{d}{dt} (\ln 750{,}000 + .6\sqrt{t})$$

$$= (.6) \left( \frac{1}{2} \right) t^{-1/2} = \frac{.3}{\sqrt{t}}.$$

When $t = 5$,

$$\frac{f'(5)}{f(5)} = \frac{.3}{\sqrt{5}} \approx .134 = 13.4\%.$$

Thus, when $t = 5$ years, the value of the investment is increasing at the relative rate of 13.4% per year. So when $t = 5$, we should expect the investment to grow by 13.4% in one year (Fig. 2). Indeed if we compute the increase in the investment from $t = 5$ to $t = 6$ and divide by its value at $t = 5$ to get the percent increase, we find

$$\text{percentage rate of change} = \frac{f(6) - f(5)}{f(5)} \approx \frac{32.6 - 28.7}{28.7} \approx .136$$

or 13.6%. This percentage is close to 13.4%, the relative rate of change of $f$ at $t = 5$.  ■

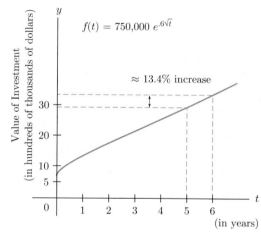

**Figure 2**

In certain mathematical models, it is assumed that for a limited period of time the percentage rate of change of a particular function is constant. The following example shows that such a function must be an exponential function.

**EXAMPLE 3**  Suppose that the function $f(t)$ has a constant relative rate of change $k$. Show that $f(t) = Ce^{kt}$ for some constant $C$.

**Solution**  We are given that

$$\frac{d}{dt} \ln f(t) = k.$$

That is,

$$\frac{f'(t)}{f(t)} = k.$$

Hence $f'(t) = kf(t)$. But this is just the differential equation satisfied by the exponential function. Therefore, we must have $f(t) = Ce^{kt}$ for some constant $C$. ∎

**Elasticity of Demand** In Section 2.7 we considered demand equations for monopolists and for entire industries. Recall that a demand equation expresses, for each quantity $q$ to be produced, the market price that will generate a demand of exactly $q$. For instance, the demand equation

$$p = 150 - .01x \tag{2}$$

says that in order to sell $x$ units, the price must be set at $150 - .01x$ dollars. To be specific, in order to sell 6000 units, the price must be set at $150 - .01(6000) = \$90$ per unit.

Equation (2) may be solved for $x$ in terms of $p$ to yield

$$x = 100(150 - p). \tag{3}$$

This last equation gives quantity in terms of price. If we let the letter $q$ represent quantity, equation (3) becomes

$$q = 100(150 - p). \tag{3'}$$

This equation is of the form $q = f(p)$, where in this case $f(p)$ is the function $f(p) = 100(150 - p)$. In what follows it will be convenient to always write our demand functions so that the quantity $q$ is expressed as a function $f(p)$ of the price $p$.

Usually, raising the price of a commodity lowers demand. Therefore, the typical demand function $q = f(p)$ is decreasing and has a negative slope everywhere, as illustrated by Fig. 3.

A demand function $q = f(p)$ relates the quantity demanded to the price. Therefore, the derivative $f'(p)$ compares the change in quantity demanded with the change in price. By way of contrast, the concept of elasticity is designed to compare the *relative* rate of change of the quantity demanded with the *relative* rate of change of price.

Let us be more explicit. Consider a particular demand function $q = f(p)$. From our interpretation of the logarithmic derivative in (1), we know that the relative rate of change of the quantity demanded with respect to $p$ is

$$\frac{\frac{d}{dp}f(p)}{f(p)} = \frac{f'(p)}{f(p)}.$$

Similarly, the relative rate of change of price with respect to $p$ is

$$\frac{\frac{d}{dp}p}{p} = \frac{1}{p}.$$

Hence the ratio of the relative rate of change of the quantity demanded to the relative rate of change of price is

$$\frac{[\text{relative rate of change of quantity}]}{[\text{relative rate of change of price}]} = \frac{f'(p)/f(p)}{1/p} = \frac{pf'(p)}{f(p)}.$$

Since $f'(p)$ is always negative for a typical demand function, the quantity $pf'(p)/f(p)$ will be negative for all values of $p$. For convenience, economists prefer to work with positive numbers and therefore the *elasticity of demand* is taken to be this quantity multiplied by $-1$.

**Figure 3**

The elasticity of demand $E(p)$ at price $p$ for the demand function $q = f(p)$ is defined to be

$$E(p) = \frac{-pf'(p)}{f(p)}.$$

**EXAMPLE 4**  **Elasticity of demand**  Suppose that the demand function for a certain metal is $q = 100 - 2p$, where $p$ is the price per pound and $q$ is the quantity demanded (in millions of pounds).

(a) What quantity can be sold at $30 per pound?

(b) Determine the function $E(p)$.

(c) Determine and interpret the elasticity of demand at $p = 30$.

(d) Determine and interpret the elasticity of demand at $p = 20$.

**Solution**  (a) In this case, $q = f(p)$, where $f(p) = 100 - 2p$. When $p = 30$, we have $q = f(30) = 100 - 2(30) = 40$. Therefore, 40 million pounds of the metal can be sold. We also say that the *demand* is 40 million pounds.

(b) $E(p) = \dfrac{-pf'(p)}{f(p)} = \dfrac{-p(-2)}{100 - 2p} = \dfrac{2p}{100 - 2p}$.

(c) The elasticity of demand at price $p = 30$ is $E(30)$.

$$E(30) = \frac{2(30)}{100 - 2(30)} = \frac{60}{40} = \frac{3}{2}$$

When the price is set at $30 per pound, a small increase in price will result in a relative rate of decrease in quantity demanded of about $\frac{3}{2}$ times the relative rate of increase in price. For example, if the price is increased from $30 by 1%, then the quantity demanded will decrease by about 1.5%.

(d) When $p = 20$, we have

$$E(20) = \frac{2(20)}{100 - 2(20)} = \frac{40}{60} = \frac{2}{3}.$$

When the price is set at $20 per pound, a small increase in price will result in a relative rate of decrease in quantity demanded of only $\frac{2}{3}$ of the relative rate of increase of price. For example, if the price is increased from $20 by 1%, the quantity demanded will decrease by $\frac{2}{3}$ of 1%.  ■

Economists say that demand is *elastic* at price $p_0$ if $E(p_0) > 1$ and *inelastic* at price $p_0$ if $E(p_0) < 1$. In Example 4, the demand for the metal is elastic at $30 per pound and inelastic at $20 per pound.

The significance of the concept of elasticity may perhaps best be appreciated by studying how revenue, $R(p)$, responds to changes in price. Let us illustrate this response with a concrete example.

**EXAMPLE 5**  Figure 4 shows the elasticity of demand for the metal in Example 4.

(a) For what values of $p$ is demand elastic? Inelastic?

(b) Find and plot the revenue function for $0 < p < 50$.

(c) How does revenue respond to an increase in price when demand is elastic, respectively, inelastic?

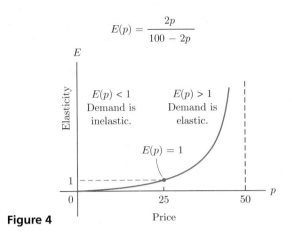

$$E(p) = \frac{2p}{100 - 2p}$$

Figure 4

**Solution** (a) In Example 4(b), we found the elasticity of demand to be

$$E(p) = \frac{2p}{100 - 2p}.$$

Let us solve $E(p) = 1$ for $p$.

$$\frac{2p}{100 - 2p} = 1$$
$$2p = 100 - 2p$$
$$4p = 100$$
$$p = 25.$$

By definition, demand is elastic at price $p$ if $E(p) > 1$ and inelastic if $E(p) < 1$. From Figure 4, we see that demand is elastic if $25 < p < 50$ and inelastic if $0 < p < 25$.

(b) Recall that

$$[\text{revenue}] = [\text{quantity}] \cdot [\text{price per unit}].$$

Using the formula for demand (in millions of pounds) from Example 4, we obtain the revenue function

$$R = (100 - 2p) \cdot p = p(100 - 2p) \quad \text{(in millions of dollars)}.$$

This is a parabola opening down with $p$-intercepts at $p = 0$ and $p = 50$. Its maximum is located at the midpoint of the $p$-intercepts, or $p = 25$ (Fig. 5).

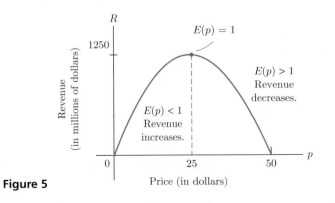

Figure 5

(c) In part (a), we determined that demand is elastic for $25 < p < 50$. For $p$ in this price range, Fig. 5 shows that an increase in price results in a decrease in revenue, and a decrease in price results in an increase in revenue. Hence we conclude that when demand is elastic, the change of revenue is in the opposite direction of the change in price. Similarly, when demand is inelastic ($0 < p < 25$), Fig. 5 shows that the change of revenue is in the same direction as the change in price. ∎

Example 5 illustrates an important application of elasticity as an indicator of the response of revenue to a change in price. As we now show, this response analysis can be applied in general.

Start by expressing revenue as a function of price

$$R(p) = f(p) \cdot p,$$

where $f(p)$ is the demand function. If we differentiate $R(p)$ using the product rule, we find that

$$R'(p) = \frac{d}{dp}[f(p) \cdot p] = f(p) \cdot 1 + p \cdot f'(p)$$

$$= f(p)\left[1 + \frac{pf'(p)}{f(p)}\right]$$

$$= f(p)[1 - E(p)]. \tag{4}$$

Now suppose that demand is elastic at some price $p_0$. Then $E(p_0) > 1$ and $1 - E(p_0)$ is negative. Since $f(p)$ is always positive, we see from (4) that $R'(p_0)$ is negative. Therefore, by the first derivative rule, $R(p)$ is decreasing at $p_0$. So an increase in price will result in a decrease in revenue, and a decrease in price will result in an increase in revenue. On the other hand, if demand is inelastic at $p_0$, then $1 - E(p_0)$ will be positive and hence $R'(p_0)$ will be positive. In this case an increase in price will result in an increase in revenue, and a decrease in price will result in a decrease in revenue. We may summarize this as follows:

> The change in revenue is in the opposite direction of the change in price when demand is elastic and in the same direction when demand is inelastic.

## PRACTICE PROBLEMS 5.3

*The current toll for the use of a certain toll road is $2.50. A study conducted by the state highway department determined that with a toll of p dollars, q cars will use the road each day, where $q = 60{,}000e^{-.5p}$.*

1. Compute the elasticity of demand at $p = 2.5$.

2. Is demand elastic or inelastic at $p = 2.5$?

3. If the state increases the toll slightly, will the revenue increase or decrease?

## EXERCISES 5.3

*Determine the percentage rate of change of the functions at the points indicated.*

1. $f(t) = t^2$ at $t = 10$ and $t = 50$

2. $f(t) = t^{10}$ at $t = 10$ and $t = 50$

3. $f(x) = e^{.3x}$ at $x = 10$ and $x = 20$

4. $f(x) = e^{-.05x}$ at $x = 1$ and $x = 10$

5. $f(t) = e^{.3t^2}$ at $t = 1$ and $t = 5$

6. $G(s) = e^{-.05s^2}$ at $s = 1$ and $s = 10$

7. $f(p) = 1/(p + 2)$ at $p = 2$ and $p = 8$

8. $g(p) = 5/(2p + 3)$ at $p = 1$ and $p = 11$

9. (*Percentage Rate of Growth*) Suppose that the annual sales $S$ (in dollars) of a company may be approximated empirically by the formula

$$S = 50{,}000\sqrt{e^{\sqrt{t}}},$$

where $t$ is the number of years beyond some fixed reference date. Use a logarithmic derivative to determine the percentage rate of growth of sales at $t = 4$.

10. (*Percentage Rate of Change*) Suppose that the price of wheat per bushel at time $t$ (in months) is approximated by

$$f(t) = 4 + .001t + .01e^{-t}.$$

What is the percentage rate of change of $f(t)$ at $t = 0$? $t = 1$? $t = 2$?

11. Suppose that an investment grows at a continuous 12% rate per year. In how many years will the value of the investment double?

12. Suppose that the value of a piece of property is growing at a continuous $r\%$ rate per year and that the value doubles in 3 years. Find $r$.

*For each demand function, find $E(p)$ and determine if demand is elastic or inelastic (or neither) at the indicated price.*

13. $q = 700 - 5p$, $p = 80$

14. $q = 600e^{-.2p}$, $p = 10$

15. $q = 400(116 - p^2)$, $p = 6$

16. $q = (77/p^2) + 3$, $p = 1$

17. $q = p^2 e^{-(p+3)}$, $p = 4$

18. $q = 700/(p + 5)$, $p = 15$

19. (*Elasticity of Demand*) Currently, 1800 people ride a certain commuter train each day and pay $4 for a ticket. The number of people $q$ willing to ride the train at price $p$ is $q = 600(5 - \sqrt{p})$. The railroad would like to increase its revenue.

   (a) Is demand elastic or inelastic at $p = 4$?

   (b) Should the price of a ticket be raised or lowered?

20. (*Elasticity of Demand*) An electronic store can sell $q = 10{,}000/(p + 50) - 30$ cellular phones at a price $p$ dollars per phone. The current price is $150.

   (a) Is demand elastic or inelastic at $p = 150$?

   (b) If the price is lowered slightly, will revenue increase or decrease?

21. (*Elasticity of Demand*) A movie theater has a seating capacity of 3000 people. The number of people attending a show at price $p$ dollars per ticket is $q = (18{,}000/p) - 1500$. Currently the price is $6 per ticket.

   (a) Is demand elastic or inelastic at $p = 6$?

   (b) If the price is lowered, will revenue increase or decrease?

22. (*Elasticity of Demand*) A subway charges 65 cents per person and has 10,000 riders each day. The demand function for the subway is $q = 2000\sqrt{90 - p}$.

   (a) Is demand elastic or inelastic at $p = 65$?

   (b) Should the price of a ride be raised or lowered in order to increase the amount of money taken in by the subway?

23. (*Elasticity of Demand*) A country that is the major supplier of a certain commodity wishes to improve its balance of trade position by lowering the price of the commodity. The demand function is $q = 1000/p^2$.

   (a) Compute $E(p)$.

   (b) Will the country succeed in raising its revenue?

24. Show that any demand function of the form $q = a/p^m$ has constant elasticity $m$.

*A cost function $C(x)$ gives the total cost of producing $x$ units of a product. The elasticity of cost at quantity $x$, $E_c(x)$, is defined to be the ratio of the relative rate of change of cost (with respect to $x$) divided by the relative rate of change of quantity (with respect to $x$).*

25. Show that $E_c(x) = x \cdot C'(x)/C(x)$.

26. Show that $E_c$ is equal to the marginal cost divided by the average cost.

27. Let $C(x) = (1/10)x^2 + 5x + 300$. Show that $E_c(50) < 1$. (Hence when producing 50 units, a small relative increase in production results in an even smaller relative increase in total cost. Also, the average cost of producing 50 units is greater than the marginal cost at $x = 50$.)

28. Let $C(x) = 1000e^{.02x}$. Determine and simplify the formula for $E_c(x)$. Show that $E_c(60) > 1$ and interpret this result.

## Technology Exercises

29. Consider the demand function $q = 60{,}000e^{-.5p}$ from the practice problems.

   (a) Determine the value of $p$ for which the value of $E(p)$ is 1. For what values of $p$ is demand inelastic?

   (b) Graph the revenue function in the window $[0, 4]$ by $[-5000, 50{,}000]$ and determine where its maximum value occurs. For what values of $p$ is the revenue an increasing function?

**1.** The demand function is $f(p) = 60{,}000e^{-.5p}$.

$$f'(p) = -30{,}000e^{-.5p}$$

$$E(p) = \frac{-pf'(p)}{f(p)} = \frac{-p(-30{,}000)e^{-.5p}}{60{,}000e^{-.5p}} = \frac{p}{2}$$

$$E(2.5) = \frac{2.5}{2} = 1.25.$$

**2.** The demand is elastic, because $E(2.5) > 1$.

**3.** Since demand is elastic at $2.50, a slight change in price causes revenue to change in the *opposite* direction. Hence revenue will decrease.

## 5.4  Further Exponential Models

A skydiver, on jumping out of an airplane, falls at an increasing rate. However, the wind rushing past the skydiver's body creates an upward force that begins to counterbalance the downward force of gravity. This air friction finally becomes so great that the skydiver's velocity reaches a limiting speed called the *terminal velocity*. If we let $v(t)$ be the downward velocity of the skydiver after $t$ seconds of free fall, then a good mathematical model for $v(t)$ is given by

$$v(t) = M(1 - e^{-kt}), \tag{1}$$

where $M$ is the terminal velocity and $k$ is some positive constant (Fig. 1). When $t$ is close to zero, $e^{-kt}$ is close to one, and the velocity is small. As $t$ increases, $e^{-kt}$ becomes small and so $v(t)$ approaches $M$.

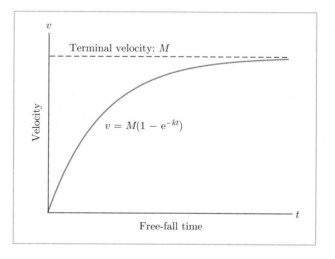

**Figure 1.** Velocity of a skydiver.

---

**EXAMPLE 1**  **Velocity of a skydiver**  Show that the velocity given in (1) satisfies the equations

$$v'(t) = k[M - v(t)], \qquad v(0) = 0. \tag{2}$$

**Solution**  From (1) we have $v(t) = M - Me^{-kt}$. Then

$$v'(t) = Mke^{-kt}.$$

However,

$$k[M - v(t)] = k[M - (M - Me^{-kt})] = kMe^{-kt},$$

so that the differential equation $v'(t) = k[M - v(t)]$ holds. Also,

$$v(0) = M - Me^0 = M - M = 0.$$ ◼

The differential equation (2) says that the rate of change in $v$ is proportional to the difference between the terminal velocity $M$ and the actual velocity $v$. It is not difficult to show that the only solution of (2) is given by the formula in (1).

The two equations (1) and (2) arise as mathematical models in a variety of situations. Some of these applications are described next.

**The Learning Curve**   Psychologists have found that in many learning situations a person's rate of learning is rapid at first and then slows down. Finally, as the task is mastered, the person's level of performance reaches a level above which it is almost impossible to rise. For example, within reasonable limits, each person seems to have a certain maximum capacity for memorizing a list of nonsense syllables. Suppose that a subject can memorize $M$ syllables in a row if given sufficient time, say an hour, to study the list but cannot memorize $M+1$ syllables in a row even if allowed several hours of study. By giving the subject different lists of syllables and varying lengths of time to study the lists, the psychologist can determine an empirical relationship between the number of nonsense syllables memorized accurately and the number of minutes of study time. It turns out that a good model for this situation is

$$f(t) = M(1 - e^{-kt})$$

for some appropriate positive constant $k$. (See Fig. 2.)

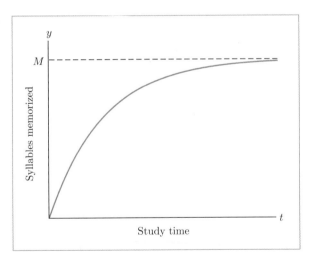

**Figure 2.**   Learning curve, $f(t) = M(1 - e^{-kt})$.

The *slope* of this learning curve at time $t$ is approximately the number of additional syllables that can be memorized if the subject is given one more minute of study time. Thus the slope is a measure of the *rate of learning*. The differential equation satisfied by the function $y = f(t)$ is

$$y' = k(M - y), \qquad f(0) = 0.$$

This equation says that if the subject is given a list of $M$ nonsense syllables, then the rate of memorization is proportional to the number of syllables remaining to be memorized.

**Diffusion of Information by Mass Media**    Sociologists have found that the differential equation (2) provides a good model for the way information is spread (or "diffused") through a population when the information is being propagated constantly by mass media, such as television or magazines.* Given a fixed population $P$, let $f(t)$ be the number of people who have already heard a certain piece of information by time $t$. Then $P - f(t)$ is the number who have not yet heard the information by time $t$. Also, $f'(t)$ is the rate of increase of the number of people who have heard the news (the "rate of diffusion" of the information). If the information is being publicized often by some mass media, then it is likely that the number of *newly informed* people per unit time is proportional to the number of people who have not yet heard the news. Therefore,

$$f'(t) = k[P - f(t)]. \qquad (3)$$

Assume that $f(0) = 0$ (i.e., there was a time $t = 0$ when nobody had heard the news). Then the remark following Example 1 shows that

$$f(t) = P(1 - e^{-kt}). \qquad (4)$$

(See Fig. 3.)

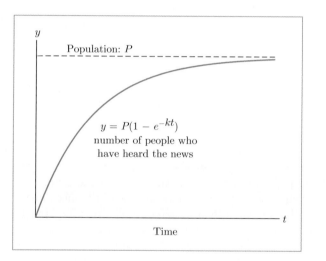

**Figure 3.**   Diffusion of information by mass media.

---

EXAMPLE 2    Suppose that the news of resignation of a public official is broadcast frequently by radio and television stations. Also suppose that one-half of the residents of a city have heard the news within 4 hours of its initial release. Use the exponential model (4) to estimate when 90% of the residents will have heard the news.

Solution    We must find the value of $k$ in (4). If $P$ is the number of residents, then the number who will have heard the news in the first four hours is given by (4) with $t = 4$. By assumption, this number is half the population. So

$$\tfrac{1}{2}P = P(1 - e^{-k \cdot 4})$$
$$.5 = 1 - e^{-4k}$$
$$e^{-4k} = 1 - .5 = .5.$$

---

*J. Coleman, *Introduction to Mathematical Sociology* (New York: The Free Press, 1964), p. 43.

Solving for $k$, we find that $k \approx .17$. So the model for this particular situation is

$$f(t) = P(1 - e^{-.17t}).$$

Now we want to find $t$ such that $f(t) = .90P$. We solve for $t$:

$$.90P = P(1 - e^{-.17t})$$
$$.90 = 1 - e^{-.17t}$$
$$e^{-.17t} = 1 - .90 = .10$$
$$-.17t = \ln .10$$
$$t = \frac{\ln .10}{-.17} \approx 14.$$

Therefore, 90% of the residents will hear the news within 14 hours of its initial release. ■

**Intravenous Infusion of Glucose**   The human body both manufactures and uses glucose ("blood sugar"). Usually, there is a balance in these two processes, so that the bloodstream has a certain "equilibrium level" of glucose. Suppose that a patient is given a single intravenous injection of glucose and let $A(t)$ be the amount of glucose (in milligrams) above the equilibrium level. Then the body will start using up the excess glucose at a rate proportional to the amount of excess glucose; that is,

$$A'(t) = -\lambda A(t), \tag{5}$$

where $\lambda$ is a positive constant called the *velocity constant of elimination*. This constant depends on how fast an individual patient's metabolic processes eliminate the excess glucose from the blood. Equation (5) describes a simple exponential decay process.

Now suppose that, instead of a single shot, the patient receives a continuous intravenous infusion of glucose. A bottle of glucose solution is suspended above the patient, and a small tube carries the glucose down to a needle that runs into a vein. In this case, there are two influences on the amount of excess glucose in the blood: the glucose being added steadily from the bottle and the glucose being removed from the body by metabolic processes. Let $r$ be the rate of infusion of glucose (often from 10 to 100 milligrams per minute). If the body did not remove any glucose, the excess glucose would increase at a constant rate of $r$ milligrams per minute; that is,

$$A'(t) = r. \tag{6}$$

Taking into account the two influences on $A'(t)$ described by (5) and (6), we can write

$$A'(t) = r - \lambda A(t). \tag{7}$$

Define $M$ to be $r/\lambda$ and note that initially there is no excess glucose; then

$$A'(t) = \lambda(M - A(t)), \qquad A(0) = 0.$$

As stated in Example 1, a solution of this differential equation is given by

$$A(t) = M(1 - e^{-\lambda t}) = \frac{r}{\lambda}(1 - e^{-\lambda t}). \tag{8}$$

Note that $M$ is the limiting value of the glucose level. Reasoning as in Example 1, we conclude that the amount of excess glucose rises until it reaches a stable level. (See Fig. 4.)

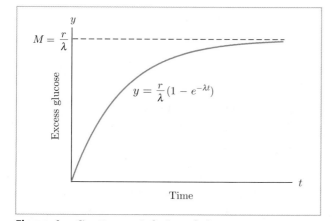

**Figure 4.** Continuous infusion of glucose.

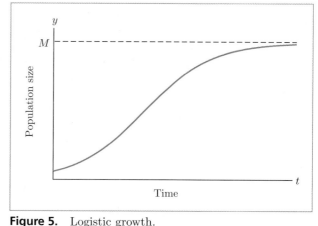

**Figure 5.** Logistic growth.

**The Logistic Growth Curve** The model for simple exponential growth discussed in Section 5.1 is adequate for describing the growth of many types of populations, but obviously a population cannot increase exponentially forever. The simple exponential growth model becomes inapplicable when the environment begins to inhibit the growth of the population. The logistic growth curve is an important exponential model that takes into account some of the effects of the environment on a population (Fig. 5). For small values of $t$, the curve has the same basic shape as an exponential growth curve. Then when the population begins to suffer from overcrowding or lack of food, the growth rate (the slope of the population curve) begins to slow down. Eventually, the growth rate tapers off to zero as the population reaches the maximum size that the environment will support. This latter part of the curve resembles the growth curves studied earlier in this section.

The equation for logistic growth has the general form

$$y = \frac{M}{1 + Be^{-Mkt}}, \tag{9}$$

where $B$, $M$, and $k$ are positive constants. We can show that $y$ satisfies the differential equation

$$y' = ky(M - y). \tag{10}$$

The factor $y$ reflects the fact that the growth rate $(y')$ depends in part on the size $y$ of the population. The factor $M - y$ reflects the fact that the growth rate also depends on how close $y$ is to the maximum level $M$.

The logistic curve is often used to fit experimental data that lie along an "S-shaped" curve. Examples are given by the growth of a fish population in a lake and the growth of a fruit fly population in a laboratory container. Also, certain enzyme reactions in animals follow a logistic law. One of the earliest applications of the logistic curve occurred in about 1840 when the Belgian sociologist P. Verhulst fit a logistic curve to six U.S. census figures, 1790 to 1840, and predicted the U.S. population for 1940. His prediction missed by less than 1 million persons (an error of about 1%).

**EXAMPLE 3**   **Logistic growth** Suppose that a lake is stocked with 100 fish. After 3 months there are 250 fish. A study of the ecology of the lake predicts that the lake can support 1000 fish. Find a formula for the number $P(t)$ of fish in the lake $t$ months after it has been stocked.

**Solution** The limiting population $M$ is 1000. Therefore, we have

$$P(t) = \frac{1000}{1 + Be^{-1000kt}}.$$

At $t = 0$ there are 100 fish, so that

$$100 = P(0) = \frac{1000}{1 + Be^0} = \frac{1000}{1 + B}.$$

Thus $1 + B = 10$, or $B = 9$. Finally, since $P(3) = 250$, we have

$$250 = \frac{1000}{1 + 9e^{-3000k}}$$

$$1 + 9e^{-3000k} = 4$$

$$e^{-3000k} = \tfrac{1}{3}$$

$$-3000k = \ln \tfrac{1}{3}$$

$$k \approx .00037.$$

Therefore,

$$P(t) = \frac{1000}{1 + 9e^{-.37t}}. \qquad \blacksquare$$

Several theoretical justifications can be given for using (9) and (10) in situations where the environment prevents a population from exceeding a certain size. A discussion of this topic may be found in *Mathematical Models and Applications* by D. Maki and M. Thompson (Englewood Cliffs, N.J.: Prentice-Hall, Inc., 1973), pp. 312–317.

**An Epidemic Model** It will be instructive to actually "build" a mathematical model. Our example concerns the spread of a highly contagious disease. We begin by making several simplifying assumptions:

1. The population is a fixed number $P$ and each member of the population is susceptible to the disease.

2. The duration of the disease is long, so that no cures occur during the time period under study.

3. All infected individuals are contagious and circulate freely among the population.

4. During each time period (such as 1 day or 1 week) each infected person makes $c$ contacts, and each contact with an uninfected person results in transmission of the disease.

Consider a short period of time from $t$ to $t + h$. Each infected person makes $c \cdot h$ contacts. How many of these contacts are with uninfected persons? If $f(t)$ is the number of infected persons at time $t$, then $P - f(t)$ is the number of uninfected persons, and $[P - f(t)]/P$ is the fraction of the population that is uninfected. Thus, of the $c \cdot h$ contacts made,

$$\left[ \frac{P - f(t)}{P} \right] \cdot c \cdot h$$

will be with uninfected persons. This is the number of new infections produced by one infected person during the time period of length $h$. The total number of *new* infections during this period is

$$f(t) \left[ \frac{P - f(t)}{P} \right] ch.$$

But this number must equal $f(t + h) - f(t)$, where $f(t + h)$ is the total number of infected persons at time $t + h$. So

$$f(t + h) - f(t) = f(t) \left[ \frac{P - f(t)}{P} \right] ch.$$

Dividing by $h$, the length of the time period, we obtain the average number of new infections per unit time (during the small time period):

$$\frac{f(t + h) - f(t)}{h} = \frac{c}{P} f(t)[P - f(t)].$$

If we let $h$ approach zero and let $y$ stand for $f(t)$, the left-hand side approaches the rate of change in the number of infected persons and we derive the following equation:

$$\frac{dy}{dt} = \frac{c}{P} y(P - y). \tag{11}$$

This is the same type of equation as that used in (10) for logistic growth, although the two situations leading to this model appear to be quite dissimilar.

Comparing (11) with (10), we see that the number of infected individuals at time $t$ is described by a logistic curve with $M = P$ and $k = c/P$. Therefore, by (9), we can write

$$f(t) = \frac{P}{1 + Be^{-ct}}.$$

$B$ and $c$ can be determined from the characteristics of the epidemic. (See Example 4.)

The logistic curve has an inflection point at that value of $t$ for which $f(t) = P/2$. The position of this inflection point has great significance for applications of the logistic curve. From inspecting a graph of the logistic curve, we see that the inflection point is the point at which the curve has greatest slope. In other words, the inflection point corresponds to the instant of fastest growth of the logistic curve. This means, for example, that in the foregoing epidemic model the disease is spreading with the greatest rapidity precisely when half the population is infected. Any attempt at disease control (through immunization, for example) must strive to reduce the incidence of the disease to as low a point as possible, but in any case at least below the inflection point at $P/2$, at which point the epidemic is spreading fastest.

---

**EXAMPLE 4**  **Spread of an epidemic**  The Public Health Service monitors the spread of an epidemic of a particularly long-lasting strain of flu in a city of 500,000 people. At the beginning of the first week of monitoring, 200 cases have been reported; during the first week 300 new cases are reported. Estimate the number of infected individuals after 6 weeks.

**Solution**  Here $P = 500{,}000$. If $f(t)$ denotes the number of cases at the end of $t$ weeks, then

$$f(t) = \frac{P}{1 + Be^{-ct}} = \frac{500{,}000}{1 + Be^{-ct}}.$$

Moreover, $f(0) = 200$, so that

$$200 = \frac{500{,}000}{1 + Be^0} = \frac{500{,}000}{1 + B},$$

and $B = 2499$. Consequently, since $f(1) = 300 + 200 = 500$, we have

$$500 = f(1) = \frac{500{,}000}{1 + 2499e^{-c}},$$

so that $e^{-c} \approx .4$ and $c \approx .92$. Finally,

$$f(t) = \frac{500{,}000}{1 + 2499e^{-.92t}}$$

and

$$f(6) = \frac{500{,}000}{1 + 2499e^{-.92(6)}} \approx 45{,}000.$$

After 6 weeks, about 45,000 individuals are infected. ■

This epidemic model is used by sociologists (who still call it an epidemic model) to describe the spread of a rumor. In economics the model is used to describe the diffusion of knowledge about a product. An "infected person" represents an individual who possesses knowledge of the product. In both cases, it is assumed that the members of the population are themselves primarily responsible for the spread of the rumor or knowledge of the product. This situation is in contrast to the model described earlier where information was spread through a population by external sources, such as radio and television.

There are several limitations to this epidemic model. Each of the four simplifying assumptions made at the outset is unrealistic in varying degrees. More complicated models can be constructed that rectify one or more of these defects, but they require more advanced mathematical tools.

**The Exponential Function in Lung Physiology**   Let us conclude this section by deriving a useful model for the pressure in a person's lungs when the air is allowed to escape passively from the lungs with no use of the person's muscles. Let $V$ be the volume of air in the lungs and let $P$ be the relative pressure in the lungs when compared with the pressure in the mouth. The *total compliance* (of the respiratory system) is defined to be the derivative $\dfrac{dV}{dP}$. For normal values of $V$ and $P$ we may assume that the total compliance is a positive constant, say $C$. That is,

$$\frac{dV}{dP} = C. \tag{12}$$

We shall assume that the airflow during the passive respiration is smooth and not turbulent. Then Poiseuille's law of fluid flow says that the rate of change of volume as a function of time (i.e., the rate of airflow) satisfies

$$\frac{dV}{dt} = -\frac{P}{R}, \tag{13}$$

where $R$ is a (positive) constant called the airway resistance. Under these conditions, we may derive a formula for $P$ as a function of time. Since the volume is a function of the pressure, and the pressure is in turn a function of time, we may use the chain rule to write

$$\frac{dV}{dt} = \frac{dV}{dP} \cdot \frac{dP}{dt}.$$

From (12) and (13),

$$-\frac{P}{R} = C \cdot \frac{dP}{dt},$$

so

$$\frac{dP}{dt} = -\frac{1}{RC} \cdot P.$$

From this differential equation we conclude that $P$ must be an exponential function of $t$. In fact,

$$P = P_0 e^{kt},$$

where $P_0$ is the initial pressure at time $t = 0$ and $k = -1/RC$. This relation between $k$ and the product $RC$ is useful to lung specialists, because they can experimentally compute $k$ and the compliance $C$, and then use the formula $k = -1/RC$ to determine the airway resistance $R$.

## PRACTICE PROBLEM 5.4

1. A sociological study* was made to examine the process by which doctors decide to adopt a new drug. The doctors were divided into two groups. The doctors in group A had little interaction with other doctors and so received most of their information through mass media. The doctors in group B had extensive interaction with other doctors and so received most of their information through word of mouth. For each group, let $f(t)$ be the number who have learned about the new drug after $t$ months. Examine the appropriate differential equations to explain why the two graphs were of the types shown in Fig. 6.

**Figure 6.** Results of a sociological study.

## EXERCISES 5.4

1. Consider the function $f(x) = 5(1 - e^{-2x})$, $x \geq 0$.

   **(a)** Show that $f(x)$ is increasing and concave down for all $x \geq 0$.

   **(b)** Explain why $f(x)$ approaches 5 as $x$ gets large.

   **(c)** Sketch the graph of $f(x)$, $x \geq 0$.

2. Consider the function $g(x) = 10 - 10e^{-.1x}$, $x \geq 0$.

   **(a)** Show that $g(x)$ is increasing and concave down for $x \geq 0$.

   **(b)** Explain why $g(x)$ approaches 10 as $x$ gets large.

   **(c)** Sketch the graph of $g(x)$, $x \geq 0$.

3. Suppose that $y = 2(1 - e^{-x})$. Compute $y'$ and show that $y' = 2 - y$.

4. Suppose that $y = 5(1 - e^{-2x})$. Compute $y'$ and show that $y' = 10 - 2y$.

5. Suppose that $f(x) = 3(1 - e^{-10x})$. Show that $y = f(x)$ satisfies the differential equation

$$y' = 10(3 - y), \qquad f(0) = 0.$$

*James S. Coleman, Eliku Katz, and Herbert Menzel, "The Diffusion of an Innovation Among Physicians," *Sociometry*, 20 (1957), 253–270.

6. (*Ebbinghaus Model for Forgetting*) Suppose that a student learns a certain amount of material for some class. Let $f(t)$ denote the percentage of the material that the student can recall $t$ weeks later. The psychologist Ebbinghaus found that this percent retention can be modeled by a function of the form

$$f(t) = (100 - a)e^{-\lambda t} + a,$$

where $\lambda$ and $a$ are positive constants and $0 < a < 100$. Sketch the graph of the function $f(t) = 85e^{-.5t} + 15$, $t \geq 0$.

7. (*Spread of News*) When a grand jury indicted the mayor of a certain town for accepting bribes, the newspaper, radio, and television immediately began to publicize the news. Within an hour, one-quarter of the citizens heard about the indictment. Estimate when three-quarters of the town heard the news.

8. Examine formula (8) for the amount $A(t)$ of excess glucose in the bloodstream of a patient at time $t$. Describe what would happen if the rate $r$ of infusion of glucose were doubled.

9. (*Glucose Elimination*) Describe an experiment that a doctor could perform in order to determine the velocity constant of elimination of glucose for a particular patient.

10. (*Concentration of Glucose in Bloodstream*) Physiologists usually describe the continuous intravenous infusion of glucose in terms of the excess *concentration* of glucose, $C(t) = A(t)/V$, where $V$ is the total volume of blood in the patient. In this case, the rate of increase in the concentration of glucose due to the continuous injection is $r/V$. Find a differential equation that gives a model for the rate of change of the excess concentration of glucose.

11. (*Spread of News*) A news item is spread by word of mouth to a potential audience of 10,000 people. After $t$ days,

$$f(t) = \frac{10,000}{1 + 50e^{-.4t}}$$

people will have heard the news. The graph of this function is shown in Fig. 7.

(a) Approximately how many people will have heard the news after 7 days?

(b) At approximately what rate will the news be spreading after 14 days?

(c) Approximately when will 7000 people have heard the news?

(d) Approximately when will the news be spreading at the rate of 600 people per day?

(e) When will the news be spreading at the greatest rate?

(f) Use (9) and (10) to determine the differential equation satisfied by $f(t)$.

(g) At what rate will the news be spreading when half of the potential audience has heard the news?

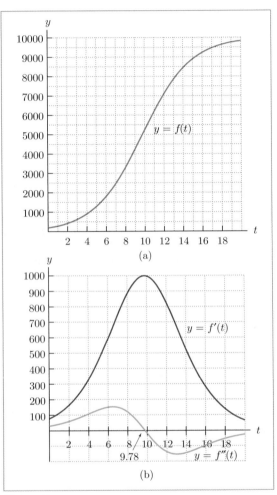

**Figure 7**

12. (*Spread of News*) A news item is broadcast by mass media to a potential audience of 50,000 people. After $t$ days,

$$f(t) = 50,000(1 - e^{-.3t})$$

people will have heard the news. The graph of this function is shown in Fig. 8.

(a) How many people will have heard the news after 10 days?

(b) At what rate is the news spreading initially?

(c) When will 22,500 people have heard the news?

(d) Approximately when will the news be spreading at the rate of 2500 people per day?

(e) Use equations (3) and (4) to determine the differential equation satisfied by $f(t)$.

(f) At what rate will the news be spreading when half of the potential audience has heard the news?

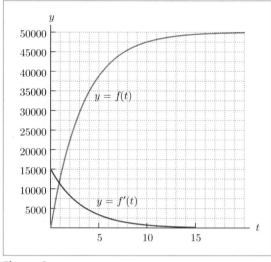

**Figure 8**

## Technology Exercises

**13.** After a drug is taken orally, the amount of the drug in the bloodstream after $t$ hours is $f(t) = 122(e^{-.2t} - e^{-t})$ units.

(a) Graph $f(t)$, $f'(t)$, and $f''(t)$ in the window $[0, 12]$ by $[-20, 75]$.

(b) How many units of the drug are in the bloodstream after 7 hours?

(c) At what rate is the level of drug in the bloodstream increasing after 1 hour?

(d) When, while the level is decreasing, is the level of the drug in the bloodstream 20 units?

(e) What is the greatest level of drug in the bloodstream and when is this level reached?

(f) When is the level of drug in the bloodstream decreasing the fastest?

**14.** A model incorporating growth restrictions for the number of bacteria in a culture after $t$ days is given by $f(t) = 5000(20 + te^{-.04t})$.

(a) Graph $f'(t)$ and $f''(t)$ in the window $[0, 100]$ by $[-700, 300]$.

(b) How fast is the culture changing after 100 days?

(c) Approximately when is the culture growing at the rate of 76.6 bacteria per day?

(d) When is the size of the culture greatest?

(e) When is the size of the culture decreasing the fastest?

## SOLUTION TO PRACTICE PROBLEM 5.4

**1.** The difference between transmission of information via mass media and via word of mouth is that in the second case the rate of transmission depends not only on the number of people who have not yet received the information, but also on the number of people who know the information and therefore are capable of spreading it. Therefore, for group A, $f'(t) = k[P - f(t)]$, and for group B, $f'(t) = kf(t)[P - f(t)]$. Note that the spread of information by word of mouth follows the same pattern as the spread of an epidemic.

# REVIEW OF FUNDAMENTAL CONCEPTS

**1.** What differential equation is key to solving exponential growth and decay problems? State a result about the solution to this differential equation.

**2.** What is a growth constant? Decay constant?

**3.** What is meant by the half-life of a radioactive element?

**4.** Explain how carbon dating works.

**5.** State the formula for each of the following quantities:

(a) the compound amount of $P$ dollars in $t$ years at interest rate $r$ compounded continuously

(b) the present value of $A$ dollars in $n$ years at interest rate $r$ compounded continuously

**6.** What is the difference between a relative rate of change and a percentage rate of change?

**7.** Define the elasticity of demand, $E(p)$, for a demand function. How is $E(p)$ used?

**8.** Describe an application of the differential equation $y' = k(M - y)$.

**9.** Describe an application of the differential equation $y' = ky(M - y)$.

# SUPPLEMENTARY EXERCISES

1. The atmospheric pressure $P(x)$ (measured in inches of mercury) at height $x$ miles above sea level satisfies the differential equation $P'(x) = -.2P(x)$. Find the formula for $P(x)$ if the atmospheric pressure at sea level is 29.92.

2. The herring gull population in North America has been doubling every 13 years since 1900. Give a differential equation satisfied by $P(t)$, the population $t$ years after 1900.

3. Find the present value of $10,000 payable at the end of 5 years if money can be invested at 12% with interest compounded continuously.

4. One thousand dollars is deposited in a savings account at 10% interest compounded continuously. How many years are required for the balance in the account to reach $3000?

5. The half-life of the radioactive element tritium is 12 years. Find its decay constant.

6. A piece of charcoal found at Stonehenge contained 63% of the level of $^{14}C$ found in living trees. Approximately how old is the charcoal?

7. From January 1, 1990 to January 1, 1997, the population of Texas grew from 17 million to 19.3 million.

   (a) Give the formula for the population $t$ years after 1990.

   (b) If this growth continues, how large will the population be in 2000?

   (c) In what year will the population reach 25 million?

8. A stock portfolio increased in value from $100,000 to $117,000 in 2 years. What rate of interest, compounded continuously, did this investment earn?

9. An investor initially invests $10,000 in a speculative venture. Suppose that the investment earns 20% interest compounded continuously for 5 years and then 6% interest compounded continuously for 5 years thereafter.

   (a) How much does the $10,000 grow to after 10 years?

   (b) Suppose that the investor has the alternative of an investment paying 14% interest compounded continuously. Which investment is superior over a 10-year period, and by how much?

10. Two different bacteria colonies are growing near a pool of stagnant water. Suppose that the first colony initially has 1000 bacteria and doubles every 21 minutes. The second colony has 710,000 bacteria and doubles every 33 minutes. How much time will elapse before the first colony becomes as large as the second?

11. The population of a city $t$ years after 1990 satisfies the differential equation $y' = .02y$. What is the growth constant? How fast will the population be growing when the population reaches 3 million people? At what level

of population will the population be growing at the rate of 100,000 people per year?

12. A colony of bacteria is growing exponentially with growth constant .4, with time measured in hours. Determine the size of the colony when the colony is growing at the rate of 200,000 bacteria per hour. Determine the rate at which the colony will be growing when its size is 1,000,000.

13. The population of a certain country is growing exponentially. The total population (in millions) in $t$ years is given by the function $P(t)$. Match each of the following answers with its corresponding question.

   *Answers:*

   **a.** Solve $P(t) = 2$ for $t$.

   **b.** $P(2)$

   **c.** $P'(2)$

   **d.** Solve $P'(t) = 2$ for $t$.

   **e.** $y' = ky$

   **f.** Solve $P(t) = 2P(0)$ for $t$.

   **g.** $P_0 e^{kt}$, $k > 0$

   **h.** $P(0)$

   *Questions:*

   **A.** How fast will the population be growing in 2 years?

   **B.** Give the general form of the function $P(t)$.

   **C.** How long will it take for the current population to double?

   **D.** What will be the size of the population in 2 years?

   **E.** What is the initial size of the population?

   **F.** When will the size of the population be 2 million?

   **G.** When will the population be growing at the rate of 2 million people per year?

   **H.** Give a differential equation satisfied by $P(t)$.

14. Suppose you have 80 grams of a certain radioactive material and the amount remaining after $t$ years is given by the function $f(t)$ shown in Fig. 1.

   (a) How much will remain after 5 years?

   (b) When will 10 grams remain?

   (c) What is the half-life of this radioactive material?

   (d) At what rate will the radioactive material be disintegrating after 1 year?

   (e) After how many years will the radioactive material be disintegrating at the rate of about 5 grams per year?

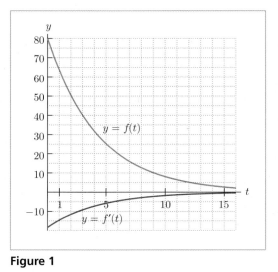

**Figure 1**

**15.** A few years after money is deposited into the bank, the compound amount is $1000 and it is growing at the rate of $60 per year. What interest rate (compounded continuously) is the money earning?

**16.** The current balance in a savings account is $1230 and the interest rate is 4.5%. At what rate is the compound amount currently growing?

**17.** Find the percentage rate of change of the function $f(t) = 50e^{.2t^2}$ at $t = 10$.

**18.** Find $E(p)$ for the demand function $q = 4000 - 40p^2$, and determine if demand is elastic or inelastic at $p = 5$.

**19.** Suppose that for a certain demand function, $E(8) = 1.5$. If the price is increased to $8.16, estimate the percentage decrease in the quantity demanded. Will the revenue increase or decrease?

**20.** Find the percentage rate of change of the function $f(p) = \dfrac{1}{3p + 1}$ at $p = 1$.

**21.** A company can sell $q = 1000p^2 e^{-.02(p+5)}$ calculators at a price of $p$ dollars per calculator. The current price is $200. If the price is decreased, will the revenue increase or decrease?

**22.** Consider a demand function of the form $q = ae^{-bp}$, where $a$ and $b$ are positive numbers. Find $E(p)$ and show that the elasticity equals 1 when $p = 1/b$.

**23.** Refer to Practice Problem 5.4. Out of 100 doctors in group A, none knew about the drug at time $t = 0$, but 66 of them were familiar with the drug after 13 months. Find the formula for $f(t)$.

**24.** The growth of the yellow nutsedge weed is described by a formula $f(t)$ of type (9) in Section 5.4. A typical weed has length 8 centimeters after 9 days, length 48 centimeters after 25 days, and reaches length 55 centimeters at maturity. Find the formula for $f(t)$.

**25.** When a rod of molten steel with a temperature of $1800°$F is placed in a large vat of water at temperature $60°$F, the temperature of the rod after $t$ seconds is

$$f(t) = 60(1 + 29e^{-.15t})$$

degrees Fahrenheit. The graph of this function is shown in Fig. 2.

**(a)** What is the temperature of the rod after 11 seconds?

**(b)** At what rate is the temperature of the rod changing after 6 seconds?

**(c)** Approximately when is the temperature of the rod 200 degrees?

**(d)** Approximately when is the rod cooling at the rate of 200 degrees per second?

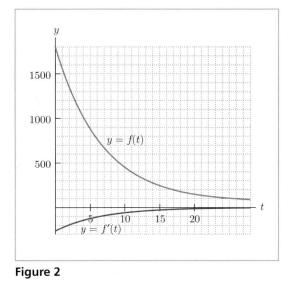

**Figure 2**

# The Definite Integral

THERE are two fundamental problems of calculus. The first is to find the slope of a curve at a point, and the second is to find the area of a region under a curve. These problems are quite simple when the curve is a straight line, as in Fig. 1. Both the slope of the line and the area of the shaded trapezoid can be calculated by geometric principles. When the graph consists of several line segments, as in Fig. 2, the slope of each line segment can be computed separately, and the area of the region can be found by adding the areas of the regions under each line segment.

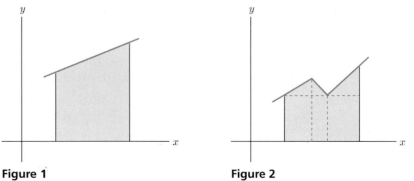

**Figure 1**          **Figure 2**

Calculus is needed when the curves are not straight lines. We have seen that the slope problem is resolved with the derivative of a function. In this chapter, we describe how the area problem is connected with the notion of the "integral" of a function. Both the slope problem and the area problem were studied by the ancient Greeks and solved in special cases. But it was not until the development of calculus in the seventeenth century that the intimate connection between the two problems was discovered. In this chapter we will discuss this connection as stated in the fundamental theorem of calculus.

## 6.1 Antidifferentiation

We have developed several techniques for calculating the derivative $F'(x)$ of a function $F(x)$. In many applications, however, it is necessary to proceed in reverse. We are given the derivative $F'(x)$ and must determine the function $F(x)$. The process of determining $F(x)$ from $F'(x)$ is called *antidifferentiation*. The next example gives a typical application involving antidifferentiation.

**EXAMPLE 1** **Federal health expenditures** Based on data from the Centers for Medicare and Medicaid Services, during the late 1990s and early 2000s, the federal health expenditures grew at an exponential rate with a growth constant of about .12. At the beginning of 2000, the rate was about 380 billion dollars per year. Let $R(t)$ denote the rate (in billions of dollars per year) of health expenditures at time $t$, where $t$ is the number of years since the beginning of 2000. Then a reasonable model for $R(t)$ is given by

$$R(t) = 380e^{.12t}. \tag{1}$$

Assuming that the federal health expenditures will continue to grow at this rate, use the formula for $R(t)$ to determine the total amount of federal health expenditures from 2000 to 2010.

**Solution** Let $T(t)$ denote the federal health expenditures from time 0 (2000) until time $t$. We wish to calculate $T(10)$, the health expenditures from 2000 to 2010. We do this by first determining a formula for $T(t)$. Since $T(t)$ is the health expenditures, the derivative, $T'(t)$, is the *rate* of health expenditures, namely, $R(t)$. Thus, although we do not yet have a formula for $T(t)$, we do know that

$$T'(t) = R(t) = 380e^{.12t}.$$

Thus the problem of determining a formula for $T(t)$ has been reduced to a problem of antidifferentiation: Find a function whose derivative is $R(t)$. We shall solve this particular problem after developing some techniques for solving antidifferentiation problems in general. ∎

**Antiderivative** Suppose that $f(x)$ is a given function and $F(x)$ is a function having $f(x)$ as its derivative—that is, $F'(x) = f(x)$. We call $F(x)$ an *antiderivative* of $f(x)$.

**EXAMPLE 2** Find an antiderivative of $f(x) = x^2$.

**Solution** The derivative of $x^3$ is $3x^2$, which is almost the same as $x^2$ except for a factor of 3.

To make this factor 1 instead of 3, take $F(x) = \frac{1}{3}x^3$. Then

$$F'(x) = \underbrace{\frac{d}{dx}\left(\frac{1}{3}x^3\right) = \frac{1}{3}\left(\frac{d}{dx}x^3\right)}_{\text{constant multiple rule}} = \frac{1}{3} \cdot 3x^2 = x^2.$$

So $F(x)$ is an antiderivative of $x^2$. Another antiderivative is $\frac{1}{3}x^3 + 5$, because the derivative of a constant function is zero:

$$\frac{d}{dx}\left(\frac{1}{3}x^3 + 5\right) = \frac{d}{dx}\left(\frac{1}{3}x^3\right) + \frac{d}{dx}(5) = x^2 + 0 = x^2.$$

In fact, if $C$ is any constant, the function $F(x) = \frac{1}{3}x^3 + C$ is also an antiderivative of $x^2$, since

$$\frac{d}{dx}\left(\frac{1}{3}x^3 + C\right) = \frac{1}{3} \cdot 3x^2 + 0 = x^2.$$

(The derivative of a constant function is zero.)　■

---

**EXAMPLE 3**　Find an antiderivative of $f(x) = e^{-2x}$.

**Solution**　Recall that the derivative of $e^{rx}$ is just a constant times $e^{rx}$. For an antiderivative of $e^{-2x}$, try a function of the form $ke^{-2x}$, where $k$ is some constant to be determined. Then

$$\frac{d}{dx}\,ke^{-2x} = k \cdot (-2e^{-2x}) = -2ke^{-2x}.$$

Choose $k$ to make $-2k = 1$; that is, choose $k = -\frac{1}{2}$. Then

$$\frac{d}{dx}\left(-\frac{1}{2}e^{-2x}\right) = \left(-\frac{1}{2}\right)(-2e^{-2x}) = 1 \cdot e^{-2x} = e^{-2x}.$$

Thus $-\frac{1}{2}e^{-2x}$ is an antiderivative of $e^{-2x}$. Also, for any constant $C$, the function $-\frac{1}{2}e^{-2x} + C$ is an antiderivative of $e^{-2x}$, because

$$\frac{d}{dx}\left(-\frac{1}{2}e^{-2x} + C\right) = e^{-2x} + 0 = e^{-2x}.$$　■

Examples 2 and 3 illustrate the fact that if $F(x)$ is an antiderivative of $f(x)$, then so is $F(x) + C$, where $C$ is any constant. (The derivative of the constant function $C$ is zero.) The next theorem says that *all* antiderivatives of $f(x)$ can be produced in this way.

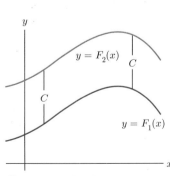

---

> **Theorem I**　Suppose that $f(x)$ is a continuous function on an interval $I$. If $F_1(x)$ and $F_2(x)$ are two antiderivatives of $f(x)$, then $F_1(x)$ and $F_2(x)$ differ by a constant on $I$. In other words, there is a constant $C$ such that
> $$F_2(x) = F_1(x) + C \quad \text{for all } x \text{ in } I.$$

---

**Figure 1.** Two antiderivatives of the same function.

Geometrically, the graph of any antiderivative $F_2(x)$ is obtained by shifting the graph of $F_1(x)$ vertically. See Fig. 1.

Our verification of Theorem I is based on the following fact, which is important in its own right.

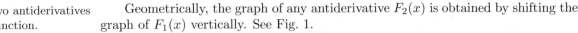

> **Theorem II** If $F'(x) = 0$ for all $x$ in an interval $I$, then there is a constant $C$ such that $F(x) = C$ for all $x$ in $I$.

It is easy to see why Theorem II is reasonable. (A formal proof of the theorem requires an important theoretical result called the mean value theorem.) If $F'(x) = 0$ for all $x$, then the curve $y = F(x)$ has slope equal to zero at every point. Thus the tangent line to $y = F(x)$ at any point is horizontal, which implies that the graph of $y = F(x)$ is a horizontal line. (Try to draw the graph of a function with a horizontal tangent everywhere. There is no choice but to keep your pencil moving on a constant, horizontal line!) If the horizontal line is $y = C$, then $F(x) = C$ for all $x$.

**Verification of Theorem I** If $F_1(x)$ and $F_2(x)$ are two antiderivatives of $f(x)$, then the function $F(x) = F_2(x) - F_1(x)$ has the derivative

$$F'(x) = F_2'(x) - F_1'(x) = f(x) - f(x) = 0.$$

So, by Theorem II, we know that $F(x) = C$ for some constant $C$. In other words, $F_2(x) - F_1(x) = C$, so that

$$F_2(x) = F_1(x) + C,$$

which is Theorem I.

Using Theorem I, we can find *all* antiderivatives of a given function once we know one antiderivative. For instance, since one antiderivative of $x^2$ is $\frac{1}{3}x^3$ (Example 2), all antiderivatives of $x^2$ have the form $\frac{1}{3}x^3 + C$, where $C$ is a constant.

> **Indefinite Integral** Suppose that $f(x)$ is a function whose antiderivatives are $F(x) + C$. The standard way to express this fact is to write
>
> $$\int f(x)\, dx = F(x) + C.$$
>
> The symbol $\int$ is called an *integral sign*. The entire notation $\int f(x)\, dx$ is called an *indefinite integral* and stands for antidifferentiation of the function $f(x)$.

We always record the variable of interest by prefacing it by the letter $d$. For example, if the variable of interest is $t$ rather than $x$, then we write $\int f(t)\, dt$ for the antiderivative.

---

**EXAMPLE 4** Determine

(a) $\displaystyle\int x^r\, dx$, $r$ a constant $\neq -1$      (b) $\displaystyle\int e^{kx}\, dx$, $k$ a constant $\neq 0$

**Solution** (a) By the constant-multiple and power rules,

$$\frac{d}{dx}\left(\frac{1}{r+1}x^{r+1}\right) = \frac{1}{r+1} \cdot \frac{d}{dx}x^{r+1} = \frac{1}{r+1} \cdot (r+1)x^r = x^r.$$

Thus $x^{r+1}/(r+1)$ is an antiderivative of $x^r$. Letting $C$ represent any constant, we have

$$\int x^r \, dx = \frac{1}{r+1} x^{r+1} + C, \qquad r \neq -1. \tag{2}$$

(b) An antiderivative of $e^{kx}$ is $e^{kx}/k$, since

$$\frac{d}{dx}\left(\frac{1}{k}e^{kx}\right) = \frac{1}{k} \cdot \frac{d}{dx}e^{kx} = \frac{1}{k}(ke^{kx}) = e^{kx}.$$

Hence

$$\int e^{kx} \, dx = \frac{1}{k}e^{kx} + C, \qquad k \neq 0. \tag{3}$$

**EXAMPLE 5**    Determine

(a) $\displaystyle\int \sqrt{x} \, dx$    (b) $\displaystyle\int \frac{1}{x^2} \, dx$

**Solution**    Both integrals follow by applying formula (2). For (a), take $r = 1/2$ in (2), then

$$\int x^{\frac{1}{2}} \, dx = \frac{1}{\frac{1}{2}+1} x^{\frac{1}{2}+1} + C = \frac{1}{3/2} x^{\frac{3}{2}} + C = \frac{2}{3} x^{\frac{3}{2}} + C.$$

For (b), take $r = -2$ in (2), then

$$\int x^{-2} \, dx = \frac{1}{-2+1} x^{-2+1} + C = -x^{-1} + C = -\frac{1}{x} + C.$$

There is a subtle point to be made concerning the integral of a discontinuous function. Consider the integral of $f(x) = \frac{1}{x^2}$ from Example 5(b). Since $f(x)$ is not defined at $x = 0$ and is continuous everywhere else, Theorem I tells us only that the antiderivatives of $f(x)$ are $-1/x + C$ on any interval that does not contain 0. So to express the antiderivatives of $f(x)$, we exclude 0 and write

$$\int \frac{1}{x^2} \, dx = \begin{cases} -\frac{1}{x} + C_1 & \text{if } x > 0, \\ -\frac{1}{x} + C_2 & \text{if } x < 0. \end{cases}$$

Note that, since we have two separate intervals ($x < 0$ and $x > 0$), we need two separate constants of integration, $C_1$ and $C_2$. For notational convenience, however, we will continue to use only one constant of integration, as we did in Example 5(b).

Formula (2) does not give an antiderivative of $x^{-1}$ because $1/(r+1)$ is undefined for $r = -1$. However, we know that for $x \neq 0$, the derivative of $\ln |x|$ is $1/x$. Hence $\ln |x|$ is an antiderivative of $1/x$, and we have

$$\int \frac{1}{x} \, dx = \ln |x| + C, \qquad x \neq 0. \tag{4}$$

Formula (4) supplies the missing case ($r = -1$) of formula (2). Because $1/x$ is not defined at $x = 0$, as we explained following Example 5, we should distinguish two intervals for the validity of formula (4) and interpret it to mean

$$\int \frac{1}{x} \, dx = \begin{cases} \ln x + C_1 & \text{if } x > 0, \\ \ln(-x) + C_2 & \text{if } x < 0. \end{cases}$$

Formulas (2), (3), and (4) are each obtained by "reversing" a familiar differentiation rule. In a similar fashion, one may use the sum rule and constant-multiple rule for derivatives to obtain corresponding rules for antiderivatives:

$$\int [f(x) + g(x)]\, dx = \int f(x)\, dx + \int g(x)\, dx \tag{5}$$

$$\int k f(x)\, dx = k \int f(x)\, dx, \quad k \text{ a constant.} \tag{6}$$

In words, (5) says that a sum of functions may be antidifferentiated term by term, and (6) says that a constant multiple may be moved through the integral sign.

**EXAMPLE 6** Compute:

$$\int \left( x^{-3} + 7e^{5x} + \frac{4}{x} \right) dx.$$

**Solution** Using the preceding rules, we have

$$\int \left( x^{-3} + 7e^{5x} + \frac{4}{x} \right) dx = \int x^{-3}\, dx + \int 7e^{5x}\, dx + \int \frac{4}{x}\, dx$$

$$= \int x^{-3}\, dx + 7 \int e^{5x}\, dx + 4 \int \frac{1}{x}\, dx$$

$$= \frac{1}{-2} x^{-2} + 7 \left( \frac{1}{5} e^{5x} \right) + 4 \ln |x| + C$$

$$= -\frac{1}{2} x^{-2} + \frac{7}{5} e^{5x} + 4 \ln |x| + C. \qquad \blacksquare$$

After some practice, most of the intermediate steps shown in the solution of Example 6 can be omitted.

A function has infinitely many different antiderivatives, corresponding to various choices of the constant $C$. In applications, it is often necessary to satisfy an additional condition, which then determines a specific value of $C$.

**EXAMPLE 7** Find the function $f(x)$ for which $f'(x) = x^2 - 2$ and $f(1) = \frac{4}{3}$.

**Solution** The unknown function $f(x)$ is an antiderivative of $x^2 - 2$. One antiderivative of $x^2 - 2$ is $\frac{1}{3}x^3 - 2x$. Therefore, by Theorem I,

$$f(x) = \frac{1}{3}x^3 - 2x + C, \quad C \text{ a constant.}$$

Figure 2 shows the graphs of $f(x)$ for several choices of $C$. We want the function whose graph passes through $(1, \frac{4}{3})$. To find the value of $C$ that makes $f(1) = \frac{4}{3}$, we set

$$\tfrac{4}{3} = f(1) = \tfrac{1}{3}(1)^3 - 2(1) + C = -\tfrac{5}{3} + C$$

and find $C = \frac{4}{3} + \frac{5}{3} = 3$. Therefore, $f(x) = \frac{1}{3}x^3 - 2x + 3$. $\qquad \blacksquare$

Having introduced the basics of antidifferentiation, let us now solve the health expenditure problem in Example 1.

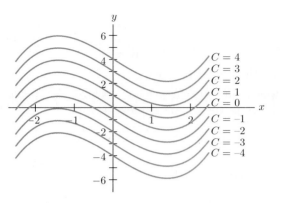

**Figure 2.** Several antiderivatives of $x^2 - 2$.

**Solution of Example 1 (Continued)** The rate of federal health expenditures at time $t$ is $R(t) = 380e^{.12t}$ billion dollars per year. Moreover, we observed that the health expenditures $T(t)$, from time 0 to time $t$, is an antiderivative of $R(t)$. Using (3) and (6), we have

$$T(t) = \int 380e^{.12t}\, dt = \frac{380}{.12}e^{.12t} + C = 3166.67e^{.12t} + C,$$

where $C$ is a constant. However, in our particular example, $T(0) = 0$, since $T(0)$ represents the expenditures from time 0 to time 0. Therefore, the constant $C$ must satisfy

$$0 = T(0) = 3166.67e^{.12(0)} + C = 3166.67 + C$$
$$C = -3166.67.$$

Thus,

$$T(t) = 3166.67e^{.12t} - 3166.67 = 3166.67(e^{.12t} - 1),$$

and so

$$T(10) = 3166.67(e^{.12(10)} - 1) \approx 7347 \text{ billion dollars.}$$

Hence the federal health expenditures from 2000 to 2010 will exceed 7 trillion dollars. ■

**EXAMPLE 8** **Position of a rocket** A rocket is fired vertically into the air. Its velocity at $t$ seconds after lift-off is $v(t) = 6t + .5$ meters per second. Before launch, the top of the rocket is 8 meters above the launch pad. Find the height of the rocket (measured from the top of the rocket to the launch pad) at time $t$.

**Solution** If $s(t)$ denotes the height of the rocket at time $t$, then $s'(t)$ is the rate at which the height is changing. That is, $s'(t) = v(t)$, and therefore $s(t)$ is an antiderivative of $v(t)$. Thus

$$s(t) = \int v(t)\, dt = \int (6t + .5)\, dt = 3t^2 + .5t + C,$$

where $C$ is a constant. When $t = 0$, the rocket's height is 8 meters. That is, $s(0) = 8$ and

$$8 = s(0) = 3(0)^2 + .5(0) + C = C.$$

Thus $C = 8$ and

$$s(t) = 3t^2 + .5t + 8.$$

■

**EXAMPLE 9** **Cost function** A company's marginal cost function is $C'(x) = .015x^2 - 2x + 80$ dollars, where $x$ denotes the number of units produced in one day. The company has fixed costs of $1000 per day.

(a) Find the cost of producing $x$ units per day.

(b) Suppose the current production level is $x = 30$. Determine the amount costs will rise if the production level is raised to $x = 60$ units.

**Solution** (a) Let $C(x)$ be the cost of producing $x$ units in one day. The derivative $C'(x)$ is the marginal cost. In other words, $C(x)$ is an antiderivative of the marginal cost function. Thus

$$C(x) = \int (.015x^2 - 2x + 80)\, dx = .005x^3 - x^2 + 80x + C.$$

The \$1000 fixed costs are the costs incurred when producing 0 units. That is, $C(0) = 1000$. So

$$1000 = C(0) = .005(0)^3 - (0)^2 + 80(0) + C.$$

Therefore, $C = 1000$, and

$$C(x) = .005x^3 - x^2 + 80x + 1000.$$

(b) The cost when $x = 30$ is $C(30)$, and the cost when $x = 60$ is $C(60)$. So the *increase* in cost when production is raised from $x = 30$ to $x = 60$ is $C(60) - C(30)$. We compute

$$C(60) = .005(60)^3 - (60)^2 + 80(60) + 1000 = 3280$$
$$C(30) = .005(30)^3 - (30)^2 + 80(30) + 1000 = 2635.$$

Thus the increase in cost is $3280 - 2635 = \$645$. ■

**Figure 3**

**Incorporating Technology** The graph of an antiderivative of a function can be obtained without actually antidifferentiating the function. See Appendices A–D for details. Figure 3 shows the graph of the solution to Example 6.

---

## PRACTICE PROBLEMS 6.1

**1.** Determine each of the following.

**(a)** $\displaystyle\int t^{7/2}\, dt$  **(b)** $\displaystyle\int \left(\frac{x^3}{3} + \frac{3}{x^3} + \frac{3}{x}\right) dx$

**2.** Find the value of $k$ that makes the following antidifferentiation formula true.

$$\int (1 - 2x)^3\, dx = k(1 - 2x)^4 + C.$$

---

## EXERCISES 6.1

*Find all antiderivatives of each of the following functions.*

**1.** $f(x) = x$  **2.** $f(x) = 9x^8$  **3.** $f(x) = e^{3x}$

**4.** $f(x) = e^{-3x}$  **5.** $f(x) = 3$  **6.** $f(x) = -4x$

*Determine the following.*

**7.** $\displaystyle\int 4x^3\, dx$  **8.** $\displaystyle\int \frac{x}{3}\, dx$  **9.** $\displaystyle\int 7\, dx$

**10.** $\displaystyle\int k^2\, dx$ ($k$ a constant)

**11.** $\displaystyle\int \frac{x}{c}\, dx$ ($c$ a constant $\neq 0$)

**12.** $\displaystyle\int x \cdot x^2\, dx$  **13.** $\displaystyle\int \left(\frac{2}{x} + \frac{x}{2}\right) dx$

**14.** $\displaystyle\int \frac{1}{7x}\, dx$  **15.** $\displaystyle\int x\sqrt{x}\, dx$

**16.** $\int \left( \frac{2}{\sqrt{x}} + 2\sqrt{x} \right) dx$    **17.** $\int \left( x - 2x^2 + \frac{1}{3x} \right) dx$

**18.** $\int \left( \frac{7}{2x^3} - \sqrt[3]{x} \right) dx$    **19.** $\int 3e^{-2x} \, dx$

**20.** $\int e^{-x} \, dx$    **21.** $\int e \, dx$

**22.** $\int \frac{1}{e^x} \, dx$    **23.** $\int -2(e^{2x} + 1) \, dx$

**24.** $\int \frac{7}{2e^{2x}} \, dx$

*In Exercises 25–36, find the value of k that makes the antidifferentiation formula true. (Note: You can check your answer without looking in the answer section. How?)*

**25.** $\int 5e^{-2t} \, dt = ke^{-2t} + C$

**26.** $\int 3e^{t/10} \, dt = ke^{t/10} + C$

**27.** $\int 2e^{4x-1} \, dx = ke^{4x-1} + C$

**28.** $\int \frac{4}{e^{3x+1}} \, dx = \frac{k}{e^{3x+1}} + C$

**29.** $\int (5x - 7)^{-2} \, dx = k(5x - 7)^{-1} + C$

**30.** $\int \sqrt{x + 1} \, dx = k(x + 1)^{3/2} + C$

**31.** $\int (4 - x)^{-1} \, dx = k \ln |4 - x| + C$

**32.** $\int \frac{7}{(8 - x)^4} \, dx = \frac{k}{(8 - x)^3} + C$

**33.** $\int (3x + 2)^4 \, dx = k(3x + 2)^5 + C$

**34.** $\int (2x - 1)^3 \, dx = k(2x - 1)^4 + C$

**35.** $\int \frac{3}{2 + x} \, dx = k \ln |2 + x| + C$

**36.** $\int \frac{5}{2 - 3x} \, dx = k \ln |2 - 3x| + C$

*Find all functions f(t) with the following property.*

**37.** $f'(t) = t^{3/2}$    **38.** $f'(t) = \frac{4}{6 + t}$

**39.** $f'(t) = 0$    **40.** $f'(t) = t^2 - 5t - 7$

*Find all functions f(x) with the following properties.*

**41.** $f'(x) = .5e^{-.2x}$, $f(0) = 0$

**42.** $f'(x) = 2x - e^{-x}$, $f(0) = 1$

**43.** $f'(x) = x$, $f(0) = 3$

**44.** $f'(x) = 8x^{1/3}$, $f(1) = 4$

**45.** $f'(x) = \sqrt{x} + 1$, $f(4) = 0$

**46.** $f'(x) = x^2 + \sqrt{x}$, $f(1) = 3$

**47.** Figure 4 shows the graphs of several functions $f(x)$ for which $f'(x) = \frac{2}{x}$. Find the expression for the function $f(x)$ whose graph passes through $(1, 2)$.

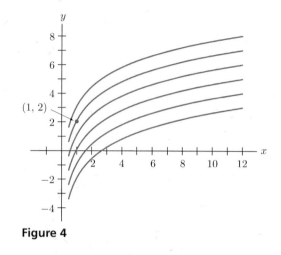

**Figure 4**

**48.** Figure 5 shows the graphs of several functions $f(x)$ for which $f'(x) = \frac{1}{3}$. Find the expression for the function $f(x)$ whose graph passes through $(6, 3)$.

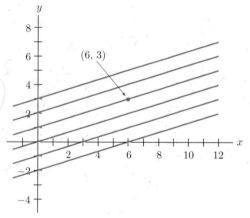

**Figure 5**

**49.** Which of the following is $\int \ln x \, dx$?

(a) $\frac{1}{x} + C$

(b) $x \cdot \ln x - x + C$

(c) $\frac{1}{2} \cdot (\ln x)^2 + C$

**50.** Which of the following is $\int x\sqrt{x + 1} \, dx$?

(a) $\frac{2}{5}(x + 1)^{5/2} - \frac{2}{3}(x + 1)^{3/2} + C$

(b) $\frac{1}{2}x^2 \cdot \frac{2}{3}(x + 1)^{3/2} + C$

**51.** Figure 6 contains the graph of a function $F(x)$. On the same coordinate system, draw the graph of the function $G(x)$ having the properties $G(0) = 0$ and $G'(x) = F'(x)$ for each $x$.

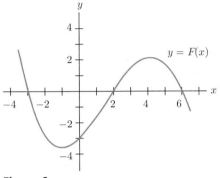

**Figure 6**

**52.** Figure 7 contains an antiderivative of the function $f(x)$. Draw the graph of another antiderivative of $f(x)$.

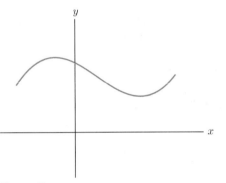

**Figure 7**

**53.** The function $g(x)$ in Fig. 8 was obtained by shifting the graph of $f(x)$ up three units. If $f'(5) = \frac{1}{4}$, what is $g'(5)$?

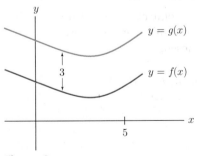

**Figure 8**

**54.** The function $g(x)$ in Fig. 9 was obtained by shifting the graph of $f(x)$ up two units. What is the derivative of $h(x) = g(x) - f(x)$?

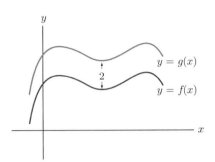

**Figure 9**

**55.** (*Position of a Ball*) A ball is thrown upward from a height of 256 feet above the ground, with an initial velocity of 96 feet per second. From physics it is known that the velocity at time $t$ is $v(t) = 96 - 32t$ feet per second.

    **(a)** Find $s(t)$, the function giving the height of the ball at time $t$.

    **(b)** How long will the ball take to reach the ground?

    **(c)** How high will the ball go?

**56.** A rock is dropped from the top of a 400-foot cliff. Its velocity at time $t$ seconds is $v(t) = -32t$ feet per second.

    **(a)** Find $s(t)$, the height of the rock above the ground at time $t$.

    **(b)** How long will the rock take to reach the ground?

    **(c)** What will be its velocity when it hits the ground?

**57.** Let $P(t)$ be the total output of a factory assembly line after $t$ hours of work. Suppose that the rate of production at time $t$ is $P'(t) = 60 + 2t - \frac{1}{4}t^2$ units per hour. Find the formula for $P(t)$.

**58.** After $t$ hours of operation a coal mine is producing coal at the rate of $C'(t) = 40 + 2t - \frac{1}{5}t^2$ tons of coal per hour. Find a formula for the total output of the coal mine after $t$ hours of operation.

**59.** (*Heat Diffusion*) A package of frozen strawberries is taken from a freezer at $-5°C$ into a room at $20°C$. At time $t$ the average temperature of the strawberries is increasing at the rate of $T'(t) = 10e^{-.4t}$ degrees Celsius per hour. Find the temperature of the strawberries at time $t$.

**60.** (*Epidemic*) A flu epidemic hits a town. Let $P(t)$ be the number of persons sick with the flu at time $t$, where time is measured in days from the beginning of the epidemic and $P(0) = 100$. Suppose that after $t$ days the flu is spreading at the rate of $P'(t) = 120t - 3t^2$ people per day. Find the formula for $P(t)$.

**61.** (*Profit*) A small tie shop finds that at a sales level of $x$ ties per day its marginal profit is $MP(x)$ dollars per tie, where $MP(x) = 1.30 + .06x - .0018x^2$. Also, the shop will lose \$95 per day at a sales level of $x = 0$. Find profit from operating the shop at a sales level of $x$ ties per day.

62. (*Cost*) A soap manufacturer estimates that its marginal cost of producing soap powder is $C'(x) = .2x + 1$ hundred dollars per ton at a production level of $x$ tons per day. Fixed costs are \$200 per day. Find the cost of producing $x$ tons of soap powder per day.

63. (*U.S. Consumption of Iron Ore*) The United States has been consuming iron ore at the rate of $R(t)$ million metric tons per year at time $t$, where $t = 0$ corresponds to 1980 and $R(t) = 94e^{.016t}$. Find a formula for the total U.S. consumption of iron ore from 1980 until time $t$.

64. (*U.S. Production of Natural Gas*) Since 1987, the rate of production of natural gas in the United States has been approximately $R(t)$ quadrillion British thermal units per year at time $t$, with $t = 0$ corresponding to 1987 and $R(t) = 17.04e^{.016t}$. Find a formula for the total U.S. production of natural gas from 1987 until time $t$.

65. (*Cost*) Suppose the drilling of an oil well has a fixed cost of \$10,000 and a marginal cost of $C'(x) = 1000 + 50x$ dollars per foot, where $x$ is the depth in feet. Find the expression for $C(x)$, the total cost of drilling $x$ feet. [*Note:* $C(0) = 10,000$.]

### Technology Exercises

*In Exercises 66–69, find an antiderivative of $f(x)$—call it $F(x)$—and then compare the graphs of $F(x)$ and $f(x)$ in the given window to check that the expression for $F(x)$ is reasonable. [That is, determine whether the two graphs are consistent. When $F(x)$ has a relative extreme point, $f(x)$ should be zero; when $F(x)$ is increasing, $f(x)$ should be positive, and so on.]*

66. $f(x) = 2x - e^{-.02x}$, $[-10, 10]$ *by* $[-20, 100]$

67. $f(x) = e^{2x} + e^{-x} + \frac{1}{2}x^2$, $[-2.4, 1.7]$ *by* $[-10, 10]$

68. $f(x) = 8x^2 - 3e^{-x} + x^3$, $[-3, 2]$ *by* $[-4, 7]$

69. $f(x) = \frac{1}{10}(x^3 - 9x^2 + 5x) + 3$, $[-4, 11]$ *by* $[-14, 7]$

## SOLUTIONS TO PRACTICE PROBLEMS 6.1

1. (a) $\displaystyle\int t^{7/2}\, dt = \frac{1}{\frac{9}{2}}t^{9/2} + C = \frac{2}{9}t^{9/2} + C$

   (b) $\displaystyle\int \left( \frac{x^3}{3} + \frac{3}{x^3} + \frac{3}{x} \right) dx$

   $\displaystyle = \int \left( \frac{1}{3}\cdot x^3 + 3x^{-3} + 3\cdot\frac{1}{x} \right) dx$

   $\displaystyle = \frac{1}{3}\left(\frac{1}{4}x^4\right) + 3\left(-\frac{1}{2}x^{-2}\right) + 3\ln|x| + C$

   $\displaystyle = \frac{1}{12}x^4 - \frac{3}{2}x^{-2} + 3\ln|x| + C.$

2. Since we are told that the antiderivative has the general form $k(1 - 2x)^4$, all we have to do is determine the value of $k$. Differentiating, we obtain

$$4k(1 - 2x)^3(-2) \quad \text{or} \quad -8k(1 - 2x)^3,$$

which is supposed to equal $(1 - 2x)^3$. Therefore, $-8k = 1$, so $k = -\frac{1}{8}$.

## 6.2 Areas and Riemann Sums

This section and the next reveal the important connection between antiderivatives and areas of regions under curves. Although the full story will have to wait until the next section, we can give a hint now by describing how "area" is related to a problem solved in Section 6.1.

Example 9 in Section 6.1 concerned a company's marginal cost function, $f(x) = .015x^2 - 2x + 80$. A calculation with an antiderivative of $f(x)$ showed that if the production level is increased from $x = 30$ to $x = 60$ units per day, the change in total costs is \$645. As we'll see, this change in cost exactly equals the area of the region under the graph of the marginal cost curve in Fig. 1 from $x = 30$ to $x = 60$. First, however, we need to learn how to find areas of such regions.

**Area Under a Graph** If $f(x)$ is a continuous nonnegative function on the interval $a \leq x \leq b$, we refer to the area of the region shown in Fig. 2 as the *area under the graph of $f(x)$ from a to b*.

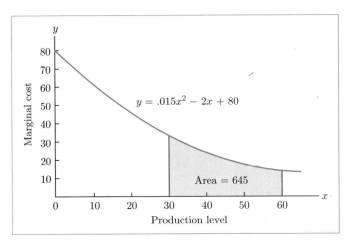

**Figure 1.** Area under a marginal cost curve.

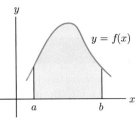

**Figure 2.** Area under a graph.

The computation of the area in Fig. 2 is not a trivial matter when the top boundary of the region is curved. However, we can *estimate* the area to any desired degree of accuracy. The basic idea is to construct rectangles whose total area is approximately the same as the area to be computed. The area of each rectangle, of course, is easy to compute.

Figure 3 shows three rectangular approximations to the area under a graph. When the rectangles are thin, the mismatch between the rectangles and the region under the graph is quite small. In general, a rectangular approximation can be made as close as desired to the exact area simply by making the width of the rectangles sufficiently small.

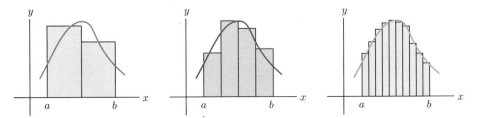

**Figure 3.** Approximating a region with rectangles.

Given a continuous nonnegative function $f(x)$ on the interval $a \leq x \leq b$, divide the $x$-axis interval into $n$ equal subintervals, where $n$ represents some positive integer. Such a subdivision is called a *partition* of the interval from $a$ to $b$. Since the entire interval is of width $b - a$, the width of each of the $n$ subintervals is $(b - a)/n$. For brevity, denote this width by $\Delta x$. That is,

$$\Delta x = \frac{b - a}{n} \quad \text{(width of one subinterval).}$$

In each subinterval, select a point. (Any point in the subinterval will do.) Let $x_1$ be the point selected from the first subinterval, $x_2$ the point from the second subinterval, and so on. These points are used to form rectangles that approximate the region under the graph of $f(x)$. Construct the first rectangle with height $f(x_1)$ and the first subinterval as base, as in Fig. 4. The top of the rectangle touches the graph directly above $x_1$. Notice that

[area of first rectangle] = [height][width] = $f(x_1) \, \Delta x$.

The second rectangle rests on the second subinterval and has height $f(x_2)$. Thus

[area of second rectangle] = [height][width] = $f(x_2) \, \Delta x$.

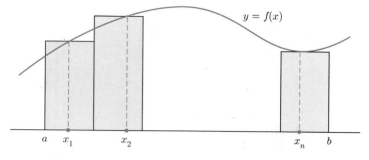

**Figure 4.**  Rectangles with heights $f(x_1), \ldots, f(x_n)$.

Continuing in this way, we construct $n$ rectangles with a combined area of

$$f(x_1)\,\Delta x + f(x_2)\,\Delta x + \cdots + f(x_n)\,\Delta x. \tag{1}$$

A sum as in (1) is called a *Riemann sum*. It provides an approximation to the area under the graph of $f(x)$ when $f(x)$ is nonnegative and continuous. In fact, as the number of subintervals increases indefinitely, the Riemann sums (1) approach a limiting value, the area under the graph.*

The value in (1) is easier to calculate when written as

$$[f(x_1) + f(x_2) + \cdots + f(x_n)]\,\Delta x.$$

This calculation requires only one multiplication.

---

**EXAMPLE 1**  Estimate the area under the graph of the marginal cost function $f(x) = .015x^2 - 2x + 80$ from $x = 30$ to $x = 60$. Use partitions of 5, 20, and 100 subintervals. Use the midpoints of the subintervals as $x_1, x_2, \ldots, x_n$ to construct the rectangles. See Fig. 5.

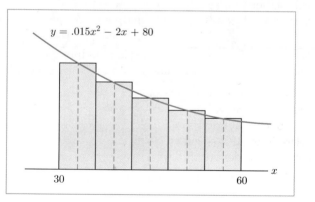

$$y = .015x^2 - 2x + 80$$

**Figure 5.**  Estimating the area under a marginal cost curve.

**Solution**  The partition of $30 \le x \le 60$ with $n = 5$ is shown in Fig. 6. The length of each subinterval is

$$\Delta x = \frac{60 - 30}{5} = 6.$$

---

*Riemann sums are named after the nineteenth-century German mathematician G. B. Riemann (pronounced "Reemahn"), who used them extensively in his work on calculus. The concept of a Riemann sum has several uses: to approximate areas under curves, to construct mathematical models in applied problems, and to give a formal definition of area. In the next section, Riemann sums are used to define the definite integral of a function.

**Figure 6.** A partition of the interval $30 \le x \le 60$.

Observe that the first midpoint is $\Delta x/2$ units from the left endpoint, and the midpoints themselves are $\Delta x$ units apart. The first midpoint is $x_1 = 30 + \Delta x/2 = 30 + 3 = 33$. Subsequent midpoints are found by successively adding $\Delta x = 6$:

$$\text{midpoints: } 33, 39, 45, 51, 57.$$

The corresponding estimate for the area under the graph of $f(x)$ is

$$f(33)\,\Delta x + f(39)\,\Delta x + f(45)\,\Delta x + f(51)\,\Delta x + f(57)\,\Delta x$$
$$= [f(33) + f(39) + f(45) + f(51) + f(57)]\,\Delta x$$
$$= [30.335 + 24.815 + 20.375 + 17.015 + 14.735]\cdot 6$$
$$= 107.275\cdot 6 = 643.65.$$

A similar calculation with 20 subintervals produces an area estimate of 644.916. With 100 subintervals the estimate is 644.997. ■

The approximations in Example 1 seem to confirm the claim made at the beginning of the section that the area under the marginal cost curve equals the change in total cost, \$645. The verification of this fact will be given in the next section.

Although the midpoints of subintervals are often selected as the $x_1, x_2, \ldots, x_n$ in a Riemann sum, left endpoints and right endpoints may also be convenient.

**EXAMPLE 2**   Use a Riemann sum with $n = 4$ to estimate the area under the graph of $f(x) = x^2$ from 1 to 3. Select the right endpoints of the subintervals as $x_1, x_2, x_3, x_4$.

**Solution**   Here $\Delta x = (3-1)/4 = .5$. The right endpoint of the first subinterval is $1 + \Delta x = 1.5$. Subsequent right endpoints are obtained by successively adding .5, as follows:

The corresponding Riemann sum is

$$f(x_1)\,\Delta x + f(x_2)\,\Delta x + f(x_3)\,\Delta x + f(x_4)\,\Delta x$$
$$= [f(x_1) + f(x_2) + f(x_3) + f(x_4)]\Delta x$$
$$= [(1.5)^2 + (2)^2 + (2.5)^2 + (3)^2](.5)$$
$$= [2.25 + 4 + 6.25 + 9](.5)$$
$$= 21.5\cdot(.5) = 10.75.$$

The rectangles used for this Riemann sum are shown in Fig. 7. The right endpoints here give an area estimate that is obviously greater than the exact area. Midpoints would work better. But if the rectangles are sufficiently narrow, even a Riemann sum using right endpoints will be close to the exact area. ■

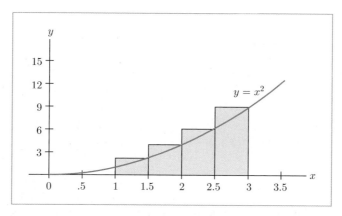

**Figure 7.** A Riemann sum using right endpoints.

---

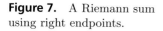

**EXAMPLE 3** To estimate the area of a 100-foot-wide waterfront lot, a surveyor measured the distance from the street to the waterline at 20-foot intervals, starting 10 feet from one corner of the lot. Use the data to construct a Riemann sum approximation to the area of the lot. See Fig. 8.

**Figure 8.** Survey of a waterfront property.

**Solution** Treat the street as the $x$-axis, and consider the waterline along the property as the graph of a function $f(x)$ over the interval from 0 to 100. The five "vertical" distances give $f(x_1), \ldots, f(x_5)$, where $x_1 = 10, \ldots, x_5 = 90$. Since there are five points $x_1, \ldots, x_5$ spread across the interval $0 \leq x \leq 100$, we partition the interval into five subintervals, with $\Delta x = \frac{100}{5} = 20$. Fortunately, each subinterval contains one $x_i$. (In fact, each $x_i$ is the midpoint of a subinterval.) Thus the area of the lot is approximated by the Riemann sum

$$
\begin{aligned}
f(x_1)\,\Delta x + \cdots + f(x_5)\,\Delta x &= [f(x_1) + \cdots + f(x_5)]\,\Delta x \\
&= [39 + 46 + 44 + 40 + 41.5] \cdot 20 \\
&= 210.5 \cdot 20 = 4210 \text{ square feet.}
\end{aligned}
$$

For a better estimate of the area, the surveyor will have to make more measurements from the street to the waterline. Note that we are able to approximate the area without ever knowing an analytic expression for the function $f(x)$. ■

The final example shows how Riemann sums arise in an application. Since the variable is time, $t$, we write $\Delta t$ in place of $\Delta x$.

---

**EXAMPLE 4** The velocity of a rocket at time $t$ is $v(t)$ feet per second. Construct a Riemann sum that estimates how far the rocket travels in the first 10 seconds. (*Note*: Don't calculate the Riemann sum. Just set it up.) What happens when the number of subintervals in the partition increases without bound?

**Solution** Partition the interval $0 \leq t \leq 10$ into $n$ subintervals of width $\Delta t$, and select points $t_1, t_2, \ldots, t_n$ from these subintervals. Although the rocket's velocity is not constant, it does not change much during a small subinterval of time. So we may use $v(t_1)$ as an approximation of the rocket's velocity during the first subinterval. Over this time subinterval of length $\Delta t = (10 - 0)/n$,

$$[\text{distance traveled}] \approx [\text{velocity}] \cdot [\text{time}]$$
$$= v(t_1) \, \Delta t. \qquad (2)$$

The distance traveled during the second time subinterval is approximately $v(t_2)\Delta t$, and so on. Thus an estimate for the total distance traveled is

$$v(t_1) \, \Delta t + v(t_2) \, \Delta t + \cdots + v(t_n) \, \Delta t. \qquad (3)$$

The sum in (3) is a Riemann sum for the velocity function on the interval $0 \leq t \leq 10$. As $n$ increases, such a Riemann sum approaches the area under the graph of the velocity function. However, from our derivation of (3), it seems reasonable that as $n$ increases, the sums become a better and better estimate of the total distance traveled. We conclude that

$$\left[ \begin{array}{c} \text{total distance rocket travels} \\ \text{during the first 10 seconds} \end{array} \right] = \left[ \begin{array}{c} \text{area under graph of} \\ v(t) \text{ over } 0 \leq t \leq 10 \end{array} \right].$$

See Fig. 9.

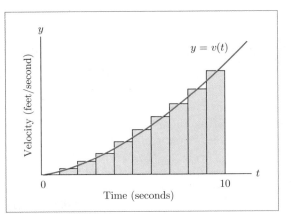

**Figure 9.** Area under a velocity curve.

It is helpful to look at Example 4 from a slightly different point of view. The height of the rocket is an increasing quantity and $v(t)$ is the *rate* of change of height at time $t$. The area under the graph of $v(t)$ from time $a = 0$ to time $b = 10$ is the *amount* of increase in the height during the first 10 seconds.

This connection between the rate of change of a function and the amount of increase of the function generalizes to a variety of situations.

> If a quantity is increasing, then the area under the rate of change function from $a$ to $b$ is the amount of increase in the quantity from $a$ to $b$.

Table 1 shows four instances of this principle. The result is justified in the next section. The first example in the table was discussed in Example 1.

| TABLE 1 | Interpretation of Areas | | |
|---|---|---|---|
| **Function** | | **a to b** | **Area under the Graph from a to b** |
| Marginal cost at production level $x$ | | 30 to 60 | Additional cost when production is increased from 30 to 60 units |
| Rate of sulfur emissions from a power plant $t$ years after 1990 | | 1 to 3 | Amount of sulfur released from 1991 to 1993 |
| Birth rate $t$ years after 1980 (in babies per year) | | 0 to 10 | Number of babies born from 1980 to 1990 |
| Rate of gas consumption $t$ years after 1985 | | 3 to 6 | Amount of gas used from 1988 to 1991 |

**Incorporating Technology**  When the points selected for a Riemann sum are all midpoints, all left endpoints, or all right endpoints, then $[f(x_1) + f(x_2) + \cdots + f(x_n)]$ is the sum of the sequence of values $f(x_1), f(x_2), \ldots, f(x_n)$ where successive numbers $x_1, x_2, \ldots, x_n$ each differ by the value $\Delta x$. If $Y_1$ represents the function $f(x)$, then this sum of a sequence can be evaluated on a graphing calculator as **sum(seq($Y_1$,X,$x_1$,$x_n$,$\Delta x$))**. Therefore, the entire Riemann sum can be evaluated as

**Figure 10.**
$Y_1 = .015X^2 - 2X + 80$
$Y_2 = X^2$

$$\text{sum(seq($Y_1$,X,$x_1$,$x_n$,$\Delta x$))} * \Delta x.$$

Figure 10 shows the computations for Examples 1 and 2. See Appendices A–D.

## PRACTICE PROBLEMS 6.2

1. Determine $\Delta x$ and the midpoints of the subintervals formed by partitioning the interval $-2 \leq x \leq 2$ into five subintervals.

2. The graph in Fig. 11 gives the rate at which new jobs were created (in millions of jobs per year) in the United States, where $t = 0$ corresponds to 1983. For instance, on January 1, 1986, jobs were being created at the rate of 2.4 million new jobs per year. Interpret the area of the shaded region.

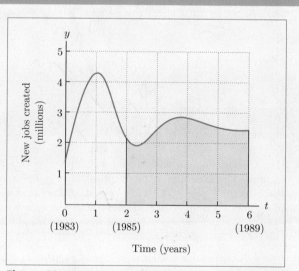

**Figure 11.**  Rate of job creation.

# EXERCISES 6.2

*Determine $\Delta x$ and the midpoints of the subintervals formed by partitioning the given interval into n subintervals. (Hint: Decimals are sometimes easier to use than fractions.)*

1. $0 \leq x \leq 2$; $n = 4$
2. $0 \leq x \leq 3$; $n = 6$
3. $1 \leq x \leq 4$; $n = 5$
4. $3 \leq x \leq 5$; $n = 5$

*In Exercises 5–10, use a Riemann sum to approximate the area under the graph of $f(x)$ on the given interval, with selected points as specified.*

5. $f(x) = x^2$; $1 \leq x \leq 3$, $n = 4$, midpoints of subintervals
6. $f(x) = x^2$; $-2 \leq x \leq 2$, $n = 4$, midpoints of subintervals

**7.** $f(x) = x^3$; $1 \le x \le 3$, $n = 5$, left endpoints

**8.** $f(x) = x^3$; $0 \le x \le 1$, $n = 5$, right endpoints

**9.** $f(x) = e^{-x}$; $2 \le x \le 3$, $n = 5$, right endpoints

**10.** $f(x) = \ln x$; $2 \le x \le 4$, $n = 5$, left endpoints

*In Exercises 11–14, use a Riemann sum to approximate the area under the graph of $f(x)$ in Fig. 12 on the given interval, with selected points as specified. Draw the approximating rectangles.*

**11.** $0 \le x \le 8$, $n = 4$, midpoints of subintervals

**12.** $3 \le x \le 7$, $n = 4$, left endpoints

**13.** $4 \le x \le 9$, $n = 5$, right endpoints

**14.** $1 \le x \le 7$, $n = 3$, midpoints of subintervals

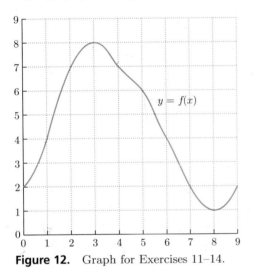

**Figure 12.**   Graph for Exercises 11–14.

**15.** Use a Riemann sum with $n = 4$ and left endpoints to estimate the area under the graph of $f(x) = 4 - x$ on the interval $1 \le x \le 4$. Then repeat with $n = 4$ and midpoints. Compare the answers with the exact answer, 4.5, which can be computed from the formula for the area of a triangle.

**16.** Use a Riemann sum with $n = 4$ and right endpoints to estimate the area under the graph of $f(x) = 2x - 4$ on the interval $2 \le x \le 3$. Then repeat with $n = 4$ and midpoints. Compare the answers with the exact answer, 1, which can be computed from the formula for the area of a triangle.

**17.** The graph of the function $f(x) = \sqrt{1 - x^2}$ on the interval $-1 \le x \le 1$ is a semicircle. The area under the graph is $\frac{1}{2}\pi(1)^2 = \pi/2 = 1.57080$, to five decimal places. Use a Riemann sum with $n = 5$ and midpoints to estimate the area under the graph. See Fig. 13. Carry out the calculations to five decimal places and compute the error (the difference between the estimate and 1.57080).

**Figure 13**

**18.** Use a Riemann sum with $n = 5$ and midpoints to estimate the area under the graph of $f(x) = \sqrt{1 - x^2}$ on the interval $0 \le x \le 1$. The graph is a quarter circle, and the area under the graph is .78540, to five decimal places. See Fig. 14. Carry out the calculations to five decimal places and compute the error. If you double the estimate from this exercise, will it be more accurate than the estimate computed in Exercise 17?

**Figure 14**

**19.** (*Airflow through the Lungs*) Lung physiologists measure the velocity of air passing through a patient's throat by having the patient breathe into a pneumotachograph. This machine produces a graph that plots airflow rate as a function of time. The graph in Fig. 15 shows the flow rate while a patient is breathing out. The area under the graph gives the total volume of air during exhalation. Estimate this volume with a Riemann sum. Use $n = 5$ and midpoints.

**Figure 15.**   Data from a pneumotachograph.

**20.** Estimate the area (in square feet) of the piece of land shown in Fig. 16.

**Figure 16.** Area of residential property.

**21.** (*Velocity*) The velocity of a car (in feet per second) is recorded from the speedometer every 10 seconds, beginning 5 seconds after the car starts to move. See Table 2. Use a Riemann sum to estimate the distance the car travels during the first 60 seconds. (*Note*: Each velocity is given at the *middle* of a 10-second interval. The first interval extends from 0 to 10, and so on.)

| TABLE 2 | A Car's Velocity | | | | | |
|---------|----|----|----|----|----|----|
| Time | 5 | 15 | 25 | 35 | 45 | 55 |
| Velocity | 20 | 44 | 32 | 39 | 65 | 80 |

**22.** Table 3 shows the velocity (in feet per second) at the end of each second for a person starting a morning jog. Make three Riemann sum estimates of the total distance jogged during the time from $t = 2$ to $t = 8$.

 (a) $n = 6$, left endpoints

 (b) $n = 6$, right endpoints

 (c) $n = 3$, midpoints

| TABLE 3 | A Jogger's Velocity | | | | | | | | |
|---------|---|---|---|---|---|---|---|---|---|
| Time | 0 | 1 | 2 | 3 | 4 | 5 | 6 | 7 | 8 |
| Velocity | 0 | 2 | 3 | 5 | 5 | 6 | 6 | 7 | 8 |

**23.** Complete the three missing entries in Table 4.

**24.** Interpret the area of the shaded region in Fig. 17.

**Figure 17.** Marginal revenue.

**25.** Interpret the area of the shaded region in Fig. 18.

**Figure 18.** Rate of soil erosion.

**26.** Interpret the area of the shaded region in Fig. 19.

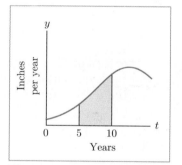

**Figure 19.** A child's growth rate.

*In Exercises 27 and 28, let A be the area under the graph of $y = f(x)$ from $x = a$ to $x = b$.*

**27.** Suppose that the function $f(x)$ is always increasing on the interval $a \leq x \leq b$. Explain why [approximation

| TABLE 4 | Interpretation of Areas | | |
|---------|-------------------------|--------|-------------------------|
| **Function** | | **a to b** | **Area under the Graph from a to b** |
| Rate of growth of a population $t$ years after 1900 (in millions of people per year) | | 10 to 50 | |
| | | 5 to 7 | Number of cigarettes smoked from 1990 to 1992 |
| Marginal profit at production level $x$ | | | Additional profit created by increasing production from 20 to 50 units |

given by the Riemann sum with left endpoints] $< A <$ [approximation given by the Riemann sum with right endpoints].

28. Suppose that the function $f(x)$ is always decreasing on the interval $a \leq x \leq b$. Explain why [approximation given by the Riemann sum with left endpoints] $> A >$ [approximation given by the Riemann sum with right endpoints].

## Technology Exercises

*Evaluate a Riemann sum to approximate the area under the graph of $f(x)$ on the given interval, with points selected as*

*specified.*

29. $f(x) = x\sqrt{1 + x^2}$; $1 \leq x \leq 3$, $n = 20$, midpoints of subintervals

30. $f(x) = \sqrt{1 - x^2}$; $-1 \leq x \leq 1$, $n = 20$, midpoints of subintervals

31. $f(x) = e^{-x^2}$; $-2 \leq x \leq 2$, $n = 100$, left endpoints

32. $f(x) = x \ln x$; $2 \leq x \leq 6$, $n = 100$, right endpoints

33. $f(x) = (\ln x)^2$; $2 \leq x \leq 3$, $n = 200$, right endpoints

34. $f(x) = (1 + x)^2 e^{-4x}$; $2 \leq x \leq 4$, $n = 200$, left endpoints

---

## SOLUTIONS TO PRACTICE PROBLEMS 6.2

1. Since $n = 5$, $\Delta x = \frac{2 - (-2)}{5} = \frac{4}{5} = .8$. The first midpoint is $x_1 = -2 + .8/2 = -1.6$. Subsequent midpoints are found by successively adding .8, to obtain

$x_2 = -.8$, $x_3 = 0$, $x_4 = .8$, and $x_5 = 1.6$.

2. The area under the curve from $t = 2$ to $t = 6$ is the number of new jobs created from 1985 to 1989.

---

## 6.3 Definite Integrals and the Fundamental Theorem

In Section 6.2 we saw that the area under the graph of a continuous nonnegative function $f(x)$ from $a$ to $b$ is the limiting value of Riemann sums of the form

$$f(x_1)\,\Delta x + f(x_2)\,\Delta x + \cdots + f(x_n)\,\Delta x$$

as $n$ increases without bound or, equivalently, as $\Delta x$ approaches zero. (Recall that $x_1, x_2, \ldots, x_n$ are selected points from a partition of $a \leq x \leq b$ and $\Delta x$ is the width of each of the $n$ subintervals.) It can be shown that even if $f(x)$ has negative values, the Riemann sums still approach a limiting value as $\Delta x \to 0$. This number is called the *definite integral of $f(x)$ from $a$ to $b$* and is denoted by

$$\int_a^b f(x)\,dx.$$

That is,

$$\int_a^b f(x)\,dx = \lim_{\Delta x \to 0} [f(x_1)\,\Delta x + f(x_2)\,\Delta x + \cdots + f(x_n)\,\Delta x]. \qquad (1)$$

If $f(x)$ is a nonnegative function, we know from Section 6.2 that the Riemann sum on the right side of (1) approaches the area under the graph of $f(x)$ from $a$ to $b$. Thus, *the definite integral of a nonnegative function $f(x)$ equals the area under the graph of $f(x)$.*

**EXAMPLE 1** Calculate

$$\int_1^4 \left(\frac{1}{3}x + \frac{2}{3}\right) dx.$$

**Solution**    Figure 1 shows the graph of the function $f(x) = \frac{1}{3}x + \frac{2}{3}$. Since $f(x)$ is nonnegative for $1 \le x \le 4$, the definite integral of $f(x)$ equals the area of the shaded region in Fig. 1. The region consists of a rectangle and a triangle. By geometry,

$$[\text{area of rectangle}] = [\text{width}] \cdot [\text{height}] = 3 \cdot 1 = 3,$$

$$[\text{area of triangle}] = \frac{1}{2}[\text{width}] \cdot [\text{height}] = \frac{1}{2} \cdot 3 \cdot 1 = \frac{3}{2}.$$

Thus the area under the graph is $4\frac{1}{2}$, and hence

$$\int_1^4 \left(\frac{1}{3}x + \frac{2}{3}\right) dx = 4\frac{1}{2}.$$

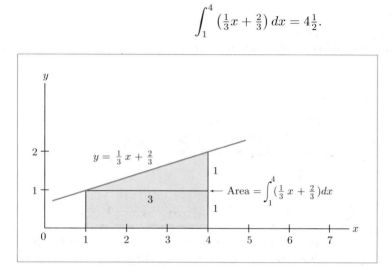

**Figure 1.**    A definite integral as the area under a curve.

In case $f(x)$ is negative at some points in the interval, we may also give a geometric interpretation of the definite integral. Consider the function $f(x)$ shown in Fig. 2. The figure shows a rectangular approximation of the region between the graph and the $x$-axis from $a$ to $b$. Consider a typical rectangle located above or below the selected point $x_i$. If $f(x_i)$ is nonnegative, the area of the rectangle equals $f(x_i)\,\Delta x$. In case $f(x_i)$ is negative, the area of the rectangle equals $(-f(x_i))\,\Delta x$. So the expression $f(x_i)\,\Delta x$ equals either the area of the corresponding rectangle or the negative of the area, according to whether $f(x_i)$ is nonnegative or negative, respectively. In particular, the Riemann sum

$$f(x_1)\,\Delta x + f(x_2)\,\Delta x + \cdots + f(x_n)\,\Delta x$$

is equal to the area of the rectangles above the $x$-axis minus the area of the rectangles below the $x$-axis. Now take the limit as $\Delta x$ approaches 0. On the one hand, the

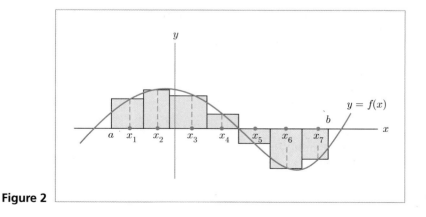

**Figure 2**

Riemann sum approaches the definite integral. On the other hand, the rectangular approximations approach the area bounded by the graph that is above the $x$-axis minus the area bounded by the graph that is below the $x$-axis. This gives us the following geometric interpretation of the definite integral.

Suppose that $f(x)$ is continuous on the interval $a \le x \le b$. Then

$$\int_a^b f(x)\,dx$$

is equal to the area above the $x$-axis bounded by the graph of $y = f(x)$ from $x = a$ to $x = b$ minus the area below the $x$-axis. Referring to Fig. 3, we have

$$\int_a^b f(x)\,dx = [\text{area of } B \text{ and } D] - [\text{area of } A \text{ and } C]$$

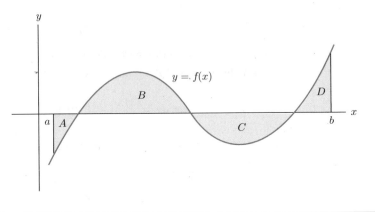

**Figure 3.** Regions above and below the $x$-axis.

**EXAMPLE 2** Calculate

$$\int_0^5 (2x - 4)\,dx.$$

**Solution** Figure 4 shows the graph of the function $f(x) = 2x - 4$ on the interval $0 \le x \le 5$. The area of the triangle above the $x$-axis is 9 and the area of the triangle below the $x$-axis is 4. Therefore, from geometry, we find that

$$\int_0^5 (2x - 4)\,dx = 9 - 4 = 5. \qquad \blacksquare$$

The values of the definite integrals in Examples 1 and 2 follow from simple geometric formulas. For integrals of more complex functions, analogous area formulas are not available. Of course, Riemann sums may always be used to estimate the value of a definite integral to any desired degree of accuracy. However, for most of the integrals in this text, the following fundamental theorem of calculus will rescue us from such calculations.

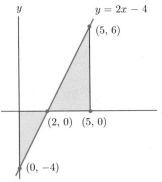

**Figure 4**

**Fundamental Theorem of Calculus** Suppose that $f(x)$ is continuous on the interval $a \le x \le b$, and let $F(x)$ be an antiderivative of $f(x)$. Then

$$\int_a^b f(x)\,dx = F(b) - F(a). \qquad (2)$$

This theorem connects the two key concepts of calculus—the integral and the derivative. An explanation of why the theorem is true is given later in this section. First we show how to use the theorem to evaluate definite integrals.

**EXAMPLE 3**   Use the fundamental theorem of calculus to evaluate the following definite integrals.

$$\text{(a)} \int_1^4 \left(\tfrac{1}{3}x + \tfrac{2}{3}\right) dx \qquad\qquad \text{(b)} \int_0^5 (2x - 4)\, dx$$

**Solution**   (a) An antiderivative of $f(x) = \tfrac{1}{3}x + \tfrac{2}{3}$ is $F(x) = \tfrac{1}{6}x^2 + \tfrac{2}{3}x$. Therefore, by the fundamental theorem,

$$
\int_1^4 \left(\tfrac{1}{3}x + \tfrac{2}{3}\right) dx = F(4) - F(1)
$$
$$
= \left[\tfrac{1}{6}(4)^2 + \tfrac{2}{3}(4)\right] - \left[\tfrac{1}{6}(1)^2 + \tfrac{2}{3}(1)\right]
$$
$$
= \left[\tfrac{16}{6} + \tfrac{8}{3}\right] - \left[\tfrac{1}{6} + \tfrac{2}{3}\right] = 4\tfrac{1}{2}.
$$

This result is the same as in Example 1.

(b) An antiderivative of $f(x) = 2x - 4$ is $F(x) = x^2 - 4x$. Therefore,

$$
\int_0^5 (2x - 4)\, dx = F(5) - F(0) = \left[5^2 - 4(5)\right] - \left[0^2 - 4(0)\right] = 5.
$$

This result is the same as in Example 2.   ■

**EXAMPLE 4**   Evaluate

$$
\int_2^5 3x^2 \, dx.
$$

**Solution**   An antiderivative of $f(x) = 3x^2$ is $F(x) = x^3 + C$, where $C$ is any constant. Then

$$
\int_2^5 3x^2 \, dx = F(5) - F(2) = \left[5^3 + C\right] - \left[2^3 + C\right] = 5^3 + C - 2^3 - C = 117.
$$

Notice how the $C$ in $F(2)$ is subtracted from the $C$ in $F(5)$. Thus the value of the definite integral does not depend on the choice of the constant $C$. For convenience, we may omit $C$ when evaluating a definite integral.   ■

The quantity $F(b) - F(a)$ is called the *net change of $F(x)$ from $x = a$ to $x = b$*. It is abbreviated by the symbol $F(x)\big|_a^b$. For instance, the net change of $F(x) = \tfrac{1}{3}e^{3x}$ from $x = 0$ to $x = 2$ is written as $\tfrac{1}{3}e^{3x}\big|_0^2$ and is evaluated as $F(2) - F(0)$.

**EXAMPLE 5**   Evaluate

$$
\int_0^2 e^{3x} \, dx.
$$

**Solution**   An antiderivative of $f(x) = e^{3x}$ is $F(x) = \tfrac{1}{3}e^{3x}$. Therefore,

$$
\int_0^2 e^{3x} \, dx = \tfrac{1}{3}e^{3x}\Big|_0^2 = \tfrac{1}{3}e^{3(2)} - \tfrac{1}{3}e^{3(0)} = \tfrac{1}{3}e^6 - \tfrac{1}{3}.
$$   ■

> **EXAMPLE 6**  Compute the area under the curve $y = x^2 - 4x + 5$ from $x = -1$ to $x = 3$.

**Solution** The graph of $f(x) = x^2 - 4x + 5$ is shown in Fig. 5. Since $f(x)$ is nonnegative for $-1 \le x \le 3$, the area under the curve is given by the definite integral

$$\int_{-1}^{3} (x^2 - 4x + 5)\, dx = \left. \left( \frac{x^3}{3} - 2x^2 + 5x \right) \right|_{-1}^{3}$$

$$= \left[ \frac{(3)^3}{3} - 2(3)^2 + 5(3) \right] - \left[ \frac{(-1)^3}{3} - 2(-1)^2 + 5(-1) \right]$$

$$= [9 - 18 + 15] - [-\tfrac{1}{3} - 2 - 5]$$

$$= [6] - [-\tfrac{22}{3}] = \tfrac{18}{3} + \tfrac{22}{3} = \tfrac{40}{3}.$$

**Figure 5**

Study this solution carefully. The calculations show how to use parentheses to avoid errors in arithmetic. It is particularly important to include the outside brackets around the value of the antiderivative at $-1$.

**Areas in Applications**  At the end of Section 6.2, we described how the area under a graph can represent the amount of change (now called the "net change") in a quantity. This interpretation of area follows immediately from the fundamental theorem of calculus, which we may rewrite as

$$\int_{a}^{b} F'(x)\, dx = F(b) - F(a) \tag{3}$$

because the $f(x)$ in (2) is the derivative of $F(x)$. If the derivative is nonnegative, then the integral in (3) is the area under the graph of the "rate of change" function $F'(x)$. Equation (3) says that this area equals the net change in $F(x)$ over the interval from $a$ to $b$.

> **EXAMPLE 7**  **Position of a rocket**  A rocket is fired vertically into the air. Its velocity at $t$ seconds after lift-off is $v(t) = 6t + .5$ meters per second.
>
> (a) Describe a region whose area represents the distance the rocket travels from time $t = 40$ to $t = 100$ seconds.
>
> (b) Compute the distance in (a).

**Solution** (a) Let $s(t)$ be the position of the rocket at time $t$, measured from some reference point. (Example 8 in Section 6.1 measured distances from the launch pad, for example.) Then the required distance is $s(100) - s(40)$, the net change in position over the interval $40 \le t \le 100$. This distance is represented by the area under the velocity curve from 40 to 100. See Fig. 6.

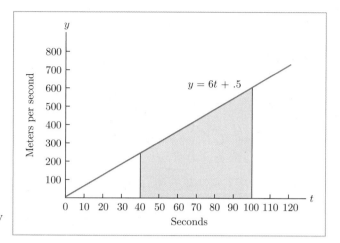

**Figure 6.** Area under velocity curve is distance traveled.

(b) Now that we know the connection between antiderivatives and area, we can easily compute the area in (a) corresponding to the distance traveled by the rocket:

$$s(100) - s(40) = \int_{40}^{100} s'(t)\, dt = \int_{40}^{100} v(t)\, dt$$

$$= \int_{40}^{100} (6t + .5)\, dt = \left(3t^2 + .5t\right)\Big|_{40}^{100}$$

$$= \left[3(100)^2 + .5(100)\right] - \left[3(40)^2 + .5(40)\right]$$

$$= 30{,}050 - 4820 = 25{,}230 \text{ meters.} \qquad \blacksquare$$

---

**EXAMPLE 8**    **Worldwide oil consumption**   During the early 1970s, the annual worldwide rate of oil consumption was $R(t) = 16.1e^{.07t}$ billion barrels of oil per year, where $t$ is the number of years since the beginning of 1970.

(a) Determine the amount of oil consumed from 1972 to 1974.

(b) Represent the answer to part (a) as an area.

**Solution**   (a) We are interested in $T(4) - T(2)$, where $T(t)$ is the total consumption of oil since 1970. This difference is the net change in $T(t)$ over the time period from $t = 2$ (1972) to $t = 4$ (1974). Now, $T(t)$ is an antiderivative of the rate function $R(t)$. Hence

$$T(4) - T(2) = \int_{2}^{4} R(t)\, dt = \int_{2}^{4} 16.1e^{.07t}\, dt = \frac{16.1}{.07}e^{.07t}\Big|_{2}^{4}$$

$$= 230e^{.07(4)} - 230e^{.07(2)} \approx 39.76 \text{ billion barrels of oil.}$$

(b) The net change in $T(t)$ is the area of the region under the graph of the rate function $T'(t) = R(t)$ from $t = 2$ to $t = 4$. See Fig. 7. $\qquad \blacksquare$

**Verification of the Fundamental Theorem of Calculus**   We shall give two explanations of why the fundamental theorem of calculus is true. Both convey important ideas about definite integrals and the fundamental theorem.

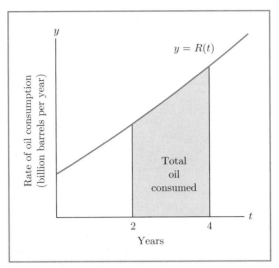

**Figure 7.**   Total oil consumed.

The first explanation of the fundamental theorem is based directly on the Riemann sum definition of the integral. As we observed earlier, the fundamental theorem may be written in the form

$$\int_a^b F'(x)\,dx = F(b) - F(a), \tag{4}$$

where $F(x)$ is any function with a continuous derivative on $a \leq x \leq b$. There are three key ideas in an explanation of why (4) is true.

**I.** If the interval $a \leq x \leq b$ is partitioned into $n$ subintervals of width $\Delta x = (b-a)/n$, then the net change in $F(x)$ over $a \leq x \leq b$ is the sum of the net changes in $F(x)$ over each subinterval.

For example, partition $a \leq x \leq b$ into three subintervals and denote the left endpoints by $x_1, x_2, x_3$, as follows:

$$\underset{a\,=\,x_1}{\vert} \quad \underset{x_2}{\vert} \quad \underset{x_3}{\vert} \quad \underset{b}{\vert}$$

Then

$$[\text{change in } F(x) \text{ over 1st interval}] = F(x_2) - F(a),$$

$$[\text{change in } F(x) \text{ over 2nd interval}] = F(x_3) - F(x_2),$$

$$[\text{change in } F(x) \text{ over 3rd interval}] = F(b) - F(x_3).$$

When these changes are summed, the intermediate terms cancel:

$$F(b) - F(x_3) + F(x_3) - F(x_2) + F(x_2) - F(a) = F(b) - F(a).$$

**II.** If $\Delta x$ is small, then the change in $F(x)$ over the $i$th subinterval is approximately $F'(x_i)\,\Delta x$.

This is the approximation of the change in a function discussed in Section 1.8. Here $x_i$ is the left endpoint of the $i$th subinterval, as follows:

**III.** The sum of the approximations in (II) is a Riemann sum for the definite integral $\int_a^b F'(x)\,dx$.

That is,

$$F(b) - F(a)$$

$$= \begin{bmatrix} \text{change over} \\ \text{1st subinterval} \end{bmatrix} + \begin{bmatrix} \text{change over} \\ \text{2nd subinterval} \end{bmatrix} + \cdots + \begin{bmatrix} \text{change over} \\ n\text{th subinterval} \end{bmatrix}$$

$$\approx F'(x_1)\,\Delta x + F'(x_2)\,\Delta x + \cdots + F'(x_n)\,\Delta x.$$

As $\Delta x \to 0$, these approximations improve. Thus, in the limit

$$F(b) - F(a) = \int_a^b F'(x)\,dx.$$

To summarize, the derivative of $F(x)$ determines the approximate change of $F(x)$ on small subintervals. The definite integral sums these approximate changes and in the limit gives the exact change of $F(x)$ over the entire interval $a \le x \le b$.

If you look back at the solution of Example 4 of Section 6.2, you will see essentially the same argument. The distance a rocket travels is the sum of the distances it travels over small intervals of time. On each subinterval, this distance is approximately the velocity (the derivative) times $\Delta t$.

**An Area Function as an Antiderivative**  The second explanation of the fundamental theorem of calculus applies only when $f(x)$ is nonnegative. We begin with the following theorem, which describes the basic relationship between area and antiderivatives. In fact, this theorem is sometimes referred to as an alternative version of the fundamental theorem of calculus.

**Theorem III**   Let $f(x)$ be a continuous nonnegative function for $a \le x \le b$. Let $A(x)$ be the area of the region under the graph of the function from $a$ to the number $x$. (See Fig. 8.) Then $A(x)$ is an antiderivative of $f(x)$.

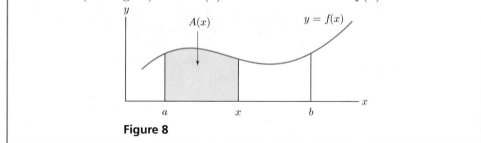

**Figure 8**

The fundamental theorem of calculus for a nonnegative function follows easily from Theorem III. Let $F(x)$ be an antiderivative of $f(x)$. Since the "area function" $A(x)$ is also an antiderivative of $f(x)$, by Theorem III, we have

$$A(x) = F(x) + C$$

for some constant $C$. Notice that $A(a)$ is 0 and $A(b)$ equals the area of the region under the graph of $f(x)$ for $a \le x \le b$. Therefore,

$$\int_a^b f(x)\,dx = A(b) = A(b) - A(a)$$

$$= [F(b) + C] - [F(a) + C]$$

$$= F(b) - F(a).$$

It is not difficult to explain why Theorem III is reasonable, although we shall not give a formal proof of the theorem. If $h$ is a small positive number, then $A(x + h) - A(x)$ is the area of the shaded region in Fig. 9. This shaded region is approximately a rectangle of width $h$, height $f(x)$, and area $h \cdot f(x)$. Thus

$$A(x + h) - A(x) \approx h \cdot f(x),$$

where the approximation becomes better as $h$ approaches zero. Dividing by $h$, we have

$$\frac{A(x + h) - A(x)}{h} \approx f(x).$$

Since the approximation improves as $h$ approaches zero, the quotient must approach $f(x)$. However, the limit definition of the derivative tells us that the quotient approaches $A'(x)$ as $h$ approaches zero. Therefore, we have $A'(x) = f(x)$. Since $x$ represented any number between $a$ and $b$, this shows that $A(x)$ is an antiderivative of $f(x)$.

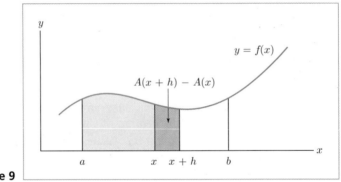

**Figure 9**

**Incorporating Technology** In Fig. 10(a) the region from Example 6 is shaded and its area evaluated, and in Fig. 10(b) the definite integral is evaluated. In both cases the values obtained are good approximations but are not exact. See Appendices A–D.

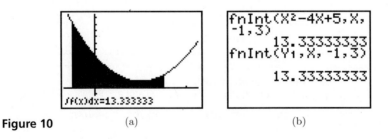

**Figure 10**      (a)                      (b)

**PRACTICE PROBLEMS 6.3**

1. Find the area under the curve $y = e^{x/2}$ from $x = -3$ to $x = 2$.

2. Let $MR(x)$ be a company's marginal revenue at production level $x$. Give an economic interpretation of the number $\int_{75}^{80} MR(x)\, dx$.

# EXERCISES 6.3

In Exercises 1–3, set up the definite integral that gives the area of the shaded region.

**1.**

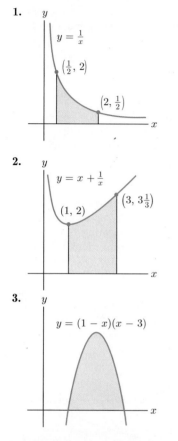

**2.**

**3.**

In Exercises 4–6, draw the region whose area is given by the definite integral.

**4.** $\displaystyle\int_2^4 x^2\,dx$    **5.** $\displaystyle\int_0^4 (8-2x)\,dx$

**6.** $\displaystyle\int_0^4 \sqrt{x}\,dx$

Calculate the following definite integrals.

**7.** $\displaystyle\int_{-1}^1 x\,dx$    **8.** $\displaystyle\int_0^1 \left(x+\frac{1}{2}\right)dx$

**9.** $\displaystyle\int_1^2 5\,dx$    **10.** $\displaystyle\int_1^4 x^2\sqrt{x}\,dx$

**11.** $\displaystyle\int_1^4 \frac{1}{2\sqrt{x}}\,dx$    **12.** $\displaystyle\int_0^1 (4x^3-1)\,dx$

**13.** $\displaystyle\int_0^1 4e^{-3x}\,dx$    **14.** $\displaystyle\int_0^1 2e^{2x}\,dx$

**15.** $\displaystyle\int_{-1}^1 \frac{e^x+e^{-x}}{2}\,dx$    **16.** $\displaystyle\int_0^1 \sqrt{e^x}\,dx$

**17.** $\displaystyle\int_0^{\ln 3} e^{-2t}\,dt$    **18.** $\displaystyle\int_{\ln 2}^{\ln 3} \frac{3}{e^{3t}}\,dt$

**19.** $\displaystyle\int_3^6 x^{-1}\,dx$    **20.** $\displaystyle\int_1^e \frac{1}{7x}\,dx$

**21.** $\displaystyle\int_{\frac{1}{5}}^5 \frac{2}{x}\,dx$

**22.** $\displaystyle\int_1^2 \left(\frac{x}{2}+\frac{2}{x}+\frac{1}{2x^2}\right)dx$

**23.** $\displaystyle\int_0^1 (3x^{1/3}-1-e^{.5x})\,dx$

**24.** $\displaystyle\int_{-1}^1 \frac{1-x-e^{-x}}{2}\,dx$

**25.** $\displaystyle\int_0^3 (x^3+x-7)\,dx$

**26.** $\displaystyle\int_0^1 \left(e^{x/3}-\frac{2x}{5}\right)dx$

**27.** $\displaystyle\int_2^3 (5-2t)^4\,dt$    **28.** $\displaystyle\int_4^9 \frac{3}{t-2}\,dt$

Find the area under each of the given curves.

**29.** $y=4x$; $x=2$ to $x=3$

**30.** $y=3x^2$; $x=-1$ to $x=1$

**31.** $y=e^{x/2}$; $x=0$ to $x=1$

**32.** $y=\sqrt{x}$; $x=0$ to $x=4$

**33.** $y=(x-3)^4$; $x=1$ to $x=4$

**34.** $y=e^{3x}$; $x=-\frac{1}{3}$ to $x=0$

**35.** Let $f(x)$ be the function pictured in Fig. 11. Determine whether $\int_0^7 f(x)\,dx$ is positive, negative, or zero.

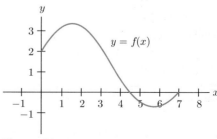

**Figure 11**

**36.** Let $g(x)$ be the function pictured in Fig. 12. Determine whether $\int_0^7 g(x)\,dx$ is positive, negative, or zero.

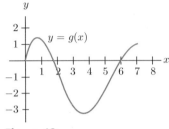

**Figure 12**

**37.** (*Cigarette Consumption*) The worldwide rate of cigarette consumption (in trillions of cigarettes per year) since 1960 is given approximately by the function $c(t) = .1t + 2.4$, where $t = 0$ corresponds to 1960. Determine the number of cigarettes sold from 1980 to 1998.

**38.** Suppose $p(t)$ is the rate (in tons per year) at which pollutants are discharged into a lake, where $t$ is the number of years since 1990. Interpret $\int_5^7 p(t)\,dt$.

**39.** A helicopter is rising straight up in the air. Its velocity at time $t$ is $v(t) = 2t + 1$ feet per second.

   **(a)** How high does the helicopter rise during the first 5 seconds?

   **(b)** Represent the answer to part (a) as an area.

**40.** (*Production*) After $t$ hours of operation, an assembly line is producing power lawn mowers at the rate of $r(t) = 21 - \frac{4}{5}t$ mowers per hour.

   **(a)** How many mowers are produced during the time from $t = 2$ to $t = 5$ hours?

   **(b)** Represent the answer to part (a) as an area.

**41.** (*Cost*) Suppose that the marginal cost function of a handbag manufacturer is $C'(x) = \frac{3}{32}x^2 - x + 200$ dollars per unit at production level $x$ (where $x$ is measured in units of 100 handbags).

   **(a)** Find the total cost of producing 6 additional units if 2 units are currently being produced.

   **(b)** Describe the answer to part (a) as an area. (Give a written description rather than a sketch.)

**42.** (*Profit*) Suppose that the marginal profit function for a company is $P'(x) = 100 + 50x - 3x^2$ at production level $x$.

   **(a)** Find the extra profit earned from the sale of 3 additional units if 5 units are currently being produced.

   **(b)** Describe the answer to part (a) as an area. (Do not make a sketch.)

**43.** Let $MP(x)$ be a company's marginal profit at production level $x$. Give an economic interpretation of the number $\int_{44}^{48} MP(x)\,dx$.

**44.** Let $MC(x)$ be a company's marginal cost at production level $x$. Give an economic interpretation of the number

$\int_0^{100} MC(x)\,dx$. (*Note:* At any production level, the total cost equals the fixed cost plus the total variable cost.)

**45.** (*Heat Diffusion*) Some food is placed in a freezer. After $t$ hours the temperature of the food is dropping at the rate of $r(t)$ degrees Fahrenheit per hour, where $r(t) = 12 + 4/(t+3)^2$.

   **(a)** Compute the area under the graph of $y = r(t)$ over the interval $0 \le t \le 2$.

   **(b)** What does the area in part (a) represent?

**46.** (*Velocity*) Suppose that the velocity of a car at time $t$ is $v(t) = 40 + 8/(t+1)^2$ kilometers per hour.

   **(a)** Compute the area under the velocity curve from $t = 1$ to $t = 9$.

   **(b)** What does the area in part (a) represent?

**47.** (*Deforestation and Fuel Wood*) Deforestation is one of the major problems facing sub-Saharan Africa. Although the clearing of land for farming has been the major cause, the steadily increasing demand for fuel wood has become a significant factor. Figure 13 summarizes projections of the World Bank. The rate of fuel wood consumption (in millions of cubic meters per year) in the Sudan $t$ years after 1980 is given approximately by the function $c(t) = 76.2e^{.03t}$. Determine the amount of fuel wood that will be consumed from 1980 to 2000.

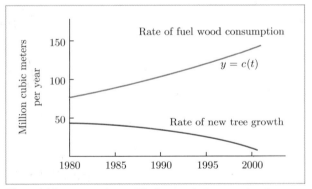

**Figure 13.** Data from "Sudan and Options in the Energy Sector," World Bank/UNDP, July 1983.

**48. (a)** Compute $\int_1^b \frac{1}{t}\,dt$, where $b > 1$.

   **(b)** Explain how the logarithm of a number greater than 1 may be interpreted as the area of a region under a curve. (What is the curve?)

**49.** For each positive number $x$, let $A(x)$ be the area of the region under the curve $y = x^2 + 1$ from 0 to $x$. Find $A'(3)$.

**50.** For each number $x > 2$, let $A(x)$ be the area of the region under the curve $y = x^3$ from 2 to $x$. Find $A'(6)$.

### Technology Exercises

*In Exercises 51–54, first use the fundamental theorem of calculus to compute the definite integral and then verify your answer numerically with a graphing utility. (For a TI calculator, use fnInt as discussed in Appendices A–D.)*

**51.** $\int_{2}^{3} \frac{1}{x}\, dx$     **52.** $\int_{1}^{3} 10x^4\, dx$

**53.** $\int_{0}^{4} 5\sqrt{x}\, dx$     **54.** $\int_{0}^{3} 6e^{2x}\, dx$

*In Exercises 55–58, sketch the region corresponding to the definite integral, and compute its area numerically to four decimal places.*

**55.** $\int_{1}^{4} \ln x\, dx$     **56.** $\int_{-1}^{1} e^{-x^2}\, dx$

**57.** $\int_{0}^{.5} \frac{1}{1+x^2}\, dx$     **58.** $\int_{0}^{3} 2^x\, dx$

---

## SOLUTIONS TO PRACTICE PROBLEMS 6.3

**1.** The desired area is

$$\int_{-3}^{2} e^{x/2}\, dx = 2e^{x/2}\Big|_{-3}^{2} = 2e - 2e^{-3/2} \approx 4.99.$$

**2.** The number $\int_{75}^{80} MR(x)\, dx$ is the net change in revenue received when the production level is raised from $x = 75$ to $x = 80$ units.

---

# 6.4  Areas in the *xy*-Plane

In this section we show how to use the definite integral to compute the area of a region that lies between the graphs of two or more functions. Three simple but important properties of the integral will be used repeatedly.

Let $f(x)$ and $g(x)$ be functions and $a$, $b$, and $k$ be any constants. Then

$$\int_{a}^{b} f(x)\, dx + \int_{a}^{b} g(x)\, dx = \int_{a}^{b} [f(x) + g(x)]\, dx \tag{1}$$

$$\int_{a}^{b} f(x)\, dx - \int_{a}^{b} g(x)\, dx = \int_{a}^{b} [f(x) - g(x)]\, dx \tag{2}$$

$$\int_{a}^{b} kf(x)\, dx = k\int_{a}^{b} f(x)\, dx. \tag{3}$$

To verify (1), let $F(x)$ and $G(x)$ be antiderivatives of $f(x)$ and $g(x)$, respectively. Then $F(x) + G(x)$ is an antiderivative of $f(x) + g(x)$. By the fundamental theorem of calculus,

$$\int_{a}^{b} [f(x) + g(x)]\, dx = [F(x) + G(x)]\Big|_{a}^{b}$$

$$= [F(b) + G(b)] - [F(a) + G(a)]$$

$$= [F(b) - F(a)] + [G(b) - G(a)]$$

$$= \int_{a}^{b} f(x)\, dx + \int_{a}^{b} g(x)\, dx.$$

The verifications of (2) and (3) are similar and use the facts that $F(x) - G(x)$ is an antiderivative of $f(x) - g(x)$ and $kF(x)$ is an antiderivative of $kf(x)$.

Let us now consider regions that are bounded both above and below by graphs of functions. Referring to Fig. 1, we would like to find a simple expression for the area of the shaded region under the graph of $y = f(x)$ and above the graph of $y = g(x)$ from $x = a$ to $x = b$. It is the region under the graph of $y = f(x)$ with the region under the graph of $y = g(x)$ taken away. Therefore,

$$[\text{area of shaded region}] = [\text{area under } f(x)] - [\text{area under } g(x)]$$

$$= \int_a^b f(x)\,dx - \int_a^b g(x)\,dx$$

$$= \int_a^b [f(x) - g(x)]\,dx \qquad [\text{by property (2)}].$$

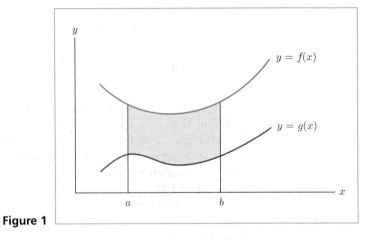

**Figure 1**

> **Area Between Two Curves**     If $y = f(x)$ lies above $y = g(x)$ from $x = a$ to $x = b$, the area of the region between $f(x)$ and $g(x)$ from $x = a$ to $x = b$ is
>
> $$\int_a^b [f(x) - g(x)]\,dx.$$

**EXAMPLE 1**   Find the area of the region between $y = 2x^2 - 4x + 6$ and $y = -x^2 + 2x + 1$ from $x = 1$ to $x = 2$.

**Solution**   Upon sketching the two graphs (Fig. 2), we see that $f(x) = 2x^2 - 4x + 6$ lies above $g(x) = -x^2 + 2x + 1$ for $1 \le x \le 2$. Therefore, our formula gives the area of the shaded region as

$$\int_1^2 [(2x^2 - 4x + 6) - (-x^2 + 2x + 1)]\,dx = \int_1^2 (3x^2 - 6x + 5)\,dx$$

$$= (x^3 - 3x^2 + 5x)\Big|_1^2 = 6 - 3 = 3. \quad \blacksquare$$

**EXAMPLE 2**   Find the area of the region between $y = x^2$ and $y = (x - 2)^2 = x^2 - 4x + 4$ from $x = 0$ to $x = 3$.

**Figure 2**

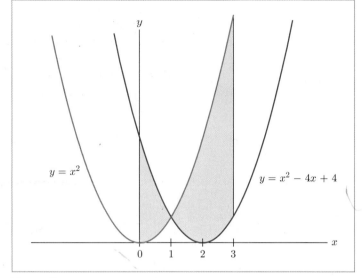

**Figure 3**

**Solution**   Upon sketching the graphs (Fig. 3), we see that the two graphs cross; by setting $x^2 = x^2 - 4x + 4$, we find that they cross when $x = 1$. Thus one graph does not always lie above the other from $x = 0$ to $x = 3$, so that we cannot directly apply our rule for finding the area between two curves. However, the difficulty is easily surmounted if we break the region into two parts, namely, the area from $x = 0$ to $x = 1$ and the area from $x = 1$ to $x = 3$. From $x = 0$ to $x = 1$, $y = x^2 - 4x + 4$ is on top; and from $x = 1$ to $x = 3$, $y = x^2$ is on top. Consequently,

$$[\text{area from } x = 0 \text{ to } x = 1] = \int_0^1 [(x^2 - 4x + 4) - (x^2)]\, dx$$

$$= \int_0^1 (-4x + 4)\, dx$$

$$= (-2x^2 + 4x)\Big|_0^1 = 2 - 0 = 2.$$

$$[\text{area from } x = 1 \text{ to } x = 3] = \int_1^3 [(x^2) - (x^2 - 4x + 4)]\, dx$$

$$= \int_1^3 (4x - 4)\, dx$$

$$= (2x^2 - 4x)\Big|_1^3 = 6 - (-2) = 8.$$

Thus the total area is $2 + 8 = 10$.  ◼

   In our derivation of the formula for the area between two curves, we examined functions that are nonnegative. However, the statement of the rule does not contain this stipulation, and rightfully so. Consider the case where $f(x)$ and $g(x)$ are not always positive. Let us determine the area of the shaded region in Fig. 4(a). Select some constant $c$ such that the graphs of the functions $f(x) + c$ and $g(x) + c$ lie completely above the $x$-axis [Fig. 4(b)]. The region between them will have the same area as the original region. Using the rule as applied to nonnegative functions,

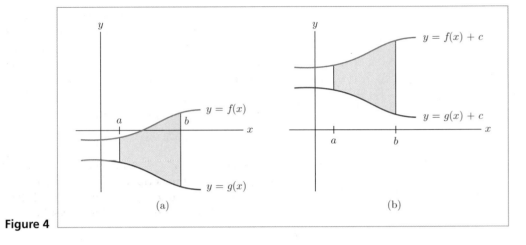

**Figure 4**

we have

$$[\text{area of the region}] = \int_a^b \left[ (f(x) + c) - (g(x) + c) \right] dx$$

$$= \int_a^b \left[ f(x) - g(x) \right] dx.$$

Therefore, we see that our rule is valid for any functions $f(x)$ and $g(x)$ as long as the graph of $f(x)$ lies above the graph of $g(x)$ for all $x$ from $x = a$ to $x = b$.

**EXAMPLE 3** Set up the integral that gives the area between the curves $y = x^2 - 2x$ and $y = -e^x$ from $x = -1$ to $x = 2$.

**Solution** Since $y = x^2 - 2x$ lies above $y = -e^x$ (Fig. 5), the rule for finding the area between two curves can be applied directly. The area between the curves is

$$\int_{-1}^2 (x^2 - 2x + e^x) \, dx.$$

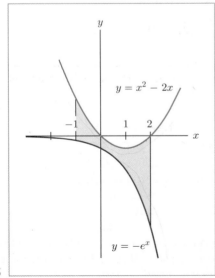

**Figure 5**

Sometimes we are asked to find the area between two curves without being given the values of $a$ and $b$. In these cases there is a region that is completely enclosed by the two curves. As the next examples illustrate, we must first find the points of intersection of the two curves in order to obtain the values of $a$ and $b$. In such problems careful curve sketching is especially important.

**EXAMPLE 4**    Set up the integral that gives the area bounded by the curves $y = x^2 + 2x + 3$ and $y = 2x + 4$.

**Solution**    The two curves are sketched in Fig. 6, and the region bounded by them is shaded. In order to find the points of intersection, we set $x^2 + 2x + 3 = 2x + 4$ and solve for $x$. We obtain $x^2 = 1$, or $x = -1$ and $x = +1$. When $x = -1$, $2x + 4 = 2(-1) + 4 = 2$. When $x = 1$, $2x + 4 = 2(1) + 4 = 6$. Thus the curves intersect at the points $(1, 6)$ and $(-1, 2)$.

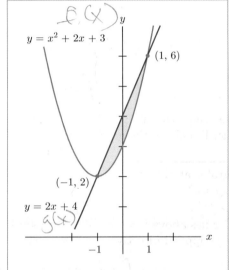

**Figure 6**

Since $y = 2x + 4$ lies above $y = x^2 + 2x + 3$ from $x = -1$ to $x = 1$, the area between the curves is given by

$$\int_{-1}^{1} [(2x + 4) - (x^2 + 2x + 3)]\, dx = \int_{-1}^{1} (1 - x^2)\, dx.$$

**EXAMPLE 5**    Set up the integral that gives the area bounded by the two curves $y = 2x^2$ and $y = x^3 - 3x$.

**Solution**    First we make a rough sketch of the two curves, as in Fig. 7. The curves intersect where $x^3 - 3x = 2x^2$, or $x^3 - 2x^2 - 3x = 0$. Note that

$$x^3 - 2x^2 - 3x = x(x^2 - 2x - 3) = x(x - 3)(x + 1).$$

So the solutions to $x^3 - 2x^2 - 3x = 0$ are $x = 0, 3, -1$, and the curves intersect at $(-1, 2)$, $(0, 0)$, and $(3, 18)$. From $x = -1$ to $x = 0$, the curve $y = x^3 - 3x$ lies above $y = 2x^2$. But from $x = 0$ to $x = 3$, the reverse is true. Thus the area between the curves is given by

$$\int_{-1}^{0} (x^3 - 3x - 2x^2)\, dx + \int_{0}^{3} (2x^2 - x^3 + 3x)\, dx.$$

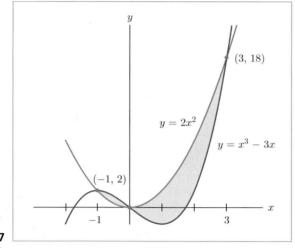

**Figure 7**

---

**EXAMPLE 6**    **Oil consumption**  Beginning in 1974, with the advent of dramatically higher oil prices, the exponential rate of growth of world oil consumption slowed down from a growth constant of 7% to a growth constant of 4% per year. A fairly good model for the annual rate of oil consumption since 1974 is given by

$$R_1(t) = 21.3e^{.04(t-4)}, \qquad t \geq 4,$$

where $t = 0$ corresponds to 1970. Determine the total amount of oil saved between 1976 and 1980 by not consuming oil at the rate predicted by the model of Example 8, Section 6.3, namely,

$$R(t) = 16.1e^{.07t}, \qquad t \geq 0.$$

**Solution**    If oil consumption had continued to grow as it did prior to 1974, then the total oil consumption between 1976 and 1980 would have been

$$\int_6^{10} R(t)\, dt. \tag{4}$$

However, taking into account the slower increase in the rate of oil consumption since 1974, we find that the total oil consumed between 1976 and 1980 was approximately

$$\int_6^{10} R_1(t)\, dt. \tag{5}$$

The integrals in (4) and (5) may be interpreted as the areas under the curves $y = R(t)$ and $y = R_1(t)$, respectively, from $t = 6$ to $t = 10$. (See Fig. 8.) By superimposing the two curves we see that the area between them from $t = 6$ to $t = 10$ represents the total oil that was saved by consuming oil at the rate given by $R_1(t)$ instead of $R(t)$. (See Fig. 9.) The area between the two curves equals

$$\int_6^{10} [R(t) - R_1(t)]\, dt = \int_6^{10} \left[16.1e^{.07t} - 21.3e^{.04(t-4)}\right] dt$$

$$= \left(\frac{16.1}{.07}e^{.07t} - \frac{21.3}{.04}e^{.04(t-4)}\right)\Bigg|_6^{10}$$

$$\approx 13.02.$$

Thus about 13 billion barrels of oil were saved between 1976 and 1980.    ■

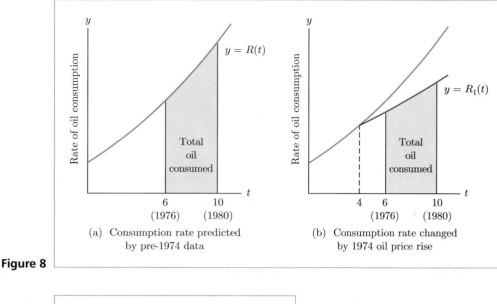

**Figure 8**

(a) Consumption rate predicted by pre-1974 data

(b) Consumption rate changed by 1974 oil price rise

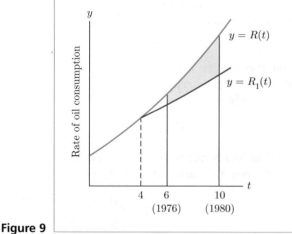

**Figure 9**

## PRACTICE PROBLEMS 6.4

1. Find the area between the curves $y = x + 3$ and $y = \frac{1}{2}x^2 + x - 7$ from $x = -2$ to $x = 1$.

2. A company plans to increase its production from 10 to 15 units per day. The present marginal cost function is $MC_1(x) = x^2 - 20x + 108$. By redesigning the production process and purchasing new equipment, the company can change the marginal cost function to $MC_2(x) = \frac{1}{2}x^2 - 12x + 75$. Determine the area between the graphs of the two marginal cost curves from $x = 10$ to $x = 15$. Interpret this area in economic terms.

# EXERCISES 6.4

**1.** Write down a definite integral or sum of definite integrals that gives the area of the shaded portion of Fig. 10.

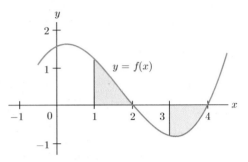

**Figure 10**

**2.** Write down a definite integral or sum of definite integrals that gives the area of the shaded portion of Fig. 11.

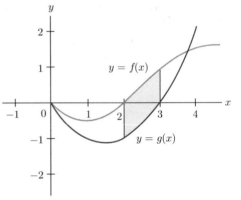

**Figure 11**

**3.** Shade the portion of Fig. 12 whose area is given by the integral

$$\int_0^2 [f(x) - g(x)]\, dx + \int_2^4 [h(x) - g(x)]\, dx.$$

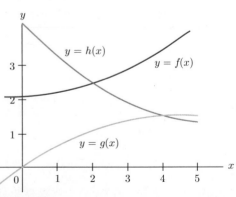

**Figure 12**

**4.** Shade the portion of Fig. 13 whose area is given by the integral

$$\int_0^1 [f(x) - g(x)] + \int_1^2 [g(x) - f(x)]\, dx.$$

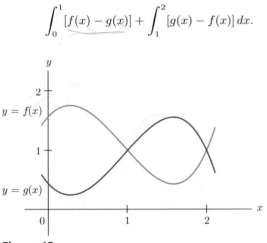

**Figure 13**

*Find the area of the region between the curves.*

**5.** $y = 2x^2$ and $y = 8$ (a horizontal line) from $x = -2$ to $x = 2$

**6.** $y = x^2 + 1$ and $y = -x^2 - 1$ from $x = -1$ to $x = 1$

**7.** $y = x^2 - 6x + 12$ and $y = 1$ from $x = 0$ to $x = 4$

**8.** $y = x(2 - x)$ and $y = 2$ from $x = 0$ to $x = 2$

**9.** $y = e^x$ and $y = \dfrac{1}{x^2}$ from $x = 1$ to $x = 2$

**10.** $y = e^{2x}$ and $y = 1 - x$ from $x = 0$ to $x = 1$

*Find the area of the region bounded by the curves.*

**11.** $y = x^2$ and $y = x$

**12.** $y = 4x(1 - x)$ and $y = \frac{3}{4}$

**13.** $y = -x^2 + 6x - 5$ and $y = 2x - 5$

**14.** $y = x^2 - 1$ and $y = 3$

**15.** $y = x(x^2 - 1)$ and the $x$-axis

**16.** $y = x^3$ and $y = 2x^2$

**17.** $y = 8x^2$ and $y = \sqrt{x}$

**18.** $y = \dfrac{4}{x}$ and $y = 5 - x$

**19.** Find the area of the region between $y = x^2 - 3x$ and the $x$-axis
  **(a)** from $x = 0$ to $x = 3$,
  **(b)** from $x = 0$ to $x = 4$,
  **(c)** from $x = -2$ to $x = 3$.

**20.** Find the area of the region between $y = x^2$ and $y = 1/x^2$
  **(a)** from $x = 1$ to $x = 4$,
  **(b)** from $x = \frac{1}{2}$ to $x = 4$.

**21.** Find the area of the region bounded by $y = 1/x^2$, $y = x$, and $y = 8x$, for $x \geq 0$. (See Fig. 14.)

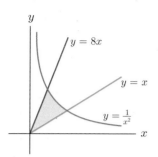

**Figure 14**

**22.** Find the area of the region bounded by $y = 1/x$, $y = 4x$, and $y = x/2$, for $x \geq 0$. (The region resembles the shaded region in Fig. 14.)

**23.** Find the area of the shaded region in Fig. 15(a) bounded by $y = 12/x$, $y = \frac{3}{2}\sqrt{x}$, and $y = x/3$.

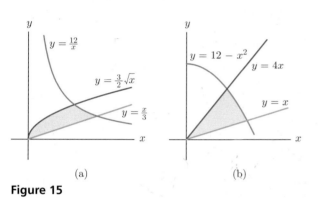

(a)                    (b)

**Figure 15**

**24.** Find the area of the shaded region in Fig. 15(b) bounded by $y = 12 - x^2$, $y = 4x$, and $y = x$.

**25.** Refer to Exercise 47 of Section 6.3. The rate of new tree growth (in millions of cubic meters per year) in the Sudan $t$ years after 1980 is given approximately by the function $g(t) = 50 - 6.03e^{.09t}$. Set up the definite integral giving the amount of depletion of the forests due to the excess of fuel wood consumption over new growth from 1980 to 2000.

**26.** Refer to the oil-consumption data in Example 6. Suppose that in 1970 the growth constant for the annual rate of oil consumption had been held to .04. What effect would this action have had on oil consumption from 1970 to 1974?

**27.** (*Profit*) The marginal profit for a certain company is $MP_1(x) = -x^2 + 14x - 24$. The company expects the daily production level to rise from $x = 6$ to $x = 8$ units. The management is considering a plan that would have the effect of changing the marginal profit to $M_2(x) = -x^2 + 12x - 20$. Should the company adopt the plan? Determine the area between the graphs of the two marginal profit functions from $x = 6$ to $x = 8$. Interpret this area in economic terms.

**28.** Two rockets are fired simultaneously straight up into the air. Their velocities (in meters per second) are $v_1(t)$ and $v_2(t)$, respectively, and $v_1(t) \geq v_2(t)$ for $t \geq 0$. Let $A$ denote the area of the region between the graphs of $y = v_1(t)$ and $y = v_2(t)$ for $0 \leq t \leq 10$. What physical interpretation may be given to the value of $A$?

**29.** (*Distance Traveled*) Cars $A$ and $B$ start at the same place and travel in the same direction, with velocities after $t$ hours given by the functions $v_A(t)$ and $v_B(t)$ in Fig. 16.

   **(a)** What does the area between the two curves from $t = 0$ to $t = 1$ represent?

   **(b)** At what time will the distance between the cars be greatest?

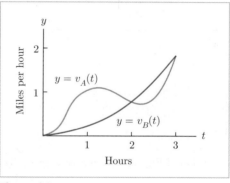

**Figure 16**

### Technology Exercises

*In Exercises 30–33, use a graphing utility to find the intersection points of the curves, and then use the utility to find the area of the region bounded by the curves.*

**30.** $y = e^x$, $y = 4x + 1$

**31.** $y = 5 - (x - 2)^2$, $y = e^x$

**32.** $y = \sqrt{x + 1}$, $y = (x - 1)^2$

**33.** $y = 1/x$, $y = 3 - x$

## SOLUTIONS TO PRACTICE PROBLEMS 6.4

1. First graph the two curves, as shown in Fig. 17. The curve $y = x + 3$ lies on top. So the area between the curves is

$$\int_{-2}^{1} \left[(x+3) - \left(\tfrac{1}{2}x^2 + x - 7\right)\right] dx$$

$$= \int_{-2}^{1} \left(-\tfrac{1}{2}x^2 + 10\right) dx$$

$$= \left(-\tfrac{1}{6}x^3 + 10x\right)\Big|_{-2}^{1} = 28.5.$$

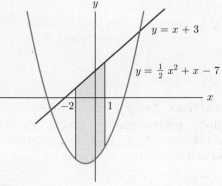

**Figure 17**

2. Graphing the two marginal cost functions yields the results shown in Fig. 18. So the area between the curves equals

$$\int_{10}^{15} [MC_1(x) - MC_2(x)]\, dx$$

$$= \int_{10}^{15} \left[(x^2 - 20x + 108) - \left(\tfrac{1}{2}x^2 - 12x + 75\right)\right] dx$$

$$= \int_{10}^{15} \left[\tfrac{1}{2}x^2 - 8x + 33\right] dx$$

$$= \left(\tfrac{1}{6}x^3 - 4x^2 + 33x\right)\Big|_{10}^{15} = 60\tfrac{5}{6}.$$

This amount represents the cost savings on the increased production (from 10 to 15) provided the new production process is used.

**Figure 18**

# 6.5   Applications of the Definite Integral

The applications in this section have two features in common. First, each example contains a quantity that is computed by evaluating a definite integral. Second, the formula for the definite integral is derived by looking at Riemann sums.

In each application, we show that a certain quantity, call it $Q$, may be approximated by dividing an interval into equal subintervals and forming an appropriate sum. We then observe that this sum is a Riemann sum for some function $f(x)$. [Sometimes $f(x)$ is not given beforehand.] The Riemann sums approach $Q$ as the number of subintervals becomes large. Since the Riemann sums also approach a definite integral, we conclude that the value of $Q$ is given by the definite integral.

Once we have expressed $Q$ as a definite integral, we may calculate its value using the fundamental theorem of calculus, which reduces the calculation to antidifferentiation. It is not necessary to calculate the value of any Riemann sum. Rather, we use the Riemann sums only as a device to express $Q$ as a definite integral. The key step is to recognize a given sum as a Riemann sum and to determine the corresponding definite integral. Example 1 illustrates this step.

**EXAMPLE 1**  Suppose that the interval $1 \le x \le 2$ is divided into 50 subintervals, each of length $\Delta x$. Let $x_1, x_2, \ldots, x_{50}$ denote points selected from these subintervals. Find an approximate value for the sum

$$(8x_1^7 + 6x_1)\,\Delta x + (8x_2^7 + 6x_2)\,\Delta x + \cdots + (8x_{50}^7 + 6x_{50})\,\Delta x.$$

**Solution**  The sum is clearly a Riemann sum for the function $f(x) = 8x^7 + 6x$ on the interval $1 \le x \le 2$. Therefore, an approximation to the sum is given by the integral

$$\int_1^2 (8x^7 + 6x)\,dx.$$

We may evaluate this integral using the fundamental theorem of calculus:

$$\int_1^2 (8x^7 + 6x)\,dx = (x^8 + 3x^2)\Big|_1^2$$

$$= [2^8 + 3(2^2)] - [1^8 + 3(1^2)]$$

$$= 268 - 4 = 264.$$

Therefore, the sum is approximately 264.  ◼

**The Average Value of a Function**  Let $f(x)$ be a continuous function on the interval $a \le x \le b$. The definite integral may be used to define the *average value* of $f(x)$ on this interval. To calculate the average of a collection of numbers $y_1, y_2, \ldots, y_n$, we add the numbers and divide by $n$ to obtain

$$\frac{y_1 + y_2 + \cdots + y_n}{n}.$$

To determine the average value of $f(x)$, we proceed similarly. Choose $n$ values of $x$, say $x_1, x_2, \ldots, x_n$, and calculate the corresponding function values $f(x_1)$, $f(x_2)$, $\ldots$, $f(x_n)$. The average of these values is

$$\frac{f(x_1) + f(x_2) + \cdots + f(x_n)}{n}. \tag{1}$$

Our goal now is to obtain a reasonable definition of the average of all the values of $f(x)$ on the interval $a \le x \le b$. If the points $x_1, x_2, \ldots, x_n$ are spread "evenly" throughout the interval, then the average (1) should be a good approximation to our intuitive concept of the average value of $f(x)$. In fact, as $n$ becomes large, the average (1) should approximate the average value of $f(x)$ to any arbitrary degree of accuracy. To guarantee that the points $x_1, x_2, \ldots, x_n$ are "evenly" spread out from $a$ to $b$, let us divide the interval from $x = a$ to $x = b$ into $n$ subintervals of equal length $\Delta x = (b-a)/n$. Then choose $x_1$ from the first subinterval, $x_2$ from the second, and so forth. The average (1) that corresponds to these points may be arranged in the form of a Riemann sum as follows:

$$\frac{f(x_1) + f(x_2) + \cdots + f(x_n)}{n}$$

$$= f(x_1)\cdot\frac{1}{n} + f(x_2)\cdot\frac{1}{n} + \cdots + f(x_n)\cdot\frac{1}{n}$$

$$= \frac{1}{b-a}\left[f(x_1)\cdot\frac{b-a}{n} + f(x_2)\cdot\frac{b-a}{n} + \cdots + f(x_n)\cdot\frac{b-a}{n}\right]$$

$$= \frac{1}{b-a}[f(x_1)\,\Delta x + f(x_2)\,\Delta x + \cdots + f(x_n)\,\Delta x].$$

The sum inside the brackets is a Riemann sum for the definite integral of $f(x)$. Thus we see that for a large number of points $x_i$, the average in (1) approaches the quantity

$$\frac{1}{b-a}\int_a^b f(x)\,dx.$$

This argument motivates the following definition.

The *average value* of a continuous function $f(x)$ over the interval $a \le x \le b$ is defined as the quantity

$$\frac{1}{b-a}\int_a^b f(x)\,dx. \tag{2}$$

---

**EXAMPLE 2**   Compute the average value of $f(x) = \sqrt{x}$ over the interval $0 \le x \le 9$.

**Solution**   Using (2) with $a = 0$ and $b = 9$, the average value of $f(x) = \sqrt{x}$ over the interval $0 \le x \le 9$ is equal to

$$\frac{1}{9-0}\int_0^9 \sqrt{x}\,dx.$$

Since $\sqrt{x} = x^{1/2}$, an antiderivative of $f(x) = \sqrt{x}$ is $F(x) = \frac{2}{3}x^{3/2}$. Therefore,

$$\frac{1}{9}\int_0^9 \sqrt{x}\,dx = \frac{1}{9}\left(\frac{2}{3}x^{3/2}\right)\Big|_0^9 = \frac{1}{9}\left(\frac{2}{3}\cdot 9^{3/2} - 0\right) = \frac{1}{9}\left(\frac{2}{3}\cdot 27\right) = 2,$$

so that the average value of $f(x) = \sqrt{x}$ over the interval $0 \le x \le 9$ is 2. The area of the shaded region is the same as the area of the rectangle pictured in Fig. 1. ■

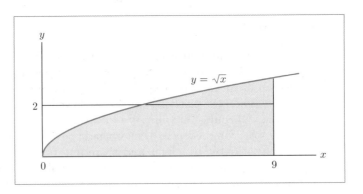

**Figure 1.**   Average value of a function.

---

**EXAMPLE 3**   On January 1, 1998 the world population was 5.9 billion and growing at the rate of 1.34% annually. Assuming this growth rate continues, the population $t$ years from then will be given by the exponential growth law

$$P(t) = 5.9e^{.0134t}.$$

Determine the average world population during the next 30 years. (This average is important in long-range planning for agricultural production and the allocation of goods and services.)

**Solution** The average value of the population $P(t)$ from $t = 0$ to $t = 30$ is

$$\frac{1}{30 - 0} \int_0^{30} P(t)\, dt = \frac{1}{30} \int_0^{30} 5.9 e^{.0134t}\, dt$$

$$= \frac{1}{30} \left( \frac{5.9}{.0134} e^{.0134t} \right) \Bigg|_0^{30} = \frac{5.9}{.402} (e^{.402} - 1)$$

$$\approx 7.26 \text{ billion.} \qquad\blacksquare$$

**Consumers' Surplus**  Using a demand curve from economics, we can derive a formula showing the amount that consumers benefit from an open system that has no price discrimination. Figure 2(a) is a *demand curve* for a commodity. It is determined by complex economic factors and gives a relationship between the quantity sold and the unit price of a commodity. Specifically, it says that, in order to sell $x$ units, the price must be set at $f(x)$ dollars per unit. Since, for most commodities, selling larger quantities requires a lowering of the price, demand curves are usually decreasing. Interactions between supply and demand determine the amount of a quantity available. Let $A$ designate the amount of the commodity currently available and $B = f(A)$ the current selling price.

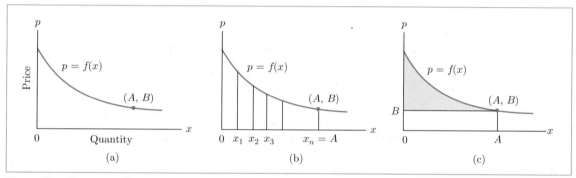

**Figure 2.**  Consumers' surplus.

Divide the interval from 0 to $A$ into $n$ subintervals, each of length $\Delta x = (A - 0)/n$, and take $x_i$ to be the right-hand endpoint of the $i$th interval. Consider the first subinterval, from 0 to $x_1$. [See Fig. 2(b).] Suppose that only $x_1$ units had been available. Then the price per unit could have been set at $f(x_1)$ dollars and these $x_1$ units sold. Of course, at this price we could not have sold any more units. However, those people who paid $f(x_1)$ dollars had a great demand for the commodity. It was extremely valuable to them, and there was no advantage in substituting another commodity at that price. They were actually paying what the commodity was worth to them. In theory, then, the first $x_1$ units of the commodity could be sold to these people at $f(x_1)$ dollars per unit, yielding

$$(\text{price per unit}) \cdot (\text{number of units}) = f(x_1) \cdot (x_1) = f(x_1) \cdot \Delta x \text{ dollars.}$$

After selling the first $x_1$ units, suppose that more units become available, so that now a total of $x_2$ units have been produced. Setting the price at $f(x_2)$, the remaining $x_2 - x_1 = \Delta x$ units can be sold, yielding $f(x_2) \cdot \Delta x$ dollars. Here, again, the second group of buyers would have paid as much for the commodity as it was worth to them. Continuing this process of price discrimination, the amount of money paid by consumers would be

$$f(x_1)\, \Delta x + f(x_2)\, \Delta x + \cdots + f(x_n)\, \Delta x.$$

Taking $n$ large, we see that this Riemann sum approaches $\int_0^A f(x)\,dx$. Since $f(x)$ is positive, this integral equals the area under the graph of $f(x)$ from $x = 0$ to $x = A$.

Of course, in an open system, everyone pays the same price, $B$, so the total amount paid is [price per unit] · [number of units] $= BA$. Since $BA$ is the area of the rectangle under the graph of the line $p = B$ from $x = 0$ to $x = A$, the amount of money saved by the consumers is the area of the shaded region in Fig. 2(c). That is, the area between the curves $p = f(x)$ and $p = B$ gives a numerical value to one benefit of a modern efficient economy.

> The *consumers' surplus* for a commodity having demand curve $p = f(x)$ is
> $$\int_0^A [f(x) - B]\,dx,$$
> where the quantity demanded is $A$ and the price is $B = f(A)$.

**EXAMPLE 4**  Find the consumers' surplus for the demand curve $p = 50 - .06x^2$ at the sales level $x = 20$.

**Solution**  Since 20 units are sold, the price must be
$$B = 50 - .06(20)^2 = 50 - 24 = 26.$$
Therefore, the consumers' surplus is
$$\int_0^{20} [(50 - .06x^2) - 26]\,dx = \int_0^{20} (24 - .06x^2)\,dx$$
$$= (24x - .02x^3)\Big|_0^{20}$$
$$= 24(20) - .02(20)^3$$
$$= 480 - 160 = 320.$$

That is, the consumers' surplus is \$320.  ■

**Future Value of an Income Stream**  The next example shows how the definite integral can be used to approximate the sum of a large number of terms.

**EXAMPLE 5**  Suppose that money is deposited daily into a savings account at an annual rate of \$1000. The account pays 6% interest compounded continuously. Approximate the amount of money in the account at the end of 5 years.

**Solution**  Divide the time interval from 0 to 5 years into daily subintervals. Each subinterval is then of duration $\Delta t = \frac{1}{365}$ years. Let $t_1, t_2, \ldots, t_n$ be points chosen from these subintervals. Since we deposit money at an annual rate of \$1000, the amount deposited during one of the subintervals is $1000\Delta t$ dollars. If this amount is deposited at time $t_i$, the $1000\Delta t$ dollars will earn interest for the remaining $5 - t_i$ years. The total amount resulting from this one deposit at time $t_i$ is then
$$1000\,\Delta t\, e^{.06(5-t_i)}.$$

Add the effects of the deposits at times $t_1, t_2, \ldots, t_n$ to arrive at the total balance in the account:
$$A = 1000e^{.06(5-t_1)}\Delta t + 1000e^{.06(5-t_2)}\Delta t + \cdots + 1000e^{.06(5-t_n)}\Delta t.$$

This is a Riemann sum for the function $f(t) = 1000e^{.06(5-t)}$ on the interval $0 \leq t \leq 5$. Since $\Delta t$ is very small when compared with the interval, the total amount in the account, $A$, is approximately

$$\int_0^5 1000e^{.06(5-t)}\, dt = \left. \frac{1000}{-.06}\, e^{.06(5-t)} \right|_0^5 = \frac{1000}{-.06}(1 - e^{.3}) \approx 5831.$$

That is, the approximate balance in the account at the end of 5 years is \$5831.  ■

▶ *Note*   The antiderivative was computed by observing that since the derivative of $e^{.06(5-t)}$ is $e^{.06(5-t)}(-.06)$, we must divide the integrand by $-.06$ to obtain an antiderivative. ◀

In Example 5, money was deposited daily into the account. If the money had been deposited several times a day, the definite integral would have given an even better approximation to the balance. Actually, the more frequently the money is deposited, the better the approximation. Economists consider a hypothetical situation where money is deposited steadily throughout the year. This flow of money is called a *continuous income stream*, and the balance in the account is given exactly by the definite integral.

> The *future value of a continuous income stream* of $K$ dollars per year for $N$ years at interest rate $r$ compounded continuously is
>
> $$\int_0^N Ke^{r(N-t)}\, dt.$$

$y = g(x)$

$a$           $b$

(a)

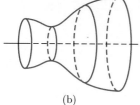

(b)

**Figure 3**

**Solids of Revolution**   When the region of Fig. 3(a) is revolved about the $x$-axis, it sweeps out a solid [Fig. 3(b)]. Riemann sums can be used to derive a formula for the volume of this *solid of revolution*. Let us break the $x$-axis between $a$ and $b$ into a large number $n$ of equal subintervals, each of length $\Delta x = (b-a)/n$. Using each subinterval as a base, we can divide the region into strips (see Fig. 4).

Let $x_i$ be a point in the $i$th subinterval. Then the volume swept out by revolving the $i$th strip is approximately the same as the volume of the cylinder swept out by revolving the rectangle of height $g(x_i)$ and base $\Delta x$ around the $x$-axis (Fig. 5). The volume of the cylinder is

$$[\text{area of circular side}] \cdot [\text{width}] = \pi[g(x_i)]^2 \cdot \Delta x.$$

The total volume swept out by all the strips is approximated by the total volume

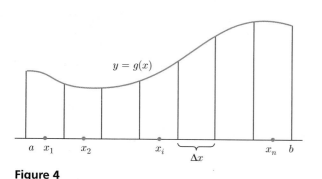

$y = g(x)$

$a$   $x_1$    $x_2$      $x_i$      $x_n$   $b$

$\Delta x$

**Figure 4**

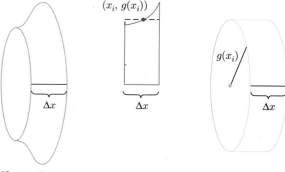

$(x_i, g(x_i))$

$g(x_i)$

$\Delta x$      $\Delta x$      $\Delta x$

**Figure 5**

swept out by the rectangles, which is

$$[\text{volume}] \approx \pi[g(x_1)]^2\Delta x + \pi[g(x_2)]^2\Delta x + \cdots + \pi[g(x_n)]^2\Delta x.$$

As $n$ gets larger and larger, this approximation becomes arbitrarily close to the true volume. The expression on the right is a Riemann sum for the definite integral of $f(x) = \pi[g(x)]^2$. Therefore, the volume of the solid equals the value of the definite integral.

> The volume of the *solid of revolution* obtained from revolving the region below the graph of $y = g(x)$ from $x = a$ to $x = b$ about the $x$-axis is
>
> $$\int_a^b \pi[g(x)]^2\,dx.$$

---

**EXAMPLE 6**   Find the volume of the solid of revolution obtained by revolving the region of Fig. 6 about the $x$-axis.

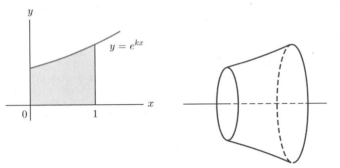

**Figure 6**

**Solution**  Here $g(x) = e^{kx}$, and

$$[\text{volume}] = \int_0^1 \pi(e^{kx})^2\,dx = \int_0^1 \pi e^{2kx}\,dx = \frac{\pi}{2k}e^{2kx}\Big|_0^1 = \frac{\pi}{2k}(e^{2k} - 1). \qquad\blacksquare$$

---

**EXAMPLE 7**   Find the volume of a right circular cone of radius $r$ and height $h$.

**Solution**  The cone [Fig. 7(a)] is the solid of revolution swept out when the shaded region in Fig. 7(b) is revolved about the $x$-axis. Using the formula developed previously, the volume of the cone is

$$\int_0^h \pi\left(\frac{r}{h}x\right)^2 dx = \frac{\pi r^2}{h^2}\int_0^h x^2\,dx = \frac{\pi r^2}{h^2}\frac{x^3}{3}\Big|_0^h = \frac{\pi r^2 h}{3}. \qquad\blacksquare$$

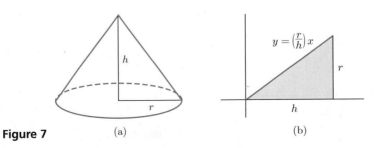

**Figure 7**      (a)                              (b)

## PRACTICE PROBLEMS 6.5

**1.** A rock dropped from a bridge has a velocity of $-32t$ feet per second after $t$ seconds. Find the average velocity of the rock during the first three seconds.

**2.** An investment yields $300 per year compounded continuously for 10 years at 10% interest. What is the (future) value of this income stream at the end of 10 years?

# Exercises 6.5

*Determine the average value of $f(x)$ over the interval from $x = a$ to $x = b$, where:*

**1.** $f(x) = x^2$; $a = 0$, $b = 3$

**2.** $f(x) = 1 - x$; $a = -1$, $b = 1$

**3.** $f(x) = 100e^{-.5x}$; $a = 0$, $b = 4$

**4.** $f(x) = 2$; $a = 0$, $b = 1$

**5.** $f(x) = 1/x$; $a = 1/3$, $b = 3$

**6.** $f(x) = \dfrac{1}{\sqrt{x}}$; $a = 1$, $b = 9$

**7.** (*Average Temperature*) During a certain 12-hour period the temperature at time $t$ (measured in hours from the start of the period) was $T(t) = 47 + 4t - \frac{1}{3}t^2$ degrees. What was the average temperature during that period?

**8.** (*Average Population*) Assuming that a country's population is now 3 million and is growing exponentially with growth constant .02, what will be the average population during the next 50 years?

**9.** (*Average Amount of Radium*) One hundred grams of radioactive radium having a half-life of 1690 years is placed in a concrete vault. What will be the average amount of radium in the vault during the next 1000 years?

**10.** (*Average Amount of Money*) One hundred dollars are deposited in the bank at 5% interest compounded continuously. What will be the average value of the money in the account during the next 20 years?

*Find the consumers' surplus for each of the following demand curves at the given sales level $x$.*

**11.** $p = 3 - \dfrac{x}{10}$; $x = 20$

**12.** $p = \dfrac{x^2}{200} - x + 50$; $x = 20$

**13.** $p = \dfrac{500}{x + 10} - 3$; $x = 40$

**14.** $p = \sqrt{16 - .02x}$; $x = 350$

*Figure 8 shows a supply curve for a commodity. It gives the relationship between the selling price of the commodity and the quantity that producers will manufacture. At a higher selling price, a greater quantity will be produced. Therefore, the curve is increasing. If $(A, B)$ is a point on the curve, then, in order to stimulate the production of $A$ units of the commodity, the price per unit must be $B$ dollars. Of course, some producers will be willing to produce the commodity even with a lower selling price. Since everyone receives the same price in an open efficient economy, most producers are receiving more than their minimal required price. The excess is called the* producers' surplus. *Using an argument analogous to that of the* consumers' surplus, *one can show that the total producers' surplus when the price is $B$ is the area of the shaded region in Fig. 8. Find the producers' surplus for each of the following supply curves at the given sales level $x$.*

**15.** $p = .01x + 3$; $x = 200$

**16.** $p = \dfrac{x^2}{9} + 1$; $x = 3$

**17.** $p = \dfrac{x}{2} + 7$; $x = 10$

**18.** $p = 1 + \frac{1}{2}\sqrt{x}$; $x = 36$

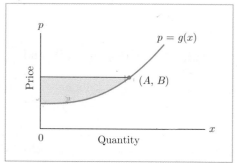

**Figure 8.**  Producers' surplus.

*For a particular commodity, the quantity produced and the unit price are given by the coordinates of the point where the supply and demand curves intersect. For each pair of supply and demand curves, determine the point of intersection $(A, B)$ and the consumers' and producers' surplus. (See Fig. 9.)*

**19.** Demand curve: $p = 12 - (x/50)$; supply curve: $p = (x/20) + 5$.

**20.** Demand curve: $p = \sqrt{25 - .1x}$; supply curve: $p = \sqrt{.1x + 9} - 2$.

**21.** (*Future Value*) Suppose that money is deposited daily into a savings account at an annual rate of $1000. If the account pays 5% interest compounded continuously, estimate the balance in the account at the end of 3 years.

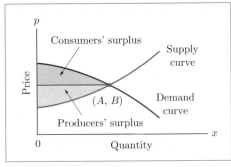

**Figure 9**

22. (*Future Value*) Suppose that money is deposited daily into a savings account at an annual rate of $2000. If the account pays 6% interest compounded continuously, approximately how much will be in the account at the end of 2 years?

23. (*Future Value*) Suppose that money is deposited steadily into a savings account at the rate of $16,000 per year. Determine the balance at the end of 4 years if the account pays 8% interest compounded continuously.

24. (*Future Value*) Suppose that money is deposited steadily into a savings account at the rate of $14,000 per year. Determine the balance at the end of 6 years if the account pays 4.5% interest compounded continuously.

25. An investment pays 10% interest compounded continuously. If money is invested steadily at the rate of $5000 per year, how much time is required until the value of the investment reaches $140,000?

26. A savings account pays 4.25% interest compounded continuously. At what rate per year must money be deposited steadily into the account in order to accumulate a balance of $100,000 after 10 years?

27. Suppose money is to be deposited daily for 5 years into a savings account at an annual rate of $1000 and the account pays 4% interest compounded continuously. Let the interval from 0 to 5 be divided into daily subintervals, with each subinterval of duration $\Delta t = \frac{1}{365}$ years. Let $t_1, \dots, t_n$ be points chosen from the subintervals.

    (a) Show that the present value of a daily deposit at time $t_i$ is $1000 \, \Delta t \, e^{-.04 t_i}$.

    (b) Find the Riemann sum corresponding to the sum of the present values of all the deposits.

    (c) What is the function and interval corresponding to the Riemann sum in (b)?

    (d) Give the definite integral that approximates the Riemann sum in (b).

    (e) Evaluate the definite integral in (d). This number is the *present value of a continuous income stream*.

28. Use the result of Exercise 27 to calculate the present value of a continuous income stream of $5000 per year for 10 years at an interest rate of 5% compounded continuously.

*Find the volume of the solid of revolution generated by revolving about the x-axis the region under each of the following curves.*

29. $y = \sqrt{4 - x^2}$ from $x = -2$ to $x = 2$ (generates a sphere of radius 2)

30. $y = \sqrt{r^2 - x^2}$ from $x = -r$ to $x = r$ (generates a sphere of radius $r$)

31. $y = x^2$ from $x = 1$ to $x = 2$

32. $y = kx$ from $x = 0$ to $x = h$ (generates a cone)

33. $y = \sqrt{x}$ from $x = 0$ to $x = 4$ (The solid generated is called a *paraboloid*.)

34. $y = 2x - x^2$ from $x = 0$ to $x = 2$

35. $y = e^{-x}$ from $x = 0$ to $x = 1$

36. $y = 2x + 1$ from $x = 0$ to $x = 1$ (The solid generated is called a *truncated cone*.)

*For the Riemann sums in Exercises 37–40, determine n, b, and f(x).*

37. $\left[ (8.25)^3 + (8.75)^3 + (9.25)^3 + (9.75)^3 \right] (.5); \; a = 8$

38. $\left[ \dfrac{3}{1} + \dfrac{3}{1.5} + \dfrac{3}{2} + \dfrac{3}{2.5} + \dfrac{3}{3} + \dfrac{3}{3.5} \right] (.5); \; a = 1$

39. $\left[ (5 + e^5) + (6 + e^6) + (7 + e^7) \right] (1); \; a = 4$

40. $\left[ 3(.3)^2 + 3(.9)^2 + 3(1.5)^2 + 3(2.1)^2 + 3(2.7)^2 \right] (.6); \; a = 0$

41. Suppose that the interval $0 \le x \le 3$ is divided into 100 subintervals of width $\Delta x = .03$. Let $x_1, x_2, \dots, x_{100}$ be points in these subintervals. Suppose that in a particular application, one needs to estimate the sum

    $$(3 - x_1)^2 \Delta x + (3 - x_2)^2 \Delta x + \cdots + (3 - x_{100})^2 \Delta x.$$

    Show that this sum is close to 9.

42. Suppose that the interval $0 \le x \le 1$ is divided into 100 subintervals of width $\Delta x = .01$. Show that the following sum is close to 5/4.

    $$\left[ 2(.01) + (.01)^3 \right] \Delta x + \left[ 2(.02) + (.02)^3 \right] \Delta x$$
    $$+ \cdots + \left[ 2(1.0) + (1.0)^3 \right] \Delta x$$

### Technology Exercises

*The following exercises ask for an unknown quantity, call it x. After setting up the appropriate formula involving a definite integral, use the fundamental theorem to evaluate the definite integral as an expression in x. Because the resulting equation will be too complicated to solve algebraically, you must use a graphing utility to obtain the solution. Note: If the quantity x is an interest rate paid by a savings account, it will most likely be between 0 and .10.*

**43.** A single deposit of $1000 is to be made into a savings account and the interest (compounded continuously) is allowed to accumulate for 3 years. Therefore, the amount at the end of $t$ years is $1000e^{rt}$.

   **(a)** Find an expression (involving $r$) that gives the average value of the money in the account during the three-year time period $0 \le t \le 3$.

   **(b)** Find the interest rate $r$ at which the average amount in the account during the three-year period is $1070.60.

**44.** A single deposit of $100 is made into a savings account paying 4% interest compounded continuously. How long must the money be held in the account so that the average amount of money during that time period will be $122.96?

**45.** Money is deposited steadily at the rate of $1000 per year into a savings account.

   **(a)** Find the expression (involving $r$) that gives the (future) balance in the account at the end of 6 years.

   **(b)** Find the interest rate that will result in a balance of $6997.18 after 6 years.

**46.** Money is deposited steadily at the rate of $3000 per year into a savings account. After 10 years the balance is $36,887. What interest rate, with interest compounded continuously, did the money earn?

---

## SOLUTIONS TO PRACTICE PROBLEMS 6.5

**1.** By definition, the average value of the function $v(t) = -32t$ for $t = 0$ to $t = 3$ is

$$\frac{1}{3-0}\int_0^3 -32t\,dt = \frac{1}{3}(-16t^2)\Big|_0^3 = \frac{1}{3}(-16 \cdot 3^2)$$

$$= -48 \text{ feet per second.}$$

   *Note:* There is another way to approach this problem:

$$[\text{average velocity}] = \frac{[\text{distance traveled}]}{[\text{time elapsed}]}.$$

As we discussed in Section 6.2, distance traveled equals the area under the velocity curve. Therefore,

$$[\text{average velocity}] = \frac{\int_0^3 -32t\,dt}{3}.$$

**2.** According to the formula developed in the text, the future value of the income stream after 10 years is equal to

$$\int_0^{10} 300e^{.1(10-t)}\,dt = -\frac{300}{.1}e^{.1(10-t)}\Big|_0^{10}$$

$$= -3000e^0 - (-3000e^1)$$

$$= 3000e - 3000$$

$$\approx \$5154.85.$$

---

# REVIEW OF FUNDAMENTAL CONCEPTS

**1.** What does it mean to antidifferentiate a function?

**2.** State the formula for $\int h(x)\,dx$ for each of the following functions.

   **(a)** $h(x) = x^r$, $r \ne -1$    **(b)** $h(x) = e^{kx}$

   **(c)** $h(x) = \dfrac{1}{x}$    **(d)** $h(x) = f(x) + g(x)$

   **(e)** $h(x) = kf(x)$

**3.** In the formula $\Delta x = \dfrac{b-a}{n}$, what do $a$, $b$, $n$, and $\Delta x$ denote?

**4.** What is a Riemann sum?

**5.** Give an interpretation of the area under a rate of change function. Give a concrete example.

**6.** What is a definite integral?

**7.** What is the difference between a definite integral and an indefinite integral?

**8.** State the fundamental theorem of calculus.

**9.** How is $F(x)\Big|_a^b$ calculated and what is it called?

**10.** Outline a procedure for finding the area of a region bounded by two curves.

**11.** State the formula for each of the following quantities:

   **(a)** average value of a function

   **(b)** consumers' surplus

   **(c)** future value of an income stream

   **(d)** volume of a solid of revolution

# SUPPLEMENTARY EXERCISES

*Calculate the following integrals.*

**1.** $\displaystyle\int 3^2\,dx$

**2.** $\displaystyle\int (x^2 - 3x + 2)\,dx$

**3.** $\displaystyle\int \sqrt{x+1}\,dx$

**4.** $\displaystyle\int \frac{2}{x+4}\,dx$

**5.** $\displaystyle 2\int (x^3 + 3x^2 - 1)\,dx$

**6.** $\displaystyle\int \sqrt[5]{x+3}\,dx$

**7.** $\displaystyle\int e^{-x/2}\,dx$

**8.** $\displaystyle\int \frac{5}{\sqrt{x-7}}\,dx$

**9.** $\displaystyle\int (3x^4 - 4x^3)\,dx$

**10.** $\displaystyle\int (2x+3)^7\,dx$

**11.** $\displaystyle\int \sqrt{4-x}\,dx$

**12.** $\displaystyle\int \left(\frac{5}{x} - \frac{x}{5}\right)dx$

**13.** $\displaystyle\int_{-1}^{1} (x+1)^2\,dx$

**14.** $\displaystyle\int_{0}^{1/8} \sqrt[3]{x}\,dx$

**15.** $\displaystyle\int_{-1}^{2} \sqrt{2x+4}\,dx$

**16.** $\displaystyle 2\int_{0}^{1} \left(\frac{2}{x+1} - \frac{1}{x+4}\right)dx$

**17.** $\displaystyle\int_{1}^{2} \frac{4}{x^5}\,dx$

**18.** $\displaystyle\frac{2}{3}\int_{0}^{8} \sqrt{x+1}\,dx$

**19.** $\displaystyle\int_{1}^{4} \frac{1}{x^2}\,dx$

**20.** $\displaystyle\int_{3}^{6} e^{2-(x/3)}\,dx$

**21.** $\displaystyle\int_{0}^{5} (5+3x)^{-1}\,dx$

**22.** $\displaystyle\int_{-2}^{2} \frac{3}{2e^{3x}}\,dx$

**23.** $\displaystyle\int_{0}^{\ln 2} (e^x - e^{-x})\,dx$

**24.** $\displaystyle\int_{\ln 2}^{\ln 3} (e^x + e^{-x})\,dx$

**25.** $\displaystyle\int_{0}^{\ln 3} \frac{e^x + e^{-x}}{e^{2x}}\,dx$

**26.** $\displaystyle\int_{0}^{1} \frac{3+e^{2x}}{e^x}\,dx$

**27.** Find the area under the curve $y = (3x-2)^{-3}$ from $x = 1$ to $x = 2$.

**28.** Find the area under the curve $y = 1 + \sqrt{x}$ from $x = 1$ to $x = 9$.

*In Exercises 29–36, find the area of the shaded region.*

**29.**

**30.**

**31.**

**32.**

**33.**

**34.**

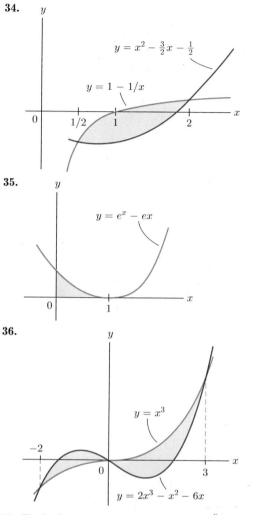

**35.**

**36.**

**37.** Find the area of the region bounded by the curves $y = x^3 - 3x + 1$ and $y = x + 1$.

**38.** Find the area of the region between the curves $y = 2x^2 + x$ and $y = x^2 + 2$ from $x = 0$ to $x = 2$.

**39.** Find the function $f(x)$ for which $f'(x) = (x - 5)^2$, $f(8) = 2$.

**40.** Find the function $f(x)$ for which $f'(x) = e^{-5x}$, $f(0) = 1$.

**41.** Describe all solutions of the following differential equations, where $y$ represents a function of $t$.

   **(a)** $y' = 4t$

   **(b)** $y' = 4y$

   **(c)** $y' = e^{4t}$

**42.** Let $k$ be a constant, and let $y = f(t)$ be a function such that $y' = kty$. Show that $y = Ce^{kt^2/2}$, for some constant $C$. [*Hint*: Use the product rule to evaluate $\frac{d}{dt}[f(t)e^{-kt^2/2}]$, and then apply Theorem II of Section 6.1.]

**43.** An airplane tire plant finds that its marginal cost of producing tires is $.04x + 150$ dollars at a production level of $x$ tires per day. If fixed costs are $500 per day, find the cost of producing $x$ tires per day.

**44.** Suppose that the marginal revenue function for a company is $400 - 3x^2$. Find the additional revenue received from doubling production if currently 10 units are being produced.

**45.** A drug is injected into a patient at the rate of $f(t)$ cubic centimeters per minute at time $t$. What does the area under the graph of $y = f(t)$ from $t = 0$ to $t = 4$ represent?

**46.** A rock thrown straight up into the air has a velocity of $v(t) = -9.8t + 20$ meters per second after $t$ seconds.

   **(a)** Determine the distance the rock travels during the first 2 seconds.

   **(b)** Represent the answer to part (a) as an area.

**47.** Use a Riemann sum with $n = 4$ and left endpoints to estimate the area under the graph in Fig. 1 for $0 \leq x \leq 2$.

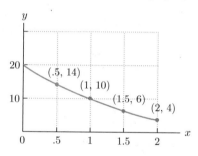

**Figure 1**

**48.** Redo Exercise 47 using right endpoints.

**49.** Use a Riemann sum with $n = 2$ and midpoints to estimate the area under the graph of

$$f(x) = \frac{1}{x + 2}$$

on the interval $0 \leq x \leq 2$. Then use a definite integral to find the exact value of the area to five decimal places.

**50.** Use a Riemann sum with $n = 5$ and midpoints to estimate the area under the graph of $f(x) = e^{2x}$ on the interval $0 \leq x \leq 1$. Then use a definite integral to find the exact value of the area to five decimal places.

**51.** Find the consumers' surplus for the demand curve $p = \sqrt{25 - .04x}$ at the sales level $x = 400$.

**52.** Three thousand dollars is deposited in the bank at 4% interest compounded continuously. What will be the average value of the money in the account during the next 10 years?

**53.** Find the average value of $f(x) = 1/x^3$ from $x = \frac{1}{3}$ to $x = \frac{1}{2}$.

**54.** Suppose that the interval $0 \le x \le 1$ is divided into 100 subintervals with a width of $\Delta x = .01$. Show that the sum

$$\left[3e^{-.01}\right]\Delta x + \left[3e^{-.02}\right]\Delta x + \left[3e^{-.03}\right]\Delta x$$
$$+ \cdots + \left[3e^{-1}\right]\Delta x$$

is close to $3(1 - e^{-1})$.

**55.** In Fig. 2, three regions are labeled with their areas. Determine $\int_a^c f(x)\,dx$ and determine $\int_a^d f(x)\,dx$.

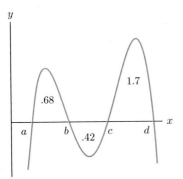

**Figure 2**

**56.** Find the volume of the solid of revolution generated by revolving about the $x$-axis the region under the curve $y = 1 - x^2$ from $x = 0$ to $x = 1$.

**57.** A store has an inventory of $Q$ units of a certain product at time $t = 0$. The store sells the product at the steady rate of $Q/A$ units per week, and exhausts the inventory in $A$ weeks.

  **(a)** Find a formula $f(t)$ for the amount of product in inventory at time $t$.

  **(b)** Find the average inventory level during the period $0 \le t \le A$.

**58.** A retail store sells a certain product at the rate of $g(t)$ units per week at time $t$, where $g(t) = rt$. At time $t = 0$, the store has $Q$ units of the product in inventory.

  **(a)** Find a formula $f(t)$ for the amount of product in inventory at time $t$.

  **(b)** Determine the value of $r$ in part (a) such that the inventory is exhausted in $A$ weeks.

  **(c)** Using $f(t)$, with $r$ as in part (b), find the average inventory level during the period $0 \le t \le A$.

**59.** Let $x$ be any positive number, and define $g(x)$ to be the number determined by the definite integral

$$g(x) = \int_0^x \frac{1}{1 + t^2}\,dt.$$

  **(a)** Give a geometric interpretation of the number $g(3)$.

  **(b)** Find the derivative $g'(x)$.

**60.** For each number $x$ satisfying $-1 \le x \le 1$, define $h(x)$ by

$$h(x) = \int_{-1}^x \sqrt{1 - t^2}\,dt.$$

  **(a)** Give a geometric interpretation of the values $h(0)$ and $h(1)$.

  **(b)** Find the derivative $h'(x)$.

**61.** Suppose that the interval $0 \le t \le 3$ is divided into 1000 subintervals of width $\Delta t$. Let $t_1, t_2, \ldots, t_{1000}$ denote the right endpoints of these subintervals. Suppose that in some problem one needs to estimate the sum

$$5000e^{-.1t_1}\,\Delta t + 5000e^{-.1t_2}\,\Delta t$$
$$+ \cdots + 5000e^{-.1t_{1000}}\,\Delta t.$$

Show that this sum is close to 13,000. (*Note:* A sum such as this would arise if one wanted to compute the present value of a continuous stream of income of $5000 per year for 3 years, with interest compounded continuously at 10%.)

**62.** What number does

$$\left[e^0 + e^{1/n} + e^{2/n} + e^{3/n} + \cdots + e^{(n-1)/n}\right] \cdot \frac{1}{n}$$

approach as $n$ gets very large?

**63.** What number does the sum

$$\left[1^3 + \left(1 + \frac{1}{n}\right)^3 + \left(1 + \frac{2}{n}\right)^3 + \left(1 + \frac{3}{n}\right)^3\right.$$
$$\left. + \cdots + \left(1 + \frac{n-1}{n}\right)^3\right] \cdot \frac{1}{n}$$

approach as $n$ gets very large?

**64.** In Fig. 3, the rectangle has the same area as the region under the graph of $f(x)$. What is the average value of $f(x)$ on the interval $2 \le x \le 6$?

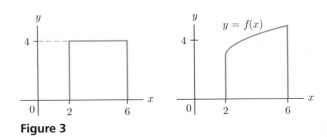

**Figure 3**

**65.** *True or false:* If $3 \le f(x) \le 4$ whenever $0 \le x \le 5$, then $3 \le \frac{1}{5}\int_0^5 f(x)\,dx \le 4$.

**66.** Suppose that water is flowing into a tank at a rate of $r(t)$ gallons per hour, where the rate depends on the time $t$ according to the formula

$$r(t) = 20 - 4t, \qquad 0 \le t \le 5.$$

(a) Consider a brief period of time, say from $t_1$ to $t_2$. The length of this time period is $\Delta t = t_2 - t_1$. During this period the rate of flow does not change much and is approximately $20 - 4t_1$ (the rate at the beginning of the brief time interval). Approximately how much water flows into the tank during the time from $t_1$ to $t_2$?

(b) Explain why the total amount of water added to the tank during the time interval from $t = 0$ to $t = 5$ is given by $\int_0^5 r(t)\, dt$.

**67.** The annual world rate of water use $t$ years after 1960, for $t \le 35$, was approximately $860e^{.04t}$ cubic kilometers per year. How much water was used between 1960 and 1995?

**68.** Suppose that money is deposited steadily into a savings account at the rate of \$4500 per year. Determine the balance at the end of 1 year if the account pays 9% interest compounded continuously.

**69.** Find a function $f(x)$ whose graph goes through the point $(1, 1)$ and whose slope at any point $(x, f(x))$ is $3x^2 - 2x + 1$.

**70.** For what value of $a$ is the shaded area in Fig. 4 equal to 1?

**71.** Show that for any positive number $b$, we have

$$\int_0^{b^2} \sqrt{x}\, dx + \int_0^b x^2\, dx = b^3.$$

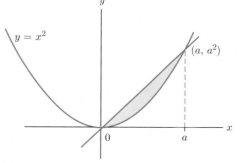

**Figure 4**

**72.** Generalize the result of Exercise 71 as follows. Let $n$ be a positive integer. Show that for any positive number $b$, we have

$$\int_0^{b^n} \sqrt[n]{x}\, dx + \int_0^b x^n\, dx = b^{n+1}.$$

**73.** Show that

$$\int_0^1 (\sqrt{x} - x^2)\, dx = \frac{1}{3}.$$

**74.** Generalize the result of Exercise 73 as follows. Let $n$ be a positive integer. Show that

$$\int_0^1 (\sqrt[n]{x} - x^n)\, dx = \frac{n-1}{n+1}.$$

# Functions of Several Variables

U NTIL now, most of our applications of calculus have involved functions of one variable. In real life, however, a quantity of interest often depends on more than one variable. For instance, the sales level of a product may depend not only on its price but also on the prices of competing products, the amount spent on advertising, and perhaps the time of year. The total cost of manufacturing the product depends on the cost of raw materials, labor, plant maintenance, and so on.

This chapter introduces the basic ideas of calculus for functions of more than one variable. Section 7.1 presents two examples that will be used throughout the chapter. Derivatives are treated in Section 7.2 and then used in Sections 7.3 and 7.4 to solve optimization problems more general than those in Chapter 2. The final two sections are devoted to least-squares problems and a brief introduction to integration of functions of two variables.

## 7.1    Examples of Functions of Several Variables

A function $f(x, y)$ of the two variables $x$ and $y$ is a rule that assigns a number to each pair of values for the variables; for instance,

$$f(x, y) = e^x(x^2 + 2y).$$

An example of a function of three variables is

$$f(x, y, z) = 5xy^2z.$$

---

**EXAMPLE 1**    A store sells butter at \$2.50 per pound and margarine at \$1.40 per pound. The revenue from the sale of $x$ pounds of butter and $y$ pounds of margarine is given by the function

$$f(x, y) = 2.50x + 1.40y.$$

Determine and interpret $f(200, 300)$.

**Solution**    $f(200, 300) = 2.50(200) + 1.40(300) = 500 + 420 = 920$. The revenue from the sale of 200 pounds of butter and 300 pounds of margarine is \$920.    ■

A function $f(x, y)$ of two variables may be graphed in a manner analogous to that for functions of one variable. It is necessary to use a three-dimensional coordinate system, where each point is identified by three coordinates $(x, y, z)$. For each choice of $x$, $y$, the graph of $f(x, y)$ includes the point $(x, y, f(x, y))$. This graph is usually a surface in three-dimensional space, with equation $z = f(x, y)$. See Fig. 1. Three graphs of specific functions are shown in Fig. 2.

**Application to Architectural Design**    When designing a building, one would like to know, at least approximately, how much heat the building loses per day. The heat loss affects many aspects of the design, such as the size of the heating plant, the size and location of duct work, and so on. A building loses heat through its sides, roof, and floor. How much heat is lost will generally differ for each face of the building and will depend on such factors as insulation, materials used in construction, exposure (north, south, east, or west), and climate. It is possible to estimate how much heat is lost per square foot of each face. Using these data, one can construct a heat-loss function as in the following example.

---

**EXAMPLE 2**    A rectangular industrial building of dimensions $x$, $y$, and $z$ is shown in Fig. 3(a). In Fig. 3(b) we give the amount of heat lost per day by each side of the building, measured in suitable units of heat per square foot. Let $f(x, y, z)$ be the total daily heat loss for such a building.

(a) Find a formula for $f(x, y, z)$.

(b) Find the total daily heat loss if the building has length 100 feet, width 70 feet, and height 50 feet.

**Solution**    (a) The total heat loss is the sum of the amount of heat loss through each face of the building. The heat loss through the roof is

[heat loss per square foot of roof] · [area of roof in square feet] $= 10xy$.

Similarly, the heat loss through the east side is $8yz$. Continuing in this way, we see that the total daily heat loss is

$$f(x, y, z) = 10xy + 8yz + 6yz + 10xz + 5xz + 1 \cdot xy.$$

**Figure 2**

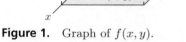

Surface
$z = f(x, y)$

$(x, y, f(x, y))$

$(x, y)$

**Figure 1.** Graph of $f(x, y)$.

We collect terms to obtain

$$f(x, y, z) = 11xy + 14yz + 15xz.$$

(b) The amount of heat loss when $x = 100$, $y = 70$, and $z = 50$ is given by $f(100, 70, 50)$, which equals

$$f(100, 70, 50) = 11(100)(70) + 14(70)(50) + 15(100)(50)$$
$$= 77{,}000 + 49{,}000 + 75{,}000 = 201{,}000.$$

In Section 7.3 we will determine the dimensions $x$, $y$, $z$ that minimize the heat loss for a building of specific volume.

**Production Functions in Economics**    The costs of a manufacturing process can generally be classified as one of two types: cost of labor and cost of capital. The meaning of the cost of labor is clear. By the cost of capital, we mean the cost of buildings, tools, machines, and similar items used in the production process. A manufacturer usually has some control over the relative portions of labor and capital utilized in his production process. He can completely automate production so that labor is at a minimum, or he can utilize mostly labor and little capital.

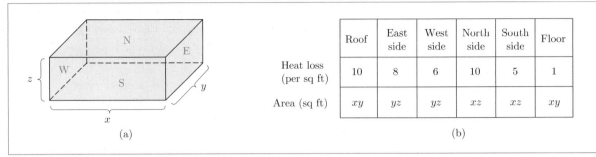

**Figure 3.** Heat loss from an industrial building.

Suppose that $x$ units of labor and $y$ units of capital are used.* Let $f(x, y)$ denote the number of units of finished product that are manufactured. Economists have found that $f(x, y)$ is often a function of the form

$$f(x, y) = Cx^A y^{1-A},$$

where $A$ and $C$ are constants, $0 < A < 1$. Such a function is called a *Cobb–Douglas production function*.

---

**EXAMPLE 3** **Production in a firm** Suppose that during a certain time period the number of units of goods produced when utilizing $x$ units of labor and $y$ units of capital is $f(x, y) = 60x^{3/4}y^{1/4}$.

(a) How many units of goods will be produced by using 81 units of labor and 16 units of capital?

(b) Show that whenever the amounts of labor and capital being used are doubled, so is the production. (Economists say that the production function has "constant returns to scale.")

**Solution** (a) $f(81, 16) = 60(81)^{3/4} \cdot (16)^{1/4} = 60 \cdot 27 \cdot 2 = 3240$. There will be 3240 units of goods produced.

(b) Utilization of $a$ units of labor and $b$ units of capital results in the production of $f(a, b) = 60a^{3/4}b^{1/4}$ units of goods. Utilizing $2a$ and $2b$ units of labor and capital, respectively, results in $f(2a, 2b)$ units produced. Set $x = 2a$ and $y = 2b$. Then we see that

$$f(2a, 2b) = 60(2a)^{3/4}(2b)^{1/4}$$

$$= 60 \cdot 2^{3/4} \cdot a^{3/4} \cdot 2^{1/4} \cdot b^{1/4}$$

$$= 60 \cdot 2^{(3/4+1/4)} \cdot a^{3/4}b^{1/4}$$

$$= 2^1 \cdot 60a^{3/4}b^{1/4}$$

$$= 2f(a, b). \qquad \blacksquare$$

**Level Curves** It is possible graphically to depict a function $f(x, y)$ of two variables using a family of curves called level curves. Let $c$ be any number. Then the graph of the equation $f(x, y) = c$ is a curve in the $xy$-plane called the *level curve of height $c$*. This curve describes all points of height $c$ on the graph of the function

---

*Economists normally use $L$ and $K$, respectively, for labor and capital. However, for simplicity, we use $x$ and $y$.

$f(x, y)$. As $c$ varies, we have a family of level curves indicating the sets of points on which $f(x, y)$ assumes various values $c$. In Fig. 4, we have drawn the graph and various level curves for the function $f(x, y) = x^2 + y^2$.

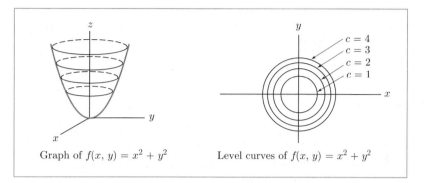

Graph of $f(x, y) = x^2 + y^2$      Level curves of $f(x, y) = x^2 + y^2$

**Figure 4.** Level curves.

Level curves often have interesting physical interpretations. For example, surveyors draw *topographic maps* that use level curves to represent points having equal altitude. Here $f(x, y) =$ the altitude at point $(x, y)$. Figure 5(a) shows the graph of $f(x, y)$ for a typical hilly region. Figure 5(b) shows the level curves corresponding to various altitudes. Note that when the level curves are closer together, the surface is steeper.

**Figure 5.** Topographic level curves show altitudes.

**EXAMPLE 4**    Determine the level curve at height 600 for the production function $f(x, y) = 60x^{3/4}y^{1/4}$ of Example 3.

**Solution**   The level curve is the graph of $f(x, y) = 600$, or

$$60x^{3/4}y^{1/4} = 600$$

$$y^{1/4} = \frac{10}{x^{3/4}}$$

$$y = \frac{10,000}{x^3}.$$

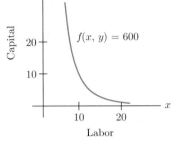

$f(x, y) = 600$

**Figure 6.**   A level curve of a production function.

Of course, since $x$ and $y$ represent quantities of labor and capital, they must both be positive. We have sketched the graph of the level curve in Fig. 6. The points on the curve are precisely those combinations of capital and labor that yield 600 units of production. Economists call this curve an *isoquant*. ∎

## PRACTICE PROBLEMS 7.1

1. Let $f(x, y, z) = x^2 + y/(x - z) - 4$. Compute $f(3, 5, 2)$.

2. Suppose that in a certain country the daily demand for coffee is given by $f(p_1, p_2) = 16p_1/p_2$ thousand pounds, where $p_1$ and $p_2$ are the respective prices of tea and coffee in dollars per pound. Compute and interpret $f(3, 4)$.

# EXERCISES 7.1

1. Let $f(x, y) = x^2 - 3xy - y^2$. Compute $f(5, 0)$, $f(5, -2)$, and $f(a, b)$.

2. Let $g(x, y) = \sqrt{x^2 + 2y^2}$. Compute $g(1, 1)$, $g(0, -1)$, and $g(a, b)$.

3. Let $g(x, y, z) = x/(y - z)$. Compute $g(2, 3, 4)$ and $g(7, 46, 44)$.

4. Let $f(x, y, z) = x^2 e^{\sqrt{y^2 + z^2}}$. Compute $f(1, -1, 1)$ and $f(2, 3, -4)$.

5. Let $f(x, y) = xy$. Show that $f(2 + h, 3) - f(2, 3) = 3h$.

6. Let $f(x, y) = xy$. Show that $f(2, 3 + k) - f(2, 3) = 2k$.

7. (*Cost*) Find a formula $C(x, y, z)$ that gives the cost of materials for the closed rectangular box in Fig. 7(a), with dimensions in feet, assuming that the material for the top and bottom costs \$3 per square foot and the material for the sides costs \$5 per square foot.

8. (*Cost*) Find a formula $C(x, y, z)$ that gives the cost of material for the rectangular enclosure in Fig. 7(b), with dimensions in feet, assuming that the material for the top costs \$3 per square foot and the material for the back and two sides costs \$5 per square foot.

9. Consider the Cobb–Douglas production function $f(x, y) = 20x^{1/3}y^{2/3}$. Compute $f(8, 1)$, $f(1, 27)$, and $f(8, 27)$. Show that for any positive constant $k$, $f(8k, 27k) = kf(8, 27)$.

10. Let $f(x, y) = 10x^{2/5}y^{3/5}$. Show that $f(3a, 3b) = 3f(a, b)$.

11. The present value of $A$ dollars to be paid $t$ years in the future (assuming a 5% continuous interest rate) is $P(A, t) = Ae^{-.05t}$. Find and interpret $P(100, 13.8)$.

12. Refer to Example 3. Suppose that labor costs \$100 per unit and capital costs \$200 per unit. Express as a function of two variables, $C(x, y)$, the cost of utilizing $x$ units of labor and $y$ units of capital.

13. (*Tax and Homeowner exemption*) The value of residential property for tax purposes is usually much lower than its actual market value. If $v$ is the market value, then the *assessed value* for real estate taxes might be only 40% of $v$. Suppose the property tax, $T$, in a community is given by the function

$$T = f(r, v, x) = \frac{r}{100}(.40v - x),$$

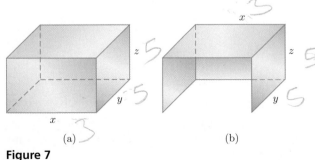

(a)          (b)

**Figure 7**

where $v$ is the estimated market value of a property (in dollars), $x$ is a *homeowner's exemption* (a number of dollars depending on the type of property), and $r$ is the tax rate (stated in dollars per hundred dollars) of net assessed value.

**(a)** Determine the real estate tax on a property valued at \$200,000 with a homeowner's exemption of \$5000, assuming a tax rate of \$2.50 per hundred dollars of net assessed value.

**(b)** Determine the tax due if the tax rate increases by 20% to \$3.00 per hundred dollars of net assessed value. Assume the same property value and homeowner's exemption. Does the tax due also increase by 20%?

**14.** Let $f(r, v, x)$ be the real estate tax function of Exercise 13.

**(a)** Determine the real estate tax on a property valued at \$100,000 with a homeowner's exemption of \$5000, assuming a tax rate of \$2.20 per hundred dollars of net assessed value.

**(b)** Determine the real estate tax when the market value rises 20% to \$120,000. Assume the same homeowner's exemption and a tax rate of \$2.20 per hundred dollars of net assessed value. Does the tax due also increase by 20%?

*Draw the level curves of heights 0, 1, and 2 for the functions in Exercises 15 and 16.*

**15.** $f(x, y) = 2x - y$       **16.** $f(x, y) = -x^2 + 2y$

**17.** Draw the level curve of the function $f(x, y) = x - y$ containing the point $(0, 0)$.

**18.** Draw the level curve of the function $f(x, y) = xy$ containing the point $(\frac{1}{2}, 4)$.

**19.** Find a function $f(x, y)$ that has the line $y = 3x - 4$ as a level curve.

**20.** Find a function $f(x, y)$ that has the curve $y = 2/x^2$ as a level curve.

**21.** Suppose that a topographic map is viewed as the graph of a certain function $f(x, y)$. What are the level curves?

**22.** A certain production process uses labor and capital. If the quantities of these commodities are $x$ and $y$, respectively, then the total cost is $100x + 200y$ dollars. Draw the level curves of height 600, 800, and 1000 for this function. Explain the significance of these curves. (Economists frequently refer to these lines as *budget lines* or *isocost lines*.)

*Match the graphs of the functions in Exercises 23–26 to the systems of level curves shown in Figs. 8(a)–(d).*

**23.**

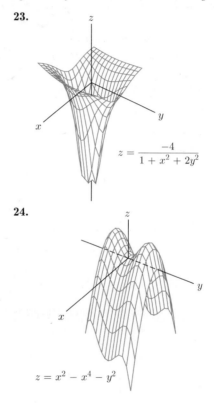

$$z = \frac{-4}{1 + x^2 + 2y^2}$$

**24.**

$$z = x^2 - x^4 - y^2$$

(a)

(b)

(c)

(d)

**Figure 8**

**25.**

$$z = \frac{15x^2y^2e^{-x^2-y^2}}{x^2+y^2}$$

**26.**

$$z = e^{-x^2} + e^{-4y^2}$$

## SOLUTIONS TO PRACTICE PROBLEMS 7.1

**1.** Substitute 3 for $x$, 5 for $y$, and 2 for $z$.

$$f(3,5,2) = 3^2 + \frac{5}{3-2} - 4 = 10.$$

**2.** To compute $f(3,4)$, substitute 3 for $p_1$ and 4 for $p_2$ into $f(p_1, p_2) = 16p_1/p_2$. Thus

$$f(3,4) = 16 \cdot \tfrac{3}{4} = 12.$$

Therefore, if the price of tea is \$3 per pound and the price of coffee is \$4 per pound, then 12,000 pounds of coffee will be sold each day. (Notice that as the price of coffee increases, the demand decreases.)

## 7.2 Partial Derivatives

In Chapter 1 we introduced the notion of a derivative to measure the rate at which a function $f(x)$ is changing with respect to changes in the variable $x$. Let us now study the analog of the derivative for functions of two (or more) variables.

Let $f(x,y)$ be a function of the two variables $x$ and $y$. Since we want to know how $f(x,y)$ changes with respect to both changes in the variable $x$ and changes in the variable $y$, we shall define two derivatives of $f(x,y)$ (to be called "partial derivatives"), one with respect to each of the variables. The *partial derivative of* $f(x,y)$ *with respect to* $x$, written $\dfrac{\partial f}{\partial x}$, is the derivative of $f(x,y)$, where $y$ is treated as a constant and $f(x,y)$ is considered as a function of $x$ alone. The *partial derivative of* $f(x,y)$ *with respect to* $y$, written $\dfrac{\partial f}{\partial y}$, is the derivative of $f(x,y)$, where $x$ is treated as a constant.

**EXAMPLE 1** Let $f(x,y) = 5x^3y^2$. Compute

$$\frac{\partial f}{\partial x} \quad \text{and} \quad \frac{\partial f}{\partial y}.$$

**Solution** To compute $\dfrac{\partial f}{\partial x}$, we think of $f(x,y)$ written as

$$f(x,y) = \left[5y^2\right]x^3,$$

where the brackets emphasize that $5y^2$ is to be treated as a constant. Therefore, when differentiating with respect to $x$, $f(x,y)$ is just a constant times $x^3$. Recall that if $k$ is any constant, then

$$\frac{d}{dx}(kx^3) = 3 \cdot k \cdot x^2.$$

Thus

$$\frac{\partial f}{\partial x} = 3 \cdot \left[5y^2\right] \cdot x^2 = 15x^2y^2.$$

After some practice, it is unnecessary to place the $y^2$ in front of the $x^3$ before differentiating.

Now, in order to compute $\dfrac{\partial f}{\partial y}$, we think of

$$f(x, y) = [5x^3]\, y^2.$$

When differentiating with respect to $y$, $f(x, y)$ is simply a constant (namely, $5x^3$) times $y^2$. Hence

$$\frac{\partial f}{\partial y} = 2 \cdot [5x^3] \cdot y = 10x^3 y.$$

&#9632;

---

**EXAMPLE 2**   Let $f(x, y) = 3x^2 + 2xy + 5y$. Compute

$$\frac{\partial f}{\partial x} \quad \text{and} \quad \frac{\partial f}{\partial y}.$$

**Solution**   To compute $\dfrac{\partial f}{\partial x}$, we think of

$$f(x, y) = 3x^2 + [2y]x + [5y].$$

Now we differentiate $f(x, y)$ as if it were a quadratic polynomial in $x$:

$$\frac{\partial f}{\partial x} = 6x + [2y] + 0 = 6x + 2y.$$

Note that $5y$ is treated as a constant when differentiating with respect to $x$, so the partial derivative of $5y$ with respect to $x$ is zero.

To compute $\dfrac{\partial f}{\partial y}$, we think of

$$f(x, y) = [3x^2] + [2x]y + 5y.$$

Then

$$\frac{\partial f}{\partial y} = 0 + [2x] + 5 = 2x + 5.$$

Note that $3x^2$ is treated as a constant when differentiating with respect to $y$, so the partial derivative of $3x^2$ with respect to $y$ is zero.

&#9632;

---

**EXAMPLE 3**   Compute

$$\frac{\partial f}{\partial x} \quad \text{and} \quad \frac{\partial f}{\partial y}$$

for each of the following.

(a) $f(x, y) = (4x + 3y - 5)^8$   (b) $f(x, y) = e^{xy^2}$   (c) $f(x, y) = y/(x + 3y)$

**Solution**   (a) To compute $\dfrac{\partial f}{\partial x}$, we think of

$$f(x, y) = (4x + [3y - 5])^8.$$

By the general power rule,

$$\frac{\partial f}{\partial x} = 8 \cdot (4x + [3y - 5])^7 \cdot 4 = 32(4x + 3y - 5)^7.$$

Here we used the fact that the derivative of $4x + 3y - 5$ with respect to $x$ is just 4.

To compute $\dfrac{\partial f}{\partial y}$, we think of

$$f(x,y) = ([4x] + 3y - 5)^8.$$

Then

$$\frac{\partial f}{\partial y} = 8 \cdot ([4x] + 3y - 5)^7 \cdot 3 = 24(4x + 3y - 5)^7.$$

(b) To compute $\dfrac{\partial f}{\partial x}$, we observe that

$$f(x,y) = e^{x[y^2]},$$

so that

$$\frac{\partial f}{\partial x} = [y^2] \, e^{x[y^2]} = y^2 e^{xy^2}.$$

To compute $\dfrac{\partial f}{\partial y}$, we think of

$$f(x,y) = e^{[x]y^2}.$$

Thus

$$\frac{\partial f}{\partial y} = e^{[x]y^2} \cdot 2[x]y = 2xy e^{xy^2}.$$

(c) To compute $\dfrac{\partial f}{\partial x}$, we use the general power rule to differentiate $[y](x + [3y])^{-1}$ with respect to $x$:

$$\frac{\partial f}{\partial x} = (-1) \cdot [y](x + [3y])^{-2} \cdot 1 = -\frac{y}{(x + 3y)^2}.$$

To compute $\dfrac{\partial f}{\partial y}$, we use the quotient rule to differentiate

$$f(x,y) = \frac{y}{[x] + 3y}$$

with respect to $y$. We find that

$$\frac{\partial f}{\partial y} = \frac{([x] + 3y) \cdot 1 - y \cdot 3}{([x] + 3y)^2} = \frac{x}{(x + 3y)^2}.$$

The use of brackets to highlight constants is helpful initially in order to compute partial derivatives. From now on we shall merely form a mental picture of those terms to be treated as constants and dispense with brackets. ■

A partial derivative of a function of several variables is also a function of several variables and hence can be evaluated at specific values of the variables. We write

$$\frac{\partial f}{\partial x}(a,b)$$

for $\dfrac{\partial f}{\partial x}$ evaluated at $x = a$, $y = b$. Similarly,

$$\frac{\partial f}{\partial y}(a,b)$$

denotes the function $\dfrac{\partial f}{\partial y}$ evaluated at $x = a$, $y = b$.

**EXAMPLE 4**   Let $f(x, y) = 3x^2 + 2xy + 5y$.

(a)  Calculate $\dfrac{\partial f}{\partial x}(1, 4)$.          (b)  Evaluate $\dfrac{\partial f}{\partial y}$ at $(x, y) = (1, 4)$.

**Solution**  (a) $\dfrac{\partial f}{\partial x} = 6x + 2y$, $\dfrac{\partial f}{\partial x}(1, 4) = 6 \cdot 1 + 2 \cdot 4 = 14$.

(b) $\dfrac{\partial f}{\partial y} = 2x + 5$, $\dfrac{\partial f}{\partial y}(1, 4) = 2 \cdot 1 + 5 = 7$. ◼

**Geometric Interpretation of Partial Derivatives**   Consider the three-dimensional surface $z = f(x, y)$ in Fig. 1. If $y$ is held constant at $b$ and $x$ is allowed to vary, the equation

$$z = f(x, b)$$

↳ constant

describes a curve on the surface. [The curve is formed by cutting the surface $z = f(x, y)$ with a vertical plane parallel to the $xz$-plane.] The value of $\dfrac{\partial f}{\partial x}(a, b)$ is the slope of the tangent line to the curve at the point where $x = a$ and $y = b$.

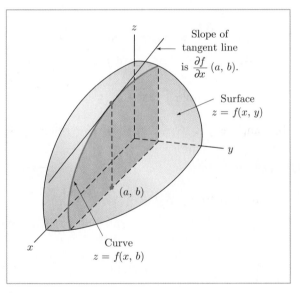

**Figure 1.** $\frac{\partial f}{\partial x}$ gives the slope of a curve formed by holding $y$ constant.

Likewise, if $x$ is held constant at $a$ and $y$ is allowed to vary, the equation

$$z = f(a, y)$$

↳ constant

describes the curve on the surface $z = f(x, y)$ shown in Fig. 2. The value of the partial derivative $\dfrac{\partial f}{\partial y}(a, b)$ is the slope of this curve at the point where $x = a$ and $y = b$.

**Partial Derivatives and Rates of Change**   Since $\dfrac{\partial f}{\partial x}$ is simply the ordinary derivative with $y$ held constant, $\dfrac{\partial f}{\partial x}$ gives the rate of change of $f(x, y)$ with respect

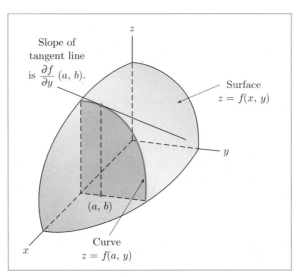

Slope of tangent line is $\frac{\partial f}{\partial y}(a, b)$.

Surface $z = f(x, y)$

$(a, b)$

Curve $z = f(a, y)$

**Figure 2.** $\frac{\partial f}{\partial y}$ gives the slope of a curve formed by holding $x$ constant.

to $x$ for $y$ held constant. In other words, keeping $y$ constant and increasing $x$ by one (small) unit produces a change in $f(x, y)$ that is approximately given by $\frac{\partial f}{\partial x}$. An analogous interpretation holds for $\frac{\partial f}{\partial y}$.

**EXAMPLE 5** Interpret the partial derivatives of $f(x, y) = 3x^2 + 2xy + 5y$ calculated in Example 4.

**Solution** We showed in Example 4 that

$$\frac{\partial f}{\partial x}(1, 4) = 14, \qquad \frac{\partial f}{\partial y}(1, 4) = 7.$$

The fact that

$$\frac{\partial f}{\partial x}(1, 4) = 14$$

means that if $y$ is kept constant at 4 and $x$ is allowed to vary near 1, then $f(x, y)$ changes at a rate 14 times the change in $x$. That is, if $x$ increases by one small unit, then $f(x, y)$ increases by approximately 14 units. If $x$ increases by $h$ units (where $h$ is small), then $f(x, y)$ increases by approximately $14 \cdot h$ units. That is,

$$f(1 + h, 4) - f(1, 4) \approx 14 \cdot h.$$

Similarly, the fact that

$$\frac{\partial f}{\partial y}(1, 4) = 7$$

means that if we keep $x$ constant at 1 and let $y$ vary near 4, then $f(x, y)$ changes at a rate equal to seven times the change in $y$. So for a small value of $k$, we have

$$f(1, 4 + k) - f(1, 4) \approx 7 \cdot k. \qquad \blacksquare$$

We can generalize the interpretations of $\frac{\partial f}{\partial x}$ and $\frac{\partial f}{\partial y}$ given in Example 5 to yield the following general fact.

Let $f(x, y)$ be a function of two variables. Then if $h$ and $k$ are small, we have

$$f(a + h, b) - f(a, b) \approx \frac{\partial f}{\partial x}(a, b) \cdot h$$

$$f(a, b + k) - f(a, b) \approx \frac{\partial f}{\partial y}(a, b) \cdot k.$$

Partial derivatives can be computed for functions of any number of variables. When taking the partial derivative with respect to one variable, we treat the other variables as constant.

**EXAMPLE 6**  Let $f(x, y, z) = x^2yz - 3z$.

(a) Compute $\dfrac{\partial f}{\partial x}, \dfrac{\partial f}{\partial y}$, and $\dfrac{\partial f}{\partial z}$.  (b) Calculate $\dfrac{\partial f}{\partial z}(2, 3, 1)$.

**Solution**  (a) $\dfrac{\partial f}{\partial x} = 2xyz, \dfrac{\partial f}{\partial y} = x^2z, \dfrac{\partial f}{\partial z} = x^2y - 3$.

(b) $\dfrac{\partial f}{\partial z}(2, 3, 1) = 2^2 \cdot 3 - 3 = 12 - 3 = 9$.  ■

**EXAMPLE 7**  Let $f(x, y, z)$ be the heat-loss function computed in Example 2 of Section 7.1. That is, $f(x, y, z) = 11xy + 14yz + 15xz$. Calculate and interpret $\dfrac{\partial f}{\partial x}(10, 7, 5)$.

**Solution**  We have

$$\frac{\partial f}{\partial x} = 11y + 15z$$

$$\frac{\partial f}{\partial x}(10, 7, 5) = 11 \cdot 7 + 15 \cdot 5 = 77 + 75 = 152.$$

The quantity $\dfrac{\partial f}{\partial x}$ is commonly referred to as the *marginal heat loss with respect to change in x*. Specifically, if $x$ is changed from 10 by $h$ units (where $h$ is small) and the values of $y$ and $z$ remain fixed at 7 and 5, then the amount of heat loss will change by approximately $152 \cdot h$ units.  ■

**EXAMPLE 8**  **Production**  Consider the production function $f(x, y) = 60x^{3/4}y^{1/4}$, which gives the number of units of goods produced when utilizing $x$ units of labor and $y$ units of capital.

(a) Find $\dfrac{\partial f}{\partial x}$ and $\dfrac{\partial f}{\partial y}$.

(b) Evaluate $\dfrac{\partial f}{\partial x}$ and $\dfrac{\partial f}{\partial y}$ at $x = 81$, $y = 16$.

(c) Interpret the numbers computed in part (b).

**Solution**  (a) $\dfrac{\partial f}{\partial x} = 60 \cdot \dfrac{3}{4}x^{-1/4}y^{1/4} = 45x^{-1/4}y^{1/4} = 45\dfrac{y^{1/4}}{x^{1/4}}$

$\dfrac{\partial f}{\partial y} = 60 \cdot \dfrac{1}{4}x^{3/4}y^{-3/4} = 15x^{3/4}y^{-3/4} = 15\dfrac{x^{3/4}}{y^{3/4}}.$

(b) $\dfrac{\partial f}{\partial x}(81, 16) = 45 \cdot \dfrac{16^{1/4}}{81^{1/4}} = 45 \cdot \dfrac{2}{3} = 30$

$\dfrac{\partial f}{\partial y}(81, 16) = 15 \cdot \dfrac{81^{3/4}}{16^{3/4}} = 15 \cdot \dfrac{27}{8} = \dfrac{405}{8} = 50\tfrac{5}{8}.$

(c) The quantities $\dfrac{\partial f}{\partial x}$ and $\dfrac{\partial f}{\partial y}$ are referred to as the *marginal productivity of labor* and the *marginal productivity of capital*. If the amount of capital is held fixed at $y = 16$ and the amount of labor increases by 1 unit, then the quantity of goods produced will increase by approximately 30 units. Similarly, an increase in capital of 1 unit (with labor fixed at 81) results in an increase in production of approximately $50\tfrac{5}{8}$ units of goods. ■

Just as we formed second derivatives in the case of one variable, we can form second partial derivatives of a function $f(x, y)$ of two variables. Since $\dfrac{\partial f}{\partial x}$ is a function of $x$ and $y$, we can differentiate it with respect to $x$ or $y$. The partial derivative of $\dfrac{\partial f}{\partial x}$ with respect to $x$ is denoted by $\dfrac{\partial^2 f}{\partial x^2}$. The partial derivative of $\dfrac{\partial f}{\partial x}$ with respect to $y$ is denoted by $\dfrac{\partial^2 f}{\partial y \partial x}$. Similarly, the partial derivative of the function $\dfrac{\partial f}{\partial y}$ with respect to $x$ is denoted by $\dfrac{\partial^2 f}{\partial x \partial y}$, and the partial derivative of $\dfrac{\partial f}{\partial y}$ with respect to $y$ is denoted by $\dfrac{\partial^2 f}{\partial y^2}$. Almost all functions $f(x, y)$ encountered in applications (and all functions $f(x, y)$ in this text) have the property that

$$\frac{\partial^2 f}{\partial y \partial x} = \frac{\partial^2 f}{\partial x \partial y}.$$

Note that in computing $\dfrac{\partial^2 f}{\partial y \partial x}$ and $\dfrac{\partial^2 f}{\partial x \partial y}$, verifying the last equation is a check that you have done the differentiation correctly.

---

**EXAMPLE 9** Let $f(x, y) = x^2 + 3xy + 2y^2$. Calculate

$$\frac{\partial^2 f}{\partial x^2}, \quad \frac{\partial^2 f}{\partial y^2}, \quad \frac{\partial^2 f}{\partial x \partial y}, \quad \text{and} \quad \frac{\partial^2 f}{\partial y \partial x}.$$

**Solution** First we compute $\dfrac{\partial f}{\partial x}$ and $\dfrac{\partial f}{\partial y}$.

$$\frac{\partial f}{\partial x} = 2x + 3y, \qquad \frac{\partial f}{\partial y} = 3x + 4y.$$

To compute $\dfrac{\partial^2 f}{\partial x^2}$, we differentiate $\dfrac{\partial f}{\partial x}$ with respect to $x$:

$$\frac{\partial^2 f}{\partial x^2} = 2.$$

Similarly, to compute $\dfrac{\partial^2 f}{\partial y^2}$, we differentiate $\dfrac{\partial f}{\partial y}$ with respect to $y$:

$$\frac{\partial^2 f}{\partial y^2} = 4.$$

To compute $\dfrac{\partial^2 f}{\partial x \partial y}$, we differentiate $\dfrac{\partial f}{\partial y}$ with respect to $x$:

$$\frac{\partial^2 f}{\partial x \partial y} = 3.$$

Finally, to compute $\dfrac{\partial^2 f}{\partial y \partial x}$, we differentiate $\dfrac{\partial f}{\partial x}$ with respect to $y$:

$$\frac{\partial^2 f}{\partial y \partial x} = 3.$$

**Incorporating Technology**  Graphing calculators can evaluate partial derivatives. See the appropriate calculator appendix for details. The function from Example 4 and its first partial derivatives are specified in Fig. 3(a) and evaluated in Fig. 3(b).

**Figure 3**   (a)    (b)

## PRACTICE PROBLEMS 7.2

1. The number of TV sets an appliance store sells per week is given by a function of two variables, $f(x, y)$, where $x$ is the price per TV set and $y$ is the amount of money spent weekly on advertising. Suppose that the current price is \$400 per set and that currently \$2000 per week is being spent for advertising.

   (a) Would you expect $\dfrac{\partial f}{\partial x}(400, 2000)$ to be positive or negative?

   (b) Would you expect $\dfrac{\partial f}{\partial y}(400, 2000)$ to be positive or negative?

2. The monthly mortgage payment for a house is a function of two variables, $f(A, r)$, where $A$ is the amount of the mortgage and the interest rate is $r\%$. For a 30-year mortgage, $f(92,000, 9) = 740.25$ and $\dfrac{\partial f}{\partial r}(92,000, 9) = 66.20$. What is the significance of the number 66.20?

## EXERCISES 7.2

*Find $\dfrac{\partial f}{\partial x}$ and $\dfrac{\partial f}{\partial y}$ for each of the following functions.*

1. $f(x, y) = 5xy$

2. $f(x, y) = x^2 - y^2$

3. $f(x, y) = 2x^2 e^y$

4. $f(x, y) = xe^{xy}$

5. $f(x, y) = \dfrac{x}{y} + \dfrac{y}{x}$

6. $f(x, y) = \dfrac{1}{x + y}$

7. $f(x, y) = (2x - y + 5)^2$

8. $f(x, y) = \dfrac{e^x}{1 + e^y}$

9. $f(x, y) = x^2 e^{3x} \ln y$

10. $f(x, y) = \ln(xy)$

11. $f(x, y) = \dfrac{x - y}{x + y}$

12. $f(x, y) = \sqrt{x^2 + y^2}$

13. Let $f(L, K) = 3\sqrt{LK}$. Find $\dfrac{\partial f}{\partial L}$.

14. Let $f(p, q) = 1 - p(1 + q)$. Find $\dfrac{\partial f}{\partial q}$ and $\dfrac{\partial f}{\partial p}$.

15. Let $f(x, y, z) = (1 + x^2 y)/z$. Find $\dfrac{\partial f}{\partial x}, \dfrac{\partial f}{\partial y}$, and $\dfrac{\partial f}{\partial z}$.

16. Let $f(x, y, z) = ze^{x/y}$. Find $\dfrac{\partial f}{\partial x}, \dfrac{\partial f}{\partial y}$, and $\dfrac{\partial f}{\partial z}$.

17. Let $f(x, y, z) = xze^{yz}$. Find $\dfrac{\partial f}{\partial x}, \dfrac{\partial f}{\partial y}$, and $\dfrac{\partial f}{\partial z}$.

18. Let $f(x, y, z) = \dfrac{xy}{z}$. Find $\dfrac{\partial f}{\partial x}, \dfrac{\partial f}{\partial y}$, and $\dfrac{\partial f}{\partial z}$.

**19.** Let $f(x,y) = x^2 + 2xy + y^2 + 3x + 5y$. Find $\dfrac{\partial f}{\partial x}(2,-3)$ and $\dfrac{\partial f}{\partial y}(2,-3)$.

**20.** Let $f(x,y) = (x+y^2)^3$. Evaluate $\dfrac{\partial f}{\partial x}$ and $\dfrac{\partial f}{\partial y}$ at $(x,y) = (1,2)$.

**21.** Let $f(x,y,z) = xy^2z + 5$. Evaluate $\dfrac{\partial f}{\partial y}$ at $(x,y,z) = (2,-1,3)$.

**22.** Let $f(x,y,z) = \dfrac{x}{y-z}$. Compute $\dfrac{\partial f}{\partial y}(2,-1,3)$.

**23.** Let $f(x,y) = x^3y + 2xy^2$. Find $\dfrac{\partial^2 f}{\partial x^2}, \dfrac{\partial^2 f}{\partial y^2}, \dfrac{\partial^2 f}{\partial x\,\partial y}$, and $\dfrac{\partial^2 f}{\partial y\,\partial x}$.

**24.** Let $f(x,y) = xe^y + x^4y + y^3$. Find $\dfrac{\partial^2 f}{\partial x^2}, \dfrac{\partial^2 f}{\partial y^2}, \dfrac{\partial^2 f}{\partial x\,\partial y}$, and $\dfrac{\partial^2 f}{\partial y\,\partial x}$.

**25.** (*Production*) A farmer can produce $f(x,y) = 200\sqrt{6x^2 + y^2}$ units of produce by utilizing $x$ units of labor and $y$ units of capital. (The capital is used to rent or purchase land, materials, and equipment.)

(a) Calculate the marginal productivities of labor and capital when $x = 10$ and $y = 5$.

(b) Let $h$ be a small number. Use the result of part (a) to determine the approximate effect on production of changing labor from 10 to $10 + h$ units while keeping capital fixed at 5 units.

(c) Use part (b) to estimate the change in production when labor decreases from 10 to 9.5 units and capital stays fixed at 5 units.

**26.** The productivity of a country is given by $f(x,y) = 300x^{2/3}y^{1/3}$, where $x$ and $y$ are the amount of labor and capital.

(a) Compute the marginal productivities of labor and capital when $x = 125$ and $y = 64$.

(b) Use part (a) to determine the approximate effect on productivity of increasing capital from 64 to 66 units, while keeping labor fixed at 125 units.

(c) What would be the approximate effect of decreasing labor from 125 to 124 units while keeping capital fixed at 64 units?

**27.** (*Demand*) In a certain suburban community, commuters have the choice of getting into the city by bus or train. The demand for these modes of transportation varies with their cost. Let $f(p_1, p_2)$ be the number of people who will take the bus when $p_1$ is the price of the bus ride and $p_2$ is the price of the train ride. For example, if $f(4.50,6) = 7000$, then 7000 commuters will take the bus when the price of a bus ticket is \$4.50 and the

price of a train ticket is \$6.00. Explain why $\dfrac{\partial f}{\partial p_1} < 0$ and $\dfrac{\partial f}{\partial p_2} > 0$.

**28.** Refer to Exercise 27. Let $g(p_1, p_2)$ be the number of people who will take the train when $p_1$ is the price of the bus ride and $p_2$ is the price of the train ride. Would you expect $\dfrac{\partial g}{\partial p_1}$ to be positive or negative? How about $\dfrac{\partial g}{\partial p_2}$?

**29.** Let $p_1$ be the average price of DVD players, $p_2$ the average price of DVDs, $f(p_1, p_2)$ the demand for DVD players, and $g(p_1, p_2)$ the demand for DVDs. Explain why $\dfrac{\partial f}{\partial p_2} < 0$ and $\dfrac{\partial g}{\partial p_1} < 0$.

**30.** The demand for a certain gas-guzzling car is given by $f(p_1, p_2)$, where $p_1$ is the price of the car and $p_2$ is the price of gasoline. Explain why $\dfrac{\partial f}{\partial p_1} < 0$ and $\dfrac{\partial f}{\partial p_2} < 0$.

**31.** The volume $(V)$ of a certain amount of a gas is determined by the temperature $(T)$ and the pressure $(P)$ by the formula $V = .08(T/P)$. Calculate and interpret $\dfrac{\partial V}{\partial P}$ and $\dfrac{\partial V}{\partial T}$ when $P = 20$, $T = 300$.

**32.** (*Beer Consumption*) Using data collected from 1929 to 1941, Richard Stone* determined that the yearly quantity $Q$ of beer consumed in the United Kingdom was approximately given by the formula $Q = f(m,p,r,s)$, where

$$f(m,p,r,s) = (1.058)m^{.136}p^{-.727}r^{.914}s^{.816}$$

and $m$ is the aggregate real income (personal income after direct taxes, adjusted for retail price changes), $p$ is the average retail price of the commodity (in this case, beer), $r$ is the average retail price level of all other consumer goods and services, and $s$ is a measure of the strength of the beer. Determine which partial derivatives are positive and which are negative and give interpretations. (For example, since $\dfrac{\partial f}{\partial r} > 0$, people buy more beer when the prices of other goods increase and the other factors remain constant.)

**33.** Richard Stone (see Exercise 32) determined that the yearly consumption of food in the United States was given by

$$f(m,p,r) = (2.186)m^{.595}p^{-.543}r^{.922}.$$

Determine which partial derivatives are positive and which are negative and give interpretations of these facts.

---

*Richard Stone, "The Analysis of Market Demand," *Journal of the Royal Statistical Society*, CVIII (1945), 286–391.

**34.** (*Distribution of Revenue*) For the production function $f(x, y) = 60x^{3/4}y^{1/4}$ considered in Example 8, think of $f(x, y)$ as the revenue when utilizing $x$ units of labor and $y$ units of capital. Under actual operating conditions, say $x = a$ and $y = b$, $\dfrac{\partial f}{\partial x}(a, b)$ is referred to as the *wage per unit of labor* and $\dfrac{\partial f}{\partial y}(a, b)$ is referred to as the *wage per unit of capital*. Show that

$$f(a, b) = a \cdot \left[\frac{\partial f}{\partial x}(a, b)\right] + b \cdot \left[\frac{\partial f}{\partial y}(a, b)\right].$$

(This equation shows how the revenue is distributed between labor and capital.)

**35.** Compute $\dfrac{\partial^2 f}{\partial x^2}$, where $f(x, y) = 60x^{3/4}y^{1/4}$, a production function (where $x$ is units of labor). Explain why $\dfrac{\partial^2 f}{\partial x^2}$ is always negative.

**36.** Compute $\dfrac{\partial^2 f}{\partial y^2}$, where $f(x, y) = 60x^{3/4}y^{1/4}$, a production function (where $y$ is units of capital). Explain why $\dfrac{\partial^2 f}{\partial y^2}$ is always negative.

**37.** Let $f(x, y) = 3x^2 + 2xy + 5y$, as in Example 5. Show that

$$f(1 + h, 4) - f(1, 4) = 14h + 3h^2.$$

Thus the error in approximating $f(1+h, 4) - f(1, 4)$ by $14h$ is $3h^2$. (If $h = .01$, for instance, the error is only .0003.)

**38.** (*Body Surface Area*) Physicians, particularly pediatricians, sometimes need to know the body surface area of a patient. For instance, the surface area is used to adjust the results of certain tests of kidney performance. Tables are available that give the approximate body surface area $A$ in square meters of a person who weighs $W$ kilograms and is $H$ centimeters tall. The following empirical formula[*] is also used:

$$A = .007W^{.425}H^{.725}.$$

Evaluate $\dfrac{\partial A}{\partial W}$ and $\dfrac{\partial A}{\partial H}$ when $W = 54$, $H = 165$, and give a physical interpretation of your answers. You may use the approximations $(54)^{.425} \approx 5.4$, $(54)^{-.575} \approx .10$, $(165)^{.725} \approx 40.5$, $(165)^{-.275} \approx .25$.

---

### SOLUTIONS TO PRACTICE PROBLEMS 7.2

**1. (a)** Negative. $\dfrac{\partial f}{\partial x}(400, 2000)$ is approximately the change in sales due to a $1 increase in $x$ (price). Since raising prices lowers sales, we would expect $\dfrac{\partial f}{\partial x}(400, 2000)$ to be negative.

**(b)** Positive. $\dfrac{\partial f}{\partial y}(400, 2000)$ is approximately the change in sales due to a $1 increase in advertising. Since spending more money on advertising brings in more customers, we would expect sales to increase; that is, $\dfrac{\partial f}{\partial y}(400, 2000)$ is most likely positive.

**2.** If the interest rate is raised from 9% to 10%, then the monthly payment will increase by about $66.20. [An increase to $9\frac{1}{2}\%$ causes an increase in the monthly payment of about $\frac{1}{2} \cdot (66.20)$ or $33.10, and so on.]

---

## 7.3 Maxima and Minima of Functions of Several Variables

Previously, we studied how to determine the maxima and minima of functions of a single variable. Let us extend that discussion to functions of several variables.

If $f(x, y)$ is a function of two variables, then we say that $f(x, y)$ has a *relative maximum* when $x = a$, $y = b$ if $f(x, y)$ is at most equal to $f(a, b)$ whenever $x$ is near $a$ and $y$ is near $b$. Geometrically, the graph of $f(x, y)$ has a peak at $(x, y) = (a, b)$. [See Fig. 1(a).] Similarly, we say that $f(x, y)$ has a *relative minimum* when $x = a$, $y = b$ if $f(x, y)$ is at least equal to $f(a, b)$ whenever $x$ is near $a$ and $y$ is near $b$. Geometrically, the graph of $f(x, y)$ has a pit whose bottom occurs at $(x, y) = (a, b)$. [See Fig. 1(b).]

Suppose the function $f(x, y)$ has a relative minimum at $(x, y) = (a, b)$, as in Fig. 2. When $y$ is held constant at $b$, $f(x, y)$ is a function of $x$ with a relative

---

[*]See J. Routh, *Mathematical Preparation for Laboratory Technicians* (Philadelphia, Pa.: W. B. Saunders Co., 1971), p. 92.

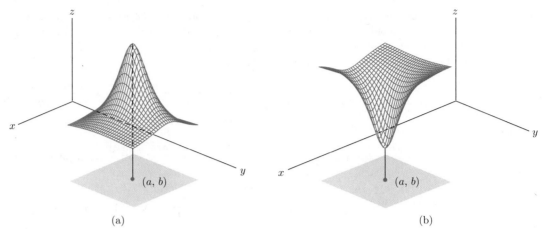

**Figure 1.** Maximum and minimum points.

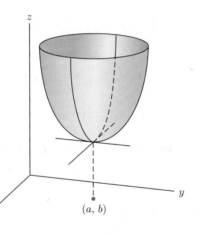

**Figure 2.** Horizontal tangent lines at a relative minimum.

minimum at $x = a$. Therefore, the tangent line to the curve $z = f(x, b)$ is horizontal at $x = a$ and hence has slope 0. That is,

$$\frac{\partial f}{\partial x}(a, b) = 0.$$

Likewise, when $x$ is held constant at $a$, $f(x, y)$ is a function of $y$ with a relative minimum at $y = b$. Therefore, its derivative with respect to $y$ is zero at $y = b$. That is,

$$\frac{\partial f}{\partial y}(a, b) = 0.$$

Similar considerations apply when $f(x, y)$ has a relative maximum at $(x, y) = (a, b)$.

---

**First-Derivative Test for Functions of Two Variables**   If $f(x, y)$ has either a relative maximum or minimum at $(x, y) = (a, b)$, then

$$\frac{\partial f}{\partial x}(a, b) = 0$$

and

$$\frac{\partial f}{\partial y}(a, b) = 0.$$

A relative maximum or minimum may or may not be an absolute maximum or minimum. However, to simplify matters in this text, the examples and exercises have been chosen so that if an absolute extremum of $f(x, y)$ exists, it will occur at a point where $f(x, y)$ has a relative extremum.

---

**EXAMPLE 1**   The function $f(x, y) = 3x^2 - 4xy + 3y^2 + 8x - 17y + 30$ has the graph pictured in Fig. 2. Find the point $(a, b)$ at which $f(x, y)$ attains its minimum value.

**Solution**   We look for those values of $x$ and $y$ at which both partial derivatives are zero. The partial derivatives are

$$\frac{\partial f}{\partial x} = 6x - 4y + 8$$

$$\frac{\partial f}{\partial y} = -4x + 6y - 17.$$

Setting $\dfrac{\partial f}{\partial x} = 0$ and $\dfrac{\partial f}{\partial y} = 0$, we obtain

$$6x - 4y + 8 = 0 \quad \text{or} \quad y = \frac{6x + 8}{4}$$

$$-4x + 6y - 17 = 0 \quad \text{or} \quad y = \frac{4x + 17}{6}.$$

By equating these two expressions for $y$, we have

$$\frac{6x + 8}{4} = \frac{4x + 17}{6}.$$

Cross-multiplying, we see that

$$36x + 48 = 16x + 68$$
$$20x = 20$$
$$x = 1.$$

When we substitute this value for $x$ into our first equation for $y$ in terms of $x$, we obtain

$$y = \frac{6x + 8}{4} = \frac{6 \cdot 1 + 8}{4} = \frac{7}{2}.$$

If $f(x, y)$ has a minimum, it must occur where $\dfrac{\partial f}{\partial x} = 0$ and $\dfrac{\partial f}{\partial y} = 0$. We have determined that the partial derivatives are zero only when $x = 1$, $y = \frac{7}{2}$. From Fig. 2 we know that $f(x, y)$ has a minimum, so it must be at $(x, y) = (1, \frac{7}{2})$.   ■

---

**EXAMPLE 2**   **Price discrimination**   A monopolist markets his product in two countries and can charge different amounts in each country. Let $x$ be the number of units to be sold in the first country and $y$ the number of units to be sold in the second country. Due to the laws of demand, the monopolist must set the price at $97 - (x/10)$ dollars in the first country and $83 - (y/20)$ dollars in the second country in order to sell all the units. The cost of producing these units is $20{,}000 + 3(x + y)$. Find the values of $x$ and $y$ that maximize the profit.

**Solution** Let $f(x, y)$ be the profit derived from selling $x$ units in the first country and $y$ in the second. Then

$$f(x, y) = [\text{revenue from first country}] + [\text{revenue from second country}] - [\text{cost}]$$

$$= \left(97 - \frac{x}{10}\right)x + \left(83 - \frac{y}{20}\right)y - [20{,}000 + 3(x + y)]$$

$$= 97x - \frac{x^2}{10} + 83y - \frac{y^2}{20} - 20{,}000 - 3x - 3y$$

$$= 94x - \frac{x^2}{10} + 80y - \frac{y^2}{20} - 20{,}000.$$

To find where $f(x, y)$ has its maximum value, we look for those values of $x$ and $y$ at which both partial derivatives are zero.

$$\frac{\partial f}{\partial x} = 94 - \frac{x}{5}$$

$$\frac{\partial f}{\partial y} = 80 - \frac{y}{10}.$$

We set $\dfrac{\partial f}{\partial x} = 0$ and $\dfrac{\partial f}{\partial y} = 0$ to obtain

$$94 - \frac{x}{5} = 0 \quad \text{or} \quad x = 470$$

$$80 - \frac{y}{10} = 0 \quad \text{or} \quad y = 800.$$

Therefore, the firm should adjust its prices to levels where it will sell 470 units in the first country and 800 units in the second country. ∎

---

**EXAMPLE 3** **Heat loss** Suppose that we want to design a rectangular building having volume 147,840 cubic feet. Assuming that the daily loss of heat is given by

$$w = 11xy + 14yz + 15xz,$$

where $x$, $y$, and $z$ are, respectively, the length, width, and height of the building, find the dimensions of the building for which the daily heat loss is minimal.

**Solution** We must minimize the function

$$w = 11xy + 14yz + 15xz, \tag{1}$$

where $x$, $y$, $z$ satisfy the constraint equation (refer to Section 2.5)

$$xyz = 147{,}840.$$

For simplicity, let us denote 147,840 by $V$. Then $xyz = V$, so that $z = V/xy$. We substitute this expression for $z$ into the objective function (1) to obtain a heat-loss function $g(x, y)$ of two variables—namely,

$$g(x, y) = 11xy + 14y\frac{V}{xy} + 15x\frac{V}{xy} = 11xy + \frac{14V}{x} + \frac{15V}{y}.$$

To minimize this function, we first compute the partial derivatives with respect to $x$ and $y$; then we equate them to zero.

$$\frac{\partial g}{\partial x} = 11y - \frac{14V}{x^2} = 0$$

$$\frac{\partial g}{\partial y} = 11x - \frac{15V}{y^2} = 0$$

These two equations yield

$$y = \frac{14V}{11x^2} \tag{2}$$

$$11xy^2 = 15V. \tag{3}$$

If we substitute the value of $y$ from (2) into (3), we see that

$$11x \left( \frac{14V}{11x^2} \right)^2 = 15V$$

$$\frac{14^2 V^2}{11x^3} = 15V$$

$$x^3 = \frac{14^2 \cdot V^2}{11 \cdot 15 \cdot V} = \frac{14^2 \cdot V}{11 \cdot 15}$$

$$= \frac{14^2 \cdot 147{,}840}{11 \cdot 15}$$

$$= 175{,}616.$$

Therefore, we see (using a calculator) that

$$x = 56.$$

From equation (2) we find that

$$y = \frac{14 \cdot V}{11x^2} = \frac{14 \cdot 147{,}840}{11 \cdot 56^2} = 60.$$

Finally,

$$z = \frac{V}{xy} = \frac{147{,}840}{56 \cdot 60} = 44.$$

Thus the building should be 56 feet long, 60 feet wide, and 44 feet high in order to minimize the heat loss.* ■

When considering a function of two variables, we find points $(x, y)$ at which $f(x, y)$ has a potential relative maximum or minimum by setting $\dfrac{\partial f}{\partial x}$ and $\dfrac{\partial f}{\partial y}$ equal to zero and solving for $x$ and $y$. However, if we are given no additional information about $f(x, y)$, it may be difficult to determine whether we have found a maximum or a minimum (or neither). In the case of functions of one variable, we studied concavity and deduced the second-derivative test. There is an analog of the second derivative test for functions of two variables, but it is much more complicated than the one-variable test. We state it without proof.

---

*For further discussion of this heat-loss problem, as well as other examples of optimization in architectural design, see L. March, "Elementary Models of Built Forms," Chapter 3 in *Urban Space and Structures*, L. Martin and L. March, eds. (Cambridge: Cambridge University Press, 1972).

**Second-Derivative Test for Functions of Two Variables** Suppose $f(x, y)$ is a function and $(a, b)$ is a point at which

$$\frac{\partial f}{\partial x}(a, b) = 0 \quad \text{and} \quad \frac{\partial f}{\partial y}(a, b) = 0,$$

and let

$$D(x, y) = \frac{\partial^2 f}{\partial x^2} \cdot \frac{\partial^2 f}{\partial y^2} - \left(\frac{\partial^2 f}{\partial x\, \partial y}\right)^2.$$

1. If

$$D(a, b) > 0 \quad \text{and} \quad \frac{\partial^2 f}{\partial x^2}(a, b) > 0,$$

then $f(x, y)$ has a relative minimum at $(a, b)$.

2. If

$$D(a, b) > 0 \quad \text{and} \quad \frac{\partial^2 f}{\partial x^2}(a, b) < 0,$$

then $f(x, y)$ has a relative maximum at $(a, b)$.

3. If

$$D(a, b) < 0,$$

then $f(x, y)$ has neither a relative maximum nor a relative minimum at $(a, b)$.

4. If $D(a, b) = 0$, then no conclusion can be drawn from this test.

The saddle-shaped graph in Fig. 3 illustrates a function $f(x, y)$ for which $D(a, b) < 0$. Both partial derivatives are zero at $(x, y) = (a, b)$ and yet the function has neither a relative maximum nor a relative minimum there. (Observe that the function has a relative maximum with respect to $x$ when $y$ is held constant and a relative minimum with respect to $y$ when $x$ is held constant.)

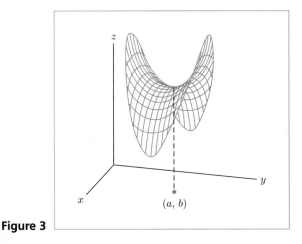

**Figure 3**

---

**EXAMPLE 4** Let $f(x, y) = x^3 - y^2 - 12x + 6y + 5$. Find all possible relative maximum and minimum points of $f(x, y)$. Use the second-derivative test to determine the nature of each such point.

**Solution**  Since

$$\frac{\partial f}{\partial x} = 3x^2 - 12, \qquad \frac{\partial f}{\partial y} = -2y + 6,$$

we find that $f(x, y)$ has a potential relative extreme point when

$$3x^2 - 12 = 0$$
$$-2y + 6 = 0.$$

From the first equation, $3x^2 = 12$, $x^2 = 4$, and $x = \pm 2$. From the second equation, $y = 3$. Thus, $\dfrac{\partial f}{\partial x}$ and $\dfrac{\partial f}{\partial y}$ are both zero when $(x, y) = (2, 3)$ and when $(x, y) = (-2, 3)$. To apply the second-derivative test, compute

$$\frac{\partial^2 f}{\partial x^2} = 6x, \qquad \frac{\partial^2 f}{\partial y^2} = -2, \qquad \frac{\partial^2 f}{\partial x\, \partial y} = 0,$$

and

$$D(x, y) = \frac{\partial^2 f}{\partial x^2} \cdot \frac{\partial^2 f}{\partial y^2} - \left( \frac{\partial^2 f}{\partial x\, \partial y} \right)^2 = (6x)(-2) - 0^2 = -12x. \tag{4}$$

Since $D(2, 3) = -12(2) = -24$, which is negative, case 3 of the second-derivative test says that $f(x, y)$ has neither a relative maximum nor a relative minimum at $(2, 3)$. However, $D(-2, 3) = -12(-2) = 24$. Since $D(-2, 3)$ is positive, the function $f(x, y)$ has either a relative maximum or a relative minimum at $(-2, 3)$. To determine which, we compute

$$\frac{\partial^2 f}{\partial x^2}(-2, 3) = 6(-2) = -12 < 0.$$

By case 2 of the second-derivative test, the function $f(x, y)$ has a relative maximum at $(-2, 3)$.  ■

In this section we have restricted ourselves to functions of two variables, but the case of three or more variables is handled in a similar fashion. For instance, here is the first-derivative test for a function of three variables.

---

If $f(x, y, z)$ has a relative maximum or minimum at $(x, y, z) = (a, b, c)$, then

$$\frac{\partial f}{\partial x}(a, b, c) = 0$$

$$\frac{\partial f}{\partial y}(a, b, c) = 0$$

$$\frac{\partial f}{\partial z}(a, b, c) = 0.$$

---

## PRACTICE PROBLEMS 7.3

1. Find all points $(x, y)$ where $f(x, y) = x^3 - 3xy + \frac{1}{2}y^2 + 8$ has a possible relative maximum or minimum.

2. Apply the second-derivative test to the function $g(x, y)$ of Example 3 to confirm that a relative minimum actually occurs when $x = 56$ and $y = 60$.

# EXERCISES 7.3

*Find all points $(x, y)$ where $f(x, y)$ has a possible relative maximum or minimum.*

1. $f(x, y) = x^2 - 3y^2 + 4x + 6y + 8$
2. $f(x, y) = \frac{1}{2}x^2 + y^2 - 3x + 2y - 5$
3. $f(x, y) = x^2 - 5xy + 6y^2 + 3x - 2y + 4$
4. $f(x, y) = -3x^2 + 7xy - 4y^2 + x + y$
5. $f(x, y) = x^3 + y^2 - 3x + 6y$
6. $f(x, y) = x^2 - y^3 + 5x + 12y + 1$
7. $f(x, y) = \frac{1}{3}x^3 - 2y^3 - 5x + 6y - 5$
8. $f(x, y) = x^4 - 8xy + 2y^2 - 3$
9. The function $f(x, y) = 2x + 3y + 9 - x^2 - xy - y^2$ has a maximum at some point $(x, y)$. Find the values of $x$ and $y$ where this maximum occurs.
10. The function $f(x, y) = \frac{1}{2}x^2 + 2xy + 3y^2 - x + 2y$ has a minimum at some point $(x, y)$. Find the values of $x$ and $y$ where this minimum occurs.

*In Exercises 11–16, both first partial derivatives of the function $f(x, y)$ are zero at the given points. Use the second-derivative test to determine the nature of $f(x, y)$ at each of these points. If the second-derivative test is inconclusive, so state.*

11. $f(x, y) = 3x^2 - 6xy + y^3 - 9y$; $(3, 3)$, $(-1, -1)$
12. $f(x, y) = 6xy^2 - 2x^3 - 3y^4$; $(0, 0)$, $(1, 1)$, $(1, -1)$
13. $f(x, y) = 2x^2 - x^4 - y^2$; $(-1, 0)$, $(0, 0)$, $(1, 0)$
14. $f(x, y) = x^4 - 4xy + y^4$; $(0, 0)$, $(1, 1)$, $(-1, -1)$
15. $f(x, y) = ye^x - 3x - y + 5$; $(0, 3)$
16. $f(x, y) = \dfrac{1}{x} + \dfrac{1}{y} + xy$; $(1, 1)$

*Find all points $(x, y)$ where $f(x, y)$ has a possible relative maximum or minimum. Then use the second-derivative test to determine, if possible, the nature of $f(x, y)$ at each of these points. If the second-derivative test is inconclusive, so state.*

17. $f(x, y) = x^2 - 2xy + 4y^2$
18. $f(x, y) = 2x^2 + 3xy + 5y^2$
19. $f(x, y) = -2x^2 + 2xy - y^2 + 4x - 6y + 5$
20. $f(x, y) = -x^2 - 8xy - y^2$
21. $f(x, y) = x^2 + 2xy + 5y^2 + 2x + 10y - 3$
22. $f(x, y) = x^2 - 2xy + 3y^2 + 4x - 16y + 22$
23. $f(x, y) = x^3 - y^2 - 3x + 4y$
24. $f(x, y) = x^3 - 2xy + 4y$
25. $f(x, y) = 2x^2 + y^3 - x - 12y + 7$
26. $f(x, y) = x^2 + 4xy + 2y^4$
27. Find the possible values of $x$, $y$, $z$ at which

$$f(x, y, z) = 2x^2 + 3y^2 + z^2 - 2x - y - z$$

assumes its minimum value.

28. Find the possible values of $x$, $y$, $z$ at which

$$f(x, y, z) = 5 + 8x - 4y + x^2 + y^2 + z^2$$

assumes its minimum value.

29. U.S. postal rules require that the length plus the girth of a package cannot exceed 84 inches in order to be mailed. Find the dimensions of the rectangular package of greatest volume that can be mailed. [*Note:* From Fig. 4 we see that $84 = $ (length)+(girth) $= l+(2x+2y)$.]

**Figure 4**

30. Find the dimensions of the rectangular box of least surface area that has a volume of 1000 cubic inches.

31. A company manufactures and sells two products, call them I and II, that sell for $10 and $9 per unit, respectively. The cost of producing $x$ units of product I and $y$ units of product II is

$$400 + 2x + 3y + .01(3x^2 + xy + 3y^2).$$

Find the values of $x$ and $y$ that maximize the company's profits. [*Note:* Profit = (revenue) − (cost).]

32. A monopolist manufactures and sells two competing products, call them I and II, that cost $30 and $20 per unit, respectively, to produce. The revenue from marketing $x$ units of product I and $y$ units of product II is $98x + 112y - .04xy - .1x^2 - .2y^2$. Find the values of $x$ and $y$ that maximize the monopolist's profits.

33. A company manufactures and sells two products, call them I and II, that sell for $p_I$ and $p_{II}$ per unit, respectively. Let $C(x, y)$ be the cost of producing $x$ units of product I and $y$ units of product II. Show that if the company's profit is maximized when $x = a$, $y = b$, then

$$\frac{\partial C}{\partial x}(a, b) = p_I \quad \text{and} \quad \frac{\partial C}{\partial y}(a, b) = p_{II}.$$

34. A monopolist manufactures and sells two competing products, call them I and II, that cost $p_I$ and $p_{II}$ per unit, respectively, to produce. Let $R(x, y)$ be the revenue from marketing $x$ units of product I and $y$ units of product II. Show that if the monopolist's profit is maximized when $x = a$, $y = b$, then

$$\frac{\partial R}{\partial x}(a, b) = p_I \quad \text{and} \quad \frac{\partial R}{\partial y}(a, b) = p_{II}.$$

1. Compute the first partial derivatives of $f(x,y)$ and solve the system of equations that results from setting the partials equal to zero.

$$\frac{\partial f}{\partial x} = 3x^2 - 3y = 0$$

$$\frac{\partial f}{\partial y} = -3x + y = 0.$$

Solve each equation for $y$ in terms of $x$.

$$\begin{cases} y = x^2 \\ y = 3x. \end{cases}$$

Equate expressions for $y$ and solve for $x$.

$$x^2 = 3x$$
$$x^2 - 3x = 0$$
$$x(x-3) = 0$$
$$x = 0 \quad \text{or} \quad x = 3.$$

When $x = 0$, $y = 0^2 = 0$. When $x = 3$, $y = 3^2 = 9$. Therefore, the possible relative maximum or minimum points are $(0,0)$ and $(3,9)$.

2. We have

$$g(x,y) = 11xy + \frac{14V}{x} + \frac{15V}{y},$$

$$\frac{\partial g}{\partial x} = 11y - \frac{14V}{x^2}, \quad \text{and} \quad \frac{\partial g}{\partial y} = 11x - \frac{15V}{y^2}.$$

Now,

$$\frac{\partial^2 g}{\partial x^2} = \frac{28V}{x^3}, \quad \frac{\partial^2 g}{\partial y^2} = \frac{30V}{y^3}, \quad \text{and} \quad \frac{\partial^2 g}{\partial x \, \partial y} = 11.$$

Therefore,

$$D(x,y) = \frac{28V}{x^3} \cdot \frac{30V}{y^3} - (11)^2$$

$$D(56,60) = \frac{28(147{,}840)}{(56)^3} \cdot \frac{30(147{,}840)}{(60)^3} - 121$$

$$= 484 - 121 = 363 > 0,$$

and

$$\frac{\partial^2 g}{\partial x^2}(56,60) = \frac{28(147{,}840)}{(56)^3} > 0.$$

It follows that $g(x,y)$ has a relative minimum at $x = 56$, $y = 60$.

# 7.4  Lagrange Multipliers and Constrained Optimization

We have seen a number of optimization problems in which we were required to minimize (or maximize) an objective function where the variables were subject to a constraint equation. For instance, in Example 4 of Section 2.5, we minimized the cost of a rectangular enclosure by minimizing the objective function $42x + 28y$, where $x$ and $y$ were subject to the constraint equation $600 - xy = 0$. In the preceding section (Example 3) we minimized the daily heat loss from a building by minimizing the objective function $11xy + 14yz + 15xz$, subject to the constraint equation $147{,}840 - xyz = 0$.

Figure 1 gives a graphical illustration of what happens when an objective function is maximized subject to a constraint. The graph of the objective function is the cone-shaped surface $z = 36 - x^2 - y^2$, and the colored curve on that surface consists of those points whose $x$- and $y$-coordinates satisfy the constraint equation $x + 7y - 25 = 0$. The constrained maximum is at the highest point on this curve. Of course, the surface itself has a higher "unconstrained maximum" at $(x,y,z) = (0,0,36)$, but these values of $x$ and $y$ do not satisfy the constraint equation.

In this section we introduce a powerful technique for solving problems of this type. Let us begin with the following general problem, which involves two variables.

**Problem**   Let $f(x,y)$ and $g(x,y)$ be functions of two variables. Find values of $x$ and $y$ that maximize (or minimize) the objective function $f(x,y)$ and that also satisfy the constraint equation $g(x,y) = 0$.

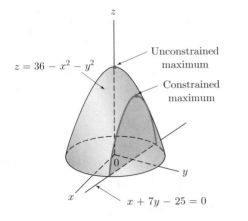

$z = 36 - x^2 - y^2$

Unconstrained maximum

Constrained maximum

$x + 7y - 25 = 0$

**Figure 1.** A constrained optimization problem.

Of course, if we can solve the equation $g(x, y) = 0$ for one variable in terms of the other and substitute the resulting expression into $f(x, y)$, we arrive at a function of a single variable that can be maximized (or minimized) by using the methods of Chapter 2. However, this technique can be unsatisfactory for two reasons. First, it may be difficult to solve the equation $g(x, y) = 0$ for $x$ or for $y$. For example, if $g(x, y) = x^4 + 5x^3y + 7x^2y^3 + y^5 - 17 = 0$, then it is difficult to write $y$ as a function of $x$ or $x$ as a function of $y$. Second, even if $g(x, y) = 0$ can be solved for one variable in terms of the other, substitution of the result into $f(x, y)$ may yield a complicated function.

One clever idea for handling the preceding problem was discovered by the eighteenth-century mathematician Lagrange, and the technique that he pioneered today bears his name—the method of *Lagrange multipliers*. The basic idea of this method is to replace $f(x, y)$ by an auxiliary function of three variables $F(x, y, \lambda)$, defined as

$$F(x, y, \lambda) = f(x, y) + \lambda g(x, y).$$

The new variable $\lambda$ (lambda) is called a *Lagrange multiplier* and always multiplies the constraint function $g(x, y)$. The following theorem is stated without proof.

---

**Theorem** Suppose that, subject to the constraint $g(x, y) = 0$, the function $f(x, y)$ has a relative maximum or minimum at $(x, y) = (a, b)$. Then there is a value of $\lambda$, say $\lambda = c$, such that the partial derivatives of $F(x, y, \lambda)$ all equal zero at $(x, y, \lambda) = (a, b, c)$.

---

The theorem implies that if we locate all points $(x, y, \lambda)$ where the partial derivatives of $F(x, y, \lambda)$ are all zero, then among the corresponding points $(x, y)$ we will find all possible places where $f(x, y)$ may have a constrained relative maximum or minimum. Thus the first step in the method of Lagrange multipliers is to set the partial derivatives of $F(x, y, \lambda)$ equal to zero and solve for $x$, $y$, and $\lambda$:

$$\frac{\partial F}{\partial x} = 0 \tag{L-1}$$

$$\frac{\partial F}{\partial y} = 0 \tag{L-2}$$

$$\frac{\partial F}{\partial \lambda} = 0. \tag{L-3}$$

From the definition of $F(x, y, \lambda)$, we see that $\dfrac{\partial F}{\partial \lambda} = g(x, y)$. Thus the third equation (L-3) is just the original constraint equation $g(x, y) = 0$. So when we find a point $(x, y, \lambda)$ that satisfies (L-1), (L-2), and (L-3), the coordinates $x$ and $y$ will automatically satisfy the constraint equation.

The first example applies this method to the problem described in Fig. 1.

**EXAMPLE 1**    Maximize $36 - x^2 - y^2$ subject to the constraint $x + 7y - 25 = 0$.

**Solution**    Here $f(x, y) = 36 - x^2 - y^2$, $g(x, y) = x + 7y - 25$, and

$$F(x, y, \lambda) = 36 - x^2 - y^2 + \lambda(x + 7y - 25).$$

Equations (L-1) to (L-3) read

$$\frac{\partial F}{\partial x} = -2x + \lambda = 0 \tag{1}$$

$$\frac{\partial F}{\partial y} = -2y + 7\lambda = 0 \tag{2}$$

$$\frac{\partial F}{\partial \lambda} = x + 7y - 25 = 0. \tag{3}$$

We solve the first two equations for $\lambda$:

$$\lambda = 2x$$

$$\lambda = \tfrac{2}{7}y. \tag{4}$$

If we equate these two expressions for $\lambda$, we obtain

$$2x = \tfrac{2}{7}y$$

$$x = \tfrac{1}{7}y. \tag{5}$$

Substituting this expression for $x$ into equation (3), we have

$$\tfrac{1}{7}y + 7y - 25 = 0$$

$$\tfrac{50}{7}y = 25$$

$$y = \tfrac{7}{2}.$$

With this value for $y$, equations (4) and (5) produce the values of $x$ and $\lambda$:

$$x = \tfrac{1}{7}y = \tfrac{1}{7}\left(\tfrac{7}{2}\right) = \tfrac{1}{2}$$

$$\lambda = \tfrac{2}{7}y = 1.$$

Therefore, the partial derivatives of $F(x, y, \lambda)$ are zero when $x = \tfrac{1}{2}$, $y = \tfrac{7}{2}$, and $\lambda = 1$. So the maximum value of $36 - x^2 - y^2$ subject to the constraint $x + 7y - 25 = 0$ is

$$36 - \left(\tfrac{1}{2}\right)^2 - \left(\tfrac{7}{2}\right)^2 = \tfrac{47}{2}. \qquad \blacksquare$$

The preceding technique for solving three equations in the three variables $x$, $y$, and $\lambda$ can usually be applied to solve Lagrange multiplier problems. Here is the basic procedure.

**1.** Solve (L-1) and (L-2) for $\lambda$ in terms of $x$ and $y$; then equate the resulting expressions for $\lambda$.

2. Solve the resulting equation for one of the variables.

3. Substitute the expression so derived into the equation (L-3) and solve the resulting equation of one variable.

4. Use the one known variable and the equations of steps 1 and 2 to determine the other two variables.

In most applications we know that an absolute (constrained) maximum or minimum exists. In the event that the method of Lagrange multipliers produces exactly one possible relative extreme value, we will assume that it is indeed the sought-after absolute extreme value. For instance, the statement of Example 1 is meant to imply that there is an absolute maximum value. Since we determined that there was just one possible relative extreme value, we concluded that it was the absolute maximum value.

---

**EXAMPLE 2** Using Lagrange multipliers, minimize $42x + 28y$, subject to the constraint $600 - xy = 0$, where $x$ and $y$ are restricted to positive values. (This problem arose in Example 4 of Section 2.5, where $42x + 28y$ was the cost of building a 600-square-foot enclosure having dimensions $x$ and $y$.)

**Solution** We have $f(x, y) = 42x + 28y$, $g(x, y) = 600 - xy$, and

$$F(x, y, \lambda) = 42x + 28y + \lambda(600 - xy).$$

The equations (L-1) to (L-3), in this case, are

$$\frac{\partial F}{\partial x} = 42 - \lambda y = 0$$

$$\frac{\partial F}{\partial y} = 28 - \lambda x = 0$$

$$\frac{\partial F}{\partial \lambda} = 600 - xy = 0.$$

From the first two equations we see that

$$\lambda = \frac{42}{y} = \frac{28}{x}. \qquad \text{(step 1)}$$

Therefore,

$$42x = 28y$$

and

$$x = \frac{2}{3}y. \qquad \text{(step 2)}$$

Substituting this expression for $x$ into the third equation, we derive

$$600 - \left(\frac{2}{3}y\right)y = 0$$

$$y^2 = \frac{3}{2} \cdot 600 = 900$$

$$y = \pm 30. \qquad \text{(step 3)}$$

We discard the case $y = -30$ because we are interested only in positive values of $x$ and $y$. Using $y = 30$, we find that

$$\left.\begin{array}{c} x = \dfrac{2}{3}(30) = 20 \\[2mm] \lambda = \dfrac{28}{20} = \dfrac{7}{5}. \end{array}\right\} \qquad \text{(step 4)}$$

So the minimum value of $42x + 28y$ with $x$ and $y$ subject to the constraint occurs when $x = 20$, $y = 30$, and $\lambda = \frac{7}{5}$. That minimum value is

$$42 \cdot (20) + 28 \cdot (30) = 1680. \qquad \blacksquare$$

---

**EXAMPLE 3**   **Production**   Suppose that $x$ units of labor and $y$ units of capital can produce $f(x, y) = 60x^{3/4}y^{1/4}$ units of a certain product. Also suppose that each unit of labor costs \$100, whereas each unit of capital costs \$200. Assume that \$30,000 is available to spend on production. How many units of labor and how many units of capital should be utilized in order to maximize production?

**Solution**   The cost of $x$ units of labor and $y$ units of capital equals $100x + 200y$. Therefore, since we want to use all the available money (\$30,000), we must satisfy the constraint equation

$$100x + 200y = 30,000$$

or

$$g(x, y) = 30,000 - 100x - 200y = 0.$$

The objective function is $f(x, y) = 60x^{3/4}y^{1/4}$. In this case, we have

$$F(x, y, \lambda) = 60x^{3/4}y^{1/4} + \lambda(30,000 - 100x - 200y).$$

The equations (L-1) to (L-3) read

$$\frac{\partial F}{\partial x} = 45x^{-1/4}y^{1/4} - 100\lambda = 0 \qquad \text{(L-1)}$$

$$\frac{\partial F}{\partial y} = 15x^{3/4}y^{-3/4} - 200\lambda = 0 \qquad \text{(L-2)}$$

$$\frac{\partial F}{\partial \lambda} = 30,000 - 100x - 200y = 0. \qquad \text{(L-3)}$$

By solving the first two equations for $\lambda$, we see that

$$\lambda = \frac{45}{100}x^{-1/4}y^{1/4} = \frac{9}{20}x^{-1/4}y^{1/4}$$

$$\lambda = \frac{15}{200}x^{3/4}y^{-3/4} = \frac{3}{40}x^{3/4}y^{-3/4}.$$

Therefore, we must have

$$\frac{9}{20}x^{-1/4}y^{1/4} = \frac{3}{40}x^{3/4}y^{-3/4}.$$

To solve for $y$ in terms of $x$, let us multiply both sides of this equation by $x^{1/4}y^{3/4}$:

$$\frac{9}{20}y = \frac{3}{40}x$$

or

$$y = \frac{1}{6}x.$$

Inserting this result in (L-3), we find that

$$100x + 200\left(\frac{1}{6}x\right) = 30{,}000$$

$$\frac{400x}{3} = 30{,}000$$

$$x = 225.$$

Hence

$$y = \frac{225}{6} = 37.5.$$

So maximum production is achieved by using 225 units of labor and 37.5 units of capital. ∎

In Example 3 it turns out that, at the optimum values of $x$ and $y$,

$$\lambda = \frac{9}{20}x^{-1/4}y^{1/4} = \frac{9}{20}(225)^{-1/4}(37.5)^{1/4} \approx .2875$$

$$\frac{\partial f}{\partial x} = 45x^{-1/4}y^{1/4} = 45(225)^{-1/4}(37.5)^{1/4} \tag{6}$$

$$\frac{\partial f}{\partial y} = 15x^{3/4}y^{-3/4} = 15(225)^{3/4}(37.5)^{-3/4}. \tag{7}$$

It can be shown that the Lagrange multiplier $\lambda$ can be interpreted as the *marginal productivity of money*. That is, if one additional dollar is available, then approximately .2875 additional units of the product can be produced.

Recall that the partial derivatives $\dfrac{\partial f}{\partial x}$ and $\dfrac{\partial f}{\partial y}$ are called the marginal productivity of labor and capital, respectively. From (6) and (7) we have

$$\frac{[\text{marginal productivity of labor}]}{[\text{marginal productivity of capital}]} = \frac{45(225)^{-1/4}(37.5)^{1/4}}{15(225)^{3/4}(37.5)^{-3/4}}$$

$$= \frac{45}{15}(225)^{-1}(37.5)^{1}$$

$$= \frac{3(37.5)}{225} = \frac{37.5}{75} = \frac{1}{2}.$$

On the other hand,

$$\frac{[\text{cost per unit of labor}]}{[\text{cost per unit of capital}]} = \frac{100}{200} = \frac{1}{2}.$$

This result illustrates the following law of economics. *If labor and capital are at their optimal levels, then the ratio of their marginal productivities equals the ratio of their unit costs.*

The method of Lagrange multipliers generalizes to functions of any number of variables. For instance, we can maximize $f(x, y, z)$, subject to the constraint equation $g(x, y, z) = 0$, by considering the Lagrange function

$$F(x, y, z, \lambda) = f(x, y, z) + \lambda g(x, y, z).$$

The analogs of equations (L-1) to (L-3) are

$$\frac{\partial F}{\partial x} = 0$$

$$\frac{\partial F}{\partial y} = 0$$

$$\frac{\partial F}{\partial z} = 0$$

$$\frac{\partial F}{\partial \lambda} = 0.$$

Let us now show how we can solve the heat-loss problem of Section 7.3 by using this method.

---

**EXAMPLE 4**  Use Lagrange multipliers to find the values of $x$, $y$, $z$ that minimize the objective function

$$f(x, y, z) = 11xy + 14yz + 15xz,$$

subject to the constraint

$$xyz = 147{,}840.$$

**Solution**  The Lagrange function is

$$F(x, y, z, \lambda) = 11xy + 14yz + 15xz + \lambda(147{,}840 - xyz).$$

The conditions for a relative minimum are

$$\frac{\partial F}{\partial x} = 11y + 15z - \lambda yz = 0$$

$$\frac{\partial F}{\partial y} = 11x + 14z - \lambda xz = 0$$

$$\frac{\partial F}{\partial z} = 14y + 15x - \lambda xy = 0$$

$$\frac{\partial F}{\partial \lambda} = 147{,}840 - xyz = 0. \tag{8}$$

From the first three equations we have

$$\left.\begin{array}{l} \lambda = \dfrac{11y + 15z}{yz} = \dfrac{11}{z} + \dfrac{15}{y} \\[2mm] \lambda = \dfrac{11x + 14z}{xz} = \dfrac{11}{z} + \dfrac{14}{x} \\[2mm] \lambda = \dfrac{14y + 15x}{xy} = \dfrac{14}{x} + \dfrac{15}{y} \end{array}\right\}. \tag{9}$$

Let us equate the first two expression for $\lambda$:

$$\frac{11}{z} + \frac{15}{y} = \frac{11}{z} + \frac{14}{x}$$

$$\frac{15}{y} = \frac{14}{x}$$

$$x = \frac{14}{15}y.$$

Next, we equate the second and third expressions for $\lambda$ in (9):

$$\frac{11}{z} + \frac{14}{x} = \frac{14}{x} + \frac{15}{y}$$

$$\frac{11}{z} = \frac{15}{y}$$

$$z = \frac{11}{15}y.$$

We now substitute the expressions for $x$ and $z$ into the constraint equation (8) and obtain

$$\frac{14}{15}y \cdot y \cdot \frac{11}{15}y = 147{,}840$$

$$y^3 = \frac{(147{,}840)(15)^2}{(14)(11)} = 216{,}000$$

$$y = 60.$$

From this, we find that

$$x = \frac{14}{15}(60) = 56 \quad \text{and} \quad z = \frac{11}{15}(60) = 44.$$

We conclude that the heat loss is minimized when $x = 56$, $y = 60$, and $z = 44$. ∎

In the solution of Example 4, we found that at the optimal values of $x$, $y$, and $z$,

$$\frac{14}{x} = \frac{15}{y} = \frac{11}{z}.$$

Referring to Example 2 of Section 7.1, we see that 14 is the combined heat loss through the east and west sides of the building, 15 is the heat loss through the north and south sides of the building, and 11 is the heat loss through the floor and roof. Thus we have that under optimal conditions

$$\frac{[\text{heat loss through east and west sides}]}{[\text{distance between east and west sides}]}$$

$$= \frac{[\text{heat loss through north and south sides}]}{[\text{distance between north and south sides}]}$$

$$= \frac{[\text{heat loss through floor and roof}]}{[\text{distance between floor and roof}]}.$$

This is a principle of optimal design: Minimal heat loss occurs when the distance between each pair of opposite sides is some fixed constant times the heat loss from the pair of sides.

The value of $\lambda$ in Example 4 corresponding to the optimal values of $x$, $y$, and $z$ is

$$\lambda = \frac{11}{z} + \frac{15}{y} = \frac{11}{44} + \frac{15}{60} = \frac{1}{2}.$$

One can show that the Lagrange multiplier $\lambda$ is the marginal heat loss with respect to volume. That is, if a building of volume slightly more than 147,840 cubic feet is optimally designed, then $\frac{1}{2}$ unit of additional heat will be lost for each additional cubic foot of volume.

## PRACTICE PROBLEMS 7.4

1. Let $F(x, y, \lambda) = 2x + 3y + \lambda(90 - 6x^{1/3}y^{2/3})$. Find $\dfrac{\partial F}{\partial x}$.

2. Refer to Exercise 29 of Section 7.3. What is the function $F(x, y, l, \lambda)$ when the exercise is solved using the method of Lagrange multipliers?

# EXERCISES 7.4

*Solve the following exercises by the method of Lagrange multipliers.*

1. Minimize $x^2 + 3y^2 + 10$, subject to the constraint $8 - x - y = 0$.

2. Maximize $x^2 - y^2$, subject to the constraint $2x + y - 3 = 0$.

3. Maximize $x^2 + xy - 3y^2$, subject to the constraint $2 - x - 2y = 0$.

4. Minimize $\frac{1}{2}x^2 - 3xy + y^2 + \frac{1}{2}$, subject to the constraint $3x - y - 1 = 0$.

5. Find the values of $x$, $y$ that maximize

$$-2x^2 - 2xy - \tfrac{3}{2}y^2 + x + 2y,$$

subject to the constraint $x + y - \frac{5}{2} = 0$.

6. Find the values of $x$, $y$ that minimize

$$x^2 + xy + y^2 - 2x - 5y,$$

subject to the constraint $1 - x + y = 0$.

7. Find the two positive numbers whose product is 25 and whose sum is as small as possible.

8. Four hundred eighty dollars are available to fence in a rectangular garden. The fencing for the north and south sides of the garden costs $10 per foot and the fencing for the east and west sides costs $15 per foot. Find the dimensions of the largest possible garden.

9. Three hundred square inches of material are available to construct an open rectangular box with a square base. Find the dimensions of the box that maximize the volume.

10. The amount of space required by a particular firm is $f(x, y) = 1000\sqrt{6x^2 + y^2}$, where $x$ and $y$ are, respectively, the number of units of labor and capital utilized. Suppose that labor costs $480 per unit and capital costs $40 per unit and that the firm has $5000 to spend. Determine the amounts of labor and capital that should be utilized in order to minimize the amount of space required.

11. Find the dimensions of the rectangle of maximum area that can be inscribed in the unit circle. [See Fig. 2(a).]

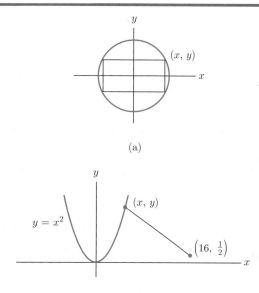

(a)

(b)

**Figure 2**

12. Find the point on the parabola $y = x^2$ that has minimal distance from the point $\left(16, \frac{1}{2}\right)$. [See Fig. 2(b).] [*Suggestion:* If $d$ denotes the distance from $(x, y)$ to $\left(16, \frac{1}{2}\right)$, then $d^2 = (x - 16)^2 + (y - \frac{1}{2})^2$. If $d^2$ is minimized, then $d$ will be minimized.]

13. Suppose that a firm makes two products A and B that use the same raw materials. Given a fixed amount of raw materials and a fixed amount of manpower, the firm must decide how much of its resources should be allocated to the production of A and how much to B. If $x$ units of A and $y$ units of B are produced, suppose that $x$ and $y$ must satisfy

$$9x^2 + 4y^2 = 18,000.$$

The graph of this equation (for $x \geq 0$, $y \geq 0$) is called a *production possibilities curve* (Fig. 3). A point $(x, y)$ on this curve represents a *production schedule* for the firm, committing it to produce $x$ units of A and $y$ units of B. The reason for the relationship between $x$ and $y$ involves the limitations on personnel and raw materials available to the firm. Suppose that each unit of A yields a $3 profit, whereas each unit of B yields a $4 profit. Then the profit of the firm is

$$P(x, y) = 3x + 4y.$$

Find the production schedule that maximizes the profit function $P(x, y)$.

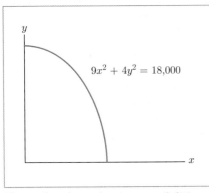

**Figure 3.** A production possibilities curve.

14. A firm makes $x$ units of product A and $y$ units of product B and has a production possibilities curve given by the equation $4x^2 + 25y^2 = 50{,}000$ for $x \geq 0$, $y \geq 0$. (See Exercise 13.) Suppose profits are \$2 per unit for product A and \$10 per unit for product B. Find the production schedule that maximizes the total profit.

15. The production function for a firm is $f(x, y) = 64x^{3/4}y^{1/4}$, where $x$ and $y$ are the number of units of labor and capital utilized. Suppose that labor costs \$96 per unit and capital costs \$162 per unit and that the firm decides to produce 3456 units of goods.

   (a) Determine the amounts of labor and capital that should be utilized in order to minimize the cost. That is, find the values of $x$, $y$ that minimize $96x + 162y$, subject to the constraint $3456 - 64x^{3/4}y^{1/4} = 0$.

   (b) Find the value of $\lambda$ at the optimal level of production.

   (c) Show that, at the optimal level of production, we have

   $$\frac{[\text{marginal productivity of labor}]}{[\text{marginal productivity of capital}]}$$
   $$= \frac{[\text{unit price of labor}]}{[\text{unit price of capital}]}.$$

16. Consider the monopolist of Example 2, Section 7.3, who sells his goods in two countries. Suppose that he must set the same price in each country. That is, $97 - (x/10) = 83 - (y/20)$. Find the values of $x$ and $y$ that maximize profits under this new restriction.

17. Find the values of $x$, $y$, and $z$ that maximize $xyz$ subject to the constraint $36 - x - 6y - 3z = 0$.

18. Find the values of $x$, $y$, and $z$ that maximize $xy + 3xz + 3yz$ subject to the constraint $9 - xyz = 0$.

19. Find the values of $x$, $y$, $z$ that maximize

$$3x + 5y + z - x^2 - y^2 - z^2,$$

subject to the constraint $6 - x - y - z = 0$.

20. Find the values of $x$, $y$, $z$ that minimize

$$x^2 + y^2 + z^2 - 3x - 5y - z,$$

subject to the constraint $20 - 2x - y - z = 0$.

21. The material for a closed rectangular box costs \$2 per square foot for the top and \$1 per square foot for the sides and bottom. Using Lagrange multipliers, find the dimensions for which the volume of the box is 12 cubic feet and the cost of the materials is minimized. [Referring to Fig. 4(a), the cost will be $3xy + 2xz + 2yz$.]

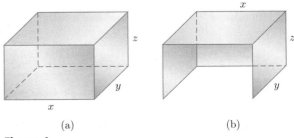

(a)                (b)

**Figure 4**

22. Use Lagrange multipliers to find the three positive numbers whose sum is 15 and whose product is as large as possible.

23. Find the dimensions of an open rectangular glass tank of volume 32 cubic feet for which the amount of material needed to construct the tank is minimized. [See Fig. 4(a).]

24. A shelter for use at the beach has a back, two sides, and a top made of canvas. [See Fig. 4(b).] Find the dimensions that maximize the volume and require 96 square feet of canvas.

25. Let $f(x, y)$ be any production function where $x$ represents labor (costing \$$a$ per unit) and $y$ represents capital (costing \$$b$ per unit). Assuming that \$$c$ is available, show that, at the values of $x$, $y$ that maximize production,

$$\frac{\dfrac{\partial f}{\partial x}}{\dfrac{\partial f}{\partial y}} = \frac{a}{b}.$$

*Note:* Let $F(x, y, \lambda) = f(x, y) + \lambda(c - ax - by)$. The result follows from (L-1) and (L-2).

26. By applying the result in Exercise 25 to the production function $f(x, y) = kx^{\alpha}y^{\beta}$, show that, for the values of $x$, $y$ that maximize production, we have

$$\frac{y}{x} = \frac{a\beta}{b\alpha}.$$

(This tells us that the ratio of capital to labor does not depend on the amount of money available nor on the level of production but only on the numbers $a$, $b$, $\alpha$, and $\beta$.)

## SOLUTIONS TO PRACTICE PROBLEMS 7.4

1. The function can be written as

$$F(x, y, \lambda) = 2x + 3y + \lambda \cdot 90 - \lambda \cdot 6x^{1/3}y^{2/3}.$$

When differentiating with respect to $x$, both $y$ and $\lambda$ should be treated as constants (so $\lambda \cdot 90$ and $\lambda \cdot 6$ are also regarded as constants).

$$\frac{\partial F}{\partial x} = 2 - \lambda \cdot 6 \cdot \frac{1}{3}x^{-2/3} \cdot y^{2/3}$$

$$= 2 - 2\lambda x^{-2/3}y^{2/3}.$$

(*Note:* It is not necessary to write out the multiplication by $\lambda$ as we did. Most people just do this mentally and then differentiate.)

2. The quantity to be maximized is the volume $xyl$. The constraint is that length plus girth is 84. This translates to $84 = l + 2x + 2y$ or $84 - l - 2x - 2y = 0$. Therefore,

$$F(x, y, l, \lambda) = xyl + \lambda(84 - l - 2x - 2y).$$

## 7.5 The Method of Least Squares

Nowadays, people compile graphs of literally thousands of different quantities: the purchasing value of the dollar as a function of time; the pressure of a fixed volume of air as a function of temperature; the average income of people as a function of their years of formal education; or the incidence of strokes as a function of blood pressure. The observed points on such graphs tend to be irregularly distributed due to the complicated nature of the phenomena underlying them as well as to errors made in observation. (For example, a given procedure for measuring average income may not count certain groups.)

In spite of the imperfect nature of the data, we are often faced with the problem of making assessments and predictions based on them. Roughly speaking, this problem amounts to filtering the sources of errors in the data and isolating the basic underlying trend. Frequently, on the basis of a suspicion or a working hypothesis, we may suspect that the underlying trend is linear—that is, the data should lie on a straight line. But which straight line? This is the problem that the *method of least squares* attempts to answer. To be more specific, let us consider the following problem:

**Problem of Fitting a Straight Line to Data**  Given observed data points $(x_1, y_1), (x_2, y_2), \ldots, (x_N, y_N)$ on a graph, find the straight line that "best" fits these points.

In order to completely understand the statement of the problem being considered, we must define what it means for a line to "best" fit a set of points. If $(x_i, y_i)$ is one of our observed points, then we will measure how far it is from a given line $y = Ax + B$ by the vertical distance from the point to the line. Since the point on the line with $x$-coordinate $x_i$ is $(x_i, Ax_i + B)$, this vertical distance is the distance between the $y$-coordinates $Ax_i + B$ and $y_i$. (See Fig. 1.) If $E_i = (Ax_i + B) - y_i$, then either $E_i$ or $-E_i$ is the vertical distance from $(x_i, y_i)$ to the line. To avoid this ambiguity, we work with the square of this vertical distance, namely,

$$E_i^2 = (Ax_i + B - y_i)^2.$$

The total error in approximating the data points $(x_1, y_1), \ldots, (x_N, y_N)$ by the line $y = Ax + B$ is usually measured by the sum $E$ of the squares of the vertical distances from the points to the line,

$$E = E_1^2 + E_2^2 + \cdots + E_N^2.$$

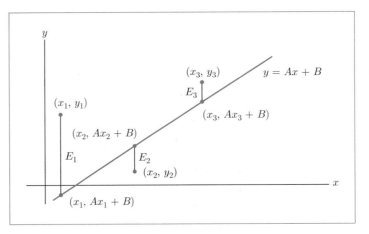

**Figure 1.** Fitting a line to data points.

$E$ is called the *least-squares error* of the observed points with respect to the line. If all the observed points lie on the line $y = Ax + B$, then all $E_i$ are zero and the error $E$ is zero. If a given observed point is far away from the line, the corresponding $E_i^2$ is large and hence makes a large contribution to the error $E$.

In general, we cannot expect to find a line $y = Ax + B$ that fits the observed points so well that the error $E$ is zero. Actually, this situation will occur only if the observed points lie on a straight line. However, we can rephrase our original problem as follows:

**Problem** Given observed data points $(x_1, y_1), (x_2, y_2), \ldots, (x_N, y_N)$, find a straight line $y = Ax + B$ for which the error $E$ is as small as possible. This line is called the *least-squares line* or *regression line*.

It turns out that this problem is a minimization problem in the two variables $A$ and $B$ and so can be solved by using the methods of Section 7.3. Let us consider an example.

**EXAMPLE 1** Find the straight line that minimizes the least-squares error for the points $(1, 4)$, $(2, 5)$, $(3, 8)$.

**Solution** Let the straight line be $y = Ax + B$. When $x = 1, 2, 3$, the $y$-coordinate of the corresponding point of the line is $A+B$, $2A+B$, $3A+B$, respectively. Therefore, the squares of the vertical distances from the points $(1, 4)$, $(2, 5)$, $(3, 8)$ are, respectively,

$$E_1^2 = (A + B - 4)^2$$
$$E_2^2 = (2A + B - 5)^2$$
$$E_3^2 = (3A + B - 8)^2.$$

(See Fig. 2.) Thus the least-squares error is

$$E = E_1^2 + E_2^2 + E_3^2 = (A + B - 4)^2 + (2A + B - 5)^2 + (3A + B - 8)^2.$$

This error obviously depends on the choice of $A$ and $B$. Let $f(A, B)$ denote this least-squares error. We want to find values of $A$ and $B$ that minimize $f(A, B)$. To do so, we take partial derivatives with respect to $A$ and $B$ and set the partial

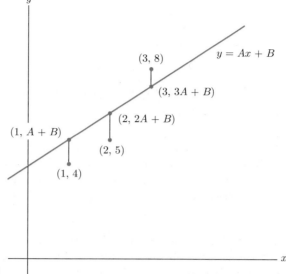

**Figure 2**

derivatives equal to zero:

$$\frac{\partial f}{\partial A} = 2(A + B - 4) + 2(2A + B - 5) \cdot 2 + 2(3A + B - 8) \cdot 3$$

$$= 28A + 12B - 76 = 0$$

$$\frac{\partial f}{\partial B} = 2(A + B - 4) + 2(2A + B - 5) + 2(3A + B - 8)$$

$$= 12A + 6B - 34 = 0.$$

In order to find $A$ and $B$, we must solve the system of simultaneous linear equations

$$28A + 12B = 76$$
$$12A + \phantom{0}6B = 34.$$

Multiplying the second equation by 2 and subtracting from the first equation, we have $4A = 8$, or $A = 2$. Therefore, $B = \frac{5}{3}$, and the straight line that minimizes the least-squares error is $y = 2x + \frac{5}{3}$.    ■

The minimization process used in Example 1 can be applied to a general set of data points $(x_1, y_1), \ldots, (x_N, y_N)$ to obtain the following algebraic formula for $A$ and $B$:

$$A = \frac{N \cdot \Sigma xy - \Sigma x \cdot \Sigma y}{N \cdot \Sigma x^2 - (\Sigma x)^2}$$

$$B = \frac{\Sigma y - A \cdot \Sigma x}{N}$$

where

$$\Sigma x = \text{sum of the } x\text{-coordinates of the data points}$$

$$\Sigma y = \text{sum of the } y\text{-coordinates of the data points}$$

$$\Sigma xy = \text{sum of the products of the coordinates of the data points}$$

$$\Sigma x^2 = \text{sum of the squares of the } x\text{-coordinates of the data points}$$

$$N = \text{number of data points}.$$

That is,

$$\Sigma x = x_1 + x_2 + \cdots + x_N$$

$$\Sigma y = y_1 + y_2 + \cdots + y_N$$

$$\Sigma xy = x_1 \cdot y_1 + x_2 \cdot y_2 + \cdots + x_N \cdot y_N$$

$$\Sigma x^2 = x_1^2 + x_2^2 + \cdots + x_N^2.$$

**EXAMPLE 2** The following table[*] gives the crude male death rate for lung cancer in 1950 and the per capita consumption of cigarettes in 1930 in various countries.

| Country | Cigarette Consumption (Per Capita) | Lung Cancer Deaths (Per Million Males) |
|---|---|---|
| Norway | 250 | 95 |
| Sweden | 300 | 120 |
| Denmark | 350 | 165 |
| Australia | 470 | 170 |

(a) Use the preceding formulas to obtain the straight line that best fits this data.

(b) In 1930 the per capita cigarette consumption in Finland was 1100. Use the straight line found in part (a) to estimate the male lung cancer death rate in Finland in 1950.

**Solution** (a) The points are plotted in Fig. 3. The sums are calculated in Table 1 and then used to determine the values of $A$ and $B$.

**Figure 3.** Lung cancer data for least-squares analysis.

| TABLE 1 Least-Squares Calculation for Smoking Data | | | |
|---|---|---|---|
| $x$ | $y$ | $xy$ | $x^2$ |
| 250 | 95 | 23,750 | 62,500 |
| 300 | 120 | 36,000 | 90,000 |
| 350 | 165 | 57,750 | 122,500 |
| 470 | 170 | 79,900 | 220,900 |
| $\Sigma x = 1370$ | $\Sigma y = 550$ | $\Sigma xy = 197{,}400$ | $\Sigma x^2 = 495{,}900$ |

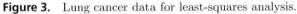

[*]These data were obtained from *Smoking and Health*, Report of the Advisory Committee to the Surgeon General of the Public Health Service, U.S. Department of Health, Education, and Welfare, Washington, D.C., Public Health Service Publication No. 1103, p. 176.

$$A = \frac{4 \cdot 197{,}400 - 1370 \cdot 550}{4 \cdot 495{,}900 - 1370^2} = \frac{36{,}100}{106{,}700} = \frac{361}{1067} \approx .338$$

$$B = \frac{550 - \frac{361}{1067} \cdot 1370}{4} = \frac{1067 \cdot 550 - 361 \cdot 1370}{1067 \cdot 4} = \frac{92{,}280}{4268} \approx 21.621.$$

Therefore, the equation of the least-squares line is $y = .338x + 21.621$.

(b) We use the straight line to estimate the lung cancer death rate in Finland by setting $x = 1100$. Then we get

$$y = .338(1100) + 21.621 = 393.421 \approx 393.$$

Therefore, we estimate the lung cancer death rate in Finland to be 393 deaths per million males. (*Note*: The actual rate was 350 deaths per million males.) ▪

**Incorporating Technology**   After the $x$- and $y$-values are placed into lists on a graphing calculator [see Fig. 4(a)], the statistical routine LinReg calculates the coefficients of the least-squares line [see Fig. 4(b)]. If desired, the equation for the line can automatically be assigned to a function and graphed along with the original points [see Fig. 4(c)]. The details for these tasks appear in Appendices A–D.

**Figure 4**    (a)    (b)    (c)

## PRACTICE PROBLEMS 7.5

**1.** Let $E = (A + B + 2)^2 + (3A + B)^2 + (6A + B - 8)^2$. What is $\dfrac{\partial E}{\partial A}$?

**2.** Find the formula (of the type in Problem 1) that gives the least-squares error for the points $(1, 10)$, $(5, 8)$, and $(7, 0)$.

## EXERCISES 7.5

**1.** Find the least-squares error $E$ for the least-squares line fit to the four points in Fig. 5.

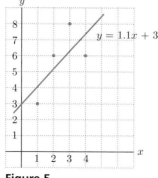

$y = 1.1x + 3$

**Figure 5**

**2.** Find the least-squares error $E$ for the least-squares line fit to the five points in Fig. 6.

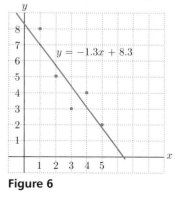

$y = -1.3x + 8.3$

**Figure 6**

3. Find the formula (of the type in Practice Problem 1) that gives the least-squares error for the points $(2, 6)$, $(5, 10)$, and $(9, 15)$.

4. Find the formula (of the type in Practice Problem 1) that gives the least-squares error for the points $(8, 4)$, $(9, 2)$, and $(10, 3)$.

*In Exercises 5–8, use partial derivatives to obtain the formula for the best least-squares fit to the data points.*

5. $(1, 2)$, $(2, 5)$, $(3, 11)$

6. $(1, 8)$, $(2, 4)$, $(4, 3)$

7. $(1, 9)$, $(2, 8)$, $(3, 6)$, $(4, 3)$

8. $(1, 5)$, $(2, 7)$, $(3, 6)$, $(4, 10)$

9. Complete Table 2 and find the values of $A$ and $B$ for the straight line that provides the best least-squares fit to the data.

**TABLE 2**

| $x$ | $y$ | $xy$ | $x^2$ |
|---|---|---|---|
| 1 | 7 | | |
| 2 | 6 | | |
| 3 | 4 | | |
| 4 | 3 | | |
| $\Sigma x =$ | $\Sigma y =$ | $\Sigma xy =$ | $\Sigma x^2 =$ |

10. Complete Table 3 and find the values of $A$ and $B$ for the straight line that provides the best least-squares fit to the data.

**TABLE 3**

| $x$ | $y$ | $xy$ | $x^2$ |
|---|---|---|---|
| 1 | 2 | | |
| 2 | 4 | | |
| 3 | 7 | | |
| 4 | 9 | | |
| 5 | 12 | | |
| $\Sigma x =$ | $\Sigma y =$ | $\Sigma xy =$ | $\Sigma x^2 =$ |

*In the remaining exercises, use one or more of the three methods discussed in this section (partial derivatives, formulas, or graphing utilities) to obtain the formula for the least-squares line.*

11. Table 4* gives the U.S. per capita health care expenditures for the years 1990–1994.

**TABLE 4    U.S. Per Capita Health Care Expenditures**

| Years (after 1990) | Dollars (in thousands) |
|---|---|
| 0 | 2.688 |
| 1 | 2.902 |
| 2 | 3.144 |
| 3 | 3.331 |
| 4 | 3.510 |

(a) Find the least-squares line for these data.

(b) Use the least-squares line to predict the per capita health care expenditures for the year 2000.

(c) Use the least-squares line to predict when per capita health care expenditures will reach $6000.

12. Table 5 shows the 1994 price of a gallon (in U.S. dollars) of fuel and the average miles driven per automobile for several countries.[†]

(a) Find the straight line that provides the best least-squares fit to these data.

(b) In 1994, the price of gas in Japan was $4.14 per gallon. Use the straight line of part (a) to estimate the average number of miles automobiles were driven in Japan.

**TABLE 5    Effect of Gas Prices on Miles Driven**

| Country | Price per Gallon | Average Miles per Auto |
|---|---|---|
| Canada | $1.57 | 10,371 |
| England | $2.86 | 10,186 |
| France | $3.31 | 8740 |
| Germany | $3.34 | 7674 |
| Sweden | $3.44 | 7456 |
| United States | $1.24 | 11,099 |

13. Table 6 gives the percent of persons 25 years and over who have completed four or more years of college.[‡]

**TABLE 6   College Completion Rates**

| Year | 1970 | 1975 | 1980 | 1985 | 1990 | 1995 |
|---|---|---|---|---|---|---|
| Percent | 10.7 | 13.9 | 16.2 | 19.4 | 21.3 | 23.0 |

(a) Use the method of least squares to obtain the straight line that best fits these data. [*Hint:* First convert *Year* to *Years after 1970.*]

(b) Estimate the percent for the year 1993.

---

*U.S. Health Care Financing Administration, *Health Care Financing Review*, Spring 1996.

[†]Source: Energy Information Administration, *International Energy Annual*. U.S. Highway Administration, *Highway Statistics*, 1994.

[‡]U.S. Bureau of the Census.

**(c)** If the trend determined by the straight line in part (a) continues, when will the percent reach 27.1?

**14.** Table 7 gives the number (in millions) of cars in use in the United States[§] for certain years.

| TABLE 7 | Automobile Population | | |
|---------|---------|------|------|
| **Year** | **Cars** | **Year** | **Cars** |
| 1980 | 104.6 | 1991 | 123.3 |
| 1985 | 114.7 | 1992 | 120.3 |
| 1989 | 122.8 | 1993 | 121.1 |
| 1990 | 123.3 | 1994 | 122 |

**(a)** Use the method of least squares to obtain the straight line that best fits these data. [*Hint:* First convert *Years* to *Years after 1980.*]

**(b)** Estimate the number of cars in use in 1983.

**(c)** If the trend determined by the straight line in part (a) continues, when will the number of cars in use reach 130 million?

**15.** An ecologist wished to know whether certain species of aquatic insects have their ecological range limited by temperature. He collected the data in Table 8, relating the average daily temperature at different portions of a creek with the elevation of that portion of the creek (above sea level).[*]

**(a)** Find the straight line that provides the best least-squares fit to these data.

**(b)** Use the linear function to estimate the average daily temperature for this creek at altitude 3.2 kilometers.

| TABLE 8 | Relationship between Elevation and Temperature in a Creek |
|---------|---------|
| **Elevation (kilometers)** | **Average Temperature (degrees Celsius)** |
| 2.7 | 11.2 |
| 2.8 | 10 |
| 3.0 | 8.5 |
| 3.5 | 7.5 |

## SOLUTIONS TO PRACTICE PROBLEMS 7.5

**1.** $\dfrac{\partial E}{\partial A} = 2(A + B + 2) \cdot 1 + 2(3A + B) \cdot 3$
$\qquad + 2(6A + B - 8) \cdot 6$

$\qquad = (2A + 2B + 4) + (18A + 6B)$
$\qquad + (72A + 12B - 96)$

$\qquad = 92A + 20B - 92.$

(Notice that we used the general power rule when differentiating and so had to always multiply by

the derivative of the quantity inside the parentheses. Also, you might be tempted to first square the terms in the expression for $E$ and then differentiate. We recommend that you resist that temptation.)

**2.** $E = (A + B - 10)^2 + (5A + B - 8)^2 + (7A + B)^2$. In general, $E$ is a sum of squares, one for each point being fitted. The point $(a, b)$ gives rise to the term $(aA + B - b)^2$.

## 7.6 Double Integrals

Up to this point, our discussion of calculus of several variables has been confined to the study of differentiation. Let us now take up the topic of integration of functions of several variables. As has been the case throughout most of this chapter, we restrict our discussion to functions $f(x, y)$ of two variables.

Let us begin with some motivation. Before we define the concept of an integral for functions of several variables, let us review the essential features of the integral in one variable.

Consider the definite integral $\int_a^b f(x)\,dx$. To write down this integral takes two pieces of information. The first is the function $f(x)$. The second is the interval over which the integration is to be performed. In this case, the interval is the portion of

---

[§]American Automobile Manufacturers Association, Inc., Detroit, Mich., *Motor Vehicle Facts and Figures.*

[*]The authors express their thanks to Dr. J. David Allen, formerly of the Department of Zoology at the University of Maryland, for providing the data for this exercise.

the $x$-axis from $x = a$ to $x = b$. The value of the definite integral is a number. In case the function $f(x)$ is nonnegative throughout the interval from $x = a$ to $x = b$, this number equals the area under the graph of $f(x)$ from $x = a$ to $x = b$. (See Fig. 1.) If $f(x)$ is negative for some values of $x$ in the interval, then the integral still equals the area bounded by the graph, but areas below the $x$-axis are counted as negative.

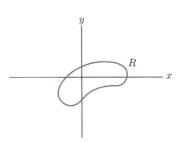

**Figure 1**

**Figure 2.** A region in the $xy$-plane.

Let us generalize the foregoing ingredients to a function $f(x, y)$ of two variables. First, we must provide a two-dimensional analog of the interval from $x = a$ to $x = b$. This is easy. We take a two-dimensional region $R$ of the plane, such as the region shown in Fig. 2. As our generalization of $f(x)$, we take a function $f(x, y)$ of two variables. Our generalization of the definite integral is denoted

$$\iint_R f(x, y) \, dx \, dy$$

and is called the *double integral of $f(x, y)$ over the region $R$*. The value of the double integral is a number defined as follows. For the sake of simplicity, let us begin by assuming that $f(x, y) \geq 0$ for all points $(x, y)$ in the region $R$. [This is the analog of the assumption that $f(x) \geq 0$ for all $x$ in the interval from $x = a$ to $x = b$.] This means that the graph of $f$ lies above the region $R$ in three-dimensional space. (See Fig. 3.) The portion of the graph over $R$ determines a solid figure. (See Fig. 4.) This figure is called the *solid bounded by $f(x, y)$ over the region $R$*. We define the double integral $\iint_R f(x, y) \, dx \, dy$ to be the volume of this solid. In case the graph of $f(x, y)$ lies partially above the region $R$ and partially below, we define the double integral to be the volume of the solid above the region minus the volume of the solid below the region. That is, we count volumes below the $xy$-plane as negative.

Now that we have defined the notion of a double integral, we must learn how to calculate its value. To do so, let us introduce the notion of an iterated integral. Let $f(x, y)$ be a function of two variables, let $g(x)$ and $h(x)$ be two functions of $x$ alone, and let $a$ and $b$ be numbers. Then an *iterated integral* is an expression of the form

$$\int_a^b \left( \int_{g(x)}^{h(x)} f(x, y) \, dy \right) dx.$$

To explain the meaning of this collection of symbols, we proceed from the inside out. We evaluate the integral

$$\int_{g(x)}^{h(x)} f(x, y) \, dy$$

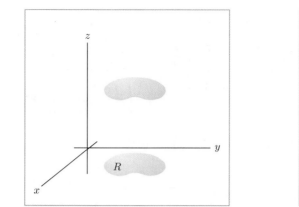

**Figure 3.** Graph of $f(x, y)$ above the region $R$.

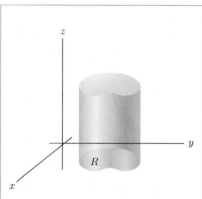

**Figure 4.** Solid bounded by $f(x, y)$ over $R$.

by considering $f(x, y)$ as a function of $y$ alone. This is indicated by the $dy$ in the inner integral. We treat $x$ as a constant in this integration. So we evaluate the integral by first finding an antiderivative $F(x, y)$ with respect to $y$. The integral above is then evaluated as

$$F(x, h(x)) - F(x, g(x)).$$

That is, we evaluate the antiderivative between the limits $y = g(x)$ and $y = h(x)$. This gives us a function of $x$ alone. To complete the evaluation of the integral, we integrate this function from $x = a$ to $x = b$. The next two examples illustrate the procedure for evaluating iterated integrals.

**EXAMPLE 1** Evaluate the iterated integral

$$\int_1^2 \left( \int_3^4 (y - x) \, dy \right) dx.$$

**Solution** Here $g(x)$ and $h(x)$ are constant functions: $g(x) = 3$ and $h(x) = 4$. We evaluate the inner integral first. The variable in this integral is $y$, so we treat $x$ as a constant.

$$\int_3^4 (y - x) \, dy = \left( \frac{1}{2} y^2 - xy \right) \Big|_3^4$$

$$= \left( \frac{1}{2} \cdot 16 - x \cdot 4 \right) - \left( \frac{1}{2} \cdot 9 - x \cdot 3 \right)$$

$$= 8 - 4x - \frac{9}{2} + 3x$$

$$= \frac{7}{2} - x$$

Now we carry out the integration with respect to $x$:

$$\int_1^2 \left( \frac{7}{2} - x \right) dx = \frac{7}{2} x - \frac{1}{2} x^2 \Big|_1^2$$

$$= \left( \frac{7}{2} \cdot 2 - \frac{1}{2} \cdot 4 \right) - \left( \frac{7}{2} - \frac{1}{2} \cdot 1 \right)$$

$$= (7 - 2) - (3) = 2.$$

So the value of the iterated integral is 2. ■

---

**EXAMPLE 2** Evaluate the iterated integral

$$\int_0^1 \left( \int_{\sqrt{x}}^{x+1} 2xy \, dy \right) dx.$$

**Solution** We evaluate the inner integral first.

$$\int_{\sqrt{x}}^{x+1} 2xy \, dy = xy^2 \Big|_{\sqrt{x}}^{x+1} = x(x+1)^2 - x(\sqrt{x})^2$$

$$= x(x^2 + 2x + 1) - x \cdot x$$

$$= x^3 + 2x^2 + x - x^2$$

$$= x^3 + x^2 + x$$

Now we evaluate the outer integral.

$$\int_0^1 (x^3 + x^2 + x) \, dx = \frac{1}{4}x^4 + \frac{1}{3}x^3 + \frac{1}{2}x^2 \Big|_0^1 = \frac{1}{4} + \frac{1}{3} + \frac{1}{2} = \frac{13}{12}$$

So the value of the iterated integral is $\frac{13}{12}$. ■

Let us now return to the discussion of the double integral $\iint\limits_R f(x,y) \, dx \, dy$. When the region $R$ has a special form, the double integral may be expressed as an iterated integral. Namely, let us suppose that $R$ is bounded by the graphs of $y = g(x)$, $y = h(x)$ and by the vertical lines $x = a$ and $x = b$. (See Fig. 5.) In this case, we have the following fundamental result, which we cite without proof.

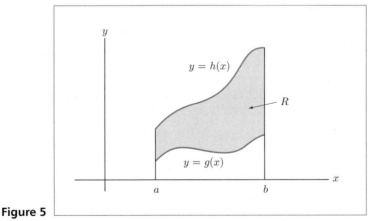

**Figure 5**

Let $R$ be the region in the $xy$-plane bounded by the graphs of $y = g(x)$, $y = h(x)$, and the vertical lines $x = a$, $x = b$. Then

$$\iint\limits_R f(x,y) \, dx \, dy = \int_a^b \left( \int_{g(x)}^{h(x)} f(x,y) \, dy \right) dx.$$

Since the value of the double integral gives the volume of the solid bounded by the graph of $f(x,y)$ over the region $R$, the preceding result may be used to calculate volumes, as the next two examples show.

---

**EXAMPLE 3**  Calculate the volume of the solid bounded above by the function $f(x,y) = y - x$ and lying over the rectangular region $R$: $1 \leq x \leq 2$, $3 \leq y \leq 4$. (See Fig. 6.)

**Solution**  The desired volume is given by the double integral $\iint\limits_{R} (y-x)\,dx\,dy$. By the result just cited, this double integral is equal to the iterated integral

$$\int_1^2 \left( \int_3^4 (y-x)\,dy \right) dx.$$

The value of this iterated integral was shown in Example 1 to be 2, so the volume of the solid shown in Fig. 6 is 2.  ■

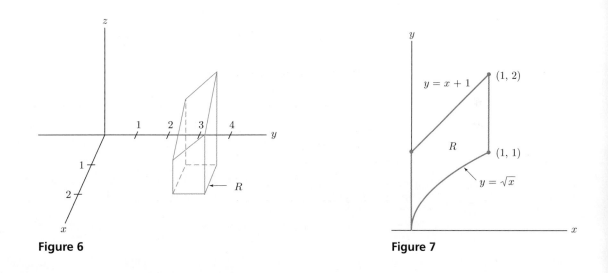

**Figure 6**                                          **Figure 7**

---

**EXAMPLE 4**  Calculate $\iint\limits_{R} 2xy\,dx\,dy$, where $R$ is the region shown in Fig. 7.

**Solution**  The region $R$ is bounded below by $y = \sqrt{x}$, above by $y = x + 1$, on the left by $x = 0$, and on the right by $x = 1$. Therefore,

$$\iint\limits_{R} 2xy\,dx\,dy = \int_0^1 \left( \int_{\sqrt{x}}^{x+1} 2xy\,dy \right) dx = \frac{13}{12} \quad \text{(by Example 2).} \quad ■$$

In our discussion, we have confined ourselves to iterated integrals in which the inner integral was with respect to $y$. In a completely analogous manner, we may treat iterated integrals in which the inner integral is with respect to $x$. Such iterated integrals may be used to evaluate double integrals over regions $R$ bounded by curves of the form $x = g(y)$, $x = h(y)$, and horizontal lines $y = a$, $y = b$. The computations are analogous to those given in this section.

## PRACTICE PROBLEMS 7.6

1. Calculate the iterated integral

$$\int_0^2 \left( \int_0^{x/2} e^{2y-x} \, dy \right) dx.$$

2. Calculate

$$\iint_R e^{2y-x} \, dx \, dy$$

where $R$ is the region in Fig. 8.

**Figure 8**

## EXERCISES 7.6

*Calculate the following iterated integrals.*

**1.** $\int_0^1 \left( \int_0^1 e^{x+y} \, dy \right) dx$

**2.** $\int_{-1}^1 \left( \int_{-1}^1 xy \, dx \right) dy$

**3.** $\int_{-2}^0 \left( \int_{-1}^1 xe^{xy} \, dy \right) dx$

**4.** $\int_0^1 \left( \int_{-1}^1 \frac{1}{3}y^3 x \, dy \right) dx$

**5.** $\int_1^4 \left( \int_x^{x^2} xy \, dy \right) dx$

**6.** $\int_0^3 \left( \int_x^{2x} y \, dy \right) dx$

**7.** $\int_{-1}^1 \left( \int_x^{2x} (x+y) \, dy \right) dx$

**8.** $\int_0^1 \left( \int_0^x e^{x+y} \, dy \right) dx$

*Let $R$ be the rectangle consisting of all points $(x, y)$ such that $0 \le x \le 2$, $2 \le y \le 3$. Calculate the following double*

*integrals. Interpret each as a volume.*

**9.** $\iint_R xy^2 \, dx \, dy$

**10.** $\iint_R (xy + y^2) \, dx \, dy$

**11.** $\iint_R e^{-x-y} \, dx \, dy$

**12.** $\iint_R e^{y-x} \, dx \, dy$

*Calculate the volumes over the following regions $R$ bounded above by the graph of $f(x, y) = x^2 + y^2$.*

**13.** $R$ is the rectangle bounded by the lines $x = 1$, $x = 3$, $y = 0$, $y = 1$.

**14.** $R$ is the region bounded by the lines $x = 0$, $x = 1$ and the curves $y = 0$ and $y = \sqrt[3]{x}$.

## SOLUTIONS TO PRACTICE PROBLEMS 7.6

1.
$$\int_0^2 \left( \int_0^{x/2} e^{2y-x} \, dy \right) dx = \int_0^2 \left( \frac{1}{2}e^{2y-x} \Big|_0^{x/2} \right) dx$$

$$= \int_0^2 \left( \frac{1}{2}e^{2(x/2)-x} - \frac{1}{2}e^{2(0)-x} \right) dx$$

$$= \int_0^2 \left( \frac{1}{2} - \frac{1}{2}e^{-x} \right) dx$$

$$= \frac{1}{2}x + \frac{1}{2}e^{-x} \Big|_0^2$$

$$= \frac{1}{2} \cdot 2 + \frac{1}{2}e^{-2} - \left( \frac{1}{2} \cdot 0 + \frac{1}{2}e^{-0} \right)$$

$$= 1 + \frac{1}{2}e^{-2} - 0 - \frac{1}{2}$$

$$= \frac{1}{2} + \frac{1}{2}e^{-2}.$$

2. The line passing through the points $(0,0)$ and $(2,1)$ has equation $y = x/2$. Hence the region $R$ is bounded below by $y = 0$, above by $y = x/2$, on the left by $x = 0$, and on the right by $x = 2$. Therefore,

$$\iint_R e^{2y-x} \, dx \, dy = \int_0^2 \left( \int_0^{x/2} e^{2y-x} \, dy \right) dx$$

$$= \frac{1}{2} + \frac{1}{2}e^{-2}$$

by Problem 1.

# REVIEW OF FUNDAMENTAL CONCEPTS

1. Give an example of a level curve of a function of two variables.

2. Explain how to find a first partial derivative of a function of two variables.

3. Explain how to find a second partial derivative of a function of two variables.

4. What expression involving a partial derivative gives an approximation to $f(a + h, b) - f(a, b)$?

5. Interpret $\dfrac{\partial f}{\partial y}(2, 3)$ as a rate of change.

6. Give an example of a Cobb–Douglas production function. What is the marginal productivity of labor? Of capital?

7. Explain how to find possible relative extreme points for a function of several variables.

8. State the second-derivative test for functions of two variables.

9. Outline how the method of Lagrange multipliers is used to solve an optimization problem.

10. What is the least-squares line approximation to a set of data points? How is the line determined?

11. Give a geometric interpretation to $\iint\limits_{R} f(x, y)\, dx\, dy$, where $f(x, y) \geq 0$.

12. Give a formula for evaluating a double integral in terms of an iterated integral.

# SUPPLEMENTARY EXERCISES

1. Let $f(x, y) = x\sqrt{y}/(1 + x)$. Compute $f(2, 9)$, $f(5, 1)$, and $f(0, 0)$.

2. Let $f(x, y, z) = x^2 e^{y/z}$. Compute $f(-1, 0, 1)$, $f(1, 3, 3)$, and $f(5, -2, 2)$.

3. If $A$ dollars are deposited in a bank (assuming a 6% continuous interest rate), the amount in the account after $t$ years is $f(A, t) = Ae^{.06t}$. Find and interpret $f(10, 11.5)$.

4. Let $f(x, y, \lambda) = xy + \lambda(5 - x - y)$. Find $f(1, 2, 3)$.

5. Let $f(x, y) = 3x^2 + xy + 5y^2$. Find $\dfrac{\partial f}{\partial x}$ and $\dfrac{\partial f}{\partial y}$.

6. Let $f(x, y) = 3x - \frac{1}{2}y^4 + 1$. Find $\dfrac{\partial f}{\partial x}$ and $\dfrac{\partial f}{\partial y}$.

7. Let $f(x, y) = e^{x/y}$. Find $\dfrac{\partial f}{\partial x}$ and $\dfrac{\partial f}{\partial y}$.

8. Let $f(x, y) = x/(x - 2y)$. Find $\dfrac{\partial f}{\partial x}$ and $\dfrac{\partial f}{\partial y}$.

9. Let $f(x, y, z) = x^3 - yz^2$. Find $\dfrac{\partial f}{\partial x}$, $\dfrac{\partial f}{\partial y}$, and $\dfrac{\partial f}{\partial z}$.

10. Let $f(x, y, \lambda) = xy + \lambda(5 - x - y)$. Find $\dfrac{\partial f}{\partial x}$, $\dfrac{\partial f}{\partial y}$, and $\dfrac{\partial f}{\partial \lambda}$.

11. Let $f(x, y) = x^3 y + 8$. Compute $\dfrac{\partial f}{\partial x}(1, 2)$ and $\dfrac{\partial f}{\partial y}(1, 2)$.

12. Let $f(x, y, z) = (x + y)z$. Evaluate $\dfrac{\partial f}{\partial y}$ at $(x, y, z) = (2, 3, 4)$.

13. Let $f(x, y) = x^5 - 2x^3 y + \frac{1}{2}y^4$. Find $\dfrac{\partial^2 f}{\partial x^2}, \dfrac{\partial^2 f}{\partial y^2}, \dfrac{\partial^2 f}{\partial x\, \partial y}$, and $\dfrac{\partial^2 f}{\partial y\, \partial x}$.

14. Let $f(x, y) = 2x^3 + x^2 y - y^2$. Compute $\dfrac{\partial^2 f}{\partial x^2}, \dfrac{\partial^2 f}{\partial y^2}$, and $\dfrac{\partial^2 f}{\partial x\, \partial y}$ at $(x, y) = (1, 2)$.

15. A dealer in a certain brand of electronic calculator finds that (within certain limits) the number of calculators she can sell per week is given by $f(p, t) = -p + 6t - .02pt$, where $p$ is the price of the calculator and $t$ is the number of dollars spent on advertising. Compute $\dfrac{\partial f}{\partial p}(25, 10{,}000)$ and $\dfrac{\partial f}{\partial t}(25, 10{,}000)$ and interpret these numbers.

16. Suppose that the crime rate in a certain city can be approximated by a function $f(x, y, z)$, where $x$ is the unemployment rate, $y$ is the amount of social services available, and $z$ is the size of the police force. Explain why $\dfrac{\partial f}{\partial x} > 0$, $\dfrac{\partial f}{\partial y} < 0$, and $\dfrac{\partial f}{\partial z} < 0$.

*In Exercises 17–20, find all points $(x, y)$ where $f(x, y)$ has a possible relative maximum or minimum.*

17. $f(x, y) = -x^2 + 2y^2 + 6x - 8y + 5$

18. $f(x, y) = x^2 + 3xy - y^2 - x - 8y + 4$

19. $f(x, y) = x^3 + 3x^2 + 3y^2 - 6y + 7$

20. $f(x, y) = \frac{1}{2}x^2 + 4xy + y^3 + 8y^2 + 3x + 2$

*In Exercises 21–23, find all points $(x, y)$ where $f(x, y)$ has a possible relative maximum or minimum. Then use the second-derivative test to determine, if possible, the nature of $f(x, y)$ at each of these points. If the second-derivative test is inconclusive, so state.*

21. $f(x, y) = x^2 + 3xy + 4y^2 - 13x - 30y + 12$

22. $f(x, y) = 7x^2 - 5xy + y^2 + x - y + 6$

23. $f(x, y) = x^3 + y^2 - 3x - 8y + 12$

24. Find the values of $x$, $y$, $z$ at which

$$f(x, y, z) = x^2 + 4y^2 + 5z^2 - 6x + 8y + 3$$

assumes its minimum value.

**25.** Using the method of Lagrange multipliers, maximize $3x^2 + 2xy - y^2$, subject to the constraint $5 - 2x - y = 0$.

**26.** Using the method of Lagrange multipliers, find the values of $x$, $y$ that minimize $-x^2 - 3xy - \frac{1}{2}y^2 + y + 10$, subject to the constraint $10 - x - y = 0$.

**27.** Using the method of Lagrange multipliers, find the values of $x$, $y$, $z$ that minimize $3x^2 + 2y^2 + z^2 + 4x + y + 3z$, subject to the constraint $4 - x - y - z = 0$.

**28.** Using the method of Lagrange multipliers, find the dimensions of the rectangular box of volume 1000 cubic inches for which the sum of the dimensions is minimized.

**29.** A person wants to plant a rectangular garden along one side of a house and put a fence on the other three sides of the garden. (See Fig. 1.) Using the method of Lagrange multipliers, find the dimensions of the garden of greatest area that can be enclosed by using 40 feet of fencing.

**Figure 1.** A garden.

**30.** The solution to Exercise 29 is $x = 10$, $y = 20$, $\lambda = 10$. Suppose that one additional foot of fencing becomes available. Compute the new optimal dimensions and the new area, and show that the increase in area (compared with the area in Exercise 29) is approximately equal to 10 (the value of $\lambda$).

*In Exercises 31–33, find the straight line that best fits the following data points, where "best" is meant in the sense of least squares.*

**31.** $(1,1)$, $(2,3)$, $(3,6)$

**32.** $(1,1)$, $(3,4)$, $(5,7)$

**33.** $(0,1)$, $(1,-1)$, $(2,-3)$, $(3,-5)$

*In Exercises 34 and 35, calculate the iterated integral.*

**34.** $\displaystyle\int_0^1 \left( \int_0^4 (x\sqrt{y} + y)\, dy \right) dx$

**35.** $\displaystyle\int_0^5 \left( \int_1^4 (2xy^4 + 3)\, dy \right) dx$

*In Exercises 36 and 37, let $R$ be the rectangle consisting of all points $(x, y)$ such that $0 \leq x \leq 4$, $1 \leq y \leq 3$, and calculate the double integral.*

**36.** $\displaystyle\iint_R (2x + 3y)\, dx\, dy$  **37.** $\displaystyle\iint_R 5\, dx\, dy$

**38.** The present value of $y$ dollars after $x$ years at 15% continuous interest is $f(x, y) = ye^{-.15x}$. Sketch some sample level curves. (Economists call this collection of level curves a *discount system*.)

# The Trigonometric Functions

IN this chapter we expand the collection of functions to which we can apply calculus by introducing the trigonometric functions. As we shall see, these functions are *periodic*. That is, after a certain point their graphs repeat themselves. This repetitive phenomenon is not displayed by any of the functions that we have considered until now. Yet many natural phenomena are repetitive or cyclical—for example, the motion of the planets in our solar system, earthquake vibrations, and the natural rhythm of the heart. Thus the functions introduced in this chapter add considerably to our capacity to describe physical processes.

## 8.1 Radian Measure of Angles

The ancient Babylonians introduced angle measurement in terms of degrees, minutes, and seconds, and these units are still generally used today for navigation and practical measurements. In calculus, however, it is more convenient to measure angles in terms of *radians*, for in this case the differentiation formulas for the trigonometric functions are easier to remember and use. Also, the radian is becoming more widely used today in scientific work because it is the unit of angle measurement in the international metric system (Système International d'Unités).

In order to define a radian, we consider a circle of radius 1 and measure angles in terms of distances around the circumference. The central angle determined by an arc of length 1 along the circumference is said to have a measure of *one radian*.

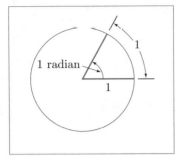

**Figure 1**

(See Fig. 1.) Since the circumference of the circle of radius 1 has length $2\pi$, there are $2\pi$ radians in one full revolution of the circle. Equivalently,

$$360° = 2\pi \text{ radians.} \tag{1}$$

The following important relations should be memorized (see Fig. 2):

$$90° = \frac{\pi}{2} \text{ radians} \qquad \text{(one quarter-revolution)}$$

$$180° = \pi \text{ radians} \qquad \text{(one half-revolution)}$$

$$270° = \frac{3\pi}{2} \text{ radians} \qquad \text{(three quarter-revolutions)}$$

$$360° = 2\pi \text{ radians} \qquad \text{(one full revolution).}$$

**Figure 2**

From formula (1) we see that

$$1° = \frac{2\pi}{360} \text{ radians} = \frac{\pi}{180} \text{ radians.}$$

If $d$ is any number, then

$$d° = d \times \frac{\pi}{180} \text{ radians.} \tag{2}$$

That is, in order to convert degrees to radians, multiply the number of degrees by $\pi/180$.

**EXAMPLE 1**  Convert $45°$, $60°$, and $135°$ to radians.

**Solution**

$$45° = \overset{1}{\cancel{45}} \times \frac{\pi}{\underset{4}{\cancel{180}}} \text{ radians} = \frac{\pi}{4} \text{ radians}$$

$$60° = \overset{1}{\cancel{60}} \times \frac{\pi}{\underset{3}{\cancel{180}}} \text{ radians} = \frac{\pi}{3} \text{ radians}$$

$$135° = \overset{3}{\cancel{135}} \times \frac{\pi}{\underset{4}{\cancel{180}}} \text{ radians} = \frac{3\pi}{4} \text{ radians}$$

These three angles are shown in Fig. 3.

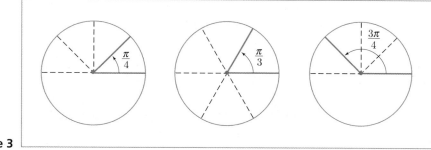

**Figure 3**

We usually omit the word "radian" when measuring angles because all our angle measurements will be in radians unless degrees are specifically indicated.

For our purposes, it is important to be able to speak of negative as well as positive angles, so let us define what we mean by a negative angle. We shall usually consider angles that are in *standard position* on a coordinate system, with the vertex of the angle at $(0,0)$ and one side, called the "initial side," along the positive $x$-axis. We measure such an angle from the initial side to the "terminal side," where a *counterclockwise angle is positive* and a *clockwise angle is negative*. Some examples are given in Fig. 4.

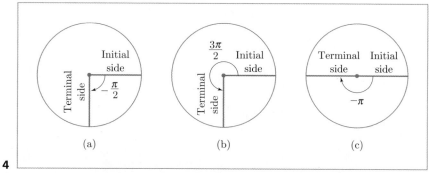

**Figure 4**

Notice in Figs. 4(a) and (b) how essentially the same picture can describe more than one angle.

By considering angles formed from more than one revolution (in the positive or negative direction), we can construct angles whose measure is of arbitrary size (i.e., not necessarily between $-2\pi$ and $2\pi$). Some examples are illustrated in Fig. 5.

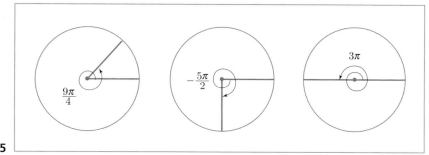

**Figure 5**

**EXAMPLE 2**   (a) What is the radian measure of the angle in Fig. 6?

(b) Construct an angle of $5\pi/2$ radians.

**Solution** (a) The angle described in Fig. 6 consists of one full revolution ($2\pi$ radians) plus three quarter-revolutions [$3 \times (\pi/2)$ radians]. That is,

$$t = 2\pi + 3 \times \frac{\pi}{2} = 4 \times \frac{\pi}{2} + 3 \times \frac{\pi}{2} = \frac{7\pi}{2}.$$

(b) Think of $5\pi/2$ radians as $5 \times (\pi/2)$ radians—that is, five quarter-revolutions of the circle. This is one full revolution plus one quarter-revolution. An angle of $5\pi/2$ radians is shown in Fig. 7. ■

**Figure 6**

**Figure 7**

## PRACTICE PROBLEMS 8.1

1. A right triangle has one angle of $\pi/3$ radians. What are the other angles?

2. How many radians are there in an angle of $-780°$? Draw the angle.

## EXERCISES 8.1

*Convert the following to radian measure.*

1. $30°$, $120°$, $315°$
2. $18°$, $72°$, $150°$
3. $450°$, $-210°$, $-90°$
4. $990°$, $-270°$, $-540°$

*Give the radian measure of each angle described.*

5.
6.

7.
8.

9.
10.

11.
12.

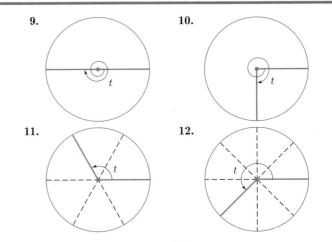

*Construct angles with the following radian measure.*

13. $3\pi/2$, $3\pi/4$, $5\pi$
14. $\pi/3$, $5\pi/2$, $6\pi$
15. $-\pi/3$, $-3\pi/4$, $-7\pi/2$
16. $-\pi/4$, $-3\pi/2$, $-3\pi$
17. $\pi/6$, $-2\pi/3$, $-\pi$
18. $2\pi/3$, $-\pi/6$, $7\pi/2$

## SOLUTIONS TO PRACTICE PROBLEMS 8.1

1. The sum of the angles of a triangle is 180° or $\pi$ radians. Since a right angle is $\pi/2$ radians and one angle is $\pi/3$ radians, the remaining angle is $\pi - (\pi/2 + \pi/3) = \pi/6$ radians.

2. $-780° = -780 \times (\pi/180)$ radians
   $= -13\pi/3$ radians.

   Since $-13\pi/3 = -4\pi - \pi/3$, we draw the angle by first making two revolutions in the negative direction and then a rotation of $\pi/3$ in the negative direction. See Fig. 8.

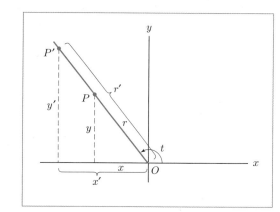

**Figure 8**

# 8.2  The Sine and the Cosine

Given a number $t$, we consider an angle of $t$ radians placed in standard position, as in Fig. 1, and we let $P$ be a point on the terminal side of this angle. Denote the coordinates of $P$ by $(x, y)$ and let $r$ be the length of the segment $OP$; that is, $r = \sqrt{x^2 + y^2}$. The *sine* and *cosine* of $t$, denoted by $\sin t$ and $\cos t$, respectively, are defined by the ratios

$$\sin t = \frac{y}{r}$$

$$\cos t = \frac{x}{r}.$$

(1)

It does not matter which point on the ray through $P$ we use to define $\sin t$ and $\cos t$. If $P' = (x', y')$ is another point on the same ray and if $r'$ is the length of $OP'$ (Fig. 2), then, by properties of similar triangles, we have

$$\frac{y'}{r'} = \frac{y}{r} = \sin t$$

$$\frac{x'}{r'} = \frac{x}{r} = \cos t.$$

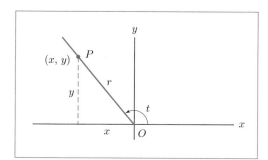

**Figure 1.**  Diagram for definitions of sine and cosine.

**Figure 2**

Three examples that illustrate the definition of $\sin t$ and $\cos t$ are shown in Fig. 3.

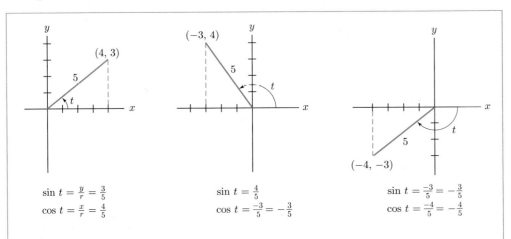

$$\sin t = \frac{y}{r} = \frac{3}{5} \qquad\qquad \sin t = \frac{4}{5} \qquad\qquad \sin t = \frac{-3}{5} = -\frac{3}{5}$$
$$\cos t = \frac{x}{r} = \frac{4}{5} \qquad\qquad \cos t = \frac{-3}{5} = -\frac{3}{5} \qquad\qquad \cos t = \frac{-4}{5} = -\frac{4}{5}$$

**Figure 3.**   Calculation of $\sin t$ and $\cos t$.

When $0 < t < \pi/2$, the values of $\sin t$ and $\cos t$ may be expressed as ratios of the lengths of the sides of a right triangle. Indeed, if we are given a right triangle as in Fig. 4, then we have

$$\sin t = \frac{\text{opposite}}{\text{hypotenuse}}, \qquad \cos t = \frac{\text{adjacent}}{\text{hypotenuse}}. \qquad (2)$$

A typical application of (2) appears in Example 1.

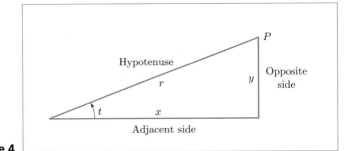

**Figure 4**

---

**EXAMPLE 1**    The hypotenuse of a right triangle is four units and one angle is .7 radian. Determine the length of the side opposite this angle.

**Solution**  See Fig. 5. Since $y/4 = \sin .7$, we have
$$y = 4\sin .7$$
$$y \approx 4(.64422) = 2.57688. \qquad \blacksquare$$

Another way to describe the sine and cosine functions is to choose the point $P$ in Fig. 1 so that $r = 1$. That is, choose $P$ on the unit circle. (See Fig. 6.) In this case

$$\sin t = \frac{y}{1} = y$$

$$\cos t = \frac{x}{1} = x.$$

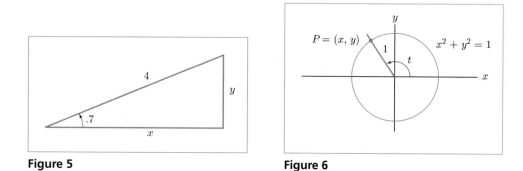

**Figure 5**    **Figure 6**

So the $y$-coordinate of $P$ is $\sin t$ and the $x$-coordinate of $P$ is $\cos t$. Thus we have the following result.

**Alternative Definition of Sine and Cosine Functions**    We can think of $\cos t$ and $\sin t$ as the $x$- and $y$-coordinates of the point $P$ on the unit circle that is determined by an angle of $t$ radians (Fig. 7).

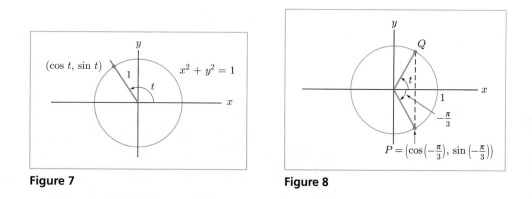

**Figure 7**    **Figure 8**

**EXAMPLE 2**    Find a value of $t$ such that $0 < t < \pi/2$ and $\cos t = \cos(-\pi/3)$.

**Solution**    On the unit circle we locate the point $P$ that is determined by an angle of $-\pi/3$ radians. The $x$-coordinate of $P$ is $\cos(-\pi/3)$. There is another point $Q$ on the unit circle with the same $x$-coordinate. (See Fig. 8.) Let $t$ be the radian measure of the angle determined by $Q$. Then

$$\cos t = \cos\left(-\frac{\pi}{3}\right)$$

because $Q$ and $P$ have the same $x$-coordinate. Also, $0 < t < \pi/2$. From the symmetry of the diagram it is clear that $t = \pi/3$.    ■

**Properties of the Sine and Cosine Functions**    Each number $t$ determines a point $(\cos t, \sin t)$ on the unit circle $x^2 + y^2 = 1$ as in Fig. 7. Therefore, $(\cos t)^2 + (\sin t)^2 = 1$. It is convenient (and traditional) to write $\sin^2 t$ instead of $(\sin t)^2$ and $\cos^2 t$ instead of $(\cos t)^2$. Thus we can write the last formula as follows:

$$\cos^2 t + \sin^2 t = 1. \tag{3}$$

The numbers $t$ and $t \pm 2\pi$ determine the same point on the unit circle (because $2\pi$ represents a full revolution of the circle). But $t + 2\pi$ and $t - 2\pi$ correspond to the points $(\cos(t + 2\pi), \sin(t + 2\pi))$ and $(\cos(t - 2\pi), \sin(t - 2\pi))$, respectively. Hence

$$\cos(t \pm 2\pi) = \cos t, \qquad \sin(t \pm 2\pi) = \sin t. \tag{4}$$

Figure 9(a) illustrates another property of the sine and cosine — namely,

$$\cos(-t) = \cos t, \qquad \sin(-t) = -\sin t. \tag{5}$$

Figure 9(b) shows that the points $P$ and $Q$ corresponding to $t$ and to $\pi/2 - t$ are reflections of each other through the line $y = x$. Consequently, the coordinates of $Q$ are obtained by interchanging the coordinates of $P$. This means that

$$\cos\left(\frac{\pi}{2} - t\right) = \sin t, \qquad \sin\left(\frac{\pi}{2} - t\right) = \cos t. \tag{6}$$

The equations in (3) to (6) are called *identities* because they hold for all values of $t$. Another identity that holds for all numbers $s$ and $t$ is

$$\sin(s + t) = \sin s \, \cos t + \cos s \, \sin t. \tag{7}$$

A proof of (7) may be found in any introductory book on trigonometry. There are a number of other identities concerning trigonometric functions, but we shall not need them here.

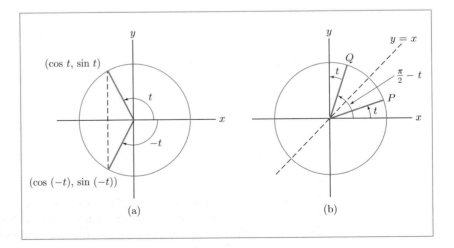

(a)  (b)

**Figure 9.**  Diagrams for two identities.

**The Graph of sin *t*** Let us analyze what happens to $\sin t$ as $t$ increases from 0 to $\pi$. When $t = 0$, the point $P = (\cos t, \sin t)$ is at $(1, 0)$, as in Fig. 10(a). As $t$ increases, $P$ moves counterclockwise around the unit circle. The $y$-coordinate of $P$—that is, $\sin t$—increases until $t = \pi/2$, where $P = (0, 1)$. [See Fig. 10(c).] As $t$ increases from $\pi/2$ to $\pi$, the $y$-coordinate of $P$—that is, $\sin t$—decreases from 1 to 0. [See Figs. 10(d) and (e).]

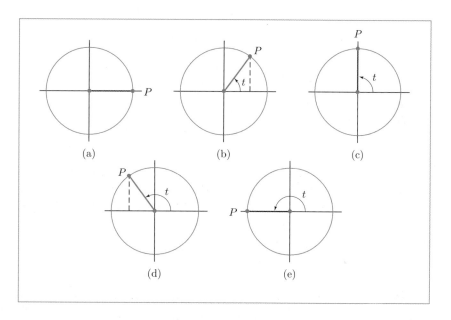

(a)　　　　　　　(b)　　　　　　　(c)

(d)　　　　　　　(e)

**Figure 10.** Movement along the unit circle.

Part of the graph of $\sin t$ is sketched in Fig. 11. Notice that, for $t$ between 0 and $\pi$, the values of $\sin t$ increase from 0 to 1 and then decrease back to 0, just as we predicted from Fig. 10. For $t$ between $\pi$ and $2\pi$, the values of $\sin t$ are negative. Can you explain why? The graph of $y = \sin t$ for $t$ between $2\pi$ and $4\pi$ is exactly the same as the graph for $t$ between 0 and $2\pi$. This result follows from formula (4). We say that the sine function is *periodic with period* $2\pi$ because the graph repeats itself every $2\pi$ units. We can use this fact to make a quick sketch of part of the graph for negative values of $t$ (Fig. 12).

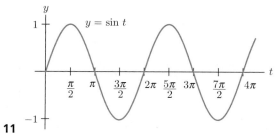

**Figure 11**

**The Graph of cos *t*** By analyzing what happens to the first coordinate of the point $(\cos t, \sin t)$ as $t$ varies, we obtain the graph of $\cos t$. Note from Fig. 13 that the graph of the cosine function is also periodic with period $2\pi$.

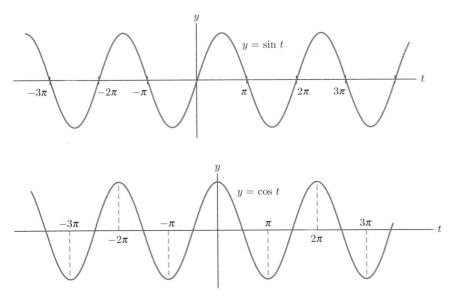

**Figure 12.** Graph of the sine function.

**Figure 13.** Graph of the cosine function.

▶ *A Remark About Notation* The sine and cosine functions assign to each number $t$ the values $\sin t$ and $\cos t$, respectively. There is nothing special, however, about the letter $t$. Although we chose to use the letters $t$, $x$, $y$, and $r$ in the *definition* of the sine and cosine, other letters could have been used as well. Now that the sine and cosine of every number are defined, we are free to use *any* letter to represent the independent variable. ◀

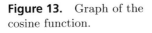

**Incorporating Technology** Many graphing calculators have a special window setting for graphing trigonometric functions. TI calculators have a ZTrig window $[-2\pi, 2\pi]$ *by* $[-4, 4]$ with $x$-scale $\pi/2$. See Fig. 14.

**Figure 14**

---

**PRACTICE PROBLEMS 8.2**

1. Find $\cos t$, where $t$ is the radian measure of the angle shown in Fig. 15.

**Figure 15**

2. Assume that $\cos(1.17) = .390$. Use properties of cosine and sine to determine $\sin(1.17)$, $\cos(1.17 + 4\pi)$, $\cos(-1.17)$, and $\sin(-1.17)$.

# EXERCISES 8.2

*In Exercises 1–12, give the values of* $\sin t$ *and* $\cos t$, *where* $t$
*is the radian measure of the angle shown.*

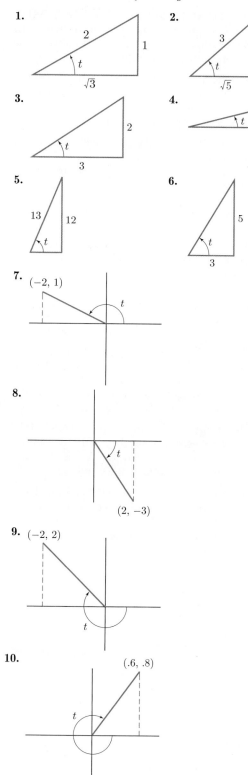

**1.**

**2.**

**3.**

**4.**

**5.**

**6.**

**7.** $(-2, 1)$

**8.**

**9.** $(-2, 2)$

**10.** $(.6, .8)$

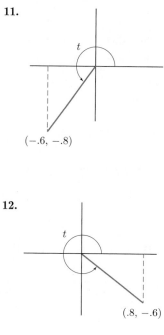

**11.**

$(-.6, -.8)$

**12.**

$(.8, -.6)$

*Exercises 13–20 refer to various right triangles whose sides
and angles are labeled as in Fig. 16. Round off all lengths
of sides to one decimal place.*

**Figure 16**

**13.** Estimate $t$ if $a = 12$, $b = 5$, and $c = 13$.

**14.** If $t = 1.1$ and $c = 10.0$, find $b$.

**15.** If $t = 1.1$ and $b = 3.2$, find $c$.

**16.** If $t = .4$ and $c = 5.0$, find $a$.

**17.** If $t = .4$ and $a = 10.0$, find $c$.

**18.** If $t = .9$ and $c = 20.0$, find $a$ and $b$.

**19.** If $t = .5$ and $a = 2.4$, find $b$ and $c$.

**20.** If $t = 1.1$ and $b = 3.5$, find $a$ and $c$.

*Find $t$ such that $0 \le t \le \pi$ and $t$ satisfy the stated condition.*

**21.** $\cos t = \cos(-\pi/6)$     **22.** $\cos t = \cos(3\pi/2)$

**23.** $\cos t = \cos(5\pi/4)$     **24.** $\cos t = \cos(-4\pi/6)$

**25.** $\cos t = \cos(-5\pi/8)$     **26.** $\cos t = \cos(-3\pi/4)$

*Find t such that $-\pi/2 \leq t \leq \pi/2$ and t satisfy the stated condition.*

**27.** $\sin t = \sin(3\pi/4)$

**28.** $\sin t = \sin(7\pi/6)$

**29.** $\sin t = \sin(-4\pi/3)$

**30.** $\sin t = -\sin(3\pi/8)$

**31.** $\sin t = -\sin(\pi/6)$

**32.** $\sin t = -\sin(-\pi/3)$

**33.** $\sin t = \cos t$

**34.** $\sin t = -\cos t$

**35.** By referring to Fig. 10, describe what happens to $\cos t$ as $t$ increases from 0 to $\pi$.

**36.** Use the unit circle to describe what happens to $\sin t$ as $t$ increases from $\pi$ to $2\pi$.

**37.** Determine the value of $\sin t$ when $t = 5\pi, -2\pi, 17\pi/2, -13\pi/2$.

**38.** Determine the value of $\cos t$ when $t = 5\pi, -2\pi, 17\pi/2, -13\pi/2$.

**39.** Assume that $\cos(.19) = .98$. Use properties of cosine and sine to determine $\sin(.19)$, $\cos(.19 - 4\pi)$, $\cos(-.19)$, and $\sin(-.19)$.

**40.** Assume that $\sin(.42) = .41$. Use properties of cosine and sine to determine $\sin(-.42)$, $\sin(6\pi - .42)$, and $\cos(.42)$.

## Technology Exercises

**41.** In any given locality, tap water temperature varies during the year. In Dallas, Texas, the tap water temperature (in degrees Fahrenheit) $t$ days after the beginning of a year is given approximately by the formula[*]

$$T = 59 + 14\cos\left[\frac{2\pi}{365}(t - 208)\right], \qquad 0 \leq t \leq 365.$$

**(a)** Graph the function in the window $[0, 365]$ *by* $[-10, 73]$.

**(b)** What is the temperature on February 14, that is, when $t = 45$?

**(c)** Use the fact that the value of the cosine function ranges from $-1$ to 1 to find the coldest and warmest tap water temperatures during the year.

**(d)** Use the **TRACE** feature or the **MINIMUM** command to estimate the day during which the tap water temperature is coldest. Find the exact day algebraically by using the fact that $\cos(-\pi) = -1$.

**(e)** Use the **TRACE** feature or the **MAXIMUM** command to estimate the day during which the tap water temperature is warmest. Find the exact day algebraically by using the fact that $\cos(0) = 1$.

**(f)** The average tap water temperature during the year is $59°$. Find the two days during which the average temperature is achieved. (*Note:* Answer this question both graphically and algebraically.)

**42.** In any given locality, the length of daylight varies during the year. In Des Moines, Iowa, the number of minutes of daylight in a day $t$ days after the beginning of a year is given approximately by the formula[†]

$$D = 720 + 200\sin\left[\frac{2\pi}{365}(t - 79.5)\right], \qquad 0 \leq t \leq 365.$$

**(a)** Graph the function in the window $[0, 365]$ *by* $[-100, 940]$.

**(b)** How many minutes of daylight are there on February 14, that is, when $t = 45$?

**(c)** Use the fact that the value of the sine function ranges from $-1$ to 1 to find the shortest and longest amounts of daylight during the year.

**(d)** Use the **TRACE** feature or the **MINIMUM** command to estimate the day with the shortest amount of daylight. Find the exact day algebraically by using the fact that $\sin(3\pi/2) = -1$.

**(e)** Use the **TRACE** feature or the **MAXIMUM** command to estimate the day with the longest amount of daylight. Find the exact day algebraically by using the fact that $\sin(\pi/2) = 1$.

**(f)** Find the two days during which the amount of daylight equals the amount of darkness. (These days are called *equinoxes*.) (*Note:* Answer this question both graphically and algebraically.)

---

### SOLUTIONS TO PRACTICE PROBLEMS 8.2

**1.** Here $P = (x, y) = (-3, -1)$. The length of the line segment $OP$ is

$$r = \sqrt{x^2 + y^2} = \sqrt{(-3)^2 + (-1)^2} = \sqrt{10}.$$

Then

$$\cos t = \frac{x}{r} = \frac{-3}{\sqrt{10}} \approx -.94868.$$

---

[*]See D. Rapp, *Solar Energy* (Englewood Cliffs, N.J.: Prentice-Hall, Inc., 1981), p. 171.

[†]See D. R. Duncan et al., "Climate Curves," *School Science and Mathematics*, Vol. LXXVI (January 1976), pp. 41–49.

## SOLUTIONS TO PRACTICE PROBLEMS 8.2 (CONTINUED)

**2.** Given $\cos(1.17) = .390$, use the relation $\cos^2 t + \sin^2 t = 1$ with $t = 1.17$ to solve for $\sin(1.17)$:

$$\cos^2(1.17) + \sin^2(1.17) = 1$$
$$\sin^2(1.17) = 1 - \cos^2(1.17)$$
$$= 1 - (.390)^2 = .8479.$$

So
$$\sin(1.17) = \sqrt{.8479} \approx .921.$$

Also, from properties (4) and (5),

$$\cos(1.17 + 4\pi) = \cos(1.17) = .390$$
$$\cos(-1.17) = \cos(1.17) = .390$$
$$\sin(-1.17) = -\sin(1.17) = -.921.$$

## 8.3  Differentiation and Integration of sin $t$ and cos $t$

In this section we study the two differentiation rules

$$\frac{d}{dt}\sin t = \cos t \tag{1}$$

$$\frac{d}{dt}\cos t = -\sin t. \tag{2}$$

It is not difficult to see why these rules might be true. Formula (1) says that the slope of the curve $y = \sin t$ at a particular value of $t$ is given by the corresponding value of $\cos t$. To check it, we draw a careful graph of $y = \sin t$ and estimate the slope at various points, as indicated in Fig. 1. Let us plot the slope as a function of $t$ (Fig. 2). As can be seen, the "slope function" (i.e., the derivative) of $\sin t$ has a graph similar to the curve $y = \cos t$. Thus formula (1) seems to be reasonable. A similar analysis of the graph of $y = \cos t$ would show why (2) might be true. Proofs of these differentiation rules are outlined in an appendix at the end of this section.

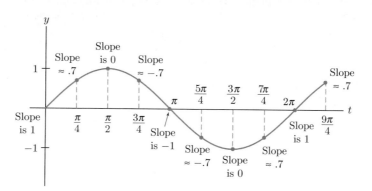

**Figure 1.** Slope estimates along the graph of $y = \sin t$.

Combining (1), (2), and the chain rule, we obtain the following general rules.

$$\frac{d}{dt}\sin g(t) = [\cos g(t)]\, g'(t) \tag{3}$$

$$\frac{d}{dt}\cos g(t) = [-\sin g(t)]\, g'(t). \tag{4}$$

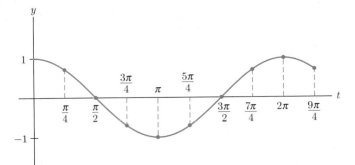

**Figure 2.** Graph of the slope function for $y = \sin t$.

---

**EXAMPLE 1** Differentiate

(a) $\sin 3t$              (b) $(t^2 + 3\sin t)^5$.

**Solution** (a) $\dfrac{d}{dt}(\sin 3t) = (\cos 3t)\dfrac{d}{dt}(3t) = (\cos 3t)\cdot 3 = 3\cos 3t$

(b) $\dfrac{d}{dt}(t^2 + 3\sin t)^5 = 5(t^2 + 3\sin t)^4 \cdot \dfrac{d}{dt}(t^2 + 3\sin t)$

$$= 5(t^2 + 3\sin t)^4(2t + 3\cos t).$$

---

**EXAMPLE 2** Differentiate

(a) $\cos(t^2 + 1)$            (b) $\cos^2 t$.

**Solution** (a) $\dfrac{d}{dt}\cos(t^2 + 1) = -\sin(t^2 + 1)\dfrac{d}{dt}(t^2 + 1) = -\sin(t^2 + 1)\cdot(2t)$

$$= -2t\sin(t^2 + 1).$$

(b) Recall that the notation $\cos^2 t$ means $(\cos t)^2$.

$$\frac{d}{dt}\cos^2 t = \frac{d}{dt}(\cos t)^2 = 2(\cos t)\frac{d}{dt}\cos t = -2\cos t\sin t.$$

---

**EXAMPLE 3** Differentiate

(a) $t^2 \cos 3t$            (b) $(\sin 2t)/t$.

**Solution** (a) From the product rule we have

$$\frac{d}{dt}(t^2\cos 3t) = t^2\frac{d}{dt}\cos 3t + (\cos 3t)\frac{d}{dt}t^2$$

$$= t^2(-3\sin 3t) + (\cos 3t)(2t)$$

$$= -3t^2\sin 3t + 2t\cos 3t.$$

(b) From the quotient rule we have

$$\frac{d}{dt}\left(\frac{\sin 2t}{t}\right) = \frac{t\dfrac{d}{dt}\sin 2t - (\sin 2t)\cdot 1}{t^2} = \frac{2t\cos 2t - \sin 2t}{t^2}.$$

**EXAMPLE 4**  A V-shaped trough is to be constructed with sides that are 200 centimeters long and 30 centimeters wide (Fig. 3). Find the angle $t$ between the sides that maximizes the capacity of the trough.

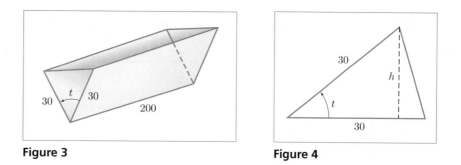

**Figure 3**                    **Figure 4**

**Solution**  The volume of the trough is its length times its cross-sectional area. Since the length is constant, it suffices to maximize the cross-sectional area. Let us rotate the diagram of a cross section so that one side is horizontal (Fig. 4). Note that $h/30 = \sin t$, so $h = 30 \sin t$. Thus the area $A$ of the cross section is

$$A = \tfrac{1}{2} \cdot \text{base} \cdot \text{height}$$

$$A = \tfrac{1}{2}(30)(h) = 15(30 \sin t) = 450 \sin t.$$

To find where $A$ is a maximum, we set the derivative equal to zero and solve for $t$.

$$\frac{dA}{dt} = 0$$

$$450 \cos t = 0$$

Physical considerations force us to consider only values of $t$ between 0 and $\pi$. From the graph of $y = \cos t$ we see that $t = \pi/2$ is the only value of $t$ between 0 and $\pi$ that makes $\cos t = 0$. So in order to maximize the volume of the trough, the two sides should be perpendicular to one another. ■

---

**EXAMPLE 5**  Calculate the following indefinite integrals.

(a) $\displaystyle\int \sin t \, dt$          (b) $\displaystyle\int \sin 3t \, dt$

**Solution**  (a) Since $\dfrac{d}{dt}(-\cos t) = \sin t$, we have

$$\int \sin t \, dt = -\cos t + C,$$

where $C$ is an arbitrary constant.

(b) From part (a) we guess that an antiderivative of $\sin 3t$ should resemble the function $-\cos 3t$. However, if we differentiate, we find that

$$\frac{d}{dt}(-\cos 3t) = (\sin 3t) \cdot \frac{d}{dt}(3t) = 3 \sin 3t,$$

which is three times too much. So we multiply this last equation by $\tfrac{1}{3}$ to derive that

$$\frac{d}{dt}\left(-\tfrac{1}{3} \cos 3t\right) = \sin 3t,$$

so

$$\int \sin 3t \, dt = -\tfrac{1}{3} \cos 3t + C.$$ ∎

---

**EXAMPLE 6**    Find the area under the curve $y = \sin 3t$ from $t = 0$ to $t = \pi/3$.

**Solution**    The area is shaded in Fig. 5.

$$[\text{shaded area}] = \int_0^{\pi/3} \sin 3t \, dt$$

$$= -\frac{1}{3} \cos 3t \Big|_0^{\pi/3}$$

$$= -\frac{1}{3} \cos \left( 3 \cdot \frac{\pi}{3} \right) - \left( -\frac{1}{3} \cos 0 \right)$$

$$= -\frac{1}{3} \cos \pi + \frac{1}{3} \cos 0$$

$$= \frac{1}{3} + \frac{1}{3} = \frac{2}{3}$$ ∎

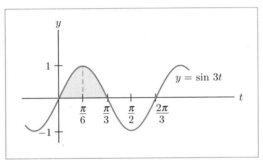

**Figure 5**

As we mentioned earlier, the trigonometric functions are required to model situations that are repetitive (or periodic). The next example illustrates such a situation.

---

**EXAMPLE 7**    In many mathematical models used to study the interaction between predators and prey, both the number of predators and the number of prey are described by periodic functions. Suppose that in one such model, the number of predators (in a particular geographical region) at time $t$ is given by the equation

$$N(t) = 5000 + 2000 \cos(2\pi t/36),$$

where $t$ is measured in months from June 1, 1990.

(a) At what rate is the number of predators changing on August 1, 1990?

(b) What is the average number of predators during the time interval from June 1, 1990, to June 1, 2002?

**Solution** (a) The date August 1, 1990, corresponds to $t = 2$. The rate of change of $N(t)$ is given by the derivative $N'(t)$:

$$N'(t) = \frac{d}{dt}\left[5000 + 2000\cos\left(\frac{2\pi t}{36}\right)\right]$$

$$= 2000\left[-\sin\left(\frac{2\pi t}{36}\right) \cdot \left(\frac{2\pi}{36}\right)\right]$$

$$= -\frac{1000\pi}{9}\sin\left(\frac{2\pi t}{36}\right),$$

$$N'(2) = -\frac{1000\pi}{9}\sin\left(\frac{\pi}{9}\right)$$

$$\approx -119.$$

Thus, on August 1, 1990, the number of predators is decreasing at the rate of 119 per month.

(b) The time interval from June 1, 1990, to June 1, 2002, corresponds to $t = 0$ to $t = 144$. (There are 144 months in 12 years!) The average value of $N(t)$ over this interval is

$$\frac{1}{144 - 0}\int_0^{144} N(t)\,dt = \frac{1}{144}\int_0^{144}\left[5000 + 2000\cos\left(\frac{2\pi t}{36}\right)\right]dt$$

$$= \frac{1}{144}\left[5000t + \frac{2000}{2\pi/36}\sin\left(\frac{2\pi t}{36}\right)\right]\Big|_0^{144}$$

$$= \frac{1}{144}\left[5000 \cdot 144 + \frac{2000}{2\pi/36}\sin(8\pi)\right]$$

$$- \frac{1}{144}\left[5000 \cdot 0 + \frac{2000}{2\pi/36}\sin(0)\right]$$

$$= 5000.$$

We have sketched the graph of $N(t)$ in Fig. 6. Note how $N(t)$ oscillates around 5000, the average value. ∎

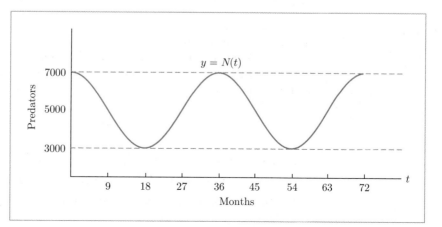

**Figure 6.** Periodic fluctuation of a predator population.

<u>**APPENDIX** Informal Justification of the Differentiation Rules for sin *t* and cos *t*</u>

First, let us examine the derivatives of $\cos t$ and $\sin t$ at $t = 0$. The function $\cos t$ has a maximum at $t = 0$; consequently, its derivative there must be zero. [See Fig. 7(a).] If we approximate the tangent line at $t = 0$ by a secant line, as in Fig. 7(b), then the slope of the secant line must approach 0 as $h \to 0$. Since the slope of the secant line is $(\cos h - 1)/h$, we conclude that

$$\lim_{h \to 0} \frac{\cos h - 1}{h} = 0. \tag{5}$$

**Figure 7**

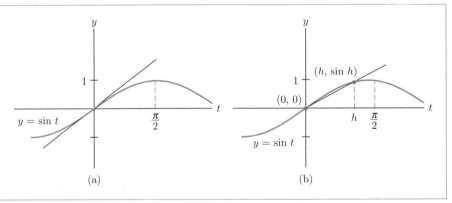

**Figure 8.** Slope of $\sin t$ at $t = 0$.

It appears from the graph of $y = \sin t$ that the tangent line at $t = 0$ has slope 1 [Fig. 8(a)]. If it does, then the slope of the approximating secant line [Fig. 8(b)] must approach 1. Since the slope of the secant line in Fig. 8(b) is $(\sin h)/h$, this would imply that

$$\lim_{h \to 0} \frac{\sin h}{h} = 1. \tag{6}$$

**Figure 9**

We can evaluate $\sin h/h$ for small values of $h$ with a calculator. See Fig. 9. The numerical evidence does not prove (6), but should be sufficiently convincing for our purposes.

To obtain the differentiation formula for $\sin t$, we approximate the slope of a tangent line by the slope of a secant line. (See Fig. 10.) The slope of a secant line

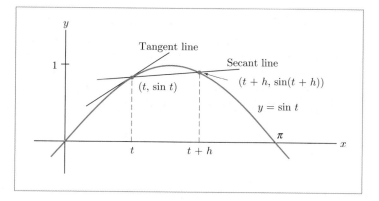

**Figure 10.** Secant line
approximation for $y = \sin t$.

is

$$\frac{\sin(t + h) - \sin t}{h}.$$

From Formula (7) of Section 8.2, we note that $\sin(t + h) = \sin t \cos h + \cos t \sin h$.
Thus,

$$[\text{slope of secant line}] = \frac{(\sin t \cos h + \cos t \sin h) - \sin t}{h}$$

$$= \frac{\sin t \, (\cos h - 1) + \cos t \sin h}{h}$$

$$= (\sin t)\frac{\cos h - 1}{h} + (\cos t)\frac{\sin h}{h}.$$

From (5) and (6) it follows that

$$\frac{d}{dt} \sin t = \lim_{h \to 0} \left[ (\sin t)\frac{\cos h - 1}{h} + (\cos t)\frac{\sin h}{h} \right]$$

$$= (\sin t) \lim_{h \to 0} \frac{\cos h - 1}{h} + (\cos t) \lim_{h \to 0} \frac{\sin h}{h}$$

$$= (\sin t) \cdot 0 + (\cos t) \cdot 1$$

$$= \cos t.$$

A similar argument may be given to verify the formula for the derivative of
$\cos t$. Here is a shorter proof that uses the chain rule and the two identities

$$\cos t = \sin \left( \frac{\pi}{2} - t \right), \qquad \sin t = \cos \left( \frac{\pi}{2} - t \right).$$

[See Formula (6) of Section 8.2.] We have

$$\frac{d}{dt} \cos t = \frac{d}{dt} \sin \left( \frac{\pi}{2} - t \right)$$

$$= \cos \left( \frac{\pi}{2} - t \right) \cdot \frac{d}{dt} \left( \frac{\pi}{2} - t \right)$$

$$= \cos \left( \frac{\pi}{2} - t \right) \cdot (-1)$$

$$= -\sin t.$$

## PRACTICE PROBLEMS 8.3

**1.** Differentiate $y - 2\sin[t^2 + (\pi/6)]$

**2.** Differentiate $y = e^t \sin 2t$.

## EXERCISES 8.3

*Differentiate (with respect to t or x):*

**1.** $y = \sin 4t$

**2.** $y = 2\cos 2t$

**3.** $y = 4\sin t$

**4.** $y = \cos(-4t)$

**5.** $y = 2\cos 3t$

**6.** $y = -\dfrac{\sin 3t}{3}$

**7.** $y = t + \cos \pi t$

**8.** $y = t\cos t$

**9.** $y = \sin(\pi - t)$

**10.** $y = \dfrac{\cos(2x+2)}{2}$

**11.** $y = \cos^3 t$

**12.** $y = \sin^3 t^2$

**13.** $y = \sin\sqrt{x-1}$

**14.** $y = \cos(e^x)$

**15.** $y = \sqrt{\sin(x-1)}$

**16.** $y = e^{\cos x}$

**17.** $y = (1+\cos t)^8$

**18.** $y = \sqrt[3]{\sin \pi t}$

**19.** $y = \cos^2 x^3$

**20.** $y = \cos^2 x + \sin^2 x$

**21.** $y = e^x \sin x$

**22.** $y = (\cos x + \sin x)^2$

**23.** $y = \sin 2x \cos 3x$

**24.** $y = \dfrac{1+x}{\cos x}$

**25.** $y = \dfrac{\sin t}{\cos t}$

**26.** $y = \cos\left[e^{2x+3}\right]$

**27.** $y = \ln(\cos t)$

**28.** $y = \ln(\sin 2t)$

**29.** $y = \sin(\ln t)$

**30.** $y = (\cos t)\ln t$

**31.** Find the slope of the line tangent to the graph of $y = \cos 3x$ at $x = 13\pi/6$.

**32.** Find the slope of the line tangent to the graph of $y = \sin 2x$ at $x = 5\pi/4$.

**33.** Find the equation of the line tangent to the graph of $y = 3\sin x + \cos 2x$ at $x = \pi/2$.

**34.** Find the equation of the line tangent to the graph of $y = 3\sin 2x - \cos 2x$ at $x = 3\pi/4$.

*Find the following indefinite integrals.*

**35.** $\displaystyle\int \cos 2x\, dx$

**36.** $\displaystyle\int 3\sin 3x\, dx$

**37.** $\displaystyle\int -\frac{1}{2}\cos\frac{x}{7}\, dx$

**38.** $\displaystyle\int 2\sin\frac{x}{2}\, dx$

**39.** $\displaystyle\int (\cos x - \sin x)\, dx$

**40.** $\displaystyle\int \left(2\sin 3x + \frac{\cos 2x}{2}\right) dx$

**41.** $\displaystyle\int (-\sin x + 3\cos(-3x))\, dx$

**42.** $\displaystyle\int \sin(-2x)\, dx$

**43.** $\displaystyle\int \sin(4x+1)\, dx$

**44.** $\displaystyle\int \cos\frac{x-2}{2}\, dx$

**45.** $\displaystyle\int 7\sin(3x-2)\, dx$

**46.** $\displaystyle\int (\cos(2x)+3)\, dx$

**47.** Suppose that a person's blood pressure $P$ at time $t$ (in seconds) is given by $P = 100 + 20\cos 6t$.

   **(a)** Find the maximum value of $P$ (called the systolic pressure) and the minimum value of $P$ (called the diastolic pressure) and give one or two values of $t$ where these maximum and minimum values of $P$ occur.

   **(b)** If time is measured in seconds, approximately how many heartbeats per minute are predicted by the equation for $P$?

**48.** The *basal metabolism* (BM) of an organism over a certain time period may be described as the total amount of heat in kilocalories (kcal) the organism produces during that period, assuming that the organism is at rest and not subject to stress. The *basal metabolic rate* (BMR) is the rate in kcal per hour at which the organism produces heat. The BMR of an animal such as a desert rat fluctuates in response to changes in temperature and other environmental factors. The BMR generally follows a *diurnal* cycle—rising at night during low temperatures and decreasing during the warmer daytime temperatures. Find the BM for 1 day if $\text{BMR}(t) = .4 + .2\cos(\pi t/12)$ kcal per hour ($t = 0$ corresponds to 3 A.M.). See Fig. 11.

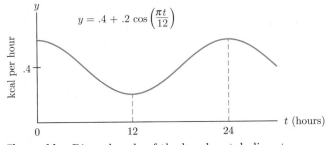

**Figure 11.** Diurnal cycle of the basal metabolic rate.

**49.** As $h$ approaches 0, what value is approached by

$$\frac{\sin\left(\frac{\pi}{2} + h\right) - 1}{h}?$$

$$\left[\textit{Hint: } \sin\frac{\pi}{2} = 1.\right]$$

**50.** As $h$ approaches 0, what value is approached by

$$\frac{\cos(\pi + h) + 1}{h}?$$

$[\textit{Hint: } \cos\pi = -1.]$

**51.** (*Average Temperature*) The average weekly temperature in Washington, D.C. $t$ weeks after the beginning of the year is

$$f(t) = 54 + 23\sin\left[\frac{2\pi}{52}(t - 12)\right].$$

The graph of this function is sketched in Fig. 12.

**(a)** What is the average weekly temperature at week 18?

**(b)** At week 20, how fast is the temperature changing?

**(c)** When is the average weekly temperature 39 degrees?

**(d)** When is the average weekly temperature falling at the rate of 1 degree per week?

**(e)** When is the average weekly temperature greatest? Least?

**(f)** When is the average weekly temperature increasing fastest? Decreasing fastest?

**52.** (*Average Daylight Hours*) The number of hours of daylight per day in Washington, D.C. $t$ weeks after the beginning of the year is

$$f(t) = 12.18 + 2.725\sin\left[\frac{2\pi}{52}(t - 12)\right].$$

The graph of this function is sketched in Fig. 13.

**(a)** How many hours of daylight are there after 42 weeks?

**Figure 12**

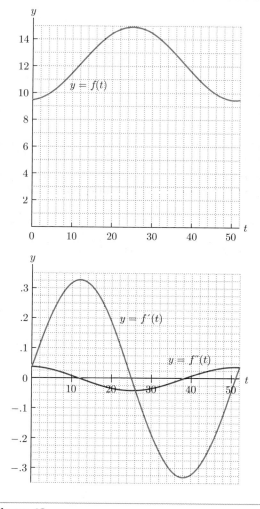

**Figure 13**

**(b)** After 32 weeks, how fast is the number of hours of daylight decreasing?

**(c)** When is there 14 hours of daylight per day?

**(d)** When is the number of hours of daylight increasing

at the rate of 15 minutes per week?

**(e)** When are the days longest? Shortest?

**(f)** When is the number of hours of daylight increasing fastest? Decreasing fastest?

## SOLUTIONS TO PRACTICE PROBLEMS 8.3

**1.** By the chain rule,

$$y' = 2\cos\left(t^2 + \frac{\pi}{6}\right) \cdot \frac{d}{dt}\left(t^2 + \frac{\pi}{6}\right)$$

$$= 2\cos\left(t^2 + \frac{\pi}{6}\right) \cdot 2t$$

$$= 4t\cos\left(t^2 + \frac{\pi}{6}\right).$$

**2.** By the product rule,

$$y' = e^t \frac{d}{dt}(\sin 2t) + (\sin 2t)\frac{d}{dt}e^t$$

$$= 2e^t \cos 2t + e^t \sin 2t.$$

## 8.4 The Tangent and Other Trigonometric Functions

Certain functions involving the sine and cosine functions occur so frequently in applications that they have been given special names. The *tangent* (tan), *cotangent* (cot), *secant* (sec), and *cosecant* (csc) are such functions and are defined as follows:

$$\tan t = \frac{\sin t}{\cos t}, \qquad \cot t = \frac{\cos t}{\sin t},$$

$$\sec t = \frac{1}{\cos t}, \qquad \csc t = \frac{1}{\sin t}.$$

They are defined only for $t$ such that the denominators in the preceding quotients are not zero. These four functions, together with the sine and cosine, are called the *trigonometric functions*. Our main interest in this section is with the tangent function. Some properties of the cotangent, secant, and cosecant are developed in the exercises.

Many identities involving the trigonometric functions can be deduced from the identities given in Section 8.2. We shall mention just one:

$$\tan^2 t + 1 = \sec^2 t. \tag{1}$$

[Here $\tan^2 t$ means $(\tan t)^2$ and $\sec^2 t$ means $(\sec t)^2$.] This identity follows from the identity $\sin^2 t + \cos^2 t = 1$ when we divide everything by $\cos^2 t$.

An important interpretation of the tangent function can be given in terms of the diagram used to define the sine and cosine. For a given $t$, let us construct an angle of $t$ radians, as in Fig. 1. Since $\sin t = y/r$ and $\cos t = x/r$, we have

$$\frac{\sin t}{\cos t} = \frac{y/r}{x/r} = \frac{y}{x},$$

where this formula holds provided that $x \neq 0$. Thus

$$\tan t = \frac{y}{x}. \tag{2}$$

**Figure 1**

**Figure 2**

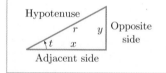

**Figure 3**

Three examples that illustrate this property of the tangent appear in Fig. 2.

When $0 < t < \pi/2$, the value of $\tan t$ is a ratio of the lengths of the sides of a right triangle. In other words, suppose that we are given a triangle as in Fig. 3. Then we have

$$\tan t = \frac{\text{opposite}}{\text{adjacent}}. \tag{3}$$

**EXAMPLE 1**  The angle of elevation from an observer to the top of a building is $29°$ (Fig. 4). If the observer is 100 meters from the base of the building, how high is the building?

**Solution**  Let $h$ denote the height of the building. Then (3) implies that

$$\frac{h}{100} = \tan 29°$$

$$h = 100 \tan 29°.$$

We convert $29°$ into radians. We find that $29° = (\pi/180) \cdot 29$ radians $\approx .5$ radian, and $\tan .5 \approx .54630$. Hence

$$h \approx 100(.54630) = 54.63 \text{ meters.}$$

**Figure 4**

**The Derivative of tan $t$**    Since $\tan t$ is defined in terms of $\sin t$ and $\cos t$, we can compute the derivative of $\tan t$ from our rules of differentiation. That is, by applying the quotient rule for differentiation, we have

$$\frac{d}{dt}(\tan t) = \frac{d}{dt}\left(\frac{\sin t}{\cos t}\right) = \frac{(\cos t)(\cos t) - (\sin t)(-\sin t)}{(\cos t)^2}$$

$$= \frac{\cos^2 t + \sin^2 t}{\cos^2 t} = \frac{1}{\cos^2 t}.$$

Now

$$\frac{1}{\cos^2 t} = \frac{1}{(\cos t)^2} = \left(\frac{1}{\cos t}\right)^2 = (\sec t)^2 = \sec^2 t.$$

So the derivative of $\tan t$ can be expressed in two equivalent ways:

$$\frac{d}{dt}(\tan t) = \frac{1}{\cos^2 t} = \sec^2 t. \tag{4}$$

Combining (4) with the chain rule, we have

$$\frac{d}{dt}(\tan g(t)) = [\sec^2 g(t)]g'(t). \tag{5}$$

**EXAMPLE 2**    Differentiate

(a) $\tan(t^3 + 1)$              (b) $\tan^3 t.$

**Solution**    (a) From (5) we find that

$$\frac{d}{dt}[\tan(t^3 + 1)] = \sec^2(t^3 + 1) \cdot \frac{d}{dt}(t^3 + 1)$$

$$= 3t^2 \sec^2(t^3 + 1).$$

(b) We write $\tan^3 t$ as $(\tan t)^3$ and use the chain rule (in this case, the general power rule):

$$\frac{d}{dt}(\tan t)^3 = (3\tan^2 t) \cdot \frac{d}{dt}\tan t = 3\tan^2 t \sec^2 t. \qquad \blacksquare$$

**The Graph of tan $t$**    Recall that $\tan t$ is defined for all $t$ except where $\cos t = 0$. (We cannot have zero in the denominator of $\sin t / \cos t$.) The graph of $\tan t$ is sketched in Fig. 5. Note that $\tan t$ is periodic with period $\pi$.

**PRACTICE PROBLEMS 8.4**

1. Show that the slope of a straight line is equal to the tangent of the angle that the line makes with the $x$-axis.

2. Calculate $\int_0^{\pi/4} \sec^2 t \, dt.$

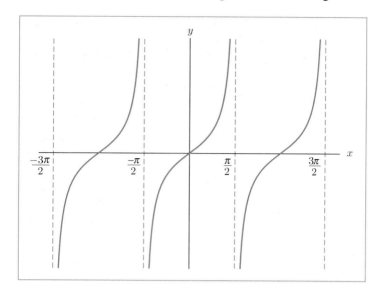

**Figure 5.** Graph of the tangent function.

# EXERCISES 8.4

**1.** Suppose that $0 < t < \pi/2$ and use Fig. 3 to describe $\sec t$ as a ratio of the lengths of the sides of a right triangle.

**2.** Describe $\cot t$ for $0 < t < \pi/2$ as a ratio of the lengths of the sides of a right triangle.

*In Exercises 3–10, give the values of* $\tan t$ *and* $\sec t$, *where $t$ is the radian measure of the angle shown.*

**3.**

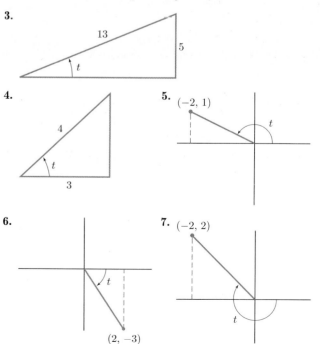

**4.**

**5.** $(-2, 1)$

**6.**

$(2, -3)$

**7.** $(-2, 2)$

**8.**

$(.6, .8)$

**9.**

$(-.6, -.8)$

**10.**

$(.8, -.6)$

11. Find the width of a river at points $A$ and $B$ if the angle $BAC$ is $90°$, the angle $ACB$ is $40°$, and the distance from $A$ to $C$ is 75 feet. See Fig. 6.

**Figure 6**

12. The angle of elevation from an observer to the top of a church is .3 radian, while the angle of elevation from the observer to the top of the church spire is .4 radian. If the observer is 70 meters from the church, how tall is the spire on top of the church?

*Differentiate (with respect to t or x):*

13. $f(t) = \sec t$

14. $f(t) = \csc t$

15. $f(t) = \cot t$

16. $f(t) = \cot 3t$

17. $f(t) = \tan 4t$

18. $f(t) = \tan \pi t$

19. $f(x) = 3 \tan(\pi - x)$

20. $f(x) = 5 \tan(2x + 1)$

21. $f(x) = 4 \tan(x^2 + x + 3)$

22. $f(x) = 3 \tan(1 - x^2)$

23. $y = \tan \sqrt{x}$

24. $y = 2 \tan \sqrt{x^2 - 4}$

25. $y = x \tan x$

26. $y = e^{3x} \tan 2x$

27. $y = \tan^2 x$

28. $y = \sqrt{\tan x}$

29. $y = (1 + \tan 2t)^3$

30. $y = \tan^4 3t$

31. $y = \ln(\tan t + \sec t)$

32. $y = \ln(\tan t)$

33. (a) Find the equation of the tangent line to the graph of $y = \tan x$ at the point $(\frac{\pi}{4}, 1)$.

    (b) Copy the portion of the graph of $y = \tan x$ for $-\frac{\pi}{2} < x < \frac{\pi}{2}$ from Fig. 5, then draw on this graph the tangent line that you found in part (a).

34. Repeat Exercise 33(a) and (b), using the point $(0, 0)$ on the graph of $y = \tan x$, instead of the point $(\frac{\pi}{4}, 1)$.

*Evaluate the following integrals.*

35. $\displaystyle\int \sec^2 3x \, dx$

36. $\displaystyle\int \sec^2(2x + 1) \, dx$

37. $\displaystyle\int_{-\pi/4}^{\pi/4} \sec^2 x \, dx$

38. $\displaystyle\int_{-\pi/8}^{\pi/8} \sec^2\left(x + \frac{\pi}{8}\right) dx$

39. $\displaystyle\int \frac{1}{\cos^2 x} \, dx$

40. $\displaystyle\int \frac{3}{\cos^2 2x} \, dx$

---

## SOLUTIONS TO PRACTICE PROBLEMS 8.4

1. A line of positive slope $m$ is shown in Fig. 7(a). Here, $\tan \theta = m/1 = m$. Suppose that $y = mx + b$ where the slope $m$ is negative. The line $y = mx$ has the same slope and makes the same angle with the $x$-axis.

[See Fig. 7(b).] We see that $\tan \theta = -m/-1 = m$.

2. $\displaystyle\int_0^{\pi/4} \sec^2 t \, dt = \tan t \Big|_0^{\pi/4} = \tan \frac{\pi}{4} - \tan 0 = 1 - 0 = 1.$

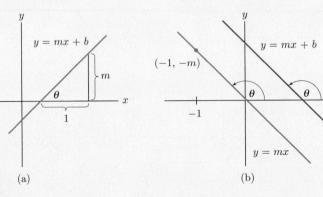

(a)          (b)

**Figure 7**

# REVIEW OF FUNDAMENTAL CONCEPTS

**1.** Explain the radian measure of an angle.

**2.** Give the formula for converting degree measure to radian measure.

**3.** Give the triangle interpretation of $\sin t$, $\cos t$, and $\tan t$ for $t$ between $0$ and $\pi/2$.

**4.** Define $\sin t$, $\cos t$, and $\tan t$ for an angle of measure $t$ for any $t$.

**5.** What does it mean when we say that the sine and cosine functions are periodic with period $2\pi$?

**6.** Give verbal descriptions of the graphs of $\sin t$, $\cos t$, and $\tan t$.

**7.** State as many identities involving the sine and cosine functions as you can recall.

**8.** Define $\cot t$, $\sec t$, and $\csc t$ for an angle of measure $t$.

**9.** State an identity involving $\tan t$ and $\sec t$.

**10.** What are the derivatives of $\sin g(t)$, $\cos g(t)$, and $\tan g(t)$?

# SUPPLEMENTARY EXERCISES

*Determine the radian measure of the angles shown in Exercises 1–3.*

**1.**

**2.**

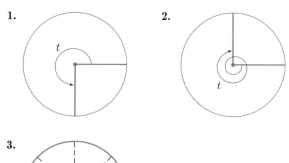

**3.**

*Construct angles with the following radian measure.*

**4.** $-\pi$

**5.** $\dfrac{5\pi}{4}$

**6.** $-\dfrac{9\pi}{2}$

*In Exercises 7–10, the point with the given coordinates determines an angle of $t$ radians, where $0 \le t \le 2\pi$. Find $\sin t$, $\cos t$, $\tan t$.*

**7.** $(3, 4)$

**8.** $(-.6, .8)$

**9.** $(-.6, -.8)$

**10.** $(3, -4)$

**11.** If $\sin t = \frac{1}{5}$, what are the possible values for $\cos t$?

**12.** If $\cos t = -\frac{2}{3}$, what are the possible values for $\sin t$?

**13.** Find the four values of $t$ between $-2\pi$ and $2\pi$ at which $\sin t = \cos t$.

**14.** Find the four values of $t$ between $-2\pi$ and $2\pi$ at which $\sin t = -\cos t$.

**15.** When $-\pi/2 < t < 0$, is $\tan t$ positive or negative?

**16.** When $\pi/2 < t < \pi$, is $\sin t$ positive or negative?

**17.** A gabled roof is to be built on a house that is 30 feet wide, so that the roof rises at a pitch of $23°$. Determine the length of the rafters needed to support the roof.

**18.** A tree casts a 60-foot shadow when the angle of elevation of the sun (measured from the horizontal) is $53°$. How tall is the tree?

*Differentiate (with respect to $t$ or $x$):*

**19.** $f(t) = 3\sin t$

**20.** $f(t) = \sin 3t$

**21.** $f(t) = \sin \sqrt{t}$

**22.** $f(t) = \cos t^3$

**23.** $g(x) = x^3 \sin x$

**24.** $g(x) = \sin(-2x)\cos 5x$

**25.** $f(x) = \dfrac{\cos 2x}{\sin 3x}$

**26.** $f(x) = \dfrac{\cos x - 1}{x^3}$

**27.** $f(x) = \cos^3 4x$

**28.** $f(x) = \tan^3 2x$

**29.** $y = \tan(x^4 + x^2)$

**30.** $y = \tan e^{-2x}$

**31.** $y = \sin(\tan x)$

**32.** $y = \tan(\sin x)$

**33.** $y = \sin x \tan x$

**34.** $y = \ln x \cos x$

**35.** $y = \ln(\sin x)$

**36.** $y = \ln(\cos x)$

**37.** $y = e^{3x}\sin^4 x$

**38.** $y = \sin^4 e^{3x}$

**39.** $f(t) = \dfrac{\sin t}{\tan 3t}$

**40.** $f(t) = \dfrac{\tan 2t}{\cos t}$

**41.** $f(t) = e^{\tan t}$

**42.** $f(t) = e^t \tan t$

**43.** If $f(t) = \sin^2 t$, find $f''(t)$.

**44.** Show that $y = 3\sin 2t + \cos 2t$ satisfies the differential equation $y'' = -4y$.

**45.** If $f(s, t) = \sin s \cos 2t$, find $\dfrac{\partial f}{\partial s}$ and $\dfrac{\partial f}{\partial t}$.

**46.** If $z = \sin wt$, find $\dfrac{\partial z}{\partial w}$ and $\dfrac{\partial z}{\partial t}$.

**47.** If $f(s, t) = t\sin st$, find $\dfrac{\partial f}{\partial s}$ and $\dfrac{\partial f}{\partial t}$.

**48.** The identity

$$\sin(s + t) = \sin s \cos t + \cos s \sin t$$

was given in Section 8.2. Compute the partial derivative of each side with respect to $t$ and obtain an identity involving $\cos(s + t)$.

**49.** Find the equation of the line tangent to the graph of $y = \tan t$ at $t = \pi/4$.

**50.** Sketch the graph of $f(t) = \sin t + \cos t$ for $-2\pi \le t \le 2\pi$, using the following steps:

**(a)** Find all $t$ (between $-2\pi$ and $2\pi$) such that $f'(t) = 0$. Plot the corresponding points on the graph of $y = f(t)$.

**(b)** Check the concavity of $f(t)$ at the points in part (a). Make sketches of the graph near these points.

**(c)** Determine any inflection points and plot them. Then complete the sketch of the graph.

**51.** Sketch the graph of $y = t + \sin t$ for $0 \le t \le 2\pi$.

**52.** Find the area under the curve $y = 2 + \sin 3t$ from $t = 0$ to $t = \pi/2$.

**53.** Find the area of the region between the curve $y = \sin t$ and the $t$-axis from $t = 0$ to $t = 2\pi$.

**54.** Find the area of the region between the curve $y = \cos t$ and the $t$-axis from $t = 0$ to $t = 3\pi/2$.

**55.** Find the area of the region bounded by the curves $y = x$ and $y = \sin x$ from $x = 0$ to $x = \pi$.

*A spirogram is a device that records on a graph the volume of air in a person's lungs as a function of time. If a person undergoes spontaneous hyperventilation, the spirogram trace will closely approximate a sine curve. A typical trace is given by*

$$V(t) = 3 + .05 \sin\left(160\pi t - \frac{\pi}{2}\right),$$

*where $t$ is measured in minutes and $V(t)$ is the lung volume in liters. (See Fig. 1.) Exercises 56–58 refer to this function.*

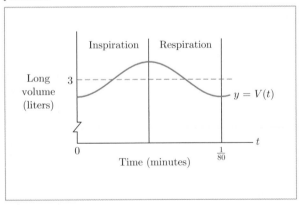

**Figure 1.** Spirogram trace.

**56. (a)** Compute $V(0)$, $V(\frac{1}{320})$, $V(\frac{1}{160})$, and $V(\frac{1}{80})$.

**(b)** What is the maximum lung volume?

**57. (a)** Find a formula for the rate of flow of air into the lungs at time $t$.

**(b)** Find the maximum rate of flow of air during inspiration (i.e., breathing in). This is called the *peak inspiratory flow*.

**(c)** Inspiration occurs during the time from $t = 0$ to $t = 1/160$. Find the average rate of flow of air during inspiration. This quantity is called the *mean inspiratory flow*.

**58.** The *minute volume* is defined to be the total amount of air inspired (breathed in) during one minute. According to a standard text on respiratory physiology, when a person undergoes spontaneous hyperventilation, the peak inspiratory flow equals $\pi$ times the minute volume, and the mean inspiratory flow equals twice the minute volume.* Verify these assertions using the data from Exercise 57.

*Evaluate the following integrals.*

**59.** $\displaystyle\int \sin(\pi - x)\, dx$

**60.** $\displaystyle\int (3\cos 3x - 2\sin 2x)\, dx$

**61.** $\displaystyle\int_0^{\pi/2} \cos 6x\, dx$

**62.** $\displaystyle\int \cos(6 - 2x)\, dx$

**63.** $\displaystyle\int_0^{\pi} (x - 2\cos(\pi - 2x))\, dx$

**64.** $\displaystyle\int_{-\pi}^{\pi} (\cos 3x + 2\sin 7x)\, dx$

**65.** $\displaystyle\int \sec^2 \frac{x}{2}\, dx$

**66.** $\displaystyle\int 2\sec^2 2x\, dx$

**67.** Find the shaded area $A_1$ in Fig. 2.

**68.** Find the shaded area $A_2$ in Fig. 2.

**69.** Find the shaded area $A_3$ in Fig. 2.

**70.** Find the shaded area $A_4$ in Fig. 2.

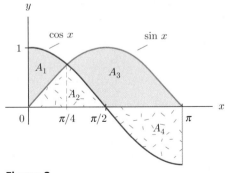

**Figure 2**

*J. F. Nunn, *Applied Respiratory Physiology*, 2nd ed. (London: Butterworths, 1977), p. 122.

In Exercises 71–74, find the average of the function $f(t)$ over the given interval.

**71.** $f(t) = 1 + \sin 2t - \frac{1}{3}\cos 2t,\ 0 \le t \le 2\pi$

**72.** $f(t) = t - \cos 2t,\ 0 \le t \le \pi$

**73.** $f(t) = 1000 + 200\sin 2\left(t - \frac{\pi}{4}\right),\ 0 \le t \le \frac{3\pi}{4}$

**74.** $f(t) = \cos t + \sin t,\ -\pi \le t \le 0$

Evaluate the given integral. [Hint: Use identity (1), Section 8.4, to transform the integral before evaluating it.]

**75.** $\displaystyle\int \tan^2 x\, dx$

**76.** $\displaystyle\int \tan^2 3x\, dx$

**77.** $\displaystyle\int (1 + \tan^2 x)\, dx$

**78.** $\displaystyle\int (2 + \tan^2 x)\, dx$

**79.** $\displaystyle\int_0^{\pi/4} \tan^2 x\, dx$

**80.** $\displaystyle\int_0^{\pi/4} (2 + 2\tan^2 x)\, dx$

# Techniques of Integration

I N this chapter we develop techniques for calculating integrals, both indefinite and definite. The need for these techniques has been justified in the preceding chapters. In addition to adding to our fund of applications, we will see even more clearly how the need to calculate integrals arises in physical problems.

As noted, integration is the reverse process of differentiation. However, integration is much harder to carry out. If a function is an expression involving elementary functions (such as $x^r$, $\sin x$, $e^x$, ...), then so is its derivative. Moreover, we were able to develop methods of calculation that enable us to differentiate, with comparative ease, almost any function that we can write down. Although many integration problems have these characteristics, some do not. There are some elementary functions (e.g., $e^{x^2}$) for which an antiderivative cannot be expressed in terms of elementary functions. Even where an elementary antiderivative exists, the techniques for finding the antiderivative are often complicated. For this reason, we must be prepared with a broad range of tools in order to cope with the problem of calculating integrals. Among the ideas to be discussed in this chapter are the following:

1. Techniques for evaluating indefinite integrals. We will concentrate on two methods—integration by substitution and integration by parts.

**2.** Evaluation of definite integrals.

**3.** Approximation of definite integrals. We will develop two new techniques for obtaining numerical approximations to $\int_a^b f(x)\,dx$. These techniques are especially useful in those cases in which we cannot find an antiderivative for $f(x)$.

Let us review the most elementary facts about integration. The indefinite integral

$$\int f(x)\,dx$$

is, by definition, a function whose derivative is $f(x)$. If $F(x)$ is one such function, then the most general function whose derivative is $f(x)$ is simply $F(x) + C$, where $C$ is any constant. We write

$$\int f(x)\,dx = F(x) + C$$

to mean that all antiderivatives of $f(x)$ are precisely the functions $F(x) + C$, where $C$ is any constant.

Each time we differentiate a function, we also derive an integration formula. For example, the fact that

$$\frac{d}{dx}(3x^2) = 6x$$

can be turned into the integration formula

$$\int 6x\,dx = 3x^2 + C.$$

Some of the formulas that follow immediately from differentiation formulas are reviewed in the following table.

| Differentiation Formula | Corresponding Integration Formula |
|---|---|
| $\dfrac{d}{dx}(x^r)\,dx = rx^{r-1}$ | $\displaystyle\int rx^{r-1}\,dx = x^r + C$ or |
| | $\displaystyle\int x^r\,dx = \dfrac{x^{r+1}}{r+1} + C,\ r \neq -1$ |
| $\dfrac{d}{dx}(e^x) = e^x$ | $\displaystyle\int e^x\,dx = e^x + C$ |
| $\dfrac{d}{dx}(\ln|x|) = \dfrac{1}{x}$ | $\displaystyle\int \dfrac{1}{x}\,dx = \ln|x| + C$ |
| $\dfrac{d}{dx}(\sin x) = \cos x$ | $\displaystyle\int \cos x\,dx = \sin x + C$ |
| $\dfrac{d}{dx}(\cos x) = -\sin x$ | $\displaystyle\int \sin x\,dx = -\cos x + C$ |
| $\dfrac{d}{dx}(\tan x) = \sec^2 x$ | $\displaystyle\int \sec^2 x\,dx = \tan x + C$ |

This table illustrates the need for techniques of integration. For although $\sin x$, $\cos x$, and $\sec^2 x$ occur as derivatives of simple trigonometric functions, the functions $\tan x$ and $\cot x$ are not on our list. In fact, if we experiment with various elementary combinations of the trigonometric functions, it is easy to convince ourselves that antiderivatives of $\tan x$ and $\cot x$ are not easy to compute. In this chapter we develop techniques for calculating such antiderivatives (among others).

## 9.1   Integration by Substitution

As noted earlier, every differentiation formula can be turned into a corresponding integration formula. This point is true even for the chain rule. The resulting formula is called *integration by substitution* and is often used to transform a complicated integral into a simpler one.

Let $f(x)$ and $g(x)$ be two given functions and let $F(x)$ be an antiderivative for $f(x)$. The chain rule asserts that

$$\frac{d}{dx}[F(g(x))] = F'(g(x))g'(x)$$

$$= f(g(x))g'(x) \qquad [\text{since } F'(x) = f(x)].$$

Turning this formula into an integration formula, we have

$$\int f(g(x))g'(x)\,dx = F(g(x)) + C, \tag{1}$$

where $C$ is any constant.

---

**EXAMPLE 1**   Determine

$$\int (x^2 + 1)^3 \cdot 2x\,dx.$$

**Solution**   If we set $f(x) = x^3$, $g(x) = x^2 + 1$, then $f(g(x)) = (x^2 + 1)^3$ and $g'(x) = 2x$. Therefore, we can apply formula (1). An antiderivative $F(x)$ of $f(x)$ is given by

$$F(x) = \frac{1}{4}x^4,$$

so that, by formula (1), we have

$$\int (x^2 + 1)^3 \cdot 2x\,dx = F(g(x)) + C = \frac{1}{4}(x^2 + 1)^4 + C. \qquad \blacksquare$$

Formula (1) can be elevated from the status of a sometimes-useful formula to a technique of integration by the introduction of a simple mnemonic device. Suppose that we are faced with integrating a function of the form $f(g(x))g'(x)$. Of course, we know the answer from formula (1). However, let us proceed somewhat differently. Replace the expression $g(x)$ by a new variable $u$, and replace $g'(x)\,dx$ by $du$. Such a substitution has the advantage that it reduces the generally complex expression $f(g(x))$ to the simpler form $f(u)$. In terms of $u$, the integration problem may be written

$$\int f(g(x))g'(x)\,dx = \int f(u)\,du.$$

However, the integral on the right is easy to evaluate, since

$$\int f(u)\,du = F(u) + C.$$

Since $u = g(x)$, we then obtain

$$\int f(g(x))g'(x)\,dx = F(u) + C = F(g(x)) + C,$$

which is the correct answer by (1). Remember, however, that replacing $g'(x)\,dx$ by $du$ only has status as a correct mathematical statement because doing so leads to the correct answers. We do not, in this book, seek to explain in any deeper way what this replacement means.

Let us rework Example 1 using this method.

**Second Solution of**    Set $u = x^2 + 1$. Then $du = \dfrac{d}{dx}(x^2 + 1)\, dx = 2x\, dx$, and
**Example 1**

$$\int (x^2 + 1)^3 \cdot 2x\, dx = \int u^3\, du$$

$$= \frac{1}{4}u^4 + C$$

$$= \frac{1}{4}(x^2 + 1)^4 + C \quad \text{(since } u = x^2 + 1\text{).}$$

---

**EXAMPLE 2**    Evaluate

$$\int 2xe^{x^2}\, dx.$$

**Solution**    Let $u = x^2$, so that $du = \dfrac{d}{dx}(x^2)\, dx = 2x\, dx$. Therefore,

$$\int 2xe^{x^2}\, dx = \int e^{x^2} \cdot 2x\, dx$$

$$= \int e^u\, du$$

$$= e^u + C$$

$$= e^{x^2} + C.$$

From Examples 1 and 2 we can deduce the following method for integration of functions of the form $f'(g(x))g'(x)$.

---

### Integration by Substitution

1. Define a new variable $u = g(x)$, where $g(x)$ is chosen in such a way that, when written in terms of $u$, the integrand is simpler than when written in terms of $x$.

2. Transform the integral with respect to $x$ into an integral with respect to $u$ by replacing $g(x)$ everywhere by $u$ and $g'(x)\, dx$ by $du$.

3. Integrate the resulting function of $u$.

4. Rewrite the answer in terms of $x$ by replacing $u$ by $g(x)$.

---

Let us try a few more examples.

---

**EXAMPLE 3**    Evaluate

$$\int 3x^2 \sqrt{x^3 + 1}\, dx.$$

**Solution**    The first problem facing us is to find an appropriate substitution that will simplify the integral. An immediate possibility is offered by setting $u = x^3 + 1$. Then $\sqrt{x^3 + 1}$ will become $\sqrt{u}$, a significant simplification. If $u = x^3 + 1$, then

$$du = \frac{d}{dx}(x^3 + 1)\, dx = 3x^2\, dx, \text{ so that}$$

$$\int 3x^2 \sqrt{x^3 + 1}\, dx = \int \sqrt{u}\, du$$

$$= \frac{2}{3} u^{3/2} + C$$

$$= \frac{2}{3}(x^3 + 1)^{3/2} + C. \qquad \blacksquare$$

**EXAMPLE 4**   Find

$$\int \frac{(\ln x)^2}{x}\, dx.$$

**Solution**   Let $u = \ln x$. Then $du = (1/x)\, dx$ and

$$\int \frac{(\ln x)^2}{x}\, dx = \int (\ln x)^2 \cdot \frac{1}{x}\, dx$$

$$= \int u^2\, du$$

$$= \frac{u^3}{3} + C$$

$$= \frac{(\ln x)^3}{3} + C \quad (\text{since } u = \ln x). \qquad \blacksquare$$

Knowing the correct substitution to make is a skill that develops through practice. Basically, we look for an occurrence of function composition, $f(g(x))$, where $f(x)$ is a function that we know how to integrate and where $g'(x)$ also appears in the integrand. Sometimes $g'(x)$ does not appear exactly, but can be obtained by multiplying by a constant. Such a shortcoming is easily remedied, as is illustrated in Examples 5 and 6.

**EXAMPLE 5**   Find

$$\int x^2 e^{x^3}\, dx.$$

**Solution**   Let $u = x^3$; then $du = 3x^2\, dx$. The integrand involves $x^2\, dx$, not $3x^2\, dx$. To introduce the missing factor "3", we write

$$\int x^2 e^{x^3}\, dx = \int \frac{1}{3} \cdot 3x^2 e^{x^3}\, dx = \frac{1}{3} \int e^{x^3} 3x^2\, dx.$$

(Recall from Section 6.1 that constant multiples may be moved through the integral sign.) Substituting, we obtain

$$\int x^2 e^{x^3}\, dx = \frac{1}{3} \int e^u\, du = \frac{1}{3} e^u + C$$

$$= \frac{1}{3} e^{x^3} + C \quad (\text{since } u = x^3).$$

Another way to handle the missing factor "3" is to write

$$u = x^3, \quad du = 3x^2\, dx, \quad \text{and} \quad \frac{1}{3}\, du = x^2\, dx.$$

Then substitution yields

$$\int x^2 e^{x^3}\,dx = \int e^{x^3}\cdot x^2\,dx = \int e^u\frac{1}{3}\,du = \frac{1}{3}\int e^u\,du$$

$$= \frac{1}{3}e^u + C = \frac{1}{3}e^{x^3} + C. \qquad \blacksquare$$

**EXAMPLE 6** Find

$$\int \frac{2-x}{\sqrt{2x^2 - 8x + 1}}\,dx.$$

**Solution** Let $u = 2x^2 - 8x + 1$; then $du = (4x - 8)\,dx$. Observe that $4x - 8 = -4(2 - x)$. So we multiply the integrand by $-4$ and compensate by placing a factor of $-\frac{1}{4}$ in front of the integral.

$$\int \frac{1}{\sqrt{2x^2 - 8x + 1}}\cdot(2 - x)\,dx = -\frac{1}{4}\int \frac{1}{\sqrt{2x^2 - 8x + 1}}\cdot(-4)(2 - x)\,dx$$

$$= -\frac{1}{4}\int \frac{1}{\sqrt{u}}\,du = -\frac{1}{4}\int u^{-1/2}\,du$$

$$= -\frac{1}{4}\cdot 2u^{1/2} + C = -\frac{1}{2}u^{1/2} + C$$

$$= -\frac{1}{2}(2x^2 - 8x + 1)^{1/2} + C. \qquad \blacksquare$$

**EXAMPLE 7** Find

$$\int \frac{2x}{x^2 + 1}\,dx.$$

**Solution** We note that the derivative of $x^2 + 1$ is $2x$. Thus we make the substitution $u = x^2 + 1$, $du = 2x\,dx$, in order to derive

$$\int \frac{2x}{x^2 + 1}\,dx = \int \frac{1}{u}\,du = \ln|u| + C = \ln(x^2 + 1) + C. \qquad \blacksquare$$

**EXAMPLE 8** Evaluate

$$\int \tan x\,dx.$$

**Solution** Since $\tan x = \dfrac{\sin x}{\cos x}$, we have

$$\int \tan x\,dx = \int \frac{\sin x}{\cos x}\,dx.$$

Let $u = \cos x$, so that $du = -\sin x\, dx$. Then

$$\int \frac{\sin x}{\cos x}\, dx = -\int \frac{-\sin x}{\cos x}\, dx$$

$$= -\int \frac{1}{u}\, du$$

$$= -\ln|u| + C$$

$$= -\ln|\cos x| + C.$$

Note that

$$-\ln|\cos x| = \ln\left|\frac{1}{\cos x}\right| = \ln|\sec x|.$$

So the preceding formula can be written

$$\int \tan x\, dx = \ln|\sec x| + C.$$

$\blacksquare$

## PRACTICE PROBLEMS 9.1

1. (*Review*) Differentiate the following functions.

   (a) $e^{(2x^3 + 3x)}$    (b) $\ln x^5$

   (c) $\ln \sqrt{x}$    (d) $\ln 5|x|$

   (e) $x \ln x$    (f) $\ln(x^4 + x^2 + 1)$

   (g) $\sin x^3$    (h) $\tan x$

2. Use the substitution $u = \dfrac{3}{x}$ to determine $\displaystyle\int \frac{e^{3/x}}{x^2}\, dx$.

## EXERCISES 9.1

*Determine the integrals in Exercises 1–36 by making appropriate substitutions.*

1. $\displaystyle\int 2x(x^2 + 4)^5\, dx$

2. $\displaystyle\int 2(2x - 1)^7\, dx$

3. $\displaystyle\int \frac{2x + 1}{\sqrt{x^2 + x + 3}}\, dx$

4. $\displaystyle\int (x^2 + 2x + 3)^6 (x + 1)\, dx$

5. $\displaystyle\int 3x^2 e^{(x^3 - 1)}\, dx$

6. $\displaystyle\int 2xe^{-x^2}\, dx$

7. $\displaystyle\int x\sqrt{4 - x^2}\, dx$

8. $\displaystyle\int \frac{(1 + \ln x)^3}{x}\, dx$

9. $\displaystyle\int \frac{1}{\sqrt{2x + 1}}\, dx$

10. $\displaystyle\int (x^3 - 6x)^7 (x^2 - 2)\, dx$

11. $\displaystyle\int xe^{x^2}\, dx$

12. $\displaystyle\int \frac{e^{\sqrt{x}}}{\sqrt{x}}\, dx$

13. $\displaystyle\int \frac{\ln(2x)}{x}\, dx$

14. $\displaystyle\int \frac{\sqrt{\ln x}}{x}\, dx$

15. $\displaystyle\int \frac{x^4}{x^5 + 1}\, dx$

16. $\displaystyle\int \frac{x}{\sqrt{x^2 + 1}}\, dx$

17. $\displaystyle\int \frac{x - 3}{(1 - 6x + x^2)^2}\, dx$

18. $\displaystyle\int x^{-2} \left(\frac{1}{x} + 2\right)^5\, dx$

19. $\displaystyle\int \frac{\ln \sqrt{x}}{x}\, dx$

20. $\displaystyle\int \frac{x^2}{3 - x^3}\, dx$

21. $\displaystyle\int \frac{x^2 - 2x}{x^3 - 3x^2 + 1}\, dx$

22. $\displaystyle\int \frac{\ln(3x)}{3x}\, dx$

23. $\displaystyle\int \frac{8x}{e^{x^2}}\, dx$

24. $\displaystyle\int \frac{3}{(2x + 1)^3}\, dx$

25. $\displaystyle\int \frac{1}{x \ln x^2}\, dx$

26. $\displaystyle\int \frac{2}{x(\ln x)^4}\, dx$

27. $\displaystyle\int (3 - x)(x^2 - 6x)^4\, dx$

28. $\displaystyle\int \frac{dx}{3 - 5x}$

29. $\displaystyle\int e^x (1 + e^x)^5\, dx$

30. $\displaystyle\int e^x \sqrt{1 + e^x}\, dx$

31. $\displaystyle\int \frac{e^x}{1 + 2e^x}\, dx$

32. $\displaystyle\int \frac{e^x + e^{-x}}{e^x - e^{-x}}\, dx$

33. $\displaystyle\int \frac{e^{-x}}{1 - e^{-x}}\, dx$

34. $\displaystyle\int \frac{(1 + e^{-x})^3}{e^x}\, dx$

[*Hint*: In Exercises 35 and 36, multiply the numerator and denominator by $e^{-x}$.]

**35.** $\displaystyle\int \frac{1}{1+e^x}\,dx$  **36.** $\displaystyle\int \frac{e^{2x}-1}{e^{2x}+1}\,dx$

**37.** Figure 1 shows graphs of several functions $f(x)$ whose slope at each $x$ is $x/\sqrt{x^2+9}$. Find the expression for the function $f(x)$ whose graph passes through $(4,8)$.

**38.** Figure 2 shows graphs of several functions $f(x)$ whose slope at each $x$ is $(2\sqrt{x}+1)/\sqrt{x}$. Find the expression for the function $f(x)$ whose graph passes through $(4,15)$.

*Determine the following integrals using the indicated substitution.*

**39.** $\displaystyle\int (x+5)^{-1/2} e^{\sqrt{x+5}}\,dx;\ u=\sqrt{x+5}$

**40.** $\displaystyle\int \frac{x^4}{x^5-7}\ln(x^5-7)\,dx;\ u=\ln(x^5-7)$

**41.** $\displaystyle\int x\sec^2 x^2\,dx;\ u=x^2$

**42.** $\displaystyle\int (1+\ln x)\sin(x\ln x)\,dx;\ u=x\ln x$

*Determine the following integrals by making an appropriate substitution.*

**43.** $\displaystyle\int \sin x\cos x\,dx$  **44.** $\displaystyle\int 2x\cos x^2\,dx$

**45.** $\displaystyle\int \frac{\cos\sqrt{x}}{\sqrt{x}}\,dx$  **46.** $\displaystyle\int \frac{\cos x}{(2+\sin x)^3}\,dx$

**47.** $\displaystyle\int \cos^3 x\sin x\,dx$  **48.** $\displaystyle\int (\sin 2x)e^{\cos 2x}\,dx$

**49.** $\displaystyle\int \frac{\cos 3x}{\sqrt{2-\sin 3x}}\,dx$  **50.** $\displaystyle\int \cot x\,dx$

**51.** $\displaystyle\int \frac{\sin x+\cos x}{\sin x-\cos x}\,dx$  **52.** $\displaystyle\int \tan x\sec^2 x\,dx$

**53.** Determine $\int 2x(x^2+5)\,dx$ by making a substitution. Then determine the integral by multiplying out the integrand and antidifferentiating. Account for the difference in the two results.

**Figure 1**

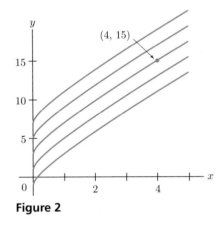

**Figure 2**

## SOLUTIONS TO PRACTICE PROBLEMS 9.1

**1. (a)** $\dfrac{d}{dx}e^{(2x^3+3x)} = e^{(2x^3+3x)}\cdot(6x^2+3)$  (chain rule)

**(b)** $\dfrac{d}{dx}\ln x^5 = \dfrac{d}{dx}5\ln x = 5\cdot\dfrac{1}{x}$  (Logarithm Property LIV)

**(c)** $\dfrac{d}{dx}\ln\sqrt{x} = \dfrac{d}{dx}\dfrac{1}{2}\ln x = \dfrac{1}{2}\cdot\dfrac{1}{x} = \dfrac{1}{2x}$  (Logarithm Property LIV)

**(d)** $\dfrac{d}{dx}\ln 5|x| = \dfrac{d}{dx}[\ln 5+\ln|x|] = 0+\dfrac{1}{x} = \dfrac{1}{x}$  (Logarithm Property LI)

**(e)** $\dfrac{d}{dx}x\ln x = x\cdot\dfrac{1}{x}+(\ln x)\cdot 1 = 1+\ln x$  (product rule)

**(f)** $\dfrac{d}{dx}\ln(x^4+x^2+1) = \dfrac{4x^3+2x}{x^4+x^2+1}$  (chain rule)

**(g)** $\dfrac{d}{dx}\sin x^3 = (\cos x^3)\cdot(3x^2)$  (chain rule)

**(h)** $\dfrac{d}{dx}\tan x = \sec^2 x$  (formula for the derivative of $\tan x$)

**2.** Let $u=3/x,\ du=(-3/x^2)\,dx$. Then

$$\int \frac{e^{3/x}}{x^2}\,dx = -\frac{1}{3}\int e^{3/x}\cdot\left(-\frac{3}{x^2}\right)dx$$

$$= -\frac{1}{3}\int e^u\,du = -\frac{1}{3}e^u+C = -\frac{1}{3}e^{3/x}+C.$$

## 9.2 Integration by Parts

In the preceding section we developed the method of integration by substitution by turning the chain rule into an integration formula. Let us do the same for the product rule. Let $f(x)$ and $g(x)$ be any two functions and let $G(x)$ be an antiderivative of $g(x)$. The product rule asserts that

$$\frac{d}{dx}[f(x)G(x)] = f(x)G'(x) + f'(x)G(x)$$

$$= f(x)g(x) + f'(x)G(x) \quad [\text{since } G'(x) = g(x)].$$

Therefore,

$$f(x)G(x) = \int f(x)g(x)\, dx + \int f'(x)G(x)\, dx.$$

This last formula can be rewritten in the following more useful form.

$$\int f(x)g(x)\, dx = f(x)G(x) - \int f'(x)G(x)\, dx. \tag{1}$$

Equation (1) is the principle of *integration by parts* and is one of the most important techniques of integration.

**EXAMPLE 1**  Evaluate

$$\int xe^x\, dx.$$

**Solution**  Set $f(x) = x$, $g(x) = e^x$. Then $f'(x) = 1$, $G(x) = e^x$, and equation (1) yields

$$\int xe^x\, dx = xe^x - \int 1 \cdot e^x\, dx = xe^x - e^x + C.$$     ■

The following principles underlie Example 1 and also illustrate general features of situations to which integration by parts may be applied:

**1.** The integrand is the product of two functions $f(x) = x$ and $g(x) = e^x$.

**2.** It is easy to compute $f'(x)$ and $G(x)$. That is, we can differentiate $f(x)$ and integrate $g(x)$.

**3.** The integral $\int f'(x)G(x)\, dx$ can be calculated.

Let us consider another example to see how these three principles work.

**EXAMPLE 2**  Evaluate

$$\int x(x+5)^8\, dx.$$

**Solution**  Our calculations can be set up as follows:

$$f(x) = x, \qquad g(x) = (x+5)^8,$$

$$f'(x) = 1, \qquad G(x) = \frac{1}{9}(x+5)^9.$$

Then

$$\int x(x+5)^8 \, dx = x \cdot \frac{1}{9}(x+5)^9 - \int 1 \cdot \frac{1}{9}(x+5)^9 \, dx$$

$$= \frac{1}{9}x(x+5)^9 - \frac{1}{9} \int (x+5)^9 \, dx$$

$$= \frac{1}{9}x(x+5)^9 - \frac{1}{9} \cdot \frac{1}{10}(x+5)^{10} + C$$

$$= \frac{1}{9}x(x+5)^9 - \frac{1}{90}(x+5)^{10} + C. \qquad \blacksquare$$

We were led to try integration by parts because our integrand is the product of two functions. We choose $f(x) = x$ [and not $(x+5)^8$] because $f'(x) = 1$, so that the new integrand has only the factor $x + 5$, thereby simplifying the integral.

**EXAMPLE 3** Evaluate

$$\int x \sin x \, dx.$$

**Solution** Let us set

$$f(x) = x, \qquad g(x) = \sin x,$$
$$f'(x) = 1, \qquad G(x) = -\cos x.$$

Then

$$\int x \sin x \, dx = -x \cos x - \int 1 \cdot (-\cos x) \, dx$$

$$= -x \cos x + \int \cos x \, dx$$

$$= -x \cos x + \sin x + C. \qquad \blacksquare$$

**EXAMPLE 4** Evaluate

$$\int x^2 \ln x \, dx.$$

**Solution** Set

$$f(x) = \ln x, \qquad g(x) = x^2$$

$$f'(x) = \frac{1}{x}, \qquad G(x) = \frac{x^3}{3}.$$

Then

$$\int x^2 \ln x \, dx = \frac{x^3}{3} \ln x - \int \frac{1}{x} \cdot \frac{x^3}{3} \, dx$$

$$= \frac{x^3}{3} \ln x - \frac{1}{3} \int x^2 \, dx$$

$$= \frac{x^3}{3} \ln x - \frac{1}{9}x^3 + C. \qquad \blacksquare$$

The next example shows how integration by parts can be used to compute a reasonably complicated integral.

**EXAMPLE 5**  Find

$$\int \frac{xe^x}{(x+1)^2}\,dx.$$

**Solution**  Let $f(x) = xe^x$, $g(x) = \dfrac{1}{(x+1)^2}$. Then

$$f'(x) = xe^x + e^x \cdot 1 = (x+1)e^x, \qquad G(x) = \frac{-1}{x+1}.$$

As a result, we have

$$\int \frac{xe^x}{(x+1)^2}\,dx = xe^x \cdot \frac{-1}{x+1} - \int (x+1)e^x \cdot \frac{-1}{x+1}\,dx$$

$$= -\frac{xe^x}{x+1} + \int e^x\,dx$$

$$= -\frac{xe^x}{x+1} + e^x + C = \frac{e^x}{x+1} + C. \qquad \blacksquare$$

Sometimes we must use integration by parts more than once.

**EXAMPLE 6**  Find

$$\int x^2 \sin x\,dx.$$

**Solution**  Let $f(x) = x^2$, $g(x) = \sin x$. Then $f'(x) = 2x$ and $G(x) = -\cos x$. Applying our formula for integration by parts, we have

$$\int x^2 \sin x\,dx = -x^2 \cos x - \int 2x \cdot (-\cos x)\,dx$$

$$= -x^2 \cos x + 2\int x \cos x\,dx. \qquad (2)$$

The integral $\int x \cos x\,dx$ can itself be handled by integration by parts. Let $f(x) = x$, $g(x) = \cos x$. Then $f'(x) = 1$ and $G(x) = \sin x$, so that

$$\int x \cos x\,dx = x \sin x - \int 1 \cdot \sin x\,dx$$

$$= x \sin x + \cos x + C. \qquad (3)$$

Combining (2) and (3), we see that

$$\int x^2 \sin x\,dx = -x^2 \cos x + 2(x \sin x + \cos x) + C$$

$$= -x^2 \cos x + 2x \sin x + 2 \cos x + C. \qquad \blacksquare$$

**EXAMPLE 7**  Evaluate

$$\int \ln x\,dx.$$

**Solution** Since $\ln x = 1 \cdot \ln x$, we may view $\ln x$ as a product $f(x)g(x)$, where $f(x) = \ln x$, $g(x) = 1$. Then

$$f'(x) = \frac{1}{x}, \qquad G(x) = x.$$

Finally,

$$\int \ln x\, dx = x \ln x - \int \frac{1}{x} \cdot x\, dx$$

$$= x \ln x - \int 1\, dx$$

$$= x \ln x - x + C.$$

## PRACTICE PROBLEMS 9.2

*Evaluate the following integrals.*

**1.** $\displaystyle\int \frac{x}{e^{3x}}\, dx$

**2.** $\displaystyle\int \ln \sqrt{x}\, dx$

## EXERCISES 9.2

*Evaluate the following integrals.*

**1.** $\displaystyle\int x e^{5x}\, dx$

**2.** $\displaystyle\int x e^{x/2}\, dx$

**3.** $\displaystyle\int x(x+7)^4\, dx$

**4.** $\displaystyle\int x(2x-3)^3\, dx$

**5.** $\displaystyle\int \frac{x}{e^x}\, dx$

**6.** $\displaystyle\int x^2 e^x\, dx$

**7.** $\displaystyle\int \frac{x}{\sqrt{x+1}}\, dx$

**8.** $\displaystyle\int \frac{x}{\sqrt{3+2x}}\, dx$

**9.** $\displaystyle\int e^{2x}(1-3x)\, dx$

**10.** $\displaystyle\int (1+x)^2 e^{2x}\, dx$

**11.** $\displaystyle\int \frac{6x}{e^{3x}}\, dx$

**12.** $\displaystyle\int \frac{x+2}{e^{2x}}\, dx$

**13.** $\displaystyle\int x\sqrt{x+1}\, dx$

**14.** $\displaystyle\int x\sqrt{2-x}\, dx$

**15.** $\displaystyle\int \sqrt{x}\ln \sqrt{x}\, dx$

**16.** $\displaystyle\int x^5 \ln x\, dx$

**17.** $\displaystyle\int x \cos x\, dx$

**18.** $\displaystyle\int x \sin 8x\, dx$

**19.** $\displaystyle\int x \ln 5x\, dx$

**20.** $\displaystyle\int x^{-3} \ln x\, dx$

**21.** $\displaystyle\int \ln x^4\, dx$

**22.** $\displaystyle\int \frac{\ln(\ln x)}{x}\, dx$

**23.** $\displaystyle\int x^2 e^{-x}\, dx$

**24.** $\displaystyle\int \ln \sqrt{x+1}\, dx$

*Evaluate the following integrals using techniques studied thus far.*

**25.** $\displaystyle\int x(x+5)^4\, dx$

**26.** $\displaystyle\int 4x \cos(x^2+1)\, dx$

**27.** $\displaystyle\int x(x^2+5)^4\, dx$

**28.** $\displaystyle\int 4x \cos(x+1)\, dx$

**29.** $\displaystyle\int (3x+1)e^{x/3}\, dx$

**30.** $\displaystyle\int \frac{(\ln x)^5}{x}\, dx$

**31.** $\displaystyle\int x \sec^2(x^2+1)\, dx$

**32.** $\displaystyle\int \frac{\ln x}{x^5}\, dx$

**33.** $\displaystyle\int (xe^{2x}+x^2)\, dx$

**34.** $\displaystyle\int (x^{3/2}+\ln 2x)\, dx$

**35.** $\displaystyle\int (xe^{x^2}-2x)\, dx$

**36.** $\displaystyle\int (x^2 - x \sin 2x)\, dx$

**37.** Figure 1 shows graphs of several functions $f(x)$ whose slope at each $x$ is $x/\sqrt{x+9}$. Find the expression for the function $f(x)$ whose graph passes through $(0, 2)$.

**38.** Figure 2 shows graphs of several functions $f(x)$ whose slope at each $x$ is $\dfrac{x}{e^{x/3}}$. Find the expression for the function $f(x)$ whose graph passes through $(0, 6)$.

**Figure 1**

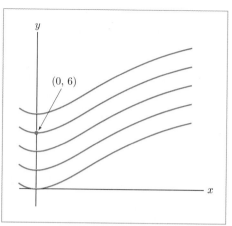

**Figure 2**

## SOLUTIONS TO PRACTICE PROBLEMS 9.2

1. $\dfrac{x}{e^{3x}}$ is the same as $xe^{-3x}$, a product of two familiar functions. Set $f(x) = x$, $g(x) = e^{-3x}$. Then

$$f'(x) = 1, \qquad G(x) = -\frac{1}{3}e^{-3x},$$

so that

$$\int \frac{x}{e^{3x}}\,dx = x \cdot \left(-\frac{1}{3}e^{-3x}\right) - \int 1 \cdot \left(-\frac{1}{3}e^{-3x}\right)dx$$

$$= -\frac{1}{3}xe^{-3x} + \frac{1}{3}\int e^{-3x}\,dx$$

$$= -\frac{1}{3}xe^{-3x} + \frac{1}{3}\left[-\frac{1}{3}e^{-3x}\right] + C$$

$$= -\frac{1}{3}xe^{-3x} - \frac{1}{9}e^{-3x} + C.$$

2. This problem is similar to Example 7, which asks for $\int \ln x\,dx$ and can be approached in the same way by letting $f(x) = \ln \sqrt{x}$ and $g(x) = 1$. Another approach is to use a property of logarithms to simplify the integrand.

$$\int \ln \sqrt{x}\,dx = \int \ln(x)^{1/2}\,dx$$

$$= \int \frac{1}{2}\ln x\,dx$$

$$= \frac{1}{2}\int \ln x\,dx$$

Since we know $\int \ln x\,dx$ from Example 7,

$$\int \ln \sqrt{x}\,dx = \frac{1}{2}\int \ln x\,dx = \frac{1}{2}(x\ln x - x) + C.$$

## 9.3 Evaluation of Definite Integrals

Earlier we discussed techniques for determining antiderivatives (indefinite integrals). One of the most important applications of such techniques concerns the computation of definite integrals. For if $F(x)$ is an antiderivative of $f(x)$, then

$$\int_a^b f(x)\,dx = F(b) - F(a).$$

Thus the techniques of the previous sections can be used to evaluate definite integrals. Here we will simplify the method of evaluating definite integrals in those cases where the antiderivative is found by integration by substitution or parts.

**EXAMPLE 1**  Evaluate

$$\int_0^1 2x(x^2 + 1)^5\,dx.$$

**Solution—First Method** Let $u = x^2 + 1$, $du = 2x\,dx$. Then

$$\int 2x(x^2+1)^5\,dx = \int u^5\,du = \frac{u^6}{6} + C = \frac{(x^2+1)^6}{6} + C.$$

Consequently,

$$\int_0^1 2x(x^2+1)^5\,dx = \frac{(x^2+1)^6}{6}\bigg|_0^1 = \frac{2^6}{6} - \frac{1^6}{6} = \frac{21}{2}. \qquad \blacksquare$$

**Solution—Second Method** Again we make the substitution $u = x^2 + 1$, $du = 2x\,dx$; however, we also apply the substitution to the limits of integration. When $x = 0$ (the lower limit of integration), we have $u = 0^2 + 1 = 1$; and when $x = 1$ (the upper limit of integration), we have $u = 1^2 + 1 = 2$. Therefore,

$$\int_0^1 2x(x^2+1)^5\,dx = \int_1^2 u^5\,du = \frac{u^6}{6}\bigg|_1^2 = \frac{2^6}{6} - \frac{1^6}{6} = \frac{21}{2}.$$

Notice that, in utilizing the second method, we did not need to reexpress the function $u^6/6$ in terms of $x$. $\qquad \blacksquare$

The foregoing computation is an example of a general computational method, which can be expressed as follows:

**Change of Limits Rule** Suppose that the integral $\int f(g(x))g'(x)\,dx$ is subjected to the substitution $u = g(x)$, so that $\int f(g(x))g'(x)\,dx$ becomes $\int f(u)\,du$. Then

$$\int_a^b f(g(x))g'(x)\,dx = \int_{g(a)}^{g(b)} f(u)\,du.$$

**Justification of Change of Limits Rule** If $F(x)$ is an antiderivative of $f(x)$, then

$$\frac{d}{dx}[F(g(x))] = F'(g(x))g'(x) = f(g(x))g'(x).$$

Therefore,

$$\int_a^b f(g(x))g'(x)\,dx = F(g(x))\bigg|_a^b = F(g(b)) - F(g(a)) = \int_{g(a)}^{g(b)} f(u)\,du.$$

**EXAMPLE 2** Evaluate

$$\int_3^5 x\sqrt{x^2 - 9}\,dx.$$

**Solution** Let $u = x^2 - 9$, then $du = 2x\,dx$. When $x = 3$, we have $u = 3^2 - 9 = 0$. When $x = 5$, we have $u = 5^2 - 9 = 16$. Thus

$$\int_3^5 x\sqrt{x^2 - 9}\,dx = \frac{1}{2}\int_3^5 2x\sqrt{x^2 - 9}\,dx$$

$$= \frac{1}{2}\int_0^{16} \sqrt{u}\,du$$

$$= \frac{1}{2} \cdot \frac{2}{3} u^{3/2}\Big|_0^{16}$$

$$= \frac{1}{3} \cdot \left[16^{3/2} - 0\right] = \frac{1}{3} \cdot 16^{3/2}$$

$$= \frac{1}{3} \cdot 64 = \frac{64}{3}.$$ ■

---

**EXAMPLE 3**   Determine the area of the ellipse $x^2/a^2 + y^2/b^2 = 1$. (See Fig. 1.)

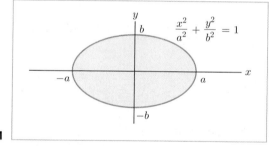

**Figure 1**

**Solution** Owing to the symmetry of the ellipse, the area is equal to twice the area of the upper half of the ellipse. Solving for $y$,

$$\frac{y^2}{b^2} = 1 - \frac{x^2}{a^2}$$

$$\frac{y}{b} = \pm\sqrt{1 - \left(\frac{x}{a}\right)^2}$$

$$y = \pm b\sqrt{1 - \left(\frac{x}{a}\right)^2}.$$

Since the area of the upper half-ellipse is the area under the curve

$$y = b\sqrt{1 - \left(\frac{x}{a}\right)^2},$$

the area of the ellipse is given by the integral

$$2\int_{-a}^a b\sqrt{1 - \left(\frac{x}{a}\right)^2}\,dx.$$

Let $u = x/a$; then $du = 1/a\,dx$. When $x = -a$, we have $u = -a/a = -1$. When $x = a$, we have $u = a/a = 1$. So

$$2 \int_{-a}^{a} b \sqrt{1 - \left(\frac{x}{a}\right)^2}\,dx = 2b \cdot a \int_{-a}^{a} \frac{1}{a} \sqrt{1 - \left(\frac{x}{a}\right)^2}\,dx$$

$$= 2ba \int_{-1}^{1} \sqrt{1 - u^2}\,du.$$

We cannot evaluate this integral by our existing techniques; we obtain its value immediately by recognizing that since the area under the curve $y = \sqrt{1 - x^2}$ from $x = -1$ to $x = 1$ is the top half of a unit circle, and we know that the area of a unit circle is $\pi$, so the area of the ellipse is $2ba \cdot (\pi/2) = \pi ab$. ∎

### Integration by Parts in Definite Integrals

**EXAMPLE 4**   Evaluate

$$\int_{0}^{\pi/2} x \cos x\,dx.$$

**Solution** We use integration by parts to find an antiderivative of $x \cos x$. Let $f(x) = x$, $g(x) = \cos x$, $f'(x) = 1$, $G(x) = \sin x$. Then

$$\int x \cos x\,dx = x \sin x - \int 1 \cdot \sin x\,dx = x \sin x + \cos x + C.$$

Hence

$$\int_{0}^{\pi/2} x \cos x\,dx = (x \sin x + \cos x)\Big|_{0}^{\pi/2}$$

$$= \left(\frac{\pi}{2} \sin \frac{\pi}{2} + \cos \frac{\pi}{2}\right) - (0 + \cos 0)$$

$$= \frac{\pi}{2} - 1.$$ ∎

**EXAMPLE 5**   Evaluate

$$\int_{0}^{5} \frac{x}{\sqrt{x + 4}}\,dx.$$

**Solution** Let $f(x) = x$, $g(x) = (x + 4)^{-1/2}$, $f'(x) = 1$, and $G(x) = 2(x + 4)^{1/2}$. Then

$$\int_{0}^{5} \frac{x}{\sqrt{x + 4}}\,dx = 2x(x + 4)^{1/2}\Big|_{0}^{5} - \int_{0}^{5} 1 \cdot 2(x + 4)^{1/2}\,dx$$

$$= 2x(x + 4)^{1/2}\Big|_{0}^{5} - \frac{4}{3}(x + 4)^{3/2}\Big|_{0}^{5}$$

$$= \left[10(9)^{1/2} - 0\right] - \left[\frac{4}{3}(9)^{3/2} - \frac{4}{3}(4)^{3/2}\right]$$

$$= [30] - \left[36 - \frac{32}{3}\right] = 4\frac{2}{3}.$$ ∎

## PRACTICE PROBLEMS 9.3

*Evaluate the following definite integrals.*

**1.** $\displaystyle\int_0^1 (2x+3)e^{x^2+3x+6}\,dx$

**2.** $\displaystyle\int_e^{e^{\pi/2}} \frac{\sin(\ln x)}{x}\,dx$

## EXERCISES 9.3

*Evaluate the following definite integrals.*

**1.** $\displaystyle\int_{5/2}^3 2(2x-5)^{14}\,dx$

**2.** $\displaystyle\int_2^6 \frac{1}{\sqrt{4x+1}}\,dx$

**3.** $\displaystyle\int_0^2 4x(1+x^2)^3\,dx$

**4.** $\displaystyle\int_0^1 \frac{2x}{\sqrt{x^2+1}}\,dx$

**5.** $\displaystyle\int_0^3 \frac{x}{\sqrt{x+1}}\,dx$

**6.** $\displaystyle\int_0^1 x(3+x)^5\,dx$

**7.** $\displaystyle\int_3^5 x\sqrt{x^2-9}\,dx$

**8.** $\displaystyle\int_0^1 \frac{1}{(1+2x)^4}\,dx$

**9.** $\displaystyle\int_{-1}^2 (x^2-1)(x^3-3x)^4\,dx$

**10.** $\displaystyle\int_0^1 (2x-1)(x^2-x)^{10}\,dx$

**11.** $\displaystyle\int_0^1 \frac{x}{x^2+3}\,dx$

**12.** $\displaystyle\int_0^4 8x(x+4)^{-3}\,dx$

**13.** $\displaystyle\int_1^3 x^2 e^{x^3}\,dx$

**14.** $\displaystyle\int_{-1}^1 2xe^x\,dx$

**15.** $\displaystyle\int_1^e \frac{\ln x}{x}\,dx$

**16.** $\displaystyle\int_1^e \ln x\,dx$

**17.** $\displaystyle\int_0^\pi e^{\sin x}\cos x\,dx$

**18.** $\displaystyle\int_0^{\pi/4} \tan x\,dx$

**19.** $\displaystyle\int_0^1 x\sin \pi x\,dx$

**20.** $\displaystyle\int_0^{\pi/2} \sin\left(2x-\frac{\pi}{2}\right)\,dx$

*Use substitutions and the fact that a circle of radius $r$ has area $\pi r^2$ to evaluate the following integrals.*

**21.** $\displaystyle\int_{-\pi/2}^{\pi/2} \sqrt{1-\sin^2 x}\,\cos x\,dx$

**22.** $\displaystyle\int_0^{\sqrt{2}} \sqrt{4-x^4}\cdot 2x\,dx$

**23.** $\displaystyle\int_{-6}^0 \sqrt{-x^2-6x}\,dx$ [Complete the square: $-x^2-6x = 9-(x+3)^2$.]

*In Exercises 24 and 25, find the area of the shaded regions.*

**24.**

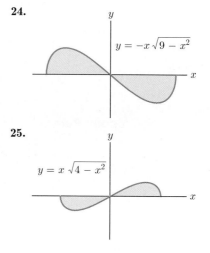

$y = -x\sqrt{9-x^2}$

**25.**

$y = x\sqrt{4-x^2}$

## SOLUTIONS TO PRACTICE PROBLEMS 9.3

**1.** Let $u = x^2+3x+6$ so that $du = (2x+3)\,dx$. When $x=0$, $u=6$; when $x=1$, $u=10$. Thus

$$\int_0^1 (2x+3)e^{x^2+3x+6}\,dx = \int_6^{10} e^u\,du = e^u\Big|_6^{10}$$

$$= e^{10} - e^6.$$

**2.** Let $u = \ln x$, $du = (1/x)\,dx$. When $x=e$, $u=\ln e =$

1; when $x = e^{\pi/2}$, $u = \ln e^{\pi/2} = \pi/2$. Thus

$$\int_e^{e^{\pi/2}} \frac{\sin(\ln x)}{x}\,dx = \int_1^{\pi/2} \sin u\,du = -\cos u\Big|_1^{\pi/2}$$

$$= -\cos\frac{\pi}{2} + \cos 1 \approx .54030.$$

# 9.4 Approximation of Definite Integrals

The definite integrals that arise in practical problems cannot always be evaluated by computing the net change in an antiderivative, as in the preceding section. Mathematicians have compiled extensive tables of antiderivatives. Moreover, there exist many excellent software programs that can determine antiderivatives. However, the form of an antiderivative may be quite complex and in some cases there may actually be no way to express an antiderivative in terms of elementary functions. In this section we discuss three methods for approximating the numerical value of a definite integral,

$$\int_a^b f(x)\,dx$$

without computing an antiderivative.

Given a positive integer $n$, divide the interval $a \le x \le b$ into $n$ equal subintervals, each of length $\Delta x = (b-a)/n$. Denote the endpoints of the subintervals by $a_0, a_1, \ldots, a_n$, and denote the midpoints of the subintervals by $x_1, x_2, \ldots, x_n$, as in Fig. 1. Recall from Chapter 6 that the definite integral is the limit of Riemann sums. When the midpoints of the subintervals in Fig. 1 are used to construct a Riemann sum, the resulting approximation to $\int_a^b f(x)\,dx$ is called the *midpoint rule*.

**Figure 1**

### Midpoint Rule

$$\int_a^b f(x)\,dx \approx f(x_1)\Delta x + f(x_2)\Delta x + \cdots + f(x_n)\Delta x \tag{1}$$

$$= [f(x_1) + f(x_2) + \cdots + f(x_n)]\Delta x.$$

**EXAMPLE 1**    Use the midpoint rule with $n = 4$ to approximate

$$\int_0^2 \frac{1}{1+e^x}\,dx.$$

**Solution**    We have $\Delta x = (b-a)/n = (2-0)/4 = .5$. The endpoints of the four subintervals begin at $a = 0$ and are spaced .5 unit apart. The first midpoint is at $a + \Delta x/2 = .25$. The midpoints are also spaced .5 unit apart. According to the midpoint rule, the integral is approximately equal to

$$\left[ \frac{1}{1 + e^{.25}} + \frac{1}{1 + e^{.75}} + \frac{1}{1 + e^{1.25}} + \frac{1}{1 + e^{1.75}} \right](.5)$$

$$\approx .5646961 \quad \text{(to seven decimal places).} \quad \blacksquare$$

A second method of approximation, the *trapezoidal rule*, uses the values of $f(x)$ at the endpoints of the subintervals of the interval $a \le x \le b$.

**Trapezoidal Rule**

$$\int_a^b f(x)\, dx \approx [f(a_0) + 2f(a_1) + \cdots + 2f(a_{n-1}) + f(a_n)]\frac{\Delta x}{2}. \qquad (2)$$

We will discuss the origin of the trapezoidal rule and why we call it by that name later in this section.

**EXAMPLE 2**   Use the trapezoidal rule with $n = 4$ to approximate

$$\int_0^2 \frac{1}{1 + e^x}\, dx.$$

**Solution**   As in Example 1, $\Delta x = .5$ and the endpoints of the subintervals are $a_0 = 0$, $a_1 = .5$, $a_2 = 1$, $a_3 = 1.5$, and $a_4 = 2$. The trapezoidal rule gives

$$\left[ \frac{1}{1 + e^0} + 2 \cdot \frac{1}{1 + e^{.5}} + 2 \cdot \frac{1}{1 + e^1} + 2 \cdot \frac{1}{1 + e^{1.5}} + \frac{1}{1 + e^2} \right]\frac{.5}{2}$$

$$\approx .5692545 \quad \text{(to seven decimal places).} \quad \blacksquare$$

When the function $f(x)$ is given explicitly, either the midpoint rule or the trapezoidal rule may be used to approximate the definite integral. However, occasionally the values of $f(x)$ may be known only at the endpoints of the subintervals. This may happen, for instance, when the values of $f(x)$ are obtained from experimental data. In this case, the midpoint rule cannot be used.

**EXAMPLE 3**   **Measuring cardiac output**[*]   Five milligrams of dye is injected into a vein leading to the heart. The concentration of the dye in the aorta, an artery leading from the heart, is determined every 2 seconds for 22 seconds. (See Table 1.) Let $c(t)$ be the concentration in the aorta after $t$ seconds. Use the trapezoidal rule to estimate $\int_0^{22} c(t)\, dt$.

**TABLE 1   Concentration of Dye in the Aorta**

| Seconds after Injection | 0 | 2 | 4 | 6 | 8 | 10 | 12 | 14 | 16 | 18 | 20 | 22 |
|---|---|---|---|---|---|---|---|---|---|---|---|---|
| Concentration (mg/liter) | 0 | 0 | .6 | 1.4 | 2.7 | 3.7 | 4.1 | 3.8 | 2.9 | 1.5 | .9 | .5 |

[*]Data from B. Horelick and S. Koont, Project UMAP, *Measuring Cardiac Output* (Newton, Mass.: Educational Development Center, Inc., 1978).

**Solution** Let $n = 11$. Then $a = 0$, $b = 22$, and $\Delta t = (22 - 0)/11 = 2$. The endpoints of the subintervals are $a_0 = 0, a_1 = 2, a_2 = 4, \ldots, a_{10} = 20$, and $a_{11} = 22$. By the trapezoidal rule,

$$\int_0^{22} c(t)\, dt \approx [c(0) + 2c(2) + 2c(4) + 2c(6) + \cdots + 2c(20) + c(22)] \left(\frac{2}{2}\right)$$

$$= [0 + 2(0) + 2(.6) + 2(1.4) + \cdots + 2(.9) + .5](1)$$

$$\approx 43.7 \text{ liters.}$$

Note that cardiac output is the rate (usually measured in liters per minute) at which the heart pumps blood, and it may be computed by the formula

$$R = \frac{60D}{\int_0^{22} c(t)\, dt},$$

where $D$ is the quantity of dye injected. For the preceding data, $R = 60(5)/43.7 \approx 6.9$ liters per minute.

Let us return to the approximations to $\displaystyle\int_0^2 \frac{1}{1 + e^x}\, dx$ found in Examples 1 and 2. These numbers are shown in Fig. 2 along with the exact value of the definite integral to seven decimal places and the error of the two approximations. [The scale is greatly enlarged. The exact value of the integral is $\ln 2 - \ln(1 + e^{-2}) \approx .5662192$. To find it, see Exercise 35, Section 9.1.] It can be shown that, in general, the error from the midpoint rule is about one-half the error from the trapezoidal rule, and the estimates from these two rules are usually on opposite sides of the actual value of the definite integral. These observations suggest that we might improve our estimate of the value of a definite integral by using a "weighted average" of these two estimates. Let $M$ and $T$ denote the estimates from the midpoint and trapezoidal rules, respectively, and define

$$S = \frac{2}{3}M + \frac{1}{3}T = \frac{2M + T}{3}. \tag{3}$$

Error:    .0015231          .0030353

.5646961    .5662192          .5692545
midpoint    actual          trapezoidal
rule        value            rule

**Figure 2**

The use of $S$ as an estimate of the value of a definite integral is called *Simpson's rule*. If we use Simpson's rule to estimate the definite integral in Example 1, we find that

$$S = \frac{2(.5646961) + .5692545}{3} \approx .5662156.$$

The error here is only .0000036. The error from the trapezoidal rule, in this example, is over 800 times as large!

As the number $n$ of subintervals increases, Simpson's rule becomes more accurate than both the midpoint rule and the trapezoidal rule. For a given definite integral, the error in the midpoint and trapezoidal rules is proportional to $1/n^2$, so doubling $n$ will divide the error by 4. However, the error in Simpson's rule is proportional to $1/n^4$, so doubling $n$ will divide the error by 16, and multiplying $n$ by a factor of 10 will divide the error by 10,000.

It is possible to combine the formulas for the midpoint and trapezoidal rules into a single formula for Simpson's rule, by using the fact that $S = (4M + 2T)/6$.

### Simpson's Rule

$$\int_a^b f(x)\,dx \approx [f(a_0) + 4f(x_1) + 2f(a_1) + 4f(x_2) + 2f(a_2)$$

$$+ \cdots + 2f(a_{n-1}) + 4f(x_n) + f(a_n)]\frac{\Delta x}{6}.$$

(4)

**Distribution of IQs**  Psychologists use various standardized tests to measure intelligence. The method most commonly used to describe the results of such tests is an intelligence quotient (or IQ). An IQ is a positive number that, in theory, indicates how a person's mental age compares with the person's chronological age. The median IQ is arbitrarily set at 100, so that half the population has an IQ less than 100 and half greater. IQs are distributed according to a bell-shaped curve called a *normal curve*, pictured in Fig. 3. The proportion of all people having IQs between $A$ and $B$ is given by the area under the curve from $A$ to $B$, that is, by the integral

$$\frac{1}{16\sqrt{2\pi}} \int_A^B e^{-(1/2)[(x-100)/16]^2}\,dx.$$

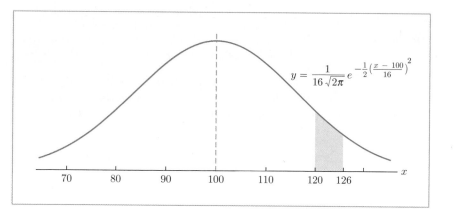

**Figure 3.**  Proportion of IQs between 120 and 126.

**EXAMPLE 4**  Estimate the proportion of all people having IQs between 120 and 126.

**Solution**  We have seen that this proportion is given by

$$\frac{1}{16\sqrt{2\pi}} \int_{120}^{126} f(x)\,dx, \quad \text{where } f(x) = e^{-(1/2)[(x-100)/16]^2}.$$

Let us approximate the definite integral by Simpson's rule with $n = 3$. Then $\Delta x = (126 - 120)/3 = 2$. The endpoints of the subintervals are 120, 122, 124, and

126; the midpoints of these subintervals are 121, 123, and 125. Simpson's rule gives

$$[f(120) + 4f(121) + 2f(122) + 4f(123) + 2f(124) + 4f(125) + f(126)]\frac{\Delta x}{6}$$

$$\approx [.4578 + 1.6904 + .7771 + 1.4235 + .6493 + 1.1801 + .2671]\left(\frac{2}{6}\right)$$

$$\approx 2.1484.$$

Multiplying this estimate by $1/(16\sqrt{2\pi})$, the constant in front of the integral, we get .0536. Thus approximately 5.36% of the population have IQs between 120 and 126. ■

**Geometric Interpretation of the Approximation Rules**   Let $f(x)$ be a continuous nonnegative function on $a \leq x \leq b$. The approximation rules discussed previously may be interpreted as methods for estimating the area under the graph of $f(x)$. The midpoint rule arises from replacing this area by a collection of $n$ rectangles, one lying over each subinterval, the first of height $f(x_1)$, the second of height $f(x_2)$, and so on. (See Fig. 4.)

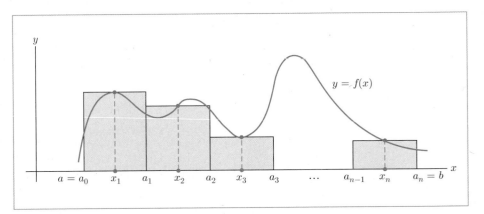

**Figure 4.**   Approximation by rectangles.

If we approximate the area under the graph of $f(x)$ by trapezoids, as in Fig. 5, then the total area of these trapezoids turns out to be the number given by the trapezoidal rule (hence the name of the rule).

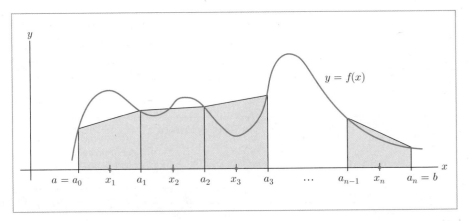

**Figure 5.**   Approximation by trapezoids.

Simpson's rule corresponds to approximating the graph of $f(x)$ on each subinterval by a parabola instead of a straight line as in the midpoint and trapezoidal rules. On each subinterval, the parabola is chosen so that it intersects the graph of $f(x)$ at the midpoint and both endpoints of the subinterval. (See Fig. 6.) It can be shown that the sum of the areas under these parabolas is the number given by Simpson's rule. Even more powerful approximation rules may be obtained by approximating the graph of $f(x)$ on each subinterval by cubic curves, or graphs of higher-order polynomials.

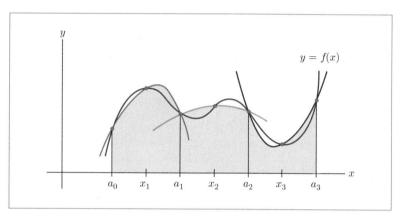

**Figure 6.** Approximation by parabolas.

**Error Analysis**  A simple measure of the error of an approximation to a definite integral is the quantity

$$\big|[\text{approximate value}] - [\text{actual value}]\big|.$$

The following theorem gives an idea of how small this error must be for the various approximation rules. In a concrete example, the actual error of an approximation may be even substantially less than the "error bound" given in the theorem.

---

**Error of Approximation Theorem**     Let $n$ be the number of subintervals used in an approximation of the definite integral

$$\int_a^b f(x)\,dx.$$

1. The error for the midpoint rule is at most $\dfrac{A(b-a)^3}{24n^2}$, where $A$ is a number such that $|f''(x)| \le A$ for all $x$ satisfying $a \le x \le b$.

2. The error for the trapezoidal rule is at most $\dfrac{A(b-a)^3}{12n^2}$, where $A$ is a number such that $|f''(x)| \le A$ for all $x$ satisfying $a \le x \le b$.

3. The error for Simpson's rule is at most $\dfrac{A(b-a)^5}{2880n^4}$, where $A$ is a number such that $|f''''(x)| \le A$ for all $x$ satisfying $a \le x \le b$.

---

**EXAMPLE 5**     Obtain a bound on the error of using the trapezoidal rule with $n = 20$ to approximate

$$\int_0^1 e^{x^2}\,dx.$$

**Solution** Here $a = 0$, $b = 1$, and $f(x) = e^{x^2}$. Differentiating twice, we find that

$$f''(x) = (4x^2 + 2)e^{x^2}.$$

**Figure 7**

How large could $|f''(x)|$ be if $x$ satisfies $0 \le x \le 1$? Since the function $(4x^2 + 2)e^{x^2}$ is clearly increasing on the interval from 0 to 1, its greatest value occurs at $x = 1$. (See Figure 7.) Therefore, its greatest value is

$$(4 \cdot 1^2 + 2)e^{1^2} = 6e,$$

so we may take $A = 6e$ in the preceding theorem. The error of approximation using the trapezoidal rule is at most

$$\frac{6e(1 - 0)^3}{12(20)^2} = \frac{e}{800} \approx \frac{2.71828}{800} \approx .003398. \qquad \blacksquare$$

**Incorporating Technology** The midpoint and trapezoidal rules can be evaluated on graphing calculators with the **sum** and **seq** functions. Specifically, suppose that $\mathbf{Y_1}$ is $f(x)$ and that $c$, $d$, and $\Delta x$ are numbers; then the value of

$$\mathbf{sum(seq(Y_1, X, c, d, \Delta x))} \qquad (5)$$

is

$$f(c) + f(c + \Delta x) + f(c + 2\Delta x) + f(c + 3\Delta x) + \cdots + f(d).$$

Appendices A–D contain the details for (5). Figures 8 and 9 calculate the values for Examples 1 and 2 and store them in the variables M and T. In Fig. 8, **sum(seq** is used to compute the entire total. In Figure 9 the first term is calculated separately, **sum(seq** is used to total the middle three terms, and then the last term is calculated separately. Figure 10 applies Simpson's rule to the same function and shows that it is very close to the highly accurate value given by **fnInt**.

**Figure 8.** Calculation for Example 1.

**Figure 9.** Calculation for Example 2.

**Figure 10.** Simpson's rule.

Graphing calculators also can be used to determine a suitable value for $A$ in the error of approximation theorem by finding the greatest value of either $f''(x)$ or $f''''(x)$. Appendices A–D show how to graph the first four derivatives of a function.

## PRACTICE PROBLEMS 9.4

*Consider* $\displaystyle\int_1^{3.4} (5x - 9)^2 \, dx$.

**1.** Divide the interval $1 \le x \le 3.4$ into three subintervals. List $\Delta x$ and the endpoints and midpoints of the subintervals.

**2.** Approximate the integral by the midpoint rule with $n = 3$.

**3.** Approximate the integral by the trapezoidal rule with $n = 3$.

**4.** Approximate the integral by Simpson's rule with $n = 3$.

**5.** Find the exact value of the integral by integration.

# EXERCISES 9.4

*In Exercises 1 and 2, divide the given interval into n subintervals and list the value of $\Delta x$ and the endpoints $a_0, a_1, \ldots, a_n$ of the subintervals.*

**1.** $3 \le x \le 5$; $n = 5$ **2.** $-1 \le x \le 2$; $n = 5$

*In Exercises 3 and 4, divide the interval into n subintervals and list the value of $\Delta x$ and the midpoints $x_1, \ldots, x_n$ of the subintervals.*

**3.** $-1 \le x \le 1$; $n = 4$ **4.** $0 \le x \le 3$; $n = 6$

**5.** Refer to the graph in Fig. 11. Draw the rectangles that approximate the area under the curve from 0 to 8 when using the midpoint rule with $n = 4$.

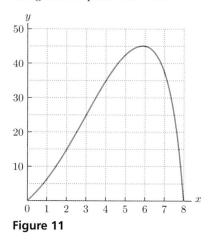

**Figure 11**

**6.** Refer to the graph in Fig. 11. Apply the trapezoidal rule with $n = 4$ to estimate the area under the curve.

*Approximate the following integrals by the midpoint rule, and then find the exact value by integration. Express your answers to five decimal places.*

**7.** $\int_0^4 (x^2 + 5)\, dx$; $n = 2, 4$

**8.** $\int_1^5 (x - 1)^2\, dx$; $n = 2, 4$

**9.** $\int_0^1 e^{-x}\, dx$; $n = 5$ **10.** $\int_1^2 \frac{1}{x+1}\, dx$; $n = 5$

*Approximate the following integrals by the trapezoidal rule, and then find the exact value by integration. Express your answers to five decimal places.*

**11.** $\int_0^1 \left(x - \frac{1}{2}\right)^2\, dx$; $n = 4$

**12.** $\int_4^9 \frac{1}{x-3}\, dx$; $n = 5$

**13.** $\int_1^5 \frac{1}{x^2}\, dx$; $n = 3$ **14.** $\int_{-1}^1 e^{2x}\, dx$; $n = 2, 4$

*Approximate the following integrals by the midpoint rule, the trapezoidal rule, and Simpson's rule. Then find the exact value by integration. Express your answers to five decimal places.*

**15.** $\int_1^4 (2x - 3)^3\, dx$; $n = 3$

**16.** $\int_{10}^{20} \frac{\ln x}{x}\, dx$; $n = 5$

**17.** $\int_0^2 2xe^{x^2}\, dx$; $n = 4$

**18.** $\int_0^3 x\sqrt{4 - x}\, dx$; $n = 5$

**19.** $\int_2^5 xe^x\, dx$; $n = 5$

**20.** $\int_1^5 (4x^3 - 3x^2)\, dx$; $n = 2$

*The following integrals cannot be evaluated in terms of elementary antiderivatives. Find an approximate value by Simpson's rule. Express your answers to five decimal places.*

**21.** $\int_0^2 \sqrt{1 + x^3}\, dx$; $n = 4$

**22.** $\int_0^1 \frac{1}{x^3 + 1}\, dx$; $n = 2$

**23.** $\int_0^2 \sqrt{\sin x}\, dx$; $n = 5$

**24.** $\int_{-1}^1 \sqrt{1 + x^4}\, dx$; $n = 4$

**25.** (*Area*) In a survey of a piece of oceanfront property, measurements of the distance to the water were made every 50 feet along a 200-foot side (Fig. 12). Use the trapezoidal rule to estimate the area of the property.

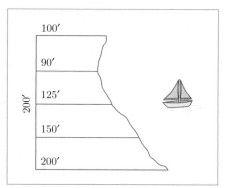

**Figure 12.** Survey of an oceanfront property.

26. (*Area*) In order to determine the amount of water flowing down a certain 100-yard-wide river, engineers need to know the area of a vertical cross section of the river. Measurements of the depth of the river were made every 20 yards from one bank to the other. The readings in fathoms were 0, 1, 2, 3, 1, 0. (One fathom equals 2 yards.) Use the trapezoidal rule to estimate the area of the cross section.

27. (*Distance Traveled*) Upon takeoff, the velocity readings of a rocket noted every second for 10 seconds were 0, 30, 75, 115, 155, 200, 250, 300, 360, 420, and 490 feet per second. Use the trapezoidal rule to estimate the distance the rocket traveled during the first 10 seconds. [*Hint*: If $s(t)$ is the distance traveled by time $t$ and $v(t)$ is the velocity at time $t$, then $s(10) = \int_0^{10} v(t)\,dt$.]

28. (*Distance Traveled*) In a drive along a country road, the speedometer readings are recorded each minute during a 5-minute interval.

| Time (minutes) | 0 | 1 | 2 | 3 | 4 | 5 |
|---|---|---|---|---|---|---|
| Velocity (mph) | 33 | 32 | 28 | 30 | 32 | 35 |

Use the trapezoidal rule to estimate the distance traveled during the 5 minutes. [*Hint*: If time is measured in minutes, then velocity should be expressed in distance per minute. For example, 35 mph is $\frac{35}{60}$ miles per minute. Also, see the hint for Exercise 27.]

29. Consider $\int_0^2 f(x)\,dx$, where $f(x) = \frac{1}{12}x^4 + 3x^2$.
    (a) Make a rough sketch of the graph of $f''(x)$ for $0 \le x \le 2$.
    (b) Find a number $A$ such that $|f''(x)| \le A$ for all $x$ satisfying $0 \le x \le 2$.
    (c) Obtain a bound on the error of using the midpoint rule with $n = 10$ to approximate the definite integral.
    (d) The exact value of the definite integral (to four decimal places) is 8.5333, and the midpoint rule with $n = 10$ gives 8.5089. What is the error for the midpoint approximation? Does this error satisfy the bound obtained in part (c)?
    (e) Redo part (c) with the number of intervals doubled to $n = 20$. Is the bound on the error halved? Quadrupled?

30. Consider $\int_1^2 f(x)\,dx$, where $f(x) = 3\ln x$.
    (a) Make a rough sketch of the graph of the fourth derivative of $f(x)$ for $1 \le x \le 2$.
    (b) Find a number $A$ such that $|f''''(x)| \le A$ for all $x$ satisfying $1 \le x \le 2$.
    (c) Obtain a bound on the error of using Simpson's rule with $n = 2$ to approximate the definite integral.
    (d) The exact value of the definite integral (to four decimal places) is 1.1589, and Simpson's rule with

$n = 2$ gives 1.1588. What is the error for the approximation by Simpson's rule? Does this error satisfy the bound obtained in part (c)?
    (e) Redo part (c) with the number of intervals tripled to $n = 6$. Is the bound on the error divided by three?

31. (a) Show that the area of the trapezoid in Fig. 13(a) is $\frac{1}{2}(h + k)l$. [*Hint*: Divide the trapezoid into a rectangle and a triangle.]
    (b) Show that the area of the first trapezoid on the left in Fig. 13(b) is $\frac{1}{2}[f(a_0) + f(a_1)]\Delta x$.
    (c) Derive the trapezoidal rule for the case $n = 4$.

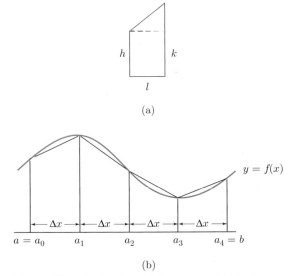

**Figure 13.**   Derivation of the trapezoidal rule.

32. Approximate the value of $\int_a^b f(x)\,dx$, where $f(x) \ge 0$, by dividing the interval $a \le x \le b$ into four subintervals and constructing five rectangles as shown in Fig. 14. Note that the width of the three inside rectangles is $\Delta x$, while the width of the two outside rectangles is $\Delta x/2$. Compute the sum of the areas of these five rectangles and compare this sum with the trapezoidal rule for $n = 4$.

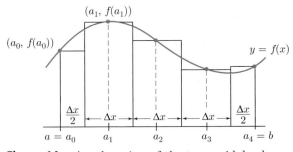

**Figure 14.**   Another view of the trapezoidal rule.

**33. (a)** Suppose that the graph of $f(x)$ is above the $x$-axis and concave down on the interval $a_0 \leq x \leq a_1$. Let $x_1$ be the midpoint of that interval, let $\Delta x = a_1 - a_0$, and construct the line tangent to the graph of $f(x)$ at $(x_1, f(x_1))$, as in Fig. 15(a). Show that the area of the shaded trapezoid in Fig. 15(a) is the same as the area of the shaded rectangle in Fig. 15(c), namely, $f(x_1)\Delta x$. [*Hint*: Look at Fig. 15(b).] This shows that the area of the rectangle in Fig. 15(c) exceeds the area under the graph of $f(x)$ on the interval $a_0 \leq x \leq a_1$.

**(b)** Suppose that the graph of $f(x)$ is above the $x$-axis and concave down for all $x$ in the interval $a \leq x \leq b$. Explain why $T \leq \int_a^b f(x)\, dx \leq M$, where $T$ and $M$ are the approximations given by the trapezoidal and midpoint rules, respectively.

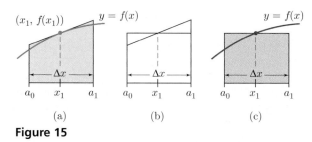

(a)  (b)  (c)

**Figure 15**

**34.** (*Riemann Sum Derivation of Formula for Cardiac Output*, see Example 3.) Subdivide the interval $0 \leq t \leq 22$ into $n$ subintervals of length $\Delta t = 22/n$ seconds. Let $t_i$ be a point in the $i$th subinterval.

**(a)** Show that $(R/60)\Delta t \approx$ [number of liters of blood flowing past the monitoring point during the $i$th time interval].

**(b)** Show that $c(t_i)(R/60)\Delta t \approx$ [quantity of dye flowing past the monitoring point during the $i$th time interval].

**(c)** Assume that basically all of the dye will have flowed past the monitoring point during the 22 seconds. Explain why $D \approx (R/60)[c(t_1) + c(t_2) + \cdots + c(t_n)]\Delta t$, where the approximation improves as $n$ gets large.

**(d)** Conclude that $D = \int_0^{22} (R/60)c(t)\, dt$, and solve for $R$.

## Technology Exercises

**35.** In Fig. 16 a definite integral of the form $\int_a^b f(x)\, dx$ is approximated by the midpoint rule. Determine $f(x)$, $a$, $b$, and $n$.

**Figure 16**

**36.** In Fig. 17 a definite integral of the form $\int_a^b f(x)\, dx$ is approximated by the trapezoidal rule. Determine $f(x)$, $a$, $b$, and $n$.

**Figure 17**

*In Exercises 37–40, approximate the integrals by the midpoint rule, the trapezoidal rule, and Simpson's rule with $n = 10$. Then find the exact value by integration and give the error for each of the approximations. Express your answers to the full accuracy given by the calculator or computer.*

**37.** $\displaystyle\int_1^{11} \frac{1}{x}\, dx$

**38.** $\displaystyle\int_0^{\pi/2} \cos x\, dx$

**39.** $\displaystyle\int_0^{\pi/4} \sec^2 x\, dx$

**40.** $\displaystyle\int_0^1 2x e^{x^2}\, dx$

*In Exercises 41 and 42, consider the definite integral $\displaystyle\int_0^1 \frac{4}{1+x^2}\, dx$, which has the value $\pi$.*

**41.** Suppose the midpoint rule is used with $n = 20$ in order to estimate $\pi$. Graph the second derivative of the function in the window $[0, 1]$ by $[-10, 10]$ and then use the graph to obtain a bound on the error of the estimate.

**42.** Suppose the trapezoidal rule is used with $n = 15$ in order to estimate $\pi$. Graph the second derivative of the function in the window $[0, 1]$ by $[-10, 10]$ and then use the graph to obtain a bound on the error of the estimate.

## SOLUTIONS TO PRACTICE PROBLEMS 9.4

1. $\Delta x = (3.4 - 1)/3 = 2.4/3 = .8$. Each subinterval will have length .8. A good way to proceed is first to draw two hatchmarks that subdivide the interval into three equal subintervals [Fig. 18(a)]. Then label the hatchmarks by successively adding .8 to the left endpoint [Fig. 18(b)]. The first midpoint can be obtained by adding one-half of .8 to the left endpoint. Then add .8 to get the next midpoint, and so on [Fig. 18(c)].

2. The midpoint rule uses only the midpoints of the subintervals:

$$\int_1^{3.4} (5x - 9)^2 \, dx \approx \left\{ (5[1.4] - 9)^2 + (5[2.2] - 9)^2 \right.$$
$$\left. + (5[3] - 9)^2 \right\} (.8)$$
$$= \left\{ (-2)^2 + 2^2 + 6^2 \right\} (.8)$$
$$= 35.2.$$

3. The trapezoidal rule uses only the endpoints of the subintervals:

$$\int_1^{3.4} (5x - 9)^2 \, dx \approx \left\{ (5[1] - 9)^2 + 2(5[1.8] - 9)^2 \right.$$
$$+ 2(5[2.6] - 9)^2$$
$$\left. + (5[3.4] - 9)^2 \right\} \left( \frac{.8}{2} \right)$$
$$= \left\{ (-4)^2 + 2(0)^2 + 2(4)^2 + 8^2 \right\} (.4)$$
$$= 44.8.$$

4. Using the formula $S = \dfrac{2M + T}{3}$, we obtain

$$\int_1^{3.4} (5x - 9)^2 \, dx \approx \frac{2(35.2) + 44.8}{3} = \frac{115.2}{3} = 38.4.$$

This approximation may also be obtained directly with formula (4), but the arithmetic requires about the same effort as calculating the midpoint and trapezoidal approximations separately and then combining them, as we did here.

5. $$\int_1^{3.4} (5x - 9)^2 \, dx = \frac{1}{15} (5x - 9)^3 \bigg|_1^{3.4}$$
$$= \frac{1}{15} \left[ 8^3 - (-4)^3 \right]$$
$$= 38.4.$$

(Notice that here Simpson's rule gives the exact answer. This was so since the function to be integrated was a quadratic polynomial. Actually, Simpson's rule gives the exact value of the definite integral of any polynomial of degree 3 or less. The reason for this can be discovered from the "error of approximation theorem.")

(a)          (b)          (c)

**Figure 18**

## 9.5 Some Applications of the Integral

Recall that the integral

$$\int_a^b f(t) \, dt$$

can be approximated by a Riemann sum as follows: We divide the $t$-axis from $a$ to $b$ into $n$ subintervals by adding intermediate points $t_0 = a, t_1, \ldots, t_{n-1}, t_n = b$.

$$t_0 = a \quad t_1 \quad t_2 \quad t_3 \quad t_4 \quad t_5 \quad \cdots \quad t_{n-1} \quad t_n = b$$

We assume that the points are equally spaced, so that each subinterval has length $\Delta t = (b - a)/n$. For large $n$, the integral is very closely approximated by the

Riemann sum

$$\int_a^b f(t)\,dt \approx f(t_1)\Delta t + f(t_2)\Delta t + \cdots + f(t_n)\Delta t. \tag{1}$$

The approximation in (1) works both ways. If we encounter a Riemann sum like the one in (1), we can approximate it by the corresponding integral. The approximation becomes better as the number of subintervals increases—that is, as $n$ gets large. Thus, as $n$ gets large, the sum approaches the value of the integral. This will be our approach in the examples that follow.

Our first two examples involve the concept of the present value of money. Suppose that we make an investment that promises to repay $A$ dollars at time $t$ (measuring the present as time 0). How much should we be willing to pay for such an investment? Clearly, we would not want to pay as much as $A$ dollars. For if we had $A$ dollars now, we could invest it at the current rate of interest and, at time $t$, we would get back our original $A$ dollars plus the accrued interest. Instead, we should only be willing to pay an amount $P$ that, if invested for $t$ years, would yield $A$ dollars. We call $P$ the *present value of $A$ dollars in $t$ years*. We shall assume continuous compounding of interest. If the current (annual) rate of interest is $r$, then $P$ dollars invested for $t$ years will yield $Pe^{rt}$ dollars (see Section 5.2). That is,

$$Pe^{rt} = A.$$

Thus the formula for the present value of $A$ dollars in $t$ years at interest rate $r$ is

$$P = Ae^{-rt}.$$

---

**EXAMPLE 1**    **Present value of an income stream**    Consider a small printing company that does most of its work on one printing press. The firm's profits are directly influenced by the amount of material that the press can produce (assuming that other factors, such as wages, are held constant). We may say that the press is producing a continuous stream of income for the company. Of course, the efficiency of the press may decline as it gets older. At time $t$, let $K(t)$ be the annual rate of income from the press. [This means that the press is producing $K(t) \cdot \frac{1}{365}$ dollars per day at time $t$.] Find a model for the present value of the income generated by the printing press over the next $T$ years, assuming an interest rate $r$ (with interest compounded continuously).

**Solution**    Let us divide the $T$-year period into $n$ small subintervals of time, each of duration $\Delta t$ years. (If each subinterval were 1 day, for example, then $\Delta t$ would equal $\frac{1}{365}$.)

We now consider the income produced by the printing press during a small time interval from $t_{j-1}$ to $t_j$. Since $\Delta t$ is small, the rate $K(t)$ of income production changes by only a negligible amount in that interval and can be considered approximately equal to $K(t_j)$. Since $K(t_j)$ gives an annual rate of income, the actual income produced during the period of $\Delta t$ years is $K(t_j)\Delta t$. This income will be produced at approximately time $t_j$ (i.e., $t_j$ years from $t = 0$), so its present value is

$$[K(t_j)\Delta t]e^{-rt_j}.$$

The present value of the total income produced over the $T$-year period is the sum of the present values of the amounts produced during each time subinterval; that is,

$$K(t_1)e^{-rt_1}\Delta t + K(t_2)e^{-rt_2}\Delta t + \cdots + K(t_n)e^{-rt_n}\Delta t. \tag{2}$$

As the number of subintervals gets large, the length $\Delta t$ of each subinterval becomes small, and the sum in (2) approaches the integral

$$\int_0^T K(t)e^{-rt}\, dt. \tag{3}$$

We call this quantity the *present value of the stream of income* produced by the printing press over the period from $t = 0$ to $t = T$ years. (The interest rate $r$ used to compute present value is often called the company's *internal rate of return*.) ■

The concept of the present value of a continuous stream of income is an important tool in management decision processes involving the selection or replacement of equipment. It is also useful when analyzing various investment opportunities. Even when $K(t)$ is a simple function, the evaluation of the integral in (3) usually requires special techniques, such as integration by parts, as we see in the next example.

**EXAMPLE 2** A company estimates that the rate of revenue produced by a machine at time $t$ will be $5000 - 100t$ dollars per year. Find the present value of this continuous stream of income over the next 4 years at a 16% interest rate.

**Solution** We use (3) with $K(t) = 5000 - 100t$, $T = 4$, and $r = .16$. The present value of this income stream is

$$\int_0^4 (5000 - 100t)e^{-.16t}\, dt.$$

Using integration by parts, with $f(t) = 5000 - 100t$ and $g(t) = e^{-.16t}$, we find that the preceding integral equals

$$(5000 - 100t)\frac{1}{-.16}e^{-.16t}\Big|_0^4 - \int_0^4 (-100)\frac{1}{-.16}e^{-.16t}\, dt$$

$$\approx 16{,}090 - \frac{100}{.16}\cdot\frac{1}{-.16}e^{-.16t}\Big|_0^4$$

$$\approx 16{,}090 - 1847 = \$14{,}243.$$ ■

A slight modification of the discussion in Example 1 gives the following general result.

> **Present Value of a Continuous Stream of Income**
>
> $$[\text{present value}] = \int_{T_1}^{T_2} K(t)e^{-rt}\, dt,$$
>
> where
> 1. $K(t)$ dollars per year is the annual rate of income at time $t$,
> 2. $r$ is the annual interest rate of invested money,
> 3. $T_1$ to $T_2$ (years) is the time period of the income stream.

**EXAMPLE 3**    **A demographic model**  It has been determined[†] that in 1940 the population density $t$ miles from the center of New York City was approximately $120e^{-.2t}$ thousand people per square mile. Estimate the number of people who lived within 2 miles of the center of New York in 1940.

Solution    Let us choose a fixed line emanating from the center of the city along which to measure distance. Subdivide this line from $t = 0$ to $t = 2$ into a large number of subintervals, each of length $\Delta t$. Each subinterval determines a ring. (See Fig. 1.)

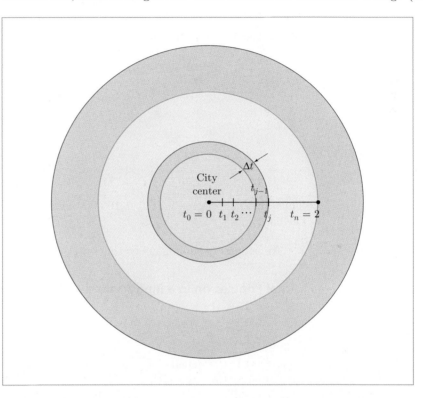

**Figure 1.**  A ring around the city center.

Let us estimate the population of each ring and then add these populations together to obtain the total population. Suppose that $j$ is an index ranging from 1 to $n$. If the outer circle of the $j$th ring is at a distance $t_j$ from the center, the inner circle of that ring is at a distance $t_{j-1} = t_j - \Delta t$ from the center. The area of the $j$th ring is

$$\pi t_j^2 - \pi t_{j-1}^2 = \pi t_j^2 - \pi (t_j - \Delta t)^2$$
$$= \pi t_j^2 - \pi [t_j^2 - 2t_j \Delta t + (\Delta t)^2]$$
$$= 2\pi t_j \Delta t - \pi (\Delta t)^2.$$

Assume that $\Delta t$ is very small. Then $\pi(\Delta t)^2$ is much smaller than $2\pi t_j \Delta t$. Hence the area of this ring is very close to $2\pi t_j \Delta t$.

Within the $j$th ring, the density of people is about $120e^{-.2t_j}$ thousand people per square mile. So the number of people in this ring is approximately

$$[\text{population density}] \cdot [\text{area of ring}] \approx 120e^{-.2t_j} \cdot 2\pi t_j \Delta t$$
$$= 240\pi t_j e^{-.2t_j} \Delta t.$$

---

[†]C. Clark, "Urban Population Densities," *Journal of the Royal Statistical Society*, Series A, 114 (1951), 490–496. See also M. J. White, "On Cumulative Urban Growth and Urban Density Functions," *Journal of Urban Economics*, 4 (1977), 104–112.

Adding up the populations of all the rings, we obtain a total of

$$240\pi t_1 e^{-.2t_1}\Delta t + 240\pi t_2 e^{-.2t_2}\Delta t + \cdots + 240\pi t_n e^{-.2t_n}\Delta t,$$

which is a Riemann sum for the function $f(t) = 240t\pi e^{-.2t}$ over the interval from $t = 0$ to $t = 2$. This approximation to the population improves as the number $n$ increases. Thus, the number of people (in thousands) who lived within 2 miles of the center of the city was

$$\int_0^2 240\pi t e^{-.2t}\,dt = 240\pi \int_0^2 t e^{-.2t}\,dt.$$

The last integral can be computed using integration by parts to obtain

$$240\pi \int_0^2 t e^{-.2t}\,dt = 240\pi \frac{t e^{-.2t}}{-.2}\Big|_0^2 - 240\pi \int_0^2 \frac{e^{-.2t}}{-.2}\,dt$$

$$= -2400\pi e^{-.4} + 1200\pi \left(\frac{e^{-.2t}}{-.2}\right)\Big|_0^2$$

$$= -2400\pi e^{-.4} + (-6000\pi e^{-.4} + 6000\pi)$$

$$\approx 1160.$$

Thus, in 1940, approximately 1,160,000 people lived within 2 miles of the center of the city. ■

An argument analogous to that in Example 3 leads to the following result.

---

**Total Population in a Ring Around the City Center**

$$[\text{population}] = \int_a^b 2\pi t D(t)\,dt,$$

where

1. $D(t)$ is the density of population (in persons per square mile) at distance $t$ miles from city center,

2. Ring includes all persons who live between $a$ and $b$ miles from the city center.

---

# EXERCISES 9.5

1. Find the present value of a continuous stream of income over 5 years when the rate of income is constant at \$35,000 per year and the interest rate is 7%.

2. A continuous stream of income is being produced at the constant rate of \$60,000 per year. Find the present value of the income generated during the time from $t = 2$ to $t = 6$ years, with a 6% interest rate.

3. Find the present value of a continuous stream of income over the time from $t = 1$ to $t = 5$ years when the interest rate is 10% and the income is produced at the rate of \$12,000 per year.

4. Find the present value of a continuous stream of income over 4 years if the rate of income is $25e^{-.02t}$ thousand dollars per year at time $t$ and the interest rate is 8%.

5. Find the present value of a continuous stream of income over 3 years if the rate of income is $80e^{-.08t}$ thousand dollars per year at time $t$ and the interest rate is 11%.

6. A continuous stream of income is produced at the rate of $20e^{1-.09t}$ thousand dollars per year at time $t$, and invested money earns 6% interest.

   (a) Write a definite integral that gives the present value of this stream of income over the time from $t = 2$ to $t = 5$ years.

   (b) Compute the present value described in part (a).

7. A "growth company" is one whose net earnings tend to increase each year. Suppose that the net earnings of a company at time $t$ are being generated at the rate of $30 + 5t$ million dollars per year.

   (a) Write a definite integral that gives the present value of the company's earnings over the next 2 years, using a 10% interest rate.

**(b)** Compute the present value described in part (a).

**8.** Find the present value of a stream of earnings generated over the next 2 years at the rate of $50 + 7t$ thousand dollars per year at time $t$, assuming a 10% interest rate.

**9.** (*Demographic Model*) In 1900 the population density of Philadelphia $t$ miles from the city center was $120e^{-.65t}$ thousand people per square mile.

**(a)** Write a definite integral whose value equals the number of people (in thousands) who lived within 5 miles of the city center.

**(b)** Calculate the definite integral in part (a).

**10.** Use the population density from Exercise 9 to calculate the number of people who lived between 3 and 5 miles of the city center.

**11.** The population density of Philadelphia in 1940 was given by the function $60e^{-.4t}$. Calculate the number of people who lived within 5 miles of the city center. Sketch the graphs of the population densities for 1900 and 1940 (see Exercise 9) on a common graph. What trend do the graphs exhibit?

**12.** (*Ecological Model*) A volcano erupts and spreads lava in all directions. The density of the deposits at a distance $t$ kilometers from the center is $D(t)$ thousand tons per square kilometer, where $D(t) = 11(t^2 + 10)^{-2}$. Find the tonnage of lava deposited between the distances of 1 and 10 kilometers from the center.

**13.** Suppose the population density function for a city is $40e^{-.5t}$ thousand people per square mile. Let $P(t)$ be the total population that lives within $t$ miles of the city center, and let $\Delta t$ be a small positive number.

**(a)** Consider the ring about the city whose inner circle is at $t$ miles and outer circle is at $t + \Delta t$ miles. The text shows that the area of this ring is approximately $2\pi t \Delta t$ square miles. Approximately how many people live within this ring? (Your answer will involve $t$ and $\Delta t$.)

**(b)** What does

$$\frac{P(t + \Delta t) - P(t)}{\Delta t}$$

approach as $\Delta t$ tends to zero?

**(c)** What does the quantity $P(5 + \Delta t) - P(5)$ represent?

**(d)** Use (a) and (c) to find a formula for

$$\frac{P(t + \Delta t) - P(t)}{\Delta t}$$

and from that obtain an approximate formula for the derivative $P'(t)$. This formula gives the rate of change of total population with respect to the distance $t$ from the city center.

**(e)** Given two positive numbers $a$ and $b$, find a formula, involving a definite integral, for the number of people who live in the city between $a$ miles and $b$ miles of the city center. [*Hint*: Use (d) and the fundamental theorem of calculus to compute $P(b) - P(a)$.]

## 9.6 Improper Integrals

In applications of calculus, especially to statistics, it is often necessary to consider the area of a region that extends infinitely far to the right or left along the $x$-axis. We have drawn several such regions in Fig. 1. Note that the area under a curve that extends infinitely far along the $x$-axis is not necessarily infinite itself. The areas of such "infinite" regions may be computed using *improper integrals*.

In order to motivate the idea of an improper integral, let us attempt to calculate the area under the curve $y = 3/x^2$ to the right of $x = 1$ (Fig. 2).

**Figure 1**

**Figure 2**

First, we shall compute the area under the graph of this function from $x = 1$ to $x = b$, where $b$ is some number greater than 1. [See Fig. 3(a).] Then we shall examine how the area increases as we let $b$ get larger, as in Figs. 3(b) and 3(c). The area from 1 to $b$ is given by

$$\int_1^b \frac{3}{x^2}\, dx = -\frac{3}{x}\bigg|_1^b = \left(-\frac{3}{b}\right) - \left(-\frac{3}{1}\right) = 3 - \frac{3}{b}.$$

When $b$ is large, $3/b$ is small and the integral nearly equals 3. That is, the area under the curve from 1 to $b$ nearly equals 3. (See Table 1.) In fact, the area gets arbitrarily close to 3 as $b$ gets larger. Thus it is reasonable to say that the region under the curve $y = 3/x^2$ for $x \geq 1$ has area 3.

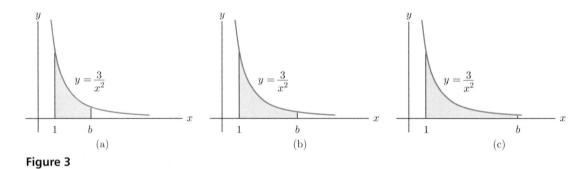

**Figure 3**

| TABLE 1 | Value of an "Infinitely Long" Area as a Limit |
|---|---|
| $b$ | $Area = \int_1^b \dfrac{3}{x^2}\, dx = 3 - \dfrac{3}{b}$ |
| 10 | 2.7000 |
| 100 | 2.9700 |
| 1,000 | 2.9970 |
| 10,000 | 2.9997 |

Recall from Chapter 1 that we write $b \to \infty$ as shorthand for "$b$ gets arbitrarily large, without bound." Then, to express the fact that the value of

$$\int_1^b \frac{3}{x^2}\, dx$$

approaches 3 as $b \to \infty$, we write

$$\int_1^\infty \frac{3}{x^2}\, dx = \lim_{b\to\infty} \int_1^b \frac{3}{x^2}\, dx = 3.$$

We call $\displaystyle\int_1^\infty \frac{3}{x^2}\, dx$ an *improper* integral because the upper limit of the integral is $\infty$ (infinity) rather than a finite number.

> **DEFINITION**  Let $a$ be fixed and suppose that $f(x)$ is a nonnegative function for $x \geq a$. If $\lim_{b \to \infty} \int_a^b f(x)\,dx = L$, we define
>
> $$\int_a^\infty f(x)\,dx = \lim_{b \to \infty} \int_a^b f(x)\,dx = L.$$
>
> We say that the improper integral $\int_a^\infty f(x)\,dx$ is *convergent* and that the region under the curve $y = f(x)$ for $x \geq a$ has area $L$. (See Fig. 4.)

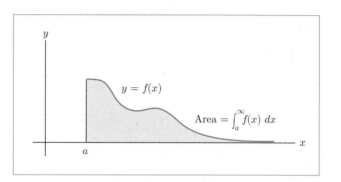

**Figure 4.**  Area defined by an improper integral.

It is possible to consider improper integrals in which $f(x)$ is both positive and negative. However, we shall consider only nonnegative functions, since this is the case occurring in most applications.

---

**EXAMPLE 1**  Find the area under the curve $y = e^{-x}$ for $x \geq 0$ (Fig. 5).

**Solution**  We must calculate the improper integral

$$\int_0^\infty e^{-x}\,dx.$$

We take $b > 0$ and compute

$$\int_0^b e^{-x}\,dx = -e^{-x}\Big|_0^b = (-e^{-b}) - (-e^0) = 1 - e^{-b} = 1 - \frac{1}{e^b}.$$

We now consider the limit as $b \to \infty$ and note that $1/e^b$ approaches zero. Thus

$$\int_0^\infty e^{-x}\,dx = \lim_{b \to \infty} \int_0^b e^{-x}\,dx = \lim_{b \to \infty}\left(1 - \frac{1}{e^b}\right) = 1.$$

**Figure 5**

Therefore, the region in Fig. 5 has area 1.  ■

---

**EXAMPLE 2**  Evaluate the improper integral

$$\int_7^\infty \frac{1}{(x-5)^2}\,dx.$$

**Solution**  $\displaystyle \int_7^b \frac{1}{(x-5)^2}\,dx = -\frac{1}{x-5}\Big|_7^b = -\frac{1}{b-5} - \left(-\frac{1}{7-5}\right) = \frac{1}{2} - \frac{1}{b-5}.$

As $b \to \infty$, the fraction $1/(b-5)$ approaches zero, so

$$\int_7^\infty \frac{1}{(x-5)^2} \, dx = \lim_{b \to \infty} \int_7^b \frac{1}{(x-5)^2} \, dx = \lim_{b \to \infty} \left( \frac{1}{2} - \frac{1}{b-5} \right) = \frac{1}{2}. \qquad \blacksquare$$

Not every improper integral is convergent. If the value $\int_a^b f(x) \, dx$ does not have a limit as $b \to \infty$, we cannot assign any numerical value to $\int_a^\infty f(x) \, dx$, and we say that the improper integral $\int_a^\infty f(x) \, dx$ is *divergent*.

---

**EXAMPLE 3** Show that

$$\int_1^\infty \frac{1}{\sqrt{x}} \, dx$$

is divergent.

**Solution** For $b > 1$ we have

$$\int_1^b \frac{1}{\sqrt{x}} \, dx = 2\sqrt{x} \Big|_1^b = 2\sqrt{b} - 2.$$

As $b \to \infty$, the quantity $2\sqrt{b} - 2$ increases without bound. That is, $2\sqrt{b} - 2$ can be made larger than any specific number. Therefore, $\int_1^b \frac{1}{\sqrt{x}} \, dx$ has no limit as $b \to \infty$, so $\int_1^\infty \frac{1}{\sqrt{x}} \, dx$ is divergent. $\qquad \blacksquare$

In some cases it is necessary to consider improper integrals of the form

$$\int_{-\infty}^b f(x) \, dx.$$

Let $b$ be fixed and examine the value of $\int_a^b f(x) \, dx$ as $a \to -\infty$, that is, as $a$ moves arbitrarily far to the left on the number line. If $\lim_{a \to -\infty} \int_a^b f(x) \, dx = L$, we say that the improper integral $\int_{-\infty}^b f(x) \, dx$ is *convergent* and we write

$$\int_{-\infty}^b f(x) \, dx = L.$$

Otherwise, the improper integral is divergent. An integral of the form $\int_{-\infty}^b f(x) \, dx$ may be used to compute the area of a region such as that shown in Fig. 1(b).

---

**EXAMPLE 4** Determine if $\int_{-\infty}^0 e^{5x} \, dx$ is convergent. If convergent, find its value.

**Solution** $$\int_{-\infty}^0 e^{5x} \, dx = \lim_{a \to -\infty} \int_a^0 e^{5x} \, dx = \lim_{a \to -\infty} \frac{1}{5} e^{5x} \Big|_a^0 = \lim_{a \to -\infty} \left( \frac{1}{5} - \frac{1}{5} e^{5a} \right)$$

As $a \to -\infty$, $e^{5a}$ approaches 0 so that $\frac{1}{5} - \frac{1}{5} e^{5a}$ approaches $\frac{1}{5}$. Thus the improper integral converges and has value $\frac{1}{5}$. $\qquad \blacksquare$

Areas of regions that extend infinitely far to the left *and* right, such as the region in Fig. 1(c), are calculated using improper integrals of the form

$$\int_{-\infty}^\infty f(x) \, dx.$$

We define such an integral to have the value

$$\int_{-\infty}^{0} f(x)\,dx + \int_{0}^{\infty} f(x)\,dx,$$

provided that both of the latter improper integrals are convergent.

An important area that arises in probability theory is the area under the so-called normal curve, whose equation is

$$y = \frac{1}{\sqrt{2\pi}} e^{-x^2/2}.$$

(See Fig. 6.) It is of fundamental importance for probability theory that this area is 1. In terms of an improper integral, this fact may be written

$$\int_{-\infty}^{\infty} \frac{1}{\sqrt{2\pi}} e^{-x^2/2}\,dx = 1.$$

The proof of this result is beyond the scope of this book.

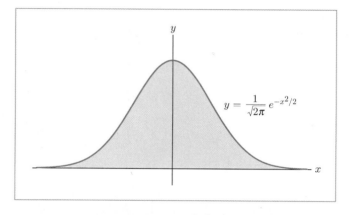

**Figure 6.**  The standard normal curve.

**Incorporating Technology**  Although graphing calculators cannot tell you whether or not an improper integral converges, you can use the calculator to obtain a reliable indication of the behavior of the integral. Just look at values of $\int_{a}^{b} f(x)\,dx$ as $b$ increases. Figures 7(a) and 7(b), which were created by setting $\mathbf{Y_1 = fnInt(e^{\wedge}(-X),X,0,X)}$, give convincing evidence that the value of the improper integral in Example 1 is 1. The final $\mathbf{X}$ in $\mathbf{fnInt(e^{\wedge}(-X),X,0,X)}$ is the upper limit of integration, that is, $b$.

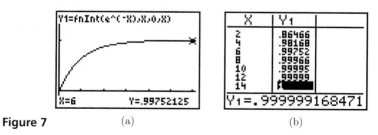

**Figure 7**            (a)                              (b)

## PRACTICE PROBLEMS 9.6

**1.** Does $1 - 2(1 - 3b)^{-4}$ approach a limit as $b \to \infty$?

**2.** Evaluate $\displaystyle\int_{1}^{\infty} \frac{x^2}{x^3 + 8}\,dx.$

**3.** Evaluate $\displaystyle\int_{-\infty}^{-2} \frac{1}{x^4}\,dx.$

# EXERCISES 9.6

*In Exercises 1–12, determine if the given expression approaches a limit as $b \to \infty$, and find that number when it does.*

**1.** $\dfrac{5}{b}$   **2.** $b^2$   **3.** $-3e^{2b}$

**4.** $\dfrac{1}{b} + \dfrac{1}{3}$   **5.** $\dfrac{1}{4} - \dfrac{1}{b^2}$   **6.** $\dfrac{1}{2}\sqrt{b}$

**7.** $2 - (b+1)^{-1/2}$   **8.** $5 - (b-1)^{-1}$   **9.** $5(b^2+3)^{-1}$

**10.** $4(1 - b^{-3/4})$   **11.** $e^{-b/2} + 5$   **12.** $2 - e^{-3b}$

**13.** Find the area under the graph of $y = 1/x^2$ for $x \geq 2$.

**14.** Find the area under the graph of $y = (x+1)^{-2}$ for $x \geq 0$.

**15.** Find the area under the graph of $y = e^{-x/2}$ for $x \geq 0$.

**16.** Find the area under the graph of $y = 4e^{-4x}$ for $x \geq 0$.

**17.** Find the area under the graph of $y = (x+1)^{-3/2}$ for $x \geq 3$.

**18.** Find the area under the graph of $y = (2x+6)^{-4/3}$ for $x \geq 1$. See Fig. 8.

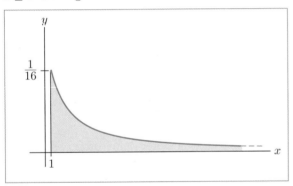

**Figure 8**

**19.** Show that the region under the graph of $y = (14x + 18)^{-4/5}$ for $x \geq 1$ cannot be assigned any finite number as its area. See Fig. 9.

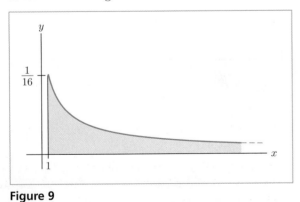

**Figure 9**

**20.** Show that the region under the graph of $y = (x-1)^{-1/3}$ for $x \geq 2$ cannot be assigned any finite number as its area.

*Evaluate the following improper integrals whenever they are convergent.*

**21.** $\displaystyle\int_1^\infty \dfrac{1}{x^3}\, dx$   **22.** $\displaystyle\int_1^\infty \dfrac{2}{x^{3/2}}\, dx$

**23.** $\displaystyle\int_0^\infty \dfrac{1}{(2x+3)^2}\, dx$   **24.** $\displaystyle\int_0^\infty e^{-3x}\, dx$

**25.** $\displaystyle\int_0^\infty e^{2x}\, dx$   **26.** $\displaystyle\int_0^\infty (x^2+1)\, dx$

**27.** $\displaystyle\int_2^\infty \dfrac{1}{(x-1)^{5/2}}\, dx$   **28.** $\displaystyle\int_2^\infty e^{2-x}\, dx$

**29.** $\displaystyle\int_0^\infty .01e^{-.01x}\, dx$   **30.** $\displaystyle\int_0^\infty \dfrac{4}{(2x+1)^3}\, dx$

**31.** $\displaystyle\int_0^\infty 6e^{1-3x}\, dx$   **32.** $\displaystyle\int_1^\infty e^{.2x}\, dx$

**33.** $\displaystyle\int_3^\infty \dfrac{x^2}{\sqrt{x^3-1}}\, dx$   **34.** $\displaystyle\int_2^\infty \dfrac{1}{x \ln x}\, dx$

**35.** $\displaystyle\int_0^\infty xe^{-x^2}\, dx$   **36.** $\displaystyle\int_0^\infty \dfrac{x}{x^2+1}\, dx$

**37.** $\displaystyle\int_0^\infty 2x(x^2+1)^{-3/2}\, dx$   **38.** $\displaystyle\int_1^\infty (5x+1)^{-4}\, dx$

**39.** $\displaystyle\int_{-\infty}^0 e^{4x}\, dx$   **40.** $\displaystyle\int_{-\infty}^0 \dfrac{8}{(x-5)^2}\, dx$

**41.** $\displaystyle\int_{-\infty}^0 \dfrac{6}{(1-3x)^2}\, dx$   **42.** $\displaystyle\int_{-\infty}^0 \dfrac{1}{\sqrt{4-x}}\, dx$

**43.** $\displaystyle\int_0^\infty \dfrac{e^{-x}}{(e^{-x}+2)^2}\, dx$   **44.** $\displaystyle\int_{-\infty}^\infty \dfrac{e^{-x}}{(e^{-x}+2)^2}\, dx$

**45.** If $k > 0$, show that $\displaystyle\int_0^\infty ke^{-kx}\, dx = 1$.

**46.** If $k > 0$, show that $\displaystyle\int_1^\infty \dfrac{k}{x^{k+1}}\, dx = 1$.

**47.** If $k > 0$, show that $\displaystyle\int_e^\infty \dfrac{k}{x(\ln x)^{k+1}}\, dx = 1$.

*The capital value of an asset such as a machine is sometimes defined as the present value of all future net earnings. (See Section 9.5.) The actual lifetime of the asset may not be known, and since some assets may last indefinitely, the capital value of the asset may be written in the form*

$$[\text{capital value}] = \int_0^\infty K(t)e^{-rt}\, dt,$$

*where $r$ is the annual rate of interest, compounded continuously.*

**48.** Find the capital value of an asset that generates income at the rate of \$5000 per year, assuming an interest rate of 10%.

**49.** Construct a formula for the capital value of a rental property that will generate a fixed income at the rate of $K$ dollars per year indefinitely, assuming an annual interest rate of $r$.

**50.** Suppose that a large farm with a known reservoir of gas beneath the ground sells the gas rights to a company for a guaranteed payment at the rate of $10,000e^{.04t}$ dollars per year. Find the present value of this perpetual stream of income, assuming an interest rate of 12%, compounded continuously.

## SOLUTIONS TO PRACTICE PROBLEMS 9.6

**1.** The expression $1 - 2(1 - 3b)^{-4}$ may also be written in the form

$$1 - \frac{2}{(1 - 3b)^4}.$$

When $b$ is large, $(1 - 3b)^4$ is very large, so $2/(1 - 3b)^4$ is very small. Thus $1 - 2(1 - 3b)^{-4}$ approaches 1 as $b \to \infty$.

**2.** The first step is to find an antiderivative of $x^2/(x^3 + 8)$. Using the substitution $u = x^3 + 8$, $du = 3x^2\, dx$, we obtain

$$\int \frac{x^2}{x^3 + 8}\, dx = \frac{1}{3} \int \frac{1}{u}\, du$$

$$= \frac{1}{3} \ln|u| + C$$

$$= \frac{1}{3} \ln|x^3 + 8| + C.$$

Now,

$$\int_1^b \frac{x^2}{x^3 + 8}\, dx = \frac{1}{3} \ln|x^3 + 8|\Big|_1^b$$

$$= \frac{1}{3} \ln(b^3 + 8) - \frac{1}{3} \ln 9.$$

Finally, we examine what happens as $b \to \infty$. Certainly, $b^3 + 8$ gets arbitrarily large, so $\ln(b^3 + 8)$ must

also get arbitrarily large. Hence

$$\int_1^b \frac{x^2}{x^3 + 8}\, dx$$

has no limit as $b \to \infty$, so the improper integral

$$\int_1^\infty \frac{x^2}{x^3 + 8}\, dx$$

is divergent.

**3.** $\displaystyle \int_a^{-2} \frac{1}{x^4}\, dx = \int_a^{-2} x^{-4}\, dx$

$$= \frac{x^{-3}}{-3}\Big|_a^{-2} = \frac{1}{-3x^3}\Big|_a^{-2}$$

$$= \frac{1}{-3(-2)^3} - \left(\frac{1}{-3 \cdot a^3}\right)$$

$$= \frac{1}{24} + \frac{1}{3a^3}$$

$$\int_{-\infty}^{-2} \frac{1}{x^4}\, dx = \lim_{a \to -\infty} \int_a^{-2} \frac{1}{x^4}\, dx$$

$$= \lim_{a \to -\infty} \left(\frac{1}{24} + \frac{1}{3a^3}\right) = \frac{1}{24}$$

## REVIEW OF FUNDAMENTAL CONCEPTS

**1.** Describe integration by substitution in your own words.

**2.** Describe integration by parts in your own words.

**3.** Describe the change of limits rule for the integration by substitution of a definite integral.

**4.** State the formula for the integration by parts of a definite integral.

**5.** State the midpoint rule. (Include the meaning of all symbols used.)

**6.** State the trapezoidal rule. (Include the meaning of all symbols used.)

**7.** Explain the formula $S = \dfrac{2M + T}{3}$.

**8.** State the error of approximation theorem for each of the three approximation rules.

**9.** State the formula for each of the following quantities:

   **(a)** present value of a continuous stream of income

   **(b)** total population in a ring around the center of a city

**10.** How do you determine whether an improper integral is convergent?

# SUPPLEMENTARY EXERCISES

*Determine the following indefinite integrals.*

**1.** $\displaystyle\int x\sin 3x^2\,dx$

**2.** $\displaystyle\int \sqrt{2x+1}\,dx$

**3.** $\displaystyle\int x(1-3x^2)^5\,dx$

**4.** $\displaystyle\int \frac{(\ln x)^5}{x}\,dx$

**5.** $\displaystyle\int \frac{(\ln x)^2}{x}\,dx$

**6.** $\displaystyle\int \frac{1}{\sqrt{4x+3}}\,dx$

**7.** $\displaystyle\int x\sqrt{4-x^2}\,dx$

**8.** $\displaystyle\int x\sin 3x\,dx$

**9.** $\displaystyle\int x^2 e^{-x^3}\,dx$

**10.** $\displaystyle\int \frac{x\ln(x^2+1)}{x^2+1}\,dx$

**11.** $\displaystyle\int x^2\cos 3x\,dx$

**12.** $\displaystyle\int \frac{\ln(\ln x)}{x\ln x}\,dx$

**13.** $\displaystyle\int \ln x^2\,dx$

**14.** $\displaystyle\int x\sqrt{x+1}\,dx$

**15.** $\displaystyle\int \frac{x}{\sqrt{3x-1}}\,dx$

**16.** $\displaystyle\int x^2\ln x^2\,dx$

**17.** $\displaystyle\int \frac{x}{(1-x)^5}\,dx$

**18.** $\displaystyle\int x(\ln x)^2\,dx$

*In Exercises 19–36, decide whether integration by parts or a substitution should be used to compute the indefinite integral. If substitution, indicate the substitution to be made. If by parts, indicate the functions $f(x)$ and $g(x)$ to be used in formula (1) of Section 9.2.*

**19.** $\displaystyle\int xe^{2x}\,dx$

**20.** $\displaystyle\int (x-3)e^{-x}\,dx$

**21.** $\displaystyle\int (x+1)^{-1/2}e^{\sqrt{x+1}}\,dx$

**22.** $\displaystyle\int x^2\sin(x^3-1)\,dx$

**23.** $\displaystyle\int \frac{x-2x^3}{x^4-x^2+4}\,dx$

**24.** $\displaystyle\int \ln\sqrt{5-x}\,dx$

**25.** $\displaystyle\int e^{-x}(3x-1)^2\,dx$

**26.** $\displaystyle\int xe^{3-x^2}\,dx$

**27.** $\displaystyle\int (500-4x)e^{-x/2}\,dx$

**28.** $\displaystyle\int x^{5/2}\ln x\,dx$

**29.** $\displaystyle\int \sqrt{x+2}\ln(x+2)\,dx$

**30.** $\displaystyle\int (x+1)^2 e^{3x}\,dx$

**31.** $\displaystyle\int (x+3)e^{x^2+6x}\,dx$

**32.** $\displaystyle\int \sin^2 x\cos x\,dx$

**33.** $\displaystyle\int x\cos(x^2-9)\,dx$

**34.** $\displaystyle\int (3-x)\sin 3x\,dx$

**35.** $\displaystyle\int \frac{2-x^2}{x^3-6x}\,dx$

**36.** $\displaystyle\int \frac{1}{x(\ln x)^{3/2}}\,dx$

*Evaluate the following definite integrals.*

**37.** $\displaystyle\int_0^1 \frac{2x}{(x^2+1)^3}\,dx$

**38.** $\displaystyle\int_0^{\pi/2} x\sin 8x\,dx$

**39.** $\displaystyle\int_0^2 xe^{-(1/2)x^2}\,dx$

**40.** $\displaystyle\int_{1/2}^1 \frac{\ln(2x+3)}{2x+3}\,dx$

**41.** $\displaystyle\int_1^2 xe^{-2x}\,dx$

**42.** $\displaystyle\int_1^2 x^{-3/2}\ln x\,dx$

*Approximate the following definite integrals by the midpoint rule, the trapezoidal rule, and Simpson's rule.*

**43.** $\displaystyle\int_1^9 \frac{1}{\sqrt{x}}\,dx; \; n=4$

**44.** $\displaystyle\int_0^{10} e^{\sqrt{x}}\,dx; \; n=5$

**45.** $\displaystyle\int_1^4 \frac{e^x}{x+1}\,dx; \; n=5$

**46.** $\displaystyle\int_{-1}^1 \frac{1}{1+x^2}\,dx; \; n=5$

*Evaluate the following improper integrals whenever they are convergent.*

**47.** $\displaystyle\int_0^\infty e^{6-3x}\,dx$

**48.** $\displaystyle\int_1^\infty x^{-2/3}\,dx$

**49.** $\displaystyle\int_1^\infty \frac{x+2}{x^2+4x-2}\,dx$

**50.** $\displaystyle\int_0^\infty x^2 e^{-x^3}\,dx$

**51.** $\displaystyle\int_{-1}^\infty (x+3)^{-5/4}\,dx$

**52.** $\displaystyle\int_{-\infty}^0 \frac{8}{(5-2x)^3}\,dx$

**53.** It can be shown that $\displaystyle\lim_{b\to\infty} be^{-b}=0$. Use this fact to compute $\int_1^\infty xe^{-x}\,dx$.

**54.** Let $k$ be a positive number. It can be shown that $\displaystyle\lim_{b\to\infty} be^{-kb}=0$. Use this fact to compute $\int_0^\infty xe^{-kx}\,dx$.

**55.** Find the present value of a continuous stream of income over the next 4 years, where the rate of income is $50e^{-.08t}$ thousand dollars per year at time $t$, and the interest rate is 12%.

**56.** Suppose that $t$ miles from the center of a certain city, the property tax revenue is approximately $R(t)$ thousand dollars per square mile, where $R(t)=50e^{-t/20}$. Use this model to predict the total property tax revenue that will be generated by property within 10 miles of the city center.

**57.** Suppose that a machine requires daily maintenance, and let $M(t)$ be the *annual* rate of maintenance expense at time $t$. Suppose that the interval $0\le t\le 2$ is divided into $n$ subintervals, with endpoints $t_0=0,t_1,\dots,t_n=2$.

**(a)** Give a Riemann sum that approximates the total maintenance expense over the next 2 years. Then write the integral that the Riemann sum approximates for large values of $n$.

**(b)** Give a Riemann sum that approximates the present value of the total maintenance expense over the next 2 years, using a 10% annual interest rate, compounded continuously. Then write the integral that the Riemann sum approximates for large values of $n$.

58. The *capitalized cost* of an asset is the total of the original cost and the present value of all future "renewals" or replacements. This concept is useful, for example, when selecting equipment that is manufactured by several different companies. Suppose that a corporation computes the present value of future expenditures using an annual interest rate $r$, with continuous compounding of interest. Assume that the original cost of an asset is $80,000 and the annual renewal expense will be $50,000, spread more or less evenly throughout each year, for a large but indefinite number of years. Find a formula, involving an integral, that gives the capitalized cost of the asset.

# Differential Equations

A *differential equation* is an equation in which derivatives of an unknown function $y = f(t)$ occur. Examples of such equations are

$$y' = 6t + 3, \quad y' = 6y, \quad y'' = 3y' - x, \quad \text{and} \quad y' + 3y + t = 0.$$

As we shall see, many physical processes can be described by differential equations. In this chapter we explore some topics in differential equations and use our resulting knowledge to study problems from many different fields, including business, genetics, and ecology.

## 10.1 Solutions of Differential Equations

A differential equation is an equation involving an unknown function $y$ and one or more of the derivatives $y'$, $y''$, $y'''$, and so on. Suppose that $y$ is a function of the variable $t$. A *solution* of a differential equation is any function $f(t)$ such that the differential equation becomes a true statement when $y$ is replaced by $f(t)$, $y'$ by $f'(t)$, $y''$ by $f''(t)$, and so forth.

**EXAMPLE 1** Show that the function $f(t) = 5e^{-2t}$ is a solution of the differential equation

$$y' + 2y = 0. \tag{1}$$

**Solution** The differential equation (1) says that $y' + 2y$ equals zero for all values of $t$. We must show that this result holds if $y$ is replaced by $5e^{-2t}$ and $y'$ is replaced by $(5e^{-2t})' = -10e^{-2t}$. Now

$$\overbrace{(5e^{-2t})'}^{y'} + 2\overbrace{(5e^{-2t})}^{y} = -10e^{-2t} + 10e^{-2t} = 0.$$

Therefore, $y = 5e^{-2t}$ is a solution of the differential equation (1). ∎

**EXAMPLE 2** Show that the function $f(t) = \frac{1}{9}t + \sin 3t$ is a solution of the differential equation

$$y'' + 9y = t. \tag{2}$$

**Solution** If $f(t) = \frac{1}{9}t + \sin 3t$, then

$$f'(t) = \frac{1}{9} + 3\cos 3t$$

$$f''(t) = -9\sin 3t.$$

Substituting $f(t)$ for $y$ and $f''(t)$ for $y''$ in the left side of (2), we obtain

$$\overbrace{-9\sin 3t}^{y''} + 9\overbrace{\left(\tfrac{1}{9}t + \sin 3t\right)}^{y} = -9\sin 3t + t + 9\sin 3t = t.$$

Therefore, $y'' + 9y = t$ if $y = \frac{1}{9}t + \sin 3t$, and hence $y = \frac{1}{9}t + \sin 3t$ is a solution to $y'' + 9y = t$. ∎

The differential equation in Example 1 is said to be of *first order*, since it involves the first derivative of the unknown function $y$. The differential equation in Example 2 is of *second order*, since it involves the second derivative of $y$. In general, the *order* of a differential equation is the order of the highest derivative that appears in the equation.

The process of determining all the functions that are solutions of a differential equation is called *solving the differential equation*. The process of antidifferentiation amounts to solving a simple type of differential equation. For example, a solution of the differential equation

$$y' = 3t^2 - 4 \tag{3}$$

is a function $y$ whose derivative is $3t^2 - 4$. Thus, solving (3) consists of finding all antiderivatives of $3t^2 - 4$. Clearly, $y$ must be of the form $y = t^3 - 4t + C$, where $C$ is a constant. The solutions of (3) corresponding to several values of $C$ are sketched in Fig. 1. Each graph is called a *solution curve*.

We encountered differential equations such as

$$y' = 2y \tag{4}$$

in our discussion of exponential functions. Unlike (3), this equation does not give a specific formula for $y'$ but instead describes a property of $y'$, namely that $y'$ is proportional to $y$ (with 2 as the constant of proportionality). At the moment, the only way we can "solve" (4) is simply to know in advance what the solutions are. Recall from Chapter 4 that the solutions of (4) have the form $y = Ce^{2t}$ for any constant $C$. Some typical solutions of (4) are sketched in Fig. 2. In the next section

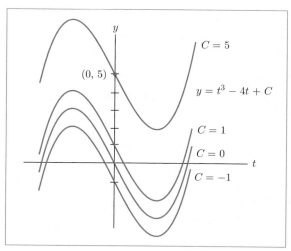

**Figure 1.** Typical solutions of $y' = 3t^2 - 4$.

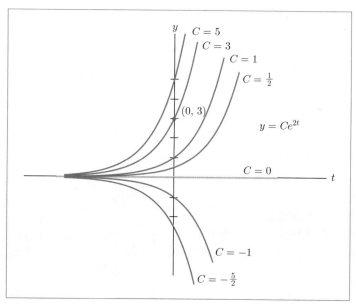

**Figure 2.** Typical solutions of $y' = 2y$.

we shall discuss a method for solving a class of differential equations that includes both (3) and (4) as special cases.

Figures 1 and 2 illustrate two important differences between differential equations and algebraic equations (e.g., $ax^2 + bx + c = 0$). First, a solution of a differential equation is a *function* rather than a number. Second, a differential equation usually has infinitely many solutions.

Sometimes we want to find a particular solution that satisfies certain additional conditions called *initial conditions*. Initial conditions specify the values of a solution and a certain number of its derivatives at some specific value of $t$, often $t = 0$. If the solution to a differential equation is $y = f(t)$, we will often write $y(0)$ for $f(0)$, $y'(0)$ for $f'(0)$, and so on. The problem of determining a solution of a differential equation that satisfies given initial conditions is called an *initial value problem*.

**EXAMPLE 3**   (a) Solve the initial value problem $y' = 3t^2 - 4$, $y(0) = 5$.

(b) Solve the initial value problem $y' = 2y$, $y(0) = 3$.

**Solution** (a) We have already noted that the general solution of $y' = 3t^2 - 4$ is $f(t) = t^3 - 4t + C$. We want the particular solution that satisfies $f(0) = 5$. Geometrically, we are looking for the curve in Fig. 1 that passes through the point $(0, 5)$. Using the general formula for $f(t)$, we have

$$5 = f(0) = (0)^3 - 4(0) + C = C$$
$$C = 5.$$

Thus $f(t) = t^3 - 4t + 5$ is the desired solution.

(b) The general solution of $y' = 2y$ is $y = Ce^{2t}$. The condition $y(0) = 3$ means that $y$ must be 3 when $t = 0$; that is, the point $(0, 3)$ must be on the graph of the solution to $y' = 2y$. (See Fig. 2.) We have

$$3 = y(0) = Ce^{2(0)} = C \cdot 1 = C$$
$$C = 3.$$

Thus $y = 3e^{2t}$ is the desired solution. ■

A constant function that satisfies a differential equation is called a *constant solution* of the differential equation. Constant solutions occur in many of the applied problems considered later in the chapter.

---

**EXAMPLE 4**  Find a constant solution of $y' = 3y - 12$.

**Solution**  Let $f(t) = c$ for all $t$. Then $f'(t)$ is zero for all $t$. If $f(t)$ satisfies the differential equation

$$f'(t) = 3 \cdot f(t) - 12,$$

then

$$0 = 3 \cdot c - 12,$$

and so $c = 4$. This is the only possible value for a constant solution. Substitution shows that the function $f(t) = 4$ is indeed a solution of the differential equation. ■

## Modeling with Differential Equations

Equations describing conditions in a physical process are often referred to as *mathematical models* and discovering these equations is called *modeling*. In our next example, we show how we can use a differential equation to model a physical process. The example should be studied carefully, for it contains the key to understanding many similar problems that will appear in exercises and in later sections.

---

**EXAMPLE 5**  **Newton's law of cooling**  Suppose that a red-hot steel rod is plunged into a bath of cool water. Let $f(t)$ be the temperature of the rod at time $t$, and suppose that the water is maintained at a constant temperature of $10°C$. According to Newton's law of cooling, the rate of change of $f(t)$ is proportional to the difference between the two temperatures $10°$ and $f(t)$. Find a differential equation that describes this physical law.

**Solution** The two key ideas are "rate of change" and "proportional." The rate of change of $f(t)$ is the derivative $f'(t)$. Since this is proportional to the difference $10 - f(t)$, there exists a constant $k$ such that

$$f'(t) = k[10 - f(t)]. \tag{5}$$

The term "proportional" does not tell us whether $k$ is positive or negative (or zero). We must decide this, if possible, from the context of the problem. In the present situation, the steel rod is hotter than the water, so $10 - f(t)$ is negative. Also, $f(t)$ will decrease as time passes, so $f'(t)$ should be negative. Thus, to make $f'(t)$ negative in (5), $k$ must be a positive number. From (5) we see that $y = f(t)$ satisfies a differential equation of the form

$$y' = k(10 - y), \quad k \text{ a positive constant.} \quad \blacksquare$$

---

**EXAMPLE 6** Suppose the constant of proportionality in Example 5 is $k = .2$ and time is measured in seconds. How fast is the temperature of the steel rod changing when the temperature is $110°C$?

**Solution** The relationship between the temperature and the rate of change of the temperature is given by the differential equation $y' = .2(10 - y)$, where $y = f(t)$ is the temperature after $t$ seconds. When $y = 110$, the rate of change is

$$y' = .2(10 - 110) = .2(-100) = -20.$$

The temperature is decreasing at the rate of 20 degrees per second. $\blacksquare$

The differential equation in Examples 5 and 6 is a special case of the differential equation

$$y' = k(M - y).$$

Differential equations of this type describe not only cooling, but also many important applications in economics, medicine, population dynamics, and engineering. Some of these applications are discussed in the exercises and in Sections 10.4 and 10.6.

## Geometric Meaning of a Differential Equation: Slope Fields

As we observed earlier, a differential equation, such as

$$y' = t - y, \tag{6}$$

does not give a specific formula for $y'$ in terms of the variable $t$; instead, it describes a property of $y'$. The key to understanding this property is to remember the geometric interpretation of the derivative as a slope formula. So if $y = f(t)$ is a solution of (6) and $(t, y)$ is a point on the graph of this solution, then equation (6) is telling us that the slope of the graph at the point $(t, y)$ (that is, $y'(t)$) is just $t - y$. For example, if a solution curve goes through the point $(1, 2)$, without knowing the formula for this solution, we can say that the slope of the solution through the point $(1, 2)$ is

$$y'(1) = 1 - 2 = -1.$$

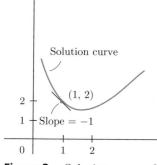

**Figure 3.** Solution curve of $y' = t - y$ and tangent line through the point $(1, 2)$.

This information gives us the direction of the solution curve through the point $(1, 2)$. In Figure 3, we show a portion of the solution curve through the point $(1, 2)$, along with the tangent line at that point. Note how the slope of the tangent lines matches with the value $y'(1) = -1$.

Imagine now repeating the previous construction of the tangent line at many points in the $ty$-plane, not just the point $(1, 2)$. This laborious process is usually done with the help of a computer or a graphing calculator that can generate a collection of small line segments, called a *slope field* or *direction field* of the differential equation [Fig. 4(a)]. Since each line segment in the slope field is tangent to a solution curve, we can visualize a solution curve by following the flow of the slope field [Fig. 4(b)].

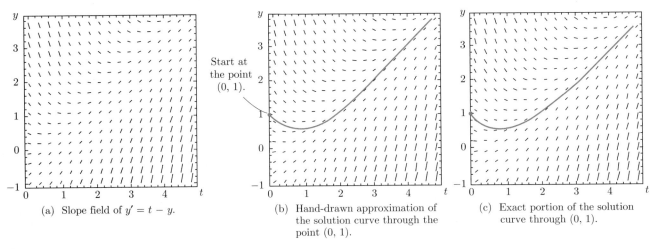

(a) Slope field of $y' = t - y$.

(b) Hand-drawn approximation of the solution curve through the point $(0, 1)$.

(c) Exact portion of the solution curve through $(0, 1)$.

**Figure 4**

---

**EXAMPLE 7**   Use the slope field of the differential equation $y' = t - y$, shown in Fig. 4(a), to draw by hand a portion of the solution curve that passes through the point $(0, 1)$.

**Solution**   We are given one point on the solution curve and the slope field. To draw the desired curve, we start at the point $(0, 1)$ and trace a curve in the $ty$-plane tangent to the line segments in the slope field. The result is a curve shown in Fig. 4(b). In Figure 4(c), we show a portion of the exact solution curve for the purpose of comparison. As you can see, the hand-drawn curve in Fig. 4(b) is a pretty good approximation of the exact solution curve.   ■

In general, we can construct a slope field for any first-order differential equation of the form

$$y' = g(t, y), \tag{7}$$

where $g(t, y)$ is a function of $t$ and $y$. The idea is that at any given point $(a, b)$ of a solution curve, the slope $y'(a)$ is given by $g(a, b)$.

As our next example shows, slope fields are useful in deriving qualitative properties of solutions.

---

**EXAMPLE 8**   Let $f(t)$ denote the number of people who have contracted a certain strain of flu after $t$ days. The function $f(t)$ satisfies the initial value problem $y' = .0002y(5000 - y)$, $y(0) = 1000$. A slope field of the differential equation is shown in Fig. 5(a). Based on the slope field, do you think that the number of infected people will ever exceed 5000?

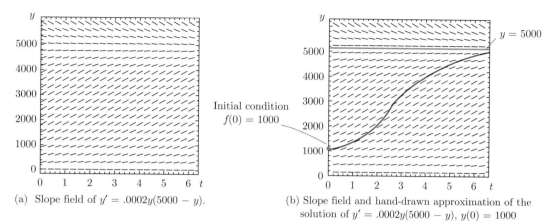

(a) Slope field of $y' = .0002y(5000 - y)$.

(b) Slope field and hand-drawn approximation of the solution of $y' = .0002y(5000 - y)$, $y(0) = 1000$

**Figure 5**

**Solution**  The initial condition $f(0) = 1000$ tells us that the point $(0, 1000)$ is on the graph of the solution curve $y = f(t)$. Starting at this point in the $ty$-plane and tracing a curve tangent to the line segments in the slope field, we obtain an approximation to a portion of the solution curve through the point $(0, 1000)$, as shown in Fig. 5(b). According to this curve, we can conclude that the solution curve will come very close to but will not exceed the value 5000. Hence the number of infected people will not exceed 5000.  ∎

The differential equation in Example 8 is a special case of the logistic differential equation, which arises frequently in the study of population growth in a restricted environment. This important equation will be studied in greater detail in later sections.

Throughout this section we have assumed that our solution functions $y$ are functions of the variable $t$, which in applications usually stands for time. In most of this chapter we will continue to use the variable $t$. However, if we use another variable occasionally, such as $x$, we will make the variable explicit by writing $\frac{dy}{dx}$ instead of $y'$.

**Incorporating Technology**  The TI-86 graphing calculator has the ability to draw slope fields and solution curves. See Appendix D for details.

## PRACTICE PROBLEMS 10.1

1. Show that any function of the form $y = Ae^{t^3/3}$, where $A$ is a constant, is a solution of the differential equation $y' - t^2 y = 0$.

2. If the function $f(t)$ is a solution of the initial value problem $y' = (t + 2)y$, $y(0) = 3$, find $f(0)$ and $f'(0)$.

3. Let $f(t)$ be the size of a population after $t$ days. Suppose that $y = f(t)$ satisfies $y' = .06y$, $y(0) = 1000$. Describe this initial value problem in words. (Include the phrase "is proportional to.") How fast is the population growing when it contains 3000 members?

## EXERCISES 10.1

1. Show that the function $f(t) = \frac{3}{2}e^{t^2} - \frac{1}{2}$ is a solution of the differential equation $y' - 2ty = t$.

2. Show that the function $f(t) = t^2 - \frac{1}{2}$ is a solution of the differential equation $(y')^2 - 4y = 2$.

3. Show that the function $f(t) = (e^{-t} + 1)^{-1}$ satisfies $y' + y^2 = y$, $y(0) = \frac{1}{2}$.

4. Show that the function $f(t) = 5e^{2t}$ satisfies $y'' - 3y' + 2y = 0$, $y(0) = 5$, $y'(0) = 10$.

*In Exercises 5 and 6, state the order of the differential equation and verify that the given function is a solution.*

5. $(1 - t^2)y'' - 2ty' + 2y = 0$, $y(t) = t$

6. $(1 - t^2)y'' - 2ty' + 6y = 0$, $y(t) = \frac{1}{2}(3t^2 - 1)$

7. Is the constant function $f(t) = 3$ a solution of the differential equation $y' = 6 - 2y$?

8. Is the constant function $f(t) = -4$ a solution of the differential equation $y' = t^2(y + 4)$?

9. Find a constant solution of $y' = t^2y - 5t^2$.

10. Find two constant solutions of $y' = 4y(y - 7)$.

11. If the function $f(t)$ is a solution of the initial value problem $y' = 2y - 3$, $y(0) = 4$, find $f(0)$ and $f'(0)$.

12. If the function $f(t)$ is a solution of the initial value problem $y' = e^t + y$, $y(0) = 0$, find $f(0)$ and $f'(0)$.

13. Let $y = v(t)$ be the downward speed (in feet per second) of a skydiver after $t$ seconds of free fall. This function satisfies the differential equation $y' = .2(160 - y)$, $y(0) = 0$. What is the skydiver's acceleration when her downward speed is 60 feet per second? (*Note:* Acceleration is the derivative of speed.)

14. A lake is stocked with 100 fish. Let $f(t)$ be the number of fish after $t$ months, and suppose that $y = f(t)$ satisfies the differential equation $y' = .0004y(1000 - y)$. Figure 6 shows the graph of the solution to this differential equation. The graph is asymptotic to the line $y = 1000$, the maximum number of fish that the lake can support. How fast is the fish population growing when it reaches one-half of its maximum population?

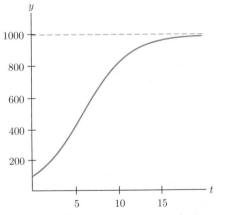

**Figure 6.** Growth of a fish population.

15. Let $f(t)$ be the balance in a savings account at the end of $t$ years and suppose $y = f(t)$ satisfies the differential equation $y' = .05y - 10,000$.

    (a) Suppose that after one year the balance is $150,000. Is the balance increasing or decreasing at that time? At what rate is it increasing or decreasing at that time?

    (b) Write the differential equation in the form $y' = k(y - M)$.

    (c) Describe this differential equation in words.

16. Let $f(t)$ be the balance in a savings account at the end of $t$ years. Suppose $y = f(t)$ satisfies the differential equation $y' = .04y + 2000$.

    (a) Suppose that after one year the balance is $10,000. Is the balance increasing or decreasing at that time? At what rate is it increasing or decreasing at that time?

    (b) Write the differential equation in the form $y' = k(y + M)$.

    (c) Describe this differential equation in words.

17. A certain piece of news is being broadcast to a potential audience of 200,000 people. Let $f(t)$ be the number of people who have heard the news after $t$ hours. Suppose that $y = f(t)$ satisfies

$$y' = .07(200,000 - y), \qquad y(0) = 10.$$

Describe this initial value problem in words.

18. Let $f(t)$ be the size of a paramecium population after $t$ days. Suppose that $y = f(t)$ satisfies the differential equation

$$y' = .003y(500 - y), \qquad y(0) = 20.$$

Describe this initial value problem in words.

19. Let $f(t)$ denote the amount of capital invested by a certain business firm at time $t$. The rate of change of invested capital, $f'(t)$, is sometimes called the *rate of net investment*. Suppose that the management of the firm decides that the optimum level of investment should be $C$ dollars and that, at any time, the rate of net investment should be proportional to the difference between $C$ and the total capital invested. Construct a differential equation that describes this situation.

20. A cool object is placed in a room that is maintained at a constant temperature of $20°C$. The rate at which the temperature of the object rises is proportional to the difference between the room temperature and the temperature of the object. Let $y = f(t)$ be the temperature of the object at time $t$, and give a differential equation that describes the rate of change of $f(t)$.

21. (*Carbon Dioxide Diffusion in Lungs During Breath Holding*) When the breath is held, carbon dioxide ($CO_2$) diffuses from the blood into the lungs at a steadily decreasing rate. Let $P_0$ and $P_b$ denote the pressure of $CO_2$ in the lungs, respectively, in the blood at the moment when the breath is held. Suppose that $P_b$ is constant during breath holding, and let $P(t)$ denote the pressure of $CO_2$ in the lungs at time $t > 0$. Experiments show that the rate of change of $P(t)$ is proportional to the difference between the two pressures $P(t)$ and $P_b$. Find an initial value problem that describes the diffusion of $CO_2$ in the lungs during breath holding.

22. The slope field in Fig. 4(a) suggests that the solution curve of the differential equation $y' = t - y$ through the point $(0, -1)$ is a straight line.

    (a) Assuming that this is true, find the equation of the line. [*Hint*: Use the differential equation to get the slope of the line through the point $(0, -1)$.]

    (b) Verify that the function that you found in part (a) is a solution by plugging its formula into the differential equation.

23. Verify that the function $f(t) = 2e^{-t} + t - 1$ is a solution of the initial value problem $y' = t - y$, $y(0) = 1$. (This is the function shown in Fig. 4(c). In Section 10.3, you will learn how to derive this solution.)

24. On the slope field in Fig. 5(a) or a copy of it, draw the solution of the initial value problem $y' = .0002y(5000 - y)$, $y(0) = 500$.

25. The health officials that studied the flu epidemic in Example 8 made an error in counting the initial number of infected people. They are now claiming that $f(t)$ (the number of infected people after $t$ days) is a solution of the initial value problem $y' = .0002y(5000 - y)$, $y(0) = 1500$. Under this new assumption, can $f(t)$ exceed 5000? [*Hint*: Since the differential equation is the same as the one in Example 8, you can use the slope field in Fig. 5(a) to answer the question.]

26. On the slope field in Fig. 4(a) (or a copy of it), draw an approximation of a portion of the solution curve of the differential equation $y' = t - y$ that goes through the point $(0, 2)$. In your opinion, based on the slope field, can this solution pass through the point $(.5, 2.2)$?

27. Figure 7 shows a slope field of the differential equation $y' = 2y(1 - y)$. With the help of this figure, determine the constant solutions, if any, of the differential equation. Verify your answer by substituting back into the equation.

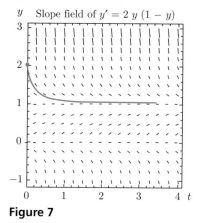

Figure 7

28. In Figure 7, we show a portion of the solution curve of the differential equation $y' = 2y(1 - y)$ through the point $(0, 2)$. On Figure 7 or a copy of it, draw an approximation of the solution curve of the differential equation $y' = 2y(1 - y)$ through the point $(0, 3)$. Use the slope field to guide your graph.

29. Suppose that $y_0 > 1$. Is it true that the solution, $y = f(t)$, of the initial value problem $y' = 2y(1 - y)$, $y(0) = y_0$, is decreasing for all $t > 0$? Answer this question based on the slope field shown in Fig. 7.

30. Answer the question in Exercise 29 by using the differential equation to determine the sign of $f'(t)$.

### Technology Exercises

31. Consider the differential equation $y' = .2(10 - y)$ from Example 6. If the initial temperature of the steel rod is $510°$, then the function $f(t) = 10 + 500e^{-.2t}$ is the solution of the differential equation.

    (a) Graph the function in the window

    $$[0, 30] \ by \ [-75, 550].$$

    (b) In the home screen, compute $.2(10 - f(5))$ and compare this value with $f'(5)$.

32. The function

    $$f(t) = \frac{5000}{1 + 49e^{-t}}$$

    is the solution of the differential equation

    $$y' = .0002y(5000 - y)$$

    from Example 8.

    (a) Graph the function in the window

    $$[0, 10] \ by \ [-750, 5750].$$

    (b) In the home screen, compute

    $$.0002f(3)(5000 - f(3))$$

    and compare this value with $f'(3)$.

**1.** If $y = Ae^{t^3/3}$, then

$$\overbrace{(Ae^{t^3/3})'}^{y'} - t^2 \overbrace{(Ae^{t^3/3})}^{y} = At^2 e^{t^3/3} - t^2 A e^{t^3/3} = 0.$$

Therefore, $y' - t^2 y = 0$ if $y = Ae^{t^3/3}$.

**2.** The initial condition $y(0) = 3$ says that $f(0) = 3$. Since $f(t)$ is a solution to $y' = (t+2)y$,

$$f'(t) = (t+2)f(t)$$

and hence

$$f'(0) = (0+2)f(0) = 2 \cdot 3 = 6.$$

**3.** Initially the population has 1000 members. At any time the rate of growth of the population is proportional to the size of the population at that time, and the constant of proportionality is .06. When $y = 3000$,

$$y' = .06y = .06 \cdot 3000 = 180.$$

Therefore, the population is growing at the rate of 180 members per day.

## 10.2 Separation of Variables

Here we describe a technique for solving an important class of differential equations—those of the form

$$y' = p(t)q(y),$$

where $p(t)$ is a function of $t$ only and $q(y)$ is a function of $y$ only. Two equations of this type are

$$y' = \frac{3t^2}{y^2} \qquad \left[p(t) = 3t^2,\, q(y) = \frac{1}{y^2}\right], \tag{1}$$

$$y' = e^{-y}(2t+1) \qquad \left[p(t) = 2t+1,\, q(y) = e^{-y}\right]. \tag{2}$$

The main feature of such equations is that we may *separate the variables*; that is, we may rewrite the equations so that $y$ occurs only on one side of the equation and $t$ on the other. For example, if we multiply both sides of equation (1) by $y^2$, the equation becomes

$$y^2 y' = 3t^2;$$

if we multiply both sides of equation (2) by $e^y$, the equation becomes

$$e^y y' = 2t + 1.$$

It should be pointed out that the differential equation

$$y' = 3t^2 - 4$$

is of the preceding type. Here $p(t) = 3t^2 - 4$ and $q(y) = 1$. The variables are already separated, however. Similarly, the differential equation

$$y' = 5y$$

is of the preceding type, with $p(t) = 5$, $q(y) = y$. We can separate the variables by writing the equation as

$$\frac{1}{y}y' = 5.$$

In the next example we present a procedure for solving differential equations in which the variables are separated.

**EXAMPLE 1**    Find all solutions of the differential equation $y^2 y' = 3t^2$.

**Solution**    (a)  Write $y'$ as $\dfrac{dy}{dt}$.

$$y^2 \frac{dy}{dt} = 3t^2$$

(b)  Integrate both sides with respect to $t$:

$$\int y^2 \frac{dy}{dt}\, dt = \int 3t^2\, dt.$$

(c)  Rewrite the left-hand side, "canceling the $dt$":

$$\int y^2\, dy = \int 3t^2\, dt.$$

(See the discussion below for an explanation of exactly what this means.)

(d)  Calculate the antiderivatives.

$$\frac{1}{3}y^3 + C_1 = t^3 + C_2.$$

(e)  Solve for $y$ in terms of $t$.

$$y^3 = 3(t^3 + C_2 - C_1)$$

$$y = \sqrt[3]{3t^3 + C}, \quad C \text{ a constant.}$$

We can check that this method works by showing that $y = \sqrt[3]{3t^3 + C}$ is a solution to $y^2 y' = 3t^2$. Since $y = (3t^3 + C)^{1/3}$, we have

$$y' = \frac{1}{3}(3t^3 + C)^{-2/3} \cdot 3 \cdot 3t^2 = 3t^2(3t^3 + C)^{-2/3}$$

$$y^2 y' = \left[(3t^3 + C)^{1/3}\right]^2 \cdot 3t^2(3t^3 + C)^{-2/3}$$

$$= 3t^2.$$

**Figure 1.** Solution curves in Example 1.

Figure 1 shows solution curves for various values of $C$. Note the linear solution that corresponds to $C = 0$.

*Discussion of Step* $(c)$    Suppose that $y = f(t)$ is a solution of the differential equation $y^2 y' = 3t^2$. Then

$$[f(t)]^2 f'(t) = 3t^2.$$

Integrating, we have

$$\int [f(t)]^2 f'(t)\, dt = \int 3t^2\, dt.$$

Let us make the substitution $y = f(t)$, $dy = f'(t)\, dt$ in the left-hand side in order to get

$$\int y^2\, dy = \int 3t^2\, dt.$$

This is just the result of step $(c)$. The process of "canceling the $dt$" and integrating with respect to $y$ is actually just equivalent to making the substitution $y = f(t)$, $dy = f'(t)\, dt$.

The technique used in Example 1 can be used for any differential equation with separated variables. Suppose that we are given such an equation:

$$h(y)y' = p(t),$$

where $h(y)$ is a function of $y$ only and $p(t)$ is a function of $t$ only. Our method of solution can be summarized as follows:

a. Write $y'$ as $\dfrac{dy}{dt}$.

$$h(y)\frac{dy}{dt} = p(t).$$

b. Integrate both sides with respect to $t$.

$$\int h(y)\frac{dy}{dt}\, dt = \int p(t)\, dt.$$

c. Rewrite the left-hand side by "canceling the $dt$."

$$\int h(y)\, dy = \int p(t)\, dt.$$

d. Calculate the antiderivatives $H(y)$ for $h(y)$ and $P(t)$ for $p(t)$.

$$H(y) = P(t) + C.$$

e. Solve for $y$ in terms of $t$.

$$y = \ldots.$$

▶ **Note**   In step (d) there is no need to write two constants of integration (as we did in Example 1), since they will be combined into one in step (e).   ◀

---

**EXAMPLE 2**   Solve $e^y y' = 2t + 1$, $y(0) = 1$.

**Solution**   (a)  $e^y \dfrac{dy}{dt} = 2t + 1$

(b)  $\displaystyle\int e^y \frac{dy}{dt}\, dt = \int (2t + 1)\, dt$

(c)  $\displaystyle\int e^y\, dy = \int (2t + 1)\, dt$

(d)  $e^y = t^2 + t + C$

(e)  $y = \ln(t^2 + t + C)$          [Take logarithms of both sides of the equation in step (d).]

If $y = \ln(t^2 + t + C)$ is to satisfy the initial condition $y(0) = 1$, then

$$1 = y(0) = \ln(0^2 + 0 + C) = \ln C,$$

so that $C = e$ and $y = \ln(t^2 + t + e)$. Figure 2 shows solutions of the differential equation for various values of $C$. The one that goes through the point $(0, 1)$ is the solution of the initial value problem.   ■

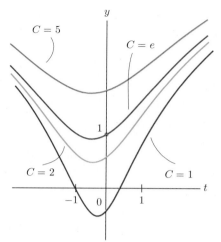

**Figure 2.** Solution curves and the solution of the initial value problem in Example 2.

---

**EXAMPLE 3**    Solve $y' = t^3 y^2 + y^2$.

**Solution**    As the equation is given, the right-hand side is not in the form $p(t)q(y)$. However, we may rewrite the equation in the form $y' = (t^3 + 1)y^2$. Now we may separate the variables, dividing both sides by $y^2$, to get

$$\frac{1}{y^2} y' = t^3 + 1. \tag{3}$$

Then we apply our method of solution:

(a)  $\dfrac{1}{y^2} \dfrac{dy}{dt} = t^3 + 1$

(b)  $\displaystyle\int \frac{1}{y^2} \frac{dy}{dt} \, dt = \int (t^3 + 1) \, dt$

(c)  $\displaystyle\int \frac{1}{y^2} \, dy = \int (t^3 + 1) \, dt$

(d)  $-\dfrac{1}{y} = \dfrac{1}{4} t^4 + t + C$, $C$ a constant

(e)  $y = -\dfrac{1}{\frac{1}{4} t^4 + t + C}$

The method that we have applied yields all the solutions of equation (3). However, we have ignored an important point. We wish to solve $y' = y^2(t^3 + 1)$ and not equation (3). Do the two equations have precisely the same solutions? We obtained equation (3) from the given equation by dividing by $y^2$. This is a permissible operation, provided that $y$ is not equal to zero for all $t$. (Of course, if $y$ is zero for some $t$, the resulting differential equation is understood to hold only for some limited range of $t$.) Thus in dividing by $y^2$, we must assume that $y$ is not the zero function. However, note that $y = 0$ is a solution of the original equation because

$$0 = (0)' = t^3 \cdot 0^2 + 0^2.$$

So, when we divided by $y^2$, we "lost" the solution $y = 0$. Finally, we see that the solutions of the differential equation $y' = t^3 y^2 + y^2$ are

$$y = -\frac{1}{\frac{1}{4}t^4 + t + C}, \quad C \text{ a constant}$$

and

$$y = 0.$$

Figure 3 shows two solutions corresponding to different values of the constant $C$, and the solution $y = 0$. It is interesting to note how the great diversity of solutions arise from a relatively simple differential equation. ∎

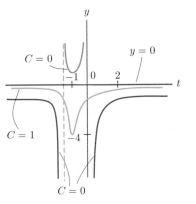

**Figure 3.** Solution curves in Example 3.

▶ **Warning** If the equation in Example 3 had been

$$y' = t^3 y^2 + 1,$$

we would not have been able to use the method of separation of variables because the expression $t^3 y^2 + 1$ cannot be written in the form $p(t)q(y)$. ◀

---

**EXAMPLE 4**  Solve the initial value problem $3y' + y^4 \cos t = 0$, $y(\frac{\pi}{2}) = \frac{1}{2}$.

**Solution**  Let us write the equation in the form

$$3y' = -y^4 \cos t. \tag{4}$$

Clearly, the constant function $y = 0$ is a solution of the differential equation because it makes both sides of (4) zero for all $t$. Now, supposing that $y \neq 0$, we may divide by $y^4$ and get

$$\frac{3}{y^4} y' = -\cos t$$

$$\int \frac{3}{y^4} \frac{dy}{dt} \, dt = -\int \cos t \, dt$$

$$-y^{-3} = -\sin t + C, \quad C \text{ a constant}$$

$$y^{-3} = \sin t + C.$$

In the last step, there is no need to change the sign of $C$, since it stands for an arbitrary constant. Solving for $y$ in terms of $t$, we find

$$y^3 = \frac{1}{\sin t + C}$$

$$y = \frac{1}{\sqrt[3]{\sin t + C}}.$$

If $y$ is to satisfy the initial condition $y(\frac{\pi}{2}) = \frac{1}{2}$, we must have

$$\frac{1}{2} = y\left(\frac{\pi}{2}\right) = \frac{1}{\sqrt[3]{\sin(\frac{\pi}{2}) + C}} = \frac{1}{\sqrt[3]{1 + C}} \qquad \text{[Recall that } \sin(\tfrac{\pi}{2}) = 1.]$$

$$2 = \sqrt[3]{1 + C} \qquad \text{[Take inverses.]}$$

$$2^3 = 1 + C$$

$$C = 7.$$

Hence the desired solution is

$$y = \frac{1}{\sqrt[3]{\sin t + 7}}.$$

**Figure 4.** Solution of the initial value problem in Example 4.

The graph of $y$, shown in Fig. 4, goes through the point $(\frac{\pi}{2}, \frac{1}{2})$. ■

---

**EXAMPLE 5**    Solve $y' = te^t/y$, $y(0) = -5$.

**Solution**    Separating the variables, we have

$$yy' = te^t$$

$$\int y \frac{dy}{dt}\, dt = \int te^t\, dt$$

$$\int y\, dy = \int te^t\, dt.$$

The integral $\int te^t\, dt$ may be found by integration by parts:

$$\int te^t\, dt = te^t - \int 1 \cdot e^t\, dt = te^t - e^t + C.$$

Therefore,

$$\tfrac{1}{2}y^2 = te^t - e^t + C$$

$$y^2 = 2te^t - 2e^t + C_1$$

$$y = \pm\sqrt{2te^t - 2e^t + C_1}.$$

Note that the $\pm$ appears because there are two square roots of $2te^t - 2e^t + C_1$, differing from one another by a minus sign. Thus the solutions are of two sorts—namely,

$$y = +\sqrt{2te^t - 2e^t + C_1}$$

$$y = -\sqrt{2te^t - 2e^t + C_1}.$$

Two of these solutions are shown in Fig. 5 with different signs in front of the square root. We must choose $C_1$ so that $y(0) = -5$. Since the values of $y$ for the first

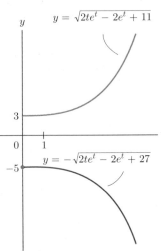

$y = \sqrt{2te^t - 2e^t + 11}$

3

0   1

$y = -\sqrt{2te^t - 2e^t + 27}$

-5

**Figure 5.**   Solution curves in Example 5.

solution are always positive, the given initial condition must correspond to the second solution, and we must have

$$-5 = y(0) = -\sqrt{2 \cdot 0 \cdot e^0 - 2e^0 + C_1} = -\sqrt{-2 + C_1}$$

$$-2 + C_1 = 25$$

$$C_1 = 27.$$

Hence the desired solution is

$$y = -\sqrt{2te^t - 2e^t + 27}.$$

Its graph is the one that goes through the point $(0, -5)$ in Fig. 5.   ■

When working the exercises at the end of the section, it is a good practice first to find the constant solution(s), if any. A constant function $y = c$ is a solution of $y' = p(t)q(y)$ if and only if $q(c) = 0$. [For $y = c$ implies that $y' = (c)' = 0$, and $p(t)q(y)$ will be zero for all $t$ if and only if $q(y) = 0$—that is, $q(c) = 0$.] After listing the constant solutions, one may assume that $q(y) \neq 0$ and go on to divide both sides of the equation $y' = p(t)q(y)$ by $q(y)$ to separate the variables.

## PRACTICE PROBLEMS 10.2

**1.** Solve the initial value problem $y' = 5y$, $y(0) = 2$, by separation of variables.

**2.** Solve $y' = \sqrt{ty}$, $y(1) = 4$.

## EXERCISES 10.2

*Solve the following differential equations.*

**1.** $\dfrac{dy}{dt} = \dfrac{5 - t}{y^2}$

**2.** $\dfrac{dy}{dt} = te^{2y}$

**3.** $\dfrac{dy}{dt} = \dfrac{e^t}{e^y}$

**4.** $\dfrac{dy}{dt} = -\dfrac{1}{t^2 y^2}$

**5.** $\dfrac{dy}{dt} = t^{1/2} y^2$

**6.** $\dfrac{dy}{dt} = \dfrac{t^2 y^2}{t^3 + 8}$

**7.** $y' = \left(\dfrac{t}{y}\right)^2 e^{t^3}$

**8.** $y' = e^{4y} t^3 - e^{4y}$

**9.** $y' = \sqrt{\dfrac{y}{t}}$

**10.** $y' = \left(\dfrac{e^t}{y}\right)^2$

**11.** $y' = 3t^2 y^2$

**12.** $(1 + t^2)y' = ty^2$

**13.** $y' e^y = te^{t^2}$

**14.** $y' = \dfrac{1}{ty + y}$

**15.** $y' = \dfrac{\ln t}{ty}$

**16.** $y^2 y' = \tan t$

**17.** $y' = (y - 3)^2 \ln t$

**18.** $yy' = t \sin(t^2 + 1)$

*Solve the following differential equations with given initial conditions.*

**19.** $y' = 2te^{-2y} - e^{-2y}$, $y(0) = 3$

**20.** $y' = y^2 - e^{3t} y^2$, $y(0) = 1$

**21.** $y^2 y' = t \cos t$, $y(0) = 2$

**22.** $y' = t^2 e^{-3y}$, $y(0) = 2$

**23.** $3y^2 y' = -\sin t$, $y(\frac{\pi}{2}) = 1$

**24.** $y' = -y^2 \sin t$, $y(\frac{\pi}{2}) = 1$

**25.** $\dfrac{dy}{dt} = \dfrac{t + 1}{ty}$, $t > 0$, $y(1) = -3$

**26.** $\dfrac{dy}{dt} = \left(\dfrac{1 + t}{1 + y}\right)^2$, $y(0) = 2$

**27.** $y' = 5ty - 2t$, $y(0) = 1$

**28.** $y' = \dfrac{t^2}{y}$, $y(0) = -5$

**29.** $\dfrac{dy}{dx} = \dfrac{\ln x}{\sqrt{xy}}$, $y(1) = 4$

**30.** $\dfrac{dN}{dt} = 2tN^2$, $N(0) = 5$

**31.** A model that describes the relationship between the price and the weekly sales of a product might have a form such as

$$\frac{dy}{dp} = -\frac{1}{2}\left(\frac{y}{p+3}\right),$$

where $y$ is the volume of sales and $p$ is the price per unit. That is, at any time the rate of decrease of sales with respect to price is directly proportional to the sales level and inversely proportional to the sales price plus a constant. Solve this differential equation. (Figure 6 shows several typical solutions.)

**32.** One problem in psychology is to determine the relation between some physical stimulus and the corresponding sensation or reaction produced in a subject. Suppose that, measured in appropriate units, the strength of a stimulus is $s$ and the intensity of the corresponding sensation is some function of $s$, say $f(s)$. Some experimental data suggests that the rate of change of intensity of the sensation with respect to the stimulus is directly proportional to the intensity of the sensation and inversely proportional to the strength of the stimulus; that is, $f(s)$ satisfies the differential equation

$$\frac{dy}{ds} = k\frac{y}{s}$$

for some positive constant $k$. Solve this differential equation. (Figure 7 shows several solutions corresponding to $k = .4$.)

**33.** Let $t$ represent the total number of hours that a truck driver spends during a year driving on a certain highway connecting two cities, and let $p(t)$ represent the probability that the driver will have at least one accident during these $t$ hours. Then $0 \leq p(t) \leq 1$, and $1 - p(t)$ represents the probability of not having an accident. Under ordinary conditions the rate of increase in the probability of an accident (as a function of $t$) is proportional to the probability of not having an accident. Construct and solve a differential equation for this situation.

**34.** In certain learning situations there is a maximum amount, $M$, of information that can be learned, and at any time the rate of learning is proportional to the amount yet to be learned. Let $y = f(t)$ be the amount of information learned up to time $t$. Construct and solve a differential equation that is satisfied by $f(t)$.

**35.** Mothballs tend to evaporate at a rate proportional to their surface area. If $V$ is the volume of a mothball, then its surface area is roughly a constant times $V^{2/3}$. So the mothball's volume decreases at a rate proportional to $V^{2/3}$. Suppose that initially a mothball has a volume of 27 cubic centimeters and 4 weeks later has a volume of 15.625 cubic centimeters. Construct and solve a differential equation satisfied by the volume at time $t$. Then determine if and when the mothball will vanish $(V = 0)$.

**36.** Some homeowner's insurance policies include automatic inflation coverage based on the U.S. Commerce Department's construction cost index (CCI). Each year the property insurance coverage is increased by an amount based on the change in the CCI. Let $f(t)$ be the CCI at time $t$ years since January 1, 1990, and let $f(0) = 100$. Suppose the construction cost index is rising at a rate proportional to the CCI and the index was 115 on January 1, 1992. Construct and solve a differential equation satisfied by $f(t)$. Then determine when the CCI will reach 200.

**37.** The Gompertz growth equation is

$$\frac{dy}{dt} = -ay\ln\frac{y}{b},$$

where $a$ and $b$ are positive constants. This equation is used in biology to describe the growth of certain populations. Find the general form of solutions to this dif-

**Figure 6.** Demand curves.

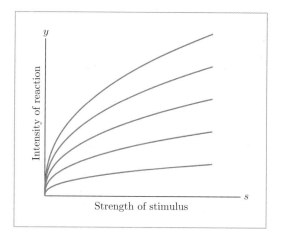

**Figure 7.** Reaction to stimuli.

ferential equation. (Figure 8 shows several solutions corresponding to $a = .04$ and $b = 90$.)

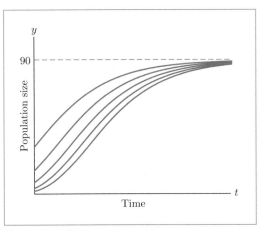

**Figure 8.**    Gompertz growth curves.

**38.** When a certain liquid substance $A$ is heated in a flask, it decomposes into a substance $B$ at a rate (measured in units of $A$ per hour) that at any time $t$ is proportional to the square of the amount of substance $A$ present. Let $y = f(t)$ be the amount of substance $A$ present at time $t$. Construct and solve a differential equation that is satisfied by $f(t)$.

**39.** Let $f(t)$ denote the number (in thousands) of fish in a lake after $t$ years, and suppose that $f(t)$ satisfies the differential equation

$$y' = 0.1y(5 - y).$$

The slope field for this equation is shown in Fig. 9.

**(a)** With the help of the slope field, discuss what happens to an initial population of 6000 fish. Does it increase or decrease?

**(b)** How about an initial population of 1000 fish? Does it increase or decrease?

**(c)** On the slope field in Fig. 9, or a copy of it, draw the solution of the initial value problem

$$y' = 0.1y(5 - y), \quad y(0) = 1.$$

What does this solution represent?

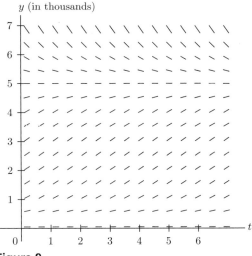

**Figure 9**

**40.** Refer to the differential equation in Exercise 39.

**(a)** Obviously if you start with zero fish, then $f(t) = 0$ for all $t$. Confirm this on the slope field. Are there any other constant solutions?

**(b)** Describe the population of fish if the initial population is greater than 5000, less than 5000. On the slope field in Fig. 9, or a copy of it, draw solution curves to illustrate your description.

## SOLUTIONS TO PRACTICE PROBLEMS 10.2

**1.** The constant function $y = 0$ is a solution of $y' = 5y$. If $y \neq 0$, we may divide by $y$ and obtain

$$\frac{1}{y} y' = 5$$

$$\int \frac{1}{y} \frac{dy}{dt} \, dt = \int 5 \, dt$$

$$\int \frac{1}{y} \, dy = \int 5 \, dt$$

$$\ln |y| = 5t + C$$

$$|y| = e^{5t+C} = e^{C} \cdot e^{5t}$$

$$y = \pm e^{C} \cdot e^{5t}.$$

These two types of solutions and the constant solution may all be written in the form

$$y = Ae^{5t},$$

where $A$ is an arbitrary constant (positive, negative, or zero). The initial condition $y(0) = 2$ implies that

$$2 = y(0) = Ae^{5(0)} = A.$$

Hence the solution of the initial value problem is $y = 2e^{5t}$.

**SOLUTIONS TO PRACTICE PROBLEMS 10.2 (CONTINUED)**

**2.** We rewrite $y' = \sqrt{ty}$ as $y' = \sqrt{t} \cdot \sqrt{y}$. The constant function $y = 0$ is one solution. To find the others, we suppose that $y \neq 0$ and divide by $\sqrt{y}$ to obtain

$$\frac{1}{\sqrt{y}} y' = \sqrt{t}$$

$$\int y^{-1/2} \frac{dy}{dt} dt = \int t^{1/2} dt$$

$$\int y^{-1/2} dy = \int t^{1/2} dt$$

$$2y^{1/2} = \frac{2}{3} t^{3/2} + C$$

$$y^{1/2} = \frac{1}{3} t^{3/2} + C_1 \qquad (5)$$

$$y = \left(\frac{1}{3} t^{3/2} + C_1\right)^2. \qquad (6)$$

We must choose $C_1$ so that $y(1) = 4$. The quickest method is to use (5) instead of (6). We have $y = 4$ when $t = 1$, so

$$4^{1/2} = \frac{1}{3}(1)^{3/2} + C_1$$

$$2 = \frac{1}{3} + C_1$$

$$C_1 = \frac{5}{3}.$$

Hence the desired solution is

$$y = \left(\frac{1}{3} t^{3/2} + \frac{5}{3}\right)^2.$$

## 10.3  First-Order Linear Differential Equations

In this section we will study first-order differential equations that can be put in the form

$$y' + a(t)y = b(t), \qquad (1)$$

where $a(t)$ and $b(t)$ are continuous functions on a given interval. Equation (1) is called a *first-order linear* differential equation in *standard form*. Section 10.4 will focus on many important applications of these equations in economics, medicine, and biology.

The following are examples of first-order linear differential equations:

$$y' - 2ty = 0 \qquad\qquad [a(t) = -2t,\ b(t) = 0]$$

$$y' + y = 2 \qquad\qquad [a(t) = 1,\ b(t) = 2]$$

$$ty' = ty + t^2 + 1 \qquad \left[y' - y = \frac{t^2 + 1}{t},\ a(t) = -1,\ b(t) = \frac{t^2 + 1}{t}\right]$$

$$e^t y' + e^t y = 5 \qquad \left[y' + y = 5e^{-t},\ a(t) = 1,\ b(t) = 5e^{-t}\right]$$

In the last two examples we first put the equation in standard form $y' + a(t)y = b(t)$ before determining the functions $a(t)$ and $b(t)$.

To solve a first-order linear differential equation, we will try to combine the terms in $y'$ and $y$ into one single term, which consists of the derivative of a product involving the function $y$. This crucial step of the solution amounts to working the product rule for differentiation backward. For example, the product rule tells us that

$$\frac{d}{dt}\left[e^t y\right] = e^t y' + e^t y,$$

and

$$\frac{d}{dt}\left[t^2 y\right] = t^2 y' + 2ty.$$

So to solve an equation in which the expression $e^t y' + e^t y$ appears, it would help to replace this expression by $\frac{d}{dt}[e^t y]$.

Similarly, to solve an equation in which the expression $t^2 y' + 2ty$ appears, it would help to combine the terms in $y'$ and $y$ and replace the expression by $\frac{d}{dt}[t^2 y]$. Here are two simple illustrations of these ideas.

---

**EXAMPLE 1**    Solve

(a) $e^t y' + e^t y = 5$        (b) $t^2 y' + 2ty = e^t + 6$

**Solution** (a)  We have already noticed that $e^t y' + e^t y = \frac{d}{dt}[e^t y]$. So the equation in (a) can be written as

$$\frac{d}{dt}[e^t y] = 5.$$

Integrating both sides with respect to $t$, we obtain

$$e^t y = \int 5 \, dt = 5t + C.$$

Multiplying both sides by $e^{-t}$ to solve for $y$, we get

$$y = 5te^{-t} + Ce^{-t}.$$

This is the solution of the equation $e^t y' + e^t y = 5$. Note how it contains one arbitrary constant $C$.

(b)  Since $\frac{d}{dt}[t^2 y] = t^2 y' + 2ty$, we can write the equation as

$$\frac{d}{dt}[t^2 y] = e^t + 6.$$

To solve, integrate both sides with respect to $t$:

$$t^2 y = \int (e^t + 6) \, dt = e^t + 6t + C.$$

Solve for $y$ by dividing both sides by $t^2$:

$$y = \frac{e^t}{t^2} + \frac{6}{t} + \frac{C}{t^2}.$$

Here again, the solution contains one arbitrary constant $C$.    ∎

The step that consists of combining the terms in $y'$ and $y$ is not always as straightforward as it was in Example 1. In general, it will be first necessary to modify the equation by multiplying it through by some (nonzero) factor. To illustrate this process, let us consider a simple example that is related to the equation in Example 1(a).

---

**EXAMPLE 2**    Solve $y' + y = -2$.

**Solution**  You can try if you want, but it is not possible to combine the terms in $y$ and $y'$ into the derivative of a product. Taking a hint from Example 1(a), if we *multiply* the equation through by the factor $e^t$ (which is never 0), we obtain the equation

$$e^t y' + e^t y = -2e^t,$$

where the terms on the left side can be combined together as $\frac{d}{dt}[e^t y]$. Now we can solve the equation as we did previously. Write the equation in the form

$$\frac{d}{dt}[e^t y] = -2e^t.$$

Integrate both sides and solve for $y$:

$$e^t y = \int -2e^t \, dt = -2e^t + C,$$

$$y = -2e^t e^{-t} + Ce^{-t} = -2 + Ce^{-t}.$$

It is instructive to verify that we have indeed found a solution of the equation $y' + y = -2$. Since $y = -2 + Ce^{-t}$, we have

$$y' = -Ce^{-t}$$

$$y' + y = \overbrace{-Ce^{-t}}^{y'} \overbrace{-2 + Ce^{-t}}^{y} = -2.$$

Thus $y = -2 + Ce^{-t}$ is a solution of $y' + y = -2$. ∎

The factor $e^t$ that we needed in Example 2 in order to combine the terms in $y'$ and $y$ was somewhat easy to guess because of our work in Example 1(a). To see what we can do with an arbitrary first-order linear equation, let us take a closer look at the equation in Example 2, $y' + y = -2$. The equation is in standard form with $a(t) = 1$. Let $A(t)$ denote an antiderivative of $a(t)$, say

$$A(t) = \int a(t) \, dt = \int 1 \, dt = t.$$

The relationship of the factor that we used in Example 2 to the equation is this: $e^t = e^{A(t)}$. That is, the factor that we needed is the exponential of $A(t)$. As we now show, this relationship holds for an arbitrary first-order linear equation.

Given equation (1), form the function $e^{A(t)}$, called an *integrating factor*, where $A(t) = \int a(t) \, dt$. Observe that, as a consequence of the chain rule and the product rule,

$$\frac{d}{dt}\left[e^{A(t)}\right] = e^{A(t)} \frac{d}{dt} A(t) = e^{A(t)} a(t) = a(t)e^{A(t)},$$

and

$$\frac{d}{dt}\left[e^{A(t)} y\right] = e^{A(t)} y' + a(t)e^{A(t)} y$$

$$= e^{A(t)}[\overbrace{y' + a(t)y}^{\text{left side of (1)}}]. \tag{2}$$

If we now multiply both sides of (1) by $e^{A(t)}$, we can simplify the resulting equation with the help of (2) as follows:

$$e^{A(t)}[y' + a(t)y] = e^{A(t)} b(t)$$

$$\frac{d}{dt}\left[e^{A(t)} y\right] = e^{A(t)} b(t) \quad \text{[by (2)]}. \tag{3}$$

In this form, equation (1) is easy to solve. Integrate both sides to get rid of the derivative on the left, and get

$$e^{A(t)} y = \int e^{A(t)} b(t) \, dt + C.$$

Solve for $y$ by multiplying both sides by $e^{-A(t)}$:

$$y = e^{-A(t)}\left[\int e^{A(t)} b(t) \, dt + C\right], \quad C \text{ a constant.} \tag{4}$$

This formula gives all the solutions of equation (1). It is called the *general solution* of (1). As our next examples illustrate, to solve a first-order linear differential equation, you can appeal directly to (4), or you can use an integrating factor and repeat the steps leading to (4).

---

**EXAMPLE 3**    Solve $y' = 3 - 2y$.

**Solution**    ***Step 1***    Put the equation in standard form: $y' + 2y = 3$.

***Step 2***    Find an integrating factor $e^{A(t)}$. We have $a(t) = 2$, so

$$A(t) = \int a(t)\, dt = \int 2\, dt = 2t.$$

Note how we picked one antiderivative of $a(t)$ by setting the constant of integration equal to zero. Hence the integrating factor is $e^{A(t)} = e^{2t}$.

***Step 3***    Multiply both sides of the differential equation by the integrating factor $e^{2t}$:

$$\overbrace{e^{2t}y' + 2e^{2t}y}^{\frac{d}{dt}[e^{2t}y]} = 3e^{2t}.$$

Recognizing the terms on the left side as a derivative of the product $e^{2t}y$, we get

$$\frac{d}{dt}\left[e^{2t}y\right] = 3e^{2t}.$$

***Step 4***    Integrating both sides then solving for $y$, we obtain

$$e^{2t}y = \int 3e^{2t}\, dt = \frac{3}{2}e^{2t} + C$$

$$y = e^{-2t}\left[\frac{3}{2}e^{2t} + C\right] = \frac{3}{2} + Ce^{-2t}.$$

Figure 1 shows solution curves for various values of $C$. Note the constant solution, $y = \frac{3}{2}$, corresponding to $C = 0$. ∎

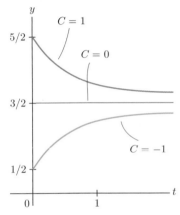

**Figure 1.** Typical solutions of the differential equation in Example 3: $y' + 2y = 3$.

From the preceding example, we can state a step-by-step process for solving a first-order linear differential equation.

---

**Solving a First-Order Linear Differential Equation**

***Step 1***    Put the equation in standard form $y' + a(t)y = b(t)$.

***Step 2***    Compute an antiderivative of $a(t)$, $A(t) = \int a(t)\, dt$. [When evaluating $\int a(t)\, dt$, it is customary to choose 0 for the constant of integration.] Form the integrating factor $e^{A(t)}$.

***Step 3***    Multiply the differential equation by the integrating factor $e^{A(t)}$. This transforms the terms on the left side of the equation into the derivative of a product, $\frac{d}{dt}\left[e^{A(t)}y\right]$, as in (3).

***Step 4***    Integrate to get rid of the derivative and then solve for $y$.

---

**EXAMPLE 4**    Solve

$$\frac{1}{3t^2}y' + y = 4, \quad t > 0.$$

**Solution** **Step 1**     To put the equation in standard form, we multiply it through by $3t^2$ and get

$$y' + 3t^2 y = 12t^2.$$

Hence $a(t) = 3t^2$.

**Step 2**     An antiderivative of $a(t)$ is

$$A(t) = \int a(t)\, dt = \int 3t^2\, dt = t^3.$$

So the integrating factor is $e^{A(t)} = e^{t^3}$.

**Step 3**     Multiply both sides of the differential equation by $e^{t^3}$ and get

$$e^{t^3} y' + 3t^2 e^{t^3} y = 12t^2 e^{t^3}$$

$$\frac{d}{dt}\left[ e^{t^3} y \right] = 12t^2 e^{t^3}.$$

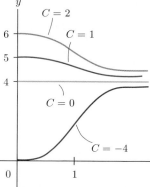

**Step 4**     Integrating both sides, we obtain

$$e^{t^3} y = \int 12t^2 e^{t^3}\, dt = 4e^{t^3} + C$$

$$y = e^{-t^3}\left[ 4e^{t^3} + C \right] = 4 + Ce^{-t^3}.$$

In evaluating the integral $\int 12t^2 e^{t^3}\, dt$, we used integration by substitution (Section 9.1): Let $u = t^3$, $du = 3t^2\, dt$. Then

$$\int 12t^2 e^{t^3}\, dt = 4\int \overbrace{e^{t^3}}^{e^u} \overbrace{3t^2\, dt}^{du} = 4\int e^u\, du = 4e^u + C = 4e^{t^3} + C.$$

**Figure 2.** Typical solutions of the differential equation in Example 4.

Figure 2 shows solution curves for various values of $C$. Note the constant solution, $y = 4$, which corresponds to $C = 0$. ■

---

**EXAMPLE 5**     Solve the initial value problem $t^2 y' + ty = 2$, $y(1) = 1$, $t > 0$.

**Solution** **Step 1**     Divide by $t^2$ to put the equation in standard form:

$$y' + \frac{1}{t}y = \frac{2}{t^2}.$$

**Step 2**     An antiderivative of $a(t) = \frac{1}{t}$ is

$$A(t) = \int \frac{1}{t}\, dt = \ln t.$$

Hence the integrating factor is $e^{A(t)} = e^{\ln t} = t$.

**Step 3**     Multiply both sides of the differential equation by $t$ and get

$$ty' + y = \frac{2}{t}$$

$$\frac{d}{dt}[ty] = \frac{2}{t}.$$

**Step 4**   Integrating both sides, then solving for $y$, we obtain

$$ly = \int \frac{2}{t}\, dt - 2\ln t + C$$

$$y = \frac{2\ln t + C}{t}.$$

In order to satisfy the initial condition, we must have

$$1 = y(1) = \frac{2\ln(1) + C}{1} = C \quad [\ln(1) = 0].$$

Hence the solution of the initial value problem is

$$y = \frac{2\ln t + 1}{t}, \quad t > 0.$$

Among all the solution curves that are shown in Fig. 3, the solution of the initial value problem is the one that goes through the point $(1, 1)$. ▪

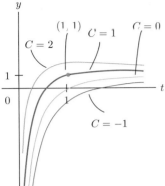

**Figure 3.**   Solution of the initial value problem $t^2 y' + ty = 2$, $y(1) = 1$, $t > 0$.

---

**PRACTICE PROBLEMS 10.3**

1. Using an integrating factor, solve $y' + y = 1 + e^{-t}$.

2. Find an integrating factor for the differential equation

$$y' = -\frac{y}{1+t} + 1, \quad t \geq 0.$$

---

# EXERCISES 10.3

*In Exercises 1–6, combine the terms in $y'$ and $y$ into the derivative of a product, as we did in Example 1.*

1. $(2t + 4)y' + 2y = 6t$
2. $te^t y' + (e^t + te^t)y = -1$
3. $t^5 y' + 5t^4 y = 2t^2 + t + 1$
4. $(t + 7)y' + y = \frac{1}{t}$, $t > 0$
5. $y' \cos t - y \sin t = 1$
6. $y' \sin t + y \cos t = 2t$

*In Exercises 7–12, combine the terms in $y'$ and $y$ into the derivative of a product, then solve the equation.*

7. $ty' + y = 1$
8. $t^3 y' + 3t^2 y = e^{2t} + 1$
9. $e^{t^2} y' + 2te^{t^2} y = \frac{2\sqrt{t}}{3}$
10. $e^{3t} y' + 3e^{3t} y = t^2$
11. $\frac{y'}{t} - \frac{y}{t^2} = 6t^2$
12. $e^t y' + e^t y = 1 + t^2$

*In Exercises 13–18, find an integrating factor for each equation. Take $t > 0$.*

13. $y' - 2y = t$
14. $y' + ty = 6t$
15. $t^3 y' + y = 0$
16. $y' + \sqrt{t}\, y = 2(t + 1)$
17. $y' - \frac{y}{10 + t} = 2$
18. $y' = t^2(y + 1)$

*In Exercises 19–32, solve the given equation using an integrating factor. Take $t > 0$.*

19. $y' + y = 1$
20. $y' + 2ty = 0$

21. $y' - 2ty = -4t$
22. $y' = 2(20 - y)$
23. $y' = .5(35 - y)$
24. $y' + y = e^{-t} + 1$
25. $y' + \frac{y}{10 + t} = 0$
26. $y' - 2y = e^{2t}$
27. $(1 + t)y' + y = -1$
28. $y' = e^{-t}(y + 1)$
29. $6y' + ty = t$
30. $e^t y' + y = 1$
31. $y' + y = 2 - e^t$
32. $\frac{1}{\sqrt{t+1}} y' + y = 1$

*In Exercises 33–40, solve the initial value problem.*

33. $y' + 2y = 1$, $y(0) = 1$
34. $ty' + y = \ln t$, $y(e) = 0$, $t > 0$
35. $y' + \frac{y}{1 + t} = 20$, $y(0) = 10$, $t \geq 0$
36. $y' = 2(10 - y)$, $y(0) = 1$
37. $y' + y = e^{2t}$, $y(0) = -1$
38. $ty' - y = -1$, $y(1) = 1$, $t > 0$
39. $y' + 2y \cos(2t) = 2\cos(2t)$, $y(\frac{\pi}{4}) = 0$
40. $ty' + y = \sin t$, $y(\frac{\pi}{2}) = 0$, $t > 0$

### Technology Exercises

41. Consider the initial value problem

$$y' = -\frac{y}{1 + t} + 10, \quad y(0) = 50.$$

**(a)** Is the solution increasing or decreasing when $t = 0$? [*Hint*: Compute $y'(0)$.]

**(b)** Find the solution and plot it for $0 \le t \le 4$.

## SOLUTIONS TO PRACTICE PROBLEMS 10.3

**1.** We follow the step-by-step method outlined in this section. The equation $y' + y = 1 + e^{-t}$ is already in standard form. We have $a(t) = 1$, $A(t) = \int 1 \, dt = t$, and so the integrating factor is $e^{A(t)} = e^t$. Multiplying the equation by the integrating factor, we transform the terms on its left side into the derivative of a product, as follows:

$e^t(y' + y) = e^t(1 + e^{-t})$    [Multiply the equation by the integrating factor $e^t$.]

$e^t y' + e^t y = e^t + 1$    [The left side is the derivative of the product $e^t y$.]

$\dfrac{d}{dt}\left[e^t y\right] = e^t + 1$

Integrating both sides, we get rid of the derivative and obtain

$$e^t y = \int (e^t + 1) \, dt = e^t + t + C.$$

To solve for $y$, we multiply both sides by $e^{-t}$ and get

$$y = e^{-t}(e^t + t + C) = 1 + te^{-t} + Ce^{-t}.$$

**2.** In standard form, the equation becomes

$$y' + \frac{y}{1+t} = 1.$$

We have $a(t) = \dfrac{1}{1+t}$. An antiderivative of $a(t)$ is

$$A(t) = \int \frac{1}{1+t} \, dt = \ln|1 + t|.$$

But since $t \ge 0$, it follows that $1 + t \ge 0$, hence $\ln|1 + t| = \ln(1 + t)$, and so the integrating factor is

$$e^{A(t)} = e^{\ln(1+t)} = 1 + t.$$

# 10.4 Applications of First-Order Linear Differential Equations

This section discusses several interesting applications that lead to first-order linear differential equations. Modeling or setting up the mathematical equations is a crucial part of the solution. You can develop your modeling skills by studying carefully the details leading to the mathematical equations in each of the following examples. Look for the key expression "the rate of change," and translate it as a derivative. Then continue your description of the rate of change in order to obtain a differential equation.

Our first examples are related to the topics of compound interest from Section 5.2 and future value of an income stream from Section 6.5.

---

**EXAMPLE 1**   **A retirement account**   You invest in a retirement account that pays 6% interest per year. You make an initial deposit of $1000 and plan on making future deposits at the rate of $2400 per year. Assume that the deposits are made continuously and that interest is compounded continuously. Let $P(t)$ denote the amount of money in the account $t$ years after the initial deposit.

(a) Set up an initial value problem that is satisfied by $P(t)$.

(b) Find $P(t)$.

---

**Solution**   (a) If no deposits or withdrawals are made, then only interest is being added to the account at a rate proportional to the amount in the account with constant of proportionality $k = .06$ or 6%. Since in this case the growth of $P(t)$ comes from interest only, it follows that $P(t)$ satisfies the equation

$$y' \qquad = \qquad .06 \qquad \times \qquad y$$

$$\begin{bmatrix} \text{rate of} \\ \text{change of } y \end{bmatrix} = \begin{bmatrix} \text{constant of} \\ \text{proportionality} \end{bmatrix} \times \begin{bmatrix} \text{amount in} \\ \text{the account} \end{bmatrix}.$$

Taking into consideration the deposits, which are adding money to the account at the rate of $2400 per year, we see that there are two influences on the way the amount of money in the account changes—the rate at which interest is added and the rate at which deposits are made. The rate of change of $P(t)$ is the *net effect* of these two influences. That is, $P(t)$ now satisfies the first-order linear differential equation

$$y' \qquad = \qquad .06y \qquad + \qquad 2400$$

$$\begin{bmatrix} \text{rate of} \\ \text{change of } y \end{bmatrix} = \begin{bmatrix} \text{rate at which} \\ \text{interest is added} \end{bmatrix} + \begin{bmatrix} \text{rate at which} \\ \text{money is deposited} \end{bmatrix}.$$

Since the initial deposit in the account was $1000, it follows that $P(t)$ satisfies the initial condition $y(0) = 1000$. Putting the equation in standard form [see Section 10.3, equation (1)], we conclude that $P(t)$ satisfies the initial value problem

$$y' - .06y = 2400, \quad y(0) = 1000.$$

(b) To solve the equation, use the step-by-step method from the previous section. We have $a(t) = -.06$. So $A(t) = -\int .06\,dt = -.06t$, and hence the integrating factor is $e^{A(t)} = e^{-.06t}$. Multiplying both sides of the equation by the integrating factor, we obtain

$$e^{-.06t}[y' - .06y] = 2400e^{-.06t}$$

$$e^{-.06t}y' - .06e^{-.06t}y = 2400e^{-.06t} \qquad \text{[The left side is a derivative of a product.]}$$

$$\frac{d}{dt}\left[e^{-.06t}y\right] = 2400e^{-.06t}.$$

Integrating both sides and then solving for $y$, we get

$$e^{-.06t}y = \int 2400e^{-.06t}\,dt$$

$$e^{-.06t}y = -40{,}000e^{-.06t} + C \qquad \left[\int 2400e^{-.06t}\,dt = -\frac{2400}{.06}e^{-.06t} + C\right]$$

$$y = e^{.06t}\left[-40{,}000e^{-.06t} + C\right] \qquad \text{[Solve for } y.\text{]}$$

$$y = -40{,}000 + Ce^{.06t} \qquad\qquad \text{[Simplify.]}$$

Thus the solution is $P(t) = -40{,}000 + Ce^{.06t}$. In order to satisfy the initial condition, we must have

$$1000 = P(0) = -40{,}000 + Ce^{.06 \cdot (0)} = -40{,}000 + C$$

$$C = 41{,}000$$

Hence the amount of money in the account at time $t$ is given by

$$P(t) = -40{,}000 + 41{,}000e^{.06t}.$$

The graph of $P(t)$ is shown in Fig. 1.

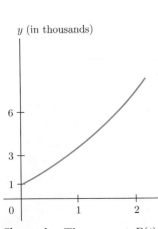

$y$ (in thousands)

**Figure 1.** The account $P(t)$ in Example 1.

Let us quickly examine the modeling ideas in Example 1. The quantity that we were looking for was the amount of money in the account at time $t$. Its rate of change was affected by two influences: the interest rate and the rate at which money was deposited. To describe the rate of change of the money in the account, we added both rates of change and obtained the differential equation $y' = .06y + 2400$.

Many interesting situations can be modeled by refining these ideas. First identify the quantity of concern (amount of money in the account). Then identify the various influences that affect the rate of change of this quantity (interest rate, rate of deposits). Finally, derive a differential equation by expressing the rate of change of the quantity of interest in terms of the rates of change of the various influences. In Example 1, we added two rates of change in order to describe $y'$. As our next example illustrates, to model problems involving amortization of consumer loans and mortgages, you may have to subtract two rates of change.

---

**EXAMPLE 2**  **Paying off a car loan**  You took a loan of $25,000 to pay for a new car. The interest rate on the loan is 5%. You arranged through your online banking to make daily payments totaling $4800 per year. This allows you to assume that your payments are flowing continuously into your account. Let $P(t)$ denote the amount that you owe on the loan at time $t$ (in years). Assume that interest is compounded continuously. Set up an initial value problem that is satisfied by $P(t)$.

**Solution**  As in Example 1, there are two influences on the way the amount of money owed changes—the rate at which interest is added to the amount owed and the rate at which payments are subtracted from the amount owed. We know that interest is being added at a rate proportional to the amount owed, with constant of proportionality $k = .05$. The effect of the payments is to subtract from the amount owed at the rate of $4800 per year. Since the rate of change of $P(t)$ is the *net effect* of these two influences, we see that $P(t)$ satisfies the first-order linear differential equation

$$y' \quad = \quad .05y \quad - \quad 4800$$

$$\begin{bmatrix} \text{rate of} \\ \text{change of } y \end{bmatrix} = \begin{bmatrix} \text{rate at which} \\ \text{interest is added} \end{bmatrix} - \begin{bmatrix} \text{rate of} \\ \text{payments} \end{bmatrix}.$$

Recalling the initial condition and rewriting the equation in standard form, we find that $P(t)$ satisfies the initial value problem

$$y' - .05y = -4800, \quad y(0) = 25,000. \qquad \blacksquare$$

Using an integrating factor and proceeding in a manner very similar to Example 1, we find the solution of the initial value problem in Example 2 to be

$$P(t) = 96,000 - 71,000e^{.05t}.$$

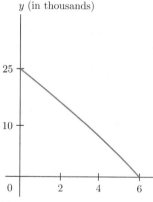

$y$ (in thousands)

25

10

0    2    4    6    $t$

**Figure 2.**  $P(t)$ is the amount owed at time $t$. The loan is paid off in approximately 6 years.

The graph of $P(t)$ is shown in Fig. 2. From the graph, we see that $P(t) = 0$ when $t \approx 6$ years. This is how long it will take to pay off the entire loan.

In the previous examples, the annual rates of deposits or loan payments were constant. If you expect your annual income to increase (which is a reasonable expectation), you may want to increase the rate of your savings deposits or you may consider paying off your loan at a faster rate. The modeling involved in these situations is similar to the previous examples, but may lead to more complicated differential equations. See the exercises for several illustrations.

In Chapter 5 we learned that simple population models are based on the assumption that the growth rate of a population is proportional to the size of the population at time $t$. The constant of proportionality is called the growth constant and is specific to the population. In reality, the rate of growth of a population may be affected by several other factors. In the next example, we consider the effect of emigration on the size of a population.

---

**EXAMPLE 3** **A population model with emigration** In 1995, people from a country suffering from serious economical problems started to emigrate to other countries in search of work and better living conditions. Let $P(t)$ denote the population of the country in millions $t$ years after 1995. Sociologists studying this population determined that for the next 30 years the number of people emigrating will gradually increase as the news of better prospects outside the country spreads. Suppose that the rate of emigration is given by $.004e^{.04t} + .04$ millions per year $t$ years after 1995. Suppose further that the growth constant of the population is $\frac{3}{125}$. Find a differential equation satisfied by $P(t)$.

**Solution** In our model, for the next 30 years (from 1995), the rate of growth of the population is affected by two influences—the rate at which the population is growing and the rate at which the population is emigrating. The rate of change of the population is the net effect of these two influences. Thus $P(t)$ satisfies the differential equation

$$ y' \quad = \quad \tfrac{3}{125}y \quad - \quad (.004e^{.04t} + .04) $$

$$ \begin{bmatrix} \text{rate of} \\ \text{change of } y \end{bmatrix} = \begin{bmatrix} \text{rate at which} \\ \text{population is growing} \end{bmatrix} - \begin{bmatrix} \text{rate at which} \\ \text{population is emigrating} \end{bmatrix}. $$

Putting the equation in standard form, we obtain

$$ y' - \frac{3}{125}y = -.004e^{.04t} - .04. $$ ∎

To obtain a formula for the population in Example 3, you need to know the initial population. For example, if in 1995 the size of the population was 2 million, then $P(0) = 2$. Solving the differential equation, subject to this initial condition, we obtain

$$ P(t) = \frac{7}{12}e^{\frac{3}{125}t} - \frac{1}{4}e^{\frac{t}{25}} + \frac{5}{3}. $$

(The details are straightforward and are left to Exercise 16.) Figure 3 shows the size of the population since 1995. Starting in 1995, the population continued to grow at a decreasing rate. According to our model, and assuming that the economical conditions remain the same, the population will peak at about 2.055 million between the years 2015 and 2020, and then it will start to decrease.

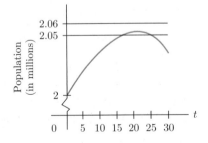

**Figure 3.** Population with emigration.

Our final topic deals with Newton's law of cooling from Example 5, Section 10.1. The modeling ideas that are involved are useful in many interesting applications such as determining the time of death of a person and the study of concentration of waste products in the body and their clearance by artificial kidneys (dialysis).

**EXAMPLE 4**    **Newton's law of cooling**    A hungry college student, in a rush to eat, turns the oven on and puts a frozen pizza in it, without preheating the oven. Let $f(t)$ denote the temperature of the pizza and $T(t)$ the oven's temperature, $t$ minutes after the oven was turned on. According to Newton's law of cooling, the rate of change of $f(t)$ is proportional to the difference between the oven's temperature and the temperature of the pizza. Find a differential equation that is satisfied by $f(t)$.

**Solution**    We reason as we did in Example 5, Section 10.1. The rate of change of the temperature of the pizza is the derivative of $f(t)$. This derivative is proportional to the difference $T(t) - f(t)$. Thus, there exists a constant $k$ such that

$$f'(t) = k[T(t) - f(t)].$$

Is $k$ positive or negative? While the pizza is heating, its temperature is rising. So $f'(t)$ is positive. Also, the temperature of the oven is always higher than the temperature of the pizza. So $T(t) - f(t)$ is positive. Thus to make $f'(t)$ positive, $k$ must be a positive number. Consequently, the differential equation satisfied by $f(t)$ is

$$y' = k[T(t) - y],$$

where $k$ is a positive constant. ∎

Example 5 treats an interesting case of the model in Example 4. In the solution, we will appeal to the integral

$$\int (at + b)e^{ct}\, dt = \frac{1}{c^2}e^{ct}(act + bc - a) + C \quad (c \neq 0). \tag{1}$$

You do not need to memorize this formula, but you should be able to evaluate integrals of this form by using integration by parts, as we now illustrate. Set $f(t) = at + b$, $g(t) = e^{ct}$. Then, $f'(t) = a$ and $G(t) = \frac{1}{c}e^{ct}$. Integrating by parts, we get

$$\int (at + b)e^{ct}\, dt = \frac{1}{c}(at + b)e^{ct} - \int \frac{a}{c}e^{ct}\, dt$$

$$= \frac{1}{c}(at + b)e^{ct} - \frac{a}{c^2}e^{ct} + C$$

$$= \frac{1}{c^2}e^{ct}(c(at + b) - a) + C,$$

and (1) follows upon simplifying.

**EXAMPLE 5**    Suppose that the oven's temperature in Example 4 is given by $T(t) = 70 + 50t$ for $0 \leq t \leq 8$. [So when the student turned on the oven ($t = 0$), the oven's temperature was $70°$, and then it started to rise at the rate of $50°$ per minute for the next 8 minutes. Presumably, the student set the oven's temperature at $470°$.] Suppose further that the constant of proportionality is $k = .1$ and that the initial temperature of the frozen pizza was $27°$. Determine the temperature of the pizza during the first 8 minutes of heating. What is the temperature of the pizza after 8 minutes of heating?

**Solution**    Substituting $T(t) = 70 + 50t$ and $k = .1$ into the differential equation in Example 4, we see that $f(t)$ satisfies

$$y' = .1[70 + 50t - y].$$

Putting the equation in standard form and recalling the initial condition $f(0) = 27$, we obtain the initial value problem

$$y' + .1y = 5t + 7, \quad y(0) = 27.$$

Multiplying both sides of the equation by the integrating factor $e^{.1t}$ and combining terms, we obtain

$$\frac{d}{dt}\left[e^{.1t}y\right] = (5t + 7)e^{.1t}.$$

Integrating both sides, we get

$$e^{.1t}y = \int (5t + 7)e^{.1t}\, dt$$

$$= 100e^{.1t}(.5t + .7 - 5) + C,$$

where we have used (1) with $a = 5$, $b = 7$, and $c = .1$. Multiplying both sides by $e^{-.1t}$, we find

$$y = 100(.5t - 4.3) + Ce^{-.1t} = 50t - 430 + Ce^{-.1t}.$$

To satisfy the initial condition, we must have

$$f(0) = 27 = -430 + C.$$

Hence $C = 457$, and so the temperature of the pizza at time $t$ is

$$f(t) = 50t - 430 + 457e^{-.1t}.$$

After 8 minutes of heating, the temperature of the pizza is

$$f(8) = 50(8) - 430 + 457e^{-.1(8)} = -30 + 457e^{-.8} \approx 175°.$$

■

The modeling techniques of this section have many interesting applications in finance, biology, medicine, and sociology. Some of these applications are presented in the exercises.

## PRACTICE PROBLEM 10.4

1. A savings account earns 4% interest per year, compounded continuously, and continuous withdrawals are made from the account at the rate of $1200 per year. Set up a differential equation that is satisfied by the amount $f(t)$ of money in the account at time $t$.

## EXERCISES 10.4

1. Refer to Example 1.

   (a) How fast was the amount in the account growing when it reached $30,000?

   (b) How much money was in the account when it was growing at twice the rate of your annual contribution?

   (c) How long do you have to wait in order for the money in the account to reach $40,000?

2. Refer to Example 2. Answer questions (a) and (b) if the interest rate is 7%. How long will it take to pay off

the $25,000 loan in this case?

3. A person planning for her retirement arranges to make continuous deposits into a savings account at the rate of $3600 per year. The savings account earns 5% interest compounded continuously.

   (a) Set up a differential equation that is satisfied by $f(t)$, the amount of money in the account at time $t$.

   (b) Solve the differential equation in part (a), assuming that $f(0) = 0$, and determine how much money will be in the account at the end of 25 years.

**4.** A person deposits $10,000 in a bank account and decides to make additional deposits at the rate of $A$ dollars per year. Suppose that the bank compounds interest continuously at the annual rate of 6% and that the deposits are made continuously into the account.

(a) Set up a differential equation that is satisfied by the amount $f(t)$ in the account at time $t$.

(b) Determine $f(t)$ (as a function of $A$).

(c) Determine $A$ if the initial deposit is to double in five years.

**5.** Twenty years ahead of her retirement, Kelly opened a savings account that earns 5% interest rate compounded continuously, and contributed to this account at the annual rate of $1200 per year for 20 years. Ten years ahead of his retirement, John opened a similar savings account that earns 5% interest rate compounded continuously, and decided to double the annual rate of contribution to $2400 per year for 10 years. Who has more money in their savings account at retirement? (Assume that the contributions are made continuously into the accounts.)

**6.** Answer the question in Exercise 5 if John contributed to his savings account at the annual rate of $3000 per year for 10 years.

**7.** A person took out a loan of $100,000 from a bank that charges 7.5% interest compounded continuously. What should be the annual rate of payments if the loan is to be paid in full in exactly 10 years? (Assume that the payments are made continuously throughout the year.)

**8.** (*Car Prices in 2001*) The National Automobile Dealers Association reported that the average retail selling price of a new vehicle* was $25,800 in 2001. A person purchased a new car at the average price and financed the entire amount. Suppose that the person can only afford to pay $500 per month. Assume that the payments are made at a continuous annual rate, and that interest is compounded continuously at the rate of 3.5%.

(a) Set up a differential equation that is satisfied by the amount $f(t)$ of money owed on the car loan at time $t$.

(b) How long will it take to pay off the car loan?

**9.** (*Home Prices in 2001*) The Federal Housing Finance Board reported that the national average one-family house† purchase price in October 2001 was $219,600. At the same time, the average interest rate on a conventional 30-year fixed-rate mortgage was 6.76%. A person purchased a home at the average price, paid a down payment equal to 10% of the purchase price, and financed the remaining balance with a 30-year fixed-rate mortgage. Assume that the person makes payments continuously at a constant annual rate $A$ and

that interest is compounded continuously at the rate of 6.76%.

(a) Set up a differential equation that is satisfied by the amount $f(t)$ of money owed on the mortgage at time $t$.

(b) Determine $A$, the rate of annual payments, that is required to pay off the loan in 30 years. What will the monthly payments be?

(c) Determine the total interest paid during the 30-year term mortgage.

**10.** Answer parts (a), (b), and (c) of Exercise 9 if the person takes a 15-year fixed-rate mortgage with a 6% interest rate and intends to pay off the entire loan in 15 years.

**11.** (*Elasticity of Demand*) Let $q = f(p)$ be the demand function for a certain commodity, where $q$ is the demand quantity and $p$ the price of one unit. In Section 5.3 we defined the elasticity of demand to be

$$E(p) = \frac{-pf'(p)}{f(p)}.$$

(a) Find a differential equation satisfied by the demand function if the elasticity of demand is a linear function of price given by $E(p) = p + 1$.

(b) Find the demand function in part (a), given $f(1) = 100$.

**12.** Find the demand function if the elasticity of demand is a linear function of price given by $E(p) = ap + b$, where $a$ and $b$ are constants.

**13.** When a red-hot steel rod is plunged in a bath of water that is kept at a constant temperature $10°C$, the temperature of the rod at time $t$, $f(t)$, satisfies the differential equation

$$y' = k[10 - y],$$

where $k > 0$ is a constant of proportionality. Determine $f(t)$ if the initial temperature of the rod is $f(0) = 350°C$ and $k = .1$.

**14.** Rework Exercise 13 for a metal with constant of proportionality $k = .2$. Which rod cools faster, the rod with constant of proportionality $k = .1$ or the rod with constant of proportionality $k = .2$? What can you say about the effect of varying the constant of proportionality in a cooling problem?

**15.** (*Determining the Time of Death*) A body was found in a room when the room's temperature was $70°F$. Let $f(t)$ denote the temperature of the body, $t$ hours from the time of death. According to Newton's law of cooling, $f$ satisfies a differential equation of the form

$$y' = k(T - y).$$

---

(a) Find $T$.

(b) After several measurements of the body's temperature, it was determined that when the temperature of the body was 80 degrees, it was decreasing at the rate of 5 degrees per hour. Find $k$.

(c) Suppose that at the time of death the body's temperature was about normal, say $98°F$. Determine $f(t)$.

(d) When the body was discovered, its temperature was $85°F$. Determine how long ago the person died.

**16.** Derive the formula for the population in Example 3, if the population in 1995 was 2 million. (The formula is given following the solution of Example 3.)

**17.** In an experiment, a certain type of bacteria was being added to a culture at the rate of $e^{.03t} + 2$ thousand bacteria per hour. Suppose that the bacteria grows at a rate proportional to the size of the culture at time $t$, with constant of proportionality $k = .45$. Let $P(t)$ denote the number of bacteria in the culture at time $t$. Find a differential equation satisfied by $P(t)$.

**18.** Find a formula for $P(t)$ in Exercise 17, if initially there were 10,000 bacteria present in the culture.

**19.** (*Dialysis and Creatinine Clearance*) According to the National Kidney Foundation, in 1997 more than 260,000 Americans suffered from chronic kidney failure and needed an artificial kidney (dialysis) to stay alive.[‡] When the kidneys fail, toxic waste products such as creatinine and urea build up in the blood. One way to remove these wastes is to use a process known as peritoneal dialysis, in which the patient's peritonium, or lining of the abdomen, is used as a filter. When the abdominal cavity is filled with a certain dialysate solution, the waste products in the blood filter through the peritonium into the solution. After a waiting period of several hours, the dialysate solution is drained out of the body along with the waste products.

In one dialysis session, the abdomen of a patient with an elevated concentration of creatinine in the blood equal to 110 grams per liter was filled with two liters of a dialysate (containing no creatinine). Let $f(t)$ denote the concentration of creatinine in the dialysate at time $t$. The rate of change of $f(t)$ is proportional to the difference between 110 (the maximum concentration that can be attained in the dialysate) and $f(t)$. Thus $f(t)$ satisfies the differential equation

$$y' = k(110 - y).$$

(a) Suppose that at the end of a four-hour dialysis session, the concentration in the dialysate was 75 grams per liter and it was rising at the rate of 10 grams per liter per hour. Find $k$.

(b) What is the rate of change of the concentration at the beginning of the dialysis session? By comparing with the rate at the end of the session, can you give a (simplistic) justification for draining and replacing the dialysate with a fresh solution after 4 hours of dialysis? [*Hint:* You do not need to solve the differential equation.]

**20.** Radium-226 is a radioactive substance with a decay constant .00043. Suppose that radium-226 is being continuously added to an initially empty container at a constant rate of 3 milligrams per year. Let $P(t)$ denote the number of grams of radium-226 remaining in the container after $t$ years.

(a) Find an initial value problem satisfied by $P(t)$.

(b) Solve the initial value problem for $P(t)$.

(c) What is the limit of the amount of radium-226 in the container as $t$ tends to infinity?

*In Exercises 21–25, solving the differential equations that arise from modeling may require using integration by parts. [See formula (1).]*

**21.** A person deposits an inheritance of $100,000 in a savings account that earns 4% interest compounded continuously. This person intends to make withdrawals that will increase gradually in size with time. Suppose that the annual rate of withdrawals is $2000 + 500t$ dollars per year, $t$ years from the time the account was opened.

(a) Assume that the withdrawals are made at a continuous rate. Set up a differential equation that is satisfied by the amount $f(t)$ in the account at time $t$.

(b) Determine $f(t)$.

(c) With the help of your calculator, plot $f(t)$ and approximate the time it will take before the account is depleted.

**22.** Suppose that you make an initial deposit of $500 in a savings account, and plan on making future deposits at a gradually increasing annual rate given by $90t + 810$ dollars per year, $t$ years after the initial deposit. Assume that the deposits are made continuously and that interest is compounded continuously at the rate of 6%. Let $P(t)$ denote the amount of money in the account.

(a) Set up an initial value problem that is satisfied by $P(t)$.

(b) Find $P(t)$.

**23.** After depositing an initial amount of $10,000 in a savings account that earns 4% interest compounded continuously, a person continued to make deposits for a certain period of time and then started to make withdrawals from the account. The annual rate of deposits was given by $3000 - 500t$ dollars per year, $t$ years from

[‡]Data obtained from the Web site of the National Kidney Foundation, www.kidney.org.

the time the account was opened. (Here negative rates of deposits correspond to withdrawals.)

(a) How many years did the person contribute to the account before starting to withdraw money from it?

(b) Let $P(t)$ denote the amount of money in the account, $t$ years after the initial deposit. Find an initial value problem satisfied by $P(t)$. (Assume that the deposits and withdrawals were made continuously.)

**24.** Figure 4 contains the solution of the initial value problem in Exercise 23.

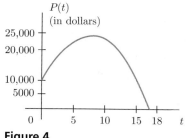

**Figure 4**

(a) With the help of the graph, approximate how long it will take before the account is depleted.

(b) Solve the initial value problem to determine $P(t)$.

(c) Use the formula for $P(t)$ to verify your answer in (a) with the help of a calculator.

**25.** (*Morphine Infusion*) Morphine is a drug that is widely used for pain management. However, morphine can be fatal by causing respiratory arrest. Since pain perception and drug tolerance vary with patients, morphine is gradually administered with small increments until pain is controlled or side effects begin to appear.

In one intravenous infusion, morphine was injected continuously at an increasing rate of $t$ milligrams per

hour. Suppose that the body removes the drug at a rate proportional to the amount of the drug present in the body, with constant of proportionality $k = .35$. Let $f(t)$ denote the amount of morphine in the body, $t$ hours from the beginning of the infusion.

(a) Find a differential equation satisfied by $f(t)$.

(b) Assuming that the infusion lasted 8 hours, determine the amount of morphine in the body during the infusion if at the beginning of the infusion the body was free of morphine.

## Technology Exercises

**26.** (*Therapeutic Level of a Drug*) A certain drug is administered intravenously to a patient at the continuous rate of $r$ milligrams per hour. The patient's body removes the drug from the bloodstream at a rate proportional to the amount of the drug in the blood, with constant of proportionality $k = .5$.

(a) Write a differential equation that is satisfied by the amount $f(t)$ of the drug in the blood at time $t$ (in hours).

(b) Find $f(t)$ assuming that $f(0) = 0$. (Give your answer in terms of $r$.)

(c) In a therapeutic two-hour infusion, the amount of drug in the body should reach 1 milligram within one hour of administration and stay above this level for another hour. However, to avoid toxicity, the amount of drug in the body should not exceed 2 milligrams at any time. Plot the graph of $f(t)$ on the interval $1 \le t \le 2$, as $r$ varies between 1 and 2 by increments of .1. That is, plot $f(t)$ for $r = 1, 1.1, 1.2, 1.3, \ldots, 2$. By looking at the graphs, pick the values of $r$ that yield a therapeutic and nontoxic two-hour infusion.

## SOLUTION TO PRACTICE PROBLEM 10.4

**1.** We reason as in Example 1. There are two influences on the way the savings account changes—the rate at which interest is added to the account and the rate at which money is withdrawn from the account. We know that interest is being added at a rate proportional to the amount in the account and withdrawals are made at a rate of $1200 per year. Since the rate of change of $f(t)$ is the net effect of these two influences, we see that $f(t)$ satisfies the first-order differential equation

$$y' = .04y - 1200$$

$$\begin{bmatrix} \text{rate of} \\ \text{change of } y \end{bmatrix} = \begin{bmatrix} \text{rate at which} \\ \text{interest is added} \end{bmatrix} - \begin{bmatrix} \text{rate at which} \\ \text{money is withdrawn} \end{bmatrix}.$$

The standard form of this first-order equation is $y' - .04y = -1200$.

## 10.5   Graphing Solutions of Differential Equations

In this section we present a technique for sketching solutions to differential equations of the form $y' = g(y)$ *without having to solve the differential equation.* This technique is based on the geometric interpretation of a differential equation that we introduced in Section 10.1 and used to construct slope fields. The technique is valuable for three reasons. First, there are many differential equations for which explicit solutions cannot be written down. Second, even when an explicit solution is available, we still face the problem of determining its behavior. For example, does the solution increase or decrease? If it increases, does it approach an asymptote or does it grow arbitrarily large? Third, and probably most significant, in many applications the explicit formula for a solution is unnecessary; only a general knowledge of the behavior of the solution is needed. That is, a qualitative understanding of the solution is sufficient.

The theory introduced in this section is part of what is called the *qualitative theory of differential equations.* We shall limit our attention to differential equations of the form $y' = g(y)$. Such differential equations are called *autonomous.* The term "autonomous" here means "independent of time" and refers to the fact that the right-hand side of $y' = g(y)$ depends only on $y$ and not on $t$. All applications studied in the next section involve autonomous differential equations.

Throughout this section we consider the values of each solution $y = f(t)$ only for $t \geq 0$. To introduce the qualitative theory, let us examine the graphs of the various typical solutions of the differential equation $y' = \frac{1}{2}(1 - y)(4 - y)$. The solution curves in Fig. 1 illustrate the following properties.

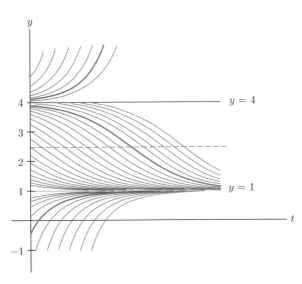

**Figure 1.**   Solutions of $y' = \frac{1}{2}(1 - y)(4 - y)$.

**Property I**   Corresponding to each zero of $g(y)$ there is a constant solution of the differential equation. Specifically, if $g(c) = 0$, the constant function $y = c$ is a solution. (The constant solutions in Fig. 1 are $y = 1$ and $y = 4$.)

**Property II**   The constant solutions divide the $ty$-plane into horizontal strips. Each nonconstant solution lies completely in one strip.

**Property III**   Each nonconstant solution is either strictly increasing or decreasing.

**Property IV**   Each nonconstant solution either is asymptotic to a constant solution or else increases or decreases without bound.

It can be shown that Properties I–IV are valid for the solutions of any autonomous differential equation $y' = g(y)$ provided that $g(y)$ is a "sufficiently well-behaved" function. We shall assume these properties in this chapter.

Using Properties I–IV, we can sketch the general shape of any solution curve by looking at the graph of the function $g(y)$ and the behavior of that graph near $y(0)$. The procedure for doing this is illustrated in the following example.

---

**EXAMPLE 1** Sketch the solution to $y' = e^{-y} - 1$ that satisfies $y(0) = -2$.

**Solution** Here $g(y) = e^{-y} - 1$. On a $yz$-coordinate system we draw the graph of the function $z = g(y) = e^{-y} - 1$ [Fig. 2(a)]. The function $g(y) = e^{-y} - 1$ has a zero when $y = 0$. Therefore, the differential equation $y' = e^{-y} - 1$ has the constant solution $y = 0$. We indicate this constant solution on a $ty$-coordinate system in Fig. 2(b). To begin the sketch of the solution satisfying $y(0) = -2$, we locate this initial value of $y$ on the (horizontal) $y$-axis in Fig. 2(a) and on the (vertical) $y$-axis in Fig. 2(b).

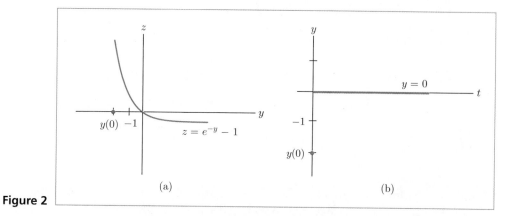

(a)         (b)

**Figure 2**

To determine whether the solution increases or decreases when it leaves the initial point $y(0)$ on the $ty$-graph, we look at the $yz$-graph and note that $z = g(y)$ is positive at $y = -2$ [Fig. 3(a)]. Consequently, since $y' = g(y)$, the derivative of the solution is positive, which implies that the solution is increasing. We indicate this by an arrow at the initial point in Fig. 3(b). Moreover, the solution $y$ will increase asymptotically to the constant solution $y = 0$, by Properties III and IV of autonomous differential equations.

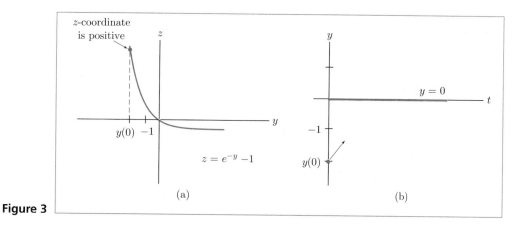

(a)         (b)

**Figure 3**

Next, we place an arrow in Fig. 4(a) to remind us that $y$ will move from $y = -2$ toward $y = 0$. As $y$ moves to the right toward $y = 0$ in Fig. 4(a), the $z$-coordinate of points on the graph of $g(y)$ becomes less positive; that is, $g(y)$ becomes less positive. Consequently, since $y' = g(y)$, the slope of the solution curve becomes less positive. Thus the solution curve is concave down, as we have shown in Fig. 4(b). ■

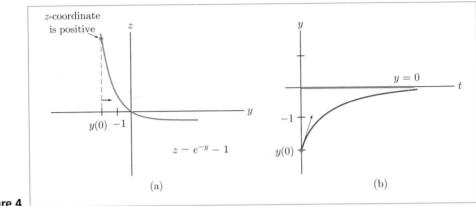

**Figure 4**

An important point to remember when sketching solutions is that $z$-coordinates on the $yz$-graph are values of $g(y)$, and since $y' = g(y)$, a $z$-coordinate gives the *slope* of the solution curve at the corresponding point on the $ty$-graph.

---

**EXAMPLE 2**   Sketch the graphs of the solutions to $y' = y + 2$ satisfying

   (a) $y(0) = 1$        (b) $y(0) = -3$

**Solution**   Here $g(y) = y + 2$. The graph of $z = g(y)$ is a straight line of slope 1 and $z$-intercept 2. [See Fig. 5(a).] This line crosses the $y$-axis only where $y = -2$. Thus the differential equation $y' = y + 2$ has one constant solution, $y = -2$. [See Fig. 5(b).]

   (a) We locate the initial value $y(0) = 1$ on the $y$-axes of both graphs in Fig. 5. The corresponding $z$-coordinate on the $yz$-graph is positive; therefore, the solution on the $ty$-graph has positive slope and is increasing as it leaves the initial point. We indicate this by an arrow in Fig. 5(b). Now, Property IV of autonomous differential equations implies that $y$ will increase without bound from its initial value. As we let $y$ increase from 1 in Fig. 6(a), we see that the $z$-coordinates

**Figure 5**

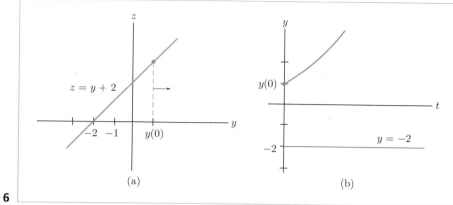

**Figure 6**

[i.e., values of $g(y)$] increase. Consequently, $y'$ is increasing, so the graph of the solution must be concave up. We have sketched the solution in Fig. 6(b).

(b) Next we graph the solution for which $y(0) = -3$. From the graph of $z = y+2$, we see that $z$ is negative when $y = -3$. This implies that the solution is decreasing as it leaves the initial point. (See Fig. 7.) It follows that the values of $y$ will continue to decrease without bound and become more and more negative. This means that on the $yz$-graph $y$ must move to the *left* [Fig. 8(a)]. We now examine what happens to $g(y)$ as $y$ moves to the left. (This is the opposite of the ordinary way to read a graph.) The $z$-coordinate becomes more negative;

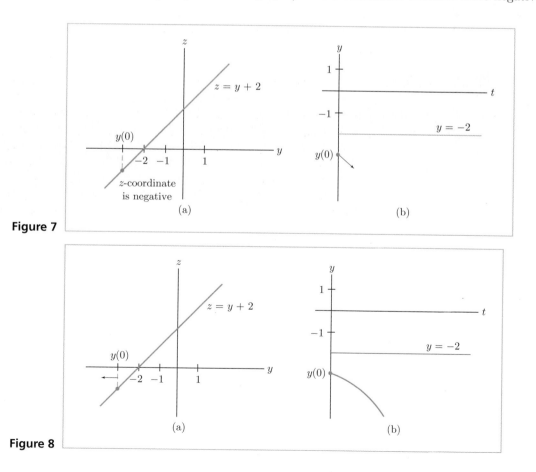

**Figure 7**

**Figure 8**

hence the slopes on the solution curve will become more negative. Thus the solution curve must be concave down, as in Fig. 8(b). ∎

From the preceding examples we can state a few rules for sketching a solution to $y' = g(y)$ with $y(0)$ given:

1. Sketch the graph of $z = g(y)$ on a $yz$-coordinate system. Find and label the zeros of $g(y)$.

2. For each zero $c$ of $g(y)$ draw the constant solution $y = c$ on the $ty$-coordinate system.

3. Plot $y(0)$ on the $y$-axes of the two coordinate systems.

4. Determine whether the value of $g(y)$ is positive or negative when $y = y(0)$. This tells us whether the solution is increasing or decreasing. On the $ty$-graph, indicate the direction of the solution through $y(0)$.

5. On the $yz$-graph, indicate which direction $y$ should move. (*Note:* If $y$ is moving *down* on the $ty$-graph, $y$ moves to the *left* on the $yz$-graph.) As $y$ moves in the proper direction on the $yz$-graph, determine whether $g(y)$ becomes more positive, less positive, more negative, or less negative. This tells us the concavity of the solution.

6. Beginning at $y(0)$ on the $ty$-graph, sketch the solution, being guided by the principle that the solution will grow (positively or negatively) without bound unless it encounters a constant solution. In this case, it will approach the constant solution asymptotically.

**EXAMPLE 3**  Sketch the solutions to $y' = y^2 - 4y$ satisfying $y(0) = 4.5$ and $y(0) = 3$.

**Solution**  Refer to Fig. 9. Since $g(y) = y^2 - 4y = y(y-4)$, the zeros of $g(y)$ are 0 and 4; hence the constant solutions are $y = 0$ and $y = 4$. The solution satisfying $y(0) = 4.5$ is increasing, because the $z$-coordinate is positive when $y = 4.5$ on the $yz$-graph. This solution continues to increase without bound. The solution satisfying $y(0) = 3$ is decreasing because the $z$-coordinate is negative when $y = 3$ on the $yz$-graph. This solution will decrease and approach asymptotically the constant solution $y = 0$.

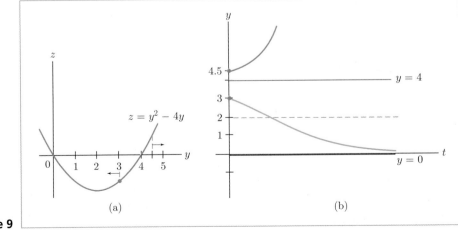

**Figure 9**

An additional piece of information about the solution satisfying $y(0) = 3$ may be obtained from the graph of $z = g(y)$. We know that $y$ decreases from 3 and approaches 0. From the graph of $z = g(y)$ in Fig. 9 it appears that at first the $z$-coordinates become more negative until $y$ reaches 2 and then becomes less negative as $y$ moves on toward 0. Since these $z$-coordinates are slopes on the solution curve, we conclude that as the solution moves downward from its initial point on the $ty$-coordinate system, its slope becomes more negative until the $y$-coordinate is 2 and then the slope becomes less negative as the $y$-coordinate approaches 0. Hence the solution is concave down until $y = 2$ and then is concave up. Thus there is an inflection point at $y = 2$, where the concavity changes. ∎

We saw in Example 3 that the inflection point at $y = 2$ was produced by the fact that $g(y)$ had a minimum at $y = 2$. A generalization of the argument in Example 3 shows that inflection points of solution curves occur at each value of $y$ where $g(y)$ has a nonzero relative maximum or minimum point. Thus we may formulate an additional rule for sketching a solution of $y' = g(y)$.

**7.** On the $ty$-coordinate system draw dashed horizontal lines at all values of $y$ at which $g(y)$ has a *nonzero* relative maximum or minimum point. A solution curve will have an inflection point whenever it crosses such a dashed line.

It is useful to note that when $g(y)$ is a quadratic function, as in Example 3, its maximum or minimum point occurs at a value of $y$ halfway between the zeros of $g(y)$. This is because the graph of a quadratic function is a parabola, which is symmetric about a vertical line through its vertex.

---

**EXAMPLE 4**   Sketch a solution to $y' = e^{-y}$ with $y(0) > 0$.

**Solution**   Refer to Fig. 10. Since $g(y) = e^{-y}$ is always positive, there are no constant solutions to the differential equation and every solution will increase without bound. When drawing solutions that asymptotically approach a horizontal line, we have no choice as to whether to draw it concave up or concave down. This decision will be obvious from its increasing or decreasing nature and from knowledge of inflection points. However, for solutions that grow without bound, we must look at $g(y)$ in order to determine concavity. In this example, as $t$ increases, the values of $y$ increase. As $y$ increases, $g(y)$ becomes less positive. Since $g(y) = y'$, we deduce that the slope of the solution curve becomes less positive; therefore, the solution curve is concave down. ∎

(a)                                    (b)

**Figure 10**

**Figure 11**

**Incorporating Technology** Any graphing calculator can be used to obtain an accurate graph of the function $g(y)$. In addition, the TI-85 and TI-86 graphing calculators can simultaneously graph several solutions to a differential equation. See Appendices C and D for details. Figure 11 shows the various types of solutions to the differential equation $y' = .5(1 - y)(4 - y)$.

## PRACTICE PROBLEMS 10.5

*Consider the differential equation $y' = g(y)$, where $g(y)$ is the function whose graph is drawn in Fig. 12.*

1. How many constant solutions are there to the differential equation $y' = g(y)$?

2. For what initial values $y(0)$ will the corresponding solution of the differential equation be an increasing function?

3. Is it true that if the initial value $y(0)$ is near 4, the corresponding solution will be asymptotic to the constant solution $y = 4$?

4. For what initial values $y(0)$ will the corresponding solution of the differential equation have an inflection point?

**Figure 12**

## EXERCISES 10.5

*Exercises 1–6 review concepts that are important in this section. In each exercise, sketch the graph of a function with the stated properties.*

1. Domain: $0 \le t \le 3$; $(0, 1)$ is on the graph; the slope is always positive, and the slope becomes less positive (as $t$ increases).

2. Domain: $0 \le t \le 4$; $(0, 2)$ is on the graph; the slope is always positive, and the slope becomes more positive (as $t$ increases).

3. Domain: $0 \le t \le 5$; $(0, 3)$ is on the graph; the slope is always negative, and the slope becomes less negative.

4. Domain: $0 \le t \le 6$; $(0, 4)$ is on the graph; the slope is always negative, and the slope becomes more negative.

5. Domain: $0 \le t \le 7$; $(0, 2)$ is on the graph; the slope is always positive, the slope becomes more positive as $t$ increases from 0 to 3, and the slope becomes less positive as $t$ increases from 3 to 7.

6. Domain: $0 \le t \le 8$; $(0, 6)$ is on the graph; the slope is always negative, the slope becomes more negative as $t$ increases from 0 to 3, and the slope becomes less negative as $t$ increases from 3 to 8.

*One or more initial conditions are given for each differential equation in the following exercises. Use the qualitative theory of autonomous differential equations to sketch the*

*graphs of the corresponding solutions. Include a $yz$-graph if one is not already provided. Always indicate the constant solutions on the $ty$-graph whether they are mentioned or not.*

7. $y' = 3 - \frac{1}{2}y$, $y(0) = 4$, $y(0) = 8$. [The graph of $z = g(y)$ is drawn in Fig. 13.]

8. $y' = \frac{2}{3}y - 3$, $y(0) = 3$, $y(0) = 6$. (See Fig. 14.)

9. $y' = y^2 - 5$, $y(0) = -4$, $y(0) = 2$, $y(0) = 3$. (See Fig. 15.)

10. $y' = 6 - y^2$, $y(0) = -3$, $y(0) = 3$. (See Fig. 16.)

11. $y' = -\frac{1}{3}(y + 2)(y - 4)$, $y(0) = -3$, $y(0) = -1$, $y(0) = 6$. (See Fig. 17.)

12. $y' = y^2 - 6y + 5$ or $y' = (y - 1)(y - 5)$, $y(0) = -2$, $y(0) = 2$, $y(0) = 4$, $y(0) = 6$. (See Fig. 18.)

13. $y' = y^3 - 9y$ or $y' = y(y^2 - 9)$, $y(0) = -4$, $y(0) = -1$, $y(0) = 2$, $y(0) = 4$. (See Fig. 19.)

14. $y' = 9y - y^3$, $y(0) = -4$, $y(0) = -1$, $y(0) = 2$, $y(0) = 4$. (See Fig. 20.)

15. Use the graph in Fig. 21 to sketch the solutions to the Gompertz growth equation

$$\frac{dy}{dt} = -\frac{1}{10}y \ln \frac{y}{100}$$

satisfying $y(0) = 10$ and $y(0) = 150$.

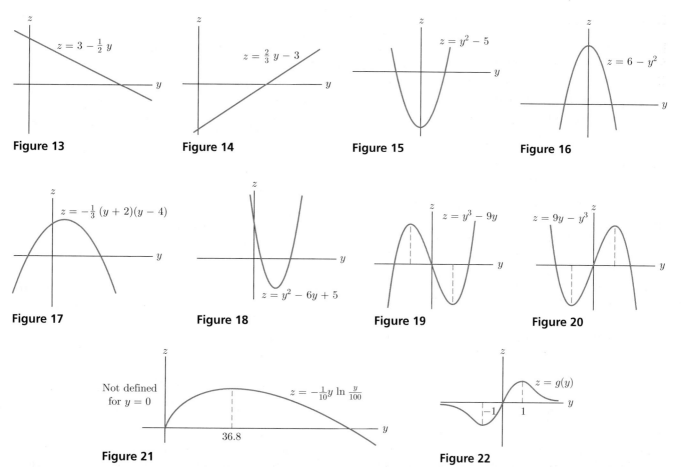

Figure 13

Figure 14

Figure 15

Figure 16

Figure 17

Figure 18

Figure 19

Figure 20

Figure 21

Figure 22

**16.** The graph of $z = -\frac{1}{2}y\ln(y/30)$ has the same general shape as the graph in Fig. 21 with relative maximum point at $y \approx 11.0364$ and $y$-intercept at $y = 30$. Sketch the solutions to the Gompertz growth equation

$$\frac{dy}{dt} = -\frac{1}{2}y\ln\frac{y}{30}$$

satisfying $y(0) = 1$, $y(0) = 20$, and $y(0) = 40$.

**17.** $y' = g(y)$, $y(0) = -.5$, $y(0) = .5$, where $g(y)$ is the function whose graph is given in Fig. 22.

**18.** $y' = g(y)$, $y(0) = 0$, $y(0) = 4$, where the graph of $g(y)$ is given in Fig. 23.

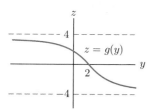

Figure 23

**19.** $y' = g(y)$, $y(0) = 0$, $y(0) = 1.2$, $y(0) = 5$, $y(0) = 7$, where the graph of $g(y)$ is given in Fig. 24.

Figure 24

**20.** $y' = g(y)$, $y(0) = 1$, $y(0) = 3$, $y(0) = 11$, where the graph of $g(y)$ is given in Fig. 25.

Figure 25

**21.** $y' = \frac{3}{4}y - 3$, $y(0) = 2$, $y(0) = 4$, $y(0) = 6$

**22.** $y' = -\frac{1}{2}y$, $y(0) = -2$, $y(0) = 0$, $y(0) = 2$

**23.** $y' = 5y - y^2$, $y(0) = 1$, $y(0) = 7$

**24.** $y' = -y^2 + 10y - 21$, $y(0) = 1$, $y(0) = 4$

25. $y' = y^2 - 3y - 4$, $y(0) = 0$, $y(0) = 3$
26. $y' = \frac{1}{2}y^2 - 3y$, $y(0) = 3$, $y(0) = 6$, $y(0) = 9$
27. $y' = y^2 + 2$, $y(0) = -1$, $y(0) = 1$
28. $y' = y - \frac{1}{4}y^2$, $y(0) = -1$, $y(0) = 1$
29. $y' = \sin y$, $y(0) = -\pi/6$, $y(0) = \pi/6$, $y(0) = 7\pi/4$
30. $y' = 1 + \sin y$, $y(0) = 0$, $y(0) = \pi$
31. $y' = 1/y$, $y(0) = -1$, $y(0) = 1$
32. $y' = y^3$, $y(0) = -1$, $y(0) = 1$
33. $y' = ky^2$, where $k$ is a negative constant, $y(0) = -2$, $y(0) = 2$.
34. $y' = ky(M - y)$, where $k > 0$, $M > 10$, and $y(0) = 1$.
35. $y' = ky - A$, where $k$ and $A$ are positive constants. Sketch solutions where $0 < y(0) < A/k$ and $y(0) > A/k$.
36. $y' = k(y - A)$, where $k < 0$ and $A > 0$. Sketch solutions where $y(0) < A$ and $y(0) > A$.
37. Suppose that once a sunflower plant has started growing, the rate of growth at any time is proportional to the product of its height and the difference between its height at maturity and its current height. Give a differential equation that is satisfied by $f(t)$, the height at time $t$, and sketch the solution.

38. A parachutist has a terminal velocity of $-176$ feet per second. That is, no matter how long a person falls, his or her speed will not exceed 176 feet per second, but it will get arbitrarily close to that value. The velocity in feet per second, $v(t)$, after $t$ seconds satisfies the differential equation $v'(t) = 32 - k \cdot v(t)$. What is the value of $k$?

### Technology Exercises

39. Draw the graph of $g(x) = (x - 2)^2(x - 6)^2$ and use the graph to sketch the solutions of the differential equation $y' = (y - 2)^2(y - 6)^2$ with initial conditions $y(0) = 1$, $y(0) = 3$, $y(0) = 5$, and $y(0) = 7$ on a $ty$-coordinate system.

40. Draw the graph of $g(x) = e^x - 100x^2 - 1$ and use the graph to sketch the solution of the differential equation $y' = e^y - 100y^2 - 1$ with initial condition $y(0) = 4$ on a $ty$-coordinate system.

## SOLUTIONS TO PRACTICE PROBLEMS 10.5

1. Three. The function $g(y)$ has zeros when $y$ is 2, 4, and 6. Therefore, $y' = g(y)$ has the constant functions $y = 2$, $y = 4$, and $y = 6$ as solutions.

2. For $2 < y(0) < 4$ and $y(0) > 6$. Since nonconstant solutions are either strictly increasing or strictly decreasing, a solution is an increasing function provided that it is increasing at time $t = 0$. This is the case when the first derivative is positive at $t = 0$. When $t = 0$, $y' = g(y(0))$. Therefore, the solution corresponding to $y(0)$ is increasing whenever $g(y(0))$ is positive.

3. Yes. If $y(0)$ is slightly to the right of 4, then $g(y(0))$ is negative, so the corresponding solution will be a decreasing function with values moving to the left closer and closer to 4. If $y(0)$ is slightly to the left of 4, then $g(y(0))$ is positive, so the corresponding solution will

be an increasing function with values moving to the right closer and closer to 4. (The constant solution $y = 4$ is referred to as a *stable* constant solution. The solution with initial value 4 stays at 4, and solutions with initial values near 4 move toward 4. The constant solution $y = 2$ is *unstable*. Solutions with initial values near 2 move away from 2.)

4. For $2 < y(0) < 3$ and $5 < y(0) < 6$. Inflection points of solutions correspond to relative maximum and relative minimum points of the function $g(y)$. If $2 < y(0) < 3$, the corresponding solution will be an increasing function. The values of $y$ will move to the right (toward 4) and therefore will cross 3, a place at which $g(y)$ has a relative maximum point. Similarly, if $5 < y(0) < 6$, the corresponding solution will be decreasing. The values of $y$ on the $yz$-graph will move to the left and cross 5.

## 10.6 Applications of Differential Equations

In this section we study real-life situations that may be modeled by an autonomous differential equation $y' = g(y)$. Here $y$ will represent some quantity that is changing with time, and the equation $y' = g(y)$ will be obtained from a description of the rate of change of $y$.

We have already encountered many situations where the rate of change of $y$ is *proportional* to some quantity. For example,

1. $y' = ky$: "the rate of change of $y$ is proportional to $y$" (exponential growth or decay).

2. $y' = k(M - y)$: "the rate of change of $y$ is proportional to the difference between $M$ and $y$" (Newton's law of cooling, for example).

Both situations involve *linear* first-order differential equations. The following example gives rise to an equation that is not linear. It concerns the rate at which a technological innovation may spread through an industry, a subject of interest to both sociologists and economists.

**EXAMPLE 1**    The by-product coke oven was first introduced into the iron and steel industry in 1894. It took about 30 years before all the major steel producers had adopted this innovation. Let $f(t)$ be the percentage of the producers that had installed the new coke ovens by time $t$. Then a reasonable model* for the way $f(t)$ increased is given by the assumption that the rate of change of $f(t)$ at time $t$ was proportional to the product of $f(t)$ and the percentage of firms that had not yet installed the new coke ovens at time $t$. Write a differential equation that is satisfied by $f(t)$.

Solution    Since $f(t)$ is the *percentage* of firms that have the new coke oven, $100 - f(t)$ is the percentage of firms that still have not installed any new coke ovens. We are told that the rate of change of $f(t)$ is proportional to the product of $f(t)$ and $100 - f(t)$. Hence there is a constant of proportionality $k$ such that

$$f'(t) = kf(t)[100 - f(t)].$$

Replacing $f(t)$ by $y$ and $f'(t)$ by $y'$, we obtain the desired differential equation,

$$y' = ky(100 - y).$$

Note that both $y$ and $100 - y$ are nonnegative quantities. Clearly, $y'$ must be positive, because $y = f(t)$ is an increasing function. Hence the constant $k$ must be positive.    ■

The differential equation obtained in Example 1 is a special case of the *logistic differential equation*,

$$y' = ky(a - y), \tag{1}$$

where $k$ and $a$ are positive constants. This equation is used as a simple mathematical model of a wide variety of physical phenomena. In Section 5.4, we described applications of the logistic equation to restricted population growth and to the spread of an epidemic. Let us use the qualitative theory of differential equations to gain more insight into this important equation.

The first step in sketching solutions of (1) is to draw the $yz$-graph. Rewriting the equation $z = ky(a - y)$ in the form

$$z = -ky^2 + kay,$$

we see that the equation is quadratic in $y$ and hence its graph will be a parabola. The parabola is concave down because the coefficient of $y^2$ is negative (since $k$ is a positive constant). The zeros of the quadratic expression $ky(a - y)$ occur where $y = 0$ and $y = a$. Since $a$ represents some positive constant, we select an arbitrary point on the positive $y$-axis and label it "$a$." With this information we can sketch a representative graph, as in Fig. 1. Note that the vertex of the parabola occurs at

---

*See E. Mansfield, "Technical Change and the Rate of Imitation," *Econometrica*, 29 (1961), 741–766.

$y = a/2$, halfway between the $y$-intercepts. (You should review how we obtained this graph, given only that $k$ and $a$ are positive constants. Similar situations will arise in the exercises.)

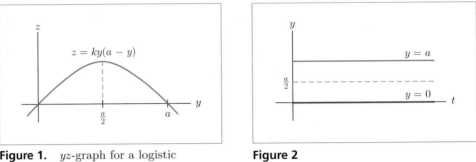

**Figure 1.** $yz$-graph for a logistic differential equation.

**Figure 2**

We begin the $ty$-graph as in Fig. 2, showing the constant solutions and placing a dashed line at $y = a/2$, where certain solution curves will have an inflection point. On either side of the constant solutions we choose initial values for $y$—say, $y_1$, $y_2$, $y_3$, $y_4$. Then we use the $yz$-graph to sketch the corresponding solution curves, as in Fig. 3.

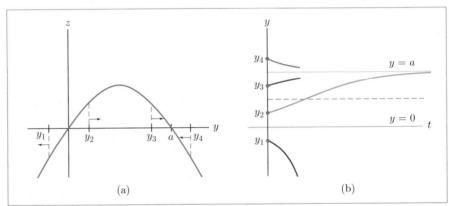

(a)

(b)

**Figure 3**

The solution in Fig. 3(b) beginning at $y_2$ has the general shape usually referred to as a *logistic curve*. This is the type of solution that would model the situation described in Example 1. The solution in Fig. 3(b) beginning at $y_1$ usually has no physical significance. The other solutions shown in Fig. 3(b) can occur in practice, particularly in the study of population growth.

In ecology, the growth of a population is often described by a logistic equation written in the form

$$\frac{dN}{dt} = rN\left(\frac{K - N}{K}\right) \tag{2}$$

or, equivalently,

$$\frac{dN}{dt} = \frac{r}{K}N(K - N),$$

where $N$ is used instead of $y$ to denote the size of the population at time $t$. Typical solutions of this equation are sketched in Fig. 4. The constant $K$ is called the *carrying capacity* of the environment. When the initial population is close to zero, the population curve has the typical S-shaped appearance, and $N$ approaches the

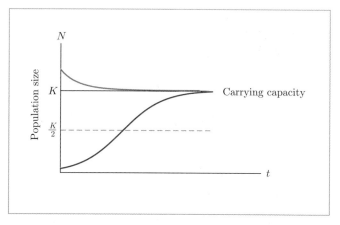

**Figure 4.** A logistic model for population change.

carrying capacity asymptotically. When the initial population is greater than $K$, the population decreases in size, again asymptotically approaching the carrying capacity.

The quantity $(K - N)/K$ in (2) is a fraction between 0 and 1. It reflects the limiting effect of the environment upon the population and is close to 1 when $N$ is close to 0. If this fraction were replaced by the constant 1, then (2) would become

$$\frac{dN}{dt} = rN.$$

This is the equation for ordinary exponential growth, where $r$ is the growth rate. For this reason the parameter $r$ in (2) is called the *intrinsic rate of growth* of the population. It expresses how the population would grow if the environment were to permit unrestricted exponential growth.

Let us now consider a concrete situation that gives rise to a logistic equation.

---

**EXAMPLE 2**   **A fish population**   A pond on a fish farm has a carrying capacity of 1000 fish. The pond was originally stocked with 100 fish. Let $N(t)$ denote the number of fish in the pond after $t$ months.

(a) Set up a logistic differential equation satisfied by $N(t)$ and plot an approximate graph of the fish population.

(b) Find the size of the population of fish with the highest rate of growth. Find this rate, given that the intrinsic rate of growth is .3.

**Solution**   (a) We are told that the equation is a logistic equation with carrying capacity $K = 1000$. Hence, from (2), the equation is

$$\frac{dN}{dt} = rN \left( \frac{1000 - N}{1000} \right) = \frac{r}{1000} N(1000 - N).$$

The fish population at time $t$ is given by the solution of this differential equation with the initial condition $N(0) = 100$. Even though we do not have a numerical value of the intrinsic rate $r$, we can still estimate the shape of the solution by using the qualitative techniques. First sketch the constant solutions, $N = 0$ and $N = 1000$, then place a dashed line at $N = 500$, where certain solutions will have inflection points. The solution starting at $N = 100$ is a typical logistic curve. It is increasing, with a horizontal asymptote $N = 1000$ and inflection point at $N = 500$, where the graph changes concavity. A solution curve satisfying these properties is shown in Fig. 5(b).

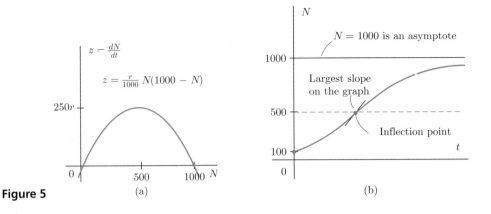

**Figure 5**

(b) Since the question concerns the rate of growth, we should look at the equation itself for answers. The equation tells us that the rate of growth is given by the quadratic function

$$\frac{dN}{dt} = \frac{r}{1000} N(1000 - N),$$

whose graph is an inverted parabola with intercepts at $N = 0$ and $N = 1000$, as shown in Fig. 5(a). Since the parabola is concave down, it has a maximum at $N = 500$, halfway between 0 and 1000. Thus the size of the population with the highest rate of growth is 500. To find the numerical value of the fastest growth rate, given that $r = .3$, we substitute $r = .3$ and $N = 500$ into the equation and get

$$\frac{dN}{dt}\bigg|_{N=500} = \frac{.3}{1000}(500)(1000 - 500) = 75 \text{ fish per month.}$$

This is the maximum rate of growth of the fish population. It is attained when 500 fish are in the pond. Note that 500 is not the maximum size of the population. In fact, we know that the fish population will approach 1000 asymptotically [see Fig. 5(b)]. ◼

We now turn to applications that involve a different sort of autonomous differential equation. The main idea is illustrated in the following familiar example of a savings account that we discussed in Section 10.4.

**EXAMPLE 3**  A savings account earns 6% interest per year, compounded continuously, and continuous withdrawals are made from the account at the rate of $900 per year. Set up a differential equation that is satisfied by the amount $f(t)$ of money in the account at time $t$. Sketch typical solutions of the differential equation.

**Solution**  At first, let us ignore the withdrawals from the account. In Section 5.2 we discussed continuous compounding of interest and showed that if no deposits or withdrawals are made, then $f(t)$ satisfies the equation

$$y' = .06y.$$

That is, the savings account grows at a rate proportional to the size of the account. Since this growth comes from the interest, we conclude that *interest is being added to the account at a rate proportional to the amount in the account.*

Now suppose that continuous withdrawals are made from this same account at the rate of $900 per year. Then there are two influences on the way the amount of money in the account changes — the rate at which interest is added and the rate at which money is withdrawn. The rate of change of $f(t)$ is the *net effect* of these two influences. That is, $f(t)$ now satisfies the equation

$$y' = .06y - 900$$

$$\begin{bmatrix} \text{rate of} \\ \text{change of } y \end{bmatrix} = \begin{bmatrix} \text{rate at which} \\ \text{interest is added} \end{bmatrix} - \begin{bmatrix} \text{rate at which} \\ \text{money is withdrawn} \end{bmatrix}.$$

The qualitative sketches for this differential equation are given in Fig. 6. The constant solution is found by solving $.06y - 900 = 0$, which gives $y = 900/.06 = 15{,}000$. If the initial amount $y(0)$ in the account is \$15,000, the balance in the account will always be \$15,000. If the initial amount is greater than \$15,000, the savings account will accumulate money without bound. If the initial amount is less than \$15,000, the account balance will decrease. Presumably, the bank will stop withdrawals when the account balance reaches zero. ■

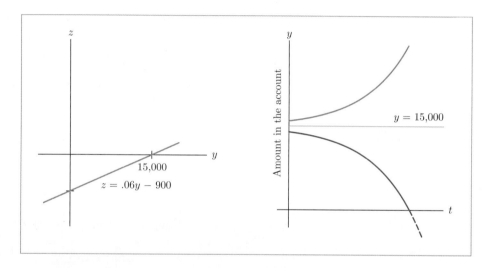

**Figure 6.** A differential equation model for a savings account: $y' = .06y - 900$.

We may think of the savings account in Example 3 as a compartment or container into which money (interest) is being steadily added and also from which money is being steadily withdrawn. (See Fig. 7.)

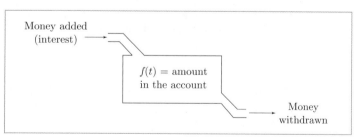

**Figure 7.** A one-compartment model in economics.

A similar situation arises frequently in physiology in what are called "one-compartment problems."* Typical examples of a compartment are a person's lungs, the digestive system, and the cardiovascular system. A common problem is to study the rate at which the amount of some substance in the compartment is changing

---

*See William Simon, *Mathematical Techniques for Physiology and Medicine* (New York: Academic Press, 1972), Chap. V.

when two or more processes act on the substance in the compartment. In many important cases, each of these processes changes the substance either at a constant rate or at a rate proportional to the amount in the compartment.

An earlier example of such a one-compartment problem, discussed in Section 5.4, concerned the continuous infusion of glucose into a patient's bloodstream. A similar situation is discussed in the next example.

---

**EXAMPLE 4** **A one-compartment mixing process** Consider a flask that contains 3 liters of salt water. Suppose that water containing 25 grams per liter of salt is pumped into the flask at the rate of 2 liters per hour, and the mixture, being steadily stirred, is pumped out of the flask at the same rate. Find a differential equation satisfied by the amount of salt $f(t)$ in the flask at time $t$.

**Solution** Let $f(t)$ be the amount of salt measured in grams. Since the volume of the mixture in the flask is being held constant at 3 liters, the concentration of salt in the flask at time $t$ is

$$[\text{concentration}] = \frac{[\text{amount of salt}]}{[\text{volume of mixture}]} = \frac{f(t) \text{ grams}}{3 \text{ liters}} = \frac{1}{3}f(t) \frac{\text{grams}}{\text{liter}}.$$

Next we compute the rates at which salt enters and leaves the flask at time $t$:

$$[\text{rate of salt entering}] = [\text{entering concentration}] \times [\text{flow rate}]$$

$$= \left[25 \frac{\text{grams}}{\text{liter}}\right] \times \left[2 \frac{\text{liters}}{\text{hour}}\right]$$

$$= 50 \frac{\text{grams}}{\text{hour}}.$$

$$[\text{rate of salt leaving}] = [\text{concentration}] \times [\text{flow rate}]$$

$$= \left[\frac{1}{3}f(t) \frac{\text{grams}}{\text{liter}}\right] \times \left[2 \frac{\text{liters}}{\text{hour}}\right]$$

$$= \frac{2}{3}f(t) \frac{\text{grams}}{\text{hour}}.$$

The *net* rate of change of salt (in grams per hour) at time $t$ is $f'(t) = 50 - \frac{2}{3}f(t)$. Hence the desired differential equation is

$$y' = 50 - \tfrac{2}{3}y. \qquad \blacksquare$$

**Differential Equations in Population Genetics** In population genetics, hereditary phenomena are studied on a populational level rather than on an individual level. Consider a particular hereditary feature of an animal, such as the length of the hair. Suppose that basically there are two types of hair for a certain animal—long hair and short hair. Also suppose that long hair is the dominant type. Let $A$ denote the gene responsible for long hair and $a$ the gene responsible for short hair. Each animal has a pair of these genes—either $AA$ ("dominant" individuals), or $aa$ ("recessive" individuals), or $Aa$ ("hybrid" individuals). If there are $N$ animals in the population, then there are $2N$ genes in the population controlling hair length. Each $Aa$ individual has one $a$ gene, and each $aa$ individual has two $a$ genes. The total number of $a$ genes in the population divided by $2N$ gives the fraction of $a$ genes. This fraction is called the *gene frequency of* $a$ in the population. Similarly, the fraction of $A$ genes is called the gene frequency of $A$. Note that

$$\begin{bmatrix} \text{gene} \\ \text{frequency} \\ \text{of } a \end{bmatrix} + \begin{bmatrix} \text{gene} \\ \text{frequency} \\ \text{of } A \end{bmatrix} = \frac{[\text{number of } a \text{ genes}]}{2N} + \frac{[\text{number of } A \text{ genes}]}{2N}$$

$$= \frac{2N}{2N} = 1. \tag{3}$$

We shall denote the gene frequency of $a$ by $q$. From (3) it follows that the gene frequency of $A$ is $1 - q$.

An important problem in population genetics involves the way the gene frequency $q$ changes as the animals in the population reproduce. If each unit on the time axis represents one "generation," we can consider $q$ as a function of time $t$ (Fig. 8). In general, many hundreds or thousands of generations are studied, so the time for one generation is small compared with the overall time period. For many purposes, $q$ is considered to be a differentiable function of $t$. In what follows we assume that the population mates at random and that the distribution of $a$ and $A$ genes is the same for males and females. In this case one can show from elementary probability theory that the gene frequency is essentially constant from one generation to the next when no "disturbing factors" are present, such as mutation or external influences on the population. We shall discuss differential equations that describe the effect of such disturbing factors on $q(t)$.*

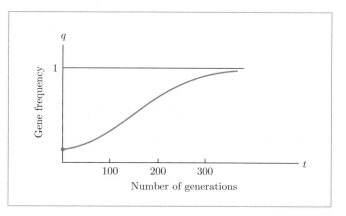

**Figure 8.**  Gene frequency in a population.

Suppose that in every generation a fraction $\nu$ of the $a$ genes mutate and become $A$ genes. Then the rate of change of the gene frequency $q$ due to this mutation is

$$\frac{dq}{dt} = -\nu q. \tag{4}$$

To understand this equation, think of $q$ as a measure of the number of $a$ genes, and think of the $a$ genes as a population that is losing members at a constant percentage rate of $100\nu\%$ per generation (i.e., per unit time). This is an exponential decay process, so $q$ satisfies the exponential decay equation (4). Now suppose, instead, that in every generation a fraction $\mu$ of the $A$ genes mutate into $a$ genes. Since the gene frequency of $A$ is $1 - q$, the decrease in the gene frequency of $A$ due to mutation will be $\mu(1 - q)$ per generation. But this must match the increase in the gene frequency of $a$, since from (3) the sum of the two gene frequencies is constant.

*See C. C. Li, *Population Genetics* (Chicago: University of Chicago Press, 1955), pp. 240–263, 283–286.

Thus the rate of change of the gene frequency of $a$ per generation, due to the mutation from $A$ to $a$, is given by

$$\frac{dq}{dt} = \mu(1 - q).$$

When mutation occurs, it frequently happens that in each generation a fraction $\mu$ of $A$ mutate to $a$ and at the same time a fraction $\nu$ of $a$ mutate to $A$. The *net effect* of these two influences on the gene frequency $q$ is described by the equation

$$\frac{dq}{dt} = \mu(1 - q) - \nu q. \tag{5}$$

(The situation here is analogous to the one-compartment problems discussed earlier.)

Let us make a qualitative analysis of equation (5). To be specific, we take $\mu = .00003$ and $\nu = .00001$. Then

$$\frac{dq}{dt} = .00003(1 - q) - .00001q = .00003 - .00004q$$

or

$$\frac{dq}{dt} = -.00004(q - .75). \tag{6}$$

Figure 9(a) shows the graph of $z = -.00004(q - .75)$ with the $z$-axis scale greatly enlarged. Typical solution curves are sketched in Fig. 9(b). We see that the gene frequency $q = .75$ is an equilibrium value. If the initial value of $q$ is smaller than .75, the value of $q$ will rise under the effect of the mutations; after many generations, it will be approximately .75. If the initial value of $q$ is between .75 and 1.00, then $q$ will eventually decrease to .75. The equilibrium value is completely determined by the magnitudes of the two opposing rates of mutation $\mu$ and $\nu$. From (6) we see that the rate of change of gene frequency is proportional to the difference between $q$ and the equilibrium value .75.

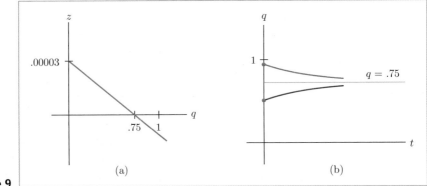

**Figure 9**

In the study of how a population adapts to an environment over a long period, geneticists assume that some hereditary types have an advantage over others in survival and reproduction. Suppose that the adaptive ability of the hybrid $(Aa)$ individuals is slightly greater than that of both the dominant $(AA)$ and the recessive $(aa)$ individuals. In this case it turns out that the rate of change of gene frequency due to this "selection pressure" is

$$\frac{dq}{dt} = q(1 - q)(c - dq), \tag{7}$$

where $c$ and $d$ are positive constants with $c < d$. On the other hand, if the adaptive ability of the hybrid individuals is slightly less than that of both the dominant and the recessive individuals, it can be shown that

$$\frac{dq}{dt} = kq(1 - q)(2q - 1), \tag{8}$$

where $k$ is a constant between 0 and 1, called the *coefficient of selection against hybrids*.

It is possible to consider the joint effects of mutation and natural selection. Suppose that mutations from $A$ to $a$ occur at a rate $\mu$ per generation and from $a$ to $A$ at a rate $\nu$ per generation. Suppose also that selection is against recessive individuals (i.e., recessive do not adapt as well as the rest of the population). Then the net rate of change in gene frequency turns out to be

$$\frac{dq}{dt} = \mu(1 - q) - \nu q - kq^2(1 - q).$$

Here $\mu(1 - q)$ represents the gain in $a$ genes from mutations $A \to a$, the term $\nu q$ is the loss in $a$ genes from mutations $a \to A$, and the term $kq^2(1 - q)$ represents the loss in $a$ genes due to natural selection pressures.

## PRACTICE PROBLEMS 10.6

1. Refer to Example 4, involving the flow of salt water through a flask. Will $f(t)$ be an increasing or a decreasing function?

2. (*Rate of Litter Accumulation*) In a certain tropical forest, "litter" (mainly dead vegetation such as leaves and vines) forms on the ground at the rate of 10 grams per square centimeter per year. At the same time, however, the litter is decomposing at the rate of 80% per year. Let $f(t)$ be the amount of litter (in grams per square centimeter) present at time $t$. Find a differential equation satisfied by $f(t)$.

## EXERCISES 10.6

*In Exercises 1–4, you are given a logistic equation with one or more initial conditions. (a) Determine the carrying capacity and intrinsic rate. (b) Sketch the graph of $\frac{dN}{dt}$ versus $N$ in a Nz-plane. (c) In the tN-plane, plot the constant solutions and place a dashed line where the concavity of certain solutions may change. (d) Sketch the solution curve corresponding to each given initial condition.*

1. $\dfrac{dN}{dt} = N(1 - N)$, $N(0) = .75$

2. $\dfrac{dN}{dt} = .3N(100 - N)$, $N(0) = 25$

3. $\dfrac{dN}{dt} = -.01N^2 + N$, $N(0) = 5$

4. $\dfrac{dN}{dt} = -N^2 + N$, $N(0) = .5$

5. Answer part (a) in Example 2 if the pond was originally stocked with 600 fish and all other data is unchanged. How does the graph of the fish population in this case differ from the one in Example 2?

6. Answer parts (a) and (b) in Example 2 if the pond has a carrying capacity of 2000 fish and all other data is unchanged.

7. (*Social Diffusion*) For information being spread by mass media, rather than through individual contact, the rate of spread of the information at any time is proportional to the percentage of the population not having the information at that time. Give the differential equation that is satisfied by $y = f(t)$, the percentage of the population having the information at time $t$. Assume that $f(0) = 1$. Sketch the solution.

8. (*Gravity*) At one point in his study of a falling body starting from rest, Galileo conjectured that its velocity at any time is proportional to the distance it has dropped. Using this hypothesis, set up the differential equation whose solution is $y = f(t)$, the distance fallen by time $t$. By making use of the initial value, show why Galileo's original conjecture is invalid.

9. (*Autocatalytic Reaction*) In an autocatalytic reaction, one substance is converted into a second substance in such a way that the second substance catalyzes its own formation. This is the process by which trypsinogen is converted into the enzyme trypsin. The reaction starts only in the presence of some trypsin, and each molecule

of trypsinogen yields one molecule of trypsin. The rate of formation of trypsin is proportional to the product of the amounts of the two substances present. Set up the differential equation that is satisfied by $y = f(t)$, the amount (number of molecules) of trypsin present at time $t$. Sketch the solution. For what value of $y$ is the reaction proceeding the fastest? [*Note:* Letting $M$ be the total amount of the two substances, the amount of trypsinogen present at time $t$ is $M - f(t)$.]

10. (*Drying*) A porous material dries outdoors at a rate that is proportional to the moisture content. Set up the differential equation whose solution is $y = f(t)$, the amount of water at time $t$ in a towel on a clothesline. Sketch the solution.

11. (*Movement of Solutes through a Cell Membrane*) Let $c$ be the concentration of a solute outside a cell that we assume to be constant throughout the process — that is, unaffected by the small influx of the solute across the membrane due to a difference in concentration. The rate of change of the concentration of the solute inside the cell at any time $t$ is proportional to the difference between the outside concentration and the inside concentration. Set up the differential equation whose solution is $y = f(t)$, the concentration of the solute inside the cell at time $t$. Sketch a solution.

12. An experimenter reports that a certain strain of bacteria grows at a rate proportional to the square of the size of the population. Set up a differential equation that describes the growth of the population. Sketch a solution.

13. (*Chemical Reaction*) Suppose that substance $A$ is converted into substance $B$ at a rate that, at any time $t$, is proportional to the square of the amount of $A$. This situation occurs, for instance, when it is necessary for two molecules of $A$ to collide in order to create one molecule of $B$. Set up the differential equation that is satisfied by $y = f(t)$, the amount of substance $A$ at time $t$. Sketch a solution.

14. (*War Fever*) L. F. Richardson proposed the following model to describe the spread of war fever.[*] If $y = f(t)$ is the percent of the population advocating war at time $t$, then the rate of change of $f(t)$ at any time is proportional to the product of the percentage of the population advocating war and the percentage not advocating war. Set up a differential equation that is satisfied by $y = f(t)$ and sketch a solution.

15. (*Capital Investment Model*) In economic theory, the following model is used to describe a possible capital investment policy. Let $f(t)$ represent the total invested capital of a company at time $t$. Additional capital is invested whenever $f(t)$ is below a certain equilibrium value $E$, and capital is withdrawn whenever $f(t)$ exceeds $E$. The rate of investment is proportional to the difference between $f(t)$ and $E$. Construct a differential

equation whose solution is $f(t)$ and sketch two or three typical solution curves.

16. (*Evans Price Adjustment Model*) Consider a certain commodity that is produced by many companies and purchased by many other firms. Over a relatively short period there tends to be an equilibrium price $p_0$ per unit of the commodity that balances the supply and the demand. Suppose that, for some reason, the price is different from the equilibrium price. The Evans price adjustment model says that the rate of change of price with respect to time is proportional to the difference between the actual market price $p$ and the equilibrium price. Write a differential equation that expresses this relation. Sketch two or more solution curves.

17. (*Continuous Annuity*) A *continuous annuity* is a steady stream of money that is paid to some person. Such an annuity may be established, for example, by making an initial deposit in a savings account and then making steady withdrawals to pay the continuous annuity. Suppose that an initial deposit of \$5400 is made into a savings account that earns $5\frac{1}{2}\%$ interest compounded continuously, and immediately continuous withdrawals are begun at the rate of \$300 per year. Set up the differential equation that is satisfied by the amount $f(t)$ of money in the account at time $t$. Sketch the solution.

18. (*A Fish Population with Harvesting*) The fish population in a pond with carrying capacity 1000 is modeled by the logistic equation

$$\frac{dN}{dt} = \frac{.4}{1000} N(1000 - N).$$

Here $N(t)$ denotes the number of fish at time $t$ in years. When the number of fish reached 250, the owner of the pond decided to remove 50 fish per year.

(a) Modify the differential equation to model the population of fish from the time it reached 250.

(b) Plot several solution curves of the new equation, including the solution curve with $N(0) = 250$.

(c) Is the practice of catching 50 fish per year sustainable or will it deplete the fish population in the pond? Will the size of the fish population ever come close to the carrying capacity of the pond?

19. A company wishes to set aside funds for future expansion and so arranges to make continuous *deposits* into a savings account at the rate of \$10,000 per year. The savings account earns 5% interest compounded continuously.

(a) Set up the differential equation that is satisfied by the amount $f(t)$ of money in the account at time $t$.

(b) Solve the differential equation in part (a), assuming that $f(0) = 0$, and determine how much money will be in the account at the end of 5 years.

---

[*]See L. F. Richardson, "War Moods I," *Psychometrica*, 1948, p. 13.

20. A company arranges to make continuous deposits into a savings account at the rate of $P$ dollars per year. The savings account earns 5% interest compounded continuously. Find the approximate value of $P$ that will make the savings account balance amount to \$50,000 in 4 years.

21. The air in a crowded room full of people contains .25% carbon dioxide ($CO_2$). An air conditioner is turned on that blows fresh air into the room at the rate of 500 cubic feet per minute. The fresh air mixes the stale air and the mixture leaves the room at the rate of 500 cubic feet per minute. The fresh air contains .01% $CO_2$, and the room has a volume of 2500 cubic feet.

    (a) Find a differential equation satisfied by the amount $f(t)$ of $CO_2$ in the room at time $t$.

    (b) The model developed in part (a) ignores the $CO_2$ produced by the respiration of the people in the room. Suppose that the people generate .08 cubic foot of $CO_2$ per minute. Modify the differential equation in (a) to take into account this additional source of $CO_2$.

22. A certain drug is administered intravenously to a patient at the continuous rate of 5 milligrams per hour. The patient's body removes the drug from the bloodstream at a rate proportional to the amount of the drug in the blood. Write a differential equation that is satisfied by the amount $f(t)$ of the drug in the blood at time $t$. Sketch a typical solution.

23. A single dose of iodine is injected intravenously into a patient. Suppose that the iodine mixes thoroughly in the blood before any is lost as a result of metabolic processes (and ignore the time required for this mixing process). Iodine will leave the blood and enter the thyroid gland at a rate proportional to the amount of iodine in the blood. Also, iodine will leave the blood and pass into the urine at a (different) rate proportional to the amount of iodine in the blood. Suppose that the iodine enters the thyroid at the rate of 4% per hour, and the iodine enters the urine at the rate of 10% per hour. Let $f(t)$ denote the amount of iodine in the blood at time $t$. Write a differential equation satisfied by $f(t)$.

24. Show that the mathematical model in Practice Problem 2 predicts that the amount of litter in the forest will eventually stabilize. What is the "equilibrium level" of litter in that problem? (*Note*: Today most forests are close to their equilibrium levels. This was not so during the Carboniferous Period when the great coal deposits were formed.)

25. In the study of the effect of natural selection on a population, one encounters the differential equation

$$\frac{dq}{dt} = -.0001q^2(1 - q),$$

where $q$ is the frequency of a gene $a$ and the selection pressure is against the recessive genotype $aa$. Sketch a solution of this equation when $q(0)$ is close to but slightly less than 1.

26. Typical values of $c$ and $d$ in equation (7) are $c = .15$, $d = .50$, and a typical value of $k$ in equation (8) is $k = .05$. Sketch representative solutions for the equations

    (a) $\frac{dq}{dt} = q(1 - q)(.15 - .50q)$ (selection favoring hybrids),

    (b) $\frac{dq}{dt} = .05q(1 - q)(2q - 1)$ (selection against hybrids).

    Consider various initial conditions with $q(0)$ between 0 and 1. Discuss possible genetic interpretations of these curves; that is, describe the effect of "selection" on the gene frequency $q$ in terms of the various initial conditions.

## SOLUTIONS TO PRACTICE PROBLEMS 10.6

1. The nature of the function $f(t)$ depends on the initial amount of salt water in the flask. Figure 10 contains solutions for three different initial amounts, $y(0)$. If the initial amount is less than 75 grams, the amount of salt in the flask will increase asymptotically to 75. If the initial concentration is greater than 75 grams, the amount of salt in the flask will decrease asymptotically to 75. Of course, if the initial concentration is exactly 75 grams, the amount of salt in the flask will remain constant.

2. This problem resembles a one-compartment problem, where the forest floor is the compartment. We have

$$\begin{bmatrix} \text{rate of change} \\ \text{of litter} \end{bmatrix} = \begin{bmatrix} \text{rate of litter} \\ \text{formation} \end{bmatrix} - \begin{bmatrix} \text{rate of litter} \\ \text{decomposition} \end{bmatrix}.$$

If $f(t)$ is the amount of litter (in grams per square centimeter) at time $t$, then the 80% decomposition rate means that at time $t$ the litter is decaying at the rate of $.80f(t)$ grams per square centimeter per year. Thus the net rate of change of litter is $f'(t) = 10 - .80f(t)$. The desired differential equation is $y' = 10 - .80y$.

**Figure 10**

# 10.7 Numerical Solution of Differential Equations

Many differential equations that arise in real-life applications cannot be solved by *any* known method. However, approximate solutions may be obtained by several different numerical techniques. In this section we describe what is known as *Euler's method* for approximating solutions to initial value problems of the form

$$y' = g(t, y), \qquad y(a) = y_0 \tag{1}$$

for values of $t$ in some interval $a \leq t \leq b$. Here $g(t, y)$ is some reasonably well-behaved function of two variables. Equations of the form $y' = p(t)q(y)$ (studied in Section 10.2) and linear equations of the form $y' = -a(t)y + b(t)$ (studied in Sections 10.3 and 10.4) are special cases of (1).

In the following discussion we assume that $f(t)$ is a solution of (1) for $a \leq t \leq b$. The basic idea on which Euler's method rests is the following: *If the graph of $y = f(t)$ passes through some given point $(t, y)$, the slope of the graph (i.e., the value of $y'$) at that point is just $g(t, y)$, because $y' = g(t, y)$. This is the same idea that we used in Section 10.1 in our discussion of slope fields.* Euler's method uses this observation to approximate the graph of $f(t)$ by a polygonal path, such as in Fig. 1.

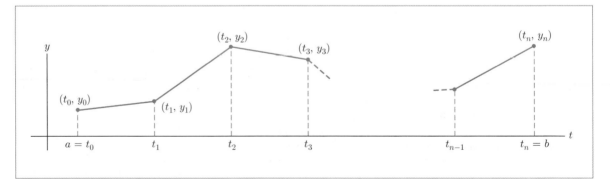

**Figure 1.** A polygonal path.

The $t$-axis from $a$ to $b$ is subdivided by the equally spaced points $t_0, t_1, \ldots, t_n$. Each of the $n$ subintervals has length $h = (b-a)/n$. The initial condition $y(a) = y_0$ in (1) implies that the graph of the solution $f(t)$ passes through the point $(t_0, y_0)$. As noted previously, the slope of this graph at $(t_0, y_0)$ must be $g(t_0, y_0)$. Thus on the first subinterval, Euler's method approximates the graph of $f(t)$ by the straight line

$$y = y_0 + g(t_0, y_0) \cdot (t - t_0),$$

which passes through $(t_0, y_0)$ and has slope $g(t_0, y_0)$. (See Fig. 2.) When $t = t_1$, the $y$-coordinate on this line is

$$y_1 = y_0 + g(t_0, y_0) \cdot (t_1 - t_0) = y_0 + g(t_0, y_0) \cdot h.$$

Since the graph of $f(t)$ is close to the point $(t_1, y_1)$ on the line, the slope of the graph of $f(t)$ when $t = t_1$ will be close to $g(t_1, y_1)$. So we draw the straight line

$$y = y_1 + g(t_1, y_1) \cdot (t - t_1) \tag{2}$$

through $(t_1, y_1)$ with slope $g(t_1, y_1)$, and we use this line to approximate $f(t)$ on the second subinterval. From (2) we determine an estimate $y_2$ for the value of $f(t)$ at $t = t_2$:

$$y_2 = y_1 + g(t_1, y_1) \cdot h.$$

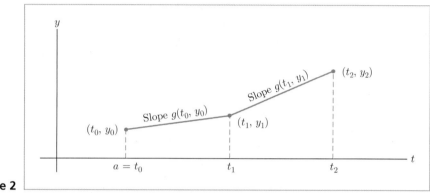

**Figure 2**

The slope of the graph of $f(t)$ at $t_2$ is now estimated by $g(t_2, y_2)$, and so on. Let us summarize this procedure:

---

**Euler's Method**    The endpoints $(t_0, y_0), \ldots, (t_n, y_n)$ of the line segments approximating the solution of (1) on the interval $a \le t \le b$ are given by the following formulas, where $h = (b - a)/n$:

$$t_0 = a \text{ (given)}, \qquad y_0 \text{ (given)}$$
$$t_1 = t_0 + h, \qquad y_1 = y_0 + g(t_0, y_0) \cdot h,$$
$$t_2 = t_1 + h, \qquad y_2 = y_1 + g(t_1, y_1) \cdot h,$$
$$\vdots \qquad\qquad \vdots$$
$$t_n = t_{n-1} + h. \qquad y_n = y_{n-1} + g(t_{n-1}, y_{n-1}) \cdot h.$$

---

**EXAMPLE 1**    Use Euler's method with $n = 4$ to approximate the solution $f(t)$ to $y' = 2t - 3y$, $y(0) = 4$, for $t$ in the interval $0 \le t \le 2$. In particular, estimate $f(2)$.

**Solution**    Here $g(t, y) = 2t - 3y$, $a = 0$, $b = 2$, $y_0 = 4$ and $h = (2 - 0)/4 = \frac{1}{2}$. Starting with $(t_0, y_0) = (0, 4)$, we find that $g(0, 4) = -12$. Thus

$$t_1 = \frac{1}{2}, \qquad y_1 = 4 + (-12) \cdot \frac{1}{2} = -2.$$

Next, $g(\frac{1}{2}, -2) = 7$, so

$$t_2 = 1, \qquad y_2 = -2 + 7 \cdot \frac{1}{2} = \frac{3}{2}.$$

Next, $g(1, \frac{3}{2}) = -\frac{5}{2}$, so

$$t_3 = \frac{3}{2}, \qquad y_3 = \frac{3}{2} + \left(-\frac{5}{2}\right) \cdot \frac{1}{2} = \frac{1}{4}.$$

Finally, $g(\frac{3}{2}, \frac{1}{4}) = \frac{9}{4}$, so

$$t_4 = 2, \qquad y_4 = \frac{1}{4} + \frac{9}{4} \cdot \frac{1}{2} = \frac{11}{8}.$$

Thus the approximation to the solution $f(t)$ is given by the polygonal path shown in Fig. 3. The last point $(2, \frac{11}{8})$ is close to the graph of $f(t)$ at $t = 2$, so $f(2) \approx \frac{11}{8}$.                                                    ■

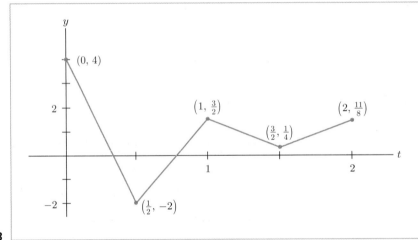

**Figure 3**

Actually, this polygonal path is somewhat misleading. The accuracy can be improved dramatically by increasing the value of $n$. Figure 4 shows the Euler approximations for $n = 8$ and $n = 20$. The graph of the exact solution is shown for comparison.

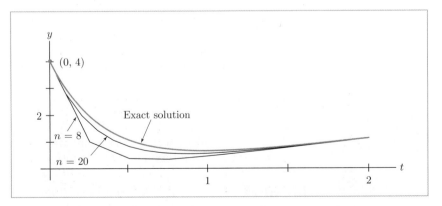

**Figure 4.** Approximating an exact solution by polygonal paths.

For many purposes, satisfactory graphs can be obtained by running Euler's method on a computer with large values of $n$. There is a limit to the accuracy obtainable, however, because each computer calculation involves a slight "round-off" error. When $n$ is very large, the cumulative round-off error can become significant.

**Incorporating Technology**  The TI-86 graphing calculator has a differential equation mode and a built-in routine for carrying out Euler's method. Figures 5 and 6 give the solution for Example 1, where the solution function is denoted by **Q1**. (The differential equation is entered into the function editor as **Q′1 = 2t − 3Q1** and the initial condition is entered into the INICT screen as **QI1 = 4**.) Figure 7 shows the Window settings used to obtain the solution. See Appendix D for details.

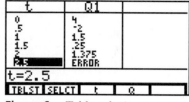

**Figure 5.**  Graphical solution to Example 1 on TI-86.

**Figure 6.**  Table solution to Example 1 on TI-86.

(a)                                    (b)

**Figure 7.**  Window settings used for Figs. 5 and 6.

The TI-85 also has a differential equation mode, but it uses a numerical technique other than Euler's method to approximate the solution of a differential equation. Graphical solutions — but not tables — can be generated. The window settings are the same as in Fig. 7, except that **EStep = 1** is replaced by **difTol = .001**. See Appendix C for details.

By using sequence mode, the TI-82 and TI-83 calculators can create tables similar to the one in Fig. 6 and the TI-83 also can display a graph of the solution. See Appendices A and B for details. Figures 8 and 9 show the solution to $y' = 2t - 3y$ with $n = 100$ as calculated by the TI-83. In Fig. 8, the variable $t$ is denoted by $x$. In Fig. 9, values of $t$ and $y$ are contained in the **u(n)** and **v(n)** columns, respectively. Figure 10 shows the settings in the sequence **Y=** editor. (Also, **nMin** was set to 0.) Figure 11 shows the Window settings for the graph.

**Figure 8.**  Graphical solution to Example 1 with $n = 100$ on TI-83.

**Figure 9.**  Table solution to Example 1 with $n = 100$ on TI-83.

**Figure 10.**  Sequence **Y=** editor.

(a)                (b)

**Figure 11.** Window settings used for the graph in Fig. 8.

## PRACTICE PROBLEMS 10.7

*Let $f(t)$ be the solution of $y' = \sqrt{ty}$, $y(1) = 4$.*

**1.** Use Euler's method with $n = 2$ on the interval $1 \le t \le 2$ to estimate $f(2)$.

**2.** Draw the polygonal path corresponding to the application of Euler's method in Problem 1.

## EXERCISES 10.7

**1.** Suppose that $f(t)$ is a solution of the differential equation $y' = ty - 5$, and the graph of $f(t)$ passes through the point $(2, 4)$. What is the slope of the graph at this point?

**2.** Suppose that $f(t)$ is a solution of $y' = t^2 - y^2$ and the graph of $f(t)$ passes through the point $(2, 3)$. Find the slope of the graph when $t = 2$.

**3.** Suppose that $f(t)$ satisfies the initial value problem $y' = y^2 + ty - 7$, $y(0) = 3$. Is $f(t)$ increasing or decreasing at $t = 0$?

**4.** Suppose that $f(t)$ satisfies the initial value problem $y' = y^2 + ty - 7$, $y(0) = 2$. Is the graph of $f(t)$ increasing or decreasing at $t = 0$?

**5.** Use Euler's method with $n = 2$ on the interval $0 \le t \le 1$ to approximate the solution $f(t)$ to $y' = t^2 y$, $y(0) = -2$. In particular, estimate $f(1)$.

**6.** Use Euler's method with $n = 2$ on the interval $2 \le t \le 3$ to approximate the solution $f(t)$ to $y' = t - 2y$, $y(2) = 3$. Estimate $f(3)$.

**7.** Use Euler's method with $n = 4$ to approximate the solution $f(t)$ to $y' = 2t - y + 1$, $y(0) = 5$ for $0 \le t \le 2$. Estimate $f(2)$.

**8.** Let $f(t)$ be the solution of $y' = y(2t - 1)$, $y(0) = 8$. Use Euler's method with $n = 4$ to estimate $f(1)$.

**9.** Let $f(t)$ be the solution of $y' = -(t + 1)y^2$, $y(0) = 1$. Use Euler's method with $n = 5$ to estimate $f(1)$. Then solve the differential equation, find an explicit formula for $f(t)$, and compute $f(1)$. How accurate is the estimated value of $f(1)$?

**10.** Let $f(t)$ be the solution of $y' = 10 - y$, $y(0) = 1$. Use Euler's method with $n = 5$ to estimate $f(1)$. Then solve the differential equation and find the exact value of $f(1)$.

**11.** Suppose that the Consumer Products Safety Commission issues new regulations that affect the toy manufacturing industry. Every toy manufacturer will have to make certain changes in its manufacturing process. Let $f(t)$ be the fraction of manufacturers that have complied with the regulations within $t$ months. Note that $0 \le f(t) \le 1$. Suppose that the rate at which new companies comply with the regulations is proportional to the fraction of companies who have not yet complied, with constant of proportionality $k = .1$.

    **(a)** Construct a differential equation satisfied by $f(t)$.

    **(b)** Use Euler's method with $n = 3$ to estimate the fraction of companies that comply with the regulations within the first 3 months.

    **(c)** Solve the differential equation in part (a) and compute $f(3)$.

    **(d)** Compare the answers in (b) and (c) and approximate the error in using Euler's method.

**12.** The Los Angeles Zoo plans to transport a California sea lion to the San Diego Zoo. The animal will be wrapped in a wet blanket during the trip. At any time $t$ the blanket will lose water (due to evaporation) at a rate proportional to the amount $f(t)$ of water in the blanket, with constant of proportionality $k = -.3$. Initially, the blanket will contain 2 gallons of seawater.

    **(a)** Set up the differential equation satisfied by $f(t)$.

    **(b)** Use Euler's method with $n = 2$ to estimate the amount of moisture in the blanket after 1 hour.

    **(c)** Solve the differential equation in part (a) and compute $f(1)$.

    **(d)** Compare the answers in (b) and (c) and approximate the error in using Euler's method.

## Technology Exercises

**13.** The differential equation $y' = .5(1 - y)(4 - y)$ has the five types of solutions labeled A–E. For each of the following initial values, graph the solution of the differential equation and identify the type of the solution. Use a small value of $h$, let $t$ range from 0 to 4, and let $y$ range from −1 to 5. Use the technique of Section 10.6 to verify your answer.

**(a)** $y(0) = -1$    **(b)** $y(0) = 1$    **(c)** $y(0) = 2$

**(d)** $y(0) = 3.9$    **(e)** $y(0) = 4.1$

   A. Constant solution

   B. Decreasing, has an inflection point, and asymptotic to the line $y = 1$

   C. Increasing, concave down, and asymptotic to the line $y = 1$

   D. Concave up and increasing indefinitely

   E. Decreasing, concave up, and asymptotic to the line $y = 1$

**14.** The differential equation $y' = .5(y - 1)(4 - y)$ has the five types of solutions labeled A–E. For each of the following initial values, graph the solution of the differential equation and identify the type of the solution. Use a small value of $h$, let $t$ range from 0 to 4, and let $y$ range from −1 to 5. Use the technique of Section 10.6 to verify your answer.

**(a)** $y(0) = .9$           **(b)** $y(0) = 1.1$

**(c)** $y(0) = 3$            **(d)** $y(0) = 4$

**(e)** $y(0) = 5$

   A. Constant solution

   B. Decreasing, concave up, and asymptotic to the line $y = 4$

   C. Increasing, has an inflection point, and asymptotic to the line $y = 4$

   D. Increasing, concave down, and asymptotic to the line $y = 4$

   E. Concave down and decreasing indefinitely

**15.** The differential equation $y' = e^t - 2y$, $y(0) = 1$, has solution $y = \frac{1}{3}(2e^{-2t} + e^t)$. In the following table, the second row should be filled in with the values obtained from the use of a numerical method and the third row should be filled in with actual values calculated from the solution. What is the greatest difference between corresponding values in the second and third rows?

| $t_i$ | 0 | .25 | .50 | .75 | 1 | 1.25 | 1.5 | 1.75 | 2 |
|---|---|---|---|---|---|---|---|---|---|
| $y_i$ | 1 | | | | | | | | |
| $y$ | 1 | .8324 | | | | | | 2.4752 | |

**16.** The differential equation $y' = 2ty + e^{t^2}$, $y(0) = 5$, has solution $y = (t + 5)e^{t^2}$. In the following table, the second row should be filled in with the values obtained from the use of a numerical method and the third row should be filled in with actual values calculated from the solution. What is the greatest difference between corresponding values in the second and third rows?

| $t_i$ | 0 | .2 | .4 | .6 | .8 | 1 | 1.2 | 1.4 | 1.6 | 1.8 | 2 |
|---|---|---|---|---|---|---|---|---|---|---|---|
| $y_i$ | 5 | | | | | | | | | | |
| $y$ | 5 | 5.412 | | | | | | | | | 382.2 |

## SOLUTIONS TO PRACTICE PROBLEMS 10.7

**1.** Here $g(t, y) = \sqrt{ty}$, $a = 1$, $b = 2$, $y_0 = 4$, and $h = (2 - 1)/2 = \frac{1}{2}$. We have

$$t_0 = 1, \quad y_0 = 4, \qquad\qquad g(1, 4) = \sqrt{1 \cdot 4}$$
$$= 2,$$
$$t_1 = \tfrac{3}{2}, \quad y_1 = 4 + 2\left(\tfrac{1}{2}\right) = 5, \quad g\left(\tfrac{3}{2}, 5\right) = \sqrt{\tfrac{3}{2} \cdot 5}$$
$$\approx 2.7386,$$

$$t_2 = 2, \quad y_2 = 5 + (2.7386)\left(\tfrac{1}{2}\right) = 6.3693.$$

Hence $f(2) \approx y_2 = 6.3693$. [In Practice Problems 10.2, we found the solution of $y' = \sqrt{ty}$, $y(1) = 4$, to be $f(t) = \left(\frac{1}{3}t^{3/2} + \frac{5}{3}\right)^2$. We find that $f(2) = 6.8094$ (to four decimal places). The error $6.8094 - 6.3693 = .4401$ in the preceding approximation is about 6.5%.]

## SOLUTIONS TO PRACTICE PROBLEMS 10.7 (CONTINUED)

2. To find the polygonal path, plot the points $(t_0, y_0)$, $(t_1, y_1)$, and $(t_2, y_2)$ and join them by line segments. See Fig. 12.

**Figure 12**

# REVIEW OF FUNDAMENTAL CONCEPTS

1. What is a differential equation?

2. What does it mean for a function to be a solution to a differential equation?

3. What is a solution curve?

4. What is a constant solution to a differential equation?

5. What is the slope field?

6. Describe the separation of variables technique for obtaining the solution to a differential equation.

7. What is a first-order linear differential equation?

8. What is an integrating factor and how does it help you solve a first-order linear differential equation?

9. What is an autonomous differential equation?

10. How do you recognize an autonomous differential equation from its slope field?

11. Outline the procedure for sketching a solution of an autonomous differential equation.

12. What is the logistic differential equation?

13. Describe Euler's method for approximating the solution of a differential equation.

# SUPPLEMENTARY EXERCISES

*Solve the differential equations in Exercises 1–10.*

1. $y^2 y' = 4t^3 - 3t^2 + 2$

2. $\dfrac{y'}{t+1} = y + 1$

3. $y' = \dfrac{y}{t} - 3y, \ t > 0$

4. $(y')^2 = t$

5. $y = 7y' + ty', \ y(0) = 3$

6. $y' = te^{t+y}, \ y(0) = 0$

7. $yy' + t = 6t^2, \ y(0) = 7$

8. $y' = 5 - 8y, \ y(0) = 1$

9. $y' - \dfrac{2}{1-t} y = (1-t)^4$

10. $y' - \dfrac{1}{2(1+t)} y = 1 + t, \ t \geq 0$

11. Find a curve in the $xy$-plane passing through the origin and whose slope at the point $(x, y)$ is $x + y$.

12. Let $P(t)$ denote the price in dollars of a certain commodity at time $t$ in days. Suppose that the rate of change of $P$ is proportional to the difference $D - S$ of the demand $D$ and supply $S$ at any time $t$. Suppose further that the demand and supply are related to the price by: $D = 10 - .3P$ and $S = -2 + 3P$.

    (a) Find a differential equation that is satisfied by $P$, given that the price was falling at the rate of one dollar a day when $D = 10$ and $S = 20$.

    (b) Find $P$ given that $P(0) = 1$.

13. Suppose that $f(t)$ is a solution of $y' = (2 - y)e^{-y}$. Is $f(t)$ increasing or decreasing at some value of $t$ where $f(t) = 3$?

14. Solve the initial value problem

$$y' = e^{y^2}(\cos y)(1 - e^{y-1}), \qquad y(0) = 1.$$

*Sketch the solutions of the differential equations in Exercises 15–24. In each case, also indicate the constant solutions.*

15. $y' = 2\cos y, \ y(0) = 0$

16. $y' = 5 + 4y - y^2, \ y(0) = 1$

17. $y' = y^2 + y, \ y(0) = -\frac{1}{3}$

18. $y' = y^2 - 2y + 1, \ y(0) = -1$

19. $y' = \ln y, \ y(0) = 2$

**20.** $y' = 1 + \cos y$, $y(0) = -\frac{3}{4}$

**21.** $y' = \dfrac{1}{y^2 + 1}$, $y(0) = -1$

**22.** $y' = \dfrac{3}{y + 3}$, $y(0) = 2$

**23.** $y' = .4y^2(1 - y)$, $y(0) = -1$, $y(0) = .1$, $y(0) = 2$

**24.** $y' = y^3 - 6y^2 + 9y$, $y(0) = -\frac{1}{4}$, $y(0) = \frac{1}{4}$, $y(0) = 4$

**25.** The birth rate in a certain city is 3.5% per year and the death rate is 2% per year. Also, there is a net movement of population out of the city at a steady rate of 3000 people per year. Let $N = f(t)$ be the city's population at time $t$.

   **(a)** Write a differential equation satisfied by $N$.

   **(b)** Use a qualitative analysis of the equation to determine if there is a size at which the population would remain constant. Is it likely that a city would have such a constant population?

**26.** Suppose that in a chemical reaction, each gram of substance $A$ combines with 3 grams of substance $B$ to form 4 grams of substance $C$. Suppose that the reaction begins with 10 grams of $A$, 15 grams of $B$, and 0 grams of $C$ present. Let $y = f(t)$ be the amount of $C$ present at time $t$. Suppose that the rate at which substance $C$ is formed is proportional to the product of the unreacted amounts of $A$ and $B$ present. That is, suppose that $f(t)$ satisfies the differential equation

$$y' = k\left(10 - \tfrac{1}{4}y\right)\left(15 - \tfrac{3}{4}y\right), \qquad y(0) = 0,$$

where $k$ is a constant.

   **(a)** What do the quantities $10 - \frac{1}{4}f(t)$ and $15 - \frac{3}{4}f(t)$ represent?

   **(b)** Should the constant $k$ be positive or negative?

   **(c)** Make a qualitative sketch of the solution of the preceding differential equation.

**27.** A bank account has $20,000 earning 5% interest compounded continuously. A pensioner uses the account to pay himself an annuity, drawing continuously at a $2000 annual rate. How long will it take for the balance in the account to drop to zero?

**28.** A continuous annuity of $12,000 per year is to be funded by steady withdrawals from a savings account that earns 6% interest compounded continuously.

   **(a)** What is the smallest initial amount in the account that will fund such an annuity forever?

   **(b)** What initial amount will fund such an annuity for exactly 20 years (at which time the savings account balance will be zero)?

**29.** Let $f(t)$ be the solution to $y' = 2e^{2t-y}$, $y(0) = 0$. Use Euler's method with $n = 4$ on $0 \le t \le 2$ to estimate $f(2)$. Then show that Euler's method gives the exact value of $f(2)$ by solving the differential equation.

**30.** Let $f(t)$ be the solution to $y' = (t+1)/y$, $y(0) = 1$. Use Euler's method with $n = 3$ on $0 \le t \le 1$ to estimate $f(1)$. Then show that Euler's method gives the exact value of $f(1)$ by solving the differential equation.

**31.** Use Euler's method with $n = 6$ on the interval $0 \le t \le 3$ to approximate the solution $f(t)$ to

$$y' = .1y(20 - y), \quad y(0) = 2.$$

**32.** Use Euler's method with $n = 5$ on the interval $0 \le t \le 1$ to approximate the solution $f(t)$ to

$$y' = \frac{1}{2}y(y - 10), \quad y(0) = 9.$$

# Taylor Polynomials and Infinite Series

IN earlier chapters we introduced the functions $e^x$, $\ln x$, $\sin x$, $\cos x$, and $\tan x$. Whenever we needed the value of one of these functions for a particular value of $x$, such as $e^{.023}$, $\ln 5.8$, or $\sin .25$, we had to use a calculator. Now we shall take up the problem of numerically computing the values of such functions for particular choices of the variable $x$. The computational methods developed have many applications—for example, to differential equations and probability theory.

## 11.1 Taylor Polynomials

A polynomial of degree $n$ is a function of the form

$$p(x) = a_0 + a_1 x + \cdots + a_n x^n$$

where $a_0, a_1, \ldots, a_n$ are given numbers and $a_n \neq 0$. There are many instances in mathematics and its applications where calculations are much simpler for polynomials than for other functions. In this section we show how to approximate a given function $f(x)$ by a polynomial $p(x)$ for all values of $x$ near some specified number, say $x = a$. To simplify matters, we begin by considering values of $x$ near $x = 0$.

In Fig. 1 we have drawn the graph of the function $f(x) = e^x$ together with the tangent line through $(0, f(0)) = (0, 1)$. The slope of the tangent line is $f'(0) = 1$.

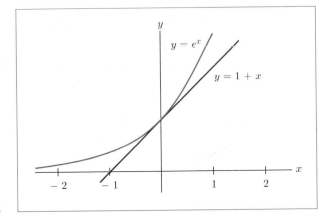

**Figure 1.** A linear approximation of $e^x$ at $x = 0$.

So the equation of the tangent line is

$$y - f(0) = f'(0)(x - 0)$$
$$y = f(0) + f'(0)x$$
$$y = 1 + x.$$

From our discussion of the derivative, we know that the tangent line at $x = 0$ closely approximates the graph of $y = e^x$ for values of $x$ near 0. Thus, if we let $p_1(x) = 1 + x$, then the values of $p_1(x)$ are close to the corresponding values of $f(x) = e^x$ for $x$ near 0.

In general, a given function $f(x)$ may be approximated for values of $x$ near 0 by the polynomial

$$p_1(x) = f(0) + f'(0)x,$$

which is called the *first Taylor polynomial of* $f(x)$ *at* $x = 0$. The graph of $p_1(x)$ is just the tangent line to $y = f(x)$ at $x = 0$.

The first Taylor polynomial "resembles" $f(x)$ near $x = 0$ in the sense that

$$\begin{aligned} p_1(0) &= f(0) \qquad \text{(both graphs go through the same point at } x = 0\text{)}, \\ p_1'(0) &= f'(0) \qquad \text{(both graphs have the same slope at } x = 0\text{)}. \end{aligned}$$

That is, $p_1(x)$ coincides with $f(x)$ in both its value at $x = 0$ and the value of its first derivative at $x = 0$. This suggests that to approximate $f(x)$ even more closely at $x = 0$, we look for a polynomial that coincides with $f(x)$ in its value at $x = 0$ and in the values of its first *and* second derivatives at $x = 0$. Yet a further approximation can be obtained by going out to the third derivative, and so on.

---

**EXAMPLE 1**    Given a function $f(x)$, suppose that $f(0) = 1$, $f'(0) = -2$, $f''(0) = 7$, and $f'''(0) = -5$. Find a polynomial of degree 3,

$$p(x) = a_0 + a_1 x + a_2 x^2 + a_3 x^3,$$

such that $p(x)$ coincides with $f(x)$ up to the third derivative at $x = 0$; that is,

$$\begin{aligned} p(0) &= f(0) = 1 \qquad &\text{(same value at } x = 0\text{)}, \\ p'(0) &= f'(0) = -2 \qquad &\text{(same first derivative at } x = 0\text{)}, \\ p''(0) &= f''(0) = 7 \qquad &\text{(same second derivative at } x = 0\text{)}, \\ p'''(0) &= f'''(0) = -5 \qquad &\text{(same third derivative at } x = 0\text{)}. \end{aligned}$$

**Solution** To find the coefficients $a_0, \ldots, a_3$ of $p(x)$, we first compute the values of $p(x)$ and its derivatives at $x = 0$:

$$
\begin{aligned}
p(x) &= a_0 + a_1 x + a_2 x^2 + a_3 x^3, & p(0) &= a_0, \\
p'(x) &= 0 + a_1 + 2a_2 x + 3a_3 x^2, & p'(0) &= a_1, \\
p''(x) &= 0 + 0 + 2a_2 + 2 \cdot 3 a_3 x, & p''(0) &= 2a_2, \\
p'''(x) &= 0 + 0 + 0 + 2 \cdot 3 a_3 & p'''(0) &= 2 \cdot 3 a_3.
\end{aligned}
$$

Since we want $p(x)$ and its derivatives to coincide with the given values of $f(x)$ and its derivatives, we must have

$$ a_0 = 1, \quad a_1 = -2, \quad 2a_2 = 7, \quad \text{and} \quad 2 \cdot 3 a_3 = -5. $$

So

$$ a_0 = 1, \quad a_1 = -2, \quad a_2 = \frac{7}{2}, \quad \text{and} \quad a_3 = \frac{-5}{2 \cdot 3}. $$

Rewriting the coefficients slightly, we have

$$ p(x) = 1 + \frac{(-2)}{1} x + \frac{7}{1 \cdot 2} x^2 + \frac{-5}{1 \cdot 2 \cdot 3} x^3. $$

The form in which we have written $p(x)$ clearly exhibits the values $1, -2, 7, -5$ of $f(x)$ and its derivatives at $x = 0$. In fact, we could also write this formula for $p(x)$ in the form

$$ p(x) = f(0) + \frac{f'(0)}{1} x + \frac{f''(0)}{1 \cdot 2} x^2 + \frac{f'''(0)}{1 \cdot 2 \cdot 3} x^3. \qquad \blacksquare $$

Given a function $f(x)$, we may use the formula in Example 1 to find a polynomial that coincides with $f(x)$ up to the third derivative at $x = 0$. To describe the general formula for higher-order polynomials, we let $f^{(n)}(x)$ denote the $n$th derivative of $f(x)$, and we let $n!$ (read "$n$ factorial") denote the product of all the integers from 1 to $n$, so that $n! = 1 \cdot 2 \cdots (n-1) \cdot n$. (Thus, $1! = 1$, $2! = 1 \cdot 2$, $3! = 1 \cdot 2 \cdot 3$, and so forth.)

---

Given a function $f(x)$, the $n$th *Taylor polynomial* of $f(x)$ at $x = 0$ is the polynomial $p_n(x)$ defined by

$$ p_n(x) = f(0) + \frac{f'(0)}{1!} x + \frac{f''(0)}{2!} x^2 + \cdots + \frac{f^{(n)}(0)}{n!} x^n. $$

This polynomial coincides with $f(x)$ up to the $n$th derivative at $x = 0$ in the sense that

$$ p_n(0) = f(0), \quad p_n'(0) = f'(0), \ldots, \quad p_n^{(n)}(0) = f^{(n)}(0). $$

---

The next example shows how Taylor polynomials are used to approximate values of $e^x$ for $x$ near 0. The choice of which polynomial to use depends on how accurate one wants the values of $e^x$.

---

**EXAMPLE 2**  Determine the first three Taylor polynomials of $f(x) = e^x$ at $x = 0$ and sketch their graphs.

**Solution**  Since all derivatives of $e^x$ are $e^x$, we see that

$$f(0) = f'(0) = f''(0) = f'''(0) = e^0 = 1.$$

Thus the desired Taylor polynomials are

$$p_1(x) = 1 + \frac{1}{1!} \cdot x = 1 + x,$$

$$p_2(x) = 1 + \frac{1}{1!}x + \frac{1}{2!}x^2 = 1 + x + \frac{1}{2}x^2,$$

$$p_3(x) = 1 + \frac{1}{1!}x + \frac{1}{2!}x^2 + \frac{1}{3!}x^3 = 1 + x + \frac{1}{2}x^2 + \frac{1}{6}x^3.$$

The relative accuracy of these approximations to $e^x$ may be seen from the graphs in Fig. 2. ∎

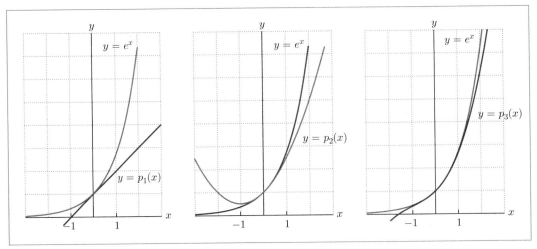

**Figure 2.**  Taylor polynomials of $e^x$ at $x = 0$.

---

**EXAMPLE 3**  Determine the $n$th Taylor polynomial of

$$f(x) = \frac{1}{1 - x}$$

at $x = 0$.

**Solution**

$$f(x) = (1 - x)^{-1}, \qquad\qquad f(0) = 1$$

$$f'(x) = 1(1 - x)^{-2}, \qquad\qquad f'(0) = 1$$

$$f''(x) = 1 \cdot 2(1 - x)^{-3} = 2!(1 - x)^{-3}, \qquad f''(0) = 2!$$

$$f'''(x) = 1 \cdot 2 \cdot 3(1 - x)^{-4} = 3!(1 - x)^{-4}, \qquad f'''(0) = 3!$$

$$f^{(4)}(x) = 1 \cdot 2 \cdot 3 \cdot 4(1 - x)^{-5} = 4!(1 - x)^{-5}, \qquad f^{(4)}(0) = 4!$$

From the pattern of calculations, it is clear that $f^{(k)}(0) = k!$ for each $k$. Therefore,

$$p_n(x) = 1 + \frac{1}{1!}x + \frac{2!}{2!}x^2 + \frac{3!}{3!}x^3 + \cdots + \frac{n!}{n!}x^n$$

$$= 1 + x + x^2 + x^3 + \cdots + x^n. \qquad\qquad ∎$$

We have already mentioned the possibility of using a polynomial to approximate the values of a function near $x = 0$. Here is another way to use a polynomial approximation.

**EXAMPLE 4**    It can be shown that the second Taylor polynomial of $\sin x^2$ at $x = 0$ is $p_2(x) = x^2$. Use this polynomial to approximate the area under the graph of $y = \sin x^2$ from $x = 0$ to $x = 1$. (See Fig. 3.)

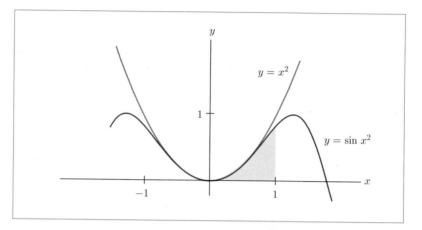

**Figure 3.**   The second Taylor polynomial of $\sin x^2$ at $x = 0$.

**Solution**    Since the graph of $p_2(x)$ is very close to the graph of $\sin x^2$ for $x$ near 0, the areas under the two graphs should be almost the same. The area under the graph of $p_2(x)$ is

$$\int_0^1 p_2(x)\,dx = \int_0^1 x^2\,dx = \left. \tfrac{1}{3}x^3 \right|_0^1 = \tfrac{1}{3}. \qquad \blacksquare$$

In Example 4, the exact area under the graph of $\sin x^2$ is given by

$$\int_0^1 \sin x^2\,dx.$$

However, this integral cannot be computed by the usual method because there is no way to construct an antiderivative of $\sin x^2$ consisting of elementary functions. Using an approximation technique from Chapter 9, one can find that the value of the integral is .3103 to four decimal places. Thus the error in using $p_2(x)$ as an approximation for $\sin x^2$ is about .023. (The error can be reduced further by using a Taylor polynomial of higher degree. In this particular example, the effort involved is far less than that required by the approximation methods of Chapter 9.)

**Taylor Polynomials at $x = a$**    Suppose now that we wish to approximate a given function $f(x)$ by a polynomial for values of $x$ near some number $a$. Since the behavior of $f(x)$ near $x = a$ is determined by the values of $f(x)$ and its derivatives at $x = a$, we should try to approximate $f(x)$ by a polynomial $p(x)$ for which the values of $p(x)$ and its derivatives at $x = a$ are the same as those of $f(x)$. This is easily done if we use a polynomial that has the form

$$p(x) = a_0 + a_1(x - a) + a_2(x - a)^2 + \cdots + a_n(x - a)^n.$$

We call this a *polynomial in $x - a$*. In this form it is easy to compute $p(a)$, $p'(a)$, and so on, because setting $x = a$ in $p(x)$ or one of its derivatives makes most of the terms equal zero. The following result is easily verified.

Given a function $f(x)$, the $n$th *Taylor polynomial of $f(x)$* at $x = a$ is the polynomial $p_n(x)$ defined by

$$p_n(x) = f(a) + \frac{f'(a)}{1!}(x - a) + \frac{f''(a)}{2!}(x - a)^2 + \cdots + \frac{f^{(n)}(a)}{n!}(x - a)^n.$$

This polynomial coincides with $f(x)$ up to the $n$th derivative at $x = a$ in the sense that

$$p_n(a) = f(a), \quad p'_n(a) = f'(a), \ldots, \ p_n^{(n)}(a) = f^{(n)}(a).$$

When $a = 0$, of course, these Taylor polynomials are just the same as those introduced earlier.

---

**EXAMPLE 5** Calculate the second Taylor polynomial of $f(x) = \sqrt{x}$ at $x = 1$ and use this polynomial to estimate $\sqrt{1.02}$.

**Solution** Here $a = 1$. Since we want the second Taylor polynomial, we must calculate the values of $f(x)$ and of its first two derivatives at $x = 1$.

$$f(x) = x^{1/2}, \qquad f'(x) = \tfrac{1}{2}x^{-1/2}, \qquad f''(x) = -\tfrac{1}{4}x^{-3/2},$$

$$f(1) = 1, \qquad f'(1) = \tfrac{1}{2}, \qquad f''(1) = -\tfrac{1}{4}$$

Therefore, the desired Taylor polynomial is

$$p_2(x) = 1 + \frac{\tfrac{1}{2}}{1!}(x - 1) + \frac{-\tfrac{1}{4}}{2!}(x - 1)^2$$

$$= 1 + \tfrac{1}{2}(x - 1) - \tfrac{1}{8}(x - 1)^2.$$

Since 1.02 is close to 1, $p_2(1.02)$ gives a good approximation to $f(1.02)$, that is, to $\sqrt{1.02}$.

$$p_2(1.02) = 1 + \tfrac{1}{2}(1.02 - 1) - \tfrac{1}{8}(1.02 - 1)^2$$

$$= 1 + \tfrac{1}{2}(.02) - \tfrac{1}{8}(.02)^2$$

$$= 1 + .01 - .00005$$

$$= 1.00995 \qquad \blacksquare$$

**Accuracy of the Approximation**  The solution to Example 5 is incomplete in a practical sense, for it offers no information about how close 1.00995 is to the true value of $\sqrt{1.02}$. In general, when we obtain an approximation to some quantity, we also want an indication of the quality of the approximation.

In order to measure the accuracy of an approximation to a function $f(x)$ by its Taylor polynomial at $x = a$, we define

$$R_n(x) = f(x) - p_n(x).$$

This difference between $f(x)$ and $p_n(x)$ is called the *$n$th remainder of $f(x)$* at $x = a$. The following formula is derived in advanced texts:

**The Remainder Formula**    Suppose that the function $f(x)$ can be differentiated $n + 1$ times on an interval containing the number $a$. Then for each $x$ in this interval, there exists a number $c$ between $a$ and $x$ such that

$$R_n(x) = \frac{f^{(n+1)}(c)}{(n+1)!}(x-a)^{n+1}. \tag{1}$$

Usually the precise value of $c$ is unknown. However, if we can find a number $M$ such that $\left|f^{(n+1)}(c)\right| \leq M$ for all $c$ between $a$ and $x$, then we don't need to know which $c$ appears in (1), because we have

$$|f(x) - p_n(x)| = |R_n(x)| \leq \frac{M}{(n+1)!}\,|x-a|^{n+1}\,.$$

**EXAMPLE 6**    Determine the accuracy of the estimate in Example 5.

**Solution**    The second remainder for a function $f(x)$ at $x = 1$ is

$$R_2(x) = \frac{f^{(3)}(c)}{3!}(x-1)^3,$$

where $c$ is between $1$ and $x$ (and where $c$ depends on $x$). Here $f(x) = \sqrt{x}$, and therefore $f^{(3)}(c) = \frac{3}{8}c^{-5/2}$. We are interested in $x = 1.02$, and so $1 \leq c \leq 1.02$. We observe that since $c^{5/2} \geq 1^{5/2} = 1$, we have $c^{-5/2} \leq 1$. Thus

$$\left|f^{(3)}(c)\right| \leq \tfrac{3}{8}\cdot 1 = \tfrac{3}{8},$$

and

$$|R_2(1.02)| \leq \frac{3/8}{3!}(1.02 - 1)^3$$

$$= \tfrac{3}{8}\cdot\tfrac{1}{6}(.02)^3$$

$$= .0000005.$$

Thus the error in using $p_2(1.02)$ as an approximation of $f(1.02)$ is at most $.0000005$. ■

**Incorporating Technology**    Graphing calculators can be used to determine how well a function is approximated by a Taylor polynomial. Figure 4 shows the (heavily shaded) graph of $\mathbf{Y_1} = \sin(x^2)$ and the graph of its sixth Taylor polynomial $\mathbf{Y_2} = x^2 - \dfrac{x^6}{6}$. The two graphs appear identical on the screen for $x$ between $-1.1$ and $1.1$. Figure 5 shows that the distance between the two functions at $x = 1.1$ is about $.02$. As $x$ increases beyond $1.1$, the goodness of fit deteriorates. For instance, the two functions are about $5.9$ units apart when $x = 2$.

## PRACTICE PROBLEMS 11.1

**1. (a)** Determine the third Taylor polynomial of $f(x) = \cos x$ at $x = 0$.

**(b)** Use the result of part (a) to estimate $\cos.12$.

**2.** Determine all Taylor polynomials of $f(x) = 3x^2 - 17$ at $x = 3$.

**Figure 4.** $\sin(x^2)$ and its sixth Taylor polynomial.

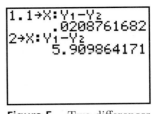

**Figure 5.** Two differences between the function and its Taylor polynomial.

# Exercises 11.1

*In Exercises 1–8, determine the third Taylor polynomial of the given function at $x = 0$.*

**1.** $f(x) = \sin x$

**2.** $f(x) = e^{-x/2}$

**3.** $f(x) = 5e^{2x}$

**4.** $f(x) = \cos(\pi - 5x)$

**5.** $f(x) = \sqrt{4x+1}$

**6.** $f(x) = \dfrac{1}{x+2}$

**7.** $f(x) = xe^{3x}$

**8.** $f(x) = \sqrt{1-x}$

**9.** Determine the fourth Taylor polynomial of $f(x) = e^x$ at $x = 0$ and use it to estimate $e^{.01}$.

**10.** Determine the fourth Taylor polynomial of $f(x) = \ln(1-x)$ at $x = 0$ and use it to estimate $\ln(.9)$.

**11.** Sketch the graphs of $f(x) = \dfrac{1}{1-x}$ and its first three Taylor polynomials at $x = 0$.

**12.** Sketch the graphs of $f(x) = \sin x$ and its first three Taylor polynomials at $x = 0$.

**13.** Determine the $n$th Taylor polynomial for $f(x) = e^x$ at $x = 0$.

**14.** Determine all Taylor polynomials for $f(x) = x^2 + 2x + 1$ at $x = 0$.

**15.** Use a second Taylor polynomial at $x = 0$ to estimate the area under the curve $y = \ln(1 + x^2)$ from $x = 0$ to $x = \frac{1}{2}$.

**16.** Use a second Taylor polynomial at $x = 0$ to estimate the area under the curve $y = \sqrt{\cos x}$ from $x = -1$ to $x = 1$. (The exact answer to three decimal places is 1.828.)

**17.** Determine the third Taylor polynomial of $\dfrac{1}{5-x}$ at $x = 4$.

**18.** Determine the fourth Taylor polynomial of $\ln x$ at $x = 1$.

**19.** Determine the third and fourth Taylor polynomials of $\cos x$ at $x = \pi$.

**20.** Determine the third and fourth Taylor polynomials of $x^3 + 3x - 1$ at $x = -1$.

**21.** Use the second Taylor polynomial of $f(x) = \sqrt{x}$ at $x = 9$ to estimate $\sqrt{9.3}$.

**22.** Use the second Taylor polynomial of $f(x) = \ln x$ at $x = 1$ to estimate $\ln .8$.

**23.** Determine all Taylor polynomials of $f(x) = x^4 + x + 1$ at $x = 2$.

**24.** Determine the $n$th Taylor polynomial of $f(x) = 1/x$ at $x = 1$.

**25.** If $f(x) = 3 + 4x - \dfrac{5}{2!}x^2 + \dfrac{7}{3!}x^3$, what are $f''(0)$ and $f'''(0)$?

**26.** If $f(x) = 2 - 6(x-1) + \dfrac{3}{2!}(x-1)^2 - \dfrac{5}{3!}(x-1)^3 + \dfrac{1}{4!}(x-1)^4$, what are $f''(1)$ and $f'''(1)$?

**27.** The third remainder for $f(x)$ at $x = 0$ is

$$R_3(x) = \frac{f^{(4)}(c)}{4!}x^4,$$

where $c$ is a number between 0 and $x$. Let $f(x) = \cos x$, as in Practice Problem 1.

  **(a)** Find a number $M$ such that $|f^{(4)}(c)| \le M$ for all values of $c$.

  **(b)** In Practice Problem 1, the error in using $p_3(.12)$ as an approximation to $f(.12) = \cos .12$ is given by $R_3(.12)$. Show that this error does not exceed $8.64 \times 10^{-6}$.

**28.** Let $p_4(x)$ be the fourth Taylor polynomial of $f(x) = e^x$ at $x = 0$. Show that the error in using $p_4(.1)$ as an approximation for $e^{.1}$ is at most $2.5 \times 10^{-7}$. [*Hint:* Observe that if $x = .1$ and if $c$ is a number between 0 and .1, then $|f^{(5)}(c)| \le f^{(5)}(.1) = e^{.1} \le e^1 \le 3$.]

**29.** Let $p_2(x)$ be the second Taylor polynomial of $f(x) = \sqrt{x}$ at $x = 9$, as in Exercise 21.

  **(a)** Give the second remainder for $f(x)$ at $x = 9$.

  **(b)** Show that $|f^{(3)}(c)| \le \frac{1}{648}$ if $c \ge 9$.

  **(c)** Show that the error in using $p_2(9.3)$ as an approximation for $\sqrt{9.3}$ is at most $\frac{1}{144} \times 10^{-3} < 7 \times 10^{-6}$.

**30.** Let $p_2(x)$ be the second Taylor polynomial of $f(x) = \ln x$ at $x = 1$, as in Exercise 22.

  **(a)** Show that $|f^{(3)}(c)| < 4$ if $c \ge .8$.

  **(b)** Show that the error in using $p_2(.8)$ as an approximation for $\ln .8$ is at most $\frac{16}{3} \times 10^{-3} < .0054$.

*Technology Exercises*

**31.** Graph the function $\mathbf{Y_1} = \dfrac{1}{1-x}$ and its fourth Taylor polynomial in the window $[-1, 1]$ *by* $[-1, 5]$. Find a number $b$ such that graphs of the two functions appear identical on the screen for $x$ between 0 and $b$. Calculate the difference between the function and its Taylor polynomial at $x = b$.

**32.** Repeat Exercise 31 for the function $\mathbf{Y_1} = \dfrac{1}{1-x}$ and its seventh Taylor polynomial.

**33.** Graph the function $\mathbf{Y_1} = e^x$ and its fourth Taylor polynomial in the window $[0, 3]$ *by* $[-2, 20]$. Find a number $b$ such that graphs of the two functions appear identical on the screen for $x$ between 0 and $b$. Calculate the difference between the function and its Taylor polynomial at $x = b$ and at $x = 3$.

**34.** Graph the function $\mathbf{Y_1} = \cos x$ and its second Taylor polynomial in the window ZDECIMAL. Find an interval of the form $[-b, b]$ over which the Taylor polynomial is a good fit to the function. What is the greatest difference between the two functions on this interval?

## SOLUTIONS TO PRACTICE PROBLEMS 11.1

**1. (a)**

$$f(x) = \cos x, \qquad f(0) = 1$$
$$f'(x) = -\sin x, \qquad f'(0) = 0$$
$$f''(x) = -\cos x, \qquad f''(0) = -1$$
$$f'''(x) = \sin x, \qquad f'''(0) = 0$$

Therefore,

$$p_3(x) = 1 + \frac{0}{1!}x + \frac{-1}{2!}x^2 + \frac{0}{3!}x^3 = 1 - \frac{1}{2}x^2.$$

[Notice that here the third Taylor polynomial is actually a polynomial of degree 2. The important thing about $p_3(x)$ is not its degree but rather the fact that it agrees with $f(x)$ at $x = 0$ up to its third derivative.]

**(b)** By part (a), $\cos x \approx 1 - \frac{1}{2}x^2$ when $x$ is near 0. Therefore,

$$\cos .12 \approx 1 - \tfrac{1}{2}(.12)^2 = .9928.$$

(*Note:* To five decimal places, $\cos .12 = .99281$.)

**2.**

$$f(x) = 3x^2 - 17, \qquad f(3) = 10$$
$$f'(x) = 6x, \qquad f'(3) = 18$$
$$f''(x) = 6, \qquad f''(3) = 6$$
$$f^{(3)}(x) = 0, \qquad f^{(3)}(3) = 0$$

The derivatives $f^{(n)}(x)$ for $n \geq 3$ are all the zero constant function. In particular, $f^{(n)}(3) = 0$ for $n \geq 3$. Therefore,

$$p_1(x) = 10 + 18(x - 3),$$

$$p_2(x) = 10 + 18(x - 3) + \frac{6}{2!}(x - 3)^2,$$

$$p_3(x) = 10 + 18(x - 3) + \frac{6}{2!}(x - 3)^2 + \frac{0}{3!}(x - 3)^3.$$

For $n \geq 3$, we have

$$p_n(x) = p_2(x) = 10 + 18(x - 3) + 3(x - 3)^2.$$

[This is the appropriate form of the Taylor polynomial at $x = 3$. However, it is instructive to multiply out the terms in $p_2(x)$ and collect the like powers of $x$:

$$p_2(x) = 10 + 18x - 18 \cdot 3 + 3(x^2 - 6x + 9)$$
$$= 10 + 18x - 54 + 3x^2 - 18x + 27 = 3x^2 - 17.$$

That is, $p_2(x)$ is $f(x)$, but written in a different form. This is not too surprising, since $f(x)$ itself is a polynomial that agrees with $f(x)$ and all of its derivatives at $x = 3$.]

# 11.2 The Newton–Raphson Algorithm

Figure 1

Many applications of mathematics involve the solution of equations. Often one has a function $f(x)$ and must find a value of $x$, say $x = r$, such that $f(r) = 0$. Such a value of $x$ is called a *zero* of the function or, equivalently, a *root* of the equation $f(x) = 0$. Graphically, a zero of $f(x)$ is a value of $x$ where the graph of $y = f(x)$ crosses the $x$-axis. (See Fig. 1.) When $f(x)$ is a polynomial, it is sometimes possible to factor $f(x)$ and quickly discover the zeros of $f(x)$. Unfortunately, in most realistic applications there is no simple way to locate zeros. However, there are several methods for finding an approximate value of a zero to any desired degree of accuracy. We shall describe one such method—the Newton–Raphson algorithm.

Suppose that we know that a zero of $f(x)$ is approximately $x_0$. The idea of the Newton–Raphson algorithm is to obtain an even better approximation of the zero by replacing $f(x)$ by its first Taylor polynomial at $x_0$—that is, by

$$p(x) = f(x_0) + \frac{f'(x_0)}{1}(x - x_0).$$

Since $p(x)$ closely resembles $f(x)$ near $x = x_0$, the zero of $f(x)$ should be close to the zero of $p(x)$. But solving the equation $p(x) = 0$ for $x$ gives

$$f(x_0) + f'(x_0)(x - x_0) = 0$$

$$xf'(x_0) = f'(x_0)x_0 - f(x_0)$$

$$x = x_0 - \frac{f(x_0)}{f'(x_0)}.$$

That is, if $x_0$ is an approximation to the zero $r$, then the number

$$x_1 = x_0 - \frac{f(x_0)}{f'(x_0)} \tag{1}$$

generally provides an improved approximation.

We may visualize the situation geometrically as in Fig. 2. The first Taylor polynomial $p(x)$ at $x_0$ has as its graph the tangent line to $y = f(x)$ at the point $(x_0, f(x_0))$. The value of $x$ for which $p(x) = 0$—that is, $x = x_1$—corresponds to the point where the tangent line crosses the $x$-axis.

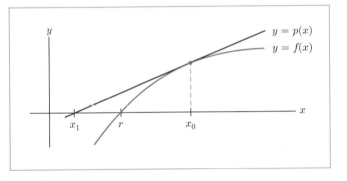

**Figure 2.** Obtaining $x_1$ from $x_0$.

Now let us use $x_1$ in place of $x_0$ as an approximation to the zero $r$. We obtain a new approximation $x_2$ from $x_1$ in the same way we obtained $x_1$ from $x_0$—namely,

$$x_2 = x_1 - \frac{f(x_1)}{f'(x_1)}.$$

We may repeat this process over and over. At each stage a new approximation $x_{\text{new}}$ is obtained from the old approximation $x_{\text{old}}$ by the following formula:

### Newton–Raphson Method

$$x_{\text{new}} = x_{\text{old}} - \frac{f(x_{\text{old}})}{f'(x_{\text{old}})}.$$

In this way we obtain a sequence of approximations $x_0, x_1, x_2, \ldots$, which usually approach as close to $r$ as desired. (See Fig. 3.)

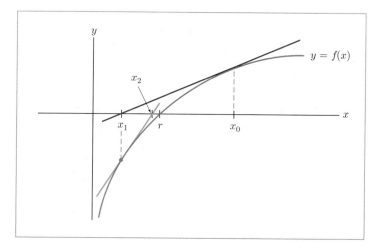

**Figure 3.** A sequence of approximations to $r$.

---

**EXAMPLE 1**  The polynomial $f(x) = x^3 - x - 2$ has a zero between 1 and 2. Let $x_0 = 1$ and find the next three approximations of the zero of $f(x)$ using the Newton–Raphson algorithm.

**Solution**  Since $f'(x) = 3x^2 - 1$, formula (1) becomes

$$x_1 = x_0 - \frac{x_0^3 - x_0 - 2}{3x_0^2 - 1}.$$

With $x_0 = 1$, we have

$$x_1 = 1 - \frac{1^3 - 1 - 2}{3(1)^2 - 1} = 1 - \frac{-2}{2} = 2,$$

$$x_2 = 2 - \frac{2^3 - 2 - 2}{3(2)^2 - 1} = 2 - \frac{4}{11} = \frac{18}{11},$$

$$x_3 = \frac{18}{11} - \frac{\left(\frac{18}{11}\right)^3 - \frac{18}{11} - 2}{3\left(\frac{18}{11}\right)^2 - 1} \approx 1.530.$$

The actual value of $r$ to three decimal places is 1.521.     ■

---

**EXAMPLE 2**  Use four repetitions of the Newton–Raphson algorithm to approximate $\sqrt{2}$.

**Solution**  $\sqrt{2}$ is a zero of the function $f(x) = x^2 - 2$. Since $\sqrt{2}$ clearly lies between 1 and 2, let us take our initial approximation as $x_0 = 1$. ($x_0 = 2$ would do just as well.) Since $f'(x) = 2x$, we have

$$x_1 = x_0 - \frac{x_0^2 - 2}{2x_0} = 1 - \frac{1^2 - 2}{2(1)} = 1 - \left(-\frac{1}{2}\right) = 1.5,$$

$$x_2 = 1.5 - \frac{(1.5)^2 - 2}{2(1.5)} \approx 1.4167,$$

$$x_3 = 1.4167 - \frac{(1.4167)^2 - 2}{2(1.4167)} \approx 1.41422,$$

$$x_4 = 1.41422 - \frac{(1.41422)^2 - 2}{2(1.41422)} \approx 1.41421.$$

This approximation to $\sqrt{2}$ is correct to five decimal places.     ■

| | EXAMPLE 3 | Approximate the zeros of the polynomial $x^3 + x + 3$. |

**Solution** By applying our curve-sketching techniques, we can make a rough sketch of the graph of $y = x^3 + x + 3$, as in Fig. 4. The graph crosses the $x$-axis between $x = -2$ and $x = -1$. So the polynomial has one zero lying between $-2$ and $-1$. Let us therefore set $x_0 = -1$. Since $f'(x) = 3x^2 + 1$, we have

$$x_1 = x_0 - \frac{x_0^3 + x_0 + 3}{3x_0^2 + 1} = -1 - \frac{(-1)^3 + (-1) + 3}{3(-1)^2 + 1} = -1.25,$$

$$x_2 = -1.25 - \frac{(-1.25)^3 + (-1.25) + 3}{3(-1.25)^2 + 1} \approx -1.21429,$$

$$x_3 \approx -1.21341,$$

$$x_4 \approx -1.21341.$$

**Figure 4**

Therefore, the zero of the given polynomial is approximately $-1.21341$. ■

---

| | EXAMPLE 4 | Approximate the positive solution of $e^x - 4 = x$. |

**Solution** The rough sketches of the two graphs in Fig. 5 indicate that the solution lies near 2. Let $f(x) = e^x - 4 - x$. Then the solution of the original equation will be a zero of $f(x)$. Let us apply the Newton–Raphson algorithm to $f(x)$ with $x_0 = 2$. Since $f'(x) = e^x - 1$,

$$x_1 = x_0 - \frac{e^{x_0} - 4 - x_0}{e^{x_0} - 1} = 2 - \frac{e^2 - 4 - 2}{e^2 - 1} \approx 2 - \frac{1.38906}{6.38906} \approx 1.78,$$

$$x_2 = 1.78 - \frac{e^{1.78} - 4 - (1.78)}{e^{1.78} - 1} \approx 1.78 - \frac{.14986}{4.92986} \approx 1.75,$$

$$x_3 = 1.75 - \frac{e^{1.75} - 4 - (1.75)}{e^{1.75} - 1} \approx 1.75 - \frac{.0046}{4.7546} \approx 1.749.$$

**Figure 5**

Therefore, an approximate solution is $x = 1.749$. ■

---

| | EXAMPLE 5 | **Internal rate of return** Suppose that an investment of $100 yields the following returns: |

$2 at the end of the first month
$15 at the end of the second month
$45 at the end of the third month
$50 at the end of the fourth (and last) month

The total of these returns is $112. This represents the initial investment of $100, plus earnings of $12 during the four months. The *internal rate of return* on this investment is the interest rate (per month) for which the sum of the present values of the returns equals the initial investment, $100. Determine the internal rate of return.

**Solution** Let $i$ be the *monthly* rate of interest. The present value of an amount $A$ to be received in $k$ months is $A(1+i)^{-k}$. Therefore, we must solve

$$\begin{bmatrix} \text{amount of initial} \\ \text{investment} \end{bmatrix} = \begin{bmatrix} \text{sum of present} \\ \text{values of returns} \end{bmatrix}$$

$$100 = 2(1+i)^{-1} + 15(1+i)^{-2} + 45(1+i)^{-3} + 50(1+i)^{-4}.$$

Multiplying both sides of the equation by $(1+i)^4$ and taking all terms to the left, we obtain

$$100(1+i)^4 - 2(1+i)^3 - 15(1+i)^2 - 45(1+i) - 50 = 0.$$

Let $x = 1 + i$, and solve the resulting equation by the Newton–Raphson algorithm with $x_0 = 1.1$.

$$100x^4 - 2x^3 - 15x^2 - 45x - 50 = 0$$

$$f(x) = 100x^4 - 2x^3 - 15x^2 - 45x - 50$$

$$f'(x) = 400x^3 - 6x^2 - 30x - 45$$

$$x_1 = x_0 - \frac{100x_0^4 - 2x_0^3 - 15x_0^2 - 45x_0 - 50}{400x_0^3 - 6x_0^2 - 30x_0 - 45}$$

$$= 1.1 - \frac{100(1.1)^4 - 2(1.1)^3 - 15(1.1)^2 - 45(1.1) - 50}{400(1.1)^3 - 6(1.1)^2 - 30(1.1) - 45}$$

$$= 1.1 - \frac{26.098}{447.14} \approx 1.042$$

$$x_2 \approx 1.035$$

$$x_3 \approx 1.035$$

Therefore, an approximate solution is $x = 1.035$. Hence $i = .035$ and the investment had an internal rate of return of 3.5% per month. ∎

In general, if an investment of $P$ dollars produces the returns

$R_1$ at the end of the first period,
$R_2$ at the end of the second period,

$\vdots$

$R_N$ at the end of the $N$th (and last) period,

then the internal rate of return, $i$, is obtained by solving[*] the equation

$$P(1+i)^N - R_1(1+i)^{N-1} - R_2(1+i)^{N-2} - \cdots - R_N = 0$$

for its positive root.

When a loan of $P$ dollars is paid back with $N$ equal periodic payments of $R$ dollars at interest rate $i$ per period, the equation to be solved for $i$ becomes

$$P(1+i)^N - R(1+i)^{N-1} - R(1+i)^{N-2} - \cdots - R = 0.$$

---

[*]We are assuming that all of the returns are nonnegative and add up to at least $P$. An analysis of the general case can be found in H. Paley, P. Colwell, and R. Cannaday, *Internal Rates of Return*, UMAP Module 640 (Lexington, MA: COMAP, Inc., 1984).

This equation can be simplified to

$$Pi + R\left[(1+i)^{-N} - 1\right] = 0.$$

(See Exercise 41 in Section 11.3.)

---

**EXAMPLE 6**    **Amortization of a loan**   A mortgage of $104,880 is repaid in 360 monthly payments of $755. Use two iterations of the Newton–Raphson algorithm to estimate the monthly rate of interest.

**Solution**   Here $P = 104{,}880$, $R = 755$, and $N = 360$. Therefore, we must solve the equation

$$104{,}880i + 755\left[(1+i)^{-360} - 1\right] = 0.$$

Let $f(i) = 104{,}880i + 755\left[(1+i)^{-360} - 1\right]$. Then

$$f'(i) = 104{,}880 - 271{,}800(1+i)^{-361}.$$

Apply the Newton–Raphson algorithm to $f(i)$ with $i_0 = .01$:

$$i_1 = i_0 - \frac{104{,}880 i_0 + 755\left[(1+i_0)^{-360} - 1\right]}{104{,}880 - 271{,}800(1+i_0)^{-361}} \approx .00676$$

$$i_2 \approx .00650.$$

Therefore, the monthly interest rate is approximately $.65\%$.    ■

### Comments

1. The values of successive approximations in the Newton–Raphson algorithm depend on the extent of round-off used during the calculation.

2. The Newton–Raphson algorithm is an excellent computational tool. However, in some cases it will not work. For instance, if $f'(x_n) = 0$ for some approximation $x_n$, then there is no way to compute the next approximation. Other instances in which the algorithm fails are presented in Exercises 25 and 26.

3. It can be shown that if $f(x)$, $f'(x)$, and $f''(x)$ are continuous near $r$ [a zero of $f(x)$] and $f'(r) \neq 0$, then the Newton–Raphson algorithm will definitely work provided that the initial approximation $x_0$ is not too far away.

### Incorporating Technology

Graphing calculators can generate a new approximation for the Newton–Raphson algorithm with each press of the ENTER key. Figure 6 shows the first five approximations for Example 1. Here $\mathbf{Y_1}$ is the function $f(x) = x^3 - x - 2$ and $\mathbf{Y_2}$ is its derivative. For details, see the section "The Newton–Raphson Algorithm" in the appropriate calculator appendix.

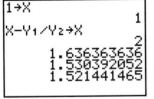

**Figure 6.**   Successive approximations for Example 1.

---

1. Use three repetitions of the Newton–Raphson algorithm to estimate $\sqrt[3]{7}$.

2. Use three repetitions of the Newton–Raphson algorithm to estimate the zeros of $f(x) = 2x^3 + 3x^2 + 6x - 3$.

# EXERCISES 11.2

*In Exercises 1–8, use three repetitions of the Newton–Raphson algorithm to approximate the following:*

1. $\sqrt{5}$

2. $\sqrt{7}$

3. $\sqrt[3]{6}$

4. $\sqrt[3]{11}$

5. The zero of $x^2 - x - 5$ between 2 and 3.

6. The zero of $x^2 + 3x - 11$ between $-5$ and $-6$.

7. The zero of $\sin x + x^2 - 1$ near $x_0 = 0$.

8. The zero of $e^x + 10x - 3$ near $x_0 = 0$.

9. Sketch the graph of $y = x^3 + 2x + 2$ and use the Newton–Raphson algorithm (three repetitions) to approximate all $x$-intercepts.

10. Sketch the graph of $y = x^3 + x - 1$ and use the Newton–Raphson algorithm (three repetitions) to approximate all $x$-intercepts.

11. Use the Newton–Raphson algorithm to find an approximate solution to $e^{-x} = x^2$.

12. Use the Newton–Raphson algorithm to find an approximate solution to $e^{5-x} = 10 - x$.

13. Suppose that an investment of $500 yields returns of $100, $200, and $300 at the end of the first, second, and third months, respectively. Determine the internal rate of return on this investment.

14. An investor buys a bond for $1000. She receives $10 at the end of each month for 2 months and then sells the bond at the end of the second month for $1040. Determine the internal rate of return on this investment.

15. A $663 television set is purchased with a down payment of $100 and a loan of $563 to be repaid in five monthly installments of $116. Determine the monthly rate of interest on the loan.

16. A mortgage of $100,050 is repaid in 240 monthly payments of $900. Determine the monthly rate of interest.

17. A function $f(x)$ has the graph given in Fig. 7. Let $x_1$ and $x_2$ be the estimates of a root of $f(x)$ obtained by applying the Newton–Raphson algorithm using an initial approximation of $x_0 = 5$. Draw the appropriate tangent lines and estimate the numerical values of $x_1$ and $x_2$.

18. Redo Exercise 17 with $x_0 = 1$.

19. Suppose the line $y = 4x + 5$ is tangent to the graph of the function $f(x)$ at $x = 3$. If the Newton–Raphson algorithm is used to find a root of $f(x) = 0$ with the initial guess $x_0 = 3$, what is $x_1$?

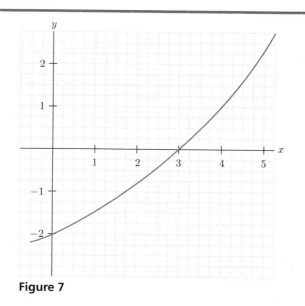

**Figure 7**

20. Suppose the graph of the function $f(x)$ has slope $-2$ at the point $(1, 2)$. If the Newton–Raphson algorithm is used to find a root of $f(x) = 0$ with the initial guess $x_0 = 1$, what is $x_1$?

21. Figure 8 contains the graph of the function $f(x) = x^2 - 2$. The function has zeros at $x = \sqrt{2}$ and $x = -\sqrt{2}$. When the Newton–Raphson algorithm is applied to find a zero, what values of $x_0$ lead to the zero $\sqrt{2}$?

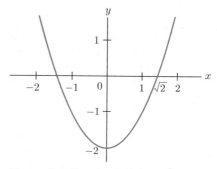

**Figure 8.** Graph of $f(x) = x^2 - 2$.

22. Figure 9 contains the graph of the function $f(x) = x^3 - 12x$. The function has zeros at $x = -\sqrt{12}$, 0, and $\sqrt{12}$. Which zero of $f(x)$ will be approximated by the Newton–Raphson method starting with $x_0 = 4$? Starting with $x_0 = 1$? Starting with $x_0 = -1.8$?

23. What special occurrence takes place when the Newton–Raphson algorithm is applied to a linear function, $f(x) = mx + b$ with $m \neq 0$?

24. What happens when the first approximation, $x_0$, is actually a zero of $f(x)$?

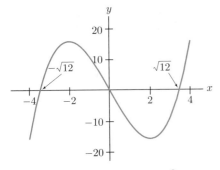

**Figure 9.**   Graph of $f(x) = x^3 - 12x$.

*Exercises 25 and 26 present two examples in which successive repetitions of the Newton–Raphson algorithm do not approach a root.*

**25.** Apply the Newton–Raphson algorithm to the function $f(x) = x^{1/3}$ whose graph is drawn in Fig. 10(a). Use $x_0 = 1$.

**26.** Apply the Newton–Raphson algorithm to the function whose graph is drawn in Fig. 10(b). Use $x_0 = 1$.

### Technology Exercises

**27.** The functions $f(x) = x^2 - 4$ and $g(x) = (x - 2)^2$ both

have a zero at $x = 2$. Apply the Newton–Raphson algorithm to each function with $x_0 = 3$ and determine the value of $n$ for which $x_n$ appears on the screen as exactly 2. Graph the two functions and explain why the sequence for $f(x)$ converges so quickly to 2, whereas the sequence for $g(x)$ converges so slowly.

**28.** Apply the Newton–Raphson algorithm to the function $f(x) = x^3 - 5x$ with $x_0 = 1$. After observing the behavior, graph the function along with the tangent lines at $x = 1$ and $x = -1$ and explain geometrically what is happening.

**29.** Draw the graph of $f(x) = x^4 - 2x^2$, $[-2, 2]$ *by* $[-2, 2]$. The function has zeros at $x = -\sqrt{2}$, $x = 0$, and $x = \sqrt{2}$. By looking at the graph, guess which zero will be approached when applying the Newton–Raphson algorithm to each of the following initial approximations:
**(a)** $x_0 = 1.1$    **(b)** $x_0 = .95$    **(c)** $x_0 = .9$
Then test your guesses by actually carrying out the computations.

**30.** Graph the function $f(x) = \dfrac{x^2}{1 + x^2}$, $[-2, 2]$ *by* $[-.5, 1]$. The function has 0 as a zero. By looking at the graph, guess at a value of $x_0$ for which $x_1$ will be exactly 0 when the Newton–Raphson algorithm is invoked. Then test your guess by actually carrying out the computation.

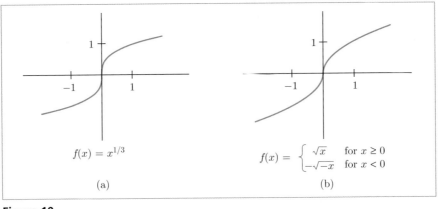

$f(x) = x^{1/3}$

(a)

$f(x) = \begin{cases} \sqrt{x} & \text{for } x \geq 0 \\ -\sqrt{-x} & \text{for } x < 0 \end{cases}$

(b)

**Figure 10**

## SOLUTIONS TO PRACTICE PROBLEMS 11.2

**1.** We wish to approximate a zero of $f(x) = x^3 - 7$. Since $f(1) = -6 < 0$ and $f(2) = 1 > 0$, the graph of $f(x)$ crosses the $x$-axis somewhere between $x = 1$ and $x = 2$. Take $x_0 = 2$ as the initial approximation to the zero. Since $f'(x) = 3x^2$, we have

$$x_1 = x_0 - \frac{x_0^3 - 7}{3x_0^2} = 2 - \frac{2^3 - 7}{3(2)^2} = \frac{23}{12} \approx 1.9167,$$

$$x_2 = 1.9167 - \frac{(1.9167)^3 - 7}{3(1.9167)^2} \approx 1.91294,$$

$$x_3 = 1.91294 - \frac{(1.91294)^3 - 7}{3(1.91294)^2} \approx 1.91293.$$

## SOLUTIONS TO PRACTICE PROBLEMS 11.2 (CONTINUED)

**2.** As a preliminary step, we use the methods of Chapter 2 to sketch the graph of $f(x)$ (Fig. 11). We see that $f(x)$ has a zero that occurs for a positive value of $x$. Since $f(0) = -3$ and $f(1) = 8$, the graph crosses the $x$-axis between 0 and 1. Let us choose $x_0 = 0$ as our initial approximation to the zero of $f(x)$. Since $f'(x) = 6x^2 + 6x + 6$, we then have

$$x_1 = x_0 - \frac{2x_0^3 + 3x_0^2 + 6x_0 - 3}{6x_0^2 + 6x_0 + 6}$$

$$= 0 - \frac{-3}{6} = \frac{1}{2},$$

$$x_2 = \frac{1}{2} - \frac{1}{\frac{21}{2}} = \frac{1}{2} - \frac{2}{21}$$

$$= \frac{17}{42} \approx .40476.$$

Continuing, we find that $x_3 \approx .39916$.

**Figure 11**

# 11.3 Infinite Series

An *infinite series* is an infinite addition of numbers

$$a_1 + a_2 + a_3 + a_4 + \cdots .$$

Here are some examples:

$$1 + \tfrac{1}{2} + \tfrac{1}{4} + \tfrac{1}{8} + \tfrac{1}{16} + \cdots , \tag{1}$$

$$1 + 1 + 1 + 1 + \cdots , \tag{2}$$

$$1 - 1 + 1 - 1 + \cdots . \tag{3}$$

Some infinite series can be associated with a "sum." To illustrate how this is done, let us consider the infinite series (1). If we add up the first two, three, four, five, and six terms of the infinite series (1), we obtain

$$1 + \tfrac{1}{2} = 1\tfrac{1}{2},$$

$$1 + \tfrac{1}{2} + \tfrac{1}{4} = 1\tfrac{3}{4},$$

$$1 + \tfrac{1}{2} + \tfrac{1}{4} + \tfrac{1}{8} = 1\tfrac{7}{8},$$

$$1 + \tfrac{1}{2} + \tfrac{1}{4} + \tfrac{1}{8} + \tfrac{1}{16} = 1\tfrac{15}{16},$$

$$1 + \tfrac{1}{2} + \tfrac{1}{4} + \tfrac{1}{8} + \tfrac{1}{16} + \tfrac{1}{32} = 1\tfrac{31}{32}.$$

Each total lies halfway between the preceding total and the number 2. It appears from the calculations above that by increasing the number of terms, we bring the total arbitrarily close to 2. Indeed, this is supported by further calculation. For example,

$$\underbrace{1 + \frac{1}{2} + \frac{1}{4} + \cdots + \frac{1}{2^9}}_{\text{10 terms}} = 2 - \frac{1}{2^9} \approx 1.998047,$$

$$\underbrace{1 + \frac{1}{2} + \frac{1}{4} + \cdots + \frac{1}{2^{19}}}_{\text{20 terms}} = 2 - \frac{1}{2^{19}} \approx 1.999998,$$

$$\underbrace{1 + \frac{1}{2} + \frac{1}{4} + \cdots + \frac{1}{2^{n-1}}}_{n \text{ terms}} = 2 - \frac{1}{2^{n-1}}.$$

Therefore, it seems reasonable to assign the infinite series (1) the "sum" 2:

$$1 + \tfrac{1}{2} + \tfrac{1}{4} + \tfrac{1}{8} + \tfrac{1}{16} + \cdots = 2. \tag{4}$$

The sum of the first $n$ terms of an infinite series is called its $n$th *partial sum* and is denoted $S_n$. In series (1) we were very fortunate that the partial sums approached a limiting value, 2. This is not always the case. For example, consider the infinite series (2). If we form the first few partial sums, we get

$$S_2 = 1 + 1 \qquad\qquad = 2,$$
$$S_3 = 1 + 1 + 1 \qquad = 3,$$
$$S_4 = 1 + 1 + 1 + 1 = 4.$$

We see that these sums do not approach any limit. Rather, they become larger and larger, eventually exceeding any specified number.

The partial sums need not grow without bound in order that an infinite series not have a sum. For example, consider the infinite series (3). Here the sums of initial terms are

$$S_2 = 1 - 1 \qquad\qquad\qquad = 0,$$
$$S_3 = 1 - 1 + 1 \qquad\qquad = 1,$$
$$S_4 = 1 - 1 + 1 - 1 \qquad = 0,$$
$$S_5 = 1 - 1 + 1 - 1 + 1 = 1,$$

and so forth. The partial sums alternate between 0 and 1 and do not approach a limit. So the infinite series (3) has no sum.

An infinite series whose partial sums approach a limit is called *convergent*. The limit is then called the *sum* of the infinite series. An infinite series whose partial sums do not approach a limit is called *divergent*. From our preceding discussion, we know that the infinite series (1) is convergent, whereas (2) and (3) are divergent.

It is often an extremely difficult task to determine whether or not a given infinite series is convergent. And intuition is not always an accurate guide. For example, one might at first suspect that the infinite series

$$1 + \tfrac{1}{2} + \tfrac{1}{3} + \tfrac{1}{4} + \tfrac{1}{5} + \cdots$$

(the so-called *harmonic series*) is convergent. However, it is not. The sums of its initial terms increase without bound, although they do so very slowly. For example, it takes about 12,000 terms before the sum exceeds 10 and about $2.7 \times 10^{43}$ terms before the sum exceeds 100. Nevertheless, the sum eventually exceeds any prescribed number. (See Exercise 42 and Section 11.4.)

There is an important type of infinite series whose convergence or divergence is easily determined. Let $a$ and $r$ be given nonzero numbers. A series of the form

$$a + ar + ar^2 + ar^3 + ar^4 + \cdots$$

is called a *geometric series with ratio* $r$. (The "ratio" of consecutive terms is $r$.)

---

**The Geometric Series**    The infinite series

$$a + ar + ar^2 + ar^3 + ar^4 + \cdots$$

converges if and only if $|r| < 1$. When $|r| < 1$, the sum of the series is

$$\frac{a}{1 - r}. \tag{5}$$

---

For example, if $a = 1$ and $r = \frac{1}{2}$, we obtain the infinite series (1). In this case

$$\frac{a}{1 - r} = \frac{1}{1 - \frac{1}{2}} = \frac{1}{\frac{1}{2}} = 2,$$

in agreement with our previous observation. Also, series (2) and (3) are divergent geometric series, with $r = 1$ and $r = -1$, respectively. A proof of the foregoing result is outlined in Exercise 41.

---

**EXAMPLE 1**    Calculate the sums of the following geometric series:

(a)   $1 + \dfrac{1}{5} + \dfrac{1}{5^2} + \dfrac{1}{5^3} + \dfrac{1}{5^4} + \cdots$

(b)   $\dfrac{2}{3^2} + \dfrac{2}{3^4} + \dfrac{2}{3^6} + \dfrac{2}{3^8} + \dfrac{2}{3^{10}} + \cdots$

(c)   $\dfrac{5}{2^2} - \dfrac{5^2}{2^5} + \dfrac{5^3}{2^8} - \dfrac{5^4}{2^{11}} + \dfrac{5^5}{2^{14}} - \cdots$

**Solution**   (a) Here $a = 1$ and $r = \frac{1}{5}$. The sum of the series is

$$\frac{a}{1 - r} = \frac{1}{1 - \frac{1}{5}} = \frac{1}{\frac{4}{5}} = \frac{5}{4}.$$

(b) We find $r$ by dividing any term by the preceding term. So

$$r = \frac{\dfrac{2}{3^4}}{\dfrac{2}{3^2}} = \frac{2}{3^4} \cdot \frac{3^2}{2} = \frac{1}{3^2} = \frac{1}{9}.$$

Since the series is a geometric series, we obtain the same result using any other pair of successive terms. For instance,

$$\frac{\dfrac{2}{3^8}}{\dfrac{2}{3^6}} = \frac{2}{3^8} \cdot \frac{3^6}{2} = \frac{1}{3^2} = \frac{1}{9}.$$

The first term of the series is $a = \dfrac{2}{3^2} = \dfrac{2}{9}$, so the sum of the series is

$$\frac{a}{1 - r} = a \cdot \frac{1}{1 - r} = \frac{2}{9} \cdot \frac{1}{1 - \frac{1}{9}} = \frac{2}{9} \cdot \frac{9}{8} = \frac{1}{4}.$$

(c) We may find $r$ as in part (b), or we may observe that the numerator of each fraction in the series (c) is increasing by a factor of 5, while the denominator is increasing by a factor of $2^3 = 8$. So the ratio of successive fractions is $\frac{5}{8}$. However, the terms in the series are alternately positive and negative, so the ratio of successive terms must be negative. Hence $r = -\frac{5}{8}$. Next, $a = \frac{5}{2^2} = \frac{5}{4}$, so the sum of series (c) is

$$a \cdot \frac{1}{1-r} = \frac{5}{4} \cdot \frac{1}{1-\left(-\frac{5}{8}\right)} = \frac{5}{4} \cdot \frac{1}{\frac{13}{8}} = \frac{5}{4} \cdot \frac{8}{13} = \frac{10}{13}. \qquad \blacksquare$$

Sometimes a rational number is expressed as an infinite repeating decimal, such as $.121\overline{212}$. The value of such a "decimal expansion" is in fact the sum of an infinite series.

---

**EXAMPLE 2**  What rational number has the decimal expansion $.121\overline{212}$?

**Solution**  This number denotes the infinite series

$$.12 + .0012 + .000012 + \cdots = \frac{12}{100} + \frac{12}{100^2} + \frac{12}{100^3} + \cdots,$$

a geometric series with $a = \frac{12}{100}$ and $r = \frac{1}{100}$. The sum of the geometric series is

$$\frac{a}{1-r} = a \cdot \frac{1}{1-r} = \frac{12}{100} \cdot \frac{1}{1-\frac{1}{100}} = \frac{12}{100} \cdot \frac{100}{99} = \frac{12}{99} = \frac{4}{33}.$$

Hence $.121\overline{212} = \frac{4}{33}$. $\qquad \blacksquare$

---

**EXAMPLE 3**  **The multiplier effect in economics**  Suppose that the federal government enacts an income tax cut of $10 billion. Assume that each person will spend 93% of all resulting extra income and save the rest. Estimate the total effect of the tax cut on economic activity.

**Solution**  Express all amounts of money in billions of dollars. Of the increase in income created by the tax cut, $(.93)(10)$ billion dollars will be spent. These dollars become extra income to someone and hence 93% will be spent again and 7% saved, so additional spending of $(.93)(.93)(10)$ billion dollars is created. The recipients of those dollars will spend 93% of them, creating yet additional spending of

$$(.93)(.93)(.93)(10) = 10(.93)^3$$

billion dollars, and so on. The total amount of new spending created by the tax cut is thus given by the infinite series

$$10(.93) + 10(.93)^2 + 10(.93)^3 + \cdots.$$

This is a geometric series with initial term $10(.93)$ and ratio $.93$. Its sum is

$$\frac{a}{1-r} = \frac{10(.93)}{1-.93} = \frac{9.3}{.07} \approx 132.86.$$

Thus a $10 billion tax cut creates new spending of about $132.86 billion. $\qquad \blacksquare$

Example 3 illustrates the *multiplier effect.* The proportion of each extra dollar that a person will spend is called the *marginal propensity to consume,* denoted MPC. In Example 3, MPC = .93. As we observed, the total new spending generated by the tax cut is

$$[\text{total new spending}] = 10 \cdot \frac{.93}{1 - .93} = [\text{tax cut}] \cdot \frac{\text{MPC}}{1 - \text{MPC}}.$$

The tax cut is multiplied by the "multiplier" $\dfrac{\text{MPC}}{1 - \text{MPC}}$ to obtain its true effect.

---

**EXAMPLE 4**   **Drug therapy**   Patients with certain heart problems are often treated with digitoxin, a derivative of the digitalis plant. The rate at which a person's body eliminates digitoxin is proportional to the amount of digitoxin present. In 1 day (24 hours) about 10% of any given amount of the drug will be eliminated. Suppose that a "maintenance dose" of .05 milligram (mg) is given daily to a patient. Estimate the total amount of digitoxin that should be present in the patient after several months of treatment.

**Solution**   For a moment, let us consider what happens to the initial dose of .05 mg, and disregard the subsequent doses. After 1 day, 10% of the .05 mg will have been eliminated and $(.90)(.05)$ mg will remain. By the end of the second day, this smaller amount will be reduced 10% to $(.90)(.90)(.05)$ mg, and so on, until after $n$ days only $(.90)^n(.05)$ mg of the original dose will remain. (See Fig. 1.) To determine the cumulative effect of all the doses of digitoxin, we observe that at the time of the second dose (one day after the first dose), the patient's body will contain the second dose of .05 mg plus $(.90)(.05)$ mg of the first dose. A day later, there will be the third dose of .05 mg, plus $(.90)(.05)$ mg of the second dose, plus $(.90)^2(.05)$ of the first dose. At the time of any new dose, the patient's body will contain that dose plus the amounts that remain of earlier doses. Let us tabulate this.

| | Total Amount (mg) of Digitoxin |
|---|---|
| **Days after Initial Dose** | 0    .05 |
| | 1    $.05 + (.90)(.05)$ |
| | 2    $.05 + (.90)(.05) + (.90)^2(.05)$ |
| | ⋮ |
| | $n$    $.05 + (.90)(.05) + (.90)^2(.05) + \cdots + (.90)^n(.05)$ |

We can see that the amounts present at the time of each new dose correspond to the partial sums of the geometric series

$$.05 + (.90)(.05) + (.90)^2(.05) + (.90)^3(.05) + \cdots ,$$

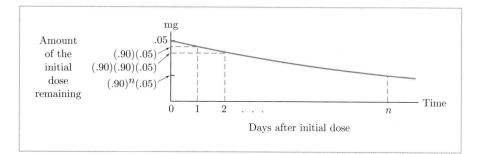

**Figure 1.**   Exponential decrease of the initial dose.

where $a = .05$ and $r = .90$. The sum of this series is

$$\frac{.05}{1 - .90} = \frac{.05}{.10} - .5.$$

Since the partial sums of the series approach the sum of .5, we may conclude that a daily maintenance dose of .05 mg will eventually raise the level of digitoxin in the patient to a "plateau" of .5 mg. Between doses the level will drop 10% down to $(.90)(.5) = .45$ mg. The use of a regular maintenance dose to sustain a certain level of a drug in a patient is an important technique in drug therapy.* ∎

**Sigma Notation** When studying series, it is often convenient to use the Greek capital letter sigma to indicate summation. For example, the sum

$$a_2 + a_3 + \cdots + a_{10}$$

is denoted by

$$\sum_{k=2}^{10} a_k$$

(read "the sum of $a$ sub $k$ from $k$ equals 2 to 10"). The $n$th partial sum of a series, $a_1 + a_2 + \cdots + a_n$, is written as $\sum_{k=1}^{n} a_k$. In these examples the letter $k$ is called the *index of summation*. Sometimes we require the index of summation to begin with 0, and sometimes with 1, but any integer value may be used for $k$. Any letter not already in use may be used as the index of summation. For instance, both

$$\sum_{i=0}^{4} a_i \quad \text{and} \quad \sum_{j=0}^{4} a_j$$

indicate the sum $a_0 + a_1 + a_2 + a_3 + a_4$.

Finally, a formal infinite series

$$a_1 + a_2 + a_3 + \cdots$$

is written as

$$\sum_{k=1}^{\infty} a_k \quad \text{or} \quad \sum_{1}^{\infty} a_k.$$

We will also write $\sum_{k=1}^{\infty} a_k$ as the symbol for the numerical value of the series when it is convergent. Using this notation (and writing $ar^0$ in place of $a$), the main result about the geometric series may be written as follows:

$$\sum_{k=0}^{\infty} ar^k = \frac{a}{1 - r} \qquad \text{if } |r| < 1,$$

$$\sum_{k=0}^{\infty} ar^k \text{ is divergent} \qquad \text{if } |r| \geq 1.$$

*See "Principles of Drug Therapy" by John A. Oates and Grant R. Wilkinson, in *Principles of Internal Medicine*, T. R. Harrison, ed., 8th ed. (New York: McGraw-Hill Book Company, 1977), pp. 334–346.

**EXAMPLE 5**   Determine the sums of the following infinite series.

(a) $\displaystyle\sum_{k=0}^{\infty} \left(\frac{2}{3}\right)^k$
(b) $\displaystyle\sum_{j=0}^{\infty} 4^{-j}$
(c) $\displaystyle\sum_{i=3}^{\infty} \frac{2}{7^i}$

**Solution**   In each case, the first step is to write out the first few terms of the series.

(a) $\displaystyle\sum_{k=0}^{\infty} \left(\frac{2}{3}\right)^k = \underset{[k=0]}{1} + \underset{[k=1]}{\frac{2}{3}} + \underset{[k=2]}{\left(\frac{2}{3}\right)^2} + \underset{[k=3]}{\left(\frac{2}{3}\right)^3} + \cdots$

This is a geometric series with initial term $a = 1$ and ratio $r = \frac{2}{3}$; its sum is

$$\frac{1}{1 - \frac{2}{3}} = \frac{1}{\frac{1}{3}} = 3.$$

(b) $\displaystyle\sum_{j=0}^{\infty} 4^{-j} = 4^0 + 4^{-1} + 4^{-2} + 4^{-3} + \cdots = 1 + \frac{1}{4} + \left(\frac{1}{4}\right)^2 + \left(\frac{1}{4}\right)^3 + \cdots = \frac{1}{1 - \frac{1}{4}} = \frac{4}{3}$

(c) $\displaystyle\sum_{i=3}^{\infty} \frac{2}{7^i} = \frac{2}{7^3} + \frac{2}{7^4} + \frac{2}{7^5} + \frac{2}{7^6} + \cdots$

This is a geometric series with $a = \dfrac{2}{7^3}$ and $r = \dfrac{1}{7}$; its sum is

$$a \cdot \frac{1}{1 - r} = \frac{2}{7^3} \cdot \frac{1}{1 - \frac{1}{7}} = \frac{2}{7^3} \cdot \frac{7}{6} = \frac{1}{147}. \qquad\blacksquare$$

**Incorporating Technology**   Graphing calculators can compute finite sums. The variable **X** can be used as the index of summation. If $f(x)$ is an expression involving $x$, then the sum

$$\sum_{x=m}^{n} f(x)$$

can be computed on TI calculators as

$$\mathbf{sum(seq(f(X), X, m, n, 1)).}$$

For instance,

$$\sum_{x=1}^{99} \frac{1}{x} = 1 + \frac{1}{2} + \frac{1}{3} + \cdots + \frac{1}{99}$$

is calculated as **sum(seq(1/X,X,1,99,1))**, and

$$\sum_{x=1}^{10} \frac{2}{3^{2x}} = \frac{2}{3^2} + \frac{2}{3^4} + \cdots + \frac{2}{3^{20}}$$

from Example 1(b) is calculated as **sum(seq(2/3^(2X), X, 1, 10, 1))**. See Fig. 2. For details, see the section "Sum a Finite Series" in the appropriate calculator appendix. Figure 3 shows the sum of two geometric series.

Also, successive partial sums can be generated by repeatedly pressing the ENTER key. In Fig. 4 the variable **S** holds the current partial sum for the series in Example 5(a).

**Figure 2**

**Figure 3**

**Figure 4**

## PRACTICE PROBLEMS 11.3

**1.** Determine the sum of the geometric series

$$8 - \frac{8}{3} + \frac{8}{9} - \frac{8}{27} + \frac{8}{81} - \cdots .$$

**2.** Find the value of $\displaystyle\sum_{k=0}^{\infty} (.7)^{-k+1}$

## EXERCISES 11.3

*Determine the sums of the following geometric series, when they are convergent.*

**1.** $1 + \dfrac{1}{6} + \dfrac{1}{6^2} + \dfrac{1}{6^3} + \dfrac{1}{6^4} \cdots$

**2.** $1 + \dfrac{3}{4} + \left(\dfrac{3}{4}\right)^2 + \left(\dfrac{3}{4}\right)^3 + \left(\dfrac{3}{4}\right)^4 + \cdots$

**3.** $1 - \dfrac{1}{3^2} + \dfrac{1}{3^4} - \dfrac{1}{3^6} + \dfrac{1}{3^8} - \cdots$

**4.** $1 + \dfrac{1}{2^3} + \dfrac{1}{2^6} + \dfrac{1}{2^9} + \dfrac{1}{2^{12}} + \cdots$

**5.** $2 + \dfrac{2}{3} + \dfrac{2}{9} + \dfrac{2}{27} + \dfrac{2}{81} + \cdots$

**6.** $3 + \dfrac{6}{5} + \dfrac{12}{25} + \dfrac{24}{125} + \dfrac{48}{625} + \cdots$

**7.** $\dfrac{1}{5} + \dfrac{1}{5^4} + \dfrac{1}{5^7} + \dfrac{1}{5^{10}} + \dfrac{1}{5^{13}} + \cdots$

**8.** $\dfrac{1}{3^2} - \dfrac{1}{3^3} + \dfrac{1}{3^4} - \dfrac{1}{3^5} + \dfrac{1}{3^6} - \cdots$

**9.** $3 - \dfrac{3^2}{7} + \dfrac{3^3}{7^2} - \dfrac{3^4}{7^3} + \dfrac{3^5}{7^4} - \cdots$

**10.** $6 - 1.2 + .24 - .048 + .0096 - \cdots$

**11.** $\dfrac{2}{5^4} - \dfrac{2^4}{5^5} + \dfrac{2^7}{5^6} - \dfrac{2^{10}}{5^7} + \dfrac{2^{13}}{5^8} - \cdots$

**12.** $\dfrac{3^2}{2^5} + \dfrac{3^4}{2^8} + \dfrac{3^6}{2^{11}} + \dfrac{3^8}{2^{14}} + \dfrac{3^{10}}{2^{17}} + \cdots$

**13.** $5 + 4 + 3.2 + 2.56 + 2.048 + \cdots$

**14.** $\dfrac{5^3}{3} - \dfrac{5^5}{3^4} + \dfrac{5^7}{3^7} - \dfrac{5^9}{3^{10}} + \dfrac{5^{11}}{3^{13}} - \cdots$

*Sum an appropriate infinite series to find the rational number whose decimal expansion is given.*

**15.** $.2727\overline{27}$

**16.** $.1731\overline{173}$

**17.** $.22\overline{2}$

**18.** $.1515\overline{15}$

**19.** $4.011\overline{011} \ (= 4 + .011\overline{011})$

**20.** $5.44\overline{4}$

**21.** Show that $.999\overline{9} = 1$.

**22.** Compute the value of $.1212\overline{1212}$ as a geometric series with $a = .1212$ and $r = .0001$. Compare your answer with the result of Example 2.

**23.** Compute the total new spending created by a \$10 billion federal income tax cut when the population's marginal propensity to consume is 95%. Compare your result with that of Example 3 and note how a small change in the MPC makes a dramatic change in the total spending generated by the tax cut.

**24.** Compute the effect of a \$20 billion federal income tax cut when the population's marginal propensity to consume is 98%. What is the "multiplier" in this case?

*A perpetuity is a periodic sequence of payments that continue forever. The capital value of the perpetuity is the sum of the present values of all future payments.*

**25.** Consider a perpetuity that promises to pay \$100 at the beginning of each month. Suppose that the interest rate is 12% compounded monthly. Then the present value of \$100 in $k$ months is $100(1.01)^{-k}$.

   **(a)** Express the capital value of the perpetuity as an infinite series.

   **(b)** Find the sum of the infinite series.

**26.** Consider a perpetuity that promises to pay $P$ dollars at the *end* of each month. (The first payment will be received in 1 month.) If the interest rate per month is $r$, the present value of $P$ dollars in $k$ months is $P(1+r)^{-k}$. Find a simple formula for the capital value of the perpetuity.

**27.** A generous corporation not only gives its CEO a \$1,000,000 bonus, but gives her enough money to cover the taxes on the bonus, the taxes on the additional taxes, the taxes on the taxes on the additional taxes, and so on. If she is in the 39.6% tax bracket, how large is her bonus?

**28.** The *coefficient of restitution* of a ball, a number between 0 and 1, specifies how much energy is conserved when the ball hits a rigid surface. A coefficient of .9, for instance, means a bouncing ball will rise to 90% of its previous height after each bounce. The coefficients of restitution for a tennis ball, basketball, super ball, and softball are .7, .75, .9, and .3, respectively. Find the total distance traveled by a tennis ball dropped from a height of 6 feet.

**29.** A patient receives 6 milligrams of a certain drug daily. Each day the body eliminates 30% of the amount of the drug present in the system. Estimate the total amount of the drug that should be present after extended treatment, immediately after a dose is given.

**30.** A patient receives 2 milligrams of a certain drug each day. Each day the body eliminates 20% of the amount of drug present in the system. Estimate the total amount of the drug present after extended treatment, immediately *before* a dose is given.

**31.** A patient receives $M$ milligrams of a certain drug each day. Each day the body eliminates 25% of the amount of drug present in the system. Determine the value of the maintenance dose $M$ such that after many days approximately 20 milligrams of the drug are present immediately after a dose is given.

**32.** A patient receives $M$ milligrams of a certain drug daily. Each day the body eliminates a fraction $q$ of the amount of the drug present in the system. Estimate the total amount of the drug that should be present after extended treatment, immediately after a dose is given.

**33.** The infinite series $a_1 + a_2 + a_3 + \cdots$ has partial sums given by $S_n = 3 - \dfrac{5}{n}$.

   **(a)** Find $\sum_{k=1}^{10} a_k$.

   **(b)** Does the infinite series converge? If so, to what value does it converge?

**34.** The infinite series $a_1 + a_2 + a_3 + \cdots$ has partial sums given by $S_n = n - \dfrac{1}{n}$.

   **(a)** Find $\sum_{k=1}^{10} a_k$.

   **(b)** Does the infinite series converge? If so, to what value does it converge?

*Determine the sums of the following infinite series.*

**35.** $\displaystyle\sum_{k=0}^{\infty} \left(\frac{5}{6}\right)^k$

**36.** $\displaystyle\sum_{k=0}^{\infty} \frac{7}{10^k}$

**37.** $\displaystyle\sum_{j=1}^{\infty} 5^{-2j}$

**38.** $\displaystyle\sum_{j=0}^{\infty} \frac{(-1)^j}{3^j}$

**39.** $\displaystyle\sum_{k=0}^{\infty} (-1)^k \frac{3^{k+1}}{5^k}$

**40.** $\displaystyle\sum_{k=1}^{\infty} \left(\frac{1}{3}\right)^{2k}$

**41.** Let $a$ and $r$ be given nonzero numbers.

   **(a)** Show that
$$(1 - r)(a + ar + ar^2 + \cdots + ar^n) = a - ar^{n+1},$$
and from this conclude that, for $r \neq 1$,
$$a + ar + ar^2 + \cdots + ar^n = \frac{a}{1 - r} - \frac{ar^{n+1}}{1 - r}.$$

   **(b)** Use the result of part (a) to explain why the geometric series $\sum_0^{\infty} ar^k$ converges to $\dfrac{a}{1 - r}$ when $|r| < 1$.

   **(c)** Use the result of part (a) to explain why the geometric series diverges for $|r| > 1$.

   **(d)** Explain why the geometric series diverges for $r = 1$ and $r = -1$.

**42.** Show that the infinite series
$$1 + \frac{1}{2} + \frac{1}{3} + \frac{1}{4} + \frac{1}{5} + \cdots$$
diverges. [*Hint:* $\frac{1}{3} + \frac{1}{4} > \frac{1}{2}$; $\frac{1}{5} + \frac{1}{6} + \frac{1}{7} + \frac{1}{8} > \frac{1}{2}$; $\frac{1}{9} + \cdots + \frac{1}{16} > \frac{1}{2}$; etc.]

### Technology Exercises

**43.** What is the exact value of the infinite geometric series whose partial sum appears at the first entry in Fig. 3?

**44.** What is the exact value of the infinite geometric series whose partial sum appears at the second entry in Fig. 3?

*In Exercises 45 and 46, the calculator screen computes a partial sum of an infinite series. Write out the first five terms of the series and determine the exact value of the infinite series.*

**45.**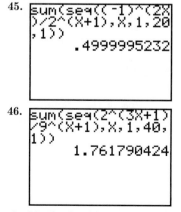

```
sum(seq((-1)^(2X
)/2^(X+1),X,1,20
,1))
         .4999995232
```

**46.**
```
sum(seq(2^(3X+1)
/9^(X+1),X,1,40,
1))
         1.761790424
```

**47.** Verify the formula
$$\sum_{x=1}^{n} x = \frac{n(n+1)}{2}$$
for $n = 10$, $50$, and $100$.

**48.** The sum of the first $n$ odd numbers is $n^2$; that is,

$$\sum_{x=1}^{n}(2x-1) = n^2.$$

Verify this formula for $n = 5$, 10, and 25.

*In Exercises 49 and 50, convince yourself that the equation is correct by summing up the first 999 terms of the infinite series and comparing it with the value on the right.*

**49.** $\displaystyle\sum_{x=1}^{\infty}\frac{1}{x^2} = \frac{\pi^2}{6}.$

**50.** $\displaystyle\sum_{x=1}^{\infty}\frac{(-1)^{x+1}}{x} = \ln 2.$

---

## SOLUTIONS TO PRACTICE PROBLEMS 11.3

**1.** Answer: 6. To obtain the sum of a geometric series, identify $a$ and $r$, and (provided that $|r| < 1$) substitute these values into the formula $\dfrac{a}{1-r}$. The initial term $a$ is just the first term of the series: $a = 8$. The ratio $r$ is often obvious by inspection. However, if in doubt, divide any term by the *preceding* term. Here, the second term divided by the first term is $\frac{-8}{3}/8 = -\frac{1}{3}$, so $r = -\frac{1}{3}$. Since $|r| = \frac{1}{3}$, the series is convergent and the sum is

$$\frac{a}{1-r} = \frac{8}{1-\left(-\frac{1}{3}\right)} = \frac{8}{\frac{4}{3}} = 8 \cdot \frac{3}{4} = 6.$$

**2.** Write out the first few terms of the series and then proceed as in Problem 1.

$$\sum_{k=0}^{\infty}(.7)^{-k+1} = \underset{[k=0]}{(.7)^1} + \underset{[k=1]}{(.7)^0} + \underset{[k=2]}{(.7)^{-1}}$$
$$+ \underset{[k=3]}{(.7)^{-2}} + \underset{[k=4]}{(.7)^{-3}}$$
$$= .7 + 1 + \frac{1}{.7} + \frac{1}{(.7)^2} + \frac{1}{(.7)^3} + \cdots$$

Here $a = .7$ and $r = 1/.7 = \frac{10}{7}$. Since $|r| = \frac{10}{7} > 1$, the series is divergent and has no sum. (The formula $\dfrac{a}{1-r}$ yields $\frac{7}{3}$; however, this value is meaningless. The formula applies only to the case in which the series is convergent.)

---

# 11.4 Series with Positive Terms

Very often, it is difficult to determine the sum of an infinite series. As a fallback to determining the sum, we can at least check that the series converges and so has a sum (even if we can't determine its exact value). Calculus includes many tests for telling if an infinite series converges. This section presents two convergence tests for infinite series consisting of positive terms. The tests are derived from geometric models of the series.

Throughout this section, we consider only series for which each term $a_k$ is positive (or zero). Suppose that $\sum_{k=1}^{\infty} ar^k$ is such a series. Consider the corresponding collection of rectangles in Fig. 1. Each rectangle is one unit wide, and the height of the $k$th rectangle is $a_k$. Hence, the area of the $k$th rectangle is $a_k$ and the area of the region consisting of the first $n$ rectangles is the $n$th partial sum $S_n = a_1 + a_2 + \cdots + a_n$. As $n$ increases, the partial sums increase and approach the area of the region consisting of all the rectangles. If this area is finite, the infinite series converges to this area. If this area is infinite, the series is divergent.

**Figure 1.** Representation of an infinite series by rectangles.

This geometric "picture" of an infinite series provides a convergence test that relates the convergence of the series to the convergence of an improper integral. For example, consider

$$\sum_{k=1}^{\infty} \frac{1}{k^2} \quad \text{and} \quad \int_{1}^{\infty} \frac{1}{x^2} \, dx.$$

Notice that the series and the integral have a similar form, namely,

$$\sum_{k=1}^{\infty} f(k) \quad \text{and} \quad \int_{1}^{\infty} f(x) \, dx, \tag{1}$$

where $f(x) = 1/x^2$.

The techniques of Section 9.6 show that the integral is convergent. Hence, the area under the graph of $y = 1/x^2$ for $x \geq 1$ is finite. Figure 2 shows the graph of $y = 1/x^2$ ($x \geq 1$) superimposed on a geometric model of the series

$$\sum_{k=1}^{\infty} \frac{1}{k^2} = 1 + \frac{1}{4} + \frac{1}{9} + \frac{1}{16} + \cdots .$$

The area of the first rectangle is 1. The region consisting of all the remaining rectangles has finite area since it is contained in the region under the graph of $y = 1/x^2$ ($x \geq 1$), which has finite area. Hence the total area of all the rectangles is finite and the series $\sum_{k=1}^{\infty} \frac{1}{k^2}$ is convergent.

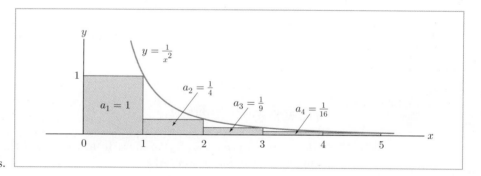

**Figure 2.** A convergent series.

As a second example, consider

$$\sum_{k=1}^{\infty} \frac{1}{k} \quad \text{and} \quad \int_{1}^{\infty} \frac{1}{x} \, dx.$$

Once again, the series and the integral have the forms in (1), where $f(x) = 1/x$. It is easy to check that the integral is divergent and hence the area under the graph of $y = 1/x$ for $x \geq 1$ is infinite. Figure 3 shows the graph of $y = 1/x$ superimposed on a geometric model of the series

$$\sum_{k=1}^{\infty} \frac{1}{k} = 1 + \frac{1}{2} + \frac{1}{3} + \frac{1}{4} + \cdots .$$

Since the area of the region formed by the rectangles clearly exceeds the infinite area of the region under the graph of $y = 1/x$ ($x \geq 1$), the series $\sum_{k=1}^{\infty} \frac{1}{k}$ is divergent.

The reasoning used for these two examples can be used to derive the following important test.

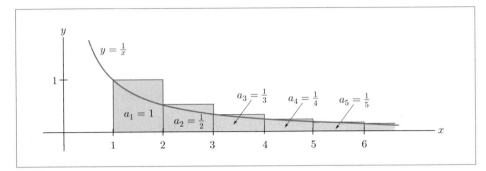

**Figure 3.** A divergent series.

**The Integral Test**   Let $f(x)$ be continuous, decreasing, and positive for $x \geq 1$. Then the infinite series

$$\sum_{k=1}^{\infty} f(k)$$

is convergent if the improper integral

$$\int_{1}^{\infty} f(x)\, dx$$

is convergent, and the infinite series is divergent if the improper integral is divergent.

**EXAMPLE 1**   Use the integral test to determine whether the infinite series

$$\sum_{k=1}^{\infty} \frac{1}{e^k}$$

is convergent or divergent.

**Solution**   Here $f(x) = 1/e^x = e^{-x}$. We know from Chapter 4 that $f(x)$ is a positive, decreasing, continuous function. Also,

$$\int_{1}^{\infty} e^{-x}\, dx = \lim_{b \to \infty} \int_{1}^{b} e^{-x}\, dx = \lim_{b \to \infty} -e^{-x} \Big|_{1}^{b}$$

$$= \lim_{b \to \infty} \left( -e^{-b} + e^{-1} \right) = e^{-1} = \frac{1}{e}.$$

Since the improper integral is convergent, so is the infinite series.   ∎

▶ **Note**   The integral test does not provide the exact value of the sum of a convergent infinite series. It only verifies convergence. Other techniques, sometimes quite advanced, must be used to find the sum. (The sum of the series in Example 1 is readily found since the series happens to be a geometric series with ratio $1/e$.)   ◀

Instead of starting our infinite series with $k = 1$, we may start with $k = N$, where $N$ is any positive integer. To test the convergence of such a series, we determine the convergence (or divergence) of the improper integral

$$\int_{N}^{\infty} f(x)\, dx.$$

**EXAMPLE 2**  Determine whether the series

$$\sum_{k=3}^{\infty} \frac{\ln k}{k}$$

is convergent.

**Solution**  We take $f(x) = (\ln x)/x$. Note that $f(x)$ is continuous and positive for $x \geq 3$. Moreover,

$$f'(x) = \frac{x \cdot \dfrac{1}{x} - \ln x \cdot 1}{x^2} = \frac{1 - \ln x}{x^2}.$$

Since $\ln x > 1$ for $x \geq 3$, we conclude that $f'(x)$ is negative and hence that $f(x)$ is a decreasing function.

To antidifferentiate $(\ln x)/x$, make the substitution $u = \ln x$, $du = (1/x)\,dx$:

$$\int \frac{\ln x}{x}\,dx = \int u\,du = \frac{u^2}{2} + C = \frac{(\ln x)^2}{2} + C.$$

Hence,

$$\int_3^{\infty} \frac{\ln x}{x}\,dx = \lim_{b \to \infty} \int_3^b \frac{\ln x}{x}\,dx = \lim_{b \to \infty} \left. \frac{(\ln x)^2}{2} \right|_3^b$$

$$= \lim_{b \to \infty} \frac{(\ln b)^2}{2} - \frac{(\ln 3)^2}{2} = \infty.$$

Thus the series $\sum_{k=3}^{\infty} \dfrac{\ln k}{k}$ is divergent.  ∎

When using the integral test, we tested the convergence of an infinite series by relating it to an improper integral. In many situations it is possible to accomplish the same result by comparing the series with another infinite series whose convergence or divergence is already known.

Suppose that $\sum_{k=1}^{\infty} a_k$ and $\sum_{k=1}^{\infty} b_k$ have the property that $0 \leq a_k \leq b_k$ for all $k$. In Fig. 4 we have superimposed the geometric models for the two series. Each rectangle for the series $\sum_{k=1}^{\infty} a_k$ lies inside (or coincides with) the corresponding rectangle for the series $\sum_{k=1}^{\infty} b_k$. Clearly, if the region formed by all the rectangles

**Figure 4.**  Term-by-term comparison of two series.

for $\sum_{k=1}^{\infty} b_k$ has finite area, so does the region for $\sum_{k=1}^{\infty} a_k$. On the other hand, if the region for $\sum_{k=1}^{\infty} a_k$ has infinite area, so does the region for $\sum_{k=1}^{\infty} b_k$. These geometric conclusions may be stated in terms of the infinite series as follows:

**Comparison Test** Suppose that $0 \le a_k \le b_k$ for $k = 1, 2, \ldots$.

If $\sum\limits_{k=1}^{\infty} b_k$ converges, so does $\sum\limits_{k=1}^{\infty} a_k$.

If $\sum\limits_{k=1}^{\infty} a_k$ diverges, so does $\sum\limits_{k=1}^{\infty} b_k$.

The comparison test also applies to two series whose terms eventually satisfy $0 \le a_k \le b_k$, say for $k \ge N$, where $N$ is some positive integer. This is so because the convergence or divergence of $\sum_{k=1}^{\infty} a_k$ and $\sum_{k=1}^{\infty} b_k$ is not affected by removing some terms from the beginning of the series.

**EXAMPLE 3** Determine the convergence or divergence of the series

$$\sum_{k=1}^{\infty} \frac{3}{1 + 5^k} = \frac{3}{6} + \frac{3}{26} + \frac{3}{126} + \frac{3}{626} + \cdots .$$

**Solution** Compare the series with the convergent geometric series

$$\sum_{k=1}^{\infty} \frac{3}{5^k} = \frac{3}{5} + \frac{3}{25} + \frac{3}{125} + \frac{3}{625} + \cdots .$$

This series converges since the ratio of successive terms is $r = \frac{1}{5}$. The $k$th terms of these two series satisfy

$$\frac{3}{1 + 5^k} < \frac{3}{5^k}$$

because the denominator of the left fraction is greater than the denominator of the right fraction. Since the series $\sum_{k=1}^{\infty} \frac{3}{5^k}$ converges, the comparison test implies that $\sum_{k=1}^{\infty} \frac{3}{1 + 5^k}$ also converges. ■

This section has considered convergence tests only for series of positive terms. Here is a version of the comparison test that works even if one of the two series has some negative terms.

Suppose that $\sum_{k=1}^{\infty} b_k$ is a convergent series of positive terms and that $|a_k| \le b_k$ for $k = 1, 2, 3, \ldots$. Then $\sum_{k=1}^{\infty} a_k$ is convergent.

## PRACTICE PROBLEMS 11.4

1. **(a)** What is the improper integral associated with the infinite series $\sum\limits_{k=1}^{\infty} \frac{k^2}{(k^3 + 6)^2}$?

   **(b)** Is the improper integral found in part (a) convergent or divergent?

   **(c)** Is the infinite series $\sum\limits_{k=1}^{\infty} \frac{k^2}{(k^3 + 6)^2}$ convergent or divergent?

2. The two series

   $$\sum_{k=1}^{\infty} \frac{1}{4k} \quad \text{and} \quad \sum_{k=1}^{\infty} \frac{1}{4k + 3}$$

   are both divergent. (This is easily established by the integral test.) Which of these series can be used in the comparison test to show that the series $\sum\limits_{k=1}^{\infty} \frac{1}{4k + 1}$ is divergent?

# EXERCISES 11.4

*In Exercises 1–16, use the integral test to determine whether the infinite series is convergent or divergent. (You may assume that the hypotheses of the integral test are satisfied.)*

**1.** $\displaystyle\sum_{k=1}^{\infty} \frac{3}{\sqrt{k}}$

**2.** $\displaystyle\sum_{k=1}^{\infty} \frac{5}{k^{3/2}}$

**3.** $\displaystyle\sum_{k=2}^{\infty} \frac{1}{(k-1)^3}$

**4.** $\displaystyle\sum_{k=0}^{\infty} \frac{7}{k+100}$

**5.** $\displaystyle\sum_{k=1}^{\infty} \frac{2}{5k-1}$

**6.** $\displaystyle\sum_{k=2}^{\infty} \frac{1}{k\sqrt{\ln k}}$

**7.** $\displaystyle\sum_{k=2}^{\infty} \frac{k}{(k^2+1)^{3/2}}$

**8.** $\displaystyle\sum_{k=1}^{\infty} \frac{1}{(2k+1)^3}$

**9.** $\displaystyle\sum_{k=2}^{\infty} \frac{1}{k(\ln k)^2}$

**10.** $\displaystyle\sum_{k=1}^{\infty} \frac{1}{(3k)^2}$

**11.** $\displaystyle\sum_{k=1}^{\infty} e^{3-k}$

**12.** $\displaystyle\sum_{k=1}^{\infty} \frac{1}{e^{2k+1}}$

**13.** $\displaystyle\sum_{k=1}^{\infty} k e^{-k^2}$

**14.** $\displaystyle\sum_{k=1}^{\infty} k^{-3/4}$

**15.** $\displaystyle\sum_{k=1}^{\infty} \frac{2k+1}{k^2+k+2}$

**16.** $\displaystyle\sum_{k=2}^{\infty} \frac{k+1}{(k^2+2k+1)^2}$

**17.** It can be shown that

$$\int_0^{\infty} \frac{3}{9+x^2}\, dx$$

is convergent. Use this fact to show that an appropriate infinite series converges. Give the series and show that the hypotheses of the integral test are satisfied.

**18.** Use the integral test to determine if $\displaystyle\sum_{k=1}^{\infty} \frac{e^{1/k}}{k^2}$ is convergent. Show that the hypotheses of the integral test are satisfied.

**19.** It can be shown that $\displaystyle\lim_{b\to\infty} be^{-b} = 0$. Use this fact and the integral test to show that $\displaystyle\sum_{k=1}^{\infty} \frac{k}{e^k}$ is convergent.

**20.** Is the series $\displaystyle\sum_{k=1}^{\infty} \frac{3^k}{4^k}$ convergent? What is the easiest way to answer this question? Can you tell if

$$\int_1^{\infty} \frac{3^x}{4^x}\, dx$$

is convergent?

*In Exercises 21–26, use the comparison test to determine whether the infinite series is convergent or divergent.*

**21.** $\displaystyle\sum_{k=2}^{\infty} \frac{1}{k^2+5}$ $\left[\text{Compare with } \displaystyle\sum_{k=2}^{\infty} \frac{1}{k^2}.\right]$

**22.** $\displaystyle\sum_{k=2}^{\infty} \frac{1}{\sqrt{k^2-1}}$ $\left[\text{Compare with } \displaystyle\sum_{k=2}^{\infty} \frac{1}{k}.\right]$

**23.** $\displaystyle\sum_{k=1}^{\infty} \frac{1}{2^k+k}$ $\left[\text{Compare with } \displaystyle\sum_{k=1}^{\infty} \frac{1}{2^k} \text{ or } \sum_{k=1}^{\infty} \frac{1}{k}.\right]$

**24.** $\displaystyle\sum_{k=1}^{\infty} \frac{1}{k3^k}$ $\left[\text{Compare with } \displaystyle\sum_{k=1}^{\infty} \frac{1}{k} \text{ or } \sum_{k=1}^{\infty} \frac{1}{3^k}.\right]$

**25.** $\displaystyle\sum_{k=1}^{\infty} \frac{1}{5^k} \cos^2\left(\frac{k\pi}{4}\right)$ $\left[\text{Compare with } \displaystyle\sum_{k=1}^{\infty} \cos^2\left(\frac{k\pi}{4}\right)\right.$ $\left.\text{or } \displaystyle\sum_{k=1}^{\infty} \frac{1}{5^k}.\right]$

**26.** $\displaystyle\sum_{k=0}^{\infty} \frac{1}{\left(\frac{3}{4}\right)^k + \left(\frac{5}{4}\right)^k}$ $\left[\text{Compare with } \displaystyle\sum_{k=0}^{\infty} \left(\frac{3}{4}\right)^{-k}\right.$ $\left.\text{or } \displaystyle\sum_{k=0}^{\infty} \left(\frac{5}{4}\right)^{-k}.\right]$

**27.** Can the comparison test be used with $\displaystyle\sum_{k=2}^{\infty} \frac{1}{k\ln k}$ and $\displaystyle\sum_{k=2}^{\infty} \frac{1}{k}$ to deduce anything about the first series?

**28.** Can the comparison test be used with $\displaystyle\sum_{k=2}^{\infty} \frac{1}{k^2\ln k}$ and $\displaystyle\sum_{k=2}^{\infty} \frac{1}{k^2}$ to deduce anything about the first series?

**29.** The following property is true for any two series (with possibly some negative terms): Let $\displaystyle\sum_{k=1}^{\infty} a_k$ and $\displaystyle\sum_{k=1}^{\infty} b_k$ be convergent series whose sums are $S$ and $T$, respectively. Then $\displaystyle\sum_{k=1}^{\infty} (a_k + b_k)$ is a convergent series whose sum is $S+T$. Make a geometric picture to illustrate why this property is true when the terms $a_k$ and $b_k$ are all positive.

**30.** Let $\displaystyle\sum_{k=1}^{\infty} a_k$ be a convergent series with sum $S$, and let $c$ be a constant. Then $\displaystyle\sum_{k=1}^{\infty} c a_k$ is a convergent series whose sum is $c \cdot S$. Make a geometric picture to illustrate why this is true when $c=2$ and the terms $a_k$ are all positive.

**31.** Use Exercise 29 to show that the series $\displaystyle\sum_{k=0}^{\infty} \frac{8^k + 9^k}{10^k}$ is convergent and determine its sum.

**32.** Use Exercise 30 to show that the series $\sum\limits_{k=1}^{\infty} \dfrac{3}{k^2}$ is convergent. Then use the comparison test to show that the

series $\sum\limits_{k=1}^{\infty} \dfrac{e^{1/k}}{k^2}$ is convergent.

## SOLUTIONS TO PRACTICE PROBLEMS 11.4

**1.** **(a)** $\displaystyle\int_1^{\infty} \dfrac{x^2}{(x^3+6)^2}\, dx$. In general, to find the function $f(x)$, replace each occurrence of $k$ by $x$; then replace the summation sign by an integral sign and adjoin $dx$.

**(b)** A substitution must first be made in order to antidifferentiate the function. Let $u = x^3 + 6$. Then $du = 3x^2\, dx$ and

$$\int \frac{x^2}{(x^3+6)^2}\, dx = \frac{1}{3}\int \frac{3x^2}{(x^3+6)^2}\, dx$$

$$= \frac{1}{3}\int \frac{1}{u^2}\, du$$

$$= -\frac{1}{3}u^{-1} + C$$

$$= -\frac{1}{3}\cdot\frac{1}{x^3+6} + C.$$

Hence,

$$\int_1^{\infty} \frac{x^2}{(x^3+6)^2}\, dx = \lim_{b\to\infty}\left[-\frac{1}{3}\cdot\frac{1}{x^3+6}\right]\Big|_1^b$$

$$= \lim_{b\to\infty}\left[-\frac{1}{3}\cdot\frac{1}{b^3+6} + \frac{1}{3}\cdot\frac{1}{7}\right]$$

$$= \frac{1}{21}.$$

Therefore, the improper integral is convergent.

**(c)** Convergent, since the infinite series is convergent if and only if the associated improper integral is convergent.

**2.** To show that a series is divergent, we need to show that its terms are *greater* than the corresponding terms of some divergent series. Since

$$\frac{1}{4k+3} < \frac{1}{4k+1},$$

the comparison should be made with

$$\sum_{k=1}^{\infty} \frac{1}{4k+3}.$$

(*Note:* If instead we had been trying to establish the *convergence* of an infinite series, we would try to show that its terms are *less* than the corresponding terms of a convergent series.)

## 11.5   Taylor Series

Consider the infinite series $1 + x + x^2 + x^3 + x^4 + \cdots$. This series is of a different type from those discussed in the preceding two sections. Its terms are not numbers but rather are powers of $x$. However, for some specific values of $x$, the series is convergent. In fact, for any value of $x$ between $-1$ and $1$, the series is a convergent geometric series with ratio $x$ and sum $\dfrac{1}{1-x}$. We write

$$\frac{1}{1-x} = 1 + x + x^2 + x^3 + x^4 + \cdots, \qquad |x| < 1. \tag{1}$$

Looking at (1) from a different point of view, we see that the function $f(x) = \dfrac{1}{1-x}$ is represented as a series involving the powers of $x$. This representation is not valid throughout the entire domain of the function $\dfrac{1}{1-x}$ but just for values of $x$ with $-1 < x < 1$.

In many important cases, a function $f(x)$ may be represented by a series of the form

$$f(x) = a_0 + a_1 x + a_2 x^2 + a_3 x^3 + \cdots, \tag{2}$$

where $a_0, a_1, a_2, \ldots$ are suitable constants, and where $x$ ranges over values that make the series converge to $f(x)$. The series is called a *power series* (because it involves powers of $x$). It may be shown that when a function $f(x)$ has a representation by a power series as in (2), the coefficients $a_0, a_1, a_2, \ldots$ are uniquely determined by $f(x)$ and its derivatives at $x = 0$. In fact, $a_0 = f(0)$, and $a_k = f^{(k)}(0)/k!$ for $k = 1, 2, \ldots$, so

$$f(x) = f(0) + \frac{f'(0)}{1!} x + \frac{f''(0)}{2!} x^2 + \cdots + \frac{f^{(k)}(0)}{k!} x^k + \cdots. \tag{3}$$

The series in (3) is often called the *Taylor series of $f(x)$ at $x = 0$* because the partial sums of the series are the Taylor polynomials of $f(x)$ at $x = 0$. The entire equation (3) is called the *Taylor series expansion of $f(x)$ at $x = 0$*.

---

**EXAMPLE 1**  Find the Taylor series expansion of

$$\frac{1}{1-x} \quad \text{at } x = 0.$$

**Solution**  We already know how to represent $\dfrac{1}{1-x}$ as a power series for $|x| < 1$. However, let us use the formula for the Taylor series to see if we get the same result.

$$f(x) = \frac{1}{1-x} = (1-x)^{-1}, \qquad f(0) = 1$$

$$f'(x) = (1-x)^{-2}, \qquad f'(0) = 1$$

$$f''(x) = 2(1-x)^{-3}, \qquad f''(0) = 2$$

$$f'''(0) = 3 \cdot 2 (1-x)^{-4}, \qquad f'''(0) = 3 \cdot 2$$

$$f^{(4)}(x) = 4 \cdot 3 \cdot 2 (1-x)^{-5}, \qquad f^{(4)}(0) = 4 \cdot 3 \cdot 2$$

$$\vdots \qquad\qquad\qquad \vdots$$

Therefore,

$$\frac{1}{1-x} = 1 + \frac{1}{1!} x + \frac{2}{2!} x^2 + \frac{3 \cdot 2}{3!} x^3 + \frac{4 \cdot 3 \cdot 2}{4!} x^4 + \cdots$$

$$= 1 + x + x^2 + x^3 + x^4 + \cdots.$$

We have verified that the Taylor series for $\dfrac{1}{1-x}$ is the familiar geometric power series. The Taylor series expansion is valid for $|x| < 1$.  ∎

---

**EXAMPLE 2**  Find the Taylor series at $x = 0$ for $f(x) = e^x$.

**Solution**

$$f(x) = e^x, \quad f'(x) = e^x, \quad f''(x) = e^x, \quad f'''(x) = e^x, \ldots$$
$$f(0) = 1, \quad f'(0) = 1, \quad f''(0) = 1, \quad f'''(0) = 1, \ldots$$

Therefore,

$$e^x = 1 + x + \frac{1}{2!} x^2 + \frac{1}{3!} x^3 + \frac{1}{4!} x^4 + \cdots.$$

It can be shown that this Taylor series expansion of $e^x$ is valid for all $x$. (*Note:* A Taylor polynomial of $e^x$ gives only an approximation of $e^x$, but the infinite Taylor series actually *equals* $e^x$ for all $x$, in the sense that for any given $x$ the sum of the series is the same as the value of $e^x$.)  ∎

**Operations on Taylor Series** It is often helpful to think of a Taylor series as a polynomial of infinite degree. Many operations on polynomials are also legitimate for Taylor series, provided that we restrict attention to values of $x$ within an appropriate interval. For example, if we have a Taylor series expansion of $f(x)$, we may differentiate the series term by term to obtain the Taylor series expansion of $f'(x)$. An analogous result holds for antiderivatives. Other permissible operations that produce Taylor series include multiplying a Taylor series expansion by a constant or a power of $x$, replacing $x$ by a power of $x$ or by a constant times a power of $x$, and adding or subtracting two Taylor series expansions. The use of such operations often makes it possible to find the Taylor series of a function without directly using the formal definition of a Taylor series. (The process of computing high-order derivatives can become quite laborious when the product or quotient rule is involved.) Once a power series expansion of a function $f(x)$ is found, that series *must* be the Taylor series of the function, since the coefficients of the series are uniquely determined by $f(x)$ and its derivatives at $x = 0$.

---

**EXAMPLE 3** Use the Taylor series at $x = 0$ for

$$\frac{1}{1-x}$$

to find the Taylor series at $x = 0$ for the following functions.

(a) $\dfrac{1}{(1-x)^2}$ (b) $\dfrac{1}{(1-x)^3}$ (c) $\ln(1-x)$

**Solution** We begin with the series expansion

$$\frac{1}{1-x} = 1 + x + x^2 + x^3 + x^4 + x^5 + \cdots, \quad |x| < 1.$$

(a) When we differentiate both sides of this equation, we obtain

$$\frac{1}{(1-x)^2} = 1 + 2x + 3x^2 + 4x^3 + 5x^4 + \cdots, \quad |x| < 1.$$

(b) Differentiating the series in part (a), we find that

$$\frac{2}{(1-x)^3} = 2 + 3 \cdot 2x + 4 \cdot 3x^2 + 5 \cdot 4x^3 + \cdots, \quad |x| < 1.$$

We may multiply a convergent series by a constant. Multiplying by $\frac{1}{2}$, we have

$$\frac{1}{(1-x)^3} = 1 + 3x + 6x^2 + 10x^3 + \cdots + \frac{(n+2)(n+1)}{2}x^n + \cdots$$

for $|x| < 1$.

(c) For $|x| < 1$, we have

$$\int \frac{1}{1-x}\, dx = \int (1 + x + x^2 + x^3 + \cdots)\, dx$$

$$-\ln(1-x) + C = x + \tfrac{1}{2}x^2 + \tfrac{1}{3}x^3 + \tfrac{1}{4}x^4 + \cdots,$$

where $C$ is the constant of integration. If we set $x = 0$ in both sides, we obtain

$$0 + C = 0,$$

so $C = 0$. Thus

$$\ln(1-x) = -x - \tfrac{1}{2}x^2 - \tfrac{1}{3}x^3 - \tfrac{1}{4}x^4 - \cdots, \qquad |x| < 1. \qquad \blacksquare$$

**EXAMPLE 4**  Use the result of Example 3(c) to compute $\ln 1.1$.

**Solution**  Take $x = -.1$ in the Taylor series expansion of $\ln(1-x)$. Then

$$\ln(1-(-.1)) = -(-.1) - \tfrac{1}{2}(-.1)^2 - \tfrac{1}{3}(-.1)^3 - \tfrac{1}{4}(-.1)^4 - \cdots$$

$$\ln 1.1 = .1 - \frac{.01}{2} + \frac{.001}{3} - \frac{.0001}{4} + \frac{.00001}{5} - \cdots.$$

This infinite series may be used to compute $\ln 1.1$ to any degree of accuracy required. For instance, the fifth partial sum gives $\ln 1.1 \approx .09531$, which is correct to five decimal places. $\qquad \blacksquare$

**EXAMPLE 5**  Use the Taylor series at $x = 0$ for $e^x$ to find the Taylor series at $x = 0$ for

(a) $x(e^x - 1)$        (b) $e^{x^2}$

**Solution**  (a) If we subtract 1 from the Taylor series for $e^x$, we obtain a series that converges to $e^x - 1$:

$$e^x - 1 = \left(1 + x + \frac{1}{2!}x^2 + \frac{1}{3!}x^3 + \frac{1}{4!}x^4 + \cdots\right) - 1$$

$$= x + \frac{1}{2!}x^2 + \frac{1}{3!}x^3 + \frac{1}{4!}x^4 + \cdots.$$

Now we multiply this series by $x$, term by term:

$$x(e^x - 1) = x^2 + \frac{1}{2!}x^3 + \frac{1}{3!}x^4 + \frac{1}{4!}x^5 + \cdots.$$

(b) To obtain the Taylor series for $e^{x^2}$, we replace every occurrence of $x$ by $x^2$ in the Taylor series for $e^x$,

$$e^{x^2} = 1 + (x^2) + \frac{1}{2!}(x^2)^2 + \frac{1}{3!}(x^2)^3 + \frac{1}{4!}(x^2)^4 + \cdots$$

$$= 1 + x^2 + \frac{1}{2!}x^4 + \frac{1}{3!}x^6 + \frac{1}{4!}x^8 + \cdots. \qquad \blacksquare$$

**EXAMPLE 6**  Find the Taylor series at $x = 0$ for

(a) $\dfrac{1}{1+x^3}$        (b) $\dfrac{x^2}{1+x^3}$

**Solution**  (a) In the Taylor series at $x = 0$ for $\dfrac{1}{1-x}$, we replace $x$ by $-x^3$, to obtain

$$\frac{1}{1-(-x^3)} = 1 + (-x^3) + (-x^3)^2 + (-x^3)^3 + (-x^3)^4 + \cdots$$

$$\frac{1}{1+x^3} = 1 - x^3 + x^6 - x^9 + x^{12} - \cdots.$$

(b) If we multiply the series in part (a) by $x^2$, we obtain

$$\frac{x^2}{1 + x^3} = x^2 - x^5 + x^8 - x^{11} + x^{14} - \cdots .$$

**Definite Integrals**  The standard normal curve of statistics has the equation

$$y = \frac{1}{\sqrt{2\pi}} e^{-x^2/2}.$$

Areas under the curve cannot be found by direct integration since there is no simple formula for an antiderivative of $e^{-x^2/2}$. However, Taylor series can be used to calculate these areas with a high degree of accuracy.

---

**EXAMPLE 7**  Find the area under the standard normal curve from $x = 0$ to $x = .8$; that is, calculate

$$\frac{1}{\sqrt{2\pi}} \int_0^{.8} e^{-x^2/2} \, dx.$$

**Solution**  A Taylor expansion for $e^x$ was obtained in Example 2.

$$e^x = 1 + x + \frac{1}{2!}x^2 + \frac{1}{3!}x^3 + \frac{1}{4!}x^4 + \cdots$$

Replace each occurrence of $x$ by $-x^2/2$. Then

$$e^{-x^2/2} = 1 + \left(-\frac{x^2}{2}\right) + \frac{1}{2!}\left(-\frac{x^2}{2}\right)^2 + \frac{1}{3!}\left(-\frac{x^2}{2}\right)^3 + \frac{1}{4!}\left(-\frac{x^2}{2}\right)^4 + \cdots$$

$$e^{-x^2/2} = 1 - \frac{1}{2 \cdot 1!}x^2 + \frac{1}{2^2 \cdot 2!}x^4 - \frac{1}{2^3 \cdot 3!}x^6 + \frac{1}{2^4 \cdot 4!}x^8 - \cdots .$$

Integrating, we obtain

$$\frac{1}{\sqrt{2\pi}} \int_0^{.8} e^{-x^2/2} \, dx$$

$$= \frac{1}{\sqrt{2\pi}} \left(x - \frac{1}{3 \cdot 2 \cdot 1!}x^3 + \frac{1}{5 \cdot 2^2 \cdot 2!}x^5 - \frac{1}{7 \cdot 2^3 \cdot 3!}x^7 + \frac{1}{9 \cdot 2^4 \cdot 4!}x^9 - \cdots \right)\Big|_0^{.8}$$

$$= \frac{1}{\sqrt{2\pi}} \left[.8 - \frac{1}{6}(.8)^3 + \frac{1}{40}(.8)^5 - \frac{1}{336}(.8)^7 + \frac{1}{3456}(.8)^9 - \cdots \right].$$

The infinite series on the right converges to the value of the definite integral. Summing up the five terms displayed gives the approximation .28815, which is accurate to four places. This approximation can be made arbitrarily accurate by summing additional terms.  ∎

**Convergence of Power Series**  When we differentiate, integrate, or perform algebraic operations on Taylor series, we are using the fact that Taylor series are *functions*. In fact, any power series in $x$ is a function of $x$, whether or not its coefficients are obtained from the derivatives of some function. The domain of a "power series function" is the set of all $x$ for which the series converges. The function value at a specific $x$ in its domain is the number to which the series converges.

For instance, the geometric series $\sum_{k=0}^{\infty} x^k$ defines a function whose domain is the set of all $x$ for which $|x| < 1$. The familiar Taylor series expansion

$$\frac{1}{1-x} = \sum_{k=0}^{\infty} x^k$$

simply states that the functions $\frac{1}{1-x}$ and $\sum_{k=0}^{\infty} x^k$ have the same value for each $x$ such that $|x| < 1$.

Given any power series $\sum_{k=0}^{\infty} a_k x^k$, one of three possibilities must occur:

(i) There is a positive constant $R$ such that the series converges for $|x| < R$ and diverges for $|x| > R$.

(ii) The series converges for all $x$.

(iii) The series converges only for $x = 0$.

In case (i) we call $R$ the *radius of convergence* of the series. The series converges for all $x$ in the interval $-R < x < R$ and may or may not converge at one or both of the endpoints of this interval. In case (ii), we say that the radius of convergence is $\infty$, and in case (iii), we say that the radius of convergence is 0.

When a power series with a positive radius of convergence is differentiated term by term, the new series will have the same radius of convergence. An analogous result holds for antiderivatives. Other operations, such as replacing $x$ by a constant times a power of $x$, may affect the radius of convergence.

Suppose we begin with a function that has derivatives of all orders, and we write down its formal Taylor series at $x = 0$. Can we conclude that the Taylor series and the function have the same values for every $x$ within the radius of convergence of the series? For all of the functions that we consider, the answer is yes. However, it is possible for the two values to differ. In this case we say that the function does not admit a power series expansion. In order to show that a function admits a power series expansion, it is necessary to show that the partial sums of the Taylor series converge to the function. The $n$th partial sum of the series is the $n$th Taylor polynomial $p_n$. Recall from Section 11.1 that we considered the $n$th remainder of $f(x)$,

$$R_n(x) = f(x) - p_n(x).$$

For a fixed $x$, the Taylor series converges to $f(x)$ if and only if $R_n(x) \to 0$ as $n \to \infty$. Exercises 45 and 46 illustrate how convergence can be verified by using the remainder formula from Section 11.1.

**Taylor Series at $x = a$** In order to simplify the discussion in this section, we have restricted our attention to series that involve powers of $x$ rather than powers of $x - a$. However, Taylor series, just like Taylor polynomials, can be formed as sums of powers of $x - a$. The Taylor expansion of $f(x)$ at $x = a$ is

$$f(x) = f(a) + \frac{f'(a)}{1!}(x-a) + \frac{f''(a)}{2!}(x-a)^2 + \frac{f'''(a)}{3!}(x-a)^3 + \cdots$$

$$+ \frac{f^{(n)}(a)}{n!}(x-a)^n + \cdots.$$

## PRACTICE PROBLEMS 11.5

1. Find the Taylor series expansion of $\sin x$ at $x = 0$.

2. Find the Taylor series expansion of $\cos x$ at $x = 0$.

3. Find the Taylor series expansion of $x^3 \cos 7x$ at $x = 0$.

4. If $f(x) = x^3 \cos 7x$, find $f^{(5)}(0)$. [*Hint*: How are the coefficients in the Taylor series of $f(x)$ related to $f(x)$ and its derivatives at $x = 0$?]

## EXERCISES 11.5

*In Exercises 1–4, find the Taylor series at $x = 0$ of the given function by computing three or four derivatives and using the definition of the Taylor series.*

**1.** $\dfrac{1}{2x + 3}$

**2.** $\ln(1 - 3x)$

**3.** $\sqrt{1 + x}$

**4.** $(1 + x)^3$

*In Exercises 5–20, find the Taylor series at $x = 0$ of the given function. Use suitable operations (differentiation, substitution, etc.) on the Taylor series at $x = 0$ of $\dfrac{1}{1 - x}$, $e^x$, or $\cos x$. These series are derived in Examples 1 and 2 and Practice Problem 2.*

**5.** $\dfrac{1}{1 - 3x}$

**6.** $\dfrac{1}{1 + x}$

**7.** $\dfrac{1}{1 + x^2}$

**8.** $\dfrac{x}{1 + x^2}$

**9.** $\dfrac{1}{(1 + x)^2}$

**10.** $\dfrac{x}{(1 - x)^3}$

**11.** $5e^{x/3}$

**12.** $x^3 e^{x^2}$

**13.** $1 - e^{-x}$

**14.** $3(e^{-2x} - 2)$

**15.** $\ln(1 + x)$

**16.** $\ln(1 + x^2)$

**17.** $\cos 3x$

**18.** $\cos x^2$

**19.** $\sin 3x$

**20.** $x \sin x^2$

**21.** Find the Taylor series of $xe^{x^2}$ at $x = 0$.

**22.** Show that $\ln\left(\dfrac{1 + x}{1 - x}\right) = 2x + \frac{2}{3}x^3 + \frac{2}{5}x^5 + \frac{2}{7}x^7 + \cdots$, $|x| < 1$. (*Hint*: Use Exercise 15 and Example 3.) This series converges much more quickly than the series for $\ln(1 - x)$ in Example 3, particularly for $x$ close to zero. The series gives a formula for $\ln y$, where $y$ is any number and $x = \dfrac{y - 1}{y + 1}$.

**23.** The *hyperbolic cosine* of $x$, denoted by $\cosh x$, is defined by
$$\cosh x = \tfrac{1}{2}(e^x + e^{-x}).$$
This function occurs often in physics and probability theory. The graph of $y = \cosh x$ is called a *catenary*.

(a) Use differentiation and the definition of a Taylor series to compute the first four nonzero terms in the Taylor series of $\cosh x$ at $x = 0$.

(b) Use the known Taylor series for $e^x$ to obtain the Taylor series for $\cosh x$ at $x = 0$.

**24.** The *hyperbolic sine* of $x$ is defined by
$$\sinh x = \tfrac{1}{2}(e^x - e^{-x}).$$
Repeat parts (a) and (b) of Exercise 23 for $\sinh x$.

**25.** Given the Taylor series expansion
$$\frac{1}{\sqrt{1 + x}} = 1 - \frac{1}{2}x + \frac{1 \cdot 3}{2 \cdot 4}x^2 - \frac{1 \cdot 3 \cdot 5}{2 \cdot 4 \cdot 6}x^3 +$$
$$\frac{1 \cdot 3 \cdot 5 \cdot 7}{2 \cdot 4 \cdot 6 \cdot 8}x^4 - \cdots,$$
find the first four terms in the Taylor series of $\dfrac{1}{\sqrt{1 - x}}$ at $x = 0$.

**26.** Find the first four terms in the Taylor series of $\dfrac{1}{\sqrt{1 - x^2}}$ at $x = 0$. (See Exercise 25.)

**27.** Use Exercise 25 and the fact that
$$\int \frac{1}{\sqrt{1 + x^2}}\, dx = \ln(x + \sqrt{1 + x^2}) + C$$
to find the Taylor series of $\ln(x + \sqrt{1 + x^2})$ at $x = 0$.

**28.** Use the Taylor series expansion for $\dfrac{x}{(1 - x)^2}$ to find the function whose Taylor series is $1 + 4x + 9x^2 + 16x^3 + 25x^4 + \cdots$.

**29.** Use the Taylor series for $e^x$ to show that $\dfrac{d}{dx}e^x = e^x$.

**30.** Use the Taylor series for $\cos x$ (see Practice Problem 2) to show that $\cos(-x) = \cos x$.

**31.** The Taylor series at $x = 0$ for
$$f(x) = \ln\left(\frac{1 + x}{1 - x}\right)$$
is given in Exercise 22. Find $f^{(5)}(0)$.

**32.** The Taylor series at $x = 0$ for $f(x) = \sec x$ is $1 + \frac{1}{2}x^2 + \frac{5}{24}x^4 + \frac{61}{720}x^6 + \cdots$. Find $f^{(4)}(0)$.

**33.** The Taylor series at $x = 0$ for $f(x) = \tan x$ is $x + \frac{1}{3}x^3 + \frac{2}{15}x^5 + \frac{17}{315}x^7 + \cdots$. Find $f^{(4)}(0)$.

**34.** The Taylor series at $x = 0$ for $\dfrac{1 + x^2}{1 - x}$ is $1 + x + 2x^2 + 2x^3 + 2x^4 + \cdots$. Find $f^{(4)}(0)$, where $f(x) = \dfrac{1 + x^4}{1 - x^2}$.

In Exercises 35–37, find the Taylor series expansion at $x = 0$ of the given antiderivative.

**35.** $\displaystyle\int e^{-x^2}\,dx$    **36.** $\displaystyle\int x e^{x^3}\,dx$    **37.** $\displaystyle\int \frac{1}{1+x^3}\,dx$

In Exercises 38–40, find an infinite series that converges to the value of the given definite integral.

**38.** $\displaystyle\int_0^1 \sin x^2\,dx$    **39.** $\displaystyle\int_0^1 e^{-x^2}\,dx$    **40.** $\displaystyle\int_0^1 x e^{x^3}\,dx$

**41.** **(a)** Use the Taylor series for $e^x$ at $x = 0$ to show that $e^x > x^2/2$ for $x > 0$.

**(b)** Deduce that $e^{-x} < 2/x^2$ for $x > 0$.

**(c)** Show that $x e^{-x}$ approaches $0$ as $x \to \infty$.

**42.** Let $k$ be a positive constant.

**(a)** Show that $e^{kx} > \dfrac{k^2 x^2}{2}$, for $x > 0$.

**(b)** Deduce that $e^{-kx} < \dfrac{2}{k^2 x^2}$, for $x > 0$.

**(c)** Show that $x e^{-kx}$ approaches $0$ as $x \to \infty$.

**43.** Show that $e^x > x^3/6$ for $x > 0$, and from this deduce that $x^2 e^{-x}$ approaches $0$ as $x \to \infty$.

**44.** If $k$ is a positive constant, show that $x^2 e^{-kx}$ approaches $0$ as $x \to \infty$.

Exercises 45 and 46 rely on the fact that

$$\lim_{n \to \infty} \frac{|x|^{n+1}}{(n+1)!} = 0.$$

The proof of this fact is omitted.

**45.** Let $R_n(x)$ be the $n$th remainder of $f(x) = \cos x$ at $x = 0$. See Section 11.1. Show that for any fixed value of $x$, $|R_n(x)| \le |x|^{n+1}/(n+1)!$, and hence conclude that $|R_n(x)| \to 0$ as $n \to \infty$. This shows that the Taylor series for $\cos x$ converges to $\cos x$ for every value of $x$.

**46.** Let $R_n(x)$ be the $n$th remainder of $f(x) = e^x$ at $x = 0$. See Section 11.1. Show that for any fixed value of $x$, $|R_n(x)| \le e^{|x|} \cdot |x|^{n+1}/(n+1)!$, and hence conclude that $|R_n(x)| \to 0$ as $n \to \infty$. This shows that the Taylor series for $e^x$ converges to $e^x$ for every value of $x$.

## SOLUTIONS TO PRACTICE PROBLEMS 11.5

**1.** Use the definition of the Taylor series as an extended Taylor polynomial.

$$f(x) = \sin x, \qquad f'(x) = \cos x,$$
$$f(0) = 0, \qquad f'(0) = 1,$$

$$f''(x) = -\sin x, \qquad f'''(x) = -\cos x,$$
$$f''(0) = 0, \qquad f'''(0) = -1,$$

$$f^{(4)}(x) = \sin x, \qquad f^{(5)}(x) = \cos x,$$
$$f^{(4)}(0) = 0, \qquad f^{(5)}(0) = 1.$$

Therefore,

$$\sin x = 0 + 1\cdot x + \frac{0}{2!}x^2 + \frac{-1}{3!}x^3 + \frac{0}{4!}x^4 + \frac{1}{5!}x^5 + \cdots$$

$$= x - \frac{1}{3!}x^3 + \frac{1}{5!}x^5 - \cdots .$$

**2.** Differentiate the Taylor series in Problem 1.

$$\frac{d}{dx}\sin x = \frac{d}{dx}\left(x - \frac{1}{3!}x^3 + \frac{1}{5!}x^5 - \cdots\right)$$

$$\cos x = 1 - \frac{1}{2!}x^2 + \frac{1}{4!}x^4 - \cdots$$

$$\left(\textit{Note: We use the fact that } \frac{3}{3!} = \frac{3}{1\cdot 2\cdot 3} = \frac{1}{1\cdot 2} = \frac{1}{2!} \text{ and } \frac{5}{5!} = \frac{1}{4!}.\right)$$

**3.** Replace $x$ by $7x$ in the Taylor series for $\cos x$ and then multiply by $x^3$.

$$\cos x = 1 - \frac{1}{2!}x^2 + \frac{1}{4!}x^4 - \cdots$$

$$\cos 7x = 1 - \frac{1}{2!}(7x)^2 + \frac{1}{4!}(7x)^4 - \cdots$$

$$= 1 - \frac{7^2}{2!}x^2 + \frac{7^4}{4!}x^4 - \cdots$$

$$x^3\cos 7x = x^3\left(1 - \frac{7^2}{2!}x^2 + \frac{7^4}{4!}x^4 - \cdots\right)$$

$$= x^3 - \frac{7^2}{2!}x^5 + \frac{7^4}{4!}x^7 - \cdots$$

**4.** The coefficient of $x^5$ in the Taylor series of $f(x)$ is $\dfrac{f^{(5)}(0)}{5!}$. By Problem 3, this coefficient is $-\dfrac{7^2}{2!}$. Therefore,

$$\frac{f^{(5)}(0)}{5!} = -\frac{7^2}{2!}$$

$$f^{(5)}(0) = -\frac{7^2}{2!}\cdot 5! = -\frac{49}{2}\cdot 120$$

$$= -(49)(60) = -2940.$$

# REVIEW OF FUNDAMENTAL CONCEPTS

1. Define the $n$th Taylor polynomial of $f(x)$ at $x = a$.

2. In what way is the $n$th Taylor polynomial of $f(x)$ at $x = a$ like $f(x)$ at $x = a$?

3. State the remainder formula for the $n$th Taylor polynomial of $f(x)$ at $x = a$.

4. Explain how the Newton–Raphson algorithm is used to approximate a zero of a function.

5. What is the $n$th partial sum of an infinite series?

6. What is a convergent infinite series? Divergent?

7. What is meant by the sum of a convergent infinite series?

8. What is a geometric series and when does it converge?

9. What is the sum of a convergent geometric series?

10. Define the Taylor series of $f(x)$ at $x = 0$.

11. Discuss the three possibilities for the radius of convergence of a Taylor series.

# SUPPLEMENTARY EXERCISES

1. Find the second Taylor polynomial of $x(x + 1)^{3/2}$ at $x = 0$.

2. Find the fourth Taylor polynomial of $(2x + 1)^{3/2}$ at $x = 0$.

3. Find the fifth Taylor polynomial of $x^3 - 7x^2 + 8$ at $x = 0$.

4. Find the $n$th Taylor polynomial of $\dfrac{2}{2 - x}$ at $x = 0$.

5. Find the third Taylor polynomial of $x^2$ at $x = 3$.

6. Find the third Taylor polynomial of $e^x$ at $x = 2$.

7. Use a second Taylor polynomial at $t = 0$ to estimate the area under the graph of $y = -\ln(\cos 2t)$ between $t = 0$ and $t = \frac{1}{2}$.

8. Use a second Taylor polynomial at $x = 0$ to estimate the value of $\tan(.1)$.

9. (a) Find the second Taylor polynomial of $\sqrt{x}$ at $x = 9$.

   (b) Use part (a) to estimate $\sqrt{8.7}$ to six decimal places.

   (c) Use the Newton–Raphson algorithm with $n = 2$ and $x_0 = 3$ to approximate the solution of the equation $x^2 - 8.7 = 0$. Express your answer to six decimal places.

10. (a) Use the third Taylor polynomial of $\ln(1 - x)$ at $x = 0$ to approximate $\ln 1.3$ to four decimal places.

   (b) Find an approximate solution of the equation $e^x = 1.3$ using the Newton–Raphson algorithm with $n = 2$ and $x_0 = 0$. Express your answer to four decimal places.

11. Use the Newton–Raphson algorithm with $n = 2$ to approximate the zero of $x^2 - 3x - 2$ near $x_0 = 4$.

12. Use the Newton–Raphson algorithm with $n = 2$ to approximate the solution of the equation $e^{2x} = 1 + e^{-x}$.

*In Exercises 13–20, find the sum of the given infinite series if it is convergent.*

13. $1 - \dfrac{3}{4} + \dfrac{9}{16} - \dfrac{27}{64} + \dfrac{81}{256} - \cdots$

14. $\dfrac{5^2}{6} + \dfrac{5^3}{6^2} + \dfrac{5^4}{6^3} + \dfrac{5^5}{6^4} + \dfrac{5^6}{6^5} + \cdots$

15. $\dfrac{1}{8} + \dfrac{1}{8^2} + \dfrac{1}{8^3} + \dfrac{1}{8^4} + \dfrac{1}{8^5} + \cdots$

16. $\dfrac{2^2}{7} - \dfrac{2^5}{7^2} + \dfrac{2^8}{7^3} - \dfrac{2^{11}}{7^4} + \dfrac{2^{14}}{7^5} - \cdots$

17. $\dfrac{1}{m + 1} + \dfrac{m}{(m + 1)^2} + \dfrac{m^2}{(m + 1)^3} + \dfrac{m^3}{(m + 1)^4} + \cdots$, where $m$ is a positive number.

18. $\dfrac{1}{m} - \dfrac{1}{m^2} + \dfrac{1}{m^3} - \dfrac{1}{m^4} + \dfrac{1}{m^5} - \cdots$, where $m$ is a positive number.

19. $1 + 2 + \dfrac{2^2}{2!} + \dfrac{2^3}{3!} + \dfrac{2^4}{4!} + \cdots$

20. $1 + \dfrac{1}{3} + \dfrac{1}{2!}\left(\dfrac{1}{3}\right)^2 + \dfrac{1}{3!}\left(\dfrac{1}{3}\right)^3 + \dfrac{1}{4!}\left(\dfrac{1}{3}\right)^4 + \cdots$

21. Use properties of convergent series to find $\displaystyle\sum_{k=0}^{\infty} \dfrac{1 + 2^k}{3^k}$.

22. Find $\displaystyle\sum_{k=0}^{\infty} \dfrac{3^k + 5^k}{7^k}$.

*In Exercises 23–26, determine if the given series is convergent.*

23. $\displaystyle\sum_{k=1}^{\infty} \dfrac{1}{k^3}$

24. $\displaystyle\sum_{k=1}^{\infty} \dfrac{1}{3^k}$

25. $\displaystyle\sum_{k=1}^{\infty} \dfrac{\ln k}{k}$

26. $\displaystyle\sum_{k=0}^{\infty} \dfrac{k^3}{(k^4 + 1)^2}$

27. For what values of $p$ is $\displaystyle\sum_{k=1}^{\infty} \dfrac{1}{k^p}$ convergent?

28. For what values of $p$ is $\displaystyle\sum_{k=1}^{\infty} \dfrac{1}{p^k}$ convergent?

*In Exercises 29–32, find the Taylor series at $x = 0$ of the given function. Use suitable operations on the Taylor series at $x = 0$ of $\dfrac{1}{1 - x}$ and $e^x$.*

**29.** $\dfrac{1}{1+x^3}$

**30.** $\ln(1+x^3)$

**31.** $\dfrac{1}{(1-3x)^2}$

**32.** $\dfrac{e^x - 1}{x}$

**33. (a)** Find the Taylor series of $\cos 2x$ at $x = 0$, either by direct calculation or by using the known series for $\cos x$.

**(b)** Use the trigonometric identity

$$\sin^2 x = \tfrac{1}{2}(1 - \cos 2x)$$

to find the Taylor series of $\sin^2 x$ at $x = 0$.

**34. (a)** Find the Taylor series of $\cos 3x$ at $x = 0$.

**(b)** Use the trigonometric identity

$$\cos^3 x = \tfrac{1}{4}(\cos 3x + 3\cos x)$$

to find the fourth Taylor polynomial of $\cos^3 x$ at $x = 0$.

**35.** Use the decomposition

$$\frac{1+x}{1-x} = \frac{1}{1-x} + \frac{x}{1-x}$$

to find the Taylor series of $\dfrac{1+x}{1-x}$ at $x = 0$.

**36.** Find an infinite series that converges to

$$\int_0^{1/2} \frac{e^x - 1}{x}\, dx.$$

[*Hint:* Use Exercise 32.]

**37.** It can be shown that the sixth Taylor polynomial of $f(x) = \sin x^2$ at $x = 0$ is $x^2 - \tfrac{1}{6}x^6$. Use this fact in parts (a), (b), and (c).

**(a)** What is the fifth Taylor polynomial of $f(x)$ at $x = 0$?

**(b)** What is $f'''(0)$?

**(c)** Estimate the area under the graph of $y = \sin x^2$ between $x = 0$ and $x = 1$. Use four decimal places, and compare your answer with the values given in Example 4 of Section 11.1.

**38.** Let $f(x) = \ln|\sec x + \tan x|$. It can be shown that $f'(0) = 1$, $f''(0) = 0$, $f'''(0) = 1$, and $f^{(4)}(0) = 0$. What is the fourth Taylor polynomial of $f(x)$ at $x = 0$?

**39.** Let $f(x) = 1 + x^2 + x^4 + x^6 + \cdots$.

**(a)** Find the Taylor series expansion of $f'(x)$ at $x = 0$.

**(b)** Find the simple formula for $f'(x)$ not involving a series. [*Hint:* First find a simple formula for $f(x)$.]

**40.** Let $f(x) = x - 2x^3 + 4x^5 - 8x^7 + 16x^9 - \cdots$.

**(a)** Find the Taylor series expansion of $\int f(x)\, dx$ at $x = 0$.

**(b)** Find a simple formula for $\int f(x)\, dx$ not involving a series. [*Hint:* First find a simple formula for $f(x)$.]

**41.** (*Fractional Reserve Banking*) Suppose that the Federal Reserve (the Fed) buys \$100 million of government debt obligations from private owners. This creates \$100 million of new money and sets off a chain reaction because of the "fractional reserve" banking system. When the \$100 million is deposited into private bank accounts, the banks keep only 15% in reserve and may loan out the remaining 85%, creating more new money—$(.85)(100)$ million dollars. The companies who borrow this money turn around and spend it, and the recipients deposit the money in their bank accounts. Assuming that all of the $(.85)(100)$ million is redeposited, the banks may again loan out 85% of this amount, creating $(.85)^2(100)$ million additional dollars. This process may be repeated indefinitely. Compute the total amount of new money that can be created theoretically by this process, beyond the original \$100 million. (In practice, only about an additional \$300 million is created, usually within a few weeks of the action of the Fed.)

**42.** Suppose that the Federal Reserve creates \$100 million of new money, as in Exercise 41, and the banks lend 85% of all new money they receive. However, suppose that out of each loan, only 80% is redeposited into the banking system. Thus, whereas the first set of loans total $(.85)(100)$ million dollars, the second set is only 85% of $(.80)(.85)(100)$, or $(.80)(.85)^2(100)$ million, and the next set is 85% of $(.80)^2(.85)^2(100)$, or $(.80)^2(.85)^3(100)$ million dollars, and so on. compute the total theoretical amount that may be loaned in this situation.

*Suppose that when you die, the proceeds of a life insurance policy will be deposited into a trust fund that will earn 8% interest, compounded continuously. According to the terms of your will, the trust fund must pay to your descendants and their heirs $c_1$ dollars (total) at the end of the first year, $c_2$ dollars at the end of the second year, $c_3$ dollars at the end of the third year, and so on, forever. The amount that must be in the trust fund initially to make the $k$th payment is $c_k e^{-.08k}$, the present value of the amount to be paid in $k$ years. So the life insurance policy should pay a total of $\sum_{k=1}^{\infty} c_k e^{-.08k}$ dollars into the trust fund in order to provide for all the payments.*

**43.** How large must the insurance policy be if $c_k = 10{,}000$ for all $k$? (Find the sum of the series.)

**44.** How large must the insurance policy be if $c_k = 10{,}000(.9)^k$ for all $k$?

**45.** Suppose that $c_k = 10{,}000(1.08)^k$ for all $k$. Find the sum of the series above if the series converges.

## CHAPTER 12

# Probability and Calculus

**12.1** Discrete Random Variables

**12.2** Continuous Random Variables

**12.3** Expected Value and Variance

**12.4** Exponential and Normal Random Variables

**12.5** Poisson and Geometric Random Variables

I N this chapter we shall survey a few applications of calculus to the theory of probability. Since we do not intend this chapter to be a self-contained course in probability, we shall select only a few salient ideas to present a taste of probability theory and provide a starting point for further study.

## 12.1 Discrete Random Variables

We will motivate the concepts of mean, variance, standard deviation, and random variable by analyzing examination grades.

Suppose that the grades on an exam taken by 10 people are 50, 60, 60, 70, 70, 90, 100, 100, 100, 100. This information is displayed in a frequency table in Fig. 1.

One of the first things we do when looking over the results of an exam is to compute the *mean* or *average* of the grades. We do this by totaling the grades and dividing by the number of people. This is the same as multiplying each distinct grade by the frequency with which it occurs, adding up those products, and dividing by the sum of the frequencies.

$$[\text{mean}] = \frac{50 \cdot 1 + 60 \cdot 2 + 70 \cdot 2 + 90 \cdot 1 + 100 \cdot 4}{10} = \frac{800}{10} = 80$$

| Grade | 50 | 60 | 70 | 90 | 100 |
|---|---|---|---|---|---|
| Frequency | 1 | 2 | 2 | 1 | 4 |

**Figure 1**

| [Grade] − [Mean] | −30 | −20 | −10 | 10 | 20 |
|---|---|---|---|---|---|
| Frequency | 1 | 2 | 2 | 1 | 4 |

**Figure 2**

In order to get an idea of how spread out the grades are, we can compute the difference between each grade and the average grade. We have tabulated these differences in Fig. 2. For example, if a person received a 50, then [grade] − [mean] is $50 - 80 = -30$. As a measure of the spread of the grades, statisticians compute the average of the squares of these differences and call it the *variance* of the grade distribution. We have

$$[\text{variance}] = \frac{(-30)^2 \cdot 1 + (-20)^2 \cdot 2 + (-10)^2 \cdot 2 + (10)^2 \cdot 1 + (20)^2 \cdot 4}{10}$$

$$= \frac{900 + 800 + 200 + 100 + 1600}{10} = \frac{3600}{10} = 360.$$

The square root of the variance is called the *standard deviation* of the grade distribution. In this case, we have

$$[\text{standard deviation}] = \sqrt{360} \approx 18.97.$$

There is another way of looking at the grade distribution and its mean and variance. This new point of view is useful because it can be generalized to other situations. We begin by converting the frequency table to a relative frequency table. (See Fig. 3.) Below each grade we list the fraction of the class receiving that grade. The grade of 50 occurred $\frac{1}{10}$ of the time, the grade of 60 occurred $\frac{2}{10}$ of the time, and so on. Note that the relative frequencies add up to 1, because they represent the various fractions of the class grouped by test scores.

| Grade | 50 | 60 | 70 | 90 | 100 |
|---|---|---|---|---|---|
| Relative frequency | $\frac{1}{10}$ | $\frac{2}{10}$ | $\frac{2}{10}$ | $\frac{1}{10}$ | $\frac{4}{10}$ |

**Figure 3**

It is sometimes helpful to display the data in the relative frequency table by constructing a *relative frequency histogram* as in Fig. 4. Over each grade we place a rectangle whose height equals the relative frequency of that grade.

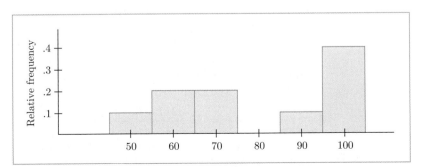

**Figure 4.**  A relative frequency histogram.

An alternative way to compute the mean grade is

$$[\text{mean}] = \frac{50\cdot1 + 60\cdot2 + 70\cdot2 + 90\cdot1 + 100\cdot4}{10}$$

$$= 50\cdot\frac{1}{10} + 60\cdot\frac{2}{10} + 70\cdot\frac{2}{10} + 90\cdot\frac{1}{10} + 100\cdot\frac{4}{10}$$

$$= 5 + 12 + 14 + 9 + 40 = 80.$$

Looking at the second line of this computation, we see that the mean is a sum of the various grades times their relative frequencies. We say that the mean is the *weighted sum* of the grades. (Grades are weighted by their relative frequencies.)

In a similar manner we see that the variance is also a weighted sum.

$$[\text{variance}] = \big[(50-80)^2\cdot1 + (60-80)^2\cdot2 + (70-80)^2\cdot2$$

$$+ (90-80)^2\cdot1 + (100-80)^2\cdot4\big]\frac{1}{10}$$

$$= (50-80)^2\cdot\frac{1}{10} + (60-80)^2\cdot\frac{2}{10} + (70-80)^2\cdot\frac{2}{10}$$

$$+ (90-80)^2\cdot\frac{1}{10} + (100-80)^2\cdot\frac{4}{10}$$

$$= 90 + 80 + 20 + 10 + 160 = 360$$

A relative frequency table such as is shown in Fig. 3 is also called a *probability table*. The reason for this terminology is as follows. Suppose that we perform an *experiment* that consists of picking an exam paper at random from among the ten papers. If the experiment is repeated many times, we expect the grade of 50 to occur about one-tenth of the time, the grade of 60 about two-tenths of the time, and so on. We say that the *probability* of the grade of 50 being chosen is $\frac{1}{10}$, the probability of the grade of 60 being chosen is $\frac{2}{10}$, and so on. In other words, the probability associated with a given grade measures the likelihood that an exam having that grade is chosen.

In this section we consider various experiments described by probability tables similar to the one in Fig. 3. The results of these experiments will be numbers (such as the preceding exam scores) called the *outcomes* of the experiment. We will also be given the probability of each outcome, indicating the relative frequency with which the given outcome is expected to occur if the experiment is repeated very often. If the outcomes of an experiment are $a_1, a_2, \ldots, a_n$, with respective probabilities $p_1, p_2, \ldots, p_n$, then we describe the experiment by a probability table as in Fig. 5. Since the probabilities indicate relative frequencies, we see that

$$0 \le p_i \le 1$$

and

$$p_1 + p_2 + \cdots + p_n = 1.$$

| Outcome | $a_1$ | $a_2$ | $a_3$ | $\cdots$ | $a_n$ |
|---|---|---|---|---|---|
| Probability | $p_1$ | $p_2$ | $p_3$ | $\cdots$ | $p_n$ |

**Figure 5**

The last equation indicates that the outcomes $a_1, \ldots, a_n$ comprise all possible results of the experiment. We will usually list the outcomes of our experiments in ascending order, so that $a_1 < a_2 < \cdots < a_n$.

We may display the data of a probability table in a histogram that has a rectangle of height $p_i$ over the outcome $a_i$. (See Fig. 6.)

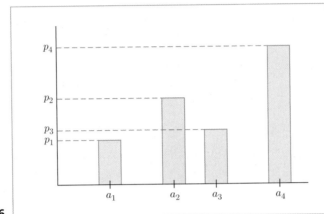

**Figure 6**

Let us define the *expected value* (or *mean*) of the probability table of Fig. 5 to be the weighted sum of the outcomes $a_1, \ldots, a_n$, each outcome weighted by the probability of its occurrence. That is,

$$[\text{expected value}] = a_1 p_1 + a_2 p_2 + \cdots + a_n p_n.$$

Similarly, let us define the *variance* of the probability table to be the weighted sum of the squares of the differences between each outcome and the expected value. That is, if $m$ denotes the expected value, then

$$[\text{variance}] = (a_1 - m)^2 p_1 + (a_2 - m)^2 p_2 + \cdots + (a_n - m)^2 p_n.$$

To keep from writing the "outcome" so many times, we shall abbreviate by $X$ the outcome of our experiment. That is, $X$ is a variable which takes on the values $a_1, a_2, \ldots, a_n$ with respective probabilities $p_1, p_2, \ldots, p_n$. We will assume that our experiment is performed many times, being repeated in an unbiased (or random) way. Then $X$ is a variable whose value depends on chance, and for this reason we say that $X$ is a *random variable*. Instead of speaking of the expected value (mean) and the variance of a probability table, let us speak of the *expected value* and the *variance of the random variable* $X$ that is associated with the probability table. We shall denote the expected value of $X$ by $\mathrm{E}(X)$ and the variance of $X$ by $\mathrm{Var}(X)$. The *standard deviation* of $X$ is defined to be $\sqrt{\mathrm{Var}(X)}$.

---

**EXAMPLE 1**     One possible bet in roulette is to wager $1 on "red." The two possible outcomes are: "lose $1" and "win $1." These outcomes and their probabilities are given in Fig. 7. (*Note:* A roulette wheel in Las Vegas has 18 red numbers, 18 black numbers, and two green numbers.) Compute the expected value and the variance of the amount won.

**Solution**  Let $X$ be the random variable "amount won." Then

$$E(X) = -1 \cdot \frac{20}{38} + 1 \cdot \frac{18}{38} = -\frac{2}{38} \approx -.0526,$$

$$Var(X) = \left[-1 - \left(-\frac{2}{38}\right)\right]^2 \cdot \frac{20}{38} + \left[1 - \left(-\frac{2}{38}\right)\right]^2 \cdot \frac{18}{38}$$

$$= \left(-\frac{36}{38}\right)^2 \cdot \frac{20}{38} + \left(\frac{40}{38}\right)^2 \cdot \frac{18}{38} \approx .997.$$

| Amount won | −1 | 1 |
|---|---|---|
| Probability | $\frac{20}{38}$ | $\frac{18}{38}$ |

**Figure 7.**  Las Vegas roulette.

The expected value of the amount won is approximately $-5\frac{1}{4}$ cents. In other words, sometimes we will win \$1 and sometimes we will lose \$1, but in the long run we can expect to lose an average of about $5\frac{1}{4}$ cents for each time we bet.  ■

**EXAMPLE 2**  An experiment consists of selecting a number at random from the set of integers $\{1, 2, 3\}$. The probabilities are given by the table in Fig. 8. Let $X$ designate the outcome. Find the expected value and the variance of $X$.

| Number | 1 | 2 | 3 |
|---|---|---|---|
| Probability | $\frac{1}{3}$ | $\frac{1}{3}$ | $\frac{1}{3}$ |

**Figure 8**

**Solution**

$$E(X) = 1 \cdot \frac{1}{3} + 2 \cdot \frac{1}{3} + 3 \cdot \frac{1}{3} = 2,$$

$$Var(X) = (1-2)^2 \cdot \frac{1}{3} + (2-2)^2 \cdot \frac{1}{3} + (3-2)^2 \cdot \frac{1}{3}$$

$$= (-1)^2 \cdot \frac{1}{3} + 0 + (1)^2 \cdot \frac{1}{3} = \frac{2}{3}$$

**EXAMPLE 3**  A cement company plans to bid on a contract for constructing the foundations of new homes in a housing development. The company is considering two bids: a high bid that will produce \$75,000 profit (if the bid is accepted), and a low bid that will produce \$40,000 profit. From past experience the company estimates that the high bid has a 30% chance of acceptance and the low bid a 50% chance. Which bid should the company make?

**Solution**  The standard method of decision is to choose the bid that has the higher expected value. Let $X$ be the amount the company makes if it submits the high bid, and let $Y$ be the amount it makes if it submits the low bid. Then the company must analyze the probability table shown in Table 1. The expected values are

$$E(X) = (75{,}000)(.30) + 0(.70) = 22{,}500,$$

$$E(Y) = (40{,}000)(.50) + 0(.50) = 20{,}000.$$

If the cement company has many opportunities to bid on similar contracts, then a "high" bid each time will be accepted sufficiently often to produce an average profit of \$22,500 per bid. A consistently "low" bid will produce an average profit of \$20,000 per bid. Thus the company should submit the high bid.  ■

| TABLE 1 Bids on a Cement Contract | | | | | | |
|---|---|---|---|---|---|---|
| | *High Bid* | | | | *Low Bid* | |
| | Accepted | Rejected | | | Accepted | Rejected |
| Value of $X$ | 75,000 | 0 | | Value of $Y$ | 40,000 | 0 |
| Probability | .30 | .70 | | Probability | .50 | .50 |

When a probability table contains a large number of possible outcomes of an experiment, the associated histogram for the random variable $X$ becomes a valuable aid for "visualizing" the data in the table. Look at Fig. 9, for example. Since the rectangles that make up the histogram all have the same width, their areas are in the same ratios as their heights. By an appropriate change of scale on the $y$-axis, we may assume that the *area* (instead of the height) of each rectangle gives the associated probability of $X$. Such a histogram is sometimes referred to as a *probability density histogram*.

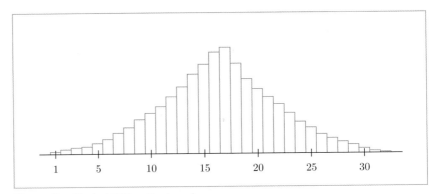

**Figure 9.** Probabilities displayed as areas.

A histogram that displays probabilities as areas is useful when one wishes to visualize the probability that $X$ has a value between two specified numbers. For example, in Fig. 9 suppose that the probabilities associated with $X = 5, X = 6, \ldots, X = 10$ are $p_5, p_6, \ldots, p_{10}$, respectively. Then the probability that $X$ lies between 5 and 10 inclusive is $p_5 + p_6 + \cdots + p_{10}$. In terms of areas, this probability is just the total area of those rectangles over the values $5, 6, \ldots, 10$. (See Fig. 10.) We will consider analogous situations in the next section.

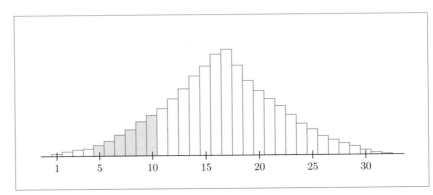

**Figure 10.** Probability that $5 \leq x \leq 10$.

## PRACTICE PROBLEMS 12.1

1. Compute the expected value and the variance of the random variable $X$ with Table 2 as its probability table.

| TABLE 2 | | | | |
|---|---|---|---|---|
| Value of $X$ | $-1$ | $0$ | $1$ | $2$ |
| Probability | $\frac{1}{8}$ | $\frac{1}{8}$ | $\frac{3}{8}$ | $\frac{3}{8}$ |

2. The production department at a radio factory sends CB radios to the inspection department in lots of 100. There an inspector examines three radios at random from each lot. If at least one of the three radios is defective and needs adjustment, the entire lot is sent back to the production department. Records of the inspection department show that the number $X$ of defective radios in a sample of three radios has Table 3 as its probability table.

| TABLE 3 | Quality Control Data | | | |
|---|---|---|---|---|
| Defectives | 0 | 1 | 2 | 3 |
| Probability | .7265 | .2477 | .0251 | .0007 |

(a) What percentage of the lots does the inspection department reject?

(b) Find the mean number of defective radios in the samples of three radios.

(c) Based on the evidence in part (b), estimate the average number of defective radios in each lot of 100 radios.

# EXERCISES 12.1

1. Table 4 is the probability table for a random variable $X$. Find $E(X)$, $Var(X)$, and the standard deviation of $X$.

| TABLE 4 | | |
|---|---|---|
| Outcome | 0 | 1 |
| Probability | $\frac{1}{5}$ | $\frac{4}{5}$ |

2. Find $E(X)$, $Var(X)$, and the standard deviation of $X$, where $X$ is the random variable whose probability table is given in Table 5.

| TABLE 5 | | | |
|---|---|---|---|
| Outcome | 1 | 2 | 3 |
| Probability | $\frac{4}{9}$ | $\frac{4}{9}$ | $\frac{1}{9}$ |

3. Compute the variances of the three random variables whose probability tables are given in Table 6. Relate the sizes of the variances to the "spread" of the values of the random variable.

| TABLE 6 | | |
|---|---|---|
| | **Outcome** | **Probability** |
| (a) | 4 | .5 |
| | 6 | .5 |
| (b) | 3 | .5 |
| | 7 | .5 |
| (c) | 1 | .5 |
| | 9 | .5 |

4. Compute the variances of the two random variables whose probability tables are given in Table 7. Relate the sizes of the variances to the "spread" of the values of the random variables.

| TABLE 7 | | |
|---|---|---|
| | **Outcome** | **Probability** |
| (a) | 2 | .1 |
| | 4 | .4 |
| | 6 | .4 |
| | 8 | .1 |
| (b) | 2 | .3 |
| | 4 | .2 |
| | 6 | .2 |
| | 8 | .3 |

5. The number of accidents per week at a busy intersection was recorded for a year. There were 11 weeks with no accidents, 26 weeks with one accident, 13 weeks with two accidents, and 2 weeks with three accidents. A week is to be selected at random and the number of accidents noted. Let $X$ be the outcome. Then $X$ is a random variable taking on the values 0, 1, 2, and 3.

(a) Write out a probability table for $X$.

(b) Compute $E(X)$.    (c) Interpret $E(X)$.

6. The number of phone calls coming into a telephone switchboard during each minute was recorded during an entire hour. During 30 of the 1-minute intervals there were no calls, during 20 intervals there was one call, and during 10 intervals there were two calls. A 1-minute interval is to be selected at random and the number of calls noted. Let $X$ be the outcome. Then $X$ is a random variable taking on the values 0, 1, and 2.

(a) Write out a probability table for $X$.

**(b)** Compute $E(X)$. **(c)** Interpret $E(X)$.

**7.** Consider a circle with radius 1.

**(a)** What percentage of the points lie within $\frac{1}{2}$ unit of the center?

**(b)** Let $c$ be a constant with $0 < c < 1$. What percentage of the points lie within $c$ units of the center?

**8.** Consider a circle with circumference 1. An arrow (i.e., a spinner) is attached at the center so that, when flicked, it spins freely. Upon stopping, it points to a particular point on the circumference of the circle. Determine the likelihood that the point is

**(a)** On the top half of the circumference.

**(b)** On the top quarter of the circumference.

**(c)** On the top one-hundredth of the circumference.

**(d)** Exactly at the top of the circumference.

**9.** A citrus grower anticipates a profit of $100,000 this year if the nightly temperatures remain mild. Unfortunately, the weather forecast indicates a 25% chance that the temperatures will drop below freezing during the next week. Such freezing weather will destroy 40% of the crop and reduce the profit to $60,000. However, the grower can protect the citrus fruit against the possible freezing (using smudge pots, electric fans, and so on) at a cost of $5000. Should the grower spend the $5000 and thereby reduce the profit to $95,000? [*Hint*: Compute $E(X)$, where $X$ is the profit the grower will get if he does nothing to protect the fruit.]

**10.** Suppose that the weather forecast in Exercise 9 indicates a 10% chance that cold weather will reduce the citrus grower's profit from $100,000 to $85,000, and a 10% chance that cold weather will reduce the profit to $75,000. Should the grower spend $5000 to protect the citrus fruit against the possible bad weather?

## SOLUTIONS TO PRACTICE PROBLEMS 12.1

**1.** $E(X) = (-1) \cdot \frac{1}{8} + 0 \cdot \frac{1}{8} + 1 \cdot \frac{3}{8} + 2 \cdot \frac{3}{8} = 1,$

$\text{Var}(X) = (-1-1)^2 \cdot \frac{1}{8} + (0-1)^2 \cdot \frac{1}{8}$

$\qquad + (1-1)^2 \cdot \frac{3}{8} + (2-1)^2 \cdot \frac{3}{8}$

$\qquad = 4 \cdot \frac{1}{8} + 1 \cdot \frac{1}{8} + 0 + 1 \cdot \frac{3}{8} = 1$

**2. (a)** There are three cases where a lot will be rejected: $X = 1$, 2, or 3. Adding the corresponding probabilities, we find that the probability of rejecting a lot is $.2477 + .0251 + .0007 = .2735$, or 27.35%. (An alternative method of solution uses the fact that the sum of the probabilities for *all* possible cases must be 1. From the table we see that the probability of accepting a lot is .7265, so the

probability of rejecting a lot is $1 - .7265 = .2735$.)

**(b)** $E(X) = 0(.7265) + 1(.2477)$

$\qquad + 2(.0251) + 3(.0007)$

$\qquad = .3000.$

**(c)** In part (b) we found that an average of .3 radio in every sample of three radios is defective. Thus about 10% of the radios in the sample are defective. Since the samples are chosen at random, we may assume that about 10% of *all* the radios are defective. Thus we estimate that on the average 10 out of each lot of 100 radios will be defective.

# 12.2 Continuous Random Variables

Consider a cell population that is growing vigorously. Suppose that when a cell is $T$ days old it divides and forms two new "daughter" cells. If the population is sufficiently large, it will contain cells of many different ages between 0 and $T$. It turns out that the proportion of cells of various ages remains constant. That is, if $a$ and $b$ are any two numbers between 0 and $T$, with $a < b$, the proportion of cells whose ages lie between $a$ and $b$ is essentially constant from one moment to the next, even though individual cells are aging and new cells are being formed all the time. In fact, biologists have found that under the ideal circumstances described, the proportion of cells whose ages are between $a$ and $b$ is given by the area under the graph of the function $f(x) = 2ke^{-kx}$ from $x = a$ to $x = b$, where $k = (\ln 2)/T$.[*] (See Fig. 1.)

---

[*] See J. R. Cook and T. W. James, "Age Distribution of Cells in Logarithmically Growing Cell Populations," in *Synchrony in Cell Division and Growth*, Erik Zeuthen, ed. (New York: John Wiley & Sons, 1964), pp. 485–495.

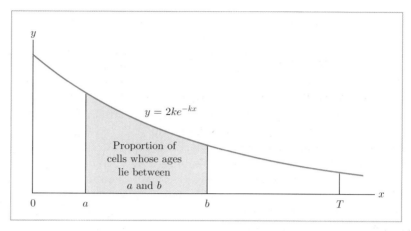

**Figure 1.** Age distribution in a cell population.

Now consider an experiment where we select a cell at random from the population and observe its age, $X$. Then the probability that $X$ lies between $a$ and $b$ is given by the area under the graph of $f(x) = 2ke^{-kx}$ from $a$ to $b$, as in Fig. 1. Let us denote this probability by $\Pr(a \leq X \leq b)$. Using the fact that the area under the graph of $f(x)$ is given by a definite integral, we have

$$\Pr(a \leq X \leq b) = \int_a^b f(x)\,dx = \int_a^b 2ke^{-kx}\,dx. \tag{1}$$

Since $X$ can assume any one of the (infinitely many) numbers in the continuous interval from 0 to $T$, we say that $X$ is a *continuous random variable*. The function $f(x)$ that determines the probability in (1) for each $a$ and $b$ is called the (*probability*) *density function* of $X$ (or of the experiment whose outcome is $X$).

More generally, suppose that we consider an experiment whose outcome may be any value between $A$ and $B$. The outcome of the experiment, denoted $X$, is called a *continuous random variable*. For the cell population described previously, $A = 0$ and $B = T$. Another typical experiment might consist of choosing a number $X$ at random between $A = 5$ and $B = 6$. Or one could observe the duration $X$ of a random telephone call passing through a given telephone switchboard. If we have no way of knowing how long a call might last, then $X$ might be any nonnegative number. In this case it is convenient to say that $X$ lies between 0 and $\infty$ and to take $A = 0$ and $B = \infty$. On the other hand, if the possible values of $X$ for some experiment include rather large negative numbers, one sometimes takes $A = -\infty$.

Given an experiment whose outcome is a continuous random variable $X$, the probability $\Pr(a \leq X \leq b)$ is a measure of the likelihood that an outcome of the experiment will lie between $a$ and $b$. If the experiment is repeated many times, the proportion of times $X$ has a value between $a$ and $b$ should be close to $\Pr(a \leq X \leq b)$. In experiments of practical interest involving a continuous random variable $X$ it is usually possible to find a function $f(x)$ such that

$$\Pr(a \leq X \leq b) = \int_a^b f(x)\,dx \tag{2}$$

for all $a$ and $b$ in the range of possible values of $X$. Such a function $f(x)$ is called a *probability density function* and satisfies the following properties:

**(I)** $f(x) \geq 0$ for $A \leq x \leq B$.

**(II)** $\displaystyle\int_A^B f(x) = 1$.

Indeed, Property I means that for $x$ between $A$ and $B$, the graph of $f(x)$ must lie on or above the $x$-axis. Property II simply says that there is probability 1 that $X$ has a value between $A$ and $B$. (Of course, if $B = \infty$ and/or $A = -\infty$, the integral in Property II is an improper integral.) Properties I and II characterize probability density functions, in the sense that any function $f(x)$ satisfying I and II is the probability density function for some continuous random variable $X$. Moreover, $\Pr(a \leq X \leq b)$ can then be calculated using equation (2).

Unlike a probability table for a discrete random variable, a density function $f(x)$ does *not* give the probability that $X$ has a certain value. Instead, $f(x)$ can be used to find the probability that $X$ is *near* a specific value in the following sense. If $x_0$ is a number between $A$ and $B$ and if $\Delta x$ is the width of a small interval centered at $x_0$, the probability that $X$ is between $x_0 - \frac{1}{2}\Delta x$ and $x_0 + \frac{1}{2}\Delta x$ is approximately $f(x_0)\Delta x$—that is, the area of the rectangle shown in Fig. 2.

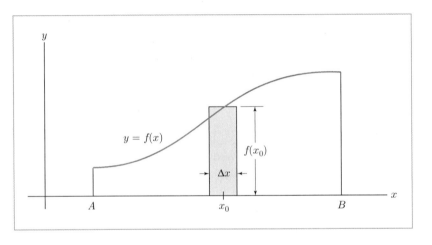

**Figure 2.**  Area of rectangle gives approximate probability that $X$ is near $x_0$.

---

**EXAMPLE 1**  Consider the cell population described earlier. Let $f(x) = 2ke^{-kx}$, where $k = (\ln 2)/T$. Show that $f(x)$ is indeed a probability density function on $0 \leq x \leq T$.

**Solution**  Clearly, $f(x) \geq 0$, since $\ln 2$ is positive and the exponential function is never negative. Thus Property I is satisfied. For Property II we check that

$$\int_0^T f(x)\, dx = \int_0^T 2ke^{-kx}\, dx = -2e^{-kx}\Big|_0^T = -2e^{-kT} + 2e^0$$

$$= -2e^{-[(\ln 2)/T]T} + 2 = -2e^{-\ln 2} + 2$$

$$= -2e^{\ln(1/2)} + 2 = -2\left(\frac{1}{2}\right) + 2 = 1. \qquad \blacksquare$$

---

**EXAMPLE 2**  Let $f(x) = kx^2$.

(a) Find the value of $k$ that makes $f(x)$ a probability density function on $0 \leq x \leq 4$.

(b) Let $X$ be a continuous random variable whose density function is $f(x)$. Compute $\Pr(1 \leq X \leq 2)$.

**Solution**    (a) We must have $k \geq 0$ so that Property I is satisfied. For Property II, we calculate

$$\int_0^4 f(x)\,dx = \int_0^4 kx^2\,dx = \left.\frac{1}{3}kx^3\right|_0^4 = \frac{1}{3}k(4)^3 - 0 = \frac{64}{3}k.$$

To satisfy Property II we must have $\frac{64}{3}k = 1$, or $k = \frac{3}{64}$. Thus $f(x) = \frac{3}{64}x^2$.

(b) $\Pr(1 \leq X \leq 2) = \int_1^2 f(x)\,dx = \int_1^2 \frac{3}{64}x^2\,dx = \left.\frac{1}{64}x^3\right|_1^2 = \frac{8}{64} - \frac{1}{64} = \frac{7}{64}$

The area corresponding to this probability is shown in Fig. 3.    ■

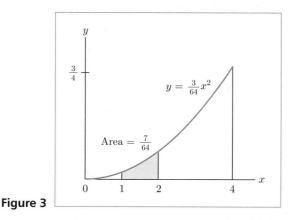

**Figure 3**

The density function in the next example is a special case of what statisticians sometimes call a *beta* probability density.

---

**EXAMPLE 3**    The parent corporation for a franchised chain of fast-food restaurants claims that the proportion of their new restaurants that make a profit during their first year of operation has the probability density function

$$f(x) = 12x(1-x)^2, \qquad 0 \leq x \leq 1.$$

(a) What is the probability that less than 40% of the restaurants opened this year will make a profit during their first year of operation?

(b) What is the probability that more than 50% of the restaurants will make a profit during their first year of operation?

**Solution**    Let $X$ be the proportion of new restaurants opened this year that make a profit during their first year of operation. Then the possible values of $X$ range between 0 and 1.

(a) The probability that $X$ is less than .4 equals the probability that $X$ is between 0 and .4. We note that $f(x) = 12x(1 - 2x + x^2) = 12x - 24x^2 + 12x^3$ and therefore,

$$\Pr(0 \leq X \leq .4) = \int_0^{.4} f(x)\,dx = \int_0^{.4} (12x - 24x^2 + 12x^3)\,dx$$

$$= \left.(6x^2 - 8x^3 + 3x^4)\right|_0^{.4} = .5248.$$

(b) The probability that $X$ is greater than .5 equals the probability that $X$ is between .5 and 1. Thus

$$\Pr(.5 \leq X \leq 1) = \int_{.5}^{1} (12x - 24x^2 + 12x^3)\, dx$$

$$= (6x^2 - 8x^3 + 3x^4)\Big|_{.5}^{1} = .3125. \quad\blacksquare$$

Each probability density function is closely related to another important function called a cumulative distribution function. To describe this relationship, let us consider an experiment whose outcome is a continuous random variable $X$, with values between $A$ and $B$, and let $f(x)$ be the associated density function. For each number $x$ between $A$ and $B$, let $F(x)$ be the probability that $X$ is less than or equal to the number $x$. Sometimes we write $F(x) = \Pr(X \leq x)$; however, since $X$ is never less than $A$, we may also write

$$F(x) = \Pr(A \leq X \leq x). \tag{3}$$

Graphically, $F(x)$ is the area under the graph of the probability density function $f(x)$ from $A$ to $x$. (See Fig. 4.) The function $F(x)$ is called the *cumulative distri-*

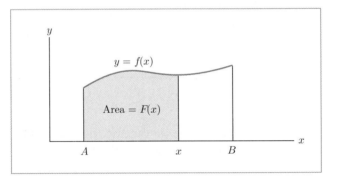

**Figure 4.**   The cumulative distribution function $F(x)$.

*bution function* of the random variable $X$ (or of the experiment whose outcome is $X$). Note that $F(x)$ also has the properties

$$F(A) = \Pr(A \leq X \leq A) = 0, \tag{4}$$
$$F(B) = \Pr(A \leq X \leq B) = 1. \tag{5}$$

Since $F(x)$ is an "area function" that gives the area under the graph of $f(x)$ from $A$ to $x$, we know from Section 6.3 that $F(x)$ is an antiderivative of $f(x)$. That is,

$$F'(x) = f(x), \qquad A \leq x \leq B. \tag{6}$$

It follows that one may use $F(x)$ to compute probabilities, since

$$\Pr(a \leq X \leq b) = \int_{a}^{b} f(x)\, dx = F(b) - F(a), \tag{7}$$

for any $a$ and $b$ between $A$ and $B$.

The relation (6) between $F(x)$ and $f(x)$ makes it possible to find one of these functions when the other is known, as we see in the following two examples.

**EXAMPLE 4**  Let $X$ be the age of a cell selected at random from the cell population described ear-lier. The probability density function for $X$ is $f(x) = 2ke^{-kx}$, where $k = (\ln 2)/T$. See Fig. 5. Find the cumulative distribution function $F(x)$ for $X$.

**Solution**  Since $F(x)$ is an antiderivative of $f(x) = 2ke^{-kx}$, we have $F(x) = -2e^{-kx} + C$ for some constant $C$. Now $F(x)$ is defined for $0 \le x \le T$. Thus (4) implies that $F(0) = 0$. Setting $F(0) = -2e^0 + C = 0$, we find that $C = 2$, so

$$F(x) = -2e^{-kx} + 2.$$

See Fig. 6.

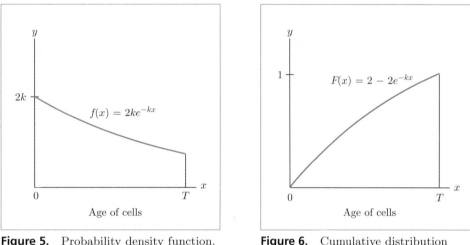

**Figure 5.**  Probability density function.

**Figure 6.**  Cumulative distribution function.

**EXAMPLE 5**  Let $X$ be the random variable associated with the experiment that consists of selecting a point at random from a circle of radius 1 and observing its distance from the center. Find the probability density function $f(x)$ and cumulative distribution function $F(x)$ of $X$.

**Solution**  The distance of a point from the center of the unit circle is a number between 0 and 1. Suppose that $0 \le x \le 1$. Let us first compute the cumulative distribution function $F(x) = \Pr(0 \le X \le x)$. That is, let us find the probability that a point selected at random lies within $x$ units of the center of the circle—in other words, lies inside the circle of radius $x$. See the shaded region in Fig. 7(b). Since the area of this shaded region is $\pi x^2$ and the area of the entire unit circle is $\pi \cdot 1^2 = \pi$, the proportion of points inside the shaded region is $\pi x^2/\pi = x^2$. Thus the probability is $x^2$ that a point selected at random will be in this shaded region. Hence

$$F(x) = x^2.$$

Differentiating, we find that the probability density function for $X$ is

$$f(x) = F'(x) = 2x.$$

Our final example involves a continuous random variable $X$ whose possible values lie between $A = 1$ and $B = \infty$; that is, $X$ is any number greater than or equal to 1.

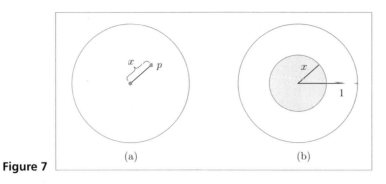

(a)          (b)

**Figure 7**

---

**EXAMPLE 6**    Let $f(x) = 3x^{-4}$, $x \geq 1$.

(a) Show that $f(x)$ is the probability density function of some random variable $X$.

(b) Find the cumulative distribution function $F(x)$ of $X$.

(c) Compute $\Pr(X \leq 4)$, $\Pr(4 \leq X \leq 5)$, and $\Pr(4 \leq X)$.

**Solution**    (a) It is clear that $f(x) \geq 0$ for $x \geq 1$. Thus Property I holds. In order to check Property II, we must compute

$$\int_1^\infty 3x^{-4}\, dx.$$

But

$$\int_1^b 3x^{-4}\, dx = -x^{-3}\Big|_1^b = -b^{-3} + 1 \to 1$$

as $b \to \infty$. Thus

$$\int_1^\infty 3x^{-4}\, dx = 1,$$

and Property II holds.

(b) Since $F(x)$ is an antiderivative of $f(x) = 3x^{-4}$, we have

$$F(x) = \int 3x^{-4}\, dx = -x^{-3} + C.$$

Since $X$ has values greater than or equal to 1, we must have $F(1) = 0$. Setting $F(1) = -1 + C = 0$, we find that $C = 1$, so

$$F(x) = 1 - x^{-3}.$$

(c) $\Pr(X \leq 4) = F(4) = 1 - 4^{-3} = 1 - \frac{1}{64} = \frac{63}{64}$

Since we know $F(x)$, we may use it to compute $\Pr(4 \leq X \leq 5)$, as follows:

$$\Pr(4 \leq X \leq 5) = F(5) - F(4) = (1 - 5^{-3}) - (1 - 4^{-3})$$

$$= \frac{1}{4^3} - \frac{1}{5^3} \approx .0076.$$

We may compute $\Pr(4 \leq X)$ directly by evaluating the improper integral

$$\int_4^\infty 3x^{-4}\, dx.$$

However, there is a simpler method. We know that

$$\int_1^4 3x^{-4}\, dx + \int_4^\infty 3x^{-4}\, dx = \int_1^\infty 3x^{-4}\, dx = 1. \tag{8}$$

In terms of probabilities, (8) may be written as

$$\Pr(X \le 4) + \Pr(4 \le X) = 1.$$

Hence

$$\Pr(4 \le X) = 1 - \Pr(X \le 4) = 1 - \frac{63}{64} = \frac{1}{64}.$$

## PRACTICE PROBLEMS 12.2

1. Suppose that in a certain farming region, and in a certain year, the number of bushels of wheat produced per acre is a random variable $X$ with a density function

$$f(x) = \frac{x - 30}{50}, \qquad 30 \le x \le 40.$$

   (a) What is the probability that an acre selected at random produced less than 35 bushels of wheat?

   (b) If the farming region had 20,000 acres of wheat, how many acres produced less than 35 bushels of wheat?

2. The density function for a continuous random variable $X$ on the interval $1 \le x \le 2$ is $f(x) = 8/(3x^3)$. Find the corresponding cumulative distribution function for $X$.

## EXERCISES 12.2

*Verify that each of the following functions is a probability density function.*

1. $f(x) = \frac{1}{18}x$, $0 \le x \le 6$
2. $f(x) = 2(x - 1)$, $1 \le x \le 2$
3. $f(x) = \frac{1}{4}$, $1 \le x \le 5$
4. $f(x) = \frac{8}{9}x$, $0 \le x \le \frac{3}{2}$
5. $f(x) = 5x^4$, $0 \le x \le 1$
6. $f(x) = \frac{3}{2}x - \frac{3}{4}x^2$, $0 \le x \le 2$

*In Exercises 7–12, find the value of $k$ that makes the given function a probability density function on the specified interval.*

7. $f(x) = kx$, $1 \le x \le 3$
8. $f(x) = kx^2$, $0 \le x \le 2$
9. $f(x) = k$, $5 \le x \le 20$
10. $f(x) = k/\sqrt{x}$, $1 \le x \le 4$
11. $f(x) = kx^2(1 - x)$, $0 \le x \le 1$
12. $f(x) = k(3x - x^2)$, $0 \le x \le 3$

13. The density function of a continuous random variable $X$ is $f(x) = \frac{1}{8}x$, $0 \le x \le 4$. Sketch the graph of $f(x)$ and shade in the areas corresponding to (a) $\Pr(X \le 1)$; (b) $\Pr(2 \le X \le 2.5)$; (c) $\Pr(3.5 \le X)$.

14. The density function of a continuous random variable $X$ is $f(x) = 3x^2$, $0 \le x \le 1$. Sketch the graph of $f(x)$ and shade in the areas corresponding to (a) $\Pr(X \le .3)$; (b) $\Pr(.5 \le X \le .7)$; (c) $\Pr(.8 \le X)$.

15. Find $\Pr(1 \le X \le 2)$ when $X$ is a random variable whose density function is given in Exercise 1.

16. Find $\Pr(1.5 \le X \le 1.7)$ when $X$ is a random variable whose density function is given in Exercise 2.

17. Find $\Pr(X \le 3)$ when $X$ is a random variable whose density function is given in Exercise 3.

18. Find $\Pr(1 \le X)$ when $X$ is a random variable whose density function is given in Exercise 4.

19. Suppose that the lifetime $X$ (in hours) of a certain type of flashlight battery is a random variable on the interval $30 \le x \le 50$ with density function $f(x) = \frac{1}{20}$, $30 \le x \le 50$. Find the probability that a battery selected at random will last at least 35 hours.

20. Suppose that at a certain supermarket the amount of time one must wait at the express lane is a random variable with density function $f(x) = 11/[10(x + 1)^2]$, $0 \le x \le 10$. See Fig. 8. Find the probability of having to wait less than 4 minutes at the express lane.

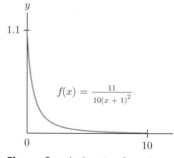

**Figure 8.** A density function.

21. The cumulative distribution function for a random variable $X$ on the interval $1 \le x \le 5$ is $F(x) = \frac{1}{2}\sqrt{x - 1}$. See Fig. 9. Find the corresponding density function.

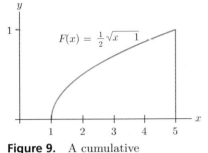

**Figure 9.** A cumulative distribution function.

22. The cumulative distribution function for a random variable $X$ on the interval $1 \le x \le 2$ is $F(x) = \frac{4}{3} - 4/(3x^2)$. Find the corresponding density function.

23. Compute the cumulative distribution function corresponding to the density function $f(x) = \frac{1}{5}$, $2 \le x \le 7$.

24. Compute the cumulative distribution function corresponding to the density function $f(x) = \frac{1}{2}(3 - x)$, $1 \le x \le 3$.

25. The time (in minutes) required to complete a certain subassembly is a random variable $X$ with density function $f(x) = \frac{1}{21}x^2$, $1 \le x \le 4$.
    (a) Use $f(x)$ to compute $\Pr(2 \le X \le 3)$.
    (b) Find the corresponding cumulative distribution function $F(x)$.
    (c) Use $F(x)$ to compute $\Pr(2 \le X \le 3)$.

26. The density function for a continuous random variable $X$ on the interval $1 \le x \le 4$ is $f(x) = \frac{4}{9}x - \frac{1}{9}x^2$.
    (a) Use $f(x)$ to compute $\Pr(3 \le X \le 4)$.
    (b) Find the corresponding cumulative distribution function $F(x)$.
    (c) Use $F(x)$ to compute $\Pr(3 \le X \le 4)$.

*An experiment consists of selecting a point at random from the square in Fig. 10(a). Let $X$ be the maximum of the coordinates of the point.*

27. Show that the cumulative distribution function of $X$ is $F(x) = x^2/4$, $0 \le x \le 2$.

28. Find the corresponding density function of $X$.

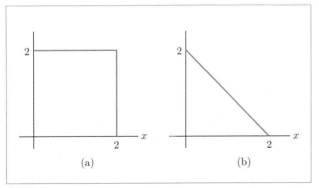

**Figure 10**

*An experiment consists of selecting a point at random from the triangle in Fig. 10(b). Let $X$ be the sum of the coordinates of the point.*

29. Show that the cumulative distribution function of $X$ is $F(x) = x^2/4$, $0 \le x \le 2$.

30. Find the corresponding density function of $X$.

*Suppose that in a certain cell population, cells divide every 10 days, and the age of a cell selected at random is a random variable $X$ with the density function $f(x) = 2ke^{-kx}$, $0 \le x \le 10$, $k = (\ln 2)/10$.*

31. Find the probability that a cell is at most 5 days old.

32. Upon examination of a slide, 10% of the cells are found to be undergoing mitosis (a change in the cell leading to division). Compute the length of time required for mitosis; that is, find the number $M$ such that
$$\int_{10-M}^{10} 2ke^{-kx}\, dx = .10.$$

33. A random variable $X$ has a density function $f(x) = \frac{1}{3}$, $0 \le x \le 3$. Find $b$ such that $\Pr(0 \le X \le b) = .6$.

34. A random variable $X$ has a density function $f(x) = \frac{2}{3}x$ on $1 \le x \le 2$. Find $a$ such that $\Pr(a \le X) = \frac{1}{3}$.

35. A random variable $X$ has a cumulative distribution function $F(x) = \frac{1}{4}x^2$ on $0 \le x \le 2$. Find $b$ such that $\Pr(X \le b) = .09$.

36. A random variable $X$ has a cumulative distribution function $F(x) = (x - 1)^2$ on $1 \le x \le 2$. Find $b$ such that $\Pr(X \le b) = \frac{1}{4}$.

37. Let $X$ be a continuous random variable with values between $A = 1$ and $B = \infty$, and with the density function $f(x) = 4x^{-5}$.
    (a) Verify that $f(x)$ is a probability density function for $x \ge 1$.
    (b) Find the corresponding cumulative distribution function $F(x)$.
    (c) Use $F(x)$ to compute $\Pr(1 \le X \le 2)$ and $\Pr(2 \le X)$.

38. Let $X$ be a continuous random variable with the density function $f(x) = 2(x + 1)^{-3}$, $x \ge 0$.
    (a) Verify that $f(x)$ is a probability density function for $x \ge 0$.
    (b) Find the cumulative distribution function for $X$.
    (c) Compute $\Pr(1 \le X \le 2)$ and $\Pr(3 \le X)$.

## SOLUTIONS TO PRACTICE PROBLEMS 12.2

**1. (a)** $\Pr(X \leq 35) = \displaystyle\int_{30}^{35} \frac{x-30}{50}\,dx = \frac{(x-30)^2}{100}\bigg|_{30}^{35}$

$$= \frac{5^2}{100} - 0 = .25$$

**(b)** Using part (a), we see that 25% of the 20,000 acres, or 5000 acres, produced less that 35 bushels of wheat.

**2.** The cumulative distribution function $F(x)$ is an antiderivative of $f(x) = 8/(3x^3) = \frac{8}{3}x^{-3}$. Thus $F(x) = -\frac{4}{3}x^{-2} + C$ for some constant $C$. Since $X$ varies over the interval $1 \leq x \leq 2$, we must have $F(1) = 0$; that is, $-\frac{4}{3}(1)^{-2} + C = 0$. Thus $C = \frac{4}{3}$, and

$$F(x) = \frac{4}{3} - \frac{4}{3}x^{-2}.$$

## 12.3 Expected Value and Variance

When studying the cell population described in Section 12.2, one might reasonably ask for the average age of the cells. In general, if one is given an experiment described by a random variable $X$ and a probability density function $f(x)$, it is often important to know the "average" outcome of the experiment, and the degree to which the experimental outcomes are spread out around the average. To provide this information in Section 12.1 we introduced the concepts of expected value and variance of a discrete random variable. Let us now examine the analogous definition for a continuous random variable.

---

**DEFINITION** Let $X$ be a continuous random variable whose possible values lie between $A$ and $B$, and let $f(x)$ be the probability density function for $X$. Then the *expected value* (or *mean*) of $X$ is the number $E(X)$ defined by

$$E(X) = \int_A^B x f(x)\,dx. \tag{1}$$

The *variance* of $X$ is the number $\mathrm{Var}(X)$ defined by

$$\mathrm{Var}(X) = \int_A^B [x - E(X)]^2 f(x)\,dx. \tag{2}$$

---

The expected value of $X$ has the same interpretation as in the discrete case—namely, if the experiment whose outcome is $X$ is performed many times, then the average of all the outcomes will approximately equal $E(X)$. As in the case of a discrete random variable, the variance of $X$ is a quantitative measure of the likely spread of the values of $X$ about the mean $E(X)$ when the experiment is performed many times.

To explain why the definition (1) of $E(X)$ is analogous to the definition in Section 12.1, let us approximate the integral in (1) by a Riemann sum of the form

$$x_1 f(x_1)\Delta x + x_2 f(x_2)\Delta x + \cdots + x_n f(x_n)\Delta x. \tag{3}$$

Here $x_1, \ldots, x_n$ are the midpoints of subintervals of the interval from $A$ to $B$, each subinterval of width $\Delta x = (B - A)/n$. (See Fig. 1.) Now recall from Section 12.2 that, for $i = 1, \ldots, n$, the quantity $f(x_i)\Delta x$ is approximately the probability that $X$ is close to $x_i$—that is, the probability that $X$ lies in the subinterval centered at $x_i$. If we write $\Pr(X \approx x_i)$ for this probability, then (3) is nearly the same as

$$x_1 \cdot \Pr(X \approx x_1) + x_2 \cdot \Pr(X \approx x_2) + \cdots + x_n \cdot \Pr(X \approx x_n). \tag{4}$$

**Figure 1**

As the number of subintervals increases, the sum becomes closer and closer to the integral in (1) defining $E(X)$. Furthermore, each approximating sum in (4) resembles the sum in the definition of the expected value of a discrete random variable, where one computes the weighted sum over all possible outcomes, with each outcome weighted by the probability of its occurrence.

A similar analysis would show that the definition (2) of variance is analogous to the definition for the discrete case.

---

**EXAMPLE 1**    Let us consider the experiment of selecting a number at random from among the numbers between 0 and $B$. Let $X$ denote the associated random variable. Determine the cumulative distribution function of $X$, the density function of $X$, and the mean and variance of $X$.

**Solution**
$$F(x) = \frac{\text{[length of the interval from 0 to } x]}{\text{[length of the interval from 0 to } B]} = \frac{x}{B}$$

Since $f(x) = F'(x)$, we see that $f(x) = 1/B$. Thus we have

$$E(X) = \int_0^B x \cdot \frac{1}{B}\, dx = \frac{1}{B}\int_0^B x\, dx = \frac{1}{B} \cdot \frac{B^2}{2} = \frac{B}{2},$$

$$\text{Var}(X) = \int_0^B \left(x - \frac{B}{2}\right)^2 \cdot \frac{1}{B}\, dx = \frac{1}{B}\int_0^B \left(x - \frac{B}{2}\right)^2 dx$$

$$= \frac{1}{B} \cdot \frac{1}{3}\left(x - \frac{B}{2}\right)^3 \Big|_0^B = \frac{1}{3B}\left[\left(\frac{B}{2}\right)^3 - \left(-\frac{B}{2}\right)^3\right] = \frac{B^2}{12}.$$

The graph of the density function $f(x)$ is shown in Fig. 2. Since the density function has a flat graph, the random variable $X$ is called the *uniform random variable* on the interval from 0 to $B$. ◼

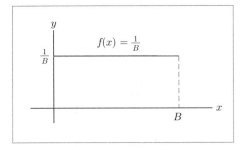

**Figure 2.** A uniform probability density function.

---

**EXAMPLE 2**    Let $X$ be the age of a cell chosen at random from the population described in Section 12.2, where the density function for $X$ was given as

$$f(x) = 2ke^{-kx}, \qquad 0 \le x \le T,$$

and $k = (\ln 2)/T$. Find the average age, $E(X)$, of the cell population.

**Solution** By definition,

$$\mathrm{E}(X) = \int_0^T x \cdot 2ke^{-kx}\, dx.$$

To calculate this integral we need integration by parts, with $f(x) = 2x$, $g(x) = ke^{-kx}$, $f'(x) = 2$, and $G(x) = -e^{-kx}$. We have

$$\int_0^T 2xke^{-kx}\, dx = -2xe^{-kx}\Big|_0^T - \int_0^T -2e^{-kx}\, dx$$

$$= -2Te^{-kT} - \left(\frac{2}{k}e^{-kx}\right)\Big|_0^T$$

$$= -2Te^{-kT} - \frac{2}{k}e^{-kT} + \frac{2}{k}.$$

This formula for $\mathrm{E}(X)$ may be simplified by noting that $e^{-kT} = e^{-\ln 2} = \frac{1}{2}$. Thus

$$\mathrm{E}(X) = -2T\left(\frac{1}{2}\right) - \frac{2}{k}\left(\frac{1}{2}\right) + \frac{2}{k} = \frac{1}{k} - T$$

$$= \frac{T}{\ln 2} - T = \left(\frac{1}{\ln 2} - 1\right)T$$

$$\approx .4427T. \qquad \blacksquare$$

---

**EXAMPLE 3** Consider the experiment of selecting a point at random in a circle of radius 1, and let $X$ be the distance from this point to the center. Compute the expected value and variance of the random variable $X$.

**Solution** We showed in Example 5 of Section 12.2 that the density function for $X$ is given by $f(x) = 2x$, $0 \le x \le 1$. Therefore, we see that

$$\mathrm{E}(X) = \int_0^1 x \cdot 2x\, dx = \int_0^1 2x^2\, dx = \frac{2x^3}{3}\Big|_0^1 = \frac{2}{3}$$

and

$$\mathrm{Var}(X) = \int_0^1 \left(x - \frac{2}{3}\right)^2 \cdot 2x\, dx \qquad \left[\text{since } \mathrm{E}(X) = \frac{2}{3}\right]$$

$$= \int_0^1 \left(x^2 - \frac{4}{3}x + \frac{4}{9}\right) \cdot 2x\, dx$$

$$= \int_0^1 \left(2x^3 - \frac{8}{3}x^2 + \frac{8}{9}x\right)dx$$

$$= \left(\frac{1}{2}x^4 - \frac{8}{9}x^3 + \frac{4}{9}x^2\right)\Big|_0^1$$

$$= \frac{1}{2} - \frac{8}{9} + \frac{4}{9} = \frac{1}{18}.$$

From our first calculation, we see that if a large number of points are chosen randomly from a circle of radius 1, their average distance to the center should be about $\frac{2}{3}$. $\qquad \blacksquare$

The following alternative formula for the variance of a random variable is usually easier to use than the actual definition of $\text{Var}(X)$.

Let $X$ be a continuous random variable whose values lie between $A$ and $B$, and let $f(x)$ be the density function for $X$. Then

$$\text{Var}(X) = \int_A^B x^2 f(x)\,dx - E(X)^2. \tag{5}$$

To prove (5), we let $m = E(X) = \int_A^B x f(x)\,dx$. Then

$$\text{Var}(X) = \int_A^B (x - m)^2 f(x)\,dx = \int_A^B (x^2 - 2xm + m^2) f(x)\,dx$$

$$= \int_A^B x^2 f(x)\,dx - 2m \int_A^B x f(x)\,dx + m^2 \int_A^B f(x)\,dx$$

$$= \int_A^B x^2 f(x)\,dx - 2m \cdot m + m^2 \cdot 1 \qquad \text{(by Property II)}$$

$$= \int_A^B x^2 f(x)\,dx - m^2.$$

**EXAMPLE 4**    A college library has found that, in any given month during a school year, the proportion of students who make some use of the library is a random variable $X$ with the cumulative distribution function

$$F(x) = 4x^3 - 3x^4, \qquad 0 \leq x \leq 1.$$

(a) Compute $E(X)$ and give an interpretation of this quantity.

(b) Compute $\text{Var}(X)$.

**Solution** (a) To compute $E(X)$ we first find the probability density function $f(x)$. From Section 12.2 we know that

$$f(x) = F'(x) = 12x^2 - 12x^3.$$

Hence

$$E(X) = \int_0^1 x f(x)\,dx = \int_0^1 (12x^3 - 12x^4)\,dx$$

$$= \left( 3x^4 - \frac{12}{5}x^5 \right)\Big|_0^1 = 3 - \frac{12}{5} = \frac{3}{5}.$$

The meaning of $E(X)$ in this example is that over a period of many months (during school years) the average proportion of students each month who make some use of the library should be close to $\frac{3}{5}$.

(b) We first compute

$$\int_0^1 x^2 f(x)\,dx = \int_0^1 (12x^4 - 12x^5)\,dx = \left( \frac{12}{5}x^5 - 2x^6 \right)\Big|_0^1$$

$$= \frac{12}{5} - 2 = \frac{2}{5}.$$

Then, from the alternative formula (5) for the variance, we find that

$$\text{Var}(X) = \frac{2}{5} - \text{E}(X)^2 = \frac{2}{5} - \left(\frac{3}{5}\right)^2 = \frac{1}{25}.$$ ■

## PRACTICE PROBLEMS 12.3

1. Find the expected value and variance of the random variable $X$ whose density function is $f(x) = 1/(2\sqrt{x})$, $1 \leq x \leq 4$.

2. An insurance company finds that the proportion $X$ of their salespeople who sell more than \$25,000 worth of insurance in a given week is a random variable with

the beta probability density function

$$f(x) = 60x^3(1-x)^2, \qquad 0 \leq x \leq 1.$$

(a) Compute $\text{E}(X)$ and give an interpretation of this quantity.

(b) Compute $\text{Var}(X)$.

## EXERCISES 12.3

*Find the expected value and variance for each random variable whose probability density function is given. When computing the variance, use formula (5).*

1. $f(x) = \frac{1}{18}x$, $0 \leq x \leq 6$
2. $f(x) = 2(x-1)$, $1 \leq x \leq 2$
3. $f(x) = \frac{1}{4}$, $1 \leq x \leq 5$
4. $f(x) = \frac{8}{9}x$, $0 \leq x \leq \frac{3}{2}$
5. $f(x) = 5x^4$, $0 \leq x \leq 1$
6. $f(x) = \frac{3}{2}x - \frac{3}{4}x^2$, $0 \leq x \leq 2$
7. $f(x) = 12x(1-x)^2$, $0 \leq x \leq 1$
8. $f(x) = \dfrac{3\sqrt{x}}{16}$, $0 \leq x \leq 4$

9. A newspaper publisher estimates that the proportion $X$ of space devoted to advertising on a given day is a random variable with the beta probability density $f(x) = 30x^2(1-x)^2$, $0 \leq x \leq 1$.
    (a) Find the cumulative distribution function for $X$.
    (b) Find the probability that less than 25% of the newspaper's space on a given day contains advertising.
    (c) Find $\text{E}(X)$ and give an interpretation of this quantity.
    (d) Compute $\text{Var}(X)$.

10. Let $X$ be the proportion of new restaurants in a given year that make a profit during their first year of operation, and suppose that the density function for $X$ is $f(x) = 20x^3(1-x)$, $0 \leq x \leq 1$.
    (a) Find $\text{E}(X)$ and give an interpretation of this quantity.
    (b) Compute $\text{Var}(X)$.

11. The useful life (in hundreds of hours) of a certain machine component is a random variable $X$ with the cumulative distribution function $F(x) = \frac{1}{9}x^2$, $0 \leq x \leq 3$.

(a) Find $\text{E}(X)$ and give an interpretation of this quantity.

(b) Compute $\text{Var}(X)$.

12. The time (in minutes) required to complete an assembly on a production line is a random variable $X$ with the cumulative distribution function $F(x) = \frac{1}{125}x^3$, $0 \leq x \leq 5$.
    (a) Find $\text{E}(X)$ and give an interpretation of this quantity.
    (b) Compute $\text{Var}(X)$.

13. Suppose that the amount of time (in minutes) a person spends reading the editorial page of the newspaper is a random variable with the density function $f(x) = \frac{1}{72}x$, $0 \leq x \leq 12$. Find the average time spent reading the editorial page.

14. Suppose that at a certain bus stop the time between buses is a random variable $X$ with the density function $f(x) = 6x(10-x)/1000$, $0 \leq x \leq 10$. Find the average time between buses.

15. When preparing a bid on a large construction project, a contractor analyzes how long each phase of the construction will take. Suppose the contractor estimates that the time required for the electrical work will be $X$ hundred worker-hours, where $X$ is a random variable with density function $f(x) = x(6-x)/18$, $3 \leq x \leq 6$. See Fig. 3.
    (a) Find the cumulative distribution function $F(x)$.
    (b) What is the likelihood that the electrical work will take less than 500 worker-hours?
    (c) Find the mean time to complete the electrical work.
    (d) Find $\text{Var}(X)$.

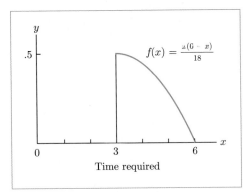

**Figure 3.** Density function for construction bid.

16. The amount of milk (in thousands of gallons) a dairy sells each week is a random variable $X$ with the density function $f(x) = 4(x - 1)^3$, $1 \le x \le 2$. See Fig. 4.

(a) What is the likelihood that the dairy will sell more than 1500 gallons?

(b) What is the average amount of milk the dairy sells each week?

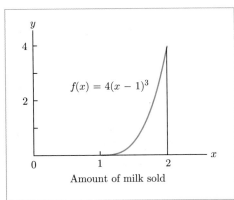

**Figure 4.** Density function for sales of milk.

17. Let $X$ be a continuous random variable with values between $A = 1$ and $B = \infty$, and with the density function $f(x) = 4x^{-5}$. Compute $E(X)$ and $Var(X)$.

18. Let $X$ be a continuous random variable with density function $f(x) = 3x^{-4}$, $x \ge 1$. Compute $E(X)$ and $Var(X)$.

*If $X$ is a random variable with density function $f(x)$ on $A \le x \le B$, the median of $X$ is that number $M$ such that*

$$\int_A^M f(x)\, dx = \frac{1}{2}.$$

*In other words,* $\Pr(X \le M) = \frac{1}{2}$.

19. Find the median of the random variable whose density function is $f(x) = \frac{1}{18}x$, $0 \le x \le 6$.

20. Find the median of the random variable whose density function is $f(x) = 2(x - 1)$, $1 \le x \le 2$.

21. The machine component described in Exercise 11 has a 50% chance of lasting at least how long?

22. In Exercise 12, find the length of time $T$ such that half of the assemblies are completed in $T$ minutes or less.

23. In Exercise 20 of Section 12.2, find the length of time $T$ such that about half of the time one waits only $T$ minutes or less in the express lane at the supermarket.

24. Find the number $M$ such that half of the time the dairy in Exercise 16 sells $M$ thousand gallons of milk or less.

25. Show that $E(X) = B - \int_A^B F(x)\, dx$, where $F(x)$ is the cumulative distribution function for $X$ on $A \le x \le B$.

26. Use the formula in Exercise 25 to compute $E(X)$ for the random variable $X$ in Exercise 12.

## SOLUTIONS TO PRACTICE PROBLEMS 12.3

1. $E(X) = \displaystyle\int_1^4 x \cdot \frac{1}{2\sqrt{x}}\, dx = \int_1^4 \frac{1}{2}x^{1/2}\, dx = \frac{1}{3}x^{3/2}\Big|_1^4$

$= \dfrac{1}{3}(4)^{3/2} - \dfrac{1}{3} = \dfrac{8}{3} - \dfrac{1}{3} = \dfrac{7}{3}$

To find $Var(X)$, we first compute

$\displaystyle\int_1^4 x^2 \cdot \frac{1}{2\sqrt{x}}\, dx = \int_1^4 \frac{1}{2}x^{3/2}\, dx = \frac{1}{5}x^{5/2}\Big|_1^4$

$= \dfrac{1}{5}(4)^{5/2} - \dfrac{1}{5} = \dfrac{32}{5} - \dfrac{1}{5} = \dfrac{31}{5}.$

Then, from formula (5),

$$Var(X) = \frac{31}{5} - \left(\frac{7}{3}\right)^2 = \frac{34}{45}.$$

**SOLUTIONS TO PRACTICE PROBLEMS 12.3 (CONTINUED)**

**2.** **(a)** First note that $f(x) = 60x^3(1-x)^2 = 60x^3(1 - 2x + x^2) = 60x^3 - 120x^4 + 60x^5$. Then

$$E(X) = \int_0^1 x f(x)\, dx$$

$$= \int_0^1 (60x^4 - 120x^5 + 60x^6)\, dx$$

$$= \left(12x^5 - 20x^6 + \frac{60}{7}x^7\right)\Big|_0^1$$

$$= 12 - 20 + \frac{60}{7} = \frac{4}{7}.$$

Thus, in an "average" week, about four-sevenths of the salespeople sell more than \$25,000 worth of insurance. More precisely, over a period of many weeks, we expect an average of four-sevenths of the salespeople each week to sell more than \$25,000 worth of insurance.

**(b)** $\displaystyle \int_0^1 x^2 f(x)\, dx = \int_0^1 (60x^5 - 120x^6 + 60x^7)\, dx$

$$= \left(10x^6 - \frac{120}{7}x^7 + \frac{60}{8}x^8\right)\Big|_0^1$$

$$= 10 - \frac{120}{7} + \frac{60}{8} = \frac{5}{14}$$

Hence

$$\operatorname{Var}(X) = \frac{5}{14} - \left(\frac{4}{7}\right)^2 = \frac{3}{98}.$$

## 12.4  Exponential and Normal Random Variables

This section is devoted to the two most important types of probability density functions—the exponential and normal density functions. These functions are associated with random variables that arise in a wide variety of applications. We will describe some typical examples.

**Exponential Density Functions**  Let $k$ be a positive constant. Then the function

$$f(x) = ke^{-kx}, \qquad x \geq 0,$$

is called an *exponential density function*. (See Fig. 1.) This function is indeed a probability density function. First, $f(x)$ is clearly greater than or equal to 0. Second,

$$\int_0^b ke^{-kx}\, dx = -e^{-kx}\Big|_0^b = 1 - e^{-kb} \to 1 \quad \text{as} \quad b \to \infty,$$

so that

$$\int_0^\infty ke^{-kx}\, dx = 1.$$

A random variable $X$ with an exponential density function is called an *exponential random variable*, and the values of $X$ are said to be *exponentially distributed*.

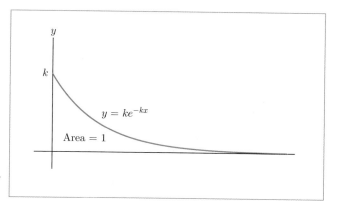

**Figure 1.**  Exponential density function.

Exponential random variables are used in reliability calculations to represent the lifetime (or "time to failure") of electronic components such as computer chips. They are used to describe the length of time between two successive random events, such as the interarrival times between successive telephone calls at a switchboard. Also, exponential random variables can arise in the study of service times (e.g., the length of time a person spends in a doctor's office or at a gas station).

Let us compute the expected value of an exponential random variable $X$, namely,

$$E(X) = \int_0^\infty x f(x)\, dx = \int_0^\infty x k e^{-kx}\, dx.$$

We may approximate this improper integral by a definite integral and use integration by parts to find that

$$\int_0^b x k e^{-kx}\, dx = -x e^{-kx}\Big|_0^b - \int_0^b -e^{-kx}\, dx$$

$$= (-be^{-kb} + 0) - \frac{1}{k} e^{-kx}\Big|_0^b$$

$$= -be^{-kb} - \frac{1}{k} e^{-kb} + \frac{1}{k}. \tag{1}$$

As $b \to \infty$, this quantity approaches $1/k$, because the numbers $-be^{-kb}$ and $-(1/k)e^{-kb}$ both approach 0. (See Section 11.5, Exercise 42.) Thus

$$E(X) = \int_0^\infty x k e^{-kx}\, dx = \frac{1}{k}.$$

Let us now compute the variance of $X$. From the alternate formula for $\text{Var}(X)$ given in Section 12.3, we have

$$\text{Var}(X) = \int_0^\infty x^2 f(x)\, dx - E(X)^2$$

$$= \int_0^\infty x^2 k e^{-kx}\, dx - \frac{1}{k^2}. \tag{2}$$

Using integration by parts, we obtain

$$\int_0^b x^2 k e^{-kx}\, dx = x^2(-e^{-kx})\Big|_0^b - \int_0^b 2x(-e^{-kx})\, dx$$

$$= (-b^2 e^{-kb} + 0) + 2\int_0^b x e^{-kx}\, dx$$

$$= -b^2 e^{-kb} + \frac{2}{k}\int_0^b x k e^{-kx}\, dx. \tag{3}$$

Now let $b \to \infty$. We know from our calculation (1) of $E(X)$ that the integral in the second term of (3) approaches $1/k$; also, it can be shown that $-b^2 e^{-kb}$ approaches 0 (see Section 11.5, Exercise 44). Therefore,

$$\int_0^\infty x^2 k e^{-kx}\, dx = \frac{2}{k} \cdot \frac{1}{k} = \frac{2}{k^2}.$$

And by equation (2), we have

$$\text{Var}(X) = \frac{2}{k^2} - \frac{1}{k^2} = \frac{1}{k^2}.$$

Let us summarize our results:

Let $X$ be a random variable with an exponential density function $f(x) = ke^{-kx}$ $(x \geq 0)$. Then

$$E(X) = \frac{1}{k} \quad \text{and} \quad \text{Var}(X) = \frac{1}{k^2}.$$

---

**EXAMPLE 1**  Suppose that the number of days of continuous use provided by a certain brand of light bulb is an exponential random variable $X$ with expected value 100 days.

(a) Find the density function of $X$.

(b) Find the probability that a randomly chosen bulb will last between 80 and 90 days.

(c) Find the probability that a randomly chosen bulb will last for more than 40 days.

**Solution**  (a) Since $X$ is an exponential random variable, its density function must be of the form $f(x) = ke^{-kx}$ for some $k > 0$. Since the expected value of such a density function is $1/k$ and is equal to 100 in this case, we see that

$$\frac{1}{k} = 100,$$

$$k = \frac{1}{100} = .01.$$

Thus $f(x) = .01e^{-.01x}$.

(b) $\Pr(80 \leq X \leq 90) = \displaystyle\int_{80}^{90} .01e^{-.01x}\, dx = -e^{-.01x} \Big|_{80}^{90} = -e^{-.9} + e^{-.8} \approx .04276$

(c) $\Pr(X \geq 40) = \displaystyle\int_{40}^{\infty} .01e^{-.01x}\, dx = 1 - \int_{0}^{40} .01e^{-.01x}\, dx$

[since $\int_0^\infty f(x)\, dx = 1$], so that

$$\Pr(X \geq 40) = 1 + (e^{-.01x})\Big|_{0}^{40} = 1 + (e^{-.4} - 1) = e^{-.4} \approx .67032. \quad \blacksquare$$

---

**EXAMPLE 2**  During a certain part of the day, the interarrival time between successive phone calls at a central telephone exchange is an exponential random variable $X$ with expected value $\frac{1}{3}$ second.

(a) Find the density function of $X$.

(b) Find the probability that between $\frac{1}{3}$ and $\frac{2}{3}$ second elapse between consecutive phone calls.

(c) Find the probability that the time between successive phone calls is more than 2 seconds.

**Solution**  (a) Since $X$ is an exponential random variable, its density function is $f(x) = ke^{-kx}$ for some $k > 0$. Since the expected value of $X$ is $1/k = \frac{1}{3}$, we have $k = 3$ and $f(x) = 3e^{-3x}$.

(b) $\Pr\left(\frac{1}{3} \leq X \leq \frac{2}{3}\right) = \displaystyle\int_{1/3}^{2/3} 3e^{-3x}\, dx = -e^{-3x}\Big|_{1/3}^{2/3} = -e^{-2} + e^{-1} \approx .23254$

(c)  $\Pr(X \geq 2) = \int_2^\infty 3e^{-3x}\,dx = 1 - \int_0^2 3e^{-3x}\,dx = 1 + \left(e^{-3x}\right)\Big|_0^2 = e^{-6} \approx .00248$

In other words, about .25% of the time, the waiting time between consecutive calls is at least 2 seconds. ■

**Normal Density Functions**  Let $\mu$, $\sigma$ be given numbers, with $\sigma > 0$. Then the function

$$f(x) = \frac{1}{\sigma\sqrt{2\pi}}\,e^{-(1/2)[(x-\mu)/\sigma]^2} \tag{4}$$

is called a *normal density function*. A random variable $X$ whose density function has this form is called a *normal random variable*, and the values of $X$ are said to be *normally distributed*. Many random variables in applications are approximately normal. For example, errors that occur in physical measurements and various manufacturing processes, as well as many human physical and mental characteristics, are all conveniently modeled by normal random variables.

The graph of the density function in (4) is called a *normal curve* (Fig. 2). A normal curve is symmetric about the line $x = \mu$ and has inflection points at $\mu - \sigma$ and $\mu + \sigma$. Figure 3 shows three normal curves corresponding to different values of $\sigma$. The parameters $\mu$ and $\sigma$ determine the shape of the curve. The value of $\mu$ determines the point where the curve reaches its maximum height, and the value of $\sigma$ determines how sharp a peak the curve has.

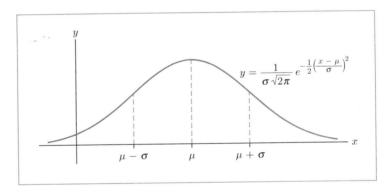

**Figure 2.**  A normal density function.

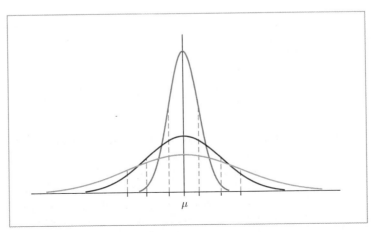

**Figure 3.**  Several normal curves.

It can be shown that the constant $1/(\sigma\sqrt{2\pi})$ in the definition (4) of a normal density function $f(x)$ is needed in order to make the area under the normal curve equal to 1 [i.e., to make $f(x)$ a probability density function]. The theoretical values of a normal random variable $X$ include all positive and negative numbers, but the normal curve approaches the horizontal axis so rapidly beyond the inflection points that the probabilities associated with intervals on the $x$-axis far to the left or right of $x = \mu$ are negligible.

Using techniques outside the scope of this book, one can verify the following basic facts about a normal random variable.

Let $X$ be a random variable with the normal density function

$$f(x) = \frac{1}{\sigma\sqrt{2\pi}}\, e^{-(1/2)[(x-\mu)/\sigma]^2}.$$

Then the expected value (mean), variance, and standard deviation of $X$ are given by

$$\mathrm{E}(X) = \mu, \quad \mathrm{Var}(X) = \sigma^2, \quad \text{and} \quad \sqrt{\mathrm{Var}(X)} = \sigma.$$

A normal random variable with expected value $\mu = 0$ and standard deviation $\sigma = 1$ is called a *standard normal random variable* and is often denoted by the letter $Z$. Using these values for $\mu$ and $\sigma$ in (4), and writing $z$ in place of the variable $x$, we see that the density function for $Z$ is

$$f(z) = \frac{1}{\sqrt{2\pi}}\, e^{-(1/2)z^2}.$$

The graph of this function is called the *standard normal curve*. (See Fig. 4.)

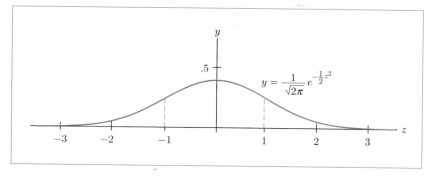

**Figure 4.** The standard normal curve.

Probabilities involving a standard normal random variable $Z$ may be written in the form

$$\Pr(a \le Z \le b) = \int_a^b \frac{1}{\sqrt{2\pi}}\, e^{-(1/2)z^2}\, dz.$$

Such an integral cannot be evaluated directly because the density function for $Z$ cannot be antidifferentiated in terms of elementary functions. However, tables of such probabilities have been compiled, using numerical approximations to the definite integrals. For $z \ge 0$, let $A(z) = \Pr(0 \le Z \le z)$ and $A(-z) = \Pr(-z \le Z \le 0)$. That is, let $A(z)$ and $A(-z)$ be the areas of the regions shown in Fig. 5. From the symmetry of the standard normal curve, it is clear that $A(-z) = A(z)$. Values of $A(z)$ for $z \ge 0$ are listed in Table 1 of Appendix E.

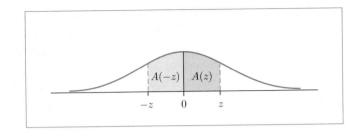

**Figure 5.** Areas under the standard normal curve.

**EXAMPLE 3** Let $Z$ be a standard normal random variable. Use Table 1 to compute the following probabilities:

(a) $\Pr(0 \leq Z \leq 1.84)$      (b) $\Pr(-1.65 \leq Z \leq 0)$

(c) $\Pr(.7 \leq Z)$      (d) $\Pr(.5 \leq Z \leq 2)$      (e) $\Pr(-.75 \leq Z \leq 1.46)$

**Solution** (a) $\Pr(0 \leq Z \leq 1.84) = A(1.84)$. In Table 1 we move down the column under "$z$" until we reach 1.8; then we move right in the same row to the column with the heading ".04." There we find that $A(1.84) = .4671$.

(b) $\Pr(-1.65 \leq Z \leq 0) = A(-1.65) = A(1.65) = .4505$ (from Table 1).

(c) Since the area under the normal curve is 1, the symmetry of the curve implies that the area to the right of the $y$-axis is .5. Now $\Pr(.7 \leq Z)$ is the area under the curve to the right of .7, and so we can find this area by subtracting from .5 the area between 0 and .7. [See Fig. 6(a).] Thus

$$\Pr(.7 \leq Z) = .5 - \Pr(0 \leq Z \leq .7)$$
$$= .5 - A(.7) = .5 - .2580 \quad \text{(from Table 1)}$$
$$= .2420.$$

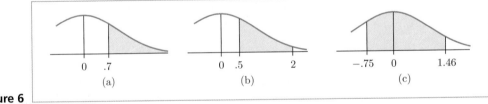

$$\begin{array}{ccc} 0 \quad .7 & 0 \quad .5 \quad 2 & -.75 \quad 0 \quad 1.46 \\ (a) & (b) & (c) \end{array}$$

**Figure 6**

(d) The area under the standard normal curve from .5 to 2 equals the area from 0 to 2 minus the area from 0 to .5. [See Fig. 6(b).] Thus we have

$$\Pr(.5 \leq Z \leq 2) = A(2) - A(.5)$$
$$= .4772 - .1915 = .2857.$$

(e) The area under the standard normal curve from $-.75$ to 1.46 equals the area from $-.75$ to 0 plus the area from 0 to 1.46. [See Fig. 6(c).] Thus

$$\Pr(-.75 \leq Z \leq 1.46) = A(-.75) + A(1.46)$$
$$= A(.75) + A(1.46)$$
$$= .2734 + .4279 = .7013. \quad \blacksquare$$

When $X$ is an *arbitrary* normal random variable, with mean $\mu$ and standard deviation $\sigma$, one may compute a probability such as $\Pr(a \leq X \leq b)$ by making the change of variable $z = (x - \mu)/\sigma$. This converts the integral for $\Pr(a \leq X \leq b)$ into an integral involving the standard normal density function. The following example illustrates this procedure.

**EXAMPLE 4**  A metal flange on a truck must be between 92.1 and 94 millimeters long in order to fit properly. Suppose that the lengths of the flanges supplied to the truck manufacturer are normally distributed with mean $\mu = 93$ millimeters and standard deviation $\sigma = .4$ millimeter.

(a) What percentage of the flanges have an acceptable length?

(b) What percentage of the flanges are too long?

**Solution**  Let $X$ be the length of a metal flange selected at random from the supply of flanges.

(a) We have

$$\Pr(92.1 \leq X \leq 94) = \int_{92.1}^{94} \frac{1}{(.4)\sqrt{2\pi}} e^{-(1/2)[(x-93)/.4]^2} \, dx.$$

Using the substitution $z = (x-93)/.4$, $dz = (1/.4)\,dx$, we note that if $x = 92.1$, then $z = (92.1-93)/.4 = -.9/.4 = -2.25$, and if $x = 94$, then $z = (94-93)/.4 = 1/.4 = 2.5$. Hence

$$\Pr(92.1 \leq X \leq 94) = \int_{-2.25}^{2.5} \frac{1}{\sqrt{2\pi}} e^{-(1/2)z^2} \, dz.$$

The value of this integral is the area under the standard normal curve from $-2.25$ to $2.5$, which equals the area from $-2.25$ to $0$ plus the area from $0$ to $2.5$. Thus

$$\Pr(92.1 \leq X \leq 94) = A(-2.25) + A(2.5)$$
$$= A(2.25) + A(2.5)$$
$$= .4878 + .4938 = .9816.$$

From this probability we conclude that about 98% of the flanges will have an acceptable length.

(b)
$$\Pr(94 \leq X) = \int_{94}^{\infty} \frac{1}{(.4)\sqrt{2\pi}} e^{-(1/2)[(x-93)/.4]^2} \, dx$$

This integral is approximated by an integral from $x = 94$ to $x = b$, where $b$ is large. If we substitute $z = (x - 93)/.4$, we find that

$$\int_{94}^{b} \frac{1}{(.4)\sqrt{2\pi}} e^{-(1/2)[(x-93)/.4]^2} \, dx = \int_{2.5}^{(b-93)/.4} \frac{1}{\sqrt{2\pi}} e^{-(1/2)z^2} \, dz. \qquad (5)$$

Now when $b \to \infty$, the quantity $(b-93)/.4$ also becomes arbitrarily large. Since the left integral in (5) approaches $\Pr(94 \leq X)$, we conclude that

$$\Pr(94 \leq X) = \int_{2.5}^{\infty} \frac{1}{\sqrt{2\pi}} e^{-(1/2)z^2} \, dz.$$

To calculate this integral we use the method of Example 3(c). The area under the standard normal curve to the right of $2.5$ equals the area to the right of $0$ minus the area from $0$ to $2.5$. That is,

$$\Pr(94 \leq X) = .5 - A(2.5)$$
$$= .5 - .4938 = .0062.$$

Approximately .6% of the flanges exceed the maximum acceptable length.  ■

**Figure 7.** Probabilities from Example 4.

**Incorporating Technology** The TI-83 graphing calculator can easily calculate normal probabilities with the **normalcdf** function from the DISTR menu. The area under the normal curve with mean $\mu$ and standard deviation $\sigma$ from $x = a$ to $x = b$ is given by **normalcdf(a,b,$\mu$,$\sigma$)**. The normal probabilities from Example 4 are calculated in Fig. 7. (*Note:* E99, which equals $10^{99}$, is generated by pressing **1** 2nd EE **99**, and is used in place of infinity. Minus infinity is represented by −E99.)

With other calculators, areas under a normal curve can be computed by evaluating a definite integral. The normal curve function for Example 4 is defined in Fig. 8, and the probabilities from Example 4 are computed in Fig. 9. (*Note:* In Fig. 9, the number 97 is ten standard deviations to the right of the mean. The function safely can be considered to have value 0 from that point on.)

**Figure 8.** Normal curve function from Example 4.

**Figure 9.** Probabilities from Example 4.

## PRACTICE PROBLEMS 12.4

1. The emergency flasher on an automobile is warranted for the first 12,000 miles that the car is driven. During that period a defective flasher will be replaced free. Suppose that the time before failure of the emergency flasher (measured in thousands of miles) is an exponential random variable $X$ with mean 50 (thousand miles). What percentage of the flashers will have to be replaced during the warranty period?

2. Suppose that the lead time between ordering furniture from a certain company and receiving delivery is a normal random variable with $\mu = 18$ weeks and $\sigma = 5$ weeks. Find the likelihood that a customer will have to wait more than 16 weeks.

## EXERCISES 12.4

*Find (by inspection) the expected values and variances of the exponential random variables with the density functions given in Exercises 1–4.*

1. $3e^{-3x}$
2. $\frac{1}{4}e^{-x/4}$
3. $.2e^{-.2x}$
4. $1.5e^{-1.5x}$

*Suppose that in a large factory there is an average of two accidents per day and the time between accidents has an exponential density function with expected value of $\frac{1}{2}$ day.*

5. Find the probability that the time between two accidents will be more than $\frac{1}{2}$ day and less than 1 day.

6. Find the probability that the time between accidents will be less than 8 hours (i.e., $\frac{1}{3}$ day).

*Suppose that the amount of time required to serve a customer at a bank has an exponential density function with mean 3 minutes.*

7. Find the probability that a customer is served in less than 2 minutes.

8. Find the probability that a customer will require more than 5 minutes.

*During a certain part of the day, the time between arrivals of automobiles at the tollgate on a turnpike is an exponential random variable with expected value 20 seconds.*

9. Find the probability that the time between successive arrivals is more than 60 seconds.

10. Find the probability that the time between successive arrivals is greater than 10 seconds and less than 30 seconds.

*Upon studying the vacancies occurring in the U.S. Supreme Court, it has been determined that the time elapsed between successive resignations is an exponential random variable with expected value 2 years.*[*]

**11.** A new president takes office at the same time a justice retires. Find the probability that the next vacancy on the court will take place during the president's 4-year term.

**12.** Find the probability that the composition of the U.S. Supreme Court will remain unchanged for a period of 5 years or more.

**13.** Suppose that the average life span of an electronic component is 72 months and that the life spans are exponentially distributed.

  **(a)** Find the probability that a component lasts for more than 24 months.

  **(b)** The *reliability function* $r(t)$ gives the probability that a component will last for more than $t$ months. Compute $r(t)$ in this case.

**14.** Consider a group of patients that have been treated for an acute disease such as cancer, and let $X$ be the number of years a person lives after receiving the treatment (the "survival time"). Under suitable conditions, the density function for $X$ will be $f(x) = ke^{-kx}$ for some constant $k$.

  **(a)** The *survival function* $S(x)$ is the probability that a person chosen at random from the group of patients survives until at least time $x$. Explain why $S(x) = 1 - F(x)$, where $F(x)$ is the cumulative distribution function for $X$, and compute $S(x)$.

  **(b)** Suppose that the probability is .90 that a patient will survive at least 5 years [i.e., $S(5) = .90$]. Find the constant $k$ in the exponential density function $f(x)$.

*Find the expected values and the standard deviations (by inspection) of the normal random variables with the density functions given in Exercises 15–18.*

**15.** $\dfrac{1}{\sqrt{2\pi}} e^{-(1/2)(x-4)^2}$

**16.** $\dfrac{1}{\sqrt{2\pi}} e^{-(1/2)(x+5)^2}$

**17.** $\dfrac{1}{3\sqrt{2\pi}} e^{-(1/18)x^2}$

**18.** $\dfrac{1}{5\sqrt{2\pi}} e^{-(1/2)[(x-3)/5]^2}$

**19.** Show that the function $f(x) = e^{-x^2/2}$ has a relative maximum at $x = 0$.

**20.** Show that the function $f(x) = e^{-(1/2)[(x-\mu)/\sigma]^2}$ has a relative maximum at $x = \mu$.

**21.** Show that the function $f(x) = e^{-x^2/2}$ has inflection points at $x = \pm 1$.

**22.** Show that the function $f(x) = e^{-(1/2)[(x-\mu)/\sigma]^2}$ has inflection points at $x = \mu \pm \sigma$.

**23.** Let $Z$ be a standard normal random variable. Calculate:

  **(a)** $\Pr(-1.3 \le Z \le 0)$    **(b)** $\Pr(.25 \le Z)$

  **(c)** $\Pr(-1 \le Z \le 2.5)$    **(d)** $\Pr(Z \le 2)$

**24.** Calculate the area under the standard normal curve for values of $z$

  **(a)** between .5 and 1.5,

  **(b)** between $-.75$ and .75

  **(c)** to the left of $-.3$

  **(d)** to the right of $-1$.

**25.** The gestation period (length of pregnancy) of pregnant females of a certain species is approximately normally distributed with a mean of 6 months and standard deviation of $\frac{1}{2}$ month.

  **(a)** Find the percentage of births that occur after a gestation period of between 6 and 7 months.

  **(b)** Find the percentage of births that occur after a gestation period of between 5 and 6 months.

**26.** Suppose that the life span of a certain automobile tire is normally distributed with $\mu = 25{,}000$ miles and $\sigma = 2000$ miles.

  **(a)** Find the probability that a tire will last between 28,000 and 30,000 miles.

  **(b)** Find the probability that a tire will last more than 29,000 miles.

**27.** Suppose that the amount of milk in a gallon container is a normal random variable with $\mu = 128.2$ ounces and $\sigma = .2$ ounce. Find the probability that a random container of milk contains less than 128 ounces.

**28.** The amount of weight required to break a certain brand of twine has a normal density function with $\mu = 43$ kilograms and $\sigma = 1.5$ kilograms. Find the probability that the breaking weight of a piece of the twine is less than 40 kilograms.

**29.** A student with an eight o'clock class at the University of Maryland commutes to school by car. She has discovered that along each of two possible routes her traveling time to school (including the time to get to class) is approximately a normal random variable. If she uses the Capital Beltway for most of her trip, $\mu = 25$ minutes and $\sigma = 5$ minutes. If she drives a longer route over local city streets, then $\mu = 28$ minutes and $\sigma = 3$ minutes. Which route should the student take if she leaves home at 7:30 A.M.? (Assume that the best route is one that minimizes the probability of being late to class.)

**30.** Which route should the student in Exercise 29 take if she leaves home at 7:26 A.M.?

---

[*]See W. A. Wallis, "The Poisson Distribution and the Supreme Court," *Journal of the American Statistical Association*, 31 (1936), 376–380.

**31.** A certain type of bolt must fit through a 20-millimeter test hole, or else it is discarded. If the diameters of the bolts are normally distributed with $\mu = 18.2$ millimeters and $\sigma = .8$ millimeters, what percentage of the bolts will be discarded?

**32.** The Math SAT scores of a recent freshman class at a university were normally distributed with $\mu = 535$ and $\sigma = 100$.

   **(a)** What percentage of the scores were between 500 and 600?

   **(b)** Find the minimum score needed to be in the top 10% of the class.

**33.** Let $X$ be the time to failure (in years) of a transistor, and suppose the transistor has been operating properly for $a$ years. Then it can be shown that the probability that the transistor will fail within the next $b$ years is

$$\frac{\Pr(a \le X \le a + b)}{\Pr(a \le X)}. \tag{6}$$

Compute this probability for the case when $X$ is an exponential random variable with density function $f(x) = ke^{-kx}$, and show that this probability equals $\Pr(0 \le X \le b)$. This means that the probability given by (6) does not depend on how long the transistor has already been operating. Exponential random variables are therefore said to be *memoryless*.

**34.** Recall that the *median* of an exponential density function is that number $M$ such that $\Pr(X \le M) = \frac{1}{2}$. Show that $M = (\ln 2)/k$. (We see that the median is less than the mean.)

**35.** Suppose that the lifetime (in weeks) of a certain brand of light bulb has an exponential density function and 80% of all light bulbs fail within the first 100 weeks. Find the average lifetime of a light bulb.

### Technology Exercises

**36.** The computations of the expected value and variance of an exponential random variable relied on the fact that

for any positive number $k$, $be^{-kb}$ and $b^2 e^{-kb}$ approach 0 as $b$ gets large. That is,

$$\lim_{x \to \infty} \frac{x}{e^{kx}} = 0 \quad \text{and} \quad \lim_{x \to \infty} \frac{x^2}{e^{kx}} = 0.$$

The validity of these limits for the case $k = 1$ is shown in Figs. 10 and 11. Convince yourself that these limits hold for all positive values of $k$ by generating graphs for $k = .1$, $k = .5$, and $k = 2$.

**Figure 10.** Graph of $Y_1 = \dfrac{x}{e^x}$.

**Figure 11.** Graph of $Y_1 = \dfrac{x^2}{e^x}$.

**37.** Use the integral routine to convince yourself that $\int_{-\infty}^{\infty} x^2 f(x)\, dx = 1$, where $f(x)$ is the standard normal density function. [*Note:* Since $f(x)$ approaches zero so rapidly as $x$ gets large in magnitude, the value of the improper integral is nearly the same as the definite integral of $x^2 f(x)$ from $x = -8$ to $x = 8$.] Conclude that the standard deviation of the standard normal random variable is 1.

---

**SOLUTIONS TO PRACTICE PROBLEMS 12.4**

**1.** The density function for $X$ is $f(x) = ke^{-kx}$, where $1/k = 50$ (thousand miles), and $k = 1/50 = .02$. Then

$$\Pr(X \le 12) = \int_0^{12} .02 e^{-.02x}\, dx = -e^{-.02x} \Big|_0^{12}$$

$$= 1 - e^{-.24} \approx .21337.$$

About 21% of the flashers will have to be replaced during the warranty period.

**2.** Let $X$ be the time between ordering and receiving the furniture. Since $\mu = 18$ and $\sigma = 5$, we have

$$\Pr(16 \le X) = \int_{16}^{\infty} \frac{1}{5\sqrt{2\pi}} e^{-(1/2)[(x-18)/5]^2} \, dx.$$

If we substitute $z = (x - 18)/5$, then $dz = \frac{1}{5}\,dx$, and $z = -.4$ when $x = 16$.

$$\Pr(16 \le X) = \int_{-.4}^{\infty} \frac{1}{\sqrt{2\pi}} e^{-(1/2)z^2} \, dz.$$

[A similar substitution was made in Example 4(b).] The preceding integral gives the area under the standard normal curve to the right of $-.4$. Since the area between $-.4$ and $0$ is $A(-.4) = A(.4)$, and the area to the right of $0$ is $.5$, we have

$$\Pr(16 \le X) = A(.4) + .5 = .1554 + .5 = .6554.$$

## 12.5  Poisson and Geometric Random Variables

Probability theory is widely applied in business, biology and the social sciences to situations that involve counting. The probability models in this section involve a random variable $X$ whose values are the discrete numbers $0, 1, 2, \dots$. Usually there is no specific upper limit to the value of $X$, even though extremely large values of $X$ are highly unlikely. Here are typical examples of such experiments. In each case, $X$ represents the outcome of the experiment.

1. At an insurance company, count the number of fire insurance claims submitted in any one month (selected at random).

2. In a microbiological study of a pond, count the number of protozoa in a drop-size random sample of water.

3. In a factory, count the number of times per month that a certain type of machine breaks down.

Suppose $X$ denotes a random variable for an experiment whose outcome is one of the values $0, 1, 2, \dots$, and for each possible value $n$, let $p_n$ be the associated probability of occurrence. That is,

$$p_0 = \Pr(X = 0)$$
$$p_1 = \Pr(X = 1)$$
$$\vdots$$
$$p_n = \Pr(X = n)$$
$$\vdots$$

Notice that since $p_0, p_1, p_2, \dots$ are probabilities, each lies between $0$ and $1$. Also, the sum of these probabilities must be $1$. (One of the outcomes $0, 1, 2, \dots$ always occurs.) That is,

$$p_0 + p_1 + p_2 + \cdots + p_n + \cdots = 1.$$

Unlike the situation in Section 12.1, this sum is an infinite series such as those studied in Sections 11.3 and 11.5.

In analogy with the case of experiments having a finite number of possible outcomes, we may define the *expected value* (or average value) of the random variable $X$ (or of the experiment whose outcome is $X$) to be the number $E(X)$ given by the following formula:

$$E(X) = 0 \cdot p_0 + 1 \cdot p_1 + 2 \cdot p_2 + 3 \cdot p_3 + \cdots$$

(provided that the infinite series converges). That is, the expected value $E(X)$ is formed by adding the products of the possible outcomes by their respective probabilities of occurrence.

In a similar fashion, letting $m$ denote $E(X)$, we define the *variance* of $X$ by

$$\text{Var}(x) = (0 - m)^2 \cdot p_0 + (1 - m)^2 \cdot p_1 + (2 - m)^2 \cdot p_2 + (3 - m)^2 \cdot p_3 + \cdots .$$

**Poisson Random Variables**  In many experiments, the probabilities $p_n$ involve a parameter $\lambda$ (depending on the particular experiment), and they have the following special form:

$$p_0 = e^{-\lambda}$$

$$p_1 = \frac{\lambda}{1}e^{-\lambda}$$

$$p_2 = \frac{\lambda^2}{2 \cdot 1}e^{-\lambda}$$

$$p_3 = \frac{\lambda^3}{3 \cdot 2 \cdot 1}e^{-\lambda}$$

$$\vdots$$

$$p_n = \frac{\lambda^n}{n!}e^{-\lambda}. \tag{1}$$

The constant $e^{-\lambda}$ in each probability is necessary to make the sum of all the probabilities equal to 1. A random variable $X$ whose probabilities are given by formula (1) is called a *Poisson random variable*, and the probabilities for $X$ are said to form a *Poisson distribution* with parameter $\lambda$. The histograms in Fig. 1 show Poisson distributions for $\lambda = 1.5$, 3, and 5.

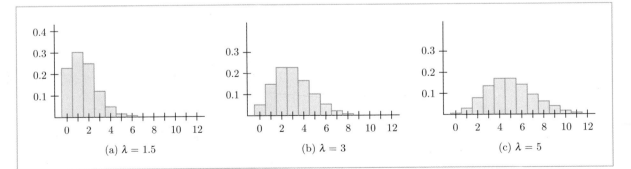

(a) $\lambda = 1.5$    (b) $\lambda = 3$    (c) $\lambda = 5$

**Figure 1**

**EXAMPLE 1**  The annual number of deaths in a certain town due to a particular disease has a Poisson distribution with parameter $\lambda = 3$. Verify that the associated probabilities sum to 1.

**Solution** The first few probabilities are (to four decimal places)

$$p_0 = e^{-3} \approx .0498$$

$$p_1 = \frac{3}{1}e^{-3} = 3 \cdot p_0 \approx .1494$$

$$p_2 = \frac{3^2}{2 \cdot 1}e^{-3} = \frac{3 \cdot 3}{2 \cdot 1}e^{-3} = \frac{3}{2}p_1 \approx .2240$$

$$p_3 = \frac{3^3}{3 \cdot 2 \cdot 1}e^{-3} = \frac{3 \cdot 3 \cdot 3}{3 \cdot 2 \cdot 1}e^{-3} = \frac{3}{3}p_2 \approx .2240$$

$$p_4 = \frac{3^4}{4 \cdot 3 \cdot 2 \cdot 1}e^{-3} = \frac{3 \cdot 3 \cdot 3 \cdot 3}{4 \cdot 3 \cdot 2 \cdot 1}e^{-3} = \frac{3}{4}p_3 \approx .1680$$

$$p_5 = \frac{3^5}{5 \cdot 4 \cdot 3 \cdot 2 \cdot 1}e^{-3} = \frac{3 \cdot 3 \cdot 3 \cdot 3 \cdot 3}{5 \cdot 4 \cdot 3 \cdot 2 \cdot 1}e^{-3} = \frac{3}{5}p_4 \approx .1008.$$

Notice how each probability $p_n$ for $n \geq 1$ is computed from the preceding probability $p_{n-1}$. In general, $p_n = (\lambda/n)p_{n-1}$.

To sum the probabilities for all $n$, use the exact values, not the decimal approximations:

$$e^{-3} + \frac{3}{1}e^{-3} + \frac{3^2}{2 \cdot 1}e^{-3} + \frac{3^3}{3 \cdot 2 \cdot 1}e^{-3} + \frac{3^4}{4 \cdot 3 \cdot 2 \cdot 1}e^{-3} + \cdots$$

$$= e^{-3}\left(1 + 3 + \frac{1}{2!}3^2 + \frac{1}{3!}3^3 + \frac{1}{4!}3^4 + \cdots\right).$$

From Section 11.5, you should recognize that the series inside the parentheses is the power series for $e^x$ evaluated at $x = 3$. The preceding sum is $e^{-3} \cdot e^3$, which equals 1. ■

The following facts about Poisson random variables provide an interpretation for the parameter $\lambda$.

Let $X$ be a random variable whose probabilities are Poisson distributed with parameter $\lambda$, that is,

$$p_0 = e^{-\lambda}$$

$$p_n = \frac{\lambda^n}{n!}e^{-\lambda} \qquad (n = 1, 2, \dots).$$

Then the expected value and variance of $X$ are given by

$$\mathrm{E}(X) = \lambda, \qquad \mathrm{Var}(X) = \lambda.$$

We shall verify only the statement about $E(X)$. The argument uses the Taylor series for $e^\lambda$. We have

$$E(X) = 0 \cdot p_0 + 1 \cdot p_1 + 2 \cdot p_2 + 3 \cdot p_3 + 4 \cdot p_4 + \cdots$$

$$= 0 \cdot e^{-\lambda} + 1 \cdot \frac{\lambda}{1} e^{-\lambda} + 2 \cdot \frac{\lambda^2}{1 \cdot 2} e^{-\lambda}$$

$$+ 3 \cdot \frac{\lambda^3}{1 \cdot 2 \cdot 3} e^{-\lambda} + 4 \cdot \frac{\lambda^4}{1 \cdot 2 \cdot 3 \cdot 4} e^{-\lambda} + \cdots$$

$$= \lambda e^{-\lambda} + \frac{\lambda^2}{1} e^{-\lambda} + \frac{\lambda^3}{1 \cdot 2} e^{-\lambda} + \frac{\lambda^4}{1 \cdot 2 \cdot 3} e^{-\lambda} + \cdots$$

$$= \lambda e^{-\lambda} \left( 1 + \frac{\lambda}{1} + \frac{\lambda^2}{1 \cdot 2} + \frac{\lambda^3}{1 \cdot 2 \cdot 3} + \cdots \right)$$

$$= \lambda e^{-\lambda} \cdot e^{\lambda}$$

$$= \lambda.$$

The next two examples illustrate some applications of Poisson random variables.

---

**EXAMPLE 2**  Suppose that we observe the number $X$ of calls received by a telephone switchboard during a 1-minute interval. Experience suggests that $X$ is Poisson distributed with $\lambda = 5$.

(a) Determine the probability that zero, one, or two calls arrive during a particular minute.

(b) Determine the probability that three or more calls arrive during a particular minute.

(c) Determine the average number of calls received per minute.

**Solution**  (a) The probability that zero, one, or two calls arrive during a given minute is $p_0 + p_1 + p_2$. Moreover,

$$p_0 = e^{-\lambda} = e^{-5} \approx .00674$$

$$p_1 = \frac{\lambda}{1} e^{-\lambda} = 5e^{-5} \approx .03369$$

$$p_2 = \frac{\lambda^2}{1 \cdot 2} e^{-\lambda} = \frac{5}{2} p_1 \approx .0842.$$

Thus $p_0 + p_1 + p_2 \approx .12465$. That is, during approximately 12% of the minutes, either zero, one, or two calls are received.

(b) The probability of receiving three or more calls is the same as the probability of *not* receiving zero, one, or two calls and so is equal to

$$1 - (p_0 + p_1 + p_2) = 1 - .12465 = .87535.$$

(c) The average number of calls received per minute is equal to $\lambda$. That is, on the average the switchboard receives five calls per minute.  ∎

---

**EXAMPLE 3**  Drop-sized water samples are drawn from a New England pond. The number of protozoa in many different samples are counted and the average number is found to be about 8.3. What is the probability that a sample chosen at random contains at most four protozoa?

**Solution** Under the assumption that the protozoa are thoroughly spread throughout the pond, without clumping, the number of protozoa per drop is a Poisson random variable, call it $X$. From the experimental data, we assume that $E(X) = 8.3$. Since $\lambda = E(X)$, the probabilities for $X$ are given by

$$p_n = \frac{8.3^n}{n!} e^{-8.3}.$$

The probability of "at most four" is $\Pr(X \leq 4)$. Using a calculator to generate the probabilities, we find that

$$
\begin{aligned}
\Pr(X \leq 4) &= p_0 + p_1 + p_2 + p_3 + p_4 \\
&\approx .00025 + .00206 + .00856 + .02368 + .04914 \\
&= .08369.
\end{aligned}
$$

The probability of at most four protozoa is about 8.4%. ∎

**Geometric Random Variables** The following two experiments give rise to discrete random variables whose values are $0, 1, \ldots$, but whose distributions are not Poisson.

- Toss a coin until a head appears and count the number of tails that precede it.

- As part of a quality control procedure, test the items coming off an assembly line. Count the number of acceptable items before the first defective item is found.

Each of these experiments involves a trial having two results (tail, head) or (acceptable, defective). In general, the two results are named *success* and *failure*, and the trial is repeated until a failure occurs. The outcome of the experiment is the number $X$ of successes $(0, 1, 2, \ldots)$ that precede the first failure. If for some number $p$ between 0 and 1 the probabilities for $X$ have the form

$$
\begin{aligned}
p_0 &= 1 - p \\
p_1 &= p(1 - p) \\
p_2 &= p^2(1 - p) \\
&\vdots \\
p_n &= p^n(1 - p) \quad\quad\quad\quad (2) \\
&\vdots
\end{aligned}
$$

then $X$ is called a *geometric random variable*, and the probabilities for $X$ are said to form a *geometric distribution* with parameter $p$. In this case, each trial of the experiment has the same probability $p$ of success. (The probability of failure is $1 - p$.) Also, the outcome of each trial is independent of the other trials. The histograms in Fig. 2 display geometric distributions for $p = .6$ and $.8$.

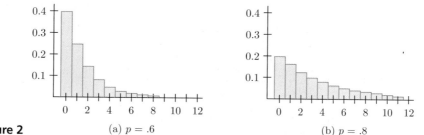

**Figure 2**  (a) $p = .6$  (b) $p = .8$

The term "geometric" is associated with formula (2) because the probabilities form a geometric series, with initial term $a = 1 - p$ and ratio $r = p$. The sum of the series is

$$p_0 + p_1 + p_2 + \cdots = \frac{a}{1 - r} = \frac{1 - p}{1 - p} = 1.$$

This is just what you would expect: The probabilities add up to 1.

---

**EXAMPLE 4**  Suppose an assembly line produces a small mechanical toy, and about 2% of the toys are defective. A quality control person selects a toy at random, inspects it, and then repeats this process until a defective toy is found. The probability of success (passing the test) is .98 for each inspection (trial).

(a) Find the probability that exactly three toys pass the test before a defective toy is found.

(b) Find the probability that at most three toys pass the test before a defective toy is found.

(c) What is the probability that at least four toys will pass the test before the first defective toy is found?

**Solution**  Let $X$ be the number of acceptable toys found before the first defective toy. A reasonable assumption is that the random variable $X$ has a geometric distribution with parameter $p = .98$.

(a) The probability that $X = 3$ is $p_3 = (.98)^3(1 - .98) \approx .0188$.

(b) $\Pr(X \leq 3) = p_0 + p_1 + p_2 + p_3$

$$= .02 + (.98)(.02) + (.98)^2(.02) + (.98)^3(.02)$$

$$\approx .02 + .0196 + .0192 + .0188 = .0776$$

(c) $\Pr(X \geq 4) = 1 - \Pr(X \leq 3) \approx 1 - .0776 = .9224$  ∎

The following properties of a geometric distribution can be established with facts about power series.

Let $X$ be a geometric random variable with parameter $p$, that is,

$$p_n = p^n(1 - p) \qquad (n = 0, 1, \dots).$$

Then the expected value and variance of $X$ are given by

$$E(X) = \frac{p}{1 - p}, \qquad \text{Var}(X) = \frac{p}{(1 - p)^2}.$$

---

**EXAMPLE 5**  For the situation described in Example 4, what is the average number of toys that pass inspection before the first defective toy is found?

**Solution**  Since $p = .98$, $E(X) = .98/(1 - .98) = 49$. If many inspections are made, then one can expect on average that 49 toys will pass inspection before a defective toy is found.  ∎

**Incorporating Technology**  The TI-83 graphing calculator has two functions that calculate Poisson probabilities. The value of **poissonpdf($\lambda$,n)** is $p_n =$

$\frac{\lambda^n}{n!}e^{-\lambda}$, and the value of **poissoncdf($\lambda$,n)** is $p_0 + p_1 + p_2 + \cdots + p_n$. (These two functions are invoked from the DISTR menu.) In Fig. 3 the functions are used to calculate two probabilities from Example 2(a). With other graphing calculators, you can find the sum of successive Poisson probabilities with **sum(seq** as shown in the last computation of Fig. 3. In Fig. 4 **sum(seq** is used to compute the geometric probabilities from parts (b) and (c) of Example 4.

**Figure 3.** Two probabilities from Example 2.

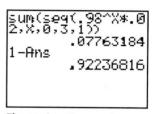

**Figure 4.** Two probabilities from Example 4.

## PRACTICE PROBLEM 12.5

1. A Public Health Officer is tracking down the source of a bacterial infection in a certain city. She analyzes the reported incidence of the infection in each city block and finds an average of three cases per block. A certain block is found to have seven cases. What is the probability that a randomly chosen block has at least seven cases, assuming that the number of cases per block is Poisson distributed?

# EXERCISES 12.5

1. Suppose a random variable $X$ has a Poisson distribution with $\lambda = 3$, as in Example 1. Compute the probabilities $p_6$, $p_7$, $p_8$.

2. Let $X$ be a Poisson random variable with parameter $\lambda = 5$. Compute the probabilities $p_0, \ldots, p_6$ to four decimal places.

3. Repeat Exercise 2 with $\lambda = .75$, and make a histogram.

4. Repeat Exercise 2 with $\lambda = 2.5$, and make a histogram.

5. The monthly number of fire insurance claims filed with the Firebug Insurance Company is Poisson distributed with $\lambda = 10$.

   (a) What is the probability that in a given month no claims are filed?

   (b) What is the probability that in a given month no more than two claims are filed? (The number of claims is either zero, one, or two.)

   (c) What is the probability that in a given month at least three claims are filed?

6. On a typical weekend evening at a local hospital, the number of persons waiting for treatment in the emergency room is Poisson distributed with $\lambda = 6.5$.

   (a) What is the likelihood that either no one or only one person is waiting for treatment?

   (b) What is the likelihood that no more than four persons are waiting?

   (c) What is the likelihood that at least five persons are waiting?

7. The number of typographical errors per page of a certain newspaper has a Poisson distribution, and there is an average of 1.5 errors per page.

   (a) What is the probability that a randomly selected page is error free?

   (b) What is the probability that a page has either two or three errors?

   (c) What is the probability that a page has at least four errors?

8. During a certain part of the day, an average of five automobiles arrives every minute at the tollgate on a turnpike. Let $X$ be the number of automobiles that arrive in any one-minute interval selected at random. Let $Y$ be the interarrival time between any two successive arrivals. (The average interarrival time is $\frac{1}{5}$ minute.) Assume that $X$ is a Poisson random variable and that $Y$ is an exponential random variable.

   (a) Find the probability that at least five cars arrive during a given one-minute interval.

**(b)** Find the probability that the time between any two successive cars is less than one-fifth of a minute.

9. A bakery makes gourmet cookies. For a batch of 4800 oatmeal and raisin cookies, how many raisins should be used so that the probability of a cookie having no raisins is .01? *Note:* A reasonable assumption is that the number of raisins in a random cookie has a Poisson distribution.

10. Suppose $X$ is a geometric random variable with parameter $p = .9$. Compute the probabilities $p_0, \ldots, p_5$ and make a histogram.

11. Repeat Exercise 10 with $p = .6$.

12. The quality control department at a sewing machine factory has determined that 1 out of 40 machines does not pass inspection. Let $X$ be the number of machines on an assembly line that pass inspection before a machine is found that fails inspection.

   **(a)** Write the formula for $\Pr(X = n)$.

   **(b)** What is the probability that of five machines coming off the assembly line, the first four pass inspection and the fifth does not?

13. In a certain town, there are two competing taxicab companies, Red Cab and Blue Cab. The taxis mix with downtown traffic in a random manner. There are three times as many Red taxis as there are Blue taxis. Suppose you stand on a downtown street corner and count the number $X$ of Red taxis that appear before the first Blue taxi appears.

   **(a)** Determine the formula for $\Pr(X = n)$.

   **(b)** What is the likelihood of observing at least three Red taxis before the first Blue taxi?

   **(c)** What is the average number of consecutive Red taxis prior to the appearance of a Blue taxi?

14. At a certain junior high school, two-thirds of the students have at least one tooth cavity. Suppose a dental survey is made of the students. What is the probability that the first student to have a cavity is the third student examined?

15. Let $X$ be a geometric random variable with parameter $p < 1$. Find a formula for $\Pr(X < n)$, for $n > 0$. *Note:* The partial sum of a geometric series with ratio $r$ is given by

$$1 + r + \cdots + r^{n-1} = \frac{1 - r^n}{1 - r}.$$

16. Whenever a document is fed into a high-speed copy machine, there is a .5% chance that a paper jam will stop the machine.

   **(a)** What is the expected number of documents that can be copied before a paper jam occurs?

   **(b)** Determine the probability that at least one hundred documents can be copied before a paper jam occurs. [*Hint:* See Exercise 15.]

17. Suppose that a large number of persons become infected by a particular strain of staphylococcus that is present in food served by a fast-food restaurant, and that the germ usually produces a certain symptom in 5% of the persons infected. What is the probability that when customers are examined, the first person to have the symptom is the fifth customer examined?

18. Suppose that you toss a fair coin until a head appears and count the number $X$ of consecutive tails that precede it.

   **(a)** Determine the probability that exactly $n$ consecutive tails occur.

   **(b)** Determine the average number of consecutive tails that occur.

   **(c)** Write down the infinite series that gives the variance for the number of consecutive tails. Use Exercise 28 in Section 11.5 to show that the variance equals 2.

*Exercises 19 and 20 illustrate a technique from statistics (called the* method of maximum likelihood*) that estimates a parameter for a probability distribution.*

19. In a production process a box of fuses is examined and found to contain two defective fuses. Suppose that the probability of having two defective fuses in a box selected at random is $(\lambda^2/2)e^{-\lambda}$ for some $\lambda$. Take first and second derivatives to determine the value of $\lambda$ for which the probability has its maximum value.

20. A person shooting at a target has five successive hits and then a miss. If $x$ is the probability of success on each shot, then the probability of having five successive hits followed by a miss is $x^5(1 - x)$. Take first and second derivatives to determine the value of $x$ for which the probability has its maximum value.

21. Let $X$ be a geometric random variable with parameter $p$. Derive the formula for $E(X)$ by using the power series formula (see Example 3 in Section 11.5):

$$1 + 2x + 3x^2 + \cdots = \frac{1}{(1 - x)^2} \quad \text{for } |x| < 1$$

22. Let $X$ be a Poisson random variable with parameter $\lambda$. Use Exercise 23 in Section 11.5 to show that the probability that $X$ is an even integer (including 0) is $e^{-\lambda} \cosh \lambda$.

### Technology Exercises

23. The number of times a printing press breaks down each month is Poisson distributed with $\lambda = 4$. What is the probability that the printing press breaks down between 2 and 8 times during a particular month?

24. The number of people arriving during a five-minute interval at supermarket checkout counter is Poisson distributed with $\lambda = 8$.

**(a)** What is the probability that exactly eight people arrive during a particular five-minute period?

**(b)** What is the probability that at most eight people arrive during a particular five-minute period?

**25.** The number of babies born each day in a certain hospital is Poisson distributed with $\lambda = 6.9$.

**(a)** During a particular day, are seven babies more likely to be born than six babies?

**(b)** What is the probability that at most 15 babies will be born during a particular day?

**26.** The number of accidents occurring each month at a certain intersection is Poisson distributed with $\lambda = 4.8$.

**(a)** During a particular month, are five accidents more likely to occur than four accidents?

**(b)** What is the probability that more than eight accidents will occur during a particular month?

---

## SOLUTION TO PRACTICE PROBLEM 12.5

**1.** The number of cases per block is Poisson distributed with $\lambda = 3$. So the probability of having at least seven cases in a given block is

$$p_7 + p_8 + p_9 + \ldots$$
$$= 1 - (p_0 + p_1 + p_2 + p_3 + p_4 + p_5 + p_6).$$

However,

$$p_n = \frac{3^n}{1 \cdot 2 \cdots n} e^{-3}$$

so that

$$p_0 = .04979, \quad p_1 = .14936, \quad p_2 = .22404$$
$$p_3 = .22404, \quad p_4 = .16803, \quad p_5 = .10082,$$
$$p_6 = .05041.$$

Therefore, the probability of at least seven cases in a given block is $1 - (.04979 + .14936 + .22404 + .22404 + .16803 + .10082 + .05041) = .03351$. See Figure 5.

**Figure 5**

---

# REVIEW OF FUNDAMENTAL CONCEPTS

**1.** What is a probability table?

**2.** What is a discrete random variable?

**3.** Make a small probability table for a discrete random variable $X$ and use it to define $\mathrm{E}(X)$, $\mathrm{Var}(X)$, and the standard deviation of $X$.

**4.** Explain how to create a probability density histogram.

**5.** What is the difference between a discrete random variable and a continuous random variable?

**6.** What are the two properties of a probability density function?

**7.** How is a probability density function used to calculate probabilities?

**8.** What is a cumulative distribution function and how is it related to the corresponding probability density function?

**9.** How is the expected value of a continuous random variable computed?

**10.** Give two ways to compute the variance of a continuous random variable.

**11.** What is an exponential density function? Give an example.

**12.** What is the expected value of an exponential random variable?

**13.** What is the density function for a normal random variable with mean $\mu$ and standard deviation $\sigma$?

**14.** What is a standard normal random variable? Write the density function.

**15.** How is an integral involving a normal density function converted to an integral involving a standard normal density function?

**16.** What is $\Pr(X = n)$ for a Poisson random variable with parameter $\lambda$? What is $\mathrm{E}(X)$ in this case?

**17.** What is $\Pr(X = n)$ for a geometric random variable with parameter $p$ (the probability of success)? What is $\mathrm{E}(X)$ in this case?

# SUPPLEMENTARY EXERCISES

1. Let $X$ be a continuous random variable on $0 \leq x \leq 2$, with the density function $f(x) = \frac{3}{8}x^2$.
   (a) Calculate $\Pr(X \leq 1)$ and $\Pr(1 \leq X \leq 1.5)$.
   (b) Find $E(X)$ and $\operatorname{Var}(X)$.

2. Let $X$ be a continuous random variable on $3 \leq x \leq 4$, with the density function $f(x) = 2(x - 3)$.
   (a) Calculate $\Pr(3.2 \leq X)$ and $\Pr(3 \leq X)$.
   (b) Find $E(X)$ and $\operatorname{Var}(X)$.

3. Verify that for any number $A$, $f(x) = e^{A-x}$, $x \geq A$, is a density function. Compute the associated cumulative distribution for $X$.

4. Verify that for any positive constants $k$ and $A$, the function $f(x) = kA^k/x^{k+1}$, $x \geq A$, is a density function. The associated cumulative distribution function $F(x)$ is called a *Pareto distribution*. Compute $F(x)$.

5. For any positive integer $n$, the function $f_n(x) = c_n x^{(n-2)/2} e^{-x/2}$, $x \geq 0$, where $c_n$ is an appropriate constant, is called the *chi-square density function* with $n$ degrees of freedom. Find $c_2$ and $c_4$ such that $f_2(x)$ and $f_4(x)$ are probability density functions.

6. Verify that for any positive number $k$, $f(x) = 1/(2k^3)x^2 e^{-x/k}$, $x \geq 0$, is a density function.

7. A medical laboratory tests many blood samples for a certain disease that occurs in about 5% of the samples. The lab collects samples from 10 persons and mixes together some blood from each sample. If a test on the mixture is positive, an additional 10 tests must be run, one on each individual sample. But if the test on the mixture is negative, no other tests are needed. It can be shown that the test of the mixture will be negative with probability $(.95)^{10} = .599$, because each of the 10 samples has a 95% chance of being free of the disease. If $X$ is the total number of tests required, then $X$ has the probability table shown in Table 1.
   (a) Find $E(X)$.
   (b) If the laboratory uses the procedure described on 200 blood samples (that is, 20 batches of 10 samples), about how many tests can it expect to run?

**TABLE 1** Blood Test Probabilities for Batches of 10 Samples

| | Test of Mixture | |
|---|---|---|
| | Negative | Positive |
| Total Tests | 1 | 11 |
| Probability | .599 | .401 |

8. Suppose that the laboratory in Exercise 7 uses batches of 5 instead of 10 samples. The probability of a negative test on the mixture of 5 samples is $(.95)^5 = .774$. Thus Table 2 gives the probabilities for the number $X$ of tests required.

**TABLE 2** Blood Test Probabilities for Batches of 5 Samples

| | Test of Mixture | |
|---|---|---|
| | Negative | Positive |
| Total Tests | 1 | 6 |
| Probability | .774 | .226 |

   (a) Find $E(X)$.
   (b) If the laboratory uses this procedure on 200 blood samples (that is, 40 batches of 5 samples), about how many tests can it expect to run?

9. A certain gas station sells $X$ thousand gallons of gas each week. Suppose that the cumulative distribution function for $X$ is $F(x) = 1 - \frac{1}{4}(2 - x)^2$, $0 \leq x \leq 2$.
   (a) If the tank contains 1.6 thousand gallons at the beginning of the week, find the probability that the gas station will have enough gas for its customers throughout the week.
   (b) How much gas must be in the tank at the beginning of the week in order to have a probability of .99 that there will be enough gasoline for the week?
   (c) Compute the density function for $X$.

10. A service contract on a computer costs \$100 per year. The contract covers all necessary maintenance and repairs on the computer. Suppose that the actual cost to the manufacturer for providing this service is a random variable $X$ (measured in hundreds of dollars) whose probability density function is $f(x) = (x - 5)^4/625$, $0 \leq x \leq 5$. Compute $E(X)$ and determine how much money the manufacturer expects to make on each service contract on the average.

11. A random variable $X$ has a uniform density function $f(x) = \frac{1}{5}$ on $20 \leq x \leq 25$.
    (a) Find $E(X)$ and $\operatorname{Var}(X)$.
    (b) Find $b$ such that $\Pr(X \leq b) = .3$.

12. A random variable $X$ has a cumulative distribution function $F(x) = (x^2 - 9)/16$ on $3 \leq x \leq 5$.
    (a) Find the density function for $X$.
    (b) Find $a$ such that $\Pr(a \leq X) = \frac{1}{4}$.

13. The annual incomes of the households in a certain community range between 5 and 25 thousand dollars. Let $X$ represent the annual income (in thousands of dollars) of a household chosen at random in this community, and suppose that the probability density function for $X$ is $f(x) = kx$, $5 \leq x \leq 25$.
    (a) Find the value of $k$ that makes $f(x)$ a density function.

**(b)** Find the fraction of households whose income exceeds $20,000.

**(c)** Find the mean annual income of the households in the community.

**14.** The density function $f(x)$ for the lifetime of a certain battery is shown in Fig. 1. Each battery lasts between 3 and 10 hours.

**(a)** Sketch the graph of the corresponding cumulative distribution function $F(x)$.

**(b)** What is the meaning of the number $F(7) - F(5)$?

**(c)** Formulate the number in (b) in terms of $f(x)$.

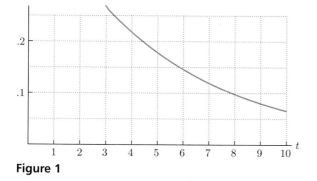

**Figure 1**

**15.** A point is selected at random from the rectangle of Fig. 2, call its coordinates $(\theta, y)$. Find the probability that $y \leq \sin \theta$.

**Figure 2**

**16.** (*Buffon Needle Problem*) A needle of length one unit is dropped on a floor that is ruled with parallel lines, one unit apart. [See Fig. 3.] Let $P$ be the lowest point of the needle, $y$ the distance of $P$ from the ruled line above it, and $\theta$ the angle the needle makes with a line parallel to the ruled lines. Show that the needle touches a ruled line if and only if $y \leq \sin \theta$. Conclude that the probability of the needle touching a ruled line is the probability found in Exercise 15.

**Figure 3**

**17.** The lifetime of a certain TV picture tube is an exponential random variable with an expected value of 5 years. Suppose that the tube manufacturer sells the tube for $100 but will give a complete refund if the tube burns out within 3 years. Then the revenue the manufacturer receives on each tube is a discrete random variable $Y$ with values 100 and 0. Determine the expected revenue per tube.

**18.** The condenser motor in an air conditioner costs $300 to replace, but a home air conditioning service will guarantee to replace it free when it burns out if you will pay an annual insurance premium of $25. Suppose that the life span of the motor is an exponential random variable with an expected life of 10 years. Should you take out the insurance for the first year? [*Hint:* Consider the random variable $Y$ such that $Y = 300$ if the motor burns out during the year and $Y = 0$ otherwise. Compare $E(Y)$ with the cost of 1 year's insurance.]

**19.** An exponential random variable $X$ has been used to model the relief times (in minutes) of arthritic patients who have taken an analgesic for their pain. Suppose that the density function for $X$ is $f(x) = ke^{-kx}$ and that a certain analgesic provides relief within 4 minutes for 75% of a large group of patients. Then one may estimate that $\Pr(X \leq 4) = .75$. Use this estimate to find an approximate value for $k$. [*Hint:* First show that $\Pr(X \leq 4) = 1 - e^{-4k}$.]

**20.** A piece of new equipment has a useful life of $X$ thousand hours, where $X$ is a random variable with the density function $f(x) = .01xe^{-x/10}$, $x \geq 0$. A manufacturer expects the machine to generate $5000 of additional income for every thousand hours of use, but the machine costs $60,000. Should the manufacturer purchase the new equipment? [*Hint:* Compute the expected value of the additional earnings generated by the machine.]

**21.** Extensive records are kept of the life spans (in months) of a certain product, and a relative frequency histogram is constructed from the data, using areas to represent relative frequencies (as in Fig. 4 in Section 12.1). It turns out that the upper boundary of the relative frequency histogram is approximated closely by the graph of the function

$$f(x) = \frac{1}{8\sqrt{2\pi}} e^{-(1/2)[(x-50)/8]^2}.$$

Determine the probability that the life span of such a product is between 30 and 50 months.

**22.** A certain machine part has a nominal length of 80 millimeters, with a tolerance of $\pm.05$ millimeter. Suppose that the actual length of the parts supplied is a normal random variable with mean 79.99 millimeters and standard deviation .02 millimeter. How many parts in a lot of 1000 should you expect to lie outside the tolerance limits?

**23.** The men hired by a certain city police department must be at least 69 inches tall. Suppose that the heights of adult men in the city are normally distributed with $\mu = 70$ inches and $\sigma = 2$ inches. What percentage of the men are tall enough to be eligible for recruitment by the police department?

**24.** Suppose that the police force in Exercise 23 maintains the same height requirements for women as men and that the heights of women in the city are normally distributed, with $\mu = 65$ inches and $\sigma = 1.6$ inches. What percentage of the women are eligible for recruitment?

**25.** Let $Z$ be a standard normal random variable. Find the number $a$ such that $\Pr(a \leq Z) = .40$.

**26.** Scores on a school's entrance exam are normally distributed, with $\mu = 500$ and $\sigma = 100$. If the school wishes to admit only the students in the top 40%, what should be the cutoff grade?

**27.** It is useful in some applications to know that about 68% of the area under the standard normal curve lies between $-1$ and $1$.

    **(a)** Verify this statement.

    **(b)** Let $X$ be a normal random variable with expected value $\mu$ and variance $\sigma^2$. Compute

$$\Pr(\mu - \sigma \leq X \leq \mu + \sigma).$$

**28. (a)** Show that about 95% of the area under the standard normal curve lies between $-2$ and $2$.

    **(b)** Let $X$ be a normal random variable with expected value $\mu$ and variance $\sigma^2$. Compute

$$\Pr(\mu - 2\sigma \leq X \leq \mu + 2\sigma).$$

**29.** The Chebychev inequality says that for any random variable $X$ with expected value $\mu$ and standard deviation $\sigma$,

$$\Pr(\mu - n\sigma \leq X \leq \mu + n\sigma) \geq 1 - \frac{1}{n^2}.$$

    **(a)** Take $n = 2$. Apply the Chebychev inequality to an exponential random variable.

    **(b)** By integrating, find the exact value of the probability in part (a).

**30.** Do the same as in Exercise 29 with a normal random variable.

*A small volume of blood is to be selected, examined under a microscope, and the number of white blood cells is counted. Suppose that for healthy people the number of white blood cells in such a specimen is Poisson distributed with $\lambda = 4$.*

**31.** What is the probability that a specimen from a healthy person has exactly four white blood cells?

**32.** What is the probability that a specimen from a healthy person has eight or more white blood cells?

**33.** What is the average number of white blood cells per specimen from a healthy person?

*A pair of dice is rolled until a "seven" or "eleven" appears, and the number of rolls preceding the final roll is observed. The probability of rolling "seven or eleven" is $\frac{2}{9}$.*

**34.** Determine the formula for $p_n$, the probability of exactly $n$ consecutive rolls preceding the final roll.

**35.** Determine the average number of consecutive rolls preceding the final roll.

**36.** What is the probability that at least three consecutive rolls precede the final roll?

# Appendices

## Appendix A
## Calculus and the TI-82 Calculator

### Functions

#### A. Define a function
- Press $\boxed{\text{Y=}}$ to obtain **Y=** edit screen.
- Cursor down to the function to be defined. (Press $\boxed{\text{CLEAR}}$ if the function is to be redefined.)
- Type in the expression for the function. (*Note*: Press $\boxed{\text{X,T,}\theta}$ to display X.)

#### B. Select or deselect a function
(Functions with highlighted equal signs are said to be *selected*. Pressing $\boxed{\text{GRAPH}}$ instructs the calculator to graph all selected functions and pressing $\boxed{\text{2nd}}$ [TABLE] creates a column of the table for each selected function.)
- Press $\boxed{\text{Y=}}$ to obtain the **Y=** edit screen.
- Cursor down to the function to be selected or deselected.
- Move the cursor to the equal sign and press $\boxed{\text{ENTER}}$ to toggle the state of the function on and off.

#### C. Display a function name, that is, $Y_1, Y_2, Y_3, \ldots$
- Press $\boxed{\text{2nd}}$ [Y-VARS] **1** to invoke the list of function names.
- Press the number of the function name, or cursor down to the function and press $\boxed{\text{ENTER}}$.

#### D. Combine functions
Suppose $Y_1$ is $f(x)$ and $Y_2$ is $g(x)$.
- If $Y_3 = Y_1 + Y_2$, then $Y_3$ is $f(x) + g(x)$. (Similarly for $-$, $\times$, and $\div$.)
- If $Y_3 = Y_1(Y_2)$, then $Y_3$ is $f(g(x))$.

### Specific Window Settings

#### A. Customize a window
- Press $\boxed{\text{WINDOW}}$ to open the window-setting screen, and edit the following values as desired.
- Xmin $=$ *the leftmost value on the x-axis*
- Xmax $=$ *the rightmost value on the x-axis*

- Xscl $=$ *the distance between tick marks on the x-axis*
- Ymin $=$ *the bottom value on the y-axis*
- Ymax $=$ *the top value on the y-axis*
- Yscl $=$ *the distance between tick marks on the y-axis*

*Note* 1: The notation $[a, b]$ *by* $[c, d]$ stands for the window settings Xmin $= a$, Xmax $= b$, Ymin $= c$, Ymax $= d$.

*Note* 2: The default values of Xscl and Yscl are 1. The value of Xscl should be made large (small) if the difference between Xmax and Xmin is large (small). For instance, with the window settings $[0, 100]$ *by* $[-1, 1]$, good scale settings are Xscl $= 10$ and Yscl $= .1$.

#### B. Use a predefined window setting
- Press $\boxed{\text{ZOOM}}$ to display the list of predefined settings.
- Either press a number or move the cursor down to an item and press $\boxed{\text{ENTER}}$.
- Select ZStandard to obtain $[-10, 10]$ *by* $[-10, 10]$, Xscl $=$ Yscl $= 1$.
- Select ZDecimal to obtain $[-4.7, 4.7]$ *by* $[-3.1, 3.1]$, Xscl $=$ Yscl $= 1$. (When TRACE is used with this setting, points have nice $x$-coordinates.)
- Select ZSquare to obtain a true-aspect window. (With such a window setting, lines that should be perpendicular actually look perpendicular, and the graph of $y = \sqrt{1 - x^2}$ looks like the top half of a circle.)
- Select ZTrig to obtain $[-2\pi, 2\pi]$ *by* $[-4, 4]$, Xscl $= \pi/2$, Yscl $= 1$, a good setting for the graphs of trigonometric functions.

#### C. Some nice window settings
(With these settings, one unit on the $x$-axis has the same length as one unit on the $y$-axis, and tracing progresses over simple values.)

- $[-4.7, 4.7]$ by $[-3.1, 3.1]$
- $[-2.35, 2.35]$ by $[-1.55, 1.55]$
- $[-7.05, 7.05]$ by $[-4.65, 4.65]$
- $[-9.4, 9.4]$ by $[-6.2, 6.2]$
- $[0, 9.4]$ by $[0, 6.2]$
- $[0, 18.8]$ by $[0, 12.4]$
- $[0, 47]$ by $[0, 31]$
- $[0, 94]$ by $[0, 62]$

*General principle*: (Xmax − Xmin) should be a number of the form $k \cdot 9.4$, where $k$ is a whole number or $\frac{1}{2}, \frac{3}{2}, \frac{5}{2}, \ldots$, then (Ymax − Ymin) should be $(31/47) \cdot$ (Xmax − Xmin).

## Derivative, Slopes, and Tangent Lines

**A. Compute $f'(a)$ from the Home screen using nDeriv($f(x)$, X, $a$)**

- Press MATH 8 to display **nDeriv(**.
- Enter either $Y_1, Y_2, \ldots$ or an expression for $f(x)$.
- Type in the remaining items and press ENTER.

**B. Define derivatives of the function $Y_1$**

- Set $Y_2 = $ **nDeriv($Y_1$, X, X)** to obtain the 1st derivative.
- Set $Y_3 = $ **nDeriv($Y_2$, X, X)** to obtain the 2nd derivative.

  *Note*: **nDeriv(** is obtained by pressing MATH 8.

**C. Compute the slope of a graph at a point**

- Press 2nd [CALC] 6.
- Use the arrow keys to move to the point of the graph.
- Press ENTER.

  *Note*: This process usually works best with a nice window setting.

**D. Draw a tangent line to a graph**

- Press 2nd [DRAW] 5 to select **Tangent**.
- Use the arrow keys to move to a point of the graph.
- Press ENTER.

  *Note* 1: This process usually works best with a nice window setting.

  *Note* 2: To remove all tangent lines, press 2nd [DRAW] 1 to execute **ClrDraw**.

## Special Points on the Graph of $Y_1$

**A. Find a point of intersection with the graph of $Y_2$, from the Home screen**
  Evaluate **solve($Y_1 - Y_2$, X, $g$)**, where $g$ is a guess, as follows:

- Press MATH 0 to display **solve(**.
- Enter either $Y_1 - Y_2$ or an expression for the difference of the two functions.

- Type in the remaining items, where $g$ is (hopefully) close to a value of $x$ for which $Y_1 = Y_2$. Press ENTER.
- A value of $x$ for which $Y_1 = Y_2$ will be displayed.

**B. Find intersection points, with graphs displayed**

- Press GRAPH to display the graphs of all selected functions.
- Press 2nd [CALC] 5 to select **intersect**.
- Reply to **"First curve?"** by using △ or ▽ (if necessary) to move the cursor to one of the two curves and then pressing ENTER.
- Reply to **"Second curve?"** by using △ or ▽ (if necessary) to move the cursor to the other curve and then pressing ENTER.
- Reply to **"Guess?"** by moving the cursor near the point of intersection and pressing ENTER.

**C. Find the second coordinate of a point with first coordinate specified, call it $a$**
  *From the Home screen:*

- Display $Y_1(a)$ and press ENTER.
  or
- Press $a$ STO ▷ X,T,θ ENTER to assign the value $a$ to the variable X.
- Display $Y_1$ and press ENTER.
  *From the Home screen or with the graph displayed:*
- Press 2nd [CALC] 1 to select **value**.
- Type in the value of $a$ and press ENTER.
- If desired, press the up-arrow key to move to points on graphs of other selected functions.
  *With the graph displayed:*
- Press TRACE.
- Move cursor with ▷ and/or ◁ until the $x$-coordinate of the cursor is as close as possible to $a$. (*Note*: This process usually works best if one of the nice window settings discussed above is used.)

**D. Find the first coordinate of a point with second coordinate specified, call it $b$**
  *From the Home screen:*

- Press MATH 0 to display **solve(**.
- Continue typing to obtain **solve($Y_1 - b$, X, $c$)** where $c$ is a guess for the value of the first coordinate. (*Note*: The expression for $Y_1$ can be used in place of $Y_1$.)
- Press ENTER.
  *With the graph displayed:*

- Set $Y_2 = b$.
- Find the point of intersection of $Y_1$ and $Y_2$ as in part B above.

**E. Find an $x$-intercept (that is, a root of $Y_1 = 0$)**

- Graph $Y_1$.
- Press 2nd [CALC] **2** to select **root**.
- Move the cursor to a point just to the left of a root and press ENTER.
- Move the cursor to a point just to the right of the root and press ENTER.
- Move the cursor to a point near the root and press ENTER.

**F. Find a relative extreme point**

- Set $Y_2 = $ **nDeriv($Y_1$, X, X)** or set $Y_2$ equal to the exact expression for the derivative of $Y_1$. [To display **nDeriv(** press MATH **8**.]
- Select $Y_2$ and deselect all other functions.
- Graph $Y_2$.
- Find an $x$-intercept of $Y_2$, call it $r$, at which the graph of $Y_2$ crosses the $x$-axis.
- The point $(r, Y_1(r))$ will be a possible relative extreme point of $Y_1$.

**G. Find an inflection point**

- Set $Y_2 = $ **nDeriv($Y_1$, X, X)** and $Y_3 = $ **nDeriv($Y_2$, X, X)**, or set $Y_3$ equal to the exact expression for the second derivative of $Y_1$.
- Select $Y_3$ and deselect all other functions.
- Graph $Y_3$.
- Find an $x$-intercept of $Y_3$, call it $r$, at which the graph of $Y_3$ crosses the $x$-axis.
- The point $(r, Y_1(r))$ will be a possible inflection point of $Y_1$.

## Tables

**A. Display values of $f(x)$ for evenly spaced values of $x$**

- Press Y=, assign the function $f(x)$ to $Y_1$.
- Press 2nd [TblSet].
- Set **TblMin** =*first value of $x$*.
- Set $\Delta$**Tbl** =*increment for values of $x$*.
- Set both **Indpnt** and **Depend** to **Auto**.

- Press 2nd [TABLE].

  *Note* 1: You can use the down- and up-arrow keys to look at function values for other values of $x$.

  *Note* 2: The table can display values of more than one function. For example, in the first step you can assign the function $g(x)$ to $Y_2$ and also select $Y_2$ to obtain a table with columns for X, $Y_1$, and $Y_2$.

**B. Display values of $f(x)$ for arbitrary values of $x$**

- Press Y= and assign the function $f(x)$ to $Y_1$.
- Press 2nd [TblSet].
- Set **Indpnt** to **Ask** by moving the cursor to Ask and pressing ENTER.
- Leave **Depend** set to **Auto**.
- Press 2nd [TABLE].
- Type in any value for X and press ENTER.
- Repeat the previous step for any other values of X.

## Riemann Sums

Suppose that $Y_1$ is $f(x)$, and $c$, $d$, and $\Delta x$ are numbers, then **sum(seq($Y_1$,X,$c$,$d$,$\Delta x$))** computes

$$f(c) + f(c + \Delta x) + f(c + 2\Delta x) + \cdots + f(d).$$

The function **sum** is in the LIST/MATH menu and the function **seq** is in the LIST/OPS menu.

**A. Compute $[f(x_1) + f(x_2) + \cdots + f(x_n)] \cdot \Delta x$**

  On the Home screen, evaluate sum(seq($f(x)$, X, $x_1$, $x_n$, $\Delta x$)) * $\Delta x$ as follows:

- Press 2nd [LIST] and move the cursor right to MATH.
- Press **5** ( to display **sum(**.
- Press 2nd [LIST] **5** to display **seq(**.
- Enter either $Y_1$ or an expression for $f(x)$.
- Type in the remaining items and press ENTER.

## Definite Integrals and Antiderivatives

**A. Compute $\int_a^b f(x)\, dx$**

  On the Home screen, evaluate fnInt($f(x), x, a, b$) as follows:

- Press MATH **9** to display **fnInt(**.
- Enter either $Y_1$ or an expression for $f(x)$.
- Type in the remaining items and press ENTER.

**B. Shade a region under the graph of a function and find its area**

- Press 2nd [CALC] **7** to select $\int$ **f(x) dx**.

- If necessary, use △ or ▽ to move the cursor to the graph.

- In response to the request for a Lower Limit, move the cursor to the left endpoint of the region and press ENTER.

- In response to the request for an Upper Limit, move the cursor to the right endpoint of the region and press ENTER.

  *Note* 1: This process often works best with a nice window setting.

  *Note* 2: To remove the shading, press 2nd [DRAW] 1 to execute the **ClrDraw** command.

C. **Obtain the graph of the solution to the differential equation** $y' = g(x)$, $y(a) = b$

  [That is, obtain the graph of the function $f(x)$ that is an antiderivative of $g(x)$ and satisfies the additional condition $f(a) = b$.]

- Set $Y_1 = g(X)$.

- Set $Y_2 = fnInt(Y_1, X, a, X) + b$. [Press MATH 9 to display **fnInt(**.] The function $Y_2$ is an antiderivative of $g(x)$ and can be evaluated and graphed.

D. **Shade the region between two curves**

  Suppose the graph of $Y_1$ lies below the graph of $Y_2$ for $a \le x \le b$ and both functions have been selected. To shade the region between these two curves execute the instructions **Shade(Y₁,Y₂,1,a,b)** as follows.

- Press 2nd [DRAW] 7 to display **Shade(**.

- Type in the remaining items and press ENTER.

  *Note* 1: Replace 1 by 2 for a striped shading.

  *Note* 2: To remove the shading, press 2nd [DRAW] 1 ENTER to execute the **ClrDraw** command.

## Functions of Several Variables

A. **Specify a function of several variables and its derivatives**

- In the **Y=** edit screen, set $Y_1 = f(X, Y)$. (The letter Y is entered by pressing ALPHA [Y].)

- Set $Y_2 = nDeriv(Y_1, X, X)$. $Y_2$ will be $\frac{\partial f}{\partial x}$.

- Set $Y_3 = nDeriv(Y_1, Y, Y)$. $Y_3$ will be $\frac{\partial f}{\partial y}$.

- Set $Y_4 = nDeriv(Y_2, X, X)$. $Y_4$ will be $\frac{\partial^2 f}{\partial x^2}$.

- Set $Y_5 = nDeriv(Y_3, Y, Y)$. $Y_5$ will be $\frac{\partial^2 f}{\partial y^2}$.

- Set $Y_6 = nDeriv(Y_3, X, X)$. $Y_6$ will be $\frac{\partial^2 f}{\partial x \, \partial y}$.

B. **Evaluate one of the functions in part A at** $x = a$ **and** $y = b$

- On the Home screen, assign the value $a$ to the variable X with $a$ STO ▷ X,T,θ.

- Press $b$ STO ▷ ALPHA [Y] to assign the value $b$ to the variable Y.

- Display the name of one of the functions, such as $Y_1, Y_2, \ldots$, and press ENTER.

## Least-Squares Approximations

A. **Obtain the equation of the least-squares line**

  Assume the points are $(x_1, y_1), \ldots, (x_n, y_n)$.

- Press STAT 1 for the EDIT screen, and obtain a table used for entering the data.

- If necessary, clear data from columns $L_1$ and $L_2$ as follows. Move the cursor to the top of column $L_1$ and press CLEAR ENTER. Repeat for column $L_2$.

- To enter the $x$-coordinates of the points, move the cursor to the first blank row of column $L_1$, enter the value of $x_1$, and press ENTER. Repeat with $x_2, \ldots, x_n$.

- Move the cursor to the first blank row of column $L_2$, and enter the values of $y_1, \ldots, y_n$.

- Press STAT ▷ for the CALC menu, and press 5 to place **LinReg(ax + b)** on the Home screen.

- Press ENTER to obtain the slope and $y$-intercept of the least-squares line.

B. **Assign the least-squares line to a function**

- Press Y=, move the cursor to the function, and press CLEAR to erase the current expression.

- Press VARS 5 to select the **Statistics** variables.

- Press ▷ ▷ to select the EQ menu, and press 7 for **RegEQ** (Regression Equation), which then assigns the least-squares line equation to the function.

C. **Display the points from part A**

- Press Y= and deselect all functions.

- Press 2nd [STAT PLOT] ENTER to select **Plot1**, press ENTER to turn **Plot1 ON**.

- Select the first plot from the four icons for the types of plots. This icon corresponds to a scatter plot.

- Select $L_1$ from Xlist and select $L_2$ from Ylist.

- Press GRAPH to display the data points.

  *Note* 1: Press ZOOM **9** to ensure that the current window setting is large enough to display the points.

  *Note* 2: When you finish using the point-plotting feature, turn it off. Press 2nd [STAT PLOT] ENTER to select **Plot1**. Then press ▷ to move to **OFF** and press ENTER.

**D. Display the line and the points from part A**

- Press Y= and deselect all functions except for the function containing the equation of the least-squares line.
- Carry out all but the first step of part C.

## The Differential Equation $y' = g(t, y)$

**A. Obtain a table for the approximation given by Euler's method, with $a$, $b$, $y_0$, $h$, and $n$ as given in Section 10.7**

*Note*: We will use $N$ to represent the value of $n$ in order to avoid confusion with the variable $n$ in the sequence **Y** = edit screen.

With the calculator in sequence mode, the values for $t_0, t_1, t_2, \ldots$ are stored as the sequence values $u(0), u(1), u(2), \ldots$, and the values for $y_0, y_1, y_2, \ldots$ are stored as the sequence values $v(0), v(1), v(2), \ldots$.

- To invoke sequence mode, press MODE, move the cursor down to the fourth line, move the cursor right to **Seq**, and press ENTER.
- Press Y= to obtain the sequence **Y=** edit screen.
- Set $U_n = U_{n-1} + h$. [In **Seq** mode, pressing 2nd [$U_{n-1}$], the second function of **7**, generates $U_{n-1}$.]
- Set $V_n = V_{n-1} + g(U_{n-1}, V_{n-1}) * h$. [In **Seq** mode, pressing 2nd [$V_{n-1}$] (the second function of **8**) generates $V_{n-1}$. To form $g(U_{n-1}, V_{n-1}) * h$, in the function $g(t, y)$ replace $t$ with $U_{n-1}$ and replace $y$ by $V_{n-1}$, and then multiply it by the value of $h$.]
- Press WINDOW and set U$n$Start to $a$, V$n$Start to $y_0$, and $n$Start to 0.
- Press 2nd [TBLSET], and set **TblMin=0**, **ΔTbl=1**, and the other items to **Auto**.
- Press 2nd [TABLE]. The successive values of $t$ and $y$ are found in the $U_n$ and $V_n$ columns, respectively.

  *Note*: After completing using Euler's method, reset the calculator to function mode by pressing MODE, moving to **Func** in the fourth line, and pressing ENTER.

## The Newton–Raphson Algorithm

**Perform the Newton–Raphson Algorithm**

- Assign the function $f(x)$ to Y$_1$ and the function $f'(x)$ to Y$_2$.
- Press 2nd [QUIT] to invoke the Home screen.
- Type in the initial approximation.
- Assign the value of the approximation to the variable X. This is accomplished with the keystrokes STO ▷ X,T,θ ENTER.
- Type in X − Y$_1$/Y$_2$ → X. (This statement calculates the value of X − Y$_1$/Y$_2$ and assigns it to X.)
- Press ENTER to display the value of this new approximation. Each time ENTER is pressed, another approximation is displayed.

  *Note*: In the first step, Y$_2$ can be set equal to **nDeriv(Y$_1$,X,X)**. In this case, the successive approximations will differ slightly from those obtained with Y$_2$ equal to $f'(x)$.

## Sum a Finite Series

**Compute the sum
$f(m) + f(m + 1) + \cdots + f(n)$**

On the Home screen, evaluate
**sum(seq($f$(X),X,$m$,$n$,1))** as follows:

- Press 2nd [LIST] and move the cursor right to MATH.
- Press **5** ( to display **sum(**.
- Press 2nd [LIST] **5** to display **seq(**.
- Enter an expression for $f(x)$.
- Type in the remaining items and press ENTER.

  *Note*: At most 99 terms can be summed.

## Miscellaneous Items and Tips

**A.** From the Home screen, if you plan to reuse a recently entered line with some minor changes, press 2nd [ENTRY] until the previous line appears. You can then make alterations to the line and press ENTER to execute the line.

**B.** If you plan to use TRACE to examine the values of various points on a graph, set Ymin to a value that is lower than is actually necessary for the graph. Then, the values of $x$ and $y$ will not obliterate the graph while you trace.

**C.** To clear the Home screen, press CLEAR twice.

# Appendix B
# Calculus and the TI-83/TI-83 Plus/TI-84 Plus Calculators

## Functions

### A. Define a function
- Press $\boxed{\text{Y=}}$ to obtain the **Y=** editor.
- Cursor down to the function to be defined. (Press $\boxed{\text{CLEAR}}$ if the function is to be redefined.)
- Type in the expression for the function. (*Note:* Press $\boxed{\text{X,T,}\theta\text{,n}}$ to display X.)

### B. Select or deselect a function
(Functions with highlighted equal signs are said to be *selected*. Pressing $\boxed{\text{GRAPH}}$ instructs the calculator to graph all selected functions and pressing $\boxed{\text{2nd}}$ $[\text{TABLE}]$ creates a column of the table for each selected function.)
- Press $\boxed{\text{Y=}}$ to obtain the **Y=** editor.
- Cursor down to the function to be selected or deselected.
- Move the cursor to the equal sign and press $\boxed{\text{ENTER}}$ to toggle the state of the function on and off.

### C. Display a function name, that is, $Y_1, Y_2, Y_3, \ldots$
- Press $\boxed{\text{VARS}}$ $\boxed{\triangleright}$ and then press **1** or $\boxed{\text{ENTER}}$ to display the list of function names.
- Press the number for the desired function, or cursor down to the function and press $\boxed{\text{ENTER}}$.

### D. Select a style [such as Line (\), Thick (\), or Dot ($\cdot\cdot$)] for the graph of a function
- Move the cursor to the symbol at the left of a function in the **Y=** editor.
- Press $\boxed{\text{ENTER}}$ repeatedly to select one of the seven styles.

### E. Combine functions
Suppose $Y_1$ is $f(x)$ and $Y_2$ is $g(x)$.
- If $Y_3 = Y_1 + Y_2$, then $Y_3$ is $f(x) + g(x)$. (Similarly for $-$, $\times$, and $\div$.)
- If $Y_3 = Y_1(Y_2)$, then $Y_3$ is $f(g(x))$.

## Specific Window Settings

### A. Customize a window
- Press $\boxed{\text{WINDOW}}$ to open the window-setting screen, and edit the following values as desired:
- Xmin = *the leftmost value on the x-axis*
- Xmax = *the rightmost value on the x-axis*
- Xscl = *the distance between tick marks on the x-axis*
- Ymin = *the bottom value on the y-axis*
- Ymax = *the top value on the y-axis*
- Yscl = *the distance between tick marks on the y-axis*
- Xres = *a whole number from 1 to 8*
  Usually Xres = 1. Higher values speed up graphing, but with a loss of resolution.

*Note* 1: The notation $[a, b]$ *by* $[c, d]$ stands for the window settings Xmin = $a$, Xmax = $b$, Ymin = $c$, Ymax = $d$.

*Note* 2: The default values of Xscl and Yscl are 1. The value of Xscl should be made large (small) if the difference between Xmax and Xmin is large (small). For instance, with the window settings $[0, 100]$ *by* $[-1, 1]$, good scale settings are Xscl = 10 and Yscl = .1.

### B. Use a predefined window setting
- Press $\boxed{\text{ZOOM}}$ to display the list of predefined settings.
- Either press a number or move the cursor down to an item and press $\boxed{\text{ENTER}}$.
- Select ZStandard to obtain $[-10, 10]$ *by* $[-10, 10]$, Xscl = Yscl = 1.
- Select ZDecimal to obtain $[-4.7, 4.7]$ *by* $[-3.1, 3.1]$, Xscl = Yscl = 1. (When TRACE is used with this setting, points have nice $x$-coordinates.)
- Select ZSquare to obtain a true-aspect window. (With such a window setting, lines that should be perpendicular actually look perpendicular, and the graph of $y = \sqrt{1 - x^2}$ looks like the top half of a circle.)
- Select ZTrig to obtain almost $[-2\pi, 2\pi]$ *by* $[-4, 4]$, Xscl = $\pi/2$, Yscl = 1, a good setting for the graphs of trigonometric functions.

### C. Some nice window settings
(With these settings, one unit on the $x$-axis has the same length as one unit on the $y$-axis, and tracing progresses over simple values.)
- $[-4.7, 4.7]$ *by* $[-3.1, 3.1]$
- $[-2.35, 2.35]$ *by* $[-1.55, 1.55]$
- $[-7.05, 7.05]$ *by* $[-4.65, 4.65]$
- $[-9.4, 9.4]$ *by* $[-6.2, 6.2]$
- $[0, 9.4]$ *by* $[0, 6.2]$
- $[0, 18.8]$ *by* $[0, 12.4]$
- $[0, 47]$ *by* $[0, 31]$
- $[0, 94]$ *by* $[0, 62]$

*General principle*: (Xmax − Xmin) should be a number of the form $k \cdot 9.4$, where $k$ is a whole number or $\frac{1}{2}, \frac{3}{2}, \frac{5}{2}, \ldots$, then (Ymax − Ymin) should be $(31/47) \cdot$ (Xmax − Xmin).

## Derivative, Slopes, and Tangent Lines

### A. Compute $f'(a)$ from the Home screen using `nDeriv(f(x), X, a)`

- Press MATH **8** to display `nDeriv(`.
- Enter either $Y_1, Y_2, \ldots$ or an expression for $f(x)$.
- Type in the remaining items and press ENTER.

### B. Define derivatives of the function $Y_1$

- Set $Y_2 = $ `nDeriv(`$Y_1$, X, X`)` to obtain the 1st derivative.
- Set $Y_3 = $ `nDeriv(`$Y_2$, X, X`)` to obtain the 2nd derivative.

  *Note*: `nDeriv(` is obtained by pressing MATH **8**.

### C. Compute the slope of a graph at a point

- Press 2nd [CALC] **6** to obtain `dy/dx`.
- Use the arrow keys to move to the point of the graph or type in the $x$-coordinate of the point.
- Press ENTER.

### D. Draw a tangent line to a graph

- Press 2nd [DRAW] **5** to select `Tangent`.
- Use the arrow keys to move to a point of the graph or type in the $x$-coordinate of a point.
- Press ENTER.

  *Note*: To remove all tangent lines, press 2nd [DRAW] **1** to execute `ClrDraw`.

## Special Points on the Graph of $Y_1$

### A. Find a point of intersection with the graph of $Y_2$, from the Home screen

- Press MATH **0** for the (equation) `Solver`.
- Press △ CLEAR to clear the edit screen of the **EQUATION SOLVER**.
- Enter either $Y_1 - Y_2$ or an expression for the difference of the two functions to the right of `"eqn:0="`.
- Press ▽ CLEAR. The equation to be solved will be on the first line of the screen and the cursor will be just to the right of `"X="`.
- Type in a guess for the $x$-coordinate of a point of intersection and then press ALPHA [SOLVE]. (The word SOLVE is above the ENTER key.) After a

delay, the value you typed in will be replaced by the value of the $x$-coordinate of a point of intersection.

*Note* 1: You needn't be concerned with the last two lines displayed.

*Note* 2: If you press ENTER by mistake, the cursor will move to the third line of the display. Press △ ALPHA [SOLVE] to continue.

- You can now insert a different guess to the right of `"X="` and press ALPHA [SOLVE] again to obtain the $x$-coordinate of another point of intersection.

### B. Find intersection points, with graphs displayed

- Press GRAPH to display the graphs of all selected functions.
- Press 2nd [CALC] **5** to select `intersect`.
- Reply to `"First curve?"` by using △ or ▽ (if necessary) to move the cursor to one of the two curves and then pressing ENTER.
- Reply to `"Second curve?"` by using △ or ▽ (if necessary) to move the cursor to the other curve and then pressing ENTER.
- Reply to `"Guess?"` by moving the cursor near the point of intersection (or typing in an approximate value of the $x$-coordinate of the point of intersection) and pressing ENTER.

### C. Find the second coordinate of the point whose first coordinate is $a$

*From the Home screen:*

- Display $Y_1(a)$ and press ENTER.
  or
- Press $a$ STO ▷ X,T,θ,n ENTER to assign the value $a$ to the variable X.
- Display $Y_1$ and press ENTER.

*From the Home screen or with the graph displayed:*

- Press 2nd [CALC] **1** to select `value`.
- Type in the value of $a$ and press ENTER.

  *Note*: The value of $a$ must be between Xmin and Xmax.

- If desired, press △ to move to points on graphs of other selected functions.

*With the graph displayed:*

- Press TRACE.
- Type in the value of $a$ and press ENTER.

### D. Find the first coordinate of a point whose second coordinate is $b$

*From the Home screen:*

- Select the **EQUATION SOLVER** screen as in part A above, and set $\text{"eqn:0"} = Y_1 - b$.
- Press $\boxed{\bigtriangledown}$ $\boxed{\text{CLEAR}}$ to switch to another screen and clear the initial value of X.
- Type in a guess for the value of X and then press $\boxed{\text{ALPHA}}$ $[\text{SOLVE}]$.

  *With the graphs displayed:*

- Set $Y_2 = b$.
- Find the point of intersection of the graphs of $Y_1$ and $Y_2$ as in part B above.

**E. Find an $x$-intercept (that is, a *zero* of $Y_1$)**

- Graph $Y_1$.
- Press $\boxed{\text{2nd}}$ $[\text{CALC}]$ **2** to select **zero**.

  *Note:* If more than one graph is displayed press $\boxed{\bigtriangleup}$ until the expression for $Y_1$ appears at the top of the screen.

- Move the cursor to a point just to the left of a zero (or type in a number less than a zero) and press $\boxed{\text{ENTER}}$.
- Move the cursor to a point just to the right of the zero (or type in a number greater than a zero) and press $\boxed{\text{ENTER}}$.
- Move the cursor to a point near the zero (or type in a number near the zero) and press $\boxed{\text{ENTER}}$.

**F. Find a relative extreme point**

- Set $Y_2 = \mathbf{nDeriv(Y_1, X, X)}$ or set $Y_2$ equal to the exact expression for the derivative of $Y_1$. [To display **nDeriv(** press $\boxed{\text{MATH}}$ **8**.]
- Select $Y_2$ and deselect all other functions.
- Graph $Y_2$.
- Find an $x$-intercept of $Y_2$, call it $r$, at which the graph of $Y_2$ crosses the $x$-axis.
- The point $(r, Y_1(r))$ will be a possible relative extreme point of $Y_1$.

**G. Find an inflection point**

- Set $Y_2 = \mathbf{nDeriv(Y_1, X, X)}$ and $Y_3 = \mathbf{nDeriv(Y_2, X, X)}$, or set $Y_3$ equal to the exact expression for the second derivative of $Y_1$.
- Select $Y_3$ and deselect all other functions.
- Graph $Y_3$.
- Find an $x$-intercept of $Y_3$, call it $r$, at which the graph of $Y_3$ crosses the $x$-axis.
- The point $(r, Y_1(r))$ will be a possible inflection point of $Y_1$.

## Tables

**A. Display values of $f(x)$ for evenly spaced values of $x$**

- Press $\boxed{Y=}$, assign the function $f(x)$ to $Y_1$.
- Press $\boxed{\text{2nd}}$ $[\text{TBLSET}]$.
- Set **TblStart** $= $*first value of $x$.*
- Set $\Delta$**Tbl** $= $*increment for values of $x$.*
- Set both **Indpnt** and **Depend** to **Auto**.
- Press $\boxed{\text{2nd}}$ $[\text{TABLE}]$.

  *Note 1:* You can use $\boxed{\bigtriangleup}$ or $\boxed{\bigtriangledown}$ to look at function values for other values of $x$.

  *Note 2:* The table can display values of more than one function. For example, in the first step you can assign the function $g(x)$ to $Y_2$ and also select $Y_2$ to obtain a table with columns for X, $Y_1$, and $Y_2$.

**B. Display values of $f(x)$ for arbitrary values of $x$**

- Press $\boxed{Y=}$ and assign the function $f(x)$ to $Y_1$, and deselect all other functions.
- Press $\boxed{\text{2nd}}$ $[\text{TblSet}]$.
- Set **Indpnt** to **Ask** by moving the cursor to Ask and pressing $\boxed{\text{ENTER}}$.
- Leave **Depend** set to **Auto**.
- Press $\boxed{\text{2nd}}$ $[\text{TABLE}]$.
- Type in any value for X and press $\boxed{\text{ENTER}}$.
- Repeat the previous step for any other values of X.

## Riemann Sums

Suppose $Y_1$ is $f(x)$, and $c$, $d$, and $\Delta x$ are numbers. Then **sum(seq($Y_1$, X, $c$, $d$, $\Delta x$))** computes

$$f(c) + f(c + \Delta x) + f(c + 2\Delta x) + \cdots + f(d).$$

The function **sum** is in the LIST/MATH menu and the function **seq** is in the LIST/OPS menu.

**Compute $[f(x_1) + f(x_2) + \cdots + f(x_n)] \cdot \Delta x$**

On the Home screen, evaluate sum(seq($f(x)$, X, $x_1$, $x_n$, $\Delta x$)) $* \Delta x$ as follows:

- Press $\boxed{\text{2nd}}$ $[\text{LIST}]$ and move the cursor right to MATH
- Press **5** to display **sum(**.
- Press $\boxed{\text{2nd}}$ $[\text{LIST}]$ and move the cursor right to OPS.
- Press **5** to display **seq(**.
- Enter either $Y_1$ or an expression for $f(x)$.
- Type in the remaining items and press $\boxed{\text{ENTER}}$.

## Definite Integrals and Antiderivatives

### A. Compute $\int_a^b f(x)\,dx$

On the Home screen, evaluate $\text{fnInt}(f(x), x, a, b)$ as follows:

- Press MATH **9** to display **fnInt(**.
- Enter either $Y_1$ or an expression for $f(x)$.
- Type in the remaining items and press ENTER.

### B. Shade a region under the graph of a function and find its area

- Press 2nd [CALC] **7** to select $\int \mathbf{f(x)\,dx}$.
- If necessary, use △ or ▽ to move the cursor to the graph.
- In response to the request for a Lower Limit, move the cursor to the left endpoint of the region (or type in the value of $a$) and press ENTER.
- In response to the request for an Upper Limit, move the cursor to the right endpoint of the region or type in the value of $b$) and press ENTER.

*Note*: To remove the shading, press 2nd [DRAW] **1** to execute the **ClrDraw** command.

### C. Obtain the graph of the solution to the differential equation $y' = g(x)$, $y(a) = b$

[That is, obtain the graph of the function $f(x)$ that is an antiderivative of $g(x)$ and satisfies the additional condition $f(a) = b$.]

- Set $Y_1 = \mathbf{g(X)}$.
- Set $Y_2 = \mathbf{fnInt(Y_1, X, a, X)} + \mathbf{b}$. [Press MATH **9** to display **fnInt(**.] The function $Y_2$ is an antiderivative of $g(x)$ and can be evaluated and graphed.

*Note*: The graphing of $Y_2$ proceeds very slowly. Graphing can be speeded up by setting xRes (in the WINDOW screen) to a high value.

### D. Shade the region between two curves

Suppose the graph of $Y_1$ lies below the graph of $Y_2$ for $a \le x \le b$ and both functions have been selected. To shade the region between these two curves execute the instructions **Shade($Y_1, Y_2, a, b$)** as follows:

- Press 2nd [DRAW] **7** to display **Shade(**.
- Type in the remaining items and press ENTER.

*Note*: To remove the shading, press 2nd [DRAW] **1** ENTER to execute the **ClrDraw** command.

## Functions of Several Variables

### A. Specify a function of several variables and its derivatives

- In the **Y=** editor, set $Y_1 = f(X, Y)$. (The letter Y is entered by pressing ALPHA [Y].)
- Set $Y_2 = \mathbf{nDeriv(Y_1, X, X)}$. $Y_2$ will be $\frac{\partial f}{\partial x}$.
- Set $Y_3 = \mathbf{nDeriv(Y_1, Y, Y)}$. $Y_3$ will be $\frac{\partial f}{\partial y}$.
- Set $Y_4 = \mathbf{nDeriv(Y_2, X, X)}$. $Y_4$ will be $\frac{\partial^2 f}{\partial x^2}$.
- Set $Y_5 = \mathbf{nDeriv(Y_3, Y, Y)}$. $Y_5$ will be $\frac{\partial^2 f}{\partial y^2}$.
- Set $Y_6 = \mathbf{nDeriv(Y_3, X, X)}$. $Y_6$ will be $\frac{\partial^2 f}{\partial x\,\partial y}$.

### B. Evaluate one of the functions in part A at $x = a$ and $y = b$

- On the Home screen, assign the value $a$ to the variable X with $a$ STO ▷ X,T,θ,n.
- Press $b$ STO ▷ ALPHA [Y] to assign the value $b$ to the variable Y.
- Display the name of one of the functions, such as $Y_1, Y_2, \ldots$, and press ENTER.

## Least-Squares Approximations

### A. Obtain the equation of the least-squares line

Assume the points are $(x_1, y_1), \ldots, (x_n, y_n)$.

- Press STAT **1** for the EDIT screen, and obtain a table used for entering the data.
- If necessary, clear data from columns $L_1$ and $L_2$ as follows. Move the cursor to the top of column $L_1$ and press CLEAR ENTER. Repeat for column $L_2$.
- To enter the $x$-coordinates of the points, move the cursor to the first blank row of column $L_1$, enter the value of $x_1$, and press ENTER. Repeat with $x_2, \ldots, x_n$.
- Move the cursor to the first blank row of column $L_2$, and enter the values of $y_1, \ldots, y_n$.
- Press STAT ▷ for the CALC menu, and press **4** to place **LinReg(ax + b)** on the Home screen.
- Press ENTER to obtain the slope and $y$-intercept of the least-squares line.

### B. Assign the least-squares line to a function

- Press Y=, move the cursor to the function, and press CLEAR to erase the current expression.
- Press VARS **5** to select the **Statistics** variables.
- Press ▷ ▷ to select the EQ menu, and press **1** for **RegEQ** (Regression Equation), which then assigns the least-squares line equation to the function.

## C. Display the points from part A

- Press [Y=] and deselect all functions.
- Press [2nd] [STAT PLOT] [ENTER] to select **Plot1**, press [ENTER] to turn **Plot1 ON**.
- Select the first plot from the six icons for the types of plots. This icon corresponds to a scatter plot.
- Press [GRAPH] to display the data points.

  *Note* 1: Press [ZOOM] **9** to ensure that the current window setting is large enough to display the points.

  *Note* 2: When you finish using the point-plotting feature, turn it off. Press [2nd] [STAT PLOT] [ENTER] to select **Plot1**. Then press [▷] to move to **OFF** and press [ENTER].

## D. Display the line and the points from part A

- Press [Y=] and deselect all functions except for the function containing the equation of the least-squares line.
- Carry out all but the first step of part C.

## The Differential Equation $y' = g(t, y)$

**Obtain a table for the approximation given by Euler's method, with $a$, $b$, $y_0$, $h$, and $n$ as given in Section 10.7**

*Note*: We will use $N$ to represent the value of $n$ in order to avoid confusion with the variable $n$ in the sequence **Y =** editor.

With the calculator in sequence mode, the values for $t_0, t_1, t_2, \ldots$ are stored as the sequence values $u(0), u(1), u(2), \ldots$, and the values for $y_0, y_1, y_2, \ldots$ are stored as the sequence values $v(0), v(1), v(2), \ldots$.

- To invoke sequence mode, press [MODE], move the cursor down to the fourth line, move the cursor right to **Seq**, and press [ENTER].
- Press [Y=] to obtain the sequence Y= editor.
- Set $n$Min $= 0$.
- Clear $u(n)$, if necessary, and set $u(n) = u(n-1) + h$. [In Seq mode, pressing [2nd] [u] (the second function of **7** key) generates $u$ and pressing [X,T,θ,n] generates $n$. Here $h$ refers to the value for the length of the subintervals.]
- Set $u(n$Min$) = a$.
- Set $v(n) = v(n-1) + g(u(n-1), v(n-1)) * h$. [In Seq mode, pressing [2nd] [v], the second function of the **8** key, generates $v$. To form $g(u(n-1), v(n-1)) * h$, in the function $g(t, y)$,

replace $t$ with $u(n-1)$ and replace $y$ by $v(n-1)$, and then multiply it by the value of $h$.]

- Set $v(n$Min$) = y_0$.
- Press [2nd][FORMAT], move on the first row to **uv** and press [ENTER].
- Press [WINDOW] and set $n$Min $= 0$, $n$Max $= N$, Xmin $= a$, and Xmax $= b$. Set the values of Xscl, Ymin, Ymax, and Yscl as you would when graphing ordinary functions. Leave PlotStart and PlotStep set at their default values of 1.
- To display a graph of the solution, press [GRAPH]. (You may use [TRACE] to examine the coordinates of points on the graph, where the first coordinate is denoted X and the second coordinate denoted Y.)
- To display a table of the points on the solution given by Euler's method, first press [2nd] [TBLSET], and set **TblStart=0**, $\Delta$**Tbl** $= 1$, and the other items to **Auto**. Then press [2nd] [TABLE]. The successive values of $t$ and $y$ are contained in the $u(n)$ and $v(n)$ columns, respectively.

  *Note*: After completing using Euler's method, reset the calculator to function mode by pressing [MODE], moving to **Func** in the fourth line, and pressing [ENTER].

## The Newton–Raphson Algorithm

**Perform the Newton–Raphson Algorithm**

- Assign the function $f(x)$ to $Y_1$ and the function $f'(x)$ to $Y_2$.
- Press [2nd] [QUIT] to invoke the Home screen.
- Type in the initial approximation.
- Assign the value of the approximation to the variable X. This is accomplished with the keystrokes [STO ▷] [X,T,θ] [ENTER].
- Type in X $-$ $Y_1/Y_2$ $\to$ X. (This statement calculates the value of X $-$ $Y_1/Y_2$ and assigns it to X.)
- Press [ENTER] to display the value of this new approximation. Each time [ENTER] is pressed, another approximation is displayed.

  *Note*: In the first step, $Y_2$ can be set equal to nDeriv($Y_1$,X,X). In this case, the successive approximations will differ slightly from those obtained with $Y_2$ equal to $f'(x)$.

## Sum a Finite Series

**Compute the sum**
$$f(m) + f(m+1) + \cdots + f(n)$$

On the Home screen, evaluate
**sum(seq(** $f$**(X),X,** $m$**,** $n$**))** as follows:

- Press 2nd [LIST] and move the cursor right to MATH.
- Press **5** to display **sum(**.
- Press 2nd [LIST] and move the cursor right to OPS.
- Press **5** to display **seq(**.
- Enter an expression for $f(x)$.
- Type in the remaining items and press ENTER.

*Note*: At most 999 terms can be summed.

## Miscellaneous Items and Tips

**A.** From the Home screen, if you plan to reuse a recently entered line with some minor changes, press 2nd [ENTRY] until the previous line appears. You can then make alterations to the line and press ENTER to execute the line.

**B.** If you plan to use TRACE to examine the values of various points on a graph, set Ymin to a value that is lower than is actually necessary for the graph. Then, the values of $x$ and $y$ will not obliterate the graph while you trace.

**C.** To clear the Home screen, press CLEAR twice.

# Appendix C
# Calculus and the TI-85 Calculator

## Functions

### A. Define a function, say y1, from the Home screen

- Press [2nd] [QUIT] to invoke the Home screen.
- Press [2nd] [alpha] [Y] **1** [ALPHA] [=] followed by an expression for the function, and press [ENTER].

### B. Define a function from the function editor

- Press [GRAPH] [F1] to select $y(x) =$ from the GRAPH menu and obtain the screen for defining functions; that is, the function editor.
- Cursor down to a function. (To delete an existing expression, press [CLEAR]. To create an additional function, cursor down to the last function and press [ENTER].)
- Type in an expression for the function. (Press [F1] or [x-VAR] to display $x$. Press [F2] or [2nd] [alpha] [Y] to display $y$.)

### C. Select or deselect a function in the function editor
(Functions with highlighted equal signs are said to be *selected*. The graph screen displays the graphs of all selected functions.)

- Press [GRAPH] [F1] to invoke the function editor.
- Cursor down to the function to be selected or deselected.
- Press [F5], that is, SELCT, to toggle the state of the function on and off.

### D. Display a function name, that is, y1, y2, y3, ...

- Press [2nd] [alpha] [Y] followed by the number.
  or
- Press [2nd] [VARS] [MORE] [F3] to invoke a list containing the function names.
- Cursor down to the desired function.
- Press [ENTER] to display the selected function name.

### E. Combine functions
Suppose y1 is $f(x)$ and y2 is $g(x)$.

- If y3 = y1 + y2, then y3 is $f(x) + g(x)$. (Similarly for $-$, $\times$, and $\div$.)
- If y3 = evalF(y1, x, y2), then y3 is $f(g(x))$. (To display evalF(, press [2nd] [CALC] [F1].)

## Specify Window Settings

### A. Customize a window

- Press [GRAPH] [F2] to invoke the RANGE screen and edit the following values as desired.
- xMin = *the leftmost value on the x-axis*
- xMax = *the rightmost value on the x-axis*
- xScl = *the distance between tick marks on the x-axis*
- yMin = *the bottom value on the y-axis*
- yMax = *the top value on the y-axis*
- yScl = *the distance between tick marks on the y-axis*

  *Note* 1: The notation $[a, b]$ *by* $[c, d]$ stands for the range settings xMin $= a$, xMax $= b$, yMin $= c$, yMax $= d$.

  *Note* 2: The default values of xScl and yScl are 1. The value of xScl should be made large (small) if the difference between xMax and xMin is large (small). For instance, with the window settings $[0, 100]$ *by* $[-1, 1]$, good scale settings are xScl $= 10$ and yScl $= .1$.

### B. Use a predefined range setting

- Press [GRAPH] [F3] to invoke the ZOOM menu.
- Press [F4], that is, ZSTD, to obtain $[-10, 10]$ *by* $[-10, 10]$, xScl $=$ yScl $= 1$.
- Press [MORE] [F2], that is, ZSQR, to obtain a true-aspect viewing rectangle. (With such a viewing rectangle, lines that should be perpendicular actually look perpendicular, and the graph of $y = \sqrt{1 - x^2}$ looks like the top half of a circle.)
- Press [MORE] [F4], that is ZDECM, to obtain $[-6.3, 6.3]$ *by* $[-3.1, 3.1]$, xScl $=$ yScl $= 1$. (When TRACE is used with this viewing rectangle, points have nice $x$-coordinates.)
- Press [MORE] [F3], that is, ZTRIG, to obtain $[-21\pi/8, 21\pi/8]$ *by* $[-4, 4]$, xScl $= \pi/2$, yScl $= 1$, a good setting for the graphs of trigonometric functions.

### C. Some nice range settings
With these settings, one unit on the $x$-axis has the same length as one unit on the $y$-axis, and tracing progresses over simple values.

- $[-6.3, 6.3]$ *by* $[-3.7, 3.7]$
- $[-3.15, 3.15]$ *by* $[-1.85, 1.85]$
- $[-9.45, 9.45]$ *by* $[-5.55, 5.55]$
- $[-12.6, 12.6]$ *by* $[-7.4, 7.4]$
- $[0, 12.6]$ *by* $[0, 7.4]$
- $[0, 25.2]$ *by* $[0, 14.8]$
- $[0, 63]$ *by* $[0, 37]$
- $[0, 126]$ *by* $[0, 74]$

*General principle*: (xMax − xMin) should be a number of the form $k \cdot 6.3$, where $k$ is a whole number or $\frac{1}{2}, \frac{3}{2}, \frac{5}{2}, \ldots$, then (yMax − yMin) should be $(37/63) \cdot ($xMax − xMin$)$.

## Derivative, Slopes, and Tangent Lines

**A. Compute $f'(a)$ from the Home screen using der1($f(x), x, a$)**

- Press 2nd [CALC] F3 to display **der1(**.
- Enter either y1, y2, … or an expression for $f(x)$.
- Type in the remaining items and press ENTER.

**B. Define derivatives of the function y5**

- Set **y1 = der1(y5, x, x)** to obtain the 1st derivative.
- Set **y2 = der2(y5, x, x)** to obtain the 2nd derivative.
- Set **y3 = nDer(y2, x, x)** to obtain the 3rd derivative.
- Set **y4 = nDer(y3, x, x)** to obtain the 4th derivative.

*Note*: **der1**, **der2**, and **nDer** are found on the menu obtained by pressing 2nd [CALC].

**C. Compute the slope of a graph at a point**

- Press GRAPH F5 to display the graph of the function.
- Press MORE F1 F4 to select dy/dx from the GRAPH/MATH menu.
- Use the arrow keys to move to the point of the graph.
- Press ENTER.

*Note*: This process usually works best with a nice range setting.

**D. Draw a tangent line to a graph**

- Press GRAPH F5 to display the graph of the function.
- Press MORE F1 MORE MORE F3 to select TANLN from the GRAPH/MATH menu.
- Move the cursor to any point on the graph.
- Press ENTER to draw the tangent line through the point and display the slope of the curve at that point. The slope is displayed at the bottom of the screen as $dy/dx = slope$.

- To draw another tangent line, press GRAPH and then repeat the previous three steps.

*Note* 1: To remove all tangent lines, press GRAPH MORE F2 MORE F5.

*Note* 2: This process usually works best with a nice range or a range in which the $x$-coordinate of the point is halfway between xMin and xMax.

## Special Points on the Graph of y1

**A. Find a point of intersection with the graph of y2, from the Home screen**

- Press 2nd [SOLVER].
- To the right of **"eqn:"** enter **y1 − y2 = 0** and press ENTER. (y1 and y2 can be entered via F keys and the equal sign is entered with ALPHA [=].)
- To the right of **"x="** type in a guess for the $x$-coordinate of the point of intersection, and then press F5. After a little delay, a value of $x$ for which y1 = y2 will be displayed in place of your guess.

**B. Find intersection points, with graphs displayed**

- Press GRAPH F5 to display the graphs of all selected functions.
- Press MORE F1 MORE F5 to select ISECT from the GRAPH/MATH menu.
- If necessary, use △ or ▽ to place the cursor on one of the two curves.
- Move the cursor close to the point of intersection and then press ENTER.
- If necessary, use △ or ▽ to place the cursor on the other curve.
- Press ENTER to display the coordinates of the point of intersection.

**C. Find the second coordinate of the point whose first coordinate is $a$**

*From the Home screen*
Compute **evalF(y1, x, $a$)** as follows:

- Press 2nd [CALC] F1 to display **evalF(**.
- Enter either y1 or an expression for the function.
- Type in the remaining items and press ENTER.
  or
- Press $a$ STO ▷ X-VAR ENTER to assign the value $a$ to the variable x.
- Display y1 and press ENTER.

*From the Home screen or with the graph displayed:*

- Press $\boxed{\text{GRAPH}}$ $\boxed{\text{MORE}}$ $\boxed{\text{MORE}}$ $\boxed{\text{F1}}$.

- Type in the value of $a$ and press $\boxed{\text{ENTER}}$. (The value of $a$ must be between xMin and xMax.)

- If desired, press the up-arrow key, $\boxed{\triangle}$, to move to points on graphs of other selected functions.

  *With the graph displayed:*

- Press $\boxed{\text{F4}}$; that is, TRACE.

- Move cursor with $\boxed{\triangleright}$ and/or $\boxed{\triangleleft}$ until the $x$-coordinate of the cursor is as close as possible to $a$.

  *Note*: Usually works best if one of the nice range settings discussed above is used.

**D. Find the first coordinate of a point whose second coordinate is $b$**

- Set $\mathbf{y2} = b$.

- Find the point of intersection of the graphs of y1 and y2 as in part B above.

**E. Find an $x$-intercept of a graph of a function**

- Press $\boxed{\text{GRAPH}}$ $\boxed{\text{MORE}}$ $\boxed{\text{F1}}$ $\boxed{\text{F3}}$ to select ROOT from the GRAPH/MATH menu.

- Move the cursor along the graph of the function close to an $x$-intercept, and press $\boxed{\text{ENTER}}$.

**F. Find a relative extreme point**

- Set $\mathbf{y2} = \mathbf{der1(y1,x,x)}$ or set y2 equal to the exact expression for the derivative of y1. [To display $\mathbf{der1(}$ press $\boxed{\text{2nd}}$ $\boxed{\text{CALC}}$ $\boxed{\text{F3}}$.]

- Select y2 and deselect all other functions.

- Graph y2.

- Find an $x$-intercept of y2, call it $r$, at which the graph of y2 crosses the $x$-axis.

- The point $(r, y1(r))$ will be a possible relative extreme point of y1.

**G. Find an inflection point**

- Set $\mathbf{y2} = \mathbf{der2(y1,x,x)}$ or set y2 equal to the exact expression for the second derivative of y1. (To display der2, press $\boxed{\text{2nd}}$ $\boxed{\text{CALC}}$ $\boxed{\text{F4}}$.)

- Select y2 and deselect all other functions.

- Graph y2.

- Find an $x$-intercept of y2, call it $r$, at which the graph of y2 crosses the $x$-axis.

- The point whose first coordinate is $r$ will be a possible inflection point of y1.

## Riemann Sums

Suppose that y1 is $f(x)$, and $c$, $d$, and $\Delta x$ are numbers, then $\mathbf{sum(seq(y1,x,}c\mathbf{,}d\mathbf{,}\Delta x\mathbf{))}$ computes

$$f(c) + f(c + \Delta x) + f(c + 2\Delta x) + \cdots + f(d).$$

The functions $\mathbf{sum}$ and $\mathbf{seq}$ are found in the LIST/OPS menu.

**Compute** $[f(x_1) + f(x_2) + \cdots + f(x_n)] \cdot \Delta x$
On the Home screen, evaluate sum(seq($f(x)$, x, $x_1$, $x_n$, $\Delta x$)) * $\Delta x$ as follows:

- Press $\boxed{\text{2nd}}$ $\boxed{\text{LIST}}$ $\boxed{\text{F5}}$ $\boxed{\text{MORE}}$ $\boxed{\text{F1}}$ $\boxed{\text{(}}$ $\boxed{\text{F3}}$ to display sum(seq(.

- Enter either y1 or an expression for $f(x)$.

- Type in the remaining items and press $\boxed{\text{ENTER}}$.

## Definite Integrals and Antiderivatives

**A. Compute $\int_a^b f(x)\, dx$**
On the Home screen, evaluate fnInt($f(x)$, x, $a$, $b$) as follows.

- Press $\boxed{\text{2nd}}$ $\boxed{\text{CALC}}$ $\boxed{\text{F5}}$ to display $\mathbf{fnInt(}$.

- Enter either y1 or an expression for $f(x)$.

- Type in the remaining items and press $\boxed{\text{ENTER}}$.

**B. Find the area of a region under the graph of a function**

- Press $\boxed{\text{GRAPH}}$ $\boxed{\text{MORE}}$ $\boxed{\text{F1}}$ $\boxed{\text{F5}}$ to select $\int f(x)$ from the GRAPH/MATH menu.

- If necessary, use $\boxed{\triangle}$ or $\boxed{\triangledown}$ to move the cursor to the graph.

- Move the cursor to the left endpoint of the region and press $\boxed{\text{ENTER}}$.

- Move the cursor to the right endpoint of the region and press $\boxed{\text{ENTER}}$.

  *Note*: This process usually works best with a nice range setting.

**C. Obtain the graph of the solution to the differential equation $y' = g(x)$, $y(a) = b$**

That is, obtain the graph of the function $f(x)$ that is an antiderivative of $g(x)$ and satisfies the additional condition $f(a) = b$.

- Set $\mathbf{y1} = \mathbf{g(x)}$.

- Set $\mathbf{y2} = \mathbf{fnInt(y1,x,a,x)} + \mathbf{b}$. (To display $\mathbf{fnInt(}$, press $\boxed{\text{2nd}}$ $\boxed{\text{CALC}}$ $\boxed{\text{F5}}$.) The function y2 is an antiderivative of $g(x)$ and can be evaluated and graphed.

  *Note*: The graphing of y2 proceeds very slowly.

## D. Shade the region between two curves

Suppose the graph of y1 lies below the graph of y2 for $a \le x \le b$ and both functions have been selected. To shade the region between these two curves, execute the instruction **Shade(y1,y2,$a$,$b$)** as follows.

- Press GRAPH MORE F2 F1 to display **Shade(** from the GRAPH/DRAW menu.
- Type in the remaining items and press ENTER.

*Note*: To remove the shading, press GRAPH MORE F2 MORE F5 to execute **ClrDraw** from the MATH/DRAW menu.

## Functions of Several Variables

### A. Specify a function of several variables and its derivatives

- In the $y(x) =$ function editor, set y1 $= f(x,y)$. (The letters $x$ and $y$ can be entered by pressing F1 and F2.)
- Set **y2 = der1(y1,x,x)**. y2 will be $\frac{\partial f}{\partial x}$.
- Set **y3 = der1(y1,y,y)**. y3 will be $\frac{\partial f}{\partial y}$.
- Set **y4 = der2(y1,x,x)**. y4 will be $\frac{\partial^2 f}{\partial x^2}$.
- Set **y5 = der2(y1,y,y)**. y5 will be $\frac{\partial^2 f}{\partial y^2}$.
- Set **y6 = nDer(y3,x,x)**. y6 will be $\frac{\partial^2 f}{\partial x\, \partial y}$.

### B. Evaluate one of the functions in part A at $x = a$ and $y = b$

- On the Home screen, assign the value $a$ to the variable $x$ with $a$ STO▷ X-VAR.
- Press $b$ STO▷ 2nd[alpha] [Y] to assign the value $b$ to the variable $y$.
- Display the name of one of the functions, such as y1, y2, . . . , and press ENTER.

## Least-Squares Approximations

### A. Obtain the equation of the least-squares line

Assume the points are $(x_1, y_1), \ldots, (x_n, y_n)$.

- Press STAT F2 ENTER ENTER to obtain a list used for entering the data.
- Press F5 to clear all previous data.
- Enter the data for the points by pressing $x_1$ ENTER $y_1$ ENTER $x_2$ ENTER $y_2$ ENTER . . . $x_n$ ENTER $y_n$.
- Press STAT F1 ENTER ENTER F2 to obtain the values of $a$ and $b$ where the least-squares line has equation $y = bx + a$.

- If desired, the straight line (and the points) can be graphed with the following steps:
  - (a) First press GRAPH F1 and deselect all functions.
  - (b) Press STAT F3 to invoke the statistical DRAW menu. (If any graphs appear, press F5 to clear them.)
  - (c) Press F4 to draw the least-squares line and press F2 to draw the $n$ points.

### B. Assign the least-squares line to a function

- Press GRAPH F1, move the cursor to the function, and press CLEAR to erase the current expression for the function.
- Press STAT F5 MORE MORE F2 to assign the equation (known as RegEq) to the function.

### C. Display the points from part A

- Press GRAPH F1 and deselect all functions.
- Press STAT F3 F2 to select SCAT from the STAT/DRAW menu.

*Note* 1: Make sure the current window setting is large enough to display the points.

*Note* 2: To also draw the least-squares line, press STAT F3 F4 to select DRREG from the STAT/DRAW menu.

*Note* 3: To erase the points, press STAT F3 F5 to select CLDRW from the STAT/DRAW menu.

## The Differential Equation $y' = g(t, y)$

The TI-85 uses a numerical approximation technique that is different from Euler's method.

### Carry out a numerical approximation with $a$, $b$, $y_0$, $h$ and $n$ as given in Section 10.7

- To invoke differential equation mode, press 2nd [MODE], move the cursor down to the fifth line, move the cursor right to DifEq, and press ENTER.
- Press GRAPH F1 and enter the differential equation. Use Q1 (or Q2, Q3, . . . ) instead of y1 (or y2, y3, . . . ). The letters t and Q can be entered with the keys F1 and F2. Up to nine differential equations, with function variables Q1, Q2, . . . , can be specified.
- Press GRAPH F4 and set $x =$ t and $y =$ Q.
- Press GRAPH F2 to invoke the range-setting screen.
- Set tMin and xMin to $a$, set tMax and xMax to $b$, and set tStep to $h$. (Leave the values of tPlot and difTol at their default settings of 0 and .001.)

- Set the values of xScl, yMin, yMax, and yScl as you would when graphing ordinary functions.
- Press GRAPH F3 and set the initial value (denoted by QI1, QI2, etc.) to $y_0$.
- Press GRAPH F5 to see a graph of a numerical solution of the differential equation.

  *Note* 1: To make the graph more accurate (and the graphing slower), decrease the value of tStep.

  *Note* 2: You can simultaneously graph a family of solutions to Q′1 = $g(t, Q1)$ by setting Q′2 = $g(t, Q2)$, Q′3 = $g(t, Q3)$, ..., and giving each of the differential equations a different initial value.

  *Note* 3: When finished with differential equation mode, reset the calculator to function mode by pressing 2nd [MODE], moving to **Func** in the fifth line, and pressing ENTER.

## The Newton–Raphson Algorithm

### Perform the Newton–Raphson Algorithm

- Assign the function $f(x)$ to y1 and the function $f'(x)$ to y2.
- Press 2nd [QUIT] to invoke the Home screen.
- Type in the initial approximation.
- Assign the value of the approximation to the variable x. This is accomplished with the keystrokes STO ▷ X-VAR ENTER.
- Type in x − y1/y2 → x. (This statement calculates the value of x − y1/y2 and assigns it to x.)
- Press ENTER to display the value of this new approximation. Each time ENTER is pressed, another approximation is displayed.

  *Note*: In the first step, y2 can be set equal to der1(y1,x,x).

## Sum a Finite Series

### Compute the sum
$$f(m) + f(m + 1) + \cdots + f(n)$$

On the Home screen, evaluate
**sum(seq(** f **(x),x,** m **,** n **,1))** as follows:

- Press 2nd [LIST] F5 MORE F1 (( F3 to display **sum(seq(**.
- Enter an expression for $f(x)$.
- Type in the remaining items and press ENTER.

  *Note*: About 2500 terms can be summed.

## Miscellaneous Items and Tips

**A.** From the Home screen, if you plan to reuse the most recently entered line with some minor changes, press 2nd [ENTRY] to display the previous line. You can then make alterations to the line and press ENTER to execute the line.

**B.** If you plan to use TRACE to examine the values of various points on a graph, set yMin to a value that is lower than is actually necessary for the graph. Then, the values of $x$ and $y$ will not obliterate the graph while you trace.

**C.** To clear the Home screen, press CLEAR twice.

**D.** When two menus are displayed at the same time, you can remove the top menu by pressing EXIT. (After that, you can remove the remaining menu with CLEAR.)

**E.** To obtain the solutions of a quadratic equation, or of any equation of the form $p(x) = 0$, where $p(x)$ is a polynomial of degree ≤ 30, press 2nd [POLY], enter the degree of the polynomial, enter the coefficients of the polynomial, and press F5. Some of the solutions might be complex numbers.

# Appendix D
# Calculus and the TI-86 Calculator

## Functions

### A. Define a function, say y1, from the Home screen

- Press $\boxed{\text{2nd}}$ [QUIT] to invoke the home screen.
- Press $\boxed{\text{2nd}}$ [alpha] [Y] **1** $\boxed{\text{ALPHA}}$ [=] followed by an expression for the function, and press $\boxed{\text{ENTER}}$.

### B. Define a function from the function editor

- Press $\boxed{\text{GRAPH}}$ $\boxed{\text{F1}}$ to select $y(x) =$ from the GRAPH menu and obtain the screen for defining functions; that is, the function editor.
- Cursor down to a function. (To delete an existing expression, press $\boxed{\text{CLEAR}}$. To create an additional function, cursor down to the last function and press $\boxed{\text{ENTER}}$.)
- Type in an expression for the function. (*Note*: Press $\boxed{\text{F1}}$ or $\boxed{\text{X-VAR}}$ to display $x$. Press $\boxed{\text{F2}}$ or $\boxed{\text{2nd}}$ [alpha] [Y] to display y.)

### C. Select or deselect a function in the function editor

Functions with highlighted equal signs are said to be *selected*. The graph screen displays the graphs of all selected functions, and tables contain a column for each selected function.

- Press $\boxed{\text{GRAPH}}$ $\boxed{\text{F1}}$ to invoke the function editor.
- Cursor down to the function to be selected or deselected.
- Press $\boxed{\text{F5}}$, that is, SELCT, to toggle the state of the function on and off.

### D. Select a style [such as Line (\\), Thick (\\), or Dot ($\cdot\cdot$.)] for the graph of a function

- Move the cursor to a function in the function editor.
- Press $\boxed{\text{MORE}}$ and then press $\boxed{\text{F3}}$ repeatedly to select one of the seven styles.

### E. Display a function name, that is, y1, y2, y3, ...

- Press $\boxed{\text{2nd}}$ [alpha] [Y] followed by the number. or
- Press $\boxed{\text{2nd}}$ [CATLG-VARS] $\boxed{\text{MORE}}$ $\boxed{\text{F4}}$ to invoke a list containing the function names.
- Cursor down to the desired function.
- Press $\boxed{\text{ENTER}}$ to display the selected function name.

### F. Combine functions

Suppose y1 is $f(x)$ and y2 is $g(x)$.

- If y3 = y1 + y2, then y3 is $f(x) + g(x)$. (Similarly for $-$, $\times$, and $\div$.)
- If y3 = y1(y2), then y3 is $f(g(x))$.

## Specify Window Settings

### A. Customize a window

- Press $\boxed{\text{GRAPH}}$ $\boxed{\text{F2}}$ to invoke the window editor and edit the following values as desired.
- xMin = *the leftmost value on the $x$-axis*
- xMax = *the rightmost value on the $x$-axis*
- xScl = *the distance between tick marks on the $x$-axis*
- yMin = *the bottom value on the $y$-axis*
- yMax = *the top value on the $y$-axis*
- yScl = *the distance between tick marks on the $y$-axis*
  Usually xRes = 1. Higher values speed up graphing, but with a loss of resolution.

*Note* 1: The notation $[a, b]$ *by* $[c, d]$ stands for the window settings xMin = $a$, xMax = $b$, yMin = $c$, yMax = $d$.

*Note* 2: The default values of xScl and yScl are 1. The value of xScl should be made large (small) if the difference between xMax and xMin is large (small). For instance, with the window settings $[0, 100]$ *by* $[-1, 1]$, good scale settings are xScl = 10 and yScl = .1.

### B. Use a predefined window setting

- Press $\boxed{\text{GRAPH}}$ $\boxed{\text{F3}}$ to invoke the ZOOM menu.
- Press $\boxed{\text{F4}}$, that is ZSTD, to obtain $[-10, 10]$ *by* $[-10, 10]$, xScl = yScl = 1.
- Press $\boxed{\text{MORE}}$ $\boxed{\text{F2}}$, that is ZSQR, to obtain a true-aspect window. (With such a window setting, lines that should be perpendicular actually look perpendicular, and the graph of $y = \sqrt{1 - x^2}$ looks like the top half of a circle.)
- Press $\boxed{\text{MORE}}$ $\boxed{\text{F4}}$, that is ZDECM, to obtain $[-6.3, 6.3]$ *by* $[-3.1, 3.1]$, xScl = yScl = 1. (When TRACE is used with this setting, points have nice $x$-coordinates.)

- Press MORE F3, that is ZTRIG, to obtain $[-21\pi/8, 21\pi/8]$ by $[-4, 4]$, xScl $= \pi/2$, yScl $= 1$, a good setting for the graphs of trigonometric functions.

### C. Some nice window settings

With these settings, one unit on the $x$-axis has the same length as one unit on the $y$-axis, and tracing progresses over simple values.

- $[-6.3, 6.3]$ by $[-3.7, 3.7]$
- $[-3.15, 3.15]$ by $[-1.85, 1.85]$
- $[-9.45, 9.45]$ by $[-5.56, 5.56]$
- $[-12.6, 12.6]$ by $[-7.4, 7.4]$
- $[0, 12.6]$ by $[0, 7.4]$
- $[0, 25.2]$ by $[0, 14.8]$
- $[0, 63]$ by $[0, 37]$
- $[0, 126]$ by $[0, 74]$

*General principle*: (xMax − xMin) should be a number of the form $k \cdot 12.6$, where $k$ is a whole number or $\frac{1}{2}, \frac{3}{2}, \frac{5}{2}, \dots$, then (yMax − yMin) should be $(37/63) \cdot$ (xMax − xMin).

## Derivative, Slopes, and Tangent Lines

### A. Compute $f'(a)$ from the Home screen using der1($f(x)$, x, a)

- Press 2nd [CALC] F3 to display der1(.
- Enter either y1, y2, ... or an expression for $f(x)$.
- Type in the remaining items and press ENTER.

### B. Define derivatives of the function y5

- Set y1 $=$ der1(y5, x, x) to obtain the 1st derivative.
- Set y2 $=$ der2(y5, x, x) to obtain the 2nd derivative.
- Set y3 $=$ nDer(y2, x, x) to obtain the 3rd derivative.
- Set y4 $=$ nDer(y3, x, x) to obtain the 4th derivative.

  *Note* 1: der1, der2, and nDer are found on the menu obtained by pressing 2nd [CALC].

  *Note* 2: To speed up the display of the 3rd and 4th derivatives, set the value of xRes in the WINDOW screen to at least 5.

### C. Compute the slope of a graph at a point

- Press GRAPH F5 to display the graph of the function.
- Press MORE F1 F2 to select dy/dx from the GRAPH/MATH menu.
- Use the arrow keys to move to the point of the graph and then press ENTER. (This process usually works best with a nice window setting.) Or, type in the value of the $x$-coordinate of a point (any number between xMin and xMax), and press ENTER.

### D. Draw a tangent line to a graph

- Press GRAPH F5 to display the graph of the function.
- Press MORE F1 MORE MORE F1 to select TANLN from the GRAPH/MATH menu.
- Move the cursor to any point on the graph or type in the first coordinate of a point.
- Press ENTER to draw the tangent line through the point and display the slope of the curve at that point. The slope is displayed at the bottom of the screen as $dy/dx = slope$.
- To draw another tangent line, press GRAPH and then repeat the previous three steps.

  *Note*: To remove all tangent lines, press GRAPH MORE F2 MORE MORE F1 to select CLDRW from the GRAPH/DRAW menu.

## Special Points on the Graph of y1

### A. Find a point of intersection with the graph of y2, from the Home screen

- Press 2nd [SOLVER].
- To the right of **"eqn:"** enter y1 − y2 = 0 and press ENTER. (*Note*: y1 and y2 can be entered via F keys and the equal sign is entered with ALPHA [=].)
- To the right of **"x="** type in a guess for the $x$-coordinate of the point of intersection, and then press F5. After a little delay, a value of $x$ for which y1 − y2 will be displayed in place of your guess.

### B. Find intersection points, with graphs displayed

- Press GRAPH F5 to display the graphs of all selected functions.
- Press MORE F1 MORE F3 to select ISECT from the GRAPH/MATH menu.
- Reply to **"First curve?"** by using △ or ▽ (if necessary) to place the cursor on one of the two curves and then pressing ENTER.
- Reply to **"Second curve?"** by using △ or ▽ (if necessary) to place the cursor on the other curve and then pressing ENTER.
- Reply to **"Guess?"** by moving the cursor close to the desired point of intersection (or typing in a guess for the first coordinate of the desired point) and pressing ENTER.

## C. Find the second coordinate of the point whose first coordinate is $a$

*From the Home screen:*

- Display **y1($a$)** and press $\boxed{\text{ENTER}}$.
  or
- Press $a$ $\boxed{\text{STO} \triangleright}$ $\boxed{\text{X-VAR}}$ $\boxed{\text{ENTER}}$ to assign the value $a$ to the variable $x$.
- Display y1 and press $\boxed{\text{ENTER}}$.

*From the Home screen or with the graph displayed:*

- Press $\boxed{\text{GRAPH}}$ $\boxed{\text{MORE}}$ $\boxed{\text{MORE}}$ $\boxed{\text{F1}}$ to display **"Eval x="**.
- Type in the value of $a$ and press $\boxed{\text{ENTER}}$. (The value of $a$ must be between xMin and xMax.)
- If desired, press the up-arrow key, $\boxed{\triangle}$, to move to points on graphs of other selected functions.

*With the graph displayed:*

- Press $\boxed{\text{F4}}$; that is, TRACE.
- Type in the value of $a$ and press $\boxed{\text{ENTER}}$.

## D. Find the first coordinate of a point whose second coordinate is $b$

- Set **y2** $= b$.
- Find the point of intersection of the graphs of y1 and y2 as in part B above.

## E. Find an $x$-intercept of a graph of a function

- Press $\boxed{\text{GRAPH}}$ $\boxed{\text{MORE}}$ $\boxed{\text{F1}}$ $\boxed{\text{F1}}$ to select ROOT from the GRAPH/MATH menu.
- If necessary, use $\boxed{\triangle}$ or $\boxed{\triangledown}$ to place the cursor on the desired graph.
- In response to the request for a Left Bound, move the cursor to a point whose first coordinate is less than the desired $x$-intercept or type in a value less than the desired $x$-intercept. Then press the $\boxed{\text{ENTER}}$ key.
- In response to the request for a Right Bound, move the cursor to a point whose first coordinate is greater than the desired $x$-intercept or type in a value greater than the desired $x$-intercept. Then press the $\boxed{\text{ENTER}}$ key.
- In response to the request for a Guess, move the cursor to a point near the desired $x$-intercept or type in a value close to the desired $x$-intercept. Then press the $\boxed{\text{ENTER}}$ key.

## F. Find a relative extreme point

- Set **y2** $=$ **der1(y1, x, x)** or set y2 equal to the exact expression for the derivative of y1. (To display **der1(**, press $\boxed{\text{2nd}}$ [CALC] $\boxed{\text{F3}}$.)

- Select y2 and deselect all other functions.
- Graph y2.
- Find an $x$-intercept of y2, call it $r$, at which the graph of y2 crosses the $x$-axis.
- The point $(r, y1(r))$ will be a possible relative extreme point of y1.

## G. Find an inflection point

- Set **y2** $=$ **der2(y1, x, x)** or set y2 equal to the exact expression for the second derivative of y1. (To display **der2**, press $\boxed{\text{2nd}}$ [CALC] $\boxed{\text{F4}}$.)
- Select y2 and deselect all other functions.
- Graph y2.
- Find an $x$-intercept of y2, call it $r$, at which the graph of y2 crosses the $x$-axis.
- The point $(r, y1(r))$ will be a possible inflection point of y1.

## Tables

### A. Display values of $f(x)$ for evenly spaced values of $x$

- Press $\boxed{\text{GRAPH}}$ $\boxed{\text{F1}}$ and assign the function $f(x)$ to y1.
- Press $\boxed{\text{TABLE}}$ $\boxed{\text{F2}}$ to invoke the TABLE SETUP window.
- Set **TblStart** $=$ *first value of $x$.*
- Set **$\Delta$Tbl** $=$ *increment for values of $x$.*
- Set **Indpnt** to **Auto**.
- Press $\boxed{\text{F1}}$ to display the table.

  *Note* 1: You can use $\boxed{\triangle}$ or $\boxed{\triangledown}$ to look at function values for other values of $x$.

  *Note* 2: The table can display values of more than one function. For example, in the first step you can assign the function $g(x)$ to another function and also select that other function to obtain a table with columns for $x$ and each of the selected functions.

### B. Display values of $f(x)$ for arbitrary values of $x$

- Press $\boxed{\text{GRAPH}}$ $\boxed{\text{F1}}$, assign the function $f(x)$ to a function, and deselect all other functions.
- Press $\boxed{\text{TABLE}}$ $\boxed{\text{F2}}$.
- Set **Indpnt** to **Ask** by moving the cursor to **Ask** and pressing $\boxed{\text{ENTER}}$.
- Press $\boxed{\text{F1}}$
- Type in any value for $x$ and press $\boxed{\text{ENTER}}$.
- Repeat the previous step for as many values as you like.

## Riemann Sums

Suppose that y1 is $f(x)$, and $c$, $d$, and $\Delta x$ are numbers, then **sum(seq(y1,x,$c$,$d$,$\Delta$x))** computes

$$f(c) + f(c + \Delta x) + f(c + 2\Delta x) + \cdots + f(d).$$

The functions **sum** and **seq** are found in the LIST/OPS menu.

### Compute $[f(x_1) + f(x_2) + \cdots + f(x_n)] \cdot \Delta x$

On the Home screen, evaluate sum(seq($f(x)$, x, $x_1$, $x_n$, $\Delta x$)) $* \Delta x$ as follows:

- Press 2nd [LIST] F5 MORE F1 ( F3 to display sum(seq(.

- Enter either y1 or an expression for $f(x)$.

- Type in the remaining items and press ENTER.

## Definite Integrals and Antiderivatives

### A. Compute $\int_a^b f(x)\, dx$

On the Home screen, evaluate fnInt($f(x)$, x, $a$, $b$) as follows:

- Press 2nd [CALC] F5 to display **fnInt(**.

- Enter either y1 or an expression for $f(x)$.

- Type in the remaining items and press ENTER.

### B. Shade a region under the graph of a function from $x = a$ to $x = b$

- Press GRAPH MORE F1 F3 to select $\int f(x)$ from the GRAPH/MATH menu.

- If necessary, use △ or ▽ to move the cursor to the graph.

- In response to the request for a Lower Limit, move the cursor to the left endpoint of the region (or type in the value of $a$) and press ENTER.

- In response to the request for an Upper Limit, move the cursor to the right endpoint of the region (or type in the value of $b$) and press ENTER.

  *Note*: To remove the shading press GRAPH MORE F2 MORE MORE F1 to select CLDRW from the GRAPH/DRAW menu.

### C. Obtain the graph of the solution to the differential equation $y' = g(x)$, $y(a) = b$

[That is, obtain the function $f(x)$ that is an antiderivative of $g(x)$ and satisfies the additional condition $f(a) = b$.]

- Set **y1 = g(x)**.

- Set **y2 = fnInt(y1, x, a, x) + b**. (To display **fnInt**, press 2nd [CALC] F5.) The function y2 is an antiderivative of $g(x)$ and can be evaluated and graphed.

  *Note*: The graphing of y2 proceeds very slowly. Graphing can be sped up by setting xRes (in the GRAPH/WIND menu) to a high value.

### D. Shade the region between two curves

Suppose the graph of y1 lies below the graph of y2 for $a \leq x \leq b$ and both functions have been selected. To shade the region between these two curves, execute the instruction **Shade(y1,y2,$a$,$b$)** as follows.

- Press GRAPH MORE F2 F1 to display **Shade(** from the GRAPH/DRAW menu.

- Type in the remaining items and press ENTER.

  *Note*: To remove the shading, press GRAPH MORE F2 MORE MORE F1 to execute **ClrDraw** from the GRAPH/DRAW menu.

## Functions of Several Variables

### A. Specify a function of several variables and its derivatives

- In the $y(x) =$ function editor, set y1 = $f(x, y)$. (The letters $x$ and $y$ can be entered by pressing F1 and F2.)

- Set **y2 = der1(y1, x, x)**. y2 will be $\frac{\partial f}{\partial x}$.

- Set **y3 = der1(y1, y, y)**. y3 will be $\frac{\partial f}{\partial y}$.

- Set **y4 = der2(y1, x, x)**. y4 will be $\frac{\partial^2 f}{\partial x^2}$.

- Set **y5 = der2(y1, y, y)**. y5 will be $\frac{\partial^2 f}{\partial y^2}$.

- Set **y6 = nDer(y3, x, x)**. y6 will be $\frac{\partial^2 f}{\partial x\, \partial y}$.

### B. Evaluate one of the functions in part A at $x = a$ and $y = b$

- On the Home screen, assign the value $a$ to the variable $x$ with $a$ STO ▷ X-VAR.

- Press $b$ STO ▷ 2nd [alpha] [Y] to assign the value $b$ to the variable $y$.

- Display the name of one of the functions, such as y1, y2, . . . , and press ENTER.

## Least-Squares Approximations

### A. Obtain the equation of the least-squares line

Assume the points are $(x_1, y_1), \ldots, (x_n, y_n)$.

- Press 2nd [STAT] F2 to obtain a table for entering the data.

- If necessary, clear data from columns xStat and yStat as follows. Move the cursor to xStat and press CLEAR ENTER. Repeat for the yStat column.

- To enter the $x$-coordinates of the points, move the cursor to the first blank row the xStat column, type in the value of $x_1$, and press ENTER. Repeat with $x_2, \ldots, x_n$.

- Move the cursor to the first blank row of the yStat column and enter the values of $y_1, \ldots, y_n$.

- Place a 1 in each of the first $n$ entries of the fStat column. (The other entries should be blank.)

- Press 2nd [QUIT] to display the home screen.

- Press 2nd STAT F1 F3 ENTER to obtain the values of $a$ and $b$ where the least-squares line has equation $y = a + bx$.

- If desired, the straight line (and the points) can be graphed with the following steps:
  (a) First press GRAPH F1 and deselect all functions.
  (b) Press 2nd STAT F4 to invoke the statistical DRAW menu. (If any graphs appear, press MORE F2 to clear them.)
  (c) Press the button for DRREG (either F1 or MORE F1) to draw the least-squares line and press MORE F2 to draw the $n$ points with SCAT.

## B. Assign the least-squares line to a function

- Press GRAPH F1, move the cursor to the function, and press CLEAR to erase the current expression for the function.

- Press 2nd [STAT] F5 MORE MORE F2 to assign the equation (known as RegEq) to the function.

## C. Display the points from part A

- Press GRAPH F1 and deselect all functions.
- Press 2nd [STAT] F3 to select **STAT PLOTS** menu.
- Press F1 to select PLOT1.
- Move the cursor to **ON** and press ENTER.
- Select SCAT as the Type, select xStat as the Xlist Name, select yStat as the Ylist Name, and select any symbol for Mark.
- Press GRAPH F5 to display the data points.

  *Note* 1: Make sure the current window setting is large enough to display the points.

  *Note* 2: When you finish using the point-plotting feature, turn it off. Press 2nd [STAT] F3 F1, move the cursor to **OFF** and press ENTER.

## D. Display the line and the points from part A

- Press GRAPH F1 and deselect all functions except for the function containing the equation of the least-squares line.

- Carry out all but the first step of part C.

## The Differential Equation $y' = g(t, y)$

### A. Carry out Euler's method, with $a$, $b$, $y_0$, and $h$ as given in Section 10.7

- To invoke differential equation mode, press 2nd [MODE], move the cursor down to the fifth line, move the cursor right to DifEq, and press ENTER.

- Press GRAPH MORE F1 and select Euler from the 4th row and **FldOff** from the 5th row.

- Press GRAPH F1 and enter the differential equation. Use Q1 (or Q2, Q3, ...) instead of y1 (or y2, y3, ...). The letters t and Q can be entered with the keys F1 and F2. Up to nine differential equations, with function variables Q1, Q2, ..., can be specified.

- Press GRAPH F4 and set $x = $ t and $y = $ Q.

- Press GRAPH F2 to invoke the window-setting screen.

- Set tMin and xMin to $a$, set tMax and xMax to $b$, and set tStep to $h$. (Leave the values of tPlot and EStep at their default settings of 0 and 1.)

- Set the values of xScl, yMin, yMax, and yScl as you would when graphing ordinary functions.

- Press GRAPH F3 and set the initial value (denoted by QI1, QI2, etc.) to $y_0$. Alternately, you can simultaneously graph solutions for several different initial values, call them $v_1, v_2, \ldots$, by setting the initial value to $\{v_1, v_2, \ldots\}$. Braces are entered via the LIST menu.

- Press GRAPH F5 to see a graph of the Euler's method of solution of the differential equation.

  *Note* 1: To make the graph more accurate (and the graphing slower), decrease the value of tStep.

  *Note* 2: A table of values can be displayed by setting **TblStart** to $a$, $\Delta$**Tbl** to $h$, and **Indpnt** to **Auto**.

  *Note* 3: The next-to-last step shows one way to obtain a family of solutions of a differential equation. Another way is presented in item B below.

## B. Graph a family of solutions of a differential equation

- To invoke differential equation mode, press 2nd [MODE], move cursor down to the fifth line, move the cursor right to DifEq, and press ENTER.

- Press GRAPH MORE F1 and select Euler from the 4th row and **FldOff** from the 5th row.

- Press GRAPH F1 and enter the differential equation. Use Q1 (or Q2, Q3, ...) instead of $y$. The letters t and Q can be entered with the keys F1 and F2.

- Press GRAPH F2 to invoke the window-setting screen.

- Set tMin and xMin to 0.

- The values of tMax and xMax must be set by trial and error. Try setting them both to values like 2, 5, or 10 and see if the solutions look like those in Section 10.6.

- Set tStep to a small value, such as .1 or the default ($\approx$ .13). The smaller the value of tStep, the better the accuracy will be and the longer the time required.

- Leave the values of tPlot and EStep at their default settings of 0 and 1.

- Set the values of xScl, yMin, yMax, and yScl as you would when graphing ordinary functions. The value of yMin should be less than the smallest constant solution, and the value of yMax should be greater than the largest constant solution.

- Press GRAPH F3 and set the initial value to $y_0$.

- Press GRAPH F5 to graph the solution of the differential equation.

- Press GRAPH MORE F5 to explore solutions with other initial values.

- Move the cursor up or down to another place on the $y$-axis and press ENTER to see the graph of the solution beginning at that point.

- Repeat the process in the previous step as often as desired.

  *Note*: To erase one of the graphs drawn, select CLDRW from the GRAPH/DRAW menu.

## C. Graph the slope field of a differential equation

- To invoke differential equation mode, press 2nd mode, move the cursor down to the fifth line, move the cursor right to **DifEq**, and press ENTER.

- Press GRAPH MORE F1 and select **SlpFld** from the sixth line.

- Press GRAPH F1 and enter the differential equation. Use Q1 (or Q2, Q3, ...) instead of y1 (or y2, y3, ...). The letter t and Q can be entered with the keys F1 and F2.

- Press GRAPH F4 and set **fldRes** = 15. To draw more line segments in the slope field, enter a larger value for **fldRes**.

- Press GRAPH F2 to invoke the window-setting screen.

- Set tMin and xMin to 0. Enter all other settings as you would when graphing solutions of a differential equation. For example, set tStep to a small value, such as .15, or use the default ($\approx$ .13). Set tMax and xMax to values like 3, 5, or 10, and see if the slope fields look like those in Section 10.1.

- Press GRAPH F3 and set an initial condition for Q1. This will add a solution curve with the specified initial condition to the slope field. If you do not specify the initial condition, then only the slope field will be graphed.

- Press GRAPH F5 to graph the slope field and a solution curve (if you have specified an initial condition) or the slope field only.

- To display the slope field with different solution curves, press GRAPH F3, change the initial value for Q1, then press GRAPH F5.

## The Newton–Raphson Algorithm

### Perform the Newton–Raphson Algorithm

- Assign the function $f(x)$ to y1 and the function $f'(x)$ to y2.

- Press 2nd [QUIT] to invoke the home screen.

- Type in the initial approximation.

- Assign the value of the approximation to the variable x. This is accomplished with the keystrokes STO ▷ X-VAR ENTER.

- Type in x − y1/y2 → x. (This statement calculates the value of x − y1/y2 and assigns it to x.)

- Press ENTER to display the value of this new approximation. Each time ENTER is pressed, another approximation is displayed.

  *Note*: In the first step, y2 can be set equal to der1(y1,x,x).

## Sum a Finite Series

**Compute the sum**
$$f(m) + f(m+1) + \cdots + f(n)$$
On the Home screen, evaluate
`sum(seq(`$f$`(x),x,`$m$`,`$n$`,1))` as follows:

- Press [2nd] [LIST] [F5] [MORE] [F1] [(] [F3] to display
  `sum(seq(`.

- Enter an expression for $f(x)$.

- Type in the remaining items and press [ENTER].
  *Note*: About 6000 terms can be summed.

## Miscellaneous Items and Tips

**A.** From the Home screen, if you plan to reuse a recently entered line with some minor changes, press [2nd] [ENTRY] until the previous line appears. You can then make alterations to the line and press [ENTER] to execute the line.

**B.** If you plan to use TRACE to examine the values of various points on a graph, set yMin to a value that is lower than is actually necessary for the graph. Then, the values of $x$ and $y$ will not obliterate the graph while you trace.

**C.** To clear the Home screen, press [CLEAR] twice.

**D.** When two menus are displayed at the same time, you can remove the top menu by pressing [EXIT]. (The remaining menu can be removed by pressing [EXIT] again.)

**E.** To obtain the solutions of a quadratic equation, or of any equation of the form $p(x) = 0$, where $p(x)$ is a polynomial of degree $\leq 30$, press [2nd] [POLY], enter the degree of the polynomial, enter the coefficients of the polynomial, and press [F5]. (Some of the solutions might be complex numbers.)

# Appendix E
# Areas under the Standard Normal Curve

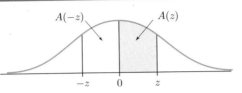

| TABLE 1 | Areas under the Standard Normal Curve | | | | | | | | |
|---|---|---|---|---|---|---|---|---|---|
| z | .00 | .01 | .02 | .03 | .04 | .05 | .06 | .07 | .08 | .09 |
| 0.0 | .0000 | .0040 | .0080 | .0120 | .0160 | .0199 | .0239 | .0279 | .0319 | .0359 |
| 0.1 | .0398 | .0438 | .0478 | .0517 | .0557 | .0596 | .0636 | .0675 | .0714 | .0754 |
| 0.2 | .0793 | .0832 | .0871 | .0910 | .0948 | .0987 | .1026 | .1064 | .1103 | .1141 |
| 0.3 | .1179 | .1217 | .1255 | .1293 | .1331 | .1368 | .1406 | .1443 | .1480 | .1517 |
| 0.4 | .1554 | .1591 | .1628 | .1664 | .1700 | .1736 | .1772 | .1808 | .1844 | .1879 |
| 0.5 | .1915 | .1950 | .1985 | .2019 | .2054 | .2088 | .2123 | .2157 | .2190 | .2224 |
| 0.6 | .2258 | .2291 | .2324 | .2357 | .2389 | .2422 | .2454 | .2486 | .2518 | .2549 |
| 0.7 | .2580 | .2612 | .2642 | .2673 | .2704 | .2734 | .2764 | .2794 | .2823 | .2852 |
| 0.8 | .2881 | .2910 | .2939 | .2967 | .2996 | .3023 | .3051 | .3078 | .3106 | .3133 |
| 0.9 | .3159 | .3186 | .3212 | .3238 | .3264 | .3289 | .3315 | .3340 | .3365 | .3389 |
| 1.0 | .3413 | .3438 | .3461 | .3485 | .3508 | .3531 | .3554 | .3577 | .3599 | .3621 |
| 1.1 | .3643 | .3665 | .3686 | .3708 | .3729 | .3749 | .3770 | .3790 | .3810 | .3820 |
| 1.2 | .3849 | .3869 | .3888 | .3907 | .3925 | .3944 | .3962 | .3980 | .3997 | .4015 |
| 1.3 | .4032 | .4049 | .4066 | .4082 | .4099 | .4115 | .4131 | .4147 | .4162 | .4177 |
| 1.4 | .4192 | .4207 | .4222 | .4236 | .4251 | .4265 | .4279 | .4292 | .4306 | .4319 |
| 1.5 | .4332 | .4345 | .4357 | .4370 | .4382 | .4394 | .4406 | .4418 | .4429 | .4441 |
| 1.6 | .4452 | .4463 | .4474 | .4484 | .4495 | .4505 | .4515 | .4525 | .4535 | .4545 |
| 1.7 | .4554 | .4564 | .4573 | .4582 | .4591 | .4599 | .4608 | .4616 | .4625 | .4633 |
| 1.8 | .4641 | .4649 | .4656 | .4664 | .4671 | .4678 | .4686 | .4693 | .4699 | .4706 |
| 1.9 | .4713 | .4719 | .4726 | .4732 | .4738 | .4744 | .4750 | .4756 | .4761 | .4767 |
| 2.0 | .4772 | .4778 | .4783 | .4788 | .4793 | .4798 | .4803 | .4808 | .4812 | .4817 |
| 2.1 | .4821 | .4826 | .4830 | .4834 | .4838 | .4842 | .4846 | .4850 | .4854 | .4857 |
| 2.2 | .4861 | .4864 | .4868 | .4871 | .4875 | .4878 | .4881 | .4884 | .4887 | .4890 |
| 2.3 | .4893 | .4896 | .4898 | .4901 | .4904 | .4906 | .4909 | .4911 | .4913 | .4916 |
| 2.4 | .4918 | .4920 | .4922 | .4925 | .4927 | .4929 | .4931 | .4932 | .4934 | .4936 |
| 2.5 | .4938 | .4940 | .4941 | .4943 | .4945 | .4946 | .4948 | .4949 | .4951 | .4952 |
| 2.6 | .4953 | .4955 | .4956 | .4957 | .4959 | .4960 | .4961 | .4962 | .4963 | .4964 |
| 2.7 | .4965 | .4966 | .4967 | .4968 | .4969 | .4970 | .4971 | .4972 | .4973 | .4974 |
| 2.8 | .4974 | .4975 | .4976 | .4977 | .4977 | .4978 | .4979 | .4979 | .4980 | .4981 |
| 2.9 | .4981 | .4982 | .4982 | .4983 | .4984 | .4984 | .4985 | .4985 | .4986 | .4986 |
| 3.0 | .4987 | .4987 | .4987 | .4988 | .4988 | .4989 | .4989 | .4989 | .4990 | .4990 |
| 3.1 | .4990 | .4991 | .4991 | .4991 | .4992 | .4992 | .4992 | .4992 | .4993 | .4993 |
| 3.2 | .4993 | .4993 | .4994 | .4994 | .4994 | .4994 | .4994 | .4995 | .4995 | .4995 |
| 3.3 | .4995 | .4995 | .4995 | .4996 | .4996 | .4996 | .4996 | .4996 | .4996 | .4997 |
| 3.4 | .4997 | .4997 | .4997 | .4997 | .4997 | .4997 | .4997 | .4997 | .4997 | .4998 |
| 3.5 | .4998 | .4998 | .4998 | .4998 | .4998 | .4998 | .4998 | .4998 | .4998 | .4998 |

# Answers

## CHAPTER 0

### EXERCISES 0.1

**1.**  $\begin{array}{c} \\ -1\ 0 \quad\quad 4 \end{array}$   **3.** $\begin{array}{c} \\ -2 \quad 0\ \sqrt{2} \end{array}$   **5.** $\begin{array}{c} \\ 0 \quad\quad 3 \end{array}$   **7.** $[2,3)$   **9.** $[-1,0)$

**11.** $(-\infty, 3)$   **13.** $0, 10, 0, 70$   **15.** $0, 0, -\frac{9}{8}, a^3 + a^2 - a - 1$   **17.** $\frac{1}{3}, 3, \dfrac{a+1}{a+2}$   **19.** $a^2 - 1, a^2 + 2a$

**21.** (a) 1990 sales   (b) 60   **23.** $x \neq 1, 2$   **25.** $x < 3$   **27.** Function   **29.** Not a function

**31.** Not a function   **33.** 1   **35.** 3   **37.** Positive   **39.** Positive   **41.** $-1, 5, 9$   **43.** .03   **45.** .04

**47.** No   **49.** Yes   **51.** $(a+1)^3$   **53.** $1, 3, 4$   **55.** $\pi, 3, 12$   **57.** $f(x) = \begin{cases} .06x & \text{for } 50 \leq x \leq 300 \\ .02x + 12 & \text{for } 300 < x \leq 600 \\ .015x + 15 & \text{for } 600 < x \end{cases}$

**59.**

**61.** $\dfrac{x^3}{4}$ is graphed   **63.**

$[-2, 4]$ by $[-8, 5]$

**65.**

$[-4, 4]$ by $[-.5, 1.5]$

**67.**

| X | Y₁ |
|---|---|
| 1 | 1.4142 |
| 2 | 2.2361 |
| 3 | 3.1623 |
| 4 | 4.1231 |
| 5 | 5.099 |
| 6 | 6.0828 |
| 7 | |

$Y_1 = 7.07106781187$

### EXERCISES 0.2

**1.**

**3.**

**5.**

**7.** $\left(-\frac{1}{3}, 0\right), (0, 3)$
**9.** $y$-intercept $(0, 5)$
**11.** $(12, 0), (0, 3)$
**13.** (a) $K = \frac{1}{250}, V = \frac{1}{50}$
(b) $\left(-\dfrac{1}{K}, 0\right), \left(0, \dfrac{1}{V}\right)$
**15.** (a) \$58
(b) $f(x) = .20x + 18$

**17.** $300x + 1500, x = $ number of days   **19.** The cost for another 5% is \$25 million. The cost for the final 5% is 21 times as much.   **21.** $a = 3, b = -4, c = 0$   **23.** $a = -2, b = 3, c = 1$   **25.** $a = -1, b = 0, c = 1$

**27.**

**29.**

**31.**

**33.** 1
**35.** $10^{-2}$
**37.** 2.5

**39.** $Y_1 = (x \leq 2) * (3x - 1) + (x > 2) * (x^2 + 1)$

$[0, 5]$ by $[0, 25]$

**41.** $Y_1 = (x \geq -2) * (x < 2) * (x \,\hat{}\, 3 + 2x)$
$Y_2 = (x \geq 2) * (\sqrt{(x - 2)})$

$[-3, 5]$ by $[-5, 5]$

**43.**

$[0, 4]$ by $[-10, 10]$

**45.**

$[0, 50]$ by $[0, 25]$

**47.**

$[-10, 10]$ by $[0, 10]$

## EXERCISES 0.3

**1.** $x^2 + 9x + 1$   **3.** $9x^3 + 9x$   **5.** $\dfrac{t}{9} + \dfrac{1}{9t}$   **7.** $\dfrac{3x + 1}{x^2 - x - 6}$   **9.** $\dfrac{4x}{x^2 - 12x + 32}$   **11.** $\dfrac{2x^2 + 5x + 50}{x^2 - 100}$

**13.** $\dfrac{2x^2 - 2x + 10}{x^2 + 3x - 10}$   **15.** $\dfrac{-x^2 + 5x}{x^2 + 3x - 10}$   **17.** $\dfrac{x^2 + 5x}{-x^2 + 7x - 10}$   **19.** $\dfrac{-x^2 + 3x + 4}{x^2 + 5x - 6}$   **21.** $\dfrac{-x^2 - 3x}{x^2 + 15x + 50}$

**23.** $\dfrac{5u - 1}{5u + 1}, u \neq 0$   **25.** $\left(\dfrac{x}{1 - x}\right)^6$   **27.** $\left(\dfrac{x}{1 - x}\right)^3 - 5\left(\dfrac{x}{1 - x}\right)^2 + 1$   **29.** $\dfrac{t^3 - 5t^2 + 1}{-t^3 + 5t^2}$   **31.** $2xh + h^2$

**33.** $4 - 2t - h$   **35. (a)** $C(A(t)) = 3000 + 1600t - 40t^2$   **(b)** \$6040   **37.** $h(x) = x + \dfrac{1}{8}; h(x)$ converts from British sizes to U.S. sizes.   **39.** The graph of $f(x) + c$ is the graph of $f(x)$ shifted up (if $c > 0$) or down (if $c < 0$) by $|c|$ units.   **41.**

**43.** $f(f(x)) = x, x \neq 1$

## EXERCISES 0.4

**1.** $2, \frac{3}{2}$    **3.** $\frac{3}{2}$    **5.** No zeros    **7.** $1, -\frac{1}{5}$    **9.** $5, 4$    **11.** $2 + \dfrac{\sqrt{6}}{3}, 2 - \dfrac{\sqrt{6}}{3}$    **13.** $(x+5)(x+3)$
**15.** $(x-4)(x+4)$    **17.** $3(x+2)^2$    **19.** $-2(x-3)(x+5)$    **21.** $x(3-x)$    **23.** $-2x(x-\sqrt{3})(x+\sqrt{3})$
**25.** $(-1,1), (5,19)$    **27.** $(-1,9), (4,4)$    **29.** $(0,0), (2,-2)$
**31.** $(0,5), (2-\sqrt{3}, 25-23\sqrt{3}/2), (2+\sqrt{3}, 25+23\sqrt{3}/2)$    **33.** $-7, 3$    **35.** $-2, 3$    **37.** $-7$
**39.** 16,667 and 78,571 subscribers    **41.** $-1, 2$    **43.** $\approx 4.56$    **45.** $\approx (-.41, -1.83), (2.41, 3.83)$
**47.** $\approx (2.14, -25.73), (4.10, -21.80)$    **49.** $[-5, 22]$ $by$ $[-1400, 100]$    **51.** $[-20, 4]$ $by$ $[-500, 2500]$

## EXERCISES 0.5

**1.** 27    **3.** 1    **5.** .0001    **7.** $-16$    **9.** 4    **11.** .01    **13.** $\frac{1}{6}$    **15.** 100    **17.** 16    **19.** 125    **21.** 1
**23.** 4    **25.** $\frac{1}{2}$    **27.** 1000    **29.** 10    **31.** 6    **33.** 16    **35.** 18    **37.** $\frac{4}{9}$    **39.** 7    **41.** $x^6 y^6$    **43.** $x^3 y^3$
**45.** $\dfrac{1}{x^{1/2}}$    **47.** $\dfrac{x^{12}}{y^6}$    **49.** $x^{12} y^{20}$    **51.** $x^2 y^6$    **53.** $16x^4$    **55.** $x^2$    **57.** $\dfrac{1}{x^7}$    **59.** $x$    **61.** $\dfrac{27 x^6}{8 y^3}$
**63.** $2\sqrt{x}$    **65.** $\dfrac{1}{8 x^6}$    **67.** $\dfrac{1}{32 x^2}$    **69.** $9x^3$    **71.** $x-1$    **73.** $1 + 6x^{1/2}$    **75.** $a^{1/2} \cdot b^{1/2} = (ab)^{1/2}$ (Law 5)
**77.** 16    **79.** $\frac{1}{4}$    **81.** 8    **83.** $\frac{1}{32}$    **85.** \$709.26    **87.** \$127,857.61    **89.** \$164.70    **91.** \$1592.75
**93.** \$3268.00    **95.** $\frac{125}{64} r^4 + \frac{125}{4} r^3 + \frac{375}{2} r^2 + 500r + 500$    **97.** $\frac{1}{20}(2x)^2 = \frac{1}{20}(4x^2) = 4(\frac{1}{20} x^2)$    **99.** .0008103
**101.** .00000823

## EXERCISES 0.6

**1.**    **3.**    **5.**    **7.** $P = 8x; 3x^2 = 25$

**9.** $A = \pi r^2; 2\pi r = 15$    **11.** $V = x^2 h; x^2 + 4xh = 65$    **13.** $\pi r^2 h = 100; C = 11\pi r^2 + 14\pi rh$
**15.** $2x + 3h = 5000; A = xh$    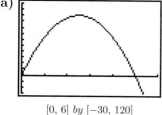    **17.** $C = 36x + 20h$    **19.** 75 cm$^2$    **21.** (a) 38    (b) \$40
**23.** (a) 200    (b) 275    (c) 25
**25.** (a) $P(x) = 12x - 800$    (b) \$640    (c) \$3150
**27.** 270 cents

**29.** A 100-in.$^3$ cylinder of radius 3 in. costs \$1.62 to construct.    **31.** \$1.08    **33.** Revenue: \$1800; cost: \$1200
**35.** 40    **37.** $C(1000) = 4000$    **39.** Find the $y$-coordinate of the point on the graph whose $x$-coordinate is 400.
**41.** The greatest profit, \$52,500, occurs when 2500 units of goods are produced.    **43.** Find the $x$-coordinate of
the point on the graph whose $y$-coordinate is 30,000.    **45.** Find $h(3)$. Find the $y$-coordinate of the point on the
graph whose $t$-coordinate is 3.    **47.** Find the maximum value of $h(t)$. Find the $y$-coordinate of the highest
point of the graph.    **49.** Solve $h(t) = 100$. Find the $t$-coordinates of the points whose $y$-coordinate is 100.
**51.** (a)    (b) 96 feet    (c) 1 and 4 seconds    (d) 5 seconds
(e) 2.5 seconds; 100 feet

$[0, 6]$ $by$ $[-30, 120]$

**53. (a)**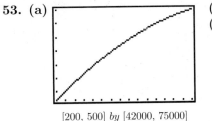

[200, 500] *by* [42000, 75000]

**(b)** 350 bicycles per year  **(c)** $68,000  **(d)** $5000
**(e)** No, revenue would only increase by $4000.

## CHAPTER 0 SUPPLEMENTARY EXERCISES

**1.** $2, 27\frac{1}{3}, -2, -2\frac{1}{8}, \dfrac{5\sqrt{2}}{2}$  **3.** $a^2 - 4a + 2$  **5.** $x \neq 0, -3$  **7.** All $x$  **9.** Yes  **11.** $5x(x-1)(x+4)$

**13.** $(-1)(x-6)(x+3)$  **15.** $-\frac{2}{5}, 1$  **17.** $\left(\dfrac{5+3\sqrt{5}}{10}, \dfrac{3\sqrt{5}}{5}\right), \left(\dfrac{5-3\sqrt{5}}{10}, -\dfrac{3\sqrt{5}}{5}\right)$  **19.** $x^2 + x - 1$

**21.** $x^{5/2} - 2x^{3/2}$  **23.** $x^{3/2} - 2x^{1/2}$  **25.** $\dfrac{x^2 - x + 1}{x^2 - 1}$  **27.** $-\dfrac{3x^2 + 1}{3x^2 + 4x + 1}$  **29.** $\dfrac{-3x^2 + 9x - 10}{3x^2 - 5x - 8}$

**31.** $\dfrac{1}{x^4} - \dfrac{2}{x^2} + 4$  **33.** $(\sqrt{x} - 1)^2$  **35.** $\dfrac{1}{(\sqrt{x}-1)^2} - \dfrac{2}{\sqrt{x}-1} + 4$  **37.** 27, 32, 4  **39.** $301 + 10t + .04t^2$

**41.** $x^2 + 2x + 1$  **43.** $x$

## CHAPTER 1

### EXERCISES 1.1

**1.** Slope is $-5$, $y$-intercept is 2  **3.** Slope is $\frac{2}{3}$, $y$-intercept is $\frac{1}{3}$  **5.** Slope is 1, $y$-intercept is $-\frac{1}{3}$
**7.** $y = 35 - 7x$  **9.** $y = 4x - 4$  **11.** $y - 5 = 6(x - \frac{2}{3})$ or $y = 6x + 1$  **13.** $y = -2x$  **15.** $y = -1$
**17.** $y = x + 1$  **19.** $y = x + 2$  **21.** $y = 2x - 4$  **23.** $y = -x + 1$  **25.** $y = -\frac{1}{2}x + 1$
**27.** Start at $(1, 0)$. To get back to the line, move one unit to the right and then one unit up.  **29.** Start at $(1, -1)$. To get back to the line, move three units to the right and then one unit down.

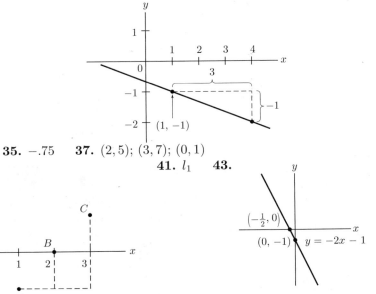

**31. (a)** C  **(b)** B  **(c)** D  **(d)** A  **33.** 2  **35.** $-.75$  **37.** $(2, 5); (3, 7); (0, 1)$
**39. (a)** Line through $A$ and $B$ has slope 1. Line through $A$ and $C$ has slope 1. The points are on one line.  **(b)**  **41.** $l_1$  **43.**

**45.** $a = 2$, $f(2) = 2$  **47. (a)** $C(10) = 1220$ dollars **(b)** Marginal cost is 12 dollars per item.
**(c)** $C(11) - C(10) =$ marginal cost $= \$12$  **49.** $P(x) = .06x + 1.69$ dollars, where $P(x)$ is the price of one gallon $x$ months since January 1st. Price of one gallon on April 1st is $P(3) = \$1.87$. Price of 15 gallons is $(15)(1.87) = \$28.05$. Price of one gallon on September 1st is $P(8) = \$2.17$. Price of 15 gallons is

$(1)(2.17) = \$32.55.$ **51.** Value of cars $t$ years after 2002 is $y(t) = -25t + 125$ thousand dollars. $t = 5$.

**53.** $G(x) = -1000x + 4100$, $G(2.70) = 1400$ gallons. **55. (a)** $C(x) = 7x + 1500$. **(b)** Marginal cost is \$7 per rod. **(c)** \$7 **57.** If the monopolist wants to sell one more unit of goods, then the price per unit must be lowered by 2 ¢. No one is willing to pay \$7 or more for a unit of goods. **59.** Let $A(x)$ be the amount in ml of the drug in the blood after $x$ minutes, then $A(x) = 6x + 1.5$. **61.** Over the interval $[0, 2]$ the rate of change is $\frac{f(2)-f(0)}{2-0} = \frac{9-7}{2} = 1$. Over the interval $[-1, 2]$ the rate of change is $\frac{f(2)-f(-1)}{2-(-1)} = \frac{9-6}{3} = 1$.

**63. (a)–(c)** **(d)** $\dfrac{f(3+h) - f(3)}{h}$

**65.** $\dfrac{L(x+h) - L(x)}{h} = \dfrac{-(x+h) + 2 - (-x + 2)}{h} = -1.$

**69. (a)** $y = .38x + 39.5$ **(b)** **(c)** Every year, .38% more of the world becomes urban. **(d)** 43.3% **(e)** 2007 **(f)** 1.9%

$[0, 30]$ *by* $[30, 60]$

**EXERCISES 1.2**

**1.** **3.** **5.** $-\frac{4}{3}$ **7.** 1 **9.** Small positive slope **11.** Zero slope **13.** Zero slope

**15.** The slope of the tangent line is approximately 1/2. Thus the federal debt rose at the rate of .5 trillion dollars per year in 1990.

**17. (a)** The debt per capita was approximately \$1000 in 1950, \$15,000 in 1990, \$21,000 in 2000, and \$24,000 in 2004.
**(b)** The slope of the tangent line is 0. Thus, in 2000, the debt per capita held steady (at about \$21,000).

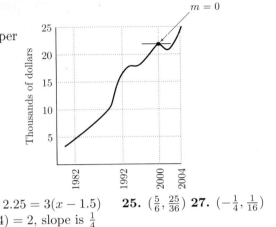

**19.** $-4$; $y - 4 = -4(x + 2)$    **21.** $\frac{8}{3}$; $y - \frac{16}{9} = \frac{8}{3}(x - \frac{4}{3})$    **23.** $y - 2.25 = 3(x - 1.5)$    **25.** $\left(\frac{5}{6}, \frac{25}{36}\right)$ **27.** $\left(-\frac{1}{4}, \frac{1}{16}\right)$
**29.** 12    **31.** $\frac{3}{4}$    **33.** $a = 1$, $f(1) = 1$, slope is 2    **35.** $a = 4$, $f(4) = 2$, slope is $\frac{1}{4}$
**37. (a)** 3, 9 **(b)** increase    **39.** $y - 2.45 = .2(x - 4)$    **41.** $y - 48 = 2.9(x - 3.5)$

## EXERCISES 1.3

**1.** 2    **3.** $5x^4$    **5.** $\frac{2}{5}x^{-3/5}$    **7.** $\frac{1}{3}x^{-2/3}$    **9.** $-2x^{-3}$    **11.** $-\frac{5}{7}x^{-\frac{12}{7}}$    **13.** 0    **15.** $-\frac{1}{3}x^{-\frac{4}{3}}$    **17.** 3
**19.** $-\frac{1}{9}$    **21.** 4    **23.** $\frac{1}{8}$    **25.** 108    **27.** 25, $-10$    **29.** 16, $\frac{8}{3}$    **31.** $\frac{1}{32}$, $-\frac{5}{64}$    **33.** $y - \frac{1}{4} = (-1)(x - (-\frac{1}{2}))$
or $y = -x - \frac{1}{4}$    **35.** Same as the line $y = 3x + 1$.    **37.** $y - 3 = \frac{1}{6}(x - 9)$ or $y = \frac{1}{6}x + \frac{3}{2}$.
**39.** $y - \frac{1}{2} = -\frac{1}{16}(x - 4)$ or $y = -\frac{1}{16}x + \frac{3}{4}$.    **41.** $f(x) = x^4$, $f'(x) = 4x^3$, $f(1) = 1$, $f'(1) = 4$, then (6) (with $a = 1$) becomes $y - 1 = 4(x - 1)$, which is the given equation of the tangent line.    **43.** $P = (\frac{1}{16}, \frac{1}{4})$, $b = \frac{1}{8}$.
**45. (a)** $(16, 4)$ **(b)**    **47. (a)**    **(b)** $f'(1) = 2$, $g'(1) = 3$. Both graphs go through the point $(1, 1)$. Since $g'(1)$ is larger than $f'(1)$, the graph of $g$ is steeper than the graph of $f$ at $(1, 1)$.

**49.** $8x^7$    **51.** $\frac{3}{4}x^{-1/4}$    **53.** 0    **55.** $\frac{1}{5}x^{-4/5}$    **57.** 4, $\frac{1}{3}$    **59.** $a = 4$, $b = 1$    **61.** 1, 1.5; 2    **63.** $2x + h - 2$
**65.** $-2x - h - 1$    **67.** $3x^2 + 3xh + h^2 + 1$    **69.** $\frac{-2}{(x+h+2)(x+2)}$

**71.** Step 1: $\frac{f(1+h)-f(1)}{h} = 2 + h$. Step 2: Let $h$ approach 0, then $2 + h$ approaches 2. Step 3: $f'(1) = 2$.

**73.** Step 1: $\frac{f(2+h)-f(2)}{h} = 12 + 6h + h^2$. Step 2: Let $h$ approach 0, then $12 + 6h + h^2$ approaches 12. Step 3: $f'(2) = 12$.

**75.** Step 1: $\frac{f(x+h)-f(x)}{h} = 2x + h$. Step 2: Let $h$ approach 0, then $2x + h$ approaches $2x$. Step 3: $f'(x) = 2x$.

**77.** Step 1: $\frac{f(x+h)-f(x)}{h} = -4x - 2h$. Step 2: Let $h$ approach 0, then $-4x - 2h$ approaches $-4x$. Step 3: $f'(x) = -4x$.

**79.** Step 1: $\frac{f(x+h)-f(x)}{h} = 3x^2 + 3xh + h^2$. Step 2: Let $h$ approach 0, then $3x^2 + 3xh + h^2$ approaches $3x^2$. Step 3: $f'(x) = 3x^2$.

**81.** Step 1: $\frac{f(x+h)-f(x)}{h} = 2$. Step 2: Let $h$ approach 0, then 2 approaches 2. Step 3: $f'(x) = 2$.

**83.** Step 1: $\frac{f(x+h)-f(x)}{h} = m$. Step 2: Let $h$ approach 0, then $m$ approaches $m$. Step 3: $f'(x) = m$.

**85.** $y - 5 = \frac{1}{2}(x-4)$ **87. (a)** and **(b)**
**(c)** Parallel lines have equal slopes:
Slope of the graph of $y = f(x)$ at the
point $(x, f(x))$ is equal to the slope of
the graph of $y = f(x) + 3$ at the point
$(x, f(x) + 3)$, which implies the given
equation.

**89.** .69315

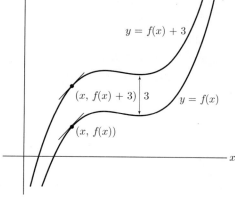

$y = f(x) + 3$

$(x,\ f(x) + 3)\ \big|\ 3$

$y = f(x)$

$(x,\ f(x))$

**91.** .70711    **93.** .11111    **95.**

**97.** $y = \dfrac{x}{6} + \dfrac{3}{2}$    **99.** $y = -16x + 12$

**101.** 2    **103.** 4

$X=2$    $Y=12$

$[0, 4]\ by\ [-5, 40]$

## EXERCISES 1.4

**1.** No limit    **3.** 1    **5.** No limit    **7.** $-5$    **9.** 5    **11.** No limit    **13.** 288    **15.** 0    **17.** 3    **19.** $-4$
**21.** $-8$    **23.** $\frac{6}{7}$    **25.** No limit    **27. (a)** 0    **(b)** $-\frac{3}{2}$    **(c)** $-\frac{1}{4}$    **(d)** $-1$    **29.** 6    **31.** 3    **33.** $-\frac{2}{121}$

**35.** $-\dfrac{\sqrt{3}}{6}$    **37.** Steps 1, 2: $\dfrac{f(3+h) - f(3)}{h} = 3$. Step 3: $f'(x) = \lim\limits_{h \to 0} 3 = 3$.

**39.** Steps 1, 2: $\dfrac{f(x+h) - f(x)}{h} = 1 + \dfrac{-1}{x(x+h)}$. Step 3: $f'(x) = \lim\limits_{h \to 0} 1 + \dfrac{-1}{x(x+h)} = 1 - \dfrac{1}{x^2}$.

**41.** Steps 1, 2: $\dfrac{f(x+h) - f(x)}{h} = \dfrac{1}{(x+1)(x+h+1)}$. Step 3: $f'(x) = \lim\limits_{h \to 0} \dfrac{1}{(x+1)(x+h+1)} = \dfrac{1}{(x+1)^2}$.

**43.** Steps 1, 2: $\dfrac{f(x+h) - f(x)}{h} = \dfrac{-2x - h}{((x+h)^2 + 1)(x^2 + 1)}$.

Step 3: $f'(x) = \lim\limits_{h \to 0} \dfrac{-2x - h}{((x+h)^2 + 1)(x^2 + 1)} = \dfrac{-2x}{(x^2 + 1)^2}$.

**45.** Steps 1, 2: $\dfrac{f(x+h) - f(x)}{h} = \dfrac{1}{\sqrt{x+h+2} + \sqrt{x+2}}$.

Step 3: $f'(x) = \lim\limits_{h \to 0} \dfrac{1}{\sqrt{x+h+2} + \sqrt{x+2}} = \dfrac{1}{2\sqrt{x+2}}$.

**47.** Steps 1, 2: $\dfrac{f(x+h) - f(x)}{h} = \dfrac{-1}{\sqrt{x}\sqrt{x+h}(\sqrt{x} + \sqrt{x+h})}$.

Step 3: $f'(x) = \lim\limits_{h \to 0} \dfrac{-1}{\sqrt{x}\sqrt{x+h}(\sqrt{x} + \sqrt{x+h})} = \dfrac{-1}{2x\sqrt{x}} = \dfrac{-1}{2x^{3/2}}$.    **49.** $f(x) = x^2; a = 1$    **51.** $f(x) = \dfrac{1}{x}$;
$a = 10$    **53.** $f(x) = \sqrt{x};\ a = 9$    **55.** Take $f(x) = x^2$, then the given limit is $f'(2) = 4$.    **57.** Take
$f(x) = \sqrt{x}$, then the given limit is $f'(2) = \frac{1}{2\sqrt{2}}$.    **59.** Take $f(x) = x^{1/3}$, then the given limit is $f'(8) = \frac{1}{12}$.
**61.** 0    **63.** 0    **65.** 0    **67.** $\frac{3}{4}$    **69.** 0    **71.** 0    **73.** 0    **75.** .5

## EXERCISES 1.5

**1.** No **3.** Yes **5.** No **7.** No **9.** Yes **11.** No **13.** Continuous, differentiable
**15.** Continuous, not differentiable **17.** Continuous, not differentiable **19.** Not continuous, not differentiable
**21.** $f(5) = 3$ **23.** Not possible **25.** $f(0) = 12$

**27. (a)** $T(x) = \begin{cases} .15x & \text{for } 0 < x \le 27{,}050 \\ .275x - 3381.25 & \text{for } 27{,}050 < x \le 65{,}550 \\ .305x - 5347.75 & \text{for } 65{,}550 < x \le 136{,}750 \end{cases}$

**(c)** $T(65{,}550) - T(27{,}050) = 10{,}587.5$ dollars

**(b)**

Taxable income in dollars

**29. (a)** $R(x) = \begin{cases} .07x + 2.5 & \text{for } 0 \le x \le 100 \\ .04x + 5.5 & \text{for } 100 < x \end{cases}$ **(b)** $P(x) = \begin{cases} .04x + 2.5 & \text{for } 0 \le x \le 100 \\ .01x + 5.5 & \text{for } 100 < x \end{cases}$

**31. (a)** 3 thousand dollars per hour **(b)** 3 thousand dollars per hour, between 8 A.M. and 10 A.M. **33.** $a = 1$

## EXERCISES 1.6

**1.** $3x^2 + 2x$ **3.** $2(\frac{1}{2})x^{-1/2}$ or $\frac{1}{\sqrt{x}}$ **5.** $\frac{1}{2} - 2(-1)x^{-2}$ or $\frac{1}{2} + \frac{2}{x^2}$ **7.** $4x^3 + 3x^2 + 1$ **9.** $3(2x+4)^2(2)$ or

$6(2x+4)^2$ **11.** $7(x^3 + x^2 + 1)^6(3x^2 + 2x)$ **13.** $-\dfrac{8}{x^3}$ **15.** $3(\frac{1}{3})(2x^2+1)^{-\frac{2}{3}}(4x)$ or $4x(2x^2+1)^{-\frac{2}{3}}$

**17.** $2 + 3(x+2)^2$ **19.** $\frac{1}{5}(-5)x^{-6}$ or $-\frac{1}{x^6}$ **21.** $(-1)(x^3+1)^{-2}(3x^2)$ or $-\dfrac{3x^2}{(x^3+1)^2}$ **23.** $1 - (x+1)^{-2}$

**25.** $\dfrac{45x^2 + 5}{2\sqrt{3x^3 + x}}$ **27.** 3 **29.** $\frac{1}{2}(1 + x + x^2)^{-\frac{1}{2}}(1 + 2x)$ **31.** $10(1 - 5x)^{-2}$

**33.** $-45(1 + x + \sqrt{x})^{-2}(1 + \frac{1}{2}x^{-\frac{1}{2}})$ **35.** $1 + \frac{1}{2}(x+1)^{-\frac{1}{2}}$ **37.** $\dfrac{3}{2}\left(\dfrac{\sqrt{x}}{2} + 1\right)^{1/2}\left(\dfrac{1}{4}x^{-1/2}\right)$ or

$\dfrac{3}{8\sqrt{x}}\left(\dfrac{\sqrt{x}}{2} + 1\right)^{1/2}$ **39.** 4 **41.** 15 **43.** $f'(4) = 48$, $y = 48x - 191$

**45.** $f'(x) = 2(3x^2 + x - 2)(6x + 1) = 36x^3 + 18x^2 - 22x - 4$ **47.** 4.8; 1.8 **49.** 14; 11 **51.** 10; $\frac{15}{4}$
**53.** $(5, \frac{161}{3})$; $(3, 49)$ **55.** $f(4) = 5$, $f'(4) = \frac{1}{2}$ **57. (a)** $S(1) = 120.560$, $S'(1) = 1.5$. **(b)** $S(3) = 80$,
$S'(3) = -6$. **59. (a)** $S(10) = \frac{372}{121} \approx 3.074$ thousand dollars, $S'(10) = -\frac{18}{11^3} \approx -.014$ thousand dollars per day.
**(b)** $S(11) \approx S(10) + S'(10) \approx 3.061$ thousand dollars. $S(11) = \frac{49}{16} = 3.0625$ thousand dollars.
**61. (a)** $A(8) = 12$, $A'(8) = .5$. **(b)** $A(9) \approx A(8) + A'(8) = 12.5$. If the company spends \$9000, it should expect
to sell 12,500 computers. **63. (a)** $D(4) \approx 5.583$ trillion dollars, $D'(4) = .082$ or 82 billion dollars per year.

## EXERCISES 1.7

**1.** $10t(t^2 + 1)^4$ **3.** $8t + \frac{11}{2}t^{-\frac{1}{2}}$ **5.** $5T^4 - 16T^3 + 6T - 1$ **7.** $6P - \frac{1}{2}$ **9.** $2a^2t + b^2$
**11.** $y' = 1$, $y'' = 0$ **13.** $y' = \frac{1}{2}x^{-1/2}$, $y'' = -\frac{1}{4}x^{-3/2}$ **15.** $y' = \frac{1}{2}(x+1)^{-\frac{1}{2}}$, $y'' = -\frac{1}{4}(x+1)^{-\frac{3}{2}}$
**17.** $f'(r) = 2\pi r$, $f''(r) = 2\pi$ **19.** $f'(P) = 15(3P + 1)^4$, $f''(P) = 180(3P + 1)^3$ **21.** 36 **23.** 0

**25.** 34    **27.** $f'(1) = -\frac{1}{9}$, $f''(1) = \frac{2}{27}$    **29.** 0    **31.** 20    **33. (a)** $f'''(x) = 60x^2 - 24x$ **(b)** $f'''(x) = \dfrac{15}{2\sqrt{x}}$

**35. (a)** $2Tx + 3P$ **(b)** $3x$ **(c)** $x^2 + 2T$    **37.** When 50 bicycles are manufactured, the cost is $5000. For every additional bicycle manufactured, there is an additional cost of $45.    **39. (a)** 2.60 dollars per unit **(b)** 100 or 200 more units    **41. (a)** $R(12) = 22$, $R'(12) = .075$ **(b)** $P(x) = R(x) - C(x)$ so $P'(x) = R'(x) - C'(x)$. When 1200 chips are produced, the marginal profit is $.75 - 1.5 = -.75$ dollars per chip.

**43.**

$[-4, 4]$ $by$ $[-2, 2]$

## EXERCISES 1.8

**1. (a)** 12; 10; 8.4 **(b)** 8    **3. (a)** 14 **(b)** 13    **5. (a)** $-\frac{7}{4}$ **(b)** $\frac{5}{6}$ **(c)** January 1, 1980
**7.** .9 g/week; .8 g/week    **9.** 63 units/hour    **11. (a)** 28 km/hour **(b)** 96 km **(c)** $\frac{1}{2}$ hr
**13. (a)** 160 ft/sec **(b)** 96 ft/sec **(c)** $-32$ ft/sec$^2$ **(d)** 10 sec **(e)** $-160$ ft/sec
**15.** A–b; B–d; C–f; D–e; E–a; F–c; G–g
**17. (a)** 15 ft/sec **(b)** No; positive velocity indicates the object is moving away from the reference point
**(c)** 5 ft/sec
**19. (a)** 5010 **(b)** 5005 **(c)** 4990 **(d)** 4980 **(e)** 4997.5
**21.** Four minutes after it has been poured, the coffee is 120°. At that time, its temperature is decreasing by 5°/min; 119.5°
**23.** When the price of a car is $10,000, 200,000 cars are sold. At that price, the number of cars sold decreases by 3 for each dollar increase in price.
**25.** When the price is set at $1200, 60,000 computers are sold. For every $100 increase in price, the sales decrease by 2000 computers. 59,000 computers.
**27.** The profit from manufacturing and selling 100 luxury cars is $90,000. Each additional car made and sold creates an additional profit of $1200; $88,800
**29. (a)** $C'(5) = 74$ dollars per item
**(b)** $C(5.25) \approx C(5) + C'(5)(.25) = 256.5$ dollars **(c)** $x = 4$ **(d)** $C'(4) = 62$, $R'(4) = 29$. If the production is increased by one unit, cost will rise by $62,000 and revenue will increase by $12,000. The company should not increase production beyond the break-even point.
**31. (a)** $500 billion **(b)** $50 billion/yr **(c)** 1994 **(d)** 1994
**33. (a)**

$y = f(t)$

$y = f'(t)$

$[.5, 6]$ $by$ $[-3, 3]$

**(b)** .85 sec
**(c)** 5 days
**(d)** $-.05$ sec/day
**(e)** 3 days

## CHAPTER 1: SUPPLEMENTARY EXERCISES

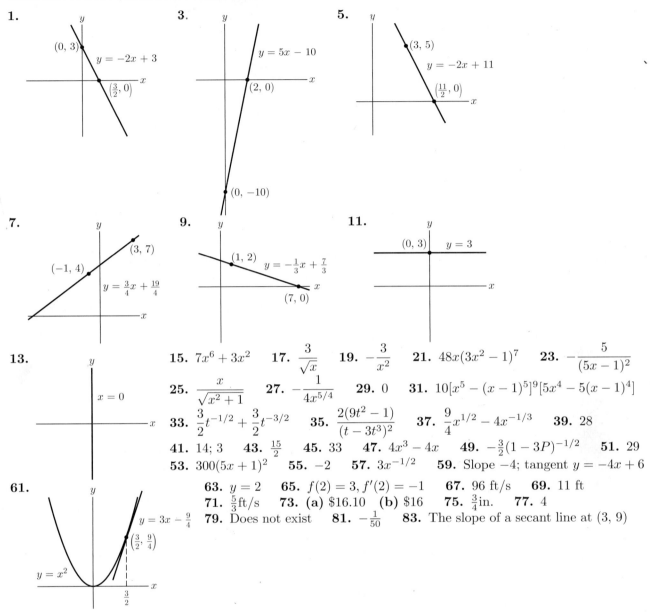

**1.** $(0, 3)$, $y = -2x + 3$, $\left(\frac{3}{2}, 0\right)$

**3.** $y = 5x - 10$, $(2, 0)$, $(0, -10)$

**5.** $(3, 5)$, $y = -2x + 11$, $\left(\frac{11}{2}, 0\right)$

**7.** $(3, 7)$, $(-1, 4)$, $y = \frac{3}{4}x + \frac{19}{4}$

**9.** $(1, 2)$, $y = -\frac{1}{3}x + \frac{7}{3}$, $(7, 0)$

**11.** $(0, 3)$, $y = 3$

**13.** $x = 0$

**15.** $7x^6 + 3x^2$ **17.** $\dfrac{3}{\sqrt{x}}$ **19.** $-\dfrac{3}{x^2}$ **21.** $48x(3x^2 - 1)^7$ **23.** $-\dfrac{5}{(5x - 1)^2}$

**25.** $\dfrac{x}{\sqrt{x^2 + 1}}$ **27.** $-\dfrac{1}{4x^{5/4}}$ **29.** $0$ **31.** $10[x^5 - (x - 1)^5]^9[5x^4 - 5(x - 1)^4]$

**33.** $\dfrac{3}{2}t^{-1/2} + \dfrac{3}{2}t^{-3/2}$ **35.** $\dfrac{2(9t^2 - 1)}{(t - 3t^3)^2}$ **37.** $\dfrac{9}{4}x^{1/2} - 4x^{-1/3}$ **39.** $28$

**41.** $14; 3$ **43.** $\frac{15}{2}$ **45.** $33$ **47.** $4x^3 - 4x$ **49.** $-\frac{3}{2}(1 - 3P)^{-1/2}$ **51.** $29$

**53.** $300(5x + 1)^2$ **55.** $-2$ **57.** $3x^{-1/2}$ **59.** Slope $-4$; tangent $y = -4x + 6$

**61.** $y = 3x - \frac{9}{4}$, $\left(\frac{3}{2}, \frac{9}{4}\right)$, $y = x^2$, $\frac{3}{2}$

**63.** $y = 2$ **65.** $f(2) = 3, f'(2) = -1$ **67.** $96\ \text{ft/s}$ **69.** $11\ \text{ft}$

**71.** $\frac{5}{3}\text{ft/s}$ **73.** (a) \$16.10 (b) \$16 **75.** $\frac{3}{4}$in. **77.** $4$

**79.** Does not exist **81.** $-\frac{1}{50}$ **83.** The slope of a secant line at $(3, 9)$

## CHAPTER 2

### EXERCISES 2.1

**1.** (a), (e), (f) **3.** (b), (c), (d) **5.** Increasing for $x < \frac{1}{2}$, relative maximum point at $x = \frac{1}{2}$, maximum value $= 1$, decreasing for $x > \frac{1}{2}$, concave down, $y$-intercept $(0,0)$, $x$- intercepts $(0, 0)$ and $(1, 0)$.
**7.** Decreasing for $x < 0$, relative minimum point at $x = 0$, relative minimum value $= 2$, increasing for $0 < x < 2$, relative maximum point at $x = 2$, relative maximum value $= 4$, decreasing for $x > 2$, concave up for $x < 1$, concave down for $x > 1$, inflection point at $(1, 3)$, $y$-intercept $(0, 2)$, $x$-intercept $(3.6, 0)$ **9.** Decreasing for $x < 2$, relative minimum at $x = 2$, minimum value $= 3$, increasing for $x > 2$, concave up for all $x$, no inflection point, defined for $x > 0$, the line $y = x$ is an asymptote, the $y$-axis is an asymptote **11.** Decreasing for $1 \le x < 3.2$, relative minimum point at $x = 3.2$, increasing for $x > 3.2$, maximum value $= 6$ (at $x = 1$), minimum value $= .9$ (at $x = 3.2$), inflection point at $x = 4$, concave up for $1 \le x < 4$, concave down for $x > 4$, the line

$y = 4$ is an asymptote.    **13.**   Slope decreases for all $x$.    **15.**   Slope decreases for $x < 1$, increases for $x > 1$. Minimum slope occurs at $x = 1$.    **17.**   (a) $C$, $F$ (b) $A$, $B$, $F$ (c) $C$

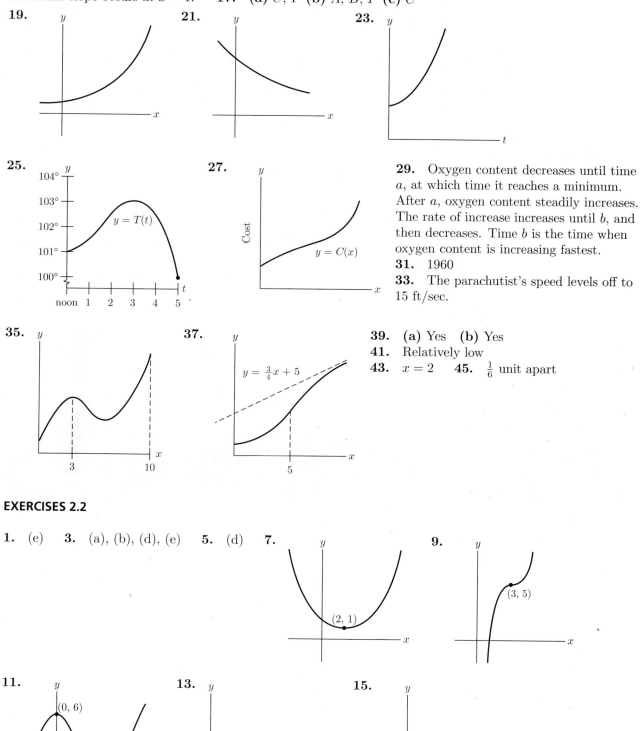

**19.**

**21.**

**23.**

**25.**

**27.**

**29.**   Oxygen content decreases until time $a$, at which time it reaches a minimum. After $a$, oxygen content steadily increases. The rate of increase increases until $b$, and then decreases. Time $b$ is the time when oxygen content is increasing fastest.
**31.**   1960
**33.**   The parachutist's speed levels off to 15 ft/sec.

**35.**

**37.**

**39.**   (a) Yes   (b) Yes
**41.**   Relatively low
**43.**   $x = 2$    **45.**   $\frac{1}{6}$ unit apart

## EXERCISES 2.2

**1.**  (e)    **3.**  (a), (b), (d), (e)    **5.**  (d)    **7.**

**9.**

**11.**

**13.**

**15.**

**17.**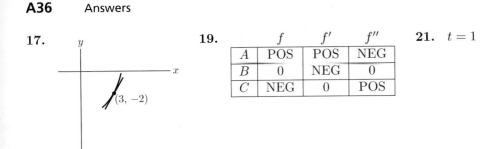

**19.**

| | $f$ | $f'$ | $f''$ |
|---|---|---|---|
| $A$ | POS | POS | NEG |
| $B$ | 0 | NEG | 0 |
| $C$ | NEG | 0 | POS |

**21.** $t = 1$

**23.** **(a)** decreasing **(b)** The function $f(x)$ is increasing for $1 \leq x < 2$ because the values of $f'(x)$ are positive. The function $f(x)$ is decreasing for $2 < x \leq 3$ because the values of $f'(x)$ are negative. Therefore, $f(x)$ has a relative maximum at $x = 2$. Coordinates: $(2, 9)$ **(c)** The function $f(x)$ is decreasing for $9 \leq x < 10$ because the values of $f'(x)$ are negative. The function $f(x)$ is increasing for $10 < x \leq 11$ because the values of $f'(x)$ are positive. Therefore, $f(x)$ has a relative minimum at $x = 10$. **(d)** concave down **(e)** at $x = 6$; coorrdinates: $(6, 5)$ **(f)** $x = 15$ **25.** The slope is positive because $f'(6) = 2$, a positive number. **27.** The slope is 0 because $f'(3) = 0$. Also, $f'(x)$ is positive for $x$ slightly less than 3, and $f'(x)$ is negative for $x$ slightly greater than 3. Hence $f(x)$ changes from increasing to decreasing at $x = 3$. **29.** $f'(x)$ is increasing at $x = 0$, so the graph of $f(x)$ is concave up. **31.** At $x = 1$, $f'(x)$ changes from increasing to decreasing, so the concavity of the graph of $f(x)$ changes from concave up to concave down. **33.** $y - 3 = 2(x - 6)$ **35.** 3.25 **37.** **(a)** $\frac{1}{6}$ in. **(b)** (ii), because the water level is falling. **39.** II **41.** I **43.** **(a)** 2 million **(b)** 30,000 farms per year **(c)** 1940 **(d)** 1945 and 1978 **(e)** 1960 **45.** rel. max: $x \approx -2.34$; rel. min: $x \approx 2.34$; inflection point: $x = 0$, $x \approx \pm 1.41$

## EXERCISES 2.3

**1.** $f'(x) = 3(x + 3)(x - 3)$; relative maximum point $(-3, 54)$; relative minimum point $(3, -54)$.

| Critical Values | | $-3$ | | $3$ | |
|---|---|---|---|---|---|
| | $x < -3$ | | $-3 < x < 3$ | | $3 < x$ |
| $3(x + 3)$ | $-$ | $0$ | $+$ | | $+$ |
| $x - 3$ | $-$ | | $-$ | $0$ | $+$ |
| $f'(x)$ | $+$ | $0$ | $-$ | $0$ | $+$ |
| $f(x)$ | Increasing on $(-\infty, -3)$ | $54$ | Decreasing on $(-3, 3)$ | $-54$ | Increasing on $(3, \infty)$ |
| | | Local maximum $(-3, 54)$ | | Local minimum $(3, -54)$ | |

**3.**  $f'(x) = -3(x-1)(x-3)$; relative maximum point $(3, 1)$; relative minimum point $(1, -3)$.

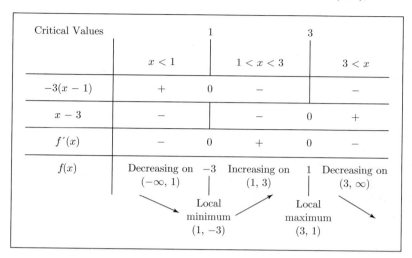

| Critical Values | | 1 | | 3 | |
|---|---|---|---|---|---|
| | $x < 1$ | | $1 < x < 3$ | | $3 < x$ |
| $-3(x-1)$ | $+$ | $0$ | $-$ | | $-$ |
| $x - 3$ | $-$ | | $-$ | $0$ | $+$ |
| $f'(x)$ | $-$ | $0$ | $+$ | $0$ | $-$ |
| $f(x)$ | Decreasing on $(-\infty, 1)$ | $-3$ | Increasing on $(1, 3)$ | $1$ | Decreasing on $(3, \infty)$ |

Local minimum $(1, -3)$   Local maximum $(3, 1)$

**5.**  $f'(x) = x(x-2)$; relative maximum point $(0, 1)$; relative minimum point $(2, -1/3)$.

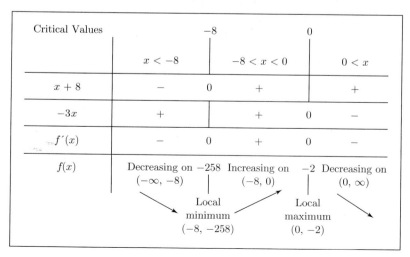

| Critical Values | | 0 | | 2 | |
|---|---|---|---|---|---|
| | $x < 0$ | | $0 < x < 2$ | | $2 < x$ |
| $x$ | $-$ | $0$ | $+$ | | $+$ |
| $x - 2$ | $-$ | | $-$ | $0$ | $+$ |
| $f'(x)$ | $+$ | $0$ | $-$ | $0$ | $+$ |
| $f(x)$ | Increasing on $(-\infty, 0)$ | $1$ | Decreasing on $(0, 2)$ | $-1/3$ | Increasing on $(2, \infty)$ |

Local maximum $(0, 1)$   Local minimum $(2, -1/3)$

**7.**  $f'(x) = -3x(x+8)$; relative maximum point $(0, -2)$; relative minimum point $(-8, -258)$.

| Critical Values | | $-8$ | | 0 | |
|---|---|---|---|---|---|
| | $x < -8$ | | $-8 < x < 0$ | | $0 < x$ |
| $x + 8$ | $-$ | $0$ | $+$ | | $+$ |
| $-3x$ | $+$ | | $+$ | $0$ | $-$ |
| $f'(x)$ | $-$ | $0$ | $+$ | $0$ | $-$ |
| $f(x)$ | Decreasing on $(-\infty, -8)$ | $-258$ | Increasing on $(-8, 0)$ | $-2$ | Decreasing on $(0, \infty)$ |

Local minimum $(-8, -258)$   Local maximum $(0, -2)$

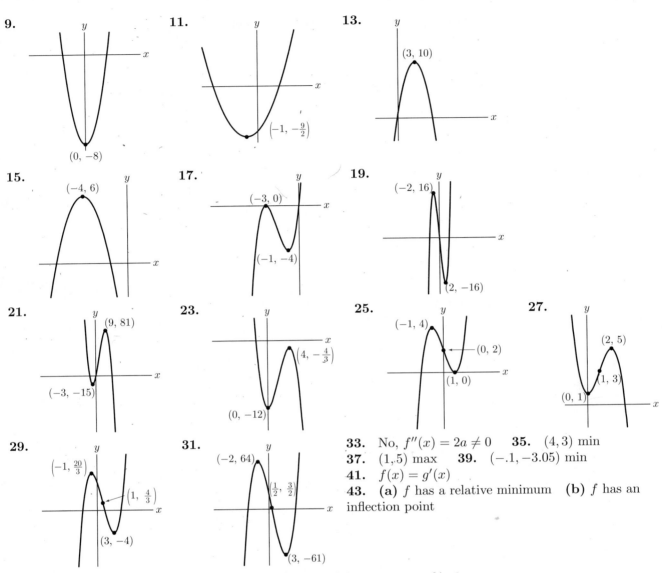

**9.** $(0, -8)$

**11.** $\left(-1, -\frac{9}{2}\right)$

**13.** $(3, 10)$

**15.** $(-4, 6)$

**17.** $(-3, 0)$, $(-1, -4)$

**19.** $(-2, 16)$, $(2, -16)$

**21.** $(9, 81)$, $(-3, -15)$

**23.** $\left(4, -\frac{4}{3}\right)$, $(0, -12)$

**25.** $(-1, 4)$, $(0, 2)$, $(1, 0)$

**27.** $(2, 5)$, $(1, 3)$, $(0, 1)$

**29.** $\left(-1, \frac{20}{3}\right)$, $\left(1, \frac{4}{3}\right)$, $(3, -4)$

**31.** $(-2, 64)$, $\left(\frac{1}{2}, \frac{3}{2}\right)$, $(3, -61)$

**33.** No, $f''(x) = 2a \neq 0$     **35.** $(4, 3)$ min
**37.** $(1, 5)$ max     **39.** $(-.1, -3.05)$ min
**41.** $f(x) = g'(x)$
**43.** **(a)** $f$ has a relative minimum   **(b)** $f$ has an
inflection point

**45.** **(a)** $A(x) = -893.103x + 460.759$ (billion dollars). **(b)** Revenue is $x\%$ of assets or
$R(x) = \frac{xA(x)}{100} = \frac{x}{100}(-893.103x + 460.759)$. $R(.3) \approx .578,484$ billion dollars or 578.484 million dollars;
$R(.1) \approx .371$, 449 billion dollars or 371.449 million dollars. **(c)** Maximum revenue when $R'(x) = 0$ or $x \approx .258$.
Maximum revenue $R(.258) \approx .594,273$ or 594.273 million dollars.     **47.** **(a)** 1984   **(b)** 70%   **(c)** 1987; 40%
**49.** $\left(5, \frac{10}{3}\right)$     **51.** $(5, 5)$

## EXERCISES 2.4

**1.** $\left(\dfrac{3 \pm \sqrt{5}}{2}, 0\right)$     **3.** $(-2, 0)$, $\left(-\frac{1}{2}, 0\right)$     **5.** $\left(\frac{1}{2}, 0\right)$     **7.** The derivative $x^2 - 4x + 5$ has no zeros.

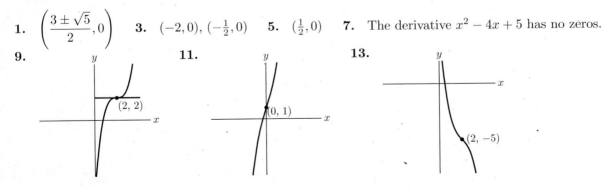

**9.** $(2, 2)$

**11.** $(0, 1)$

**13.** $(2, -5)$

**15.**

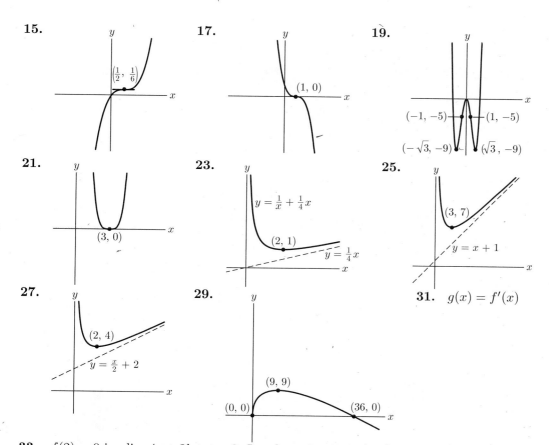

**17.**

**19.**

**21.**

**23.**

**25.**

**27.**

**29.**

**31.** $g(x) = f'(x)$

**33.** $f(2) = 0$ implies $4a + 2b + c = 0$. Local maximum at $(0, 1)$ implies $f'(0) = 0$ and $f(0) = 1$. $a = -1/4$, $b = 0$, $c = 1$, $f(x) = -1/4x^2 + 1$. **35.** If $f'(a) = 0$ and $f'$ is increasing at $x = a$, then $f'(x) < 0$ for $x_r < a$ and $f'(x) > 0$ for $x > a$. By the first derivative test (case (b)), $f$ has a local minimum at $x = a$.

**37.** (a)

$[0, 20]$ by $[-12, 50]$

(b) 15.0 g
(c) after 12.0 days
(d) 1.6 g/day
(e) after 6.0 days and after 17.6 days
(f) after 11.8 days

## EXERCISES 2.5

**1.** 20    **3.** $t = 4$, $f(4) = 8$    **5.** $x = 1, y = 1$, maximum $= 1$    **7.** $x = 3, y = 3$, minimum $= 18$
**9.** $x = 6, y = 6$, minimum $= 12$    **11.** (a) Objective: $A = xy$, constraint: $8x + 4y = 320$ (b) $A = -2x^2 + 80x$
(c) $x = 20$ ft, $y = 40$ ft    **13.** (a)

(b) $h + 4x$
(c) Objective: $V = x^2h$; constraint: $h + 4x = 84$
(d) $V = -4x^3 + 84x^2$
(e) $x = 14$ in., $h = 28$ in.

**15.** Let $x$ be the length of the fence and $y$ the other dimension. Objective: $C = 15x + 20y$; constraint: $xy = 75$; $x = 10$ ft, $y = 7.5$ ft
**17.** Let $x$ be the length of each edge of the base and $h$ the height. Objective: $A = 2x^2 + 4xh$; constraint: $x^2h = 8000$; 20 cm by 20 cm by 20 cm

**19.** Let $x$ be the length of the fence parallel to the river and $y$ the length of each section perpendicular to the river. Objective: $A = xy$; constraint: $6x + 15y = 1500$; $x = 125$ ft, $y = 50$ ft

**21.** Objective: $P = xy$; constraint: $x + y = 100$; $x = 50$, $y = 50$    **23.** Objective: $A = \dfrac{\pi x^2}{2} + 2xh$;

constraint: $(2 + \pi)x + 2h = 14$; $x = \dfrac{14}{4 + \pi}$ ft    **25.**    $w = 20$ ft, $x = 10$ ft

**27.** **(a)** $C(x) = 6x + 10\sqrt{(20 - x)^2 + 24^2}$; $C'(x) = 6 - \dfrac{10(20-x)}{\sqrt{(20-x)^2+24^2}}$; $C'(x) = 0$ $(0 \leq x \leq 20)$ implies $x = 2$.

Use the first derivative test to conclude that the minimum cost is $C(2) = 312$ dollars.    **29.** $\left(\frac{3}{2}, \sqrt{\frac{3}{2}}\right)$

**31.** $x = 2$, $y = 1$

## EXERCISES 2.6

**1.** **(a)** 90  **(b)** 180  **(c)** 6  **(d)** 1080 pounds    **3.** **(a)** $C = 16r + 2x$  **(b)** Constraint $rx = 800$
**(c)** $x = 80$, $r = 10$, minimum inventory cost $= \$320$    **5.** Let $x$ be the number of cases per order and $r$ the number of orders per year. Objective: $C = 80r + 5x$; constraint: $rx = 10{,}000$ **(a)** $\$4100$  **(b)** 400 cases
**7.** Let $r$ be the number of production runs and $x$ the number of microscopes manufactured per run. Objective: $C = 2500r + 25x$; constraint: $rx = 1600$; 4 runs    **11.** Objective: $A = (100 + x)w$; constraint: $2x + 2w = 300$; $x = 25$ ft, $w = 125$ ft    **13.** Objective: $F = 2x + 3w$; constraint: $xw = 54$; $x = 9$ m, $w = 6$ m    **15.** **(a)** $A(x) = 100x + 1000$. **(b)** $R(x) = A(x) \cdot (\text{Price}) = (100x + 1000)(18 - x)$ $(0 \leq x \leq 18)$. The graph of $R(x)$ is a parabola looking downward, with a maximum at $x = 4$. **(c)** $A(x)$ does not change, $R(x) = (100x + 1000)(9 - x)$ $(0 \leq x \leq 9)$. Maximum value when $x = 0$.    **17.** Let $x$ be the length of each edge of the base and $h$ the height. Objective: $C = 6x^2 + 10xh$; constraint: $x^2 h = 150$; 5 ft by 5 ft by 6 ft    **19.** Let $x$ be the length of each edge of the end and $h$ the length. Objective: $V = x^2 h$; constraint: $2x + h = 120$; 40 cm by 40 cm by 40 cm    **21.** Objective: $V = w^2 x$; constraint: $2x + w = 16$; $\frac{8}{3}$ in.    **23.** After 20 days
**25.** $2\sqrt{3}$ by 6    **27.** 10 in. by 10 in. by 4 in.    **29.** $\approx 3.77$ cm

## EXERCISES 2.7

**1.** $\$1$    **3.** 32    **5.** 5    **7.** $x = 20$ units, $p = \$133.33$    **9.** 2 million tons, $\$156$ per ton    **11.** **(a)** $\$2.00$
**(b)** $\$2.30$    **13.** Let $x$ be the number of prints and $p$ the price per print. Demand equation: $p = 650 - 5x$; revenue: $R(x) = (650 - 5x)x$; 65 prints    **15.** Let $x$ be the number of tables and $p$ the profit per table.
$p = 16 - .5x$; profit from the cafe: $R = (16 - .5x)x$; 16 tables    **17.** **(a)** $x = 15 \cdot 10^5$, $p = \$45$. **(b)** No. Profit is maximized when price is increased to $\$50$.    **19.** 5%    **21.** **(a)** $\$75{,}000$  **(b)** $\$3200$ per unit  **(c)** 15 units
**(d)** 32.5 units  **(e)** 35 units

## CHAPTER 2: SUPPLEMENTARY EXERCISES

**1.** **(a)** increasing: $-3 < x < 1, x > 5$; decreasing: $x < -3, 1 < x < 5$   **(b)** concave up: $x < -1, x > 3$; concave down: $-1 < x < 3$   **3.**   **5.**   **7.** d, e   **9.** c, d   **11.** e

**13.** Graph goes through $(1, 2)$, increasing at $x = 1$.   **15.** Increasing and concave up at $x = 3$.
**17.** $(10, 2)$ is a relative minimum point.   **19.** Graph goes through $(5, -1)$, decreasing at $x = 5$.
**21.** **(a)** after 2 hours   **(b)** .8   **(c)** after 3 hours   **(d)** $-.02$ units per hour

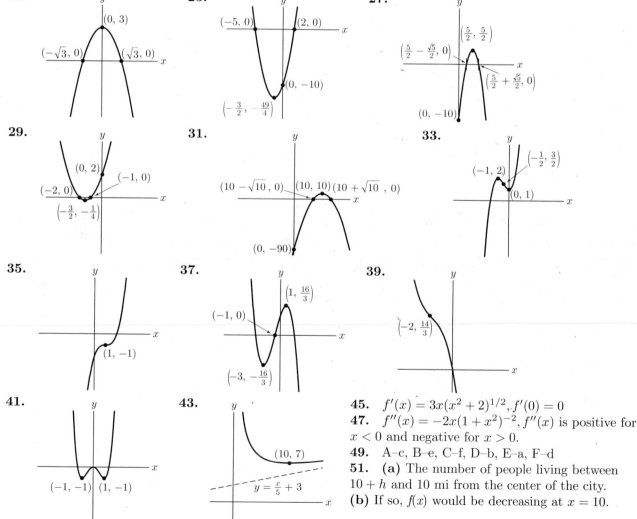

**45.** $f'(x) = 3x(x^2 + 2)^{1/2}, f'(0) = 0$
**47.** $f''(x) = -2x(1 + x^2)^{-2}, f''(x)$ is positive for $x < 0$ and negative for $x > 0$.
**49.** A–c, B–e, C–f, D–b, E–a, F–d
**51.** **(a)** The number of people living between $10 + h$ and 10 mi from the center of the city.
**(b)** If so, $f(x)$ would be decreasing at $x = 10$.

**53.** The endpoint maximum value of 2 occurs at $x = 0$.   **55.** Let $x$ be the width and $h$ the height.
Objective: $A = 4x + 2xh + 8h$; constraint: $4xh = 200$; 4 ft by 10 ft by 5 ft   **57.** $\frac{15}{2}$ in.
**59.** Let $r$ be the number of production runs and $x$ the number of books manufactured per run.
Objective: $C = 1000r + (.25)x$; constraint: $rx = 400,000$; $x = 40,000$   **61.** $x = 3500$   **63.** Let $x$ be the number of people and $c$ the cost. Objective: $R = xc$; constraint: $c = 1040 - 20x$; 25 people.

## CHAPTER 3

### EXERCISES 3.1

**1.** $(x+1) \cdot (3x^2+5) + (x^3+5x+2) \cdot 1$, or $4x^3 + 3x^2 + 10x + 7$

**3.** $(2x^4 - x + 1) \cdot (-5x^4) + (-x^5 + 1) \cdot (8x^3 - 1)$, or $-18x^8 + 6x^5 - 5x^4 + 8x^3 - 1$

**5.** $x(4)(x^2+1)^3(2x) + (x^2+1)^4(1)$ or $(x^2+1)^3(9x^2+1)$

**7.** $(x^2+3) \cdot 10(x^2-3)^9(2x) + (x^2-3)^{10} \cdot 2x$, or $2x(x^2-3)^9(11x^2+27)$

**9.** $(5x+1) \cdot 2x + (x^2-1) \cdot 5 + \frac{2}{3}$, or $15x^2 + 2x - \frac{13}{3}$

**11.** $\dfrac{(x+1) \cdot 1 - (x-1) \cdot 1}{(x+1)^2}$, or $\dfrac{2}{(x+1)^2}$  **13.** $\dfrac{(x^2+1) \cdot 2x - (x^2-1) \cdot 2x}{(x^2+1)^2}$ or $\dfrac{4x}{(x^2+1)^2}$

**15.** $\dfrac{(2x+1)^2 - 2(2x+1)(2)(x+3)}{(2x+1)^4}$ or $\dfrac{-2x-11}{(2x+1)^3}$  **17.** $\dfrac{-4x}{(x^2+1)^2}$

**19.** $\dfrac{(3-x^2) \cdot (6x+5) - (3x^2+5x+1) \cdot (-2x)}{(3-x^2)^2}$, or $\dfrac{5(x+1)(x+3)}{(3-x^2)^2}$

**21.** $2[(3x^2+2x+2)(x-2)][(3x^2+2x+2) \cdot 1 + (x-2) \cdot (6x+2)]$, or $2(3x^2+2x+2)(x-2)(9x^2-8x-2)$

**23.** $\dfrac{-1}{2\sqrt{x}(\sqrt{x}+1)^2}$  **25.** $\dfrac{3x^4 - 4x^2 - 3}{x^2}$

**27.** $\sqrt{x+2}\,2\,(2x+1)(2) + (2x+1)^2\dfrac{1}{2\sqrt{x+2}}$ or $\dfrac{(2x+1)(10x+17)}{2\sqrt{x+2}}$

**29.** $y - 16 = 88(x-3)$  **31.** $2, 7$  **33.** $0, \pm 2, \pm\frac{5}{4}$  **35.** $(\frac{1}{2}, \frac{3}{2}), (-\frac{1}{2}, \frac{9}{2})$  **37.** $\dfrac{dy}{dx} = 8x(x^2+1)^3$,

$\dfrac{d^2y}{dx^2} = 8(x^2+1)^2(7x^2+1)$  **39.** $\dfrac{dy}{dx} = \dfrac{x}{2\sqrt{x+1}} + \sqrt{x+1}$ or $\dfrac{dy}{dx} = \dfrac{3x+2}{2\sqrt{x+1}}, \dfrac{d^2y}{dx^2} = \dfrac{3x+4}{4(x+1)^{3/2}}$

**41.** $x \cdot f'(x) + f(x)$  **43.** $\dfrac{(x^2+1) \cdot f'(x) - 2x \cdot f(x)}{(x^2+1)^2}$  **45.** $2 \times 3 \times 1$

**47.** $150; AC(150) = 35 = C'(150)$  **49.** $AR$ is maximized where $0 = \dfrac{d}{dx}(AR) = \dfrac{x \cdot R'(x) - R(x) \cdot 1}{x^2}$. This happens when the production level $x$ satisfies $xR'(x) - R(x) = 0$, and hence $R'(x) = R(x)/x = AR$.

**51.** $38 \text{ in.}^2/\text{s}$  **53.** $150,853,600 \text{ gal/yr}$  **55.** $(2, 10)$  **59.** $\dfrac{1 - 2x \cdot f(x)}{(1+x^2)^2}$  **61.** $\frac{1}{8}$

**65.** $f(x)g(x)h'(x) + f(x)g'(x)h(x) + f'(x)g(x)h(x)$  **67.** $\dfrac{h(t)w'(t) - 2w(t)h'(t)}{[h(t)]^3}$

**69.** (a)

[0, 6] by [−5, 20]

(b) $10.8 \text{ mm}^2$ (c) $2.61$ units of light (d) $-.55 \text{ mm}^2/\text{unit of light}$

### EXERCISES 3.2

**1.** $\dfrac{x^3}{x^3+1}$  **3.** $\sqrt{x}(x+1)$  **5.** $f(x) = x^5, g(x) = x^3 + 8x - 2$  **7.** $f(x) = \sqrt{x}, g(x) = 4 - x^2$

**9.** $f(x) = \dfrac{1}{x}, g(x) = x^3 - 5x^2 + 1$  **11.** $30x(x^2+5)^{14}$

**13.** $6x^2 \cdot 3(x-1)^2(1) + (x-1)^3 \cdot 12x$, or $6x(x-1)^2(5x-2)$

**15.** $2(x^3-1) \cdot 4(3x^2+1)^3(6x) + (3x^2+1)^4 \cdot 2(3x^2)$, or $6x(3x^2+1)^3(11x^3 + x - 8)$

**17.** $\dfrac{d}{dx}[4^3(1-x)^{-3}] = 192(1-x)^{-4}$  **19.** $3\left(\dfrac{4x-1}{3x+1}\right)^2 \cdot \dfrac{(3x+1) \cdot 4 - (4x-1) \cdot 3}{(3x+1)^2}$, or $\dfrac{21(4x-1)^2}{(3x+1)^4}$

**21.** $2xf'(x^2)$   **23.** $f'(-x)$   **25.** $\dfrac{2x^2 f'(x^2) - f(x^2)}{x^2}$   **27.**   **29.** $30(6x-1)^4$

$(1, 1)$   $\left(2, \frac{8}{9}\right)$

$y = \dfrac{4x}{(x+1)^2}$

**31.** $-(1-x^2)^{-2} \cdot (-2x)$, or $2x(1-x^2)^{-2}$   **33.** $[4(x^2-4)^3 - 2(x^2-4)] \cdot (2x)$, or $8x(x^2-4)^3 - 4x(x^2-4)$

**35.** $2[(x^2+5)^3+1]3(x^2+5)^2 \cdot (2x)$, or $12x[(x^2+5)^3+1](x^2+5)^2$   **37.** $6(4x+1)^{1/2}$

**39.** $\dfrac{dy}{dx} = \left(\dfrac{1}{2} - \dfrac{2}{u^2}\right)(1-2x)$ or $\dfrac{dy}{dx} = \left(\dfrac{1}{2} - \dfrac{2}{(x-x^2)^2}\right)(1-2x)$

**41.** $\dfrac{dy}{dt} = (2x-3)(2t)$ when $t=0$, $x=3$, $\dfrac{dy}{dt} = (3)(0) = 0$

**43.** $\dfrac{dy}{dt} = \dfrac{-2}{(x-1)^2} \cdot \dfrac{t}{2}$ when $t=3$, $x=\dfrac{9}{4}$, $\dfrac{dy}{dt} = -\dfrac{48}{25}$   **45.** $y = 62x - 300$   **47.** $1; 2; 3$

**49.** **(a)** $\dfrac{dV}{dt} = \dfrac{dV}{dx} \cdot \dfrac{dx}{dt}$   **(b)** 2   **51.** **(a)** $\dfrac{dy}{dt}, \dfrac{dP}{dy}, \dfrac{dP}{dt}$   **(b)** $\dfrac{dP}{dt} = \dfrac{dP}{dy} \cdot \dfrac{dy}{dt}$

**53.** **(a)** $\dfrac{200(100-x^2)}{(100+x^2)^2}$   **(b)** $\dfrac{400[100-(4+2t)^2]}{[100+(4+2t)^2]^2}$   **(c)** Falling at the rate of \$480 per week

**55.** **(a)** $.4 + .0002x$   **(b)** increasing at the rate of 25 thousand persons per year   **(c)** rising at the rate of 14 ppm per year   **57.** $x^3 + 1$   **59.** 24

**61.** **(a)** $t = 1.5$, $x = 40$, $W \approx 77.209$ million dollars; $t = 3.5$, $x = 30$, $W \approx 76.364$ million dollars

**(b)** $\dfrac{dx}{dt}\Big|_{t=1.5} = 20$; the price of one share is \$40 and is rising at the rate of \$20 per month. $\dfrac{dx}{dt}\Big|_{t=3.5} = 0$; the price of one share is steady at \$30 a share.

**63.** **(a)** $t = 2.5$, $x = 40$, $\dfrac{dx}{dt}\Big|_{t=2.5} = -20$; the price of one share is \$40 and is falling at the rate of \$20 per month. $t = 4$, $x = 30$, $\dfrac{dx}{dt}\Big|_{t=4} = 0$; the price of one share is steady at \$30 a share.

**(b)** $\dfrac{dW}{dt}\Big|_{t=2.5} = \dfrac{dW}{dx}\Big|_{x=40} \dfrac{dx}{dt}\Big|_{t=2.5} = \dfrac{120}{1849} \cdot (-20) \approx -1.3$; the total value of the company is falling at the rate of 1.3 million dollars per month. $\dfrac{dW}{dt} = \dfrac{dW}{dx}\dfrac{dx}{dt} = 0$; when $x = 30$, the total value of the company was steady at $W \approx 76.364$ million dollars.

**65.** The derivative of the composite function $f(g(x))$ is the derivative of the outer function evaluated at the inner function and then multiplied by the derivative of the inner function.

## EXERCISES 3.3

**1.** $\dfrac{x}{y}$   **3.** $\dfrac{1+6x}{5y^4}$   **5.** $\dfrac{2x^3-x}{2y^3-y}$   **7.** $\dfrac{1-6x^2}{1-6y^2}$   **9.** $-\dfrac{y}{x}$   **11.** $-\dfrac{y+2}{5x}$   **13.** $\dfrac{8-3xy^2}{2x^2y}$   **15.** $\dfrac{x^2(y^3-1)}{y^2(1-x^3)}$

**17.** $-\dfrac{y^2+2xy}{x^2+2xy}$   **19.** $\dfrac{1}{2}$   **21.** $-\dfrac{8}{3}$   **23.** $-\dfrac{2}{15}$   **25.** $y - \dfrac{1}{2} = -\dfrac{1}{16}(x-4), y + \dfrac{1}{2} = \dfrac{1}{16}(x-4)$

**27. (a)** $\dfrac{2x-x^3-xy^2}{2y+y^3+x^2y}$   **(b)** 0   **29. (a)** $\dfrac{-y}{2x}$   **(b)** $\dfrac{-27}{16}$   **31.** $-\dfrac{x^3}{y^3}\dfrac{dx}{dt}$   **33.** $\dfrac{2x-y}{x}\dfrac{dx}{dt}$   **35.** $\dfrac{2x+2y}{3y^2-2x}\dfrac{dx}{dt}$

**37.** $-\dfrac{15}{8}$ units per second   **39.** Rising at 3 thousand units per week   **41.** Increasing at \$20 thousand per month   **43.** Decreasing at $\dfrac{1}{14}$ L per second   **45. (a)** $x^2 + y^2 = 100$   **(b)** $\dfrac{dy}{dt} = -4$, so the top of the ladder is falling at the rate of 4 ft/sec.   **47.** Decreasing at $\dfrac{22}{\sqrt{5}}$ ft/sec (or 9.84 ft/sec)

## CHAPTER 3: SUPPLEMENTARY EXERCISES

**1.** $(4x - 1) \cdot 4(3x + 1)^3(3) + (3x + 1)^4 \cdot 4$, or $4(3x + 1)^3(15x - 2)$

**3.** $x \cdot 3(x^5 - 1)^2 \cdot 5x^4 + (x^5 - 1)^3 \cdot 1$, or $(x^5 - 1)^2(16x^5 - 1)$

**5.** $5(x^{1/2} - 1)^4 \cdot 2(x^{1/2} - 2)(\frac{1}{2}x^{-1/2}) + (x^{1/2} - 2)^2 \cdot 20(x^{1/2} - 1)^3(\frac{1}{2}x^{-1/2})$, or $5x^{-1/2}(x^{1/2} - 1)^3(x^{1/2} - 2)(3x^{1/2} - 5)$

**7.** $3(x^2 - 1)^3 \cdot 5(x^2 + 1)^4(2x) + (x^2 + 1)^5 \cdot 9(x^2 - 1)^2(2x)$, or $12x(x^2 - 1)^2(x^2 + 1)^4(4x^2 - 1)$

**9.** $\dfrac{(x - 2) \cdot (2x - 6) - (x^2 - 6x) \cdot 1}{(x - 2)^2}$, or $\dfrac{x^2 - 4x + 12}{(x - 2)^2}$

**11.** $2\left(\dfrac{3 - x^2}{x^3}\right) \cdot \dfrac{x^3 \cdot (-2x) - (3 - x^2) \cdot 3x^2}{x^6}$, or $\dfrac{2(3 - x^2)(x^2 - 9)}{x^7}$

**13.** $-\frac{1}{3}, 3, \frac{31}{27}$    **15.** $y + 32 = 176(x + 1)$    **17.** $x = 44$ m, $y = 22$ m    **19.** $\dfrac{dC}{dt} = \dfrac{dC}{dx} \cdot \dfrac{dx}{dt} = 40 \cdot 3 = 120$. Costs are rising $120 per day.    **21.** $0; -\frac{7}{2}$    **23.** $\frac{3}{2}; -\frac{7}{8}$    **25.** $1; -\frac{3}{2}$    **27.** $\dfrac{3x^2}{x^6 + 1}$    **29.** $\dfrac{2x}{(x^2 + 1)^2 + 1}$

**31.** $\frac{1}{2}\sqrt{1 - x}$    **33.** $\dfrac{3x^2}{2(x^3 + 1)}$    **35.** $-\dfrac{25}{x(25 + x^2)}$    **37.** $\dfrac{x^{1/2}}{(1 + x^2)^{1/2}} \cdot \frac{1}{2}(x^{-1/2})$, or $\dfrac{1}{2\sqrt{1 + x^2}}$

**39. (a)** $\dfrac{dR}{dA}, \dfrac{dA}{dt}, \dfrac{dR}{dx}$, and $\dfrac{dx}{dA}$    **(b)** $\dfrac{dR}{dt} = \dfrac{dR}{dx}\dfrac{dx}{dA}\dfrac{dA}{dt}$    **41. (a)** $-y^{1/3}/x^{1/3}$    **(b)** $1$    **43.** $-3$    **45.** $\frac{3}{5}$

**47. (a)** $\dfrac{dy}{dx} = \dfrac{15x^2}{2y}$    **(b)** $\frac{20}{3}$ thousand dollars per thousand unit increase in production    **(c)** $\dfrac{dy}{dt} = \dfrac{15x^2}{2y}\dfrac{dx}{dt}$    **(d)** 2 thousand dollars per week    **49.** Increasing at the rate of 2.5 units per unit time    **51.** 1.89 m²/year

## CHAPTER 4

### EXERCISES 4.1

**1.** $2^{2x}, 3^{(1/2)x}, 3^{-2x}$    **3.** $2^{2x}, 3^{3x}, 2^{-3x}$    **5.** $2^{-4x}, 2^{9x}, 3^{-2x}$    **7.** $2^x, 3^x, 3^x$    **9.** $3^{2x}, 2^{6x}, 3^{-x}$    **11.** $2^{(1/2)x}, 3^{(4/3)x}$    **13.** $2^{-2x}, 3^x$    **15.** $\frac{1}{9}$    **17.** $1$    **19.** $2$    **21.** $-1$    **23.** $\frac{1}{5}$    **25.** $\frac{5}{2}$    **27.** $-1$    **29.** $4$    **31.** $1$    **33.** $1$ or $2$    **35.** $1$ or $2$    **37.** $2^h$    **39.** $2^h - 1$    **41.** $3^x + 1$    **43.**

0.6931    **45.** 2.7

dy/dx=.69314724

$[-1, 2]$ by $[-1, 4]$

### EXERCISES 4.2

**1.** 1.16, 1.11, 1.10    **3. (a)** $2m$ **(b)** $m/4$ where $m \approx .693$    **5. (a)** $e$ **(b)** $1/e$    **7.** 1.005, 1.001, 1.000

**9.** $10e^{10x}$    **11. (a)** $4e^{4x}$ **(b)** $ke^{kx}$    **13.** $e^{2x}, e^{-x}$    **15.** $e^{-6x}, e^{2x}$    **17.** $e^{6x}, e^{2x}$    **19.** $x = 4$

**21.** $x = 4, -2$    **23.** $1$ or $-1$    **25.** $\dfrac{e^x}{2}$    **27.** $2(1 + 2e^x)(2e^x)$ or $4e^x(1 + 2e^x)$    **29.** $-\dfrac{1}{e^x}$

**31.** $xe^x + e^x$ or $e^x(x + 1)$    **33.** $\dfrac{e^x(1 + e^x) - e^x \cdot e^x}{(1 + e^x)^2}$ or $\dfrac{e^x}{(1 + e^x)^2}$    **35.** $20(1 + 5e^x)^3$    **37.** $y' = e^x(1 + x)^2$, $y' = 0$ when $x = -1$. The point is $(-1, 2/e)$.    **39.** $1$    **41.** $y - \frac{1}{3} = \frac{1}{9}x$    **43.** $f'(x) = e^x(x^2 + 4x + 3)$, $f''(x) = e^x(x^2 + 6x + 7)$    **45. (a)** 800 g/cm² **(b)** 14 km **(c)** $-50$ g/cm² per km **(d)** 2 km

**47.** $y = x + 1$

**51.** $\left.\dfrac{d}{dx}(10^x)\right|_{x=0} \approx 2.3026; \quad \dfrac{d}{dx}(10^x) = m10^x$ where $m = \left.\dfrac{d}{dx}(10^x)\right|_{x=0}$

$[-1, 3]$ *by* $[-3, 20]$

## EXERCISES 4.3

**1.** $8e^{2x}$ **3.** $4 - 2e^{-2x}$ **5.** $e^{t^2}(2t^2 + 2t + 1)$ **7.** $3(e^x + e^{-x})^2(e^x - e^{-x})$ **9.** $\frac{1}{8}e^{\frac{t}{2}+1}$ **11.** $-\frac{3}{t^2}e^{\frac{3}{t}}$
**13.** $e^{x^2-5x+4}(2x - 5)$ **15.** $\frac{1-x+e^{2x}}{e^x}$ **17.** $2e^{2t} - 4e^{-4t}$ **19.** $-xe^{-x+2}$ **21.** $x^3(2e^{2x}) + e^{2x}(3x^2)$ or
$x^2e^{2x}(2x + 3)$ **23.** $\dfrac{(15 + 4x)e^{4x}}{(4 + x)^2}$ **25.** $(-x^{-2} + 2x^{-1} + 6)e^{2x}$ **27.** Max at $x = -2/3$ **29.** Min at
$x = 5/4$ **31.** Max at $x = 9/10$ **33.** \$54,366 per year **35.** **(a)** 45 m/sec **(b)** 10 m/sec$^2$ **(c)** 4 sec
**(d)** 4 sec **37.** 2 in./week **39.** $.02e^{-2e^{-.01x}}e^{-.01x}$ **41.** $y = Ce^{-4x}$ **43.** $y = e^{-.5x}$ **47.** 1
**49.** **(a)** Volume seems to stabilize near 6 ml **(b)** 3.2 ml **(c)** 7.7 wk **(d)** .97 ml/wk **(e)** 3.7 wk
**(f)** 1.13 ml/wk

## EXERCISES 4.4

**1.** $\frac{1}{2}$ **3.** $\ln 5$ **5.** $\frac{1}{e}$ **7.** $-3$ **9.** $e$ **11.** $0$ **13.** $x^2$ **15.** $\frac{1}{49}$ **17.** $2x$ **19.** $\frac{1}{2}\ln 5$
**21.** $4 - e^{1/2}$ **23.** $\pm e^{9/2}$ **25.** $-\dfrac{\ln .5}{.00012}$ **27.** $\frac{5}{3}$ **29.** $\dfrac{e}{3}$ **31.** $3\ln\frac{9}{2}$ **33.** $\frac{1}{2}e^{8/5}$ **35.** $\frac{1}{2}\ln 4$
**37.** $-\ln\frac{3}{2}$ **39.** $x = \ln 5, y = 5(1 - \ln 5) \approx -3.07$ **41.** **(a)** $y' = -e^{-x}, y' = -2$ when $x = -\ln 2 = \ln(\frac{1}{2})$.
The point is $(\ln(\frac{1}{2}), 2)$. **(b)** **43.** $(-\ln 3, 3 - 3\ln 3)$, minimum

**45.** Max at $t = 2\ln 51$ **47.** $\ln 2$ **49.** The graph of $y = e^{\ln x}$ is the same as the graph of $y = x$ for $x > 0$.
**51.** $\frac{1}{5}e^2 \approx 1.4778$

$[-1, 3]$ *by* $[-3, 3]$

## EXERCISES 4.5

**1.** $\dfrac{1}{x}$ **3.** $\dfrac{1}{x + 3}$ **5.** $\dfrac{e^x - e^{-x}}{e^x + e^{-x}}$ **7.** $\left(\dfrac{1}{x} + 1\right)e^{\ln x + x}$ **9.** $\dfrac{1}{x}(\ln x + \ln 2x)$ **11.** $\dfrac{2\ln x}{x} + \dfrac{2}{x}$ or
$\dfrac{2}{x}(\ln x + 1)$ **13.** $\dfrac{1}{x}$ **15.** $2x(1 + \ln(x^2 + 1))$ **17.** $\dfrac{2}{x^2 - 1}$ **19.** $\dfrac{5e^{5x}}{e^{5x} + 1}$ **21.** $2(\ln 4)t$

**23.** $\dfrac{6\ln t - 3(\ln t)^2}{t^2}$ **25.** $y = 1$ **27.** $\left(e^2, \dfrac{2}{e}\right)$ **29.** $y' = x(1 + 2\ln x)$, $y'' = 3 + 2\ln x$, $y' = 0$ when $x = e^{-1/2}$, $y''(e^{-1/2}) > 0$, relative minimum at $x = e^{-1/2}$, $y = -e^{-1}/2$. **31.** $1 + 3\ln 10$

**33.** $R'(x) = \dfrac{45(\ln x - 1)}{(\ln x)^2}$; $R'(20) \approx 10$ **35.** $\frac{1}{7}$ **37.**

$[-5, 5]$ $by$ $[-2, 2]$

## EXERCISES 4.6

**1.** $\ln 5x$ **3.** $\ln 3$ **5.** $\ln 2$ **7.** $x^2$ **9.** $\ln \dfrac{x^5 z^3}{y^{1/2}}$ **11.** $3\ln x$ **13.** $3\ln 3$ **15.** (a) $2\ln 2 = 2(.69) = 1.38$
(b) $\ln 2 + \ln 3 = 1.79$ (c) $3\ln 3 + \ln 2 = 3.99$ **17.** (a) $-\ln 2 - \ln 3 = -1.79$ (b) $\ln 2 - 2\ln 3 = -1.51$
(c) $-\frac{1}{2}\ln 2 = -.345$ **19.** d **21.** d **23.** 3 **25.** $\sqrt{e}$ **27.** $e^4$ **29.** $e$ or $e^{-1}$ **31.** $\dfrac{1 + 2e}{e - 1}$

**33.** $\dfrac{1}{x + 5} + \dfrac{2}{2x - 1} - \dfrac{1}{4 - x}$ **35.** $\dfrac{2}{1 + x} + \dfrac{3}{2 + x} + \dfrac{4}{3 + x}$

**37.** $5 + \dfrac{1}{x + 4} + \dfrac{3}{3x - 2} - \dfrac{1}{x + 1}$ **39.** $\dfrac{5}{5x + 1} + \dfrac{4}{4x + 1} + \dfrac{1}{x \ln x} - \dfrac{1}{2x + 1}$

**41.** $\ln(3x + 1)\dfrac{5}{5x + 1} + \ln(5x + 1)\dfrac{3}{3x + 1}$ **43.** $(x + 1)^4(4x - 1)^2 \left(\dfrac{4}{x + 1} + \dfrac{8}{4x - 1}\right)$

**45.** $\dfrac{(x + 1)(2x + 1)(3x + 1)}{\sqrt{4x + 1}} \left(\dfrac{1}{x + 1} + \dfrac{2}{2x + 1} + \dfrac{3}{3x + 1} - \dfrac{2}{4x + 1}\right)$

**47.** $2^x \ln 2$ **49.** $x^x[1 + \ln x]$ **51.** $y = cx^k$ **53.** $h = 3, k = \ln 2$

## CHAPTER 4: SUPPLEMENTARY EXERCISES

**1.** 81 **3.** $\frac{1}{25}$ **5.** 4 **7.** 9 **9.** $e^{3x^2}$ **11.** $e^{2x}$ **13.** $e^{11x} + 7e^x$ **15.** $x = 4$ **17.** $x = -5$
**19.** $70e^{7x}$ **21.** $e^{x^2} + 2x^2 e^{x^2}$ **23.** $e^x \cdot e^{e^x} = e^{x + e^x}$ **25.** $\dfrac{(e^{3x} + 3)(2x - 1) - (x^2 - x + 5)3e^{3x}}{(e^{3x} + 3)^2}$

**27.** $y = Ce^{-x}$ **29.** $y = 2e^{1.5x}$ **31.** **33.**

**35.**

**37.** $y = \frac{1}{4}x + \frac{1}{2}$ **39.** $\sqrt{5}$ **41.** $\frac{2}{3}$ **43.** 1 **45.** $e, \dfrac{1}{e}$

**47.** $\frac{1}{2}\ln 5$ **49.** $e^{5/2}$ **51.** $\dfrac{6x^5 + 12x^3}{x^6 + 3x^4 + 1}$ **53.** $\dfrac{5}{5x - 7}$

**55.** $\dfrac{2\ln x}{x}$ **57.** $\dfrac{1}{x} + 1 - \dfrac{1}{2(1+x)}$ **59.** $\ln x$ **61.** $\dfrac{1}{x\ln x}$

**63.** $\dfrac{e^x}{x} + e^x \ln x$ **65.** $\dfrac{x}{x^2 + 1} - \dfrac{1}{2x + 3}$ **67.** $2x - \dfrac{1}{x}$

**69.** $\ln 2$ **71.** $\dfrac{1}{x - 1}$ **73.** $-\frac{1}{2}x^{-\frac{1}{2}}$ or $\dfrac{-1}{2\sqrt{x}}$

**75.** $\sqrt[5]{\dfrac{x^5 + 1}{x^5 + 5x + 1}}\left[\dfrac{x^4}{x^5 + 1} - \dfrac{x^4 + 1}{x^5 + 5x + 1}\right]$ **77.** $x^{\sqrt{x} - \frac{1}{2}}\left[1 + \frac{1}{2}\ln x\right]$

**79.** $(x^2 + 5)^6(x^3 + 7)^8(x^4 + 9)^{10}\left[\dfrac{12x}{x^2 + 5} + \dfrac{24x^2}{x^3 + 7} + \dfrac{40x^3}{x^4 + 9}\right]$ **81.** $10^x \ln 10$

**83.** $\frac{1}{2}\sqrt{\dfrac{xe^x}{x^3 + 3}}\left[\dfrac{1}{x} + 1 - \dfrac{3x^2}{x^3 + 3}\right]$ **85.** $e^{x+1}(x^2 + 1)x\left[1 + \dfrac{2x}{x^2 + 1} + \dfrac{1}{x}\right]$ or $e^{x+1}(x^3 + 3x^2 + x + 1)$

**87.**

(0, ln 2)

**89.**

(0, ln 2)

**91.**

(e, 1)

(1, 0)

**93.**

(1, 1)

**95.** $\dfrac{e^x}{3 + e^x}$ **97.** 1

## CHAPTER 5

### EXERCISES 5.1

**1. (a)** $P(t) = 3e^{.02t}$ **(b)** 3 million **(c)** .02 **(d)** 3.52 million
**(e)** 80,000 people per year **(f)** 3.5 million **3. (a)** 5000 **(b)** $P'(t) = .2P(t)$ **(c)** 3.5 h **(d)** 6.9 h
**5.** .017 **7.** 22 yr **9.** 27 million cells **11.** 34.0 million **13. (a)** $P(t) = 8e^{-.021t}$ **(b)** 8 g **(c)** .021
**(d)** 6.5 g **(e)** .021 g/yr **(f)** 5 g **(g)** 4 g; 2 g; 1 g **15. (a)** $f'(t) = -.6f(t)$ **(b)** 14.9 mg **(c)** 1.2 h
**17.** 30.1 yr **19.** 176 days **21.** $f(t) = 8e^{-.014t}$ **23. (a)** 8 g **(b)** 3.5 h **(c)** .6 g/h **(d)** 8 h
**25.** 13,412 years **27.** 58.3% **29.** 10,900 years ago **31.** a–D, b–G, c–E, d–B, e–H, f–F, g–A, h–C

### EXERCISES 5.2

**1. (a)** \$5000 **(b)** 4% **(c)** \$7459.12 **(d)** $A'(t) = .04A(t)$ **(e)** \$298.36 per year **(f)** \$7000
**3. (a)** $A(t) = 4000e^{.035t}$ **(b)** $A'(t) = .035A(t)$ **(c)** \$4290.03 **(d)** 6.4 yr **(e)** \$175 per year
**5.** \$378 per year **7.** 15.3 yr **9.** 29.3% **11.** 17.3 years **13.** 7.3% **15.** 2002 **17.** 2006

**19.** $786.63     **21.** $7985.16     **23.** 15.7 yr     **25.** a–B, b–D, c–G, d–A, e–F, f–E, g–H, h–C     **27. (a)** $200
**(b)** $8 per year **(c)** 4% **(d)** 30 yr **(e)** 30 yr **(f)** $A'(t)$ is a constant multiple of $A(t)$ since $A'(t) = rA(t)$.
**29.**     **31.** .06

$[-.5, .5]$ $by$ $[-.5, 4]$

## EXERCISES 5.3

**1.** 20%, 4%     **3.** 30%, 30%     **5.** 60%, 300%     **7.** $-25\%, -10\%$     **9.** 12.5%     **11.** 5.8 yr
**13.** $p/(140 - p)$, elastic     **15.** $2p^2/(116 - p^2)$, inelastic     **17.** $p - 2$, elastic     **19. (a)** inelastic     **(b)** raised
**21. (a)** elastic     **(b)** increase     **23. (a)** 2     **(b)** yes     **29. (a)** $p < 2$     **(b)** $p < 2$

## EXERCISES 5.4

**1. (a)** $f'(x) = 10e^{-2x} > 0$, $f(x)$ increasing; $f''(x) = -20e^{-2x} < 0$, $f(x)$ concave down

**(b)** As $x$ becomes large, $e^{-2x} = \dfrac{1}{e^{2x}}$ approaches 0.     **(c)**

$y = 5$

$y = 5(1 - e^{-2x})$

**3.** $y' = 2e^{-x} = 2 - (2 - 2e^{-x}) = 2 - y$
**5.** $y' = 30e^{-10x} = 30 - (30 - 30e^{-10x}) = 30 - 10y = 10(3 - y)$, $f(0) = 3(1 - 1) = 0$     **7.** 4.8 h
**11. (a)** 2500     **(b)** 500 people/day     **(c)** day 12     **(d)** day 6 and day 13.5     **(e)** at time 9.78 days
**(f)** $f'(t) = .00004f(t)(10,000 - f(t))$ **(g)** 1000 people per day
**13. (a)**     **(b)** 30 units     **(c)** 25 units per hour     **(d)** 9 hr
**(e)** 65.3 units after 2 hr     **(f)** 4 hr

$y = f(t)$

$y = f'(t)$

$y = f''(t)$

$[0, 12]$ $by$ $[-20, 75]$

## CHAPTER 5: SUPPLEMENTARY EXERCISES

**1.** $29.92e^{-.2x}$     **3.** $5488.12     **5.** .058     **7. (a)** $17e^{.018t}$     **(b)** 20.4 million     **(c)** 2011     **9. (a)** $36,693
**(b)** The alternative investment is superior by $3859.     **11.** .02; 60,000 people per year; 5 million people
**13.** a–F, b–D, c–A, d–G, e–H, f–C, g–B, h–E     **15.** 6%     **17.** 400%     **19.** 3%, decrease     **21.** increase
**23.** $100(1 - e^{-.083t})$     **25. (a)** 400°     **(b)** decreasing at a rate of 100°/sec     **(c)** 17 sec     **(d)** 2 sec

## CHAPTER 6

### EXERCISES 6.1

**1.** $\frac{1}{2}x^2 + C$  **3.** $\frac{1}{3}e^{3x} + C$  **5.** $3x + C$  **7.** $x^4 + C$  **9.** $7x + C$  **11.** $\dfrac{x^2}{2c} + C$  **13.** $2\ln|x| + \dfrac{x^2}{4} + C$

**15.** $\frac{2}{5}x^{5/2} + C$  **17.** $\frac{1}{2}x^2 - \frac{2}{3}x^3 + \frac{1}{3}\ln|x| + C$  **19.** $-\frac{3}{2}e^{-2x} + C$  **21.** $ex + C$  **23.** $-e^{2x} - 2x + C$

**25.** $-\frac{5}{2}$  **27.** $\frac{1}{2}$  **29.** $-\frac{1}{5}$  **31.** $-1$  **33.** $\frac{1}{15}$  **35.** $3$  **37.** $\frac{2}{5}t^{5/2} + C$  **39.** $C$  **41.** $-\frac{5}{2}e^{-.2x} + \frac{5}{2}$

**43.** $\dfrac{x^2}{2} + 3$  **45.** $\frac{2}{3}x^{3/2} + x - \frac{28}{3}$  **47.** $2\ln|x| + 2$

**49.** Testing all three functions reveals that (b) is the only one that works.

**51.**

**53.** $\frac{1}{4}$  **55. (a)** $-16t^2 + 96t + 256$  **(b)** 8 sec  **(c)** 400 ft

**57.** $P(t) = 60t + t^2 - \frac{1}{12}t^3$  **59.** $20 - 25e^{-.4t}$ °C

**61.** $-95 + 1.3x + .03x^2 - .0006x^3$  **63.** $5875(e^{.016t} - 1)$

**65.** $C(x) = 25x^2 + 1000x + 10{,}000$

**67.** $F(x) = \frac{1}{2}e^{2x} - e^{-x} + \frac{1}{6}x^3$

$[-2.4, 1.7]$ by $[-10, 10]$

**69.** $F(x) = \frac{1}{10}\left(\frac{1}{4}x^4 - 3x^3 + \frac{5}{2}x^2\right) + 3x$

$[-4, 11]$ by $[-14, 7]$

### EXERCISES 6.2

**1.** .5; .25, .75, 1.25, 1.75  **3.** .6; 1.3, 1.9, 2.5, 3.1, 3.7  **5.** 8.625  **7.** 15.12  **9.** .077278  **11.** 40  **13.** 15

**15.** 5.625; 4.5  **17.** 1.61321; error = .04241  **19.** 1.08 L  **21.** 2800 ft  **23.** Increase in the population (in millions) from 1910 to 1950; rate of cigarette consumption $t$ years after 1985; 20 to 50  **25.** Tons of soil eroded during a 5-day period  **29.** 9.5965  **31.** 1.7641  **33.** .8427

### EXERCISES 6.3

**1.** $\displaystyle\int_{1/2}^{2} \frac{1}{x}\,dx$  **3.** $\displaystyle\int_{1}^{3}(1-x)(x-3)\,dx$  **5.**

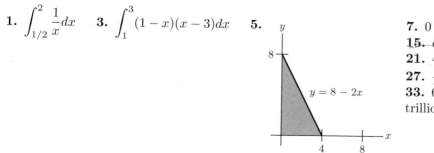

$y = 8 - 2x$

**7.** 0  **9.** 5  **11.** 1  **13.** $\frac{4}{3}(1 - e^{-3})$

**15.** $e - e^{-1}$  **17.** $\frac{4}{9}$  **19.** $\ln 2$

**21.** $4\ln 5$  **23.** $\frac{13}{4} - 2\sqrt{e}$  **25.** $3\frac{3}{4}$

**27.** $\frac{1}{5}$  **29.** 10  **31.** $2(e^{1/2} - 1)$

**33.** $6\frac{3}{5}$  **35.** Positive  **37.** 95.4 trillion

**39. (a)** 30 ft    **(b)**

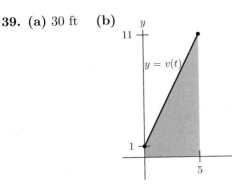

**41. (a)** $1185.75    **(b)** The area under the marginal cost curve from $x = 2$ to $x = 8$.    **43.** The increase in profits resulting from increasing the production level from 44 to 48 units    **45. (a)** $368/15 \approx 24.5$    **(b)** The amount the temperature falls during the first 2 h    **47.** 2088 million m³    **49.** 10    **51.** $\ln \frac{3}{2}$    **53.** $\frac{80}{3}$    **55.** 2.5452    **57.** .4636

## EXERCISES 6.4

**1.** $\int_1^2 f(x)dx + \int_3^4 -f(x)dx$    **3.**

**5.** $\frac{64}{3}$    **7.** $\frac{52}{3}$    **9.** $e^2 - e - \frac{1}{2}$    **11.** $\frac{1}{6}$    **13.** $\frac{32}{3}$    **15.** $\frac{1}{2}$    **17.** $\frac{1}{24}$    **19. (a)** $\frac{9}{2}$    **(b)** $\frac{19}{3}$    **(c)** $\frac{79}{6}$    **21.** $\frac{3}{2}$    **23.** $2 + 12\ln\frac{3}{2}$

**25.** $\int_0^{20} (76.2e^{.03t} - 50 + 6.03e^{.09t})dt$

**27.** No; 20; the additional profit from using the original plan

**29. (a)** The distance between the two cars after 1 h    **(b)** after 2 h    **31.** 1.4032    **33.** 1.4293

## EXERCISES 6.5

**1.** 3    **3.** $50(1 - e^{-2})$    **5.** $\frac{3}{4}\ln 3$    **7.** 55°    **9.** $\approx 82$ g    **11.** $20    **13.** $404.72    **15.** $200    **17.** $25
**19.** Intersection (100, 10), consumers' surplus = $100, producers' surplus = $250    **21.** $3236.68    **23.** $75,426
**25.** 13.35 y    **27. (b)** $1000e^{-.04t_1}\Delta t + 1000e^{-.04t_2}\Delta t + \cdots + 1000e^{-.04t_n}\Delta t$    **(c)** $f(t) = 1000e^{-.04t}; 0 \le t \le 5$
**(d)** $\int_0^5 1000e^{-.04t}dt$    **(e)** $4531.73    **29.** $\frac{32}{3}\pi$    **31.** $\frac{31\pi}{5}$    **33.** $8\pi$    **35.** $\frac{\pi}{2}(1 - e^{-2})$
**37.** $n = 4$, $b = 10$, $f(x) = x^3$    **39.** $n = 3$, $b = 7$, $f(x) = x + e^x$    **41.** The sum is approximated by
$\int_0^3 (3 - x)^2 dx = 9$.    **43. (a)** $\frac{1000}{3r}(e^{3r} - 1)$    **(b)** 4.5%    **45. (a)** $\frac{1000}{r}(e^{6r} - 1)$    **(b)** 5%

## CHAPTER 6: SUPPLEMENTARY EXERCISES

**1.** $9x + C$    **3.** $\frac{2}{3}(x + 1)^{3/2} + C$    **5.** $2(\frac{1}{4}x^4 + x^3 - x) + C$    **7.** $-2e^{-x/2} + C$    **9.** $\frac{3}{5}x^5 - x^4 + C$
**11.** $-\frac{2}{3}(4 - x)^{3/2} + C$    **13.** $\frac{8}{3}$    **15.** $\frac{14}{3}\sqrt{2}$    **17.** $\frac{15}{16}$    **19.** $\frac{3}{4}$    **21.** $\frac{1}{3}\ln 4$    **23.** $\frac{1}{2}$    **25.** $\frac{80}{81}$
**27.** $\frac{5}{32}$    **29.** $\frac{1}{3}$    **31.** $\frac{1}{2}$    **33.** $\frac{28}{3}$    **35.** $\frac{e}{2} - 1$    **37.** 8    **39.** $\frac{1}{3}(x - 5)^3 - 7$    **41. (a)** $2t^2 + C$
**(b)** $Ce^{4t}$    **(c)** $\frac{1}{4}e^{4t} + C$    **43.** $.02x^2 + 150x + 500$ dollars    **45.** The total quantity of drug (in cubic centimeters) injected during the first 4 min    **47.** 25    **49.** .68571; .69315    **51.** $433.33    **53.** 15
**55.** .26; 1.96    **57. (a)** $f(t) = Q - \frac{Q}{A}t$    **(b)** $\frac{Q}{2}$    **59. (a)** The area under the curve $y = \frac{1}{1 + t^2}$ from $t = 0$
to $t = 3$    **(b)** $\frac{1}{1 + x^2}$    **63.** $\frac{15}{4}$    **65.** True    **67.** 65,000 km³    **69.** $f(x) = x^3 - x^2 + x$

## CHAPTER 7

## EXERCISES 7.1

**1.** $f(5, 0) = 25$, $f(5, -2) = 51$, $f(a, b) = a^2 - 3ab - b^2$    **3.** $g(2, 3, 4) = -2$, $g(7, 46, 44) = \frac{7}{2}$
**7.** $C(x, y, z) = 6xy + 10xz + 10yz$    **9.** $f(8, 1) = 40$, $f(1, 27) = 180$, $f(8, 27) = 360$    **11.** $\approx $50$. $50 invested at 5% continuously compounded interest will yield $100 in 13.8 years    **13. (a)** $1875    **(b)** $2250; yes

**15.**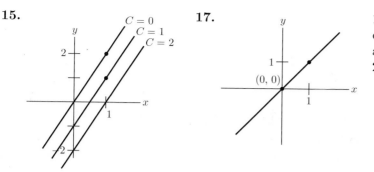

**17.**

**19.** $f(x, y) = y - 3x$ **21.** They correspond to the points having the same altitude above sea level. **23.** d **25.** c

## EXERCISES 7.2

**1.** $5y, 5x$ **3.** $4xe^y, 2x^2 e^y$ **5.** $\dfrac{1}{y} - \dfrac{y}{x^2}; \dfrac{-x}{y^2} + \dfrac{1}{x}$ **7.** $4(2x - y + 5), -2(2x - y + 5)$

**9.** $(2xe^{3x} + 3x^2 e^{3x}) \ln y, x^2 e^{3x}/y$ **11.** $\dfrac{2y}{(x + y)^2}, -\dfrac{2x}{(x + y)^2}$ **13.** $\dfrac{3}{2}\sqrt{\dfrac{K}{L}}$ **15.** $\dfrac{2xy}{z}, \dfrac{x^2}{z}, -\dfrac{1 + x^2 y}{z^2}$

**17.** $ze^{yz}, xz^2 e^{yz}, x(yz + 1)e^{yz}$ **19.** 1, 3 **21.** $-12$

**23.** $\dfrac{\partial f}{\partial x} = 3x^2 y + 2y^2, \dfrac{\partial^2 f}{\partial x^2} = 6xy, \dfrac{\partial f}{\partial y} = x^3 + 4xy, \dfrac{\partial^2 f}{\partial y^2} = 4x, \dfrac{\partial^2 f}{\partial y \partial x} = \dfrac{\partial^2 f}{\partial x \partial y} = 3x^2 + 4y$

**25. (a)** Marginal productivity of labor = 480; of capital = 40 **(b)** $480h$ **(c)** Production decreases by 240 units. **27.** If the price of a bus ride increases and the price of a train ticket remains constant, fewer people will ride the bus. An increase in train-ticket prices coupled with constant bus fare should cause more people to ride the bus. **29.** If the average price of DVDs increases and the average price of a DVD player remains constant, people will purchase fewer DVDs. An increase in average DVD player prices coupled with constant DVD prices should cause a decline in the number of DVD players purchased. **31.** $\dfrac{\partial V}{\partial P}(20, 300) = -.06, \dfrac{\partial V}{\partial T}(20, 300) = .004$

**33.** $\dfrac{\partial f}{\partial r} > 0, \dfrac{\partial f}{\partial m} > 0, \dfrac{\partial f}{\partial p} < 0$ **35.** $\dfrac{\partial^2 f}{\partial x^2} = -\dfrac{45}{4}x^{-5/4}y^{1/4}$. Marginal productivity of labor is decreasing.

## EXERCISES 7.3

**1.** $(-2, 1)$ **3.** $(26, 11)$ **5.** $(1, -3), (-1, -3)$ **7.** $(\sqrt{5}, 1), (\sqrt{5}, -1), (-\sqrt{5}, 1), (-\sqrt{5}, -1)$ **9.** $\left(\frac{1}{3}, \frac{4}{3}\right)$
**11.** Relative minimum; neither relative maximum nor relative minimum **13.** Relative maximum; neither relative maximum nor relative minimum; relative maximum **15.** Neither relative maximum nor relative minimum **17.** $(0, 0)$ min **19.** $(-1, -4)$ max **21.** $(0, -1)$ min **23.** $(-1, 2)$ max; $(1, 2)$ neither max nor min **25.** $\left(\frac{1}{4}, 2\right)$ min; $\left(\frac{1}{4}, -2\right)$ neither max nor min **27.** $\left(\frac{1}{2}, \frac{1}{6}, \frac{1}{2}\right)$ **29.** 14 in. × 14 in. × 28 in.
**31.** $x = 120, y = 80$

## EXERCISES 7.4

**1.** 58 at $x = 6, y = 2, \lambda = 12$ **3.** 13 at $x = 8, y = -3, \lambda = 13$ **5.** $x = \frac{1}{2}, y = 2$ **7.** 5, 5

**9.** Base 10 in., height 5 in. **11.** $F(x, y, \lambda) = 4xy + \lambda(1 - x^2 - y^2); \dfrac{\sqrt{2}}{2} \times \dfrac{\sqrt{2}}{2}$
**13.** $F(x, y, \lambda) = 3x + 4y + \lambda(18{,}000 - 9x^2 - 4y^2); x = 20, y = 60$
**15. (a)** $F(x, y, \lambda) = 96x + 162y + \lambda(3456 - 64x^{3/4}y^{1/4}); x = 81, y = 16$ **(b)** $\lambda = 3$ **17.** $x = 12, y = 2, z = 4$
**19.** $x = 2, y = 3, z = 1$ **21.** $F(x, y, z, \lambda) = 3xy + 2xz + 2yz + \lambda(12 - xyz); x = 2, y = 2, z = 3$
**23.** $F(x, y, z, \lambda) = xy + 2xz + 2yz + \lambda(32 - xyz); x = y = 4, z = 2$

## EXERCISES 7.5

**1.** $E = 6.7$ **3.** $E = (2A + B - 6)^2 + (5A + B - 10)^2 + (9A + B - 15)^2$ **5.** $y = 4.5x - 3$ **7.** $y = -2x + 11.5$
**9.** $y = -1.4x + 8.5$ **11. (a)** $y = .2073x + 2.7$ **(b)** \$4773 **(c)** 2006 **13. (a)** $y = .497x + 11.2$
**(b)** 22.6 percent **(c)** 2002 **15. (a)** $y = -4.24x + 22.01$ **(b)** $y = 8.442$ degrees Celsius

## EXERCISES 7.6

**1.** $e^2 - 2e + 1$    **3.** $2 - e^{-2} - e^2$    **5.** $309\frac{3}{8}$    **7.** $\frac{5}{3}$    **9.** $\frac{38}{3}$    **11.** $e^{-5} + e^{-2} - e^{-3} - e^{-4}$    **13.** $9\frac{1}{3}$

## CHAPTER 7: SUPPLEMENTARY EXERCISES

**1.** $2, \frac{5}{6}, 0$    **3.** $\approx 19.94$. Ten dollars increases to 20 dollars in 11.5 y.    **5.** $6x + y, x + 10y$    **7.** $\frac{1}{y}e^{x/y}, -\frac{x}{y^2}e^{x/y}$
**9.** $3x^2, -z^2, -2yz$    **11.** 6, 1    **13.** $20x^3 - 12xy, 6y^2, -6x^2, -6x^2$    **15.** $-201, 5.5$. At the level $p = 25$,
$t = 10,000$, an increase in price of \$1 will result in a loss in sales of approximately 201 calculators, and an
increase in advertising of \$1 will result in the sale of approximately 5.5 additional calculators.    **17.** $(3, 2)$
**19.** $(0, 1), (-2, 1)$    **21.** Min at $(2, 3)$    **23.** Min at $(1, 4)$; neither max nor min at $(-1, 4)$
**25.** $20; x = 3, y = -1$    **27.** $x = \frac{1}{2}, y = \frac{3}{2}, z = 2$    **29.** $F(x, y, \lambda) = xy + \lambda(40 - 2x - y); x = 10, y = 20$
**31.** $y = \frac{5}{2}x - \frac{5}{3}$    **33.** $y = -2x + 1$    **35.** 5160    **37.** 40

## CHAPTER 8
### EXERCISES 8.1

**1.** $\frac{\pi}{6}, \frac{2\pi}{3}, \frac{7\pi}{4}$    **3.** $\frac{5\pi}{2}, -\frac{7\pi}{6}, -\frac{\pi}{2}$    **5.** $4\pi$    **7.** $\frac{7\pi}{2}$    **9.** $-3\pi$    **11.** $\frac{2\pi}{3}$
**13.**

**15.**

**17.**

### EXERCISES 8.2

**1.** $\sin t = \frac{1}{2}, \cos t = \frac{\sqrt{3}}{2}$    **3.** $\sin t = \frac{2}{\sqrt{13}}, \cos t = \frac{3}{\sqrt{13}}$    **5.** $\sin t = \frac{12}{13}, \cos t = \frac{5}{13}$

**7.** $\sin t = \frac{1}{\sqrt{5}}, \cos t = -\frac{2}{\sqrt{5}}$    **9.** $\sin t = \frac{\sqrt{2}}{2}, \cos t = -\frac{\sqrt{2}}{2}$    **11.** $\sin t = -.8, \cos t = -.6$    **13.** .4    **15.** 3.6

**17.** 10.9    **19.** $b = 1.3, c = 2.7$    **21.** $\frac{\pi}{6}$    **23.** $\frac{3\pi}{4}$    **25.** $\frac{5\pi}{8}$    **27.** $\frac{\pi}{4}$    **29.** $\frac{\pi}{3}$    **31.** $-\frac{\pi}{6}$    **33.** $\frac{\pi}{4}$
**35.** Here $\cos t$ decreases from 1 to $-1$.    **37.** $0, 0, 1, -1$    **39.** $.2, .98, .98, -.2$

**41. (a)**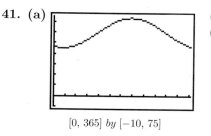

[0, 365] by [−10, 75]

**(b)** 46°  **(c)** 45° coldest, 73° warmest  **(d)** January 26  **(e)** July 27
**(f)** October 27 and April 27

[0, 365] by [−10, 75]

## EXERCISES 8.3

**1.** $4\cos 4t$  **3.** $4\cos t$  **5.** $-6\sin 3t$  **7.** $1 - \pi\sin \pi t$  **9.** $-\cos(\pi - t)$  **11.** $-3\cos^2 t \sin t$

**13.** $\dfrac{\cos\sqrt{x-1}}{2\sqrt{x-1}}$  **15.** $\dfrac{\cos(x-1)}{2\sqrt{\sin(x-1)}}$  **17.** $-8\sin t(1+\cos t)^7$  **19.** $-6x^2\cos x^3\sin x^3$  **21.** $e^x(\sin x + \cos x)$

**23.** $2\cos(2x)\cos(3x) - 3\sin(2x)\sin(3x)$  **25.** $\cos^{-2}t$  **27.** $-\dfrac{\sin t}{\cos t}$  **29.** $\dfrac{\cos(\ln t)}{t}$  **31.** $-3$  **33.** $y=2$

**35.** $\frac{1}{2}\sin 2x + C$  **37.** $-\dfrac{7}{2}\sin\dfrac{x}{7} + C$  **39.** $\sin x + \cos x + C$  **41.** $\cos x + \sin 3x + C$

**43.** $-\frac{1}{4}\cos(4x+1) + C$  **45.** $-\frac{7}{3}\cos(3x-2) + C$  **47. (a)** max = 120 at 0, $\dfrac{\pi}{3}$; min = 80 at $\dfrac{\pi}{6}, \dfrac{\pi}{2}$  **(b)** 57
**49.** 0  **51. (a)** 69°  **(b)** increasing 1.6°/wk  **(c)** weeks 6 and 44  **(d)** weeks 28 and 48  **(e)** week 25, week
51  **(f)** week 12, week 38

## EXERCISES 8.4

**1.** $\sec t = \dfrac{\text{hypotenuse}}{\text{adjacent}}$  **3.** $\tan t = \frac{5}{12}, \sec t = \frac{13}{12}$  **5.** $\tan t = -\dfrac{1}{2}, \sec t = -\dfrac{\sqrt{5}}{2}$

**7.** $\tan t = -1, \sec t = -\sqrt{2}$  **9.** $\tan t = \frac{4}{3}, \sec t = -\frac{5}{3}$  **11.** $75\tan(.7) \approx 63$ ft  **13.** $\tan t \sec t$

**15.** $-\csc^2 t$  **17.** $4\sec^2(4t)$  **19.** $-3\sec^2(\pi - x)$  **21.** $4(2x+1)\sec^2(x^2+x+3)$  **23.** $\dfrac{\sec^2\sqrt{x}}{2\sqrt{x}}$

**25.** $\tan x + x\sec^2 x$  **27.** $2\tan x\sec^2 x$  **29.** $6[1+\tan(2t)]^2\sec^2(2t)$  **31.** $\sec t$
**33. (a)** $y - 1 = 2(x - \frac{\pi}{4})$  **(b)**

**35.** $\frac{1}{3}\tan 3x + C$  **37.** 2
**39.** $\tan x + C$

## CHAPTER 8: SUPPLEMENTARY EXERCISES

**1.** $\dfrac{3\pi}{2}$  **3.** $-\dfrac{3\pi}{4}$  **5.**

**7.** $\frac{4}{5}, \frac{3}{5}, \frac{4}{3}$  **9.** $-.8, -.6, \frac{4}{3}$  **11.** $\pm\dfrac{2\sqrt{6}}{5}$

**13.** $\dfrac{\pi}{4}, \dfrac{5\pi}{4}, -\dfrac{3\pi}{4}, -\dfrac{7\pi}{4}$  **15.** Negative  **17.** 16.3 ft

**19.** $3\cos t$  **21.** $(\cos\sqrt{t})\cdot\frac{1}{2}t^{-1/2}$
**23.** $x^3\cos x + 3x^2\sin x$

**25.** $-\dfrac{2\sin(3x)\sin(2x) + 3\cos(2x)\cos(3x)}{\sin^2(3x)}$

**27.** $-12\cos^2(4x)\sin(4x)$  **29.** $[\sec^2(x^4+x^2)](4x^3+2x)$  **31.** $\cos(\tan x)\sec^2 x$  **33.** $\sin x\sec^2 x + \sin x$

**35.** $\cot x$  **37.** $4e^{3x}\sin^3 x\cos x + 3e^{3x}\sin^4 x$  **39.** $\dfrac{\tan(3t)\cos t - 3\sin t\sec^2(3t)}{\tan^2(3t)}$  **41.** $e^{\tan t}\sec^2 t$

**43.** $2(\cos^2 t - \sin^2 t)$  **45.** $\dfrac{\partial f}{\partial s} = \cos s\cos(2t),\ \dfrac{\partial f}{\partial t} = -2\sin s\sin(2t)$

**47.** $\dfrac{\partial f}{\partial s} = t^2\cos(st),\ \dfrac{\partial f}{\partial t} = \sin(st) + st\cos(st)$  **49.** $y - 1 = 2\left(t - \dfrac{\pi}{4}\right)$  **51.**

**53.** $4$  **55.** $\dfrac{\pi^2}{2} - 2$  **57.** (a) $V'(t) = 8\pi\cos\left(160\pi t - \dfrac{\pi}{2}\right)$  (b) $8\pi$ L/min  (c) $16$ L/min

**59.** $\cos(\pi - x) + C$  **61.** $0$  **63.** $\dfrac{\pi^2}{2}$  **65.** $2\tan\dfrac{x}{2} + C$  **67.** $\sqrt{2} - 1$  **69.** $\sqrt{2}$  **71.** $1$

**73.** $1000 + \dfrac{400}{3\pi}$  **75.** $\tan x - x + C$  **77.** $\tan x + C$  **79.** $1 - \dfrac{\pi}{4}$

# CHAPTER 9

## EXERCISES 9.1

**1.** $\frac{1}{6}(x^2+4)^6 + C$  **3.** $2\sqrt{x^2+x+3} + C$  **5.** $e^{(x^3-1)} + C$  **7.** $-\frac{1}{3}(4-x^2)^{3/2} + C$  **9.** $(2x+1)^{1/2} + C$

**11.** $\frac{1}{2}e^{x^2} + C$  **13.** $\frac{1}{2}(\ln 2x)^2 + C$  **15.** $\frac{1}{5}\ln|x^5+1| + C$  **17.** $-(2-12x+2x^2)^{-1} + C$

**19.** $\frac{1}{4}(\ln x)^2 + C = (\ln\sqrt{x})^2 + C$  **21.** $\frac{1}{3}\ln|x^3-3x^2+1| + C$  **23.** $-4e^{-x^2} + C$  **25.** $\frac{1}{2}\ln|\ln x^2| + C$

**27.** $-\frac{1}{10}(x^2-6x)^5 + C$  **29.** $\frac{1}{6}(1+e^x)^6 + C$  **31.** $\frac{1}{2}\ln(1+2e^x) + C$  **33.** $\ln|1-e^{-x}| + C$

**35.** $-\ln(e^{-x}+1) + C$  **37.** $f(x) = (x^2+9)^{1/2} + 3$  **39.** $2e^{\sqrt{x+5}} + C$  **41.** $\frac{1}{2}\tan x^2 + C$

**43.** $\frac{1}{2}(\sin x)^2 + C$  **45.** $2\sin\sqrt{x} + C$  **47.** $-\frac{1}{4}\cos^4 x + C$  **49.** $-\frac{2}{3}\sqrt{2-\sin 3x} + C$

**51.** $\ln|\sin x - \cos x| + C$  **53.** $\frac{1}{2}(x^2+5)^2 + C = \frac{1}{2}x^4 + 5x^2 + \frac{25}{2} + C; \frac{1}{2}x^4 + 5x^2 + C_1$

## EXERCISES 9.2

**1.** $\frac{1}{5}xe^{5x} - \frac{1}{25}e^{5x} + C$  **3.** $\frac{x}{5}(x+7)^5 - \frac{1}{30}(x+7)^6 + C$  **5.** $-xe^{-x} - e^{-x} + C$

**7.** $2x(x+1)^{1/2} - \frac{4}{3}(x+1)^{3/2} + C$  **9.** $(1-3x)(\frac{1}{2}e^{2x}) + \frac{3}{4}e^{2x} + C$  **11.** $-2xe^{-3x} - \frac{2}{3}e^{-3x} + C$

**13.** $\frac{2}{3}x(x+1)^{3/2} - \frac{4}{15}(x+1)^{5/2} + C$  **15.** $\frac{1}{3}x^{3/2}\ln x - \frac{2}{9}x^{3/2} + C$  **17.** $x\sin x + \cos x + C$

**19.** $\dfrac{x^2}{2}\ln 5x - \dfrac{1}{4}x^2 + C$  **21.** $4x\ln x - 4x + C$  **23.** $-e^{-x}(x^2+2x+2) + C$

**25.** $\frac{1}{5}x(x+5)^5 - \frac{1}{30}(x+5)^6 + C$  **27.** $\frac{1}{10}(x^2+5)^5 + C$  **29.** $3(3x+1)e^{x/3} - 27e^{x/3} + C$

**31.** $\frac{1}{2}\tan(x^2+1) + C$  **33.** $\frac{1}{2}xe^{2x} - \frac{1}{4}e^{2x} + \frac{1}{3}x^3 + C$  **35.** $\frac{1}{2}e^{x^2} - x^2 + C$  **37.** $2x\sqrt{x+9} - \frac{4}{3}(x+9)^{3/2} + 38$

## EXERCISES 9.3

**1.** $\frac{1}{15}$  **3.** $312$  **5.** $\frac{8}{3}$  **7.** $\frac{64}{3}$  **9.** $0$  **11.** $\frac{1}{2}\ln(\frac{4}{3})$  **13.** $\frac{1}{3}(e^{27}-e)$  **15.** $\frac{1}{2}$  **17.** $0$  **19.** $\dfrac{1}{\pi}$  **21.** $\dfrac{\pi}{2}$

**23.** $\dfrac{9\pi}{2}$  **25.** $\dfrac{16}{3}$

## EXERCISES 9.4

**1.** $\Delta x = .4$; 3, 3.4, 3.8, 4.2, 4.6, 5
**3.** $\Delta x = .5$; $-.75, -.25, .25, .75$

**5.**

**7.** $(n = 2)$ 40, $(n = 4)$ 41, exact: $41\frac{1}{3}$
**9.** .63107, exact: .63212
**11.** .09375, exact: .08333
**13.** 1.03740, exact: .8
**15.** $M = 72$, $T = 90$, $S = 78$, exact: 78
**17.** $M = 44.96248$, $T = 72.19005$,
$S = 54.03834$, exact: 53.59815

**19.** $M = 573.41797, T = 612.10806, S = 586.31466$, exact: 586.26358   **21.** 3.24124   **23.** 1.61347
**25.** 25,750 ft$^2$   **27.** 2150 ft   **29.** (a) $f''(x) = x^2 + 6$   (b) $A = 10$   (c) .0333

(d) $-.0244$, satisfies the bound in (c)   (e) quartered

**35.** $f(x) = \dfrac{1}{25 - x^2}, a = 3, b = 5, n = 10$

**37.** Midpoint rule: 2.361749156; trapezoidal rule: 2.474422799; Simpson's rule: 2.399307037; Exact value: 2.397895273; Error using midpoint rule: $-.036146117$; Error using trapezoidal rule: .076527526; Error using Simpson's rule: .001411764

**39.** Midpoint rule: .9989755866; trapezoidal rule: 1.00205197; Simpson's rule: 1.000001048; Exact value: 1; Error using midpoint rule: $-.0010244134$; Error using trapezoidal rule: .00205197; Error using Simpson's rule: .000001048

**41.**
bound on the error is .0008333

## EXERCISES 9.5

**1.** $147,656   **3.** $35,797   **5.** $182,937   **7.** (a) $\displaystyle\int_0^2 (30 + 5t)e^{-.10t}\,dt$   (b) $63.1 million

**9. (a)** $240\pi \displaystyle\int_0^5 te^{-.65t}dt$   **(b)** 1,490,623    **11.** $\approx 1,400,000;$

**13. (a)** $80\pi te^{-.5t}\Delta t$ thousand people   **(b)** $P'(t)$   **(c)** the number of people who live between 5 mi and $5 + \Delta t$ mi from the city center   **(d)** $P'(t) = 80\pi te^{-.5t}$   **(e)** $P(b) - P(a) = \displaystyle\int_a^b P'(t)dt = \int_a^b 80\pi te^{-.5t}dt$

## EXERCISES 9.6

**1.** 0    **3.** No limit    **5.** $\frac{1}{4}$    **7.** 2    **9.** 0    **11.** 5    **13.** $\frac{1}{2}$    **15.** 2    **17.** 1    **19.** Area under the graph from 1 to $b$ is $\frac{5}{14}(14b + 18)^{1/5} - \frac{5}{7}$. This has no limit as $b \to \infty$.    **21.** $\frac{1}{2}$    **23.** $\frac{1}{6}$    **25.** Divergent    **27.** $\frac{2}{3}$
**29.** 1    **31.** $2e$    **33.** Divergent    **35.** $\frac{1}{2}$    **37.** 2    **39.** $\frac{1}{4}$    **41.** 2    **43.** $\frac{1}{6}$    **49.** $\dfrac{K}{r}$

## CHAPTER 9: SUPPLEMENTARY EXERCISES

**1.** $-\frac{1}{6}\cos(3x^2) + C$    **3.** $-\frac{1}{36}(1 - 3x^2)^6 + C$    **5.** $\frac{1}{3}[\ln x]^3 + C$    **7.** $-\frac{1}{3}(4 - x^2)^{3/2} + C$    **9.** $-\frac{1}{3}e^{-x^3} + C$
**11.** $\frac{1}{3}x^2\sin(3x) - \frac{2}{27}\sin(3x) + \frac{2}{9}x\cos(3x) + C$    **13.** $2(x\ln x - x) + C$    **15.** $\frac{2}{3}x(3x - 1)^{1/2} - \frac{4}{27}(3x - 1)^{3/2} + C$
**17.** $\frac{1}{4}\left[\dfrac{x}{(1 - x)^4}\right] - \frac{1}{12}\left[\dfrac{1}{(1 - x)^3}\right] + C$    **19.** $f(x) = x, g(x) = e^{2x}$    **21.** $u = \sqrt{x + 1}$    **23.** $u = x^4 - x^2 + 4$
**25.** $f(x) = (3x - 1)^2, g(x) = e^{-x}$; then integrate by parts again.    **27.** $f(x) = 500 - 4x, g(x) = e^{-x/2}$
**29.** $f(x) = \ln(x + 2), g(x) = \sqrt{x + 2}$    **31.** $u = x^2 + 6x$    **33.** $u = x^2 - 9$    **35.** $u = x^3 - 6x$    **37.** $\frac{3}{8}$
**39.** $1 - e^{-2}$    **41.** $\frac{3}{4}e^{-2} - \frac{5}{4}e^{-4}$    **43.** $M = 3.93782, T = 4.13839, S = 4.00468$
**45.** $M = 12.84089, T = 13.20137, S = 12.96105$    **47.** $\frac{1}{3}e^6$    **49.** Divergent    **51.** $2^{7/4}$    **53.** $2e^{-1}$
**55.** \$137,668    **57. (a)** $M(t_1)\Delta t + \cdots + M(t_n)\Delta t \approx \displaystyle\int_0^2 M(t)dt$

**(b)** $M(t_1)e^{-.1t_1}\Delta t + \cdots + M(t_n)e^{-.1t_n}\Delta t \approx \displaystyle\int_0^2 M(t)e^{-.1t}dt$

## CHAPTER 10

## EXERCISES 10.1

**5.** Order $= 2$    **7.** Yes    **9.** $y = 5$    **11.** $f(0) = 4, f'(0) = 5$    **13.** 20 feet per second per second
**15. (a)** Decreasing at \$2500 per year   **(b)** $y' = .05(y - 200,000)$   **(c)** The rate of change of the savings account balance is proportional to the difference between the balance at the end of $t$ years and \$200,000.
**17.** The number of people who have heard the news broadcast after $t$ hours is increasing at a rate that is proportional to the difference between that number and 200,000. At the beginning of the broadcast there are 10 people tuned in.    **19.** $y' = k(C - y), k > 0$    **21.** $y' = k(P_b - y), y(0) = P_0$, where $k$ is a positive constant.
**25.** No, it will come very close to 5000 but will not attain or exceed this value.    **27.** $y = 0$ and $y = 1$
**29.** All solution curves with $y(0) > 1$ will decrease and tend to 1. A typical solution curve is shown in Fig. 7.

**31.** **(a)**      **(b)** $.2(10 - f(5)) = -36.78794412$, using nDeriv $f'(5) = -36.78794436$

[0, 30] *by* [−75, 550]

## EXERCISES 10.2

**1.** $y = \sqrt[3]{15t - \frac{3}{2}t^2 + C}$     **3.** $y = \ln|e^t + C|$     **5.** $y = (-\frac{2}{3}t^{3/2} + C)^{-1}$ or $y = 0$     **7.** $y = (e^{t^3} + C)^{1/3}$

**9.** $y = (\sqrt{t} + C)^2$ or $y = 0$     **11.** $y = \dfrac{-1}{t^3 + C}$ or $y = 0$     **13.** $y = \ln(\frac{1}{2}e^{t^2} + C)$     **15.** $y = \pm\sqrt{(\ln t)^2 + C}$

**17.** $y = \dfrac{1}{t - t\ln t + C} + 3$ or $y = 3$     **19.** $y = \frac{1}{2}\ln(2t^2 - 2t + e^6)$     **21.** $y = \sqrt[3]{3t\sin t + 3\cos t + 5}$

**23.** $y = \sqrt[3]{\cos t + 1}$     **25.** $y = -\sqrt{2t + 2\ln t + 7}$     **27.** $y = \frac{2}{5} + \frac{3}{5}e^{(5/2)t^2}$

**29.** $y = (3x^{1/2}\ln x - 6x^{1/2} + 14)^{2/3}$     **31.** $y = A(p + 3)^{-1/2}$, $A > 0$

**33.** $y' = k(1 - y)$, where $y = p(t)$, $y(0) = 0$; $y = 1 - e^{-kt}$

**35.** $\dfrac{dV}{dt} = kV^{2/3}$, where $k < 0$; $V = (3 - \frac{1}{8}t)^3$, $V = 0$ when $t = 24$ weeks.

**37.** $y = be^{Ce^{-at}}$, $C$ any number

**39.** **(a)** The population will decrease and approach 5000. **(b)** The population will increase and approach 5000.

**(c)**

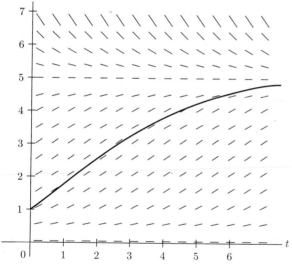

*y* (in thousands)

The solution curve represents the number of fish in the pond, if we start with an initial population of 1000 fish.

## EXERCISES 10.3

**1.** $\dfrac{d}{dt}[(2t + 4)y] = 6t$     **3.** $\dfrac{d}{dt}[t^5 y] = 2t^2 + t + 1$     **5.** $\dfrac{d}{dt}[(\cos t)y] = 1$     **7.** $\dfrac{d}{dt}[t\,y] = 1, y = 1 + \dfrac{C}{t}$

**9.** $\dfrac{d}{dt}[e^{t^2} y] = \frac{2}{3}\sqrt{t}, y = \frac{4}{9}t^{3/2}e^{-t^2} + Ce^{-t^2}$     **11.** $\dfrac{d}{dt}\left[\dfrac{1}{t}y\right] = 6t^2, y = 2t^4 + Ct$     **13.** $e^{-2t}$     **15.** $e^{-\frac{1}{2t^2}}$

**17.** $\dfrac{1}{10 + t}$     **19.** $y = 1 + Ce^{-t}$     **21.** $y = 2 + Ce^{t^2}$     **23.** $y = 35 + Ce^{-.5t}$     **25.** $y = \dfrac{C}{10 + t}$

**27.** $y = \dfrac{C - t}{1 + t}$     **29.** $y = 1 + Ce^{-\frac{1}{12}t^2}$     **31.** $y = 2 - \frac{1}{2}e^t + Ce^{-t}$     **33.** $y = \frac{1}{2} + \frac{1}{2}e^{-2t}$

**35.** $y = 10 + 10t$     **37.** $y = \frac{1}{3}e^{2t} - \frac{4}{3}e^{-t}$     **39.** $y = 1 - e^{-\sin 2t}$

**41.** (a) $y'(0) = -40$, $y(t)$ is decreasing at $t = 0$. (b) $y = \dfrac{50 + 10t + 5t^2}{1+t}$

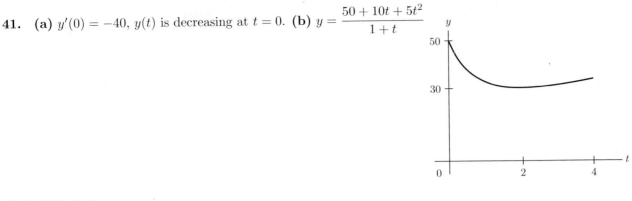

## EXERCISES 10.4

**1.** (a) 4200 dollars per year   (b) 40,000 dollars   (c) $t = \frac{1}{.06}\ln(\frac{80}{41}) \approx 11.1$ years   **3.** (a) $y' = .05y + 3600$
(b) $f(t) = -72,000 + 72,000e^{.05t}$, $f(25) = 179,305$ dollars.   **5.** At retirement, Kelly has \$41,239 in her
savings account, and John has \$31,139 in his savings account.   **7.** Approximately \$14,214 per year.
**9.** (a) $y' = .0676y - A$, $y(0) = 197,640$   (b) $A = 15,385$ dollars per year, or 1282.1 dollars per month
(c) \$263,914   **11.** (a) $py' + (p+1)y = 0$   (b) $f(p) = 100\,\frac{e^{-p+1}}{p}$   **13.** $f(t) = 10 + 340e^{-.1t}$
**15.** (a) $T = 70$   (b) $k = .5$   (c) $f(t) = 70 + 28e^{-.5t}$ (d) Approximately one hour and 15 minutes
**17.** $y' = .45y + e^{.03t} + 2$   **19.** (a) $k = \frac{2}{7} \approx .286$   (b) $\frac{220}{7} \approx 31.43$ grams per liter per hour. Replacing the
dialysate with a fresh solution after 4 hours triples the rate at which waste products are filtered from the body.
**21.** (a) $y' - .04y = -500\,t - 2000$, $y(0) = 100,000$   (b) $f(t) = 362,500 - 262,500e^{.04t} + 12,500t$
(c)                                                   The account is depleted in approximately 22.4 years.

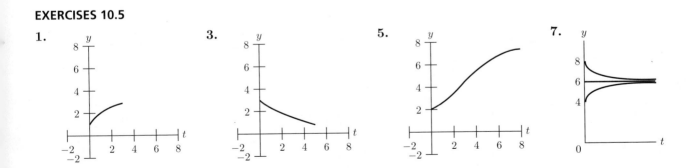

**23.** (a) 6 years   (b) $y' - .04y = 3000 - 500t$, $y(0) = 10,000$
**25.** (a) $y' + .35y = t$ (b) $f(t) = \frac{1}{(.35)^2}(.35t - 1 + e^{-.35t})$, $f(8) \approx 15.2$ milligrams

## EXERCISES 10.5

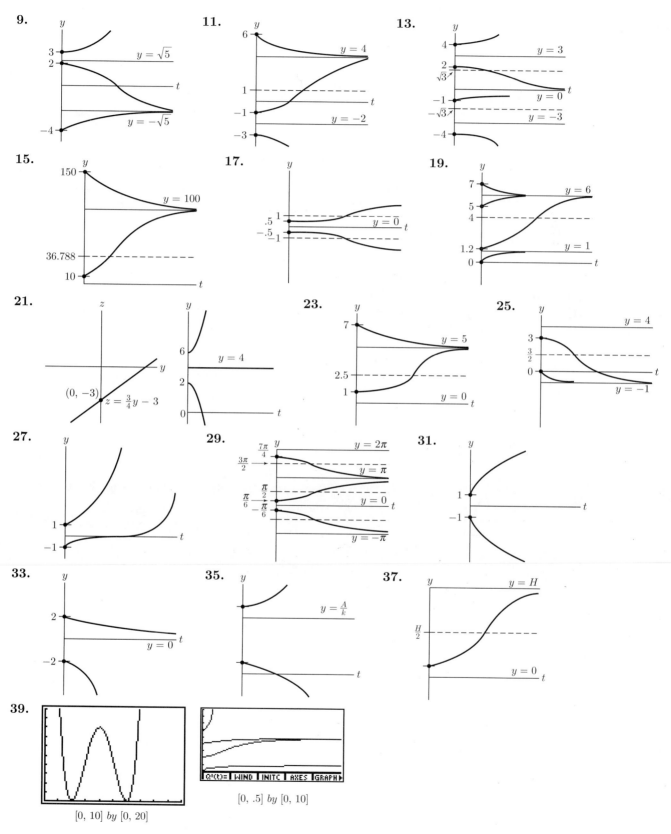

**9.**

**11.**

**13.**

**15.**

**17.**

**19.**

**21.**

**23.**

**25.**

**27.**

**29.**

**31.**

**33.**

**35.**

**37.**

**39.**

$[0, 10]$ *by* $[0, 20]$

$[0, .5]$ *by* $[0, 10]$

**EXERCISES 10.6**

1.  **(a)** Carrying capacity $= 1$,   **(b)**
    intrinsic rate $= 1$

**(c), (d)**

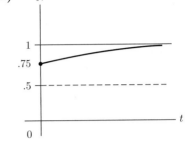

3.  **(a)** Carrying capacity $= 100$,  **(b)**
    intrinsic rate $= 1$

**(c), (d)**

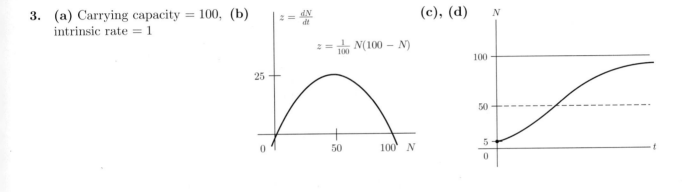

5.  $\dfrac{dN}{dt} = \dfrac{r}{1000} N(1000 - N)$, $N(0) = 600$;     **7.**  $y' = k(100 - y)$, $k > 0$     **9.**  $y' = ky(M - y)$, $k > 0$

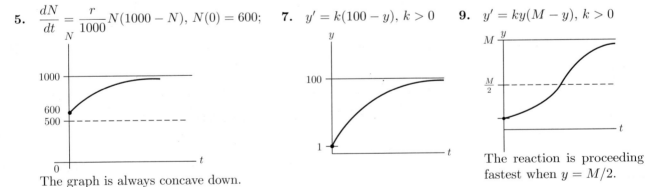

The graph is always concave down.

The reaction is proceeding
fastest when $y = M/2$.

**11.**  $y' = k(c - y)$, $k > 0$     **13.**  $y' = ky^2$, $k < 0$     **15.**  $y' = k(E - y)$, $k > 0$

$0 < y(0) < c$

$y(0) > 0$

$y(0) > E$

$y = E$

$y(0) < E$

**17.** $y' = .055y - 300,$
$y(0) = 5400$

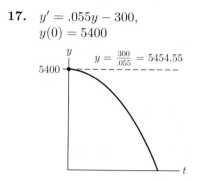

$y = \frac{300}{.055} = 5454.55$

5400

**19.** (a) $y' = .05y + 10,000, \ y(0) = 0$
(b) $y = 200,000(e^{.05t} - 1), \ \$56,805$
**21.** (a) $y' = .05 - .2y, \ y(0) = 6.25$
(b) $y' = .13 - .2y, \ y(0) = 6.25$
**23.** $y' = -.14y$

**25.**

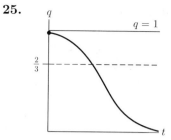

$q = 1$

$\frac{2}{3}$

**EXERCISES 10.7**

**1.** 3   **3.** Increasing   **5.** $f(1) \approx -\frac{9}{4}$   **7.** $f(2) \approx \frac{27}{8}$

$(0, -2)$   $\left(\frac{1}{2}, -2\right)$   $\left(1, -\frac{9}{4}\right)$

$(0, 5)$   $\left(1, \frac{5}{2}\right)$   $\left(\frac{1}{2}, 3\right)$   $\left(2, \frac{27}{8}\right)$   $\left(\frac{3}{2}, \frac{11}{4}\right)$

**9.** Euler's method yields $f(1) \approx .37011$; solution: $f(t) = \dfrac{1}{\frac{1}{2}t^2 + t + 1}$; $f(1) = .4$; error $= .02989$

**11.** (a) $y' = .1(1 - y), y(0) = 0$   (b) .271   (c) $y = 1 - e^{-.1t}, y(3) = 1 - e^{-.3} \approx .25918$   (d) error $\approx .01182$

**13.** (a) C

(b) A

(c) E

[0, 4] by [−1, 5]

[0, 4] by [−1, 5]

[0, 4] by [−1, 5]

**(d)** B

**(e)** D

[0, 4] by [−1, 5]

[0, 4] by [−1, 5]

**15.** $y_i = 1, .75, .6960, .7602, .9093, 1.1342,$
$1.4397, 1.8403, 2.3588; \ y = 1, .8324, .7948,$
$.8544, .9963, 1.2182, 1.5271, 1.9383, 2.4752,$
greatest difference $= .1163$

**CHAPTER 10: SUPPLEMENTARY EXERCISES**

**1.** $y = \sqrt[3]{3t^4 - 3t^3 + 6t + C}$   **3.** $y = Ate^{-3t}$   **5.** $y = \frac{3}{7}t + 3$   **7.** $y = \sqrt{4t^3 - t^2 + 49}$

**9.** $y(t) = \frac{1}{7}(t-1)^5 + \frac{C}{(t-1)^2}$    **11.** $y = -1 - x + e^x$    **13.** Decreasing    **15.**

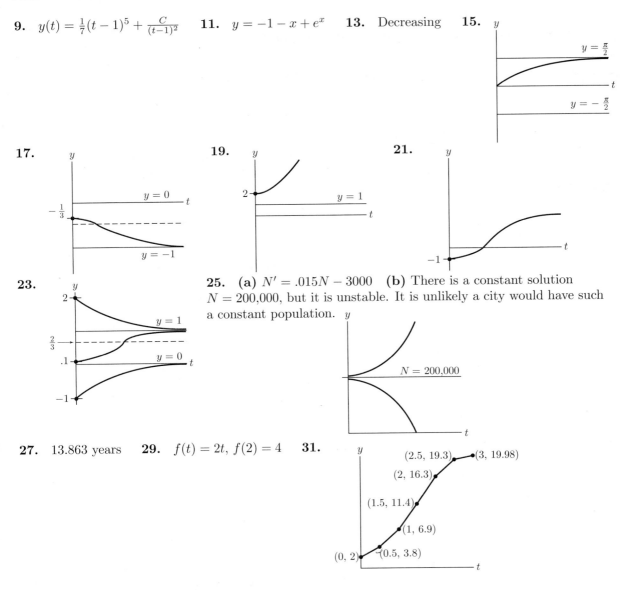

**17.**

**19.**

**21.**

**23.**

**25.** **(a)** $N' = .015N - 3000$    **(b)** There is a constant solution $N = 200{,}000$, but it is unstable. It is unlikely a city would have such a constant population.

**27.** 13.863 years    **29.** $f(t) = 2t$, $f(2) = 4$    **31.**

# CHAPTER 11

## EXERCISES 11.1

**1.** $x - \frac{1}{6}x^3$    **3.** $5 + 10x + 10x^2 + \frac{20}{3}x^3$    **5.** $1 + 2x - 2x^2 + 4x^3$    **7.** $x + 3x^2 + \frac{9}{2}x^3$

**9.** $p_4(x) = 1 + x + \frac{1}{2}x^2 + \frac{1}{6}x^3 + \frac{1}{24}x^4$, $e^{.01} \approx p_4(.01) = 1.01005$

**11.**

$$f(x) = \frac{1}{1-x}$$

$$p_1(x) = 1 + x$$    $$p_2(x) = 1 + x + x^2$$    $$p_3(x) = 1 + x + x^2 + x^3$$

**13.** $p_n(x) = 1 + x + \frac{1}{2}x^2 + \frac{1}{3!}x^3 + \cdots + \frac{1}{n!}x^n$    **15.** $p_2(x) = x^2$, area $\approx .0417$

**17.** $1 + (x-4) + (x-4)^2 + (x-4)^3$   **19.** $p_3(x) = -1 + \frac{1}{2}(x-\pi)^2$, $p_4(x) = -1 + \frac{1}{2}(x-\pi)^2 - \frac{1}{24}(x-\pi)^4$
**21.** $3.04958$   **23.** $p_1(x) = 19 + 33(x-2)$, $p_2(x) = 19 + 33(x-2) + 24(x-2)^2$,
$p_3(x) = 19 + 33(x-2) + 24(x-2)^2 + 8(x-2)^3$, $p_n(x) = 19 + 33(x-2) + 24(x-2)^2 + 8(x-2)^3 + (x-2)^4$, $n \geq 4$
**25.** $f''(0) = -5$, $f'''(0) = 7$   **27.** **(a)** 1   **(b)** $|R_3(.12)| \leq \frac{1}{4!}(.12)^4 = 8.64 \times 10^{-6}$
**29.** **(a)** $R_2(x) = \dfrac{f'''(c)}{3!}(x-9)^3$, where $c$ is between 9 and $x$   **(b)** $f^{(3)}(c) = \frac{3}{8}c^{-5/2} \leq \frac{3}{8}9^{-5/2} = \frac{1}{648}$
**(c)** $|R_2(x)| \leq \dfrac{1}{648} \cdot \dfrac{1}{3!}(.3)^3 = \dfrac{1}{144} \times 10^{-3} < 7 \times 10^{-6}$
**31.** When $b = .55$ difference is approximately .11;   **33.** When $b = 1.85$ difference is approximately .2552.
when $b = -.68$ difference is approximately .087.   When $x = 3$ difference is approximately 3.7105.

$[-1, 1]$ *by* $[-1, 5]$          $[0, 3]$ *by* $[-2, 20]$

## EXERCISES 11.2

**1.** Let $f(x) = x^2 - 5$, $x_0 = 2$; then $x_1 = 2.25$, $x_2 \approx 2.2361$, $x_3 \approx 2.23607$.   **3.** Let $f(x) = x^3 - 6$, $x_0 = 2$;
then $x_1 \approx 1.8333$, $x_2 \approx 1.81726$, $x_3 \approx 1.81712$.   **5.** If $x_0 = 2$, then $x_3 \approx 2.79130$.   **7.** $x_3 \approx .63707$
**9.** $x_0 = -1$, $x_1 = -.8$,   **11.** .703   **13.** 8.21% per month   **15.** 1% per month
$x_2 \approx -.77143$, $x_3 \approx -.77092$

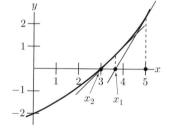

**17.** $x_1 \approx 3.5$, $x_2 \approx 3.0$   **19.** $-\frac{5}{4}$   **21.** $x_0 > 0$   **23.** $x_1$ will be the exact root.

**25.** $x_0 = 1$, $x_1 = -2$, $x_2 = 4$, $x_3 = -8$   **27.** 4; 31   **29.** **(a)** $\sqrt{2}$   **(b)** $-\sqrt{2}$   **(c)** 0

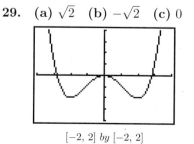

$[-6, 6]$ *by* $[-5, 10]$          $[-2, 2]$ *by* $[-2, 2]$

## EXERCISES 11.3

**1.** $\frac{6}{5}$   **3.** $\frac{9}{10}$   **5.** 3   **7.** $\frac{25}{124}$   **9.** $\frac{21}{10}$   **11.** Divergent $(r = -\frac{8}{5})$   **13.** 25   **15.** $\frac{3}{11}$   **17.** $\frac{2}{9}$

**19.** $\frac{4007}{999}$   **21.** $.99\overline{9}\ldots = (.9)\dfrac{1}{1 - .1} = \dfrac{9}{10} \cdot \dfrac{10}{9} = 1$   **23.** \$190 billion

**25.** (a) $100 + 100(1.01)^{-1} + 100(1.01)^{-2} + \cdots = \sum_{k=0}^{\infty} 100(1.01)^{-k}$   (b) \$10,100   **27.** \$1,655,629

**29.** 20 mg   **31.** 5 mg   **33.** (a) 2.5   (b) yes; 3   **35.** 6   **37.** $\frac{1}{24}$   **39.** $\frac{15}{8}$   **43.** 12

**45.** $\frac{1}{4} + \frac{1}{8} + \frac{1}{16} + \frac{1}{32} + \frac{1}{64}; \frac{1}{2}$

## EXERCISES 11.4

**1.** Divergent   **3.** Convergent   **5.** Divergent   **7.** Convergent   **9.** Convergent   **11.** Convergent

**13.** Convergent   **15.** Divergent   **17.** $\sum_{k=0}^{\infty} \dfrac{3}{9 + k^2}$   **21.** Convergent   **23.** Convergent

**25.** Convergent   **27.** No   **29.**

## EXERCISES 11.5

**1.** $\dfrac{1}{3} - \dfrac{2}{9}x + \dfrac{2^2}{3^3}x^2 - \dfrac{2^3}{3^4}x^3 + \cdots$   **3.** $1 + \dfrac{1}{2}x - \dfrac{1}{2^2 \cdot 2!}x^2 + \dfrac{1 \cdot 3}{2^3 \cdot 3!}x^3 - \dfrac{1 \cdot 3 \cdot 5}{2^4 \cdot 4!}x^4 + \cdots$

**5.** $1 + 3x + 3^2x^2 + 3^3x^3 + \cdots$   **7.** $1 - x^2 + x^4 - x^6 + \cdots$   **9.** $1 - 2x + 3x^2 - 4x^3 + 5x^4 - \cdots$

**11.** $5 + \dfrac{5}{3}x + \dfrac{5}{3^2 \cdot 2!}x^2 + \dfrac{5}{3^3 \cdot 3!}x^3 + \cdots$   **13.** $x - \dfrac{1}{2!}x^2 + \dfrac{1}{3!}x^3 - \dfrac{1}{4!}x^4 + \cdots$

**15.** $x - \dfrac{1}{2}x^2 + \dfrac{1}{3}x^3 - \dfrac{1}{4}x^4 + \cdots$   **17.** $1 - \dfrac{3^2}{2!}x^2 + \dfrac{3^4}{4!}x^4 - \dfrac{3^6}{6!}x^6 + \cdots$   **19.** $3x - \dfrac{3^3}{3!}x^3 + \dfrac{3^5}{5!}x^5 - \cdots$

**21.** $x + x^3 + \dfrac{1}{2!}x^5 + \dfrac{1}{3!}x^7 + \cdots$   **23.** $1 + \dfrac{1}{2!}x^2 + \dfrac{1}{4!}x^4 + \dfrac{1}{6!}x^6 + \cdots$

**25.** $1 + \dfrac{1}{2}x + \dfrac{1 \cdot 3}{2 \cdot 4}x^2 + \dfrac{1 \cdot 3 \cdot 5}{2 \cdot 4 \cdot 6}x^3 + \cdots$   **27.** $x - \dfrac{1}{2 \cdot 3}x^3 + \dfrac{1 \cdot 3}{2 \cdot 4 \cdot 5}x^5 - \dfrac{1 \cdot 3 \cdot 5}{2 \cdot 4 \cdot 6 \cdot 7}x^7 + \cdots$

**31.** 48   **33.** 0   **35.** $\left[ x - \dfrac{1}{3}x^3 + \dfrac{1}{5 \cdot 2!}x^5 - \dfrac{1}{7 \cdot 3!}x^7 + \cdots \right] + C$

**37.** $\left[ x - \dfrac{1}{4}x^4 + \dfrac{1}{7}x^7 - \dfrac{1}{10}x^{10} + \cdots \right] + C$   **39.** $1 - \dfrac{1}{3} + \dfrac{1}{5 \cdot 2!} - \dfrac{1}{7 \cdot 3!} + \cdots$

## CHAPTER 11: SUPPLEMENTARY EXERCISES

**1.** $x + \frac{3}{2}x^2$   **3.** $8 - 7x^2 + x^3$   **5.** $9 + 6(x - 3) + (x - 3)^2$   **7.** $p_2(t) = 2t^2; \frac{1}{12}$

**9.** (a) $3 + \frac{1}{6}(x - 9) - \frac{1}{216}(x - 9)^2$   (b) 2.949583   (c) 2.949576   **11.** 3.5619   **13.** $\frac{4}{7}$   **15.** $\frac{1}{7}$   **17.** 1

**19.** $e^2$   **21.** $\frac{9}{2}$   **23.** Convergent   **25.** Divergent   **27.** $p > 1$   **29.** $1 - x^3 + x^6 - x^9 + x^{12} - \cdots$

**31.** $1 + 6x + 27x^2 + 108x^3 + \cdots$

**33.** (a) $1 - \dfrac{2^2}{2!}x^2 + \dfrac{2^4}{4!}x^4 - \dfrac{2^6}{6!}x^6 + \cdots$   (b) $x^2 - \dfrac{2^3}{4!}x^4 + \dfrac{2^5}{6!}x^6 - \dfrac{2^7}{8!}x^8 + \cdots$   **35.** $1 + 2x + 2x^2 + 2x^3 + \cdots$

**37.** (a) $x^2$   (b) 0   (c) .3095   **39.** (a) $f'(x) = 2x + 4x^3 + 6x^5 + \cdots$   (b) $f'(x) = \dfrac{2x}{(1 - x^2)^2}$

**41.** \$566,666,667   **43.** \$120,066.66   **45.** \$3,285,603.18

## CHAPTER 12

## EXERCISES 12.1

**1.** $E(X) = \frac{4}{5}$, $Var(X) = .16$, standard deviation $= .4$   **3.** (a) $Var(X) = 1$   (b) $Var(X) = 4$   (c) $Var(X) = 16$

**5.** (a)

| Accidents | 0 | 1 | 2 | 3 |
|---|---|---|---|---|
| Probability | .21 | .5 | .25 | .04 |

(b) $E(X) \approx 1.12$   (c) Average of 1.12 accidents per week during the year   **7.** (a) 25%   (b) $100c^2$%   **9.** $E(X) = \$90,000$. The grower should spend the \$5000.

## EXERCISES 12.2

**7.** $\frac{1}{4}$   **9.** $\frac{1}{15}$   **11.** 12   **13.**

$y = \frac{1}{8}x$

**15.** $\frac{1}{12}$   **17.** $\frac{1}{2}$   **19.** $\frac{3}{4}$
**21.** $f(x) = \frac{1}{4}(x-1)^{-1/2}$   **23.** $f(x) = \frac{1}{5}x - \frac{2}{5}$
**25.** (a) $\frac{19}{63}$   (b) $f(x) = (x^3 - 1)/63$   (c) $\frac{19}{63}$
**31.** $2 - \sqrt{2} \approx .59$   **33.** 1.8   **35.** .6
**37.** (b) $f(x) = 1 - x^{-4}$   (c) $\frac{15}{16}, \frac{1}{16}$

## EXERCISES 12.3

**1.** $E(X) = 4$, $Var(X) = 2$   **3.** $E(X) = 3$, $Var(X) = \frac{4}{3}$   **5.** $E(X) = \frac{5}{6}$, $Var(X) = \frac{5}{252}$
**7.** $E(X) = \frac{2}{5}$, $Var(X) = \frac{1}{25}$   **9.** (a) $F(x) = 10x^3 - 15x^4 + 6x^5$   (b) $\frac{53}{512}$   (c) $\frac{1}{2}$. On the average about half of the newspaper's space is devoted to advertising.   (d) $\frac{1}{28}$   **11.** (a) 2. The average useful life of the component is 200 h.   (b) $\frac{1}{2}$   **13.** 8 min   **15.** (a) $F(x) = \frac{1}{6}x^2 - \frac{1}{54}x^3 - 1$   (b) $\frac{23}{27}$   (c) 412.5 worker-hours
(d) .5344   **17.** $E(X) = \frac{4}{3}$, $Var(X) = \frac{2}{9}$   **19.** $3\sqrt{2}$   **21.** $\dfrac{3\sqrt{2}}{2}$ hundred hours   **23.** $\frac{5}{6}$ min

**25.** *Hint*: Compute $\displaystyle\int_A^B xf(x)\,dx$ using integration by parts.

## EXERCISES 12.4

**1.** $E(X) = \frac{1}{3}$, $Var(X) = \frac{1}{9}$   **3.** $E(X) = 5$, $Var(X) = 25$   **5.** $e^{-1} - e^{-2}$   **7.** $1 - e^{-2/3}$   **9.** $e^{-3}$
**11.** $1 - e^{-2}$   **13.** (a) $e^{-1/3}$   (b) $r(t) = e^{-t/72}$   **15.** $\mu = 4$, $\sigma = 1$   **17.** $\mu = 0$, $\sigma = 3$   **23.** (a) .4032
(b) .4013   (c) .8351   (d) .9772   **25.** (a) 47.72%   (b) 47.72%   **27.** .1587   **29.** The Capital Beltway
**31.** 1.22%   **33.** *Hint*: First show that $Pr(a \le X \le a+b) = e^{-ka}(1 - e^{-kb})$.   **35.** 62.15 weeks

## EXERCISES 12.5

**1.** .0504, .0216, .0081   **3.** .4724, .3543, .1329, .0332, .0062, .0009, .0001

**5.** (a) .0000454   (b) .0027694   (c) .9972306   **7.** (a) .2231302   (b) .3765321   (c) .0656425   **9.** 22,105
**11.**

.4, .24, .144, .0864, .05184, .031104   **13.** (a) $\left(\frac{3}{4}\right)^n\left(\frac{1}{4}\right)$   (b) .4219   (c) 3

**15.** $1 - p^n$   **17.** .04073   **19.** $\lambda = 2$   **23.** .74053326   **25.** (a) No   (b) .9979061

## CHAPTER 12: SUPPLEMENTARY EXERCISES

**1.** (a) $\frac{1}{8}$, $\frac{19}{64}$ (b) $E(X) = \frac{3}{2}$, $Var(X) = \frac{3}{20}$   **3.** $f(x) = 1 - e^{A-x}$   **5.** $c_2 = \frac{1}{2}$, $c_4 = \frac{1}{4}$
**7.** (a) 5.01   (b) 100   **9.** (a) .96   (b) 1.8 thousand gallons   (c) $f(x) = 1 - x/2$, $0 \le x \le 2$

**11.** (a) $E(X) = 22.5$, $\mathrm{Var}(X) = 2.0833$ (b) 21.5     **13.** (a) $\frac{1}{300}$ (b) $\frac{3}{8}$ (c) \$17,222

**15.** $\frac{2}{\pi}$     **17.** \$54.88     **19.** $k \approx .35$     **21.** 4938     **23.** 69.15%     **25.** $a \approx .25$     **27.** (b) .6826

**29.** (a) $\Pr\left(-\frac{1}{k} \le X \le \frac{3}{k}\right) = \Pr\left(0 \le X \le \frac{3}{k}\right) \ge \frac{3}{4}$     (b) $1 - e^{-3} \approx .9502$

**31.** .1953668     **33.** 4     **35.** $\frac{2}{7}$

# Index